Springer Geology

For further volumes:
http://www.springer.com/series/10172

Mansour Ghorbani

The Economic Geology of Iran

Mineral Deposits and Natural Resources

Springer

Mansour Ghorbani
Faculty of Geoscience
Shahid Beheshti University
Tehran, Iran

ISBN 978-94-007-5624-3 ISBN 978-94-007-5625-0 (eBook)
DOI 10.1007/978-94-007-5625-0
Springer Dordrecht Heidelberg New York London

Library of Congress Control Number: 2012951116

Printed on acid-free paper

Springer is part of Springer Science+Business Media (www.springer.com)

Preface

If the heart does not recognize the value of the passing time and stall
Our only harvest will be embarrassment through the lifetime for all

Hafez (Iranian Poet, 1325–1389 A.D.)

Mineral exploration, extraction and production in Iran dates back to about 4000 B.C. Ever since not only our knowledge about minerals and developing mines has improved, but the diversity of the discovered mineral deposits has increased and their utilization potentials have been amplified. Taking into account the ancient history of mining and the multiplicity of mineral deposits in the country, one expects that the mining sector of the country plays a pivotal role in the economy; nevertheless, it is not so because of the dominance of the petroleum industry in the country's macro economy.

In geology and mining circles and among many foreigners, the common belief is that Iran is undiscovered and untouched in terms of exploration and mining. However, existence of 5,000 dynamic mineral production units extracting over 60 different types of minerals and rocks, some dating back to thousands of years, point to the contrary.

Notwithstanding, what remains unknown about the active 5,000 mining sites of the country are their exact composition, mode and time of formation, host rocks, associated minerals, configuration and dimensions, and the amount of reserves. Frustration increases when we learn that comprehensive data on mining activities carried out in world-famous mineral reserves. Most countries that are poor in terms of natural riches invest heavily in mineral exploration and, following a minor find, invest even more to determine its potentials and feasibility.

Knowledge of surface deposits in our country is relatively good and acceptable due to a long history of mining activity, lack of dense vegetation cover, extensive surface exploration studies carried out in the past 50 years, and the role of non-experts in claiming rights of mineral territories; the last factor being the main reason for the need of information about mineral deposits. Withal, this does not mean that all surface reserves are known. The most immediate need of the country is to acknowledge the small and scattered surface indications and estimate their amount of reserves.

Few countries of the world have such a variety of surface exposure of natural resources as Iran. The country has an important position within many domains such as mineral reserves, energy resources and natural attractions. This natural richness and climatic-morphological diversity on one hand and the ancient civilization on the other form the foundation of Iran's social and economic activities leading to a socio-cultural diversity.

This book manuscript is an attempt to prioritize Iran's natural riches and is intended to present a wide-spectrum illustration of its mineral resources in order to provide the international scientific and academic community, all real and legal persons of economic interests, investors, and global enterprises, with helpful information on the economic geology of Iran. Although the audience is considered to comprise of geologists, miners, and perhaps international mining and economic institutions, the main objective is to introduce Iran's mineral reserves and natural resources to the international community with vested interest in Iran's resources and its mining industry.

Iran's Economic Geology book is primarily a synopsis of the country's geological, natural and climatic features, its mining history, mineral resources, mining potentials as well as an overview of its other natural resources as energy and water.

Chapter 1 covers the general characteristics of Iran's nature, climate and its biodiversity; a summary of the geology of Iran is presented in Chap. 2; the third chapter refers to history of mining and early mining skills; Chap. 4 denotes different metallogenic and mineralization phases through geological periods in Iran; Chap. 5 states metallogeny and distribution of mineral resources; metallogenic and mining provinces, belts, and zones of Iran are described in Chap. 6 while the next chapter provides a description about the position of Iranian mining industry in the world; energy resources, production, and consumption in Iran vis-à-vis those in the world are presented in Chap. 8; the final chapter points out a complete list of mineral deposits and indications of Iran with their metallogenic attributes.

One of my lasting wishes was to create an encyclopedia on Iran's natural resources in order to demonstrate at the national and international level the country's natural mineral resources, energy resources, state of forests and rangelands, water resources, and natural attractions. Typically such a book should be written by a team of skillful experts and scholars and one or more governmental organization should sponsor it, by providing logistics and references, thus helping the expert team in its formulation. Unfortunately, I did not find such support by any organization or agency willing to sponsor it and, therefore, decided to write the book *The Economic Geology of Iran* singlehandedly with my own limited scientific expertise and funds by carrying out surveys and field visits across Iran over the years collecting the results of former studies, bearing all costs personally. Consequently, the book might be rife with many shortcomings, the most prominent being that it has been developed by only one person who was guilty on his neck for wishing and still having such a wish to carry on such work. The passion and love that drove me to a point as described by Hafez:

I am the servant of the house of wisdom for so lengthy
Even though I have nothing comparing to the wealthy

Apart from noted imperfections, this book presents comprehensive information about Iran's general geography, climate and geology, its mineral, energy and water resources, and their historical applications; it is a true reflection of Iran's economic geology, its mineral and natural resources. The book is the harvest of 25 years of my study and research in field, travelling through mineral and natural resources across my country. This book is a small gift to all who promote a prosperous, free, and peaceful world for all mankind.

In preparing the manuscript I confronted countless difficulties that include:

- Diversity, abundance, and scattered distribution of minerals and mining indications in Iran. Though they are endowments and should be considered as valuable assets, their survey necessitates collaboration from all stakeholder individuals and organizations, the opportunity which the author was denied of.
- Sporadic distribution of information, especially the metallogenic data and the results of recent explorations.
- Lack of descriptive information even about the most known and important deposits.
- Lack of a repository where all mining related reports and researches could be found.
- Unavailability of reports on exploration carried out by mining engineers consulting offices.
- Archiving of information in organizations and places unfamiliar to experts.

In order to overcome the above difficulties I tried to keep track of reports on mining exploration, during visits to the mining indications and deposits as much as possible, and later on compare and integrate them with my own findings. It is hoped that with thoughtful future teamwork the unknowns about Iran's deposits will be further clarified.

I would like to take this opportunity to express my cordial gratitude and appreciation to those individuals who have helped me during different stages of this endeavor. I am indebted to Dr. A. Kani, Assistant Professor of Geology, Shahid Beheshti University (Tehran, Iran) and Peyman Tajbakhsh, project manager at Pars Geological Research Center (Arian Zamin) for translation of the Persian text of the book into English. Special thanks are due to Mohsen Iranmanesh, administrative manager and head of publications section at Pars Geological Research Center (Arian Zamin) who has tirelessly worked on typing the manuscript and drafting of the figures.

Lastly, I would like to end this introduction by a translation of a piece from contemporary Iranian poet Houshang Ebtehaj, which best describes my emotions:

What do you think?
The world is like a broken mirror, where even a straight pine tree, looks broken to you;

The sitting mountains ambushing sunset narrow valleys makes the roads seem close;
Don't measure Time's infinite shorelines with our lives' steps!
Our pains and sorrows are only an instant for it.
Like river that towards the downhill, strike the stone with his head, flow forward;
There is no hope that a dead accomplish a miracle;
Be alive!

Mansour Ghorbani

Contents

1 Nature of Iran and Its Climate 1
1.1 Location, Boundaries, and Surface Area 1
1.2 Administrative Division 2
1.3 Iranian Plateau 2
1.3.1 Trend of Iranian Plateau Transformation 5
1.4 Physiography 8
1.4.1 Physiographic Division 8
1.5 Vegetation 13
1.5.1 Caspian Floral Province 14
1.5.2 Zagros Floral Province 14
1.5.3 Iran's Central Plateau Floral Province 14
1.5.4 Southern Coastal Plains Floral Province 15
1.6 Summarized Facts About Vegetation Cover 15
1.7 Weather Systems 16
1.7.1 Air Pressure and Wind Flow in Different Season 16
1.7.2 Precipitation 17
1.7.3 Average Annual Rainfall 19
1.7.4 Characteristics of Rainfall in Iranian Plateau 19
1.7.5 Genesis of Dry and Wet Weather Cycles 19
1.8 Aves 20
1.9 Mammals 21
1.10 Mountains 21
1.10.1 Northern Mountain Ranges 22
1.10.2 Zagros Mountains 23
1.10.3 Central Mountains 23
1.10.4 Eastern Mountains 24
1.11 Seas 24
1.11.1 Caspian Sea 24
1.11.2 Persian Gulf 25
1.11.3 Oman Sea (Makran Sea) 26

1.12 Lakes 26
1.13 Watersheds 29
1.14 Rivers 29
1.14.1 Caspian Sea Category 30
1.14.2 Persian Gulf Category 31
1.15 Deserts 33

2 A Summary of Geology of Iran 45
2.1 Structural Divisions of Iran 45
2.1.1 Central Iran 47
2.1.2 Sanandaj–Sirjan 47
2.1.3 Zagros 47
2.1.4 Alborz 50
2.1.5 Azerbaijan 51
2.1.6 Eastern Iran 52
2.1.7 Southeastern Iran (Makran) 53
2.1.8 Kopet Dagh 53
2.2 Ophiolite Series and Ultramafic Rocks of Iran 53
2.3 Basement Rocks 54
2.4 Summary of Stratigraphy of Iran 55
2.4.1 Precambrian 55
2.4.2 Paleozoic 55
2.4.3 Mesozoic in Iran 56
2.4.4 Tertiary 57
2.5 Stratigraphic Succession in Various Structural Divisions 57
2.6 Overview of Distribution and Nature of Magmatic Rocks in Iran 57
2.6.1 Magmatic Rocks of Upper Precambrian–Lower Cambrian 60
2.6.2 Lower Paleozoic 60
2.6.3 Upper Paleozoic 60
2.6.4 Mesozoic 61
2.6.5 Tertiary 61
2.7 Structural and Orogenic Events in Iran 62
2.7.1 Pan-African 62
2.7.2 Caledonian 62
2.7.3 Hercynian 62
2.7.4 Early Cimmerian 63
2.7.5 Middle Cimmerian 63
2.7.6 Late Cimmerian 63
2.7.7 Laramide 63
2.7.8 Pyrenean 63
2.7.9 Pasadenian 64

3 **History of Mining** 65
3.1 Introduction 65
3.2 History of Mining 67
3.2.1 Hakhamaneshian Period (550 B.C.–330 B.C.) 67
3.2.2 Seleucid Period (320 B.C.–63 B.C.) 68
3.2.3 Arsacid Period (247 B.C.–228 A.D.) 68
3.2.4 Sassanid Period (224 A.D.–670 A.D.) 69
3.2.5 Islamic Era (651 A.D.–1230 A.D.) 69
3.2.6 Mughal Era (1230 A.D.–1357 A.D.) 70
3.2.7 Safavid Period (1501 A.D.–1786 A.D.) 70
3.2.8 Qajar Period (1794 A.D.–1925 A.D.) 71
3.2.9 Pahlavi Period (1925 A.D.–1979 A.D.) 72
3.3 Mining Techniques in Ancient Iran 74
3.3.1 Exploration 74
3.3.2 Exploitation 74
3.3.3 Processing and Metallurgy 75
3.4 Mining Provinces in Ancient Iran 77

4 **Metallogenic and Mineralization Phases of Iran** 79
4.1 Introduction 79
4.1.1 Late Proterozoic–Early Cambrian 80
4.1.2 Early Paleozoic 81
4.1.3 Late Paleozoic–Triassic 82
4.1.4 Jurassic–Early Cretaceous 83
4.1.5 Late Cretaceous–Paleogene (Laramide), Chromite 84
4.1.6 Tertiary–Quaternary (Middle–Late Alpine) 85

5 **Metallogeny and Distribution of Minerals** 87
5.1 Iron 87
5.1.1 Metallogenic Phases of Iron 87
5.1.2 Distribution of Iron Deposits 91
5.2 Manganese 95
5.2.1 Mineralization Phases of Manganese 95
5.2.2 Distribution of Manganese Deposits 96
5.3 Chromite 97
5.3.1 Metallogenic Phases of Chromite 97
5.3.2 Distribution of Chromite Deposits 99
5.4 Copper 103
5.4.1 Introduction 103
5.4.2 Copper Alloys 105
5.4.3 Copper Metallogeny in Iran 105
5.4.4 Phases of Copper Mineralization 107
5.4.5 Distribution of Copper Deposits in Iran 108
5.5 Lead and Zinc 113
5.5.1 Introduction 113
5.5.2 Metallogenic Phases of Lead and Zinc 114

5.5.3 Spatial (Geographical) Distribution of Lead–Zinc Deposits and Indications 115
5.5.4 Economically Important Lead–Zinc Deposits 119
5.6 Aluminum 119
5.6.1 Introduction 119
5.6.2 Temporal Settings of Bauxite Deposits 124
5.6.3 Distribution of Bauxite Deposits 125
5.7 Gold 132
5.7.1 Introduction 132
5.7.2 Geological Distribution of Gold Reserves 135
5.7.3 Geographic Distribution of Gold Deposits 137
5.8 Antimony, Arsenic, and Mercury 141
5.8.1 Mineralization Phases of Antimony, Arsenic, and Mercury 141
5.8.2 Distribution of Antimony, Arsenic, and Mercury Deposits 143
5.9 Nickel and Cobalt 146
5.9.1 Mineralization Phases of Nickel and Cobalt 146
5.9.2 Distribution of Nickel and Cobalt Deposits in Iran 146
5.10 Titanium 147
5.10.1 Mineralization Phases of Titanium in Iran 147
5.10.2 Distribution of Titanium Deposits in Iran 148
5.11 Tungsten 149
5.11.1 Mineralization Phases of Tungsten 149
5.11.2 Distribution of Tungsten Deposits in Iran 150
5.12 Uranium 152
5.12.1 Mineralization Phases of Uranium in Iran 152
5.12.2 Distribution of Uranium Deposits in Iran 152
5.13 Phosphate 153
5.13.1 Introduction 153
5.13.2 Mineralization Phases of Phosphate Resources in Iran 154
5.13.3 Distribution of Phosphate Deposits in Iran 155
5.14 Talc 160
5.14.1 Mineralization Phases of Talc in Iran 160
5.14.2 Distribution of Talc Deposits in Iran 160
5.15 Mica 161
5.15.1 Mineralization Phases of Mica in Iran 161
5.15.2 Distribution of Mica Deposits in Iran 162
5.16 Boron and Borax 165
5.16.1 Mineralization Phases of Borax in Iran 165
5.16.2 Distribution of Borax Deposits in Iran 165
5.17 Magnesite 165
5.17.1 Mineralization Phases of Magnesite 166
5.17.2 Distribution of Magnesite Deposits 166

5.18 Celestine (Strontium Sulfate) 166
5.18.1 Mineralization Phases of Celestine Deposits in Iran 166
5.18.2 Distribution of Celestine Deposits in Iran 166
5.19 Barite 169
5.19.1 Mineralization Phases of Barite Deposits in Iran 169
5.19.2 Distribution of Barite Deposits in Iran 170
5.20 Feldspar Mineralization 170
5.20.1 Mineralization Phases of Feldspar Deposits 170
5.20.2 Distribution of Feldspar Deposits of Iran 172
5.21 Rock Salt 172
5.21.1 Mineralization Phases of Rock Salt Deposits in Iran 172
5.21.2 Distribution of Rock Salt Deposits in Iran 172
5.22 Potash Mineralization in Iran 174
5.22.1 Mineralization Phases of Potash Deposits in Iran 174
5.22.2 Distribution of Potash Deposits in Iran 174
5.23 Silica Mineralization in Iran 174
5.23.1 Mineralization Phases of Silica Deposits in Iran 174
5.23.2 Distribution of Silica Deposits in Iran 175
5.24 Kaolin and Fireclay 177
5.24.1 Mineralization Phases of Kaolin and Fireclay Deposits in Iran 177
5.24.2 Distribution of Kaolin and Fireclay Deposits in Iran 178
5.25 Bentonite 181
5.25.1 Introduction 181
5.25.2 Distribution of Bentonite Deposits in Iran 181
5.25.3 Specifications of Bentonite Deposits in Iran and Their Genesis 181
5.25.4 Chemical Composition of Bentonite Deposits in Iran 182
5.25.5 Age of Bentonite Deposits in Iran 184
5.25.6 Environment of Bentonite Deposition 185
5.25.7 Origin and Genesis of Bentonite Deposits in Iran 186
5.25.8 Formation of Bentonite Deposits 187
5.26 Diatomite 188
5.26.1 Distribution of Diatomite Deposits 188
5.27 Perlite 189
5.27.1 Mineralization Phases of Perlite Deposits in Iran 189
5.27.2 Distribution of Perlite Deposits in Iran 189
5.28 Dolomite 189
5.28.1 Mineralization Phases of Dolomite Deposits in Iran 189
5.28.2 Distribution of Dolomite Deposits in Iran 190
5.29 Tar and Natural Asphalt 191
5.29.1 Origin of Tar and Natural Asphalt 191
5.29.2 Distribution of Tar and Natural Asphalt Deposits in Iran 192

5.30 Dimension and Ornamental (Decorative) Stones ... 192
5.30.1 Standards of Dimension Stones ... 192
5.30.2 Distribution of Dimension Stones ... 195
5.31 Rare Earth Elements (REEs) Resources in Iran ... 196

6 Metallogenic and Mining Provinces, Belts and Zones of Iran ... 199
6.1 Introduction ... 199
6.2 Categorization of Metallogenic and Mining Areas in Iran ... 200
6.3 Metallogenic Provinces of Iran ... 202
6.3.1 Central Iran Metallogenic Province ... 202
6.3.2 Urumiyeh–Dokhtar Metallogenic Province ... 243
6.3.3 Sanandaj–Sirjan Metallogenic Province ... 249
6.3.4 Northeast Metallogenic Province (Taknar and Kavir–Sabzevar Metallogenic Belts) ... 257
6.3.5 Alborz Metallogenic Province ... 266
6.3.6 Southeast and East of Iran Metallogenic Province (Science and Technology Research Institute 1999) ... 286
6.3.7 Zagros Oil and Gas Province (Motiei 1993) ... 294
6.3.8 Kopet–Dagh Natural Gas Province ... 294

7 The Position of Iranian Mining Industry in the World ... 297
7.1 Introduction ... 297
7.2 Brief History ... 298
7.3 Mining Activities in Iran ... 299
7.4 Share of Mining Sector in Iranian Economy ... 299
7.5 Mineral Resources of Iran ... 307
7.5.1 Iron ... 314
7.5.2 Manganese ... 314
7.5.3 Chromium (Chromite) ... 314
7.5.4 Molybdenum ... 316
7.5.5 Copper ... 316
7.5.6 Lead–Zinc ... 318
7.5.7 Aluminum (Bauxite) ... 318
7.5.8 Arsenic ... 319
7.5.9 Antimony ... 320
7.5.10 Rhenium ... 321
7.5.11 Gold ... 321
7.5.12 Phosphate ... 322
7.5.13 Fluorite ... 322
7.5.14 Rock Salt ... 324
7.5.15 Boron (Borax) ... 325
7.5.16 Bentonite ... 326
7.5.17 Vermiculite ... 326
7.5.18 Mica ... 326
7.5.19 Strontium ... 326
7.5.20 Barite ... 327

7.5.21 Diatomite 327
7.5.22 Feldspars 327
7.5.23 Perlite 328
7.5.24 Dimension Stones 328
7.5.25 Magnesite 329
7.5.26 Fossil Fuels (Petroleum, Natural Gas, and Coal) 330
7.5.27 Water Resources (Budget) 332

8 Energy Resources, Production, and Consumption in Iran Compared with the World 333
8.1 Types of Energy Resources 333
8.2 Nonrenewable Energy Resources 334
8.2.1 Crude Oil 334
8.2.2 Gas 334
8.2.3 Gas Condensates 334
8.2.4 Tar Sands 334
8.2.5 Kerogens 335
8.2.6 Other Nonrenewable Resources 336
8.3 Fossil Fuel Reserves of Iran 336
8.3.1 Crude oil 336
8.3.2 Gas Condensates 339
8.3.3 Gas 340
8.3.4 Coal 340
8.4 Fossilized Fuel Production of Iran and the World 342
8.4.1 Crude Oil 342
8.4.2 Gas 344
8.4.3 Coal 344
8.5 Fossilized Fuel Consumption in Iran and the World 345
8.5.1 Consumption Pattern 346
8.6 Global Status of Iran in Terms of Energy Resources 346
8.7 Nuclear Energy in Iran and the World 346
8.8 Renewable Energy Resources of Iran 348
8.8.1 Hydroelectric Energy 350
8.8.2 Solar Energy 350
8.8.3 Wind Energy 350
8.8.4 Geothermal Energy 351
8.8.5 Tidal Energy 353

9 List of Mineral Deposits and Indications of Iran 355
9.1 Introduction 355

References 543

Knowledge Sources 549

Index 567

Chapter 1
Nature of Iran and Its Climate

Abstract This chapter briefly describes Iran's physical geography, climatic characteristics, morphology, faunal and floral diversity, and water bodies inside and along its borders. The chapter is written in such a fashion that it gives the reader an idea about Iran's natural features and attractions even if one has never visited Iran. The information provided herein lays ground for the study of physical features of Iran in any discipline.

Keywords Iranian Plateau • Climate of Iran • Mountains of Iran • Fauna and Flora of Iran • Sea, lakes and rivers of Iran

1.1 Location, Boundaries, and Surface Area

Iran is positioned on the southwest of Asia forming a bridge connecting three continents of Asia, Europe, and Africa. It is bordered by Azerbaijan, Armenia, Turkmenistan, and the Caspian Sea on the north, Afghanistan and Pakistan on the east, Oman Sea and Persian Gulf on the south, and Turkey and Iraq on the west (Road Atlas of Iran 2004). The country is part of the Iranian Plateau that constitutes a vast and high terrane, which is bordered by the Caspian Sea on the north, Amu Darya, Syr Darya, and Kora River plains on the northeast, Sindh and Punjab Rivers plain on the southeast, the Oman Sea and the Persian Gulf on the south, and the Tigris river plain on the southwest (Comprehensive Geography of Iran 1987).

The total areal extent of the plateau is about 2,600,000 km^2, of which 1,648,195 km^2 is accounted for by Iran, and the rest covers Afghanistan, Pakistan, and former Soviet republics. The area of Iran is 3.7% of Asia and 1.09% of the total land on the Earth's surface (Comprehensive Geography of Iran 1987).

The Iranian Plateau is a triangular-shaped piece of land between the Persian Gulf and the Oman Sea on the south and the Caspian Sea on the north and plays the role of a bridge between Central Asia and other plateaus in western Asia and Europe

M. Ghorbani, *The Economic Geology of Iran: Mineral Deposits and Natural Resources*, Springer Geology, DOI 10.1007/978-94-007-5625-0_1,

(Darehshouri and Kasraian 1998). The longest stretch of Iran runs from the Ararat Mountains northwest to the Goater Port on southwest and measures 2,210 km, while the widest stretch is between Sarakhs on the northeast and the Arvand River on the southwest, measuring about 1,400 km (Comprehensive Geography of Iran 1987). Half of Iran's land surface is mountainous, 1/4 covered by fertile and productive plains and the other 1/4 covered with salty arid deserts (Darehshouri and Kasraian 1998).

The southernmost point of Iran is the Goater Port located on 25° N latitude, and the northernmost point is the Ararat foothills 40° N. The easternmost point is Kuhak on the border with Pakistan, while the westernmost point is Bazargan on the border with Turkey. Iran's geographic coordinates are between 44° and 63° 5′ 30″ E longitude and 25–40° N latitude (Comprehensive Geography of Iran 1987).

The time difference between the easternmost and westernmost points is about 1 h and 18 min. The perimeter of Iran is about 8,700 km of which 2,700 km (or nearly one-third) is marine (Persian Gulf, Oman or Makran Sea and Caspian Sea) and the rest is terrestrial. The marine border line from the Arvand River estuary to the Goater Port stretches about 2,000 km (Comprehensive Geography of Iran 1987).

1.2 Administrative Division

In the old days, Iran was governed by a totalitarian monarchy system, and the political subdivisions were ruled by a local ruler. This system continued until 1907 when a constitutional monarchy was established. In this year, a law was ratified based on which Iran was divided into four provinces: Azerbaijan, Fars, Khorasan, and Kerman and Baluchistan. This system remained in force until 1937 when, based on a new administrative division system, Iran was divided into 10 provinces and 49 counties. In the year 1971, the divisions were increased to 14 provinces, 8 districts, and 161 counties (National Historical Atlas of Iran 1999). Based on the latest administrative division system provided by the Interior Ministry of Iran Website, Iran consists of 31 provinces, 385 counties, 961 districts, 1,120 cities, and 2,473 rural districts (Fig. 1.1). The general statistics of various provinces is presented in Table 1.1.

1.3 Iranian Plateau

The Iranian Plateau that covers a major part of Iran, Afghanistan, and Pakistan seems to have experienced similar developments throughout its extent during the evolution of the Earth.

Following the breakup of Gondwana supercontinent into several pieces, the Iranian Plateau, like the other pieces, began to drift away from the supercontinent, thus forming the Tethys Sea that lasted for tens of millions of years until it finally

Fig. 1.1 Administrative division map of Iran (http://en.wikipedia.org/wiki/Template:Provinces_of_Iran_Labelled_Map http://wikimediafoundation.org/wiki/Terms_of_Use)

reached the southern coasts of Laurasia at the beginning of Cenozoic. The collision point was at the coastal plains of Touran (Ghorbani 2002b).

Undoubtedly the geological history of the Iranian Plateau is one of the most interesting, complicated, and adventurous on the face of the Earth because the Iranian Plateau alone experienced numerous and various geological events and phenomena.

Different theories about the history of the Iranian Plateau from the moment it separated from Gondwana till its collision with Laurasia have been presented by many geologists.

Some significant and important geological events occurred following the accretion of the Iranian Plateau to Laurasia. Due to the general moving direction of Africa toward Laurasia and also the continuous pressure applied by the Iranian Plateau onto this supercontinent, several orogenic phases have occurred (Ghorbani 2002a, b, c).

Table 1.1 General statistics about the provinces of Iran (http://fa.wikipedia.org – for detailed URLs, please refer to the reference list)

No.	Province name	Area (sq. km)	Population
1	Alborz	5,833	2,412,513
2	Ardabil	17,800	1,248,488
3	Bushehr	27,653	1,032,949
4	Chahar-mahal and Bakhtiyari	16,332	895,263
5	East Azerbaijan	45,650	3,724,620
6	Fars	122,608	4,596,685
7	Golestan	20,367	1,777,014
8	Guilan	14,042	2,480,874
9	Hamedan	19,368	1,758,268
10	Hormozgan	70,697	1,578,183
11	Ilam	20,133	557,599
12	Isfahan	107,029	4,879,312
13	Kerman	181,785	2,938,988
14	Kermanshah	24,998	1,945,227
15	Khuzestan	64,055	4,531,720
16	Kohkilooyeh and Boyer-ahmad	15,504	658,629
17	Kurdistan	29,137	1,493,645
18	Lurestan	24,294	1,754,243
19	Markazi	29,127	1,413,959
20	Mazandaran	23,842	3,070,943
20	Northern Khorasan	28,434	867,727
22	Qazvin	15,567	1,201,565
23	Qom	11,526	1,151,672
24	Razavi Khorasan	128,949	5,994,402
25	Semnan	97,491	631,218
26	Sistan and Baluchestan	180,726	2,534,327
27	Southern Khorasan	85,290	622,534
28	Tehran	12,981	12,183,391
29	West Azerbaijan	37,411	3,080,576
30	Yazd	129,285	1,074,428
31	Zanjan	21,773	1,015,734
Total area (as of 2010):		1,629,687	
Total population (as of 2011):			75,106,696

The separation of the Indian subcontinent from Gondwana and its collision with Asia (which led to the formation of the great Himalayan mountains) caused the closure of the Tethys Sea on the east of the Iranian Plateau (Ghorbani 1999f).

Another factor that played a significant role in the final shaping of Iran was the spreading of the Red Sea (Darehshouri and Kasraian 1998).

The northward movement of the Arabian Plate and the pressure exerted against Iran and Turkey caused new geological phenomena to occur (Ref).

As the Arabian Plate collided with the Iranian Plate, it plunged beneath the Iranian Plate and resulted in various geological and topographical episodes, such as the formation of the Zagros Mountains and other ranges parallel to it (Darehshouri and Kasraian 1998).

This collision should be considered as a continent–continent type. Geologists believe that the very thick Iranian continental crust in the north Zagros area is the result of plunging the Arabian Plate underneath the Iranian Plate (Berberian 1984).

Currently the Arabian Plate is still pushing against the Iranian Plate, and recent earthquakes in the southwest of the country are the results of such movement (Alavi 1994).

The Iranian Plateau collision with Asia probably occurred in the early Mesozoic, which was followed by numerous geological and biological developments (Darehshouri and Kasraian 1998).

Further folding in Alborz and Central Iran as well as folding of the Zagros Mountains occurred in this period of time (Darehshouri and Kasraian 1998).

Prior to the detachment of the Iranian Plateau, Gondwana was closer to the Equator and experienced a warm climate (Darehshouri and Kasraian 1998).

The characteristics of phenomena experienced by the Iranian Plateau caused looming of numerous morphological and climatological diversities with each of them having a specific mechanism. Each of these phenomena resulted in the formation of a variety of natural resources and biological diversities.

The Iranian Plateau entered a new climatic condition once attached to Asia. Under the newly established circumstance, many biological species that could not adjust themselves to the new condition gradually vanished, and the susceptible Iranian terrane was occupied by the fauna and flora of the neighboring Touran and Europe.

Many of the immigrant fauna and flora from northeastern Touran entered the Iranian Plateau and spread all the way to the south; the European fauna and flora were also spread through northwest gateway and settled in the areas suitable for them.

Although the Iranian Plateau managed to preserve some of its biota throughout these incidents, it was strongly affected by the incoming communities of Touran, and, to some extent, the European ones. While preserving some of its previous fauna and flora, the natural appearance of the biota eventually transformed to be more similar to that of Laurasia.

1.3.1 Trend of Iranian Plateau Transformation

The appearance of the Iranian Plateau changed following the changes in geology and climate of southern Asia. One of the major factors in this transformation was the Alpine Orogeny, whose final phases changed the outlook of the Iranian Plateau (Ghorbani 1999a, b, c, d, e, f). From Late Pliocene to Quaternary, that is, during the Pasadenian orogeny, the Iranian Plateau underwent significant changes. Following this orogeny phase, the distribution of land and water bodies in the Iranian Plateau approached that of today, and some inland lakes vanished. At the end of Pliocene and beginning of Quaternary, three significant phenomena were dominant on the Iranian Plateau (Ref):

- Volcanism
- Seismic activities and large earthquakes
- Fluvial inundation and flooding

These phenomena were so significant that the earthquakes and flooding happening today cannot be compared to those occurring from Pliocene until Quaternary. Some of these are briefly described next.

At the said time, the Alborz, Zagros, Sahand, and Sabalan Mountains were much taller than they are today. Steep topography, on the one hand, and climate characteristics of the early Quaternary, on the otherhand, caused torrential rains and river inundations, which led to severe mechanical weathering and erosion. In fact, the large volume and thick unsorted sediments and alluvium bespeak of such activities (Hezaar-Darreh Formation in north of Iran, and Bakhtiari conglomerate and loose alluviums in the south of Iran).

Displacement and disorder in river terraces in places where they are positioned perpendicular to Quaternary faults, or the effects of faults within young terraces, as well as fracturing, sliding, and slumping of multiton boulders in some areas indicate the occurrence of numerous large and strong earthquakes at the beginning of Quaternary.

A comprehensive review of the characteristics of Quaternary sediments and the phenomena occurred in this period, and it can be assumed that incidents such as floods, earthquakes, and volcanism were severely strong during Late Pliocene-Quaternary throughout Iran, but they gradually weakened. Volcanic activities totally ceased, and seismic activities lost their strength, with their intensity and amplitude decreasing.

This climate change led to the shrinkage of woodlands causing the extinction of jungle-dwelling animals whereas increase of ungulates and grazers in plains and prairies. Some of the extinct animals were giraffes, mastodons, marals (ghazals), and mammals like boars. Monkeys, carnivorous mammals, and predators belonging to the Canidae family were abundant in this period of time (Darehshouri and Kasraian 1998).

Tropical and subtropical flora gradually vanished, and other plant species replaced them as seen today in the woodlands in the northern flanks of the Alborz Mountains (Darehshouri and Kasraian 1998).

Solid evidence has been found on the habitats of flora and fauna for the period of 4–10 million years ago in north Maragheh, northwest of Iran (Darehshouri and Kasraian 1998). These evidences include a unique collection of animal fossils buried under volcanic ash released by the Sahand Volcano.

This valuable collection of vertebrate fossils in Maragheh, which is believed to have been formed during the last ten million years, is very important for understanding the paleoclimatic and paleoecologic conditions of Iran as well as the evolutionary studies of animal and human populations of Iran.

There are remnants of animals such as rams, ewes, marals, mountain goats (capras), giraffes, beavers, felidae, canidae, hyperion, and hyenas found in different sizes, in some cases gigantic, in the fossil-bearing Maragheh Formation (Darehshouri and Kasraian 1998). Rams and ewes are very sensitive to environmental changes and are therefore of paleontological importance. The presence of such animals in the past few million years points to the presence of grazelands and hilly prairies with moderate and humid climate. These rams and ewes had somehow limited distribution, and their remnants are only found in younger Cenozoic sediments, and therefore they are very helpful in understanding the environmental and climatic condition of the said times.

The Maragheh fossiliferous basin must be considered as a natural heritage and protected as part of Iran's cultural heritage.

Based on the excavation works in the Maragheh area and identification of vertebrate fossils as well as their comparison to their counterparts in the same time period (Darehshouri and Kasraian 1998), it can be stated that the northwestern part of Iran has been totally influenced by the animal species from Europe, and, with the exception of hyperion in northwest, Iran never had common animal species with India. Excavation works in the Maragheh area confirm that Azerbaijan Province was the gateway for European fauna and flora species to enter Iran (Darehshouri and Kasraian 1998).

Although excavation works did not yield any evidence on the very early and remote ancestors of mankind in Iran, they provided comprehensive information about the vertebrates. Two million years ago, most of the mammals became extinct as they could not bear the cold weather caused by the Ice Age. The Ice Age lasted until 10,000 years ago and strongly affected the northern hemisphere. The floral and faunal assemblages of the northern hemisphere were immensely hit by this cold ice age (Darehshouri and Kasraian 1998).

Most of the deciduous forests of previous time periods vanished, and those situated away from the glaciers (like the forests in the Caspian area) survived as the remnants of the Cenozoic forests that once covered all of Laurasia.

The surviving animals swarmed and migrated toward the southern part of Asia, Touran, and Iran; the surviving animals included marals, elephants, wild goats, bears, and boars. The few mammals that lived to see the final stage of the Ice Age had to adjust to tolerate the severe cold by growing thick hair and fatty layers underneath their skin (Darehshouri and Kasraian 1998); these mammals included mammoths, rhinoceros, cows, yaks, marals, saber-tooth tigers, and foxes.

Although glaciers did not fully cover the Iranian Plateau, they significantly affected vegetation coverage and animal variety in Iran.

The movement of Iranian Plateau from north of Africa toward South Asia was the most important event in the geological history of Iran. It caused the extinction of part of flora and fauna species as well as the migration of numerous species from neighboring lands so that, besides the endemic plants and animals specific to Iran, many other species that had originated in Touran, Europe, Mediterranean, Africa, and India are still living on this land (Darehshouri and Kasraian 1998).

1.4 Physiography

Iran features a very diverse climatic and geographic characteristic. The highest temperature in the Persian Gulf area can reach up to 53°C in summer, and the lowest temperature in the northwestern part of Iran can reach up to –40°C in winter. The average precipitation in lush forests on southwest of the Caspian Sea is about 2,400 mm per year, whereas the Lut Desert may experience no rainfall for some successive years (Darehshouri and Kasraian 1998).

The elevation at the Caspian Sea coasts is 25 m below the mean sea level, whereas there are over 100 peaks with heights of more than 4,000 m, among which Mount Damavand stands at 5,610 m (Jafari 2005).

The biodiversity of Iran is as contrasting as its geography and climate. While the northern flanks of Alborz are covered with deciduous forests, no trees can be seen in the millions of hectares of deserts. Mangroves forests cover the coasts of the Persian Gulf and the Oman Sea (Darehshouri and Kasraian 1998).

1.4.1 Physiographic Division

Physiographically, Iran can be divided into four regions, namely, Caspian, Zagros, central plateau, and southern coastal plains. This division corresponds to the climatic conditions, vegetation, faunal distribution, and, to a great extent, geological and morphological characteristics. These four regions are immensely affected by various natural and physiographic factors of the northern hemisphere (Climate Atlas of Iran; University of Tehran, Institute of Geography 1965).

1. *Caspian*: Some scientists define the Caspian region as a stretch of land with the Golestan forests at its eastern ends and Astara at its western ends, while some others stretch the Caspian region from Bojnourd on the east to Arasbaran on the west. This region is a long and narrow land with a width of 20–70 km and a length of 800 km. The climatic conditions in this region are moderate to warm with high humidity and plenty of precipitation. This region is covered with deciduous forests. The maximum annual rainfall in this region occurs at the Anzali area and exceeds 2,000 mm sometimes; however, moving toward the east, the annual rainfall drops down to 600 mm as in the Gorgan area. The temperature tremendously varies from the coastal plains to the high mountains. The different floral communities found in the Caspian region are the result of such climatic diversity (Climate Atlas of Iran; University of Tehran, Institute of Geography 1965).

2. *Zagros*: It is a wide, extensive mountain range stretching from Azerbaijan Province on the northwest toward south and southeast extending all the way parallel to the Persian Gulf till it reaches the Strait of Hormuz. The Zagros Mountains are very young, and in fact they are the most recent mountains formed in Iran (Alavi 1994). Currently this mountain range is still rising as the Arabian Plate is subducting under the Iranian Plate; consequently, this range has not yet stabilized.

Zagros consists of a series of mountain ranges with their crests parallel to each other. This series runs from Azerbaijan southwardly to Lorestan and Bakhtiari, and then to Boyer Ahmad and Fars; it finally ends at the Strait of Hormuz and Minab Fault.

The width of the Zagros Mountain range sometimes reaches 300 km, and there is no gap seen along this mountain range. The highest point along the Zagros Mountain range is Mount Dena with a height of 4,447 m.

Rain clouds from the Atlantic Ocean travel eastwardly across the Mediterranean and either come to a halt as they reach the Zagros Mountain range descending as precipitation or ascend to higher altitudes in atmosphere and scatter. Zagros prevents this weather system from entering into Central Iran.

The temperature and amount of precipitation vary extremely along the Zagros Mountain range. In general, the temperature and rainfall drop when moving from north of Zagros toward the southern parts.

The rainfall over the eastern flanks is less than that of the western flanks. The average precipitation varies from 200 to more than 1,000 mm in different areas of Zagros. Most of the rainfall occurs during the winter. Summer is dry just like in the Mediterranean area. The climate is drier than the western part of the Mediterranean area but eventually follows the same pattern of the Mediterranean climate contour lines (Climate Atlas of Iran; University of Tehran, Institute of Geography 1965).

Since the rainfall is less on the eastern flanks as compared to the western parts, the vegetation cover of the Zagros area is of dry Mediterranean (Darehshouri and Kasraian 1998) type.

In Fars Province, moving toward the east, Zagros loses height gradually and hence the annual rainfall drops. As a result, the temperature rises, thus leading to a significant diversity in climate, vegetation, and animal habitats in the eastern terminus of the Zagros Mountain range. Corresponding to elevation and climate, various plant communities have inhibited the southern end of Zagros. Relatively homogeneous forests of Boneh, Peanuts, and Aras are seen in places like Arsanjan, Qir Mountains, and Mount Shab in Larestan, and remain relatively intact and unharmed.

3. Central plateau: This is a very vast plateau spreading out from Alborz to Zagros and eastwardly to Afghanistan and Pakistan. The average elevation of this plateau is 1,300 m, but it drops to 700 m inside Dasht-e Kavir, and further to 300 m in the central part of Lut. Alborz and Zagros Mountains block the Mediterranean rain-bearing winds from entering into Iran's central plateau, thus lowering the average annual rainfall within this plateau to less than 300 mm (Darehshouri and Kasraian 1998).

The plateau has no streams. Those originating from Alborz and Zagros run down into salt marshes and salty wetlands and vaporize (Darehshouri and Kasraian 1998). The plateau features desert characteristics.

Dasht-e Kavir and Lut Kavir are two large arid regions that cover most of this central plateau. In addition to these two deserts, some small to large basins are seen in between or alongside them (e.g., Gav-khooni Swamp). With the exception of central parts of Dasht-e Kavir and Lut Kavir, scarce mountains are seen almost everywhere throughout this plateau (e.g., Toroud Chah-shirin Mountains, Shotori Mountains).

Iran's central plateau region is located almost in the middle of the Iran-Touran block. Its climate is of continental type that features hot summers and cold winters. The shortage of rainfall and lengthy dry season differentiates this plateau from other areas. The lowland and highland, soil type, salinity, and aridness in this area are very diverse (Darehshouri and Kasraian 1998).

The vastest sandy desert of Iran is located east of Lut Kavir, which is one of the largest of its kind in the world. The second-largest sandy desert of Iran is located north of Kashan. There are a few large sandy deserts south of Dasht-e Kavir, including Choopanan, Rig-Zarrin, Mesr, and Aroosan (Darehshouri and Kasraian 1998).

4. Southern coastal plains: This region is bounded between mountains, Persian Gulf and Oman Sea, and its width varies at different places. This region stretches from Ghasr-e Shirin to the Goater Port and is characterized by low rainfall and hot weather. The islands in the Persian Gulf are parts of this region.

The existence of numerous trees and plants locally affects the climate of this region. The average temperature is lower in woodlands and summers are cooler. The woodlands block direct sunlight from reaching the ground and therefore diminish heat radiation. On the other hand, moisture in the ground absorbs the heat to some extent, thereby leading to moderation of climate and a decrease in temperature. This effect is by and large sensed up to elevation of 1,500 m (Comprehensive Geography of Iran 1987).

The forests in the mountainous areas play a significant role in increasing the rainfall because due to lower temperatures, the rain clouds are saturated faster and precipitate. However, the adverse consequence on the climate of the region due to deforestation is obvious.

Based on physiographical characteristics and the geographical and climatological literatures and maps (Comprehensive Geography of Iran 1987; Darehshouri and Kasraian 1998; Climatic Atlas of Iran 1965; Shahrabi 1994; Nabavi 1994), Iran can be divided into ten climatic areas.

1.4.1.1 Arid Climate

Much of the Iranian Plateau, especially the central part, consists of dry and barren deserts that are bounded by Alborz on the north, Zagros on the west, Persian Gulf and Oman Sea on the south, and Afghanistan and Pakistan on the east. This area features arid climate, and the temperature difference between day and night is extreme with very hot days and relatively colder nights.

The general characteristics of this climate are summarized as a long hot season, extreme aridness, and shortage of rainfall. The aridity index[1] in this area is 300–350 days. Rain falls in only a few days and mostly in the mountainous parts. Arid climate dominates the basins within Iran's central plateau, and the annual rainfall is negligible, as low as 75 mm.

Besides Namak and Lut deserts, areas such as Jazmurian, Khur and Jandagh, Farrokhi, Biyabanak, Anarak, Yazd, Bafgh, Tabas, Nehbandan to Nosrat-abad, Bam, Zahedan, Zabol, Saravan, Bahu-kalat, and Abar-kuh can be included in this type of climate.

[1] Dryness coefficient reflects the number of days with no rainfall.

1.4.1.2 Severe Semiarid Climate

This climate prevails in surrounding areas with arid climate and covers the edge of the deserts. It features hot and dry weather. Cloud systems coming from the north and the west of Iran can hardly affect the climate of this area. As a result, the average annual rainfall is between 100 and 280 mm. The aridity index in this type of climate is 250–300 days. Parts of Khuzestan plain, Mah-shahr, Abadan, Ahwaz, Hamidiyeh, Andimeshk, Aghajari, Khash, Jask, southern parts of Fars Province, and most of Iran's central plateau feature characteristics of semiarid climate.

The average annual temperature in these areas is between 16 and 20°C. Extreme dryness and severe heat are characteristics of this climate. The weather becomes moderate in winter with occasional rainfall that usually leads to flash floods.

The southern parts of Khuzestan plain and Persian Gulf coast are often affected by dry and hot weather systems from the Arabian Peninsula and show features of such climate.

1.4.1.3 Moderate Semiarid Climate

The climate also encircles areas with an aridity index of 200–250 days. Since the Zagros and Alborz Mountain ranges prevent the entry of enough moisture into these areas, the rainfall is low and the temperature is high in areas with this type of climate. The moist weather systems lose their moisture as they hit and pass through the Zagros and Alborz Regions, and therefore the rainfall drops and temperature rises away from southern flanks of Alborz and eastern flanks of the Zagros toward Central Iran.

Masjed Soleiman, Gachsaran, Behbahan, Bushehr, Fasa, Jahrom, Varamin, Saveh, Naein, Esfahan, Najaf-abad, Rafsanjan, Sabzevar, Bandarabbas, Birjand, Gonabad, Kashmar, Shahrood, Sirjan, and Kerman feature this type of climate.

1.4.1.4 Hot and Dry Mediterranean Climate

In this weather condition, the number of dry days drops down, and the aridity index is between 150 and 200 days. This type of climate includes the western parts of Iran (Ghasr-e Shirin to Khoram-abad), parts of the plains in the southern flanks of Alborz, and areas from Tehran to north of Khorasan Province (i.e., Mashhad, Ghoochan, Fariman, Torbat-e Jam).

Much of the precipitation falls on the southern slopes of Alborz and north of Khorasan because of the presence of low-pressure cold weather systems coming in from the north or the west of Iran. In general, the average annual rainfall in these areas is between 200 and 500 mm.

The Zagros Mountain range provides suitable condition for precipitation in some parts within this type of climate. The weather systems coming from the Mediterranean area hit the Zagros Mountain range, causing them to lose temperature and therefore produce relatively significant rainfalls. The rainfall rate in Zagros varies from 400 to 1,200 mm and mostly falls on the western flanks.

1.4.1.5 Warm Mediterranean Climate

This type of weather condition is mostly dominant in northwestern and western Iran, including Ravansar, Kermanshah, Gorgan plain, Qazvin plain, Rudbar, Abhar, Takestan, and southern parts of Arak. Relatively high elevation of the aforementioned areas and their placement in higher geographic latitude cause the rain to fall in the form of snow in the winter and in the form of torrential showers in the spring. The average annual rainfall is between 300 and 600 mm, and the aridity index is 100–150 days.

1.4.1.6 Moderate Mediterranean Climate

This type of climate with an aridity index of 40–100 days prevails over parts of the Caspian Sea coast, which is from Anzali to Nowshahr. The Alborz Mountain range in the southern and western parts of this region acts as a barrier blocking moisture systems coming from the Caspian Sea. The average annual rainfall in this region varies from 700 to 1,800 mm with maximum rainfall pouring down on the coastal areas and northern slopes of Alborz. The rainfall rate recorded in Anzali is 1,781 mm. From the temperature standpoint, this region is considered as mild or moderate as its average annual temperature is 15–18°C.

1.4.1.7 Semi-Mediterranean Climate

The area affected by this climate is not vast and is limited between the Caspian Sea and the Alborz Mountains. Alborz separates this plain from areas inside Iran and acts as a barrier that disconnects two very different climates (i.e., warm and humid climate and dry and arid climate).

This type of climate covers the southern and eastern parts of the Caspian Sea coast, which is from Babolsar to Hosseingholi Gulf. The aridity index is less than 40 days, and the average annual rainfall is 700–1,000 mm. This region and its climate are separated from the contrasting dry and arid region by the Alborz Mountain range. The moisture systems coming from the Caspian Sea are drawn toward the coast and then up in the mountain slopes, where they condense due to the cold weather. Such condensations usually lead to precipitation. Some of the characteristics of this region are frequent rain, high humidity, and small seasonal temperature variation. The rainfall rate decreases when moving from the west to the east. The coastal lands never freeze because of high humidity.

1.4.1.8 Exerophire Climate

This type of climate does not experience dry season. It rains over part of coastal plains, including Astara, Fooman, Lakan, Bibalani, Koran Talar, and eastern slopes of the Talesh Mountains. The average annual rainfall in this region varies from 1,000 to 2,000 mm.

1.4.1.9 Cold Exerophire Climate

This type of climate prevails over heights in Alborz and Zagros. Freezing weather is dominant in these areas for 5–11 months of the year. The average annual rainfall is over 2,000 mm.

1.4.1.10 Cold Steppe Climate

This climate is seen in some parts of the Alborz and Zagros Mountains with 5–8 months of freezing weather. The annual precipitation is measured around 200 mm.

1.5 Vegetation

Naturalists divide northern hemisphere into four major provinces from a botanic point of view. Iran stands as a link between all these botanic provinces. Numerous factors from each of these provinces have affected the floral community of Iran. The four botanic provinces are

1. Europo-Siberian
2. Mediterranean
3. Irano-Touranian
4. Arid-Sandy

The aforementioned botanic provinces, which are located next to Iran's four geographical regions of Khazar, Zagros, central plateau, and southern coastal plains, respectively, have clearly affected and made their marks on the floral community of Iran. Obviously these provinces cannot be considered independently. Environmental intertwining based on distribution patterns of species in these geographic regions has always been possible. However, deserts and mountains are the main barriers against such distributions. Migration of mountainous floral species toward arid areas is hampered as they reach the edges of deserts, and the same blockade happens to the arid species as they reach foothills on their way toward the mountainous areas.

Extreme hot and cold weather as well as other climatic factors (e.g., elevation difference) played significant roles in the distribution and propagation of the floral species. The folding and major orogenic incidents in various geological eras and the migration time of fauna and flora species into Iran also played a clear role in the distribution of the species.

Although only 5% of the plants in Iran are of European-Siberian origin, this small portion includes 80 tree species and more than 50 shrub species. Most trees in the Caspian region have managed to survive as they were situated at a far distance from the glaciers, but some trees that once existed there (Europe-Siberia) have not survived. Due to the lack of sufficient knowledge and research works about the precise distribution pattern of these species, as well as the absence of data on fossil flora and fauna of Iran, there are many complicating questions. In the following paragraphs, a brief description of the four floral provinces is presented.

1.5.1 Caspian Floral Province

The northern plain forests, which are now mostly wiped out, were covered with box trees. Oak and carpinus trees form plant communities along the foothills, wherein single trees such as Azad, Shabe Hasab, Namdar, Lilaki, elm, and ash are seen.

Besides the oak trees, beech groves are also seen on slopes with higher altitude within the Caspian Province, which could appear along with carpinus trees at elevations around 2,000 m. When moving to higher elevations, more species are found, with trees being replaced with brushes and bushes.

Forests in the Caspian Province are truly a live museum displaying the preglacial European plants. Unfortunately, a large part of these forests has been transformed into rice fields, tea gardens, and citrus groves, and, more regretfully, into residential areas, where the greedy developers have destroyed the natural vegetal cover to build residential complexes.

1.5.2 Zagros Floral Province

The vegetation cover in Zagros varies with elevation. At greater heights and high peaks, the vegetation cover is more grassy and short. When moving down below the elevation of 4,000 m, plant species such as astragalus bushes are abundant, and below 2,800 m juniper (oros) trees can be seen. These juniper trees are not in the form of forest but are scattered single trees rising up through the rocks. However, in the southern parts of Zagros, there are vast forests of junipers mostly accompanied by wild olive trees. In these places, junipers grow in lower elevation as opposed to the northern part of Zagros, where these trees are only found at higher elevations.

1.5.3 Iran's Central Plateau Floral Province

This province is covered with a type of vegetation considered to be nurturing ground for many species of the northern hemisphere. Iran's central plateau covers about 3/4 of Iran's surface area and 70% of flora species of Iran are reported from this province. The plants in this area are well adopted with severe aridity, salinity, heat, long summers, and cold winters. Parts of mountains and foothills are covered with xerophytes. There are patches of junipers in the southern and northern part as well as in higher altitudes, and scarce small plant communities of turpentine and almond trees seen at higher elevations within central deserts. When moving from foothills toward the plains, the number of trees shrinks and brushes replace the trees. Rolling sand dunes in the form of small and large patches are scattered throughout this area, which displays a special vegetation coverage. The concentration of brushes and bushes in Iran's central plateau is remarkable. This vegetation cover is very rich in species diversity, and its abundance depends on elevation, rainfall, and type of soil. Vital activities come to a halt twice: (1) when it is extremely cold in winters and (2) due to long dry summers.

1.5.4 Southern Coastal Plains Floral Province

The vegetation in this province includes warmth-loving plants whose origin goes back to Africa or tropical Asia such as palm (date) and lotus trees. Their need for warmth and poor resistance to cold weather makes their distribution limited to the southern warm coastal belt.

Various types of acacia trees are found in this province, which cover the coastal plains and rolling hills uniformly at some places along with prosopis trees in other places. Vast forests of acacia and prosopis still remain in Tangeh Khiyal, Port of Magham to Port of Charak, near Port of Lengeh, and also between Minab and Jask.

From Jask to Chabahar and alongside Jagin and Gabrik, rivers and other seasonal rivers prosopis forests are seen, which unfortunately are being excessively exploited putting them on the verge of extinction.

Lotus trees have invaded the Irano-Touran and Zagros areas more than other species have. At the bottom of valleys and low plains on the south of Zagros, there are thick lotus forests, which are accompanied by nerium brushes in humid areas. Wild pomegranate trees also grow scarcely in some parts. In plains north of Chabahar and in the Dashte-yari Plain, various trees such as acacia and prosopis along with wild palm trees form a diverse vegetation cover.

1.6 Summarized Facts About Vegetation Cover

Much of the vegetation cover consists of forests, prairies, and fruit groves and orchards. Not very far back, Iran had 30,000,000 ha of forests, which accounted for 1/5 of its surface area; however, the number has been shrinking due to uncontrolled timbering. In general, 13.86 million hectares of land is presently covered with forests and 90 million hectares with prairies (9.3 million hectares with high quality, 37.3 with moderate to poor quality, and 43.3 with low quality).

The forest areas can be categorized as follows:

1. Northern forests that spread over 2.1 million hectares, also known as Hirkani forests
2. Other forests, which are divided into four groups:
   (a) Zagros (six million hectares)
   (b) Arasbaran (160,000 ha)
   (c) Irano-Touranian (4.5 million hectares)
   (d) Gulf (1.1 million hectares)

1.7 Weather Systems

1.7.1 Air Pressure and Wind Flow in Different Season

During winter, a high-pressure weather system forms over eastern Turkestan, Aral, and Caspian Seas to the north of the Black Sea and stretches to Carpathian and Pyrenean, and even close to the Azure Islands in the Atlantic Ocean. Woikof (quoted in Comprehensive Geography of Iran 1987) called this huge weather system the axis of the continents, which is the common line separating the southern and northern wind systems. On the north of this axis, the winds blow in the southwestern direction, while on its south the winds blow toward the northeast.

Winter Season: During winter, a high-pressure system forms over the Iranian Plateau whose effects are not as significant as those of the Central Asian wind systems. This weather system is of low-pressure type over the Black Sea, the Persian Gulf, the Oman Sea, and the Indian Ocean.

In general, it can be said that the direction of winds during winter is from land toward sea; therefore, a large front of cold weather comes from Central Asia and moves toward the west, crossing over Turkestan and Kyrgyzstan. A branch off of this weather system shifts toward the Caspian Sea but the mountains on the south and southwest block its flow. Only part of the moisture being carried by this branch gets through into Iran via the Manjil Valley.

The direction of the winds is usually perpendicular to the general trend of the mountains within the plateau, thus lowering its speed dramatically. The speed of winds in winter is about 1–3 m/s, that is, 3.6–10.8 km/h.

Spring Season: Spring is in fact a transitional period between the winter weather systems and summer weather systems. The length of spring is not the same in all locations; it is longer in the northwestern parts of the Iranian Plateau and Afghanistan as compared to the central and southern parts.

During spring, the high-pressure systems usually detour toward the west but do not totally disappear. In May, a series of low-pressure systems, which originate from the Azure Islands, appear in such a way that the slope direction of the system is not toward the Mediterranean Sea but the pressure (if any) flows from the west of Mediterranean to its east.

Summer Season: Low-pressure weather system dominates over southern Asia in summer. To the north of this low-pressure system, there is a high-pressure system originating from the Azure Islands. One of the effects of this high-pressure system is that a significant volume of oceanic weather starts moving toward central Europe, the Mediterranean, and Alpine Mountains.

At the same time, high-pressure over the Indian Ocean covers a large area, and weather systems flow toward the northeast.

The winds flowing toward the east come from two different oceanic sources. The weather systems flow from the northwest direction, and therefore the direction of major winds in summer shifts between the north and west.

As opposed to winter, all weather systems in summer flow from oceans toward lands, and since the direction of the summer winds almost parallels the trend of mountains in the Iranian Plateau, the mountains do not block the winds and hence the speed of summer winds is higher than that of the winter winds.

Winds blowing from the northwest are known as "North Wind" in the Persian Gulf coast and mostly flow at the speed of 10 m/s in the middle of summer. Although these winds lose their strength in August, they retain their direction. The strongest winds in summer blow in the Sistan area, which are known as the 120-day winds. The winds blowing from the north flow perpendicular to the northern mountain range and enter into the plateau via the deep Harrir-rud Valley and Morghab River Valley. This wind system starts flowing from late May and continues until late August or early September and reigns all over the Sistan area.

The average speed of the 120-day winds has been measured as much as 10 m/s or 36 km/h. The highest speed is usually between 72 and 90 km/h, but in one exceptional case a speed of 126 km/h was recorded. At the time these winds blow, the land becomes so dusty that all human activities come to a standstill.

The winds blowing from the northwest are of remarkable speed and effects, and the sandy deserts in Lut Kavir confirm this.

Based on the aforesaid facts, the Atlantic Ocean is the source for the summer winds in Iran, which can be termed as the seasonal Atlantic Ocean winds that proceed to almost Near East region.

The northwest winds are of basically oceanic origin but as they lose their moisture while crossing over the hot deserts in Syria and Iraq, they turn into dry winds.

The effect of seasonal Atlantic Ocean winds on the Near East and Middle East regions is more than that of the seasonal Indian Ocean winds.

The Seasonal Indian Ocean winds bring about torrential rains during summer over Pakistan and India, and since their direction is SW-NE, they just brush up the southeast corner of Iran (Chabahar) and no sign of the winds is seen in the other parts of Iran.

Autumn Season: Autumn is also a transitional period between the summer weather systems and winter weather systems, meaning the weather systems of summer transform gradually toward wintery systems.

1.7.2 Precipitation

Winter Season: Those winds blowing from low-pressure zones in Central Asia toward west remain dry if they do not cross over lakes and therefore bring no rainfall to the Near East region. Upon reviewing the rainfall statistics, it is realized that much of the precipitation in Iran occurs in winter, so the source of rainfall in Iran cannot be Central Asia. Then, what is the source for the winter rainfall in Iran?

Wiickmann (1922) described the low-pressure weather systems in Asia Minor and concluded:

> Since the nature of the rainfalls in Asia Minor and Iranian Plateau are completely similar to each other, it is most likely reasonable to accept that a few rain cloud systems come from low pressure system in Asia Minor that reach the Iranian Plateau and even India after crossing over north, south, and center parts of Iranian Plateau and produce winter rainfalls in India.

Of course, there might be other factors to produce such rainfall, and one of these is associated with the Caspian Sea.

There is no doubt that rainfalls over the Caspian Sea coast originate from this very large sea. As mentioned before, in winter the cold front from Central Asia crosses over the Aral and Caspian Seas and absorb significant volumes of moisture off of these large water bodies. Once these moist winds from the east-northeast hit the Alborz Mountains, they become condensed and produce significant rainfall.

During the winter, precipitation on most parts of the Iranian Plateau is in the form of snow, which remains on the grounds and peaks higher than 3,000 m and forms the water source for permanent rivers. It is to be noted that there are few places in Iran with permanent snow reserves or glaciers and thus the limited number of permanent rivers in Iran.

Spring Season: The average rainfall on the Caspian coasts during winter and spring is almost the same. For example, the rainfall in the Port of Anzali in winter and spring is 17 and 13% of the total annual rainfall, respectively. Although the rainfall in Central Iran and the surrounding mountains is noticeable in spring, the precipitation in local basins within the plateau in winter is much more than that of spring. Spring in the southern part of the plateau is shorter than in the northern, northwestern, and eastern parts.

Summer Season: Only on the Caspian coasts remarkable volume of rain pours down in summer, and there is almost no rainfall in the other parts of Iran during this season.

The seasonal Indian Ocean winds bring rain clouds to a very limited area in the southeast of Iran as the direction of these winds is from the southwest to northeast.

The extension of the summer weather system of the Azure Islands (which brings significant rainfall to Europe and even causes river inundation) reaches the Iranian Plateau. Although this system is of oceanic origin and is expected to bring rainfall to Iran as well, as it crosses over the hot deserts before reaching Iran, it loses its moisture due to the high temperature along the eastern coast of the Mediterranean (about 16–17°C) and hot deserts on the east of Iran (26–28°C).

The thickness of the Indian Ocean clouds moving toward Iran does not exceed a few hundred meters, and therefore they do not bring much rain; especially when they rise up to the altitude of 1,000 m, they lose their rain-producing potentials due to mixing with the northwestern weather systems.

Only 1/10 of the Iranian Plateau is affected by the seasonal Indian Ocean winds in summers, and the remaining 9/10 is influenced by the passage of the northwestern weather systems. In general, summer is a dry and arid season for the Iranian Plateau.

Autumn Season: The weather system in this season features a transitional state between the summer and winter seasonal winds, meaning the seasonal Indian Ocean winds gradually recede toward the southeast allowing procession of the Atlantic Ocean weather system into Iran, which produces 15% of the average annual rainfall.

The Caspian coast gets more rainfall in autumn than any other season, which in some places accounts for 60% of the total annual rainfall. The climate situation in the Iranian Plateau in autumn seems to be more or less the continuation of the summer climate.

1.7.3 Average Annual Rainfall

A brief overview of Iranian Plateau's Annual Rainfall map (Nabavi 1994) makes us realize that the Caspian coast is indeed where the most rainfall occurs. On the northern flanks of the Alborz Mountain range (Talesh area), the amount of rainfall can even reach 2,000 mm per year.

Next to the Caspian coast are areas in the western and northwestern part of Iran where the average annual rainfall is 500 mm. Moreover, the peaks of Sahand and Sabalan receive annually as much as 1,000 mm of precipitation.

The average annual rainfall in Iran decreases, moving from the northwest to the southwest, and matching the direction of rain-bearing winds blowing from the northwest to the southeast. The more these winds approach the southeastern parts, the more they lose their moisture, and the reduction of rainfall from the northwest to the southeast is explained.

1.7.4 Characteristics of Rainfall in the Iranian Plateau

Rainfall in the Iranian Plateau is as much noticeable as in the other arid and semiarid areas. Torrential rain could result in significant flash floods. Weather change may be accompanied by hailstorms (up to the size of a walnut) in some areas, which could damage orchards and agricultural plants.

1.7.5 Genesis of Dry and Wet Weather Cycles

The Iranian Plateau has experienced many droughts, which severely damaged agricultural production. Unfortunately not enough statistical data are available to help figure out the pattern(s) of these wet and dry weather cycles.

Andreas and Stolze tried to analyze these cycles based on the famines that happened in Iran during 1860–1880, and also in comparison with the weather pattern in India and Iran. They concluded that drought would happen every 10–11 years in Iran, which matches the 11-year solar eclipse cycle. It is to be noted that droughts were not the only reason for the loss of agricultural production but the swarms of locust and other pests were the other factors causing such losses.

The elevation of the Caspian Sea is presently 28 m below the mean sea level, but it experienced several fluctuations in the past. Bruckner (quoted in Seas and Lakes of Iran, Shahrabi 1994) studied and analyzed the writings of historians, explorers, and travelers between the years 880 and 1915 that documented these fluctuations and concluded that the climate in the Iranian Plateau underwent changes every 30 years. This 30-year cycle has been named in his honor in meteorological literature. Bruckner believed the fluctuation of the Caspian Sea is a function of precipitation onto this large body of water.

Some experts believe that the past climate of Iran suddenly turned arid. Therefore, the source of rainfall in the Iranian Plateau is attributed to effective factors from a distance of thousands of kilometers. The climatic conditions of Iran are not isolated or local, and just like other parts of the world, they are governed by intracontinental weather systems.

In general, the precipitation in central plateau is limited to 5–6 months and no sign of rain is seen for the rest of the year.

Since the rainfall is sometimes in the form of torrential showers, the resulting water either vaporizes or runs into marshy basins rapidly if no measure is taken to collect it. Ancient Iranians were aware of the significance of such supplies and built storage dams to collect the water to utilize over the dry season (Ministry of Energy 1973).

1.8 Aves

Iran has a wide range of bird species, and based on the statistical data, about 500 species have so far been identified, of which more than 350 species are reproducing in Iran (Darehshouri and Kasraian 1998). This diversity is due to the geographic location of Iran as it is situated between several biogeographic areas with specific bird species. Out of these 350 species, 250 belong to other geographic provinces where they evolved and then migrated to Iran and found habitat similar to their original ones to survive and reproduce.

In fact, close to 50 bird species that live in relatively cool plains and forest in the Caspian and Azerbaijan areas belong to the Asian and mild European vast woodlands and steppes (Darehshouri and Kasraian 1998). This part of Northern Iran constitutes the southern fringes of their habitats. Some of these birds include green woodpecker, dotted woodpecker, Divar-khazak, and Elikaei. About 40 bird species from the mild European deciduous forests are seen on northern and western Iran, and in forests within the Alborz, Azerbaijan, and Zagros areas. Some of these birds include wild pigeon, starling, blue-headed cyanistes, and bullfinch.

Another 30 bird species related to the warm Mediterranean bushlands live in scarce forests in Zagros and southern flanks of Alborz, which they have found to be suitable for survival. Some of these birds include Syrian woodpecker (garden woodpecker), black-headed cyanistes, and Cinereous bunting (Darehshouri and Kasraian 1998).

Only a few bird species belonging to warm steppes of Turkistan and southern Russia live in steppe-like environment in north and northeast of Iran.

Out of the 350 bird species that are reproducing in Iran, only 45 belong to these vast areas. The abundance and diversity of these species in dry and arid habitats are limited. Alborz heights, and Zagros to some extent, are the habitat for some birds, which are native of the mountain ranges in south of Eurasia. About 15 species belong to a widely spread faunal group that spreads from the southern Europe to the Himalayan Mountains. There are birds in the southern part of Iran with African or Indian origin whose distribution area is bordered with the coastal plains and southern flanks of Zagros.

1.9 Mammals

Due to the large extent of Iran and its location between Saudi Arabia, Caucasus, Turkistan, and India, and more importantly because of the unique nature of the Iranian Plateau, a wide variety of fauna is seen in Iran. Mammals around the world are categorized into 20 orders or 4,000 species. The mammals in Iran belong to 160 species related to 10 orders, which is high considering the area of Iran. The continent of Europe, with an area four times larger than Iran, has less mammal species (Darehshouri and Kasraian 1998).

Some of these mammals play a significant role in preserving the ecosystem, for example, Persian squirrel, Iranian fallow deer, zebra, wild goat (capra), and gerbil play important roles in reviving the oak forests of Zagros.

About 20% of the mammalian population of Iran are native to this land, among which Iranian fallow deer, zebras, capras, and Iranian squirrels can be cited. Some mammals and rodents have migrated to Iran from other places like Touran, and some mammals with origin of Touran entered Iran from the northeast including *Roobah Sardom Siah, Payka, Jard-e Nimrooz and some other rodents. Roobah Sardom Siah* exclusively settled on the southeast plains of Caspian, whereas *Payka* and other rodents spread furthermore inside Iran.

Baluchistan was the passageway for different mammals moving from India to Iran, but they could not spread far due to the cold weather and high elevation that acted as a barrier on the way of migration.

There are also some mammals living in the south and southeastern parts of Iran that are of African origin.

It can be stated that more than 50% of mammals in Iran are of Euroasian origin, 20% are native, 15% Indian, and the rest are African and those common in all of the aforementioned areas (Darehshouri and Kasraian 1998).

1.10 Mountains

Except the narrow plains along the Caspian Sea and the Persian Gulf and Khouzestan Region, Iran is considered as a highland with an average elevation of 1,000 m. The lowest point in Iran is located in salt plain northeast of Shahdad on the margin of Lut Kavir with an elevation of 350 m, and the highest point is Mount Damavand with a height of 5,774 m in the Alborz Mountain range. In fact, Iran is a land that consists of a central plateau with mountainous margin, and this mountainous margin includes four mountainous areas; northern (Alborz) mountains, Zagros Mountains, central mountains, and eastern mountains.

1.10.1 Northern Mountain Ranges

These mountains are divided into three sections: Alborz, Khorasan, and Azerbaijan mountains.

Alborz Mountains: This mountain range stretches for 950 km, and its width varies from 15 to 110 km (Orohydrography Map of Iran 1994). It covers an area of approximately 51,500 km^2. The Alborz Mountains begin at Ali-abad Mountains in Gorgan, extends like a bow along the south of the Caspian Sea and ends in Astara.

The Alborz Mountains act as a barrier between the Caspian Sea and the Iranian Plateau, obstructing the humidity from Caspian. This range includes several mountains such as Talesh, Guilan, Lahijan, Deylaman, and the major Alborz Mountains. The peaks of these mountains are covered with snow most of the year, where several permanent rivers originate (especially on the northern flank). The valleys of these rivers are narrow and deep. The highest peak in these mountains is Damavand with a maximum elevation of 5,774 m, which is the highest peak in Iran. There are many hot springs around this peak. Besides Damavand, there is another high peak in these mountains called Alam-kuh, which stands at 4,650 m (Jafari 2005).

The Alborz Mountains can be divided into three sections: western Alborz (from Astara to Sefid-rud valley), central Alborz (from Sefid-rud valley to Talar River and Firuzkuh), and eastern Alborz (from Talar River to Gorgan River and all the way to the state border between Khorasan and Mazandaran Provinces).

Azerbaijan Mountains (Northwestern Iran): These highlands start from the Ararat Mountains and extends through the Ghezel Ozan River valley, ending in Talesh Mountains. These mountain ranges include several peaks such as Arasbaran, Sabalan, Sahand and Boz-ghoosh, Mishou-dagh and Gharah-dash, or Takab-Shahindezh, which together cover an area of 63,000 km^2. These mountains are the source for several permanent rivers such as Aras, Talkheh-rud, and some tributaries of Ghezel Ozan. The highest peak within this mountain range is Sabalan with a height of 4,811 m that is permanently covered with snow (Jafar 2005).

The highest peaks in Arasbaran, Sahand and Boz-ghoosh, Mishou-dagh and Takab, or Gharah-dash are Kiyamaki-dagh (3,347 m), Sahand (3,707 m), Alamdar (3,155 m), and Takhte-belgheys and Salim-khan (3,332 m), respectively.

Khorasan Mountains (Northeastern Iran): These mountains are situated between the Gorgan and Tajan valleys, and their most important peaks are Kopeh-dagh and Hezar-masjed in north of Khorasan, and Aladagh, Binaloud and Shah-jahan in south of Khorasan. These mountains are mostly lower in height and are easy to pass through, which made it easy for the invaders from Central Asia.

This mountain range covers an area of 48,000 km^2 and extends as long as 531 km in the northern part (Comprehensive Geography of Iran 1987). The Hezar-masjed and Binaloud heights are separated by the Atrak and Kashaf-rud Rivers. The highest peak in the Khoransan mountain range is Binaloud (3,249 m). The mountains in northern Khorasan are the source of important rivers such as Atrak, Daroungar, Kashaf-rud, Kal-shoor, and Jajarm.

Due to the polar winds blowing from the north, these mountains experience cold winters and mild summers. There are scarce forests seen at the border of Khoransan with Turkmenistan. These mountains keep off Turkmenistan's dry summer and cold winter weather from entering into northern Khorasan.

1.10.2 *Zagros Mountains*

These mountains begin in Kordestan and extend to Khouzestan, Fars, and the southern coast. The length of this mountain range is about 1,000 km and its width around 200 km. These mountains are much more expanded, well ordered, and simpler (from stratigraphical standpoint) than Alborz. Most geographers call Zagros as "Great Jura" because there are well-ordered and compressed anticlines and synclines than those seen in Alborz, and look like what is seen in the Jura Mountains of Europe (Comprehensive Geography of Iran 1987).

Western and Southwestern Mountain Ranges: The highest mountains in Lorestan are Oshtoran-kuh (4,050 m) and Garin (3,645 m). In middle Zagros, the highest peak is Dena (5,200 m), which is the highest in all Zagros Mountain range. In northern Zagros (Kordestan and Kermanshahan), the highest mountains are Chehel-cheshmeh (3,173 m), Shahou (3,390 m), and Parou (3,357 m). The Alvand Peak in Hamedan Province with a height of 3,580 m is one of most beautiful mountains in Iran.

The Zagros Mountain blocks the moisture coming from the Mediterranean and Atlantic Ocean on their western flanks, resulting in the formation of huge reserves of snow and ice. The water produced due to melting of these reserves runs down through numerous valleys, such as Karoon, Karkheh, and Zayandeh-rud. The rivers originating from Zagros erode the mountains, especially in the northern and central parts. In some places, rivers like Zab and Sirvan cut the mountains sharply, while in the southern part, due to orderly folding of Zagros, rivers carve their valley parallel to the axis of the mountains (Orohydrography Map of Iran 1994).

The Zagros Mountain range is among the most magnificent mountains of the world. They gradually lose height moving from north to south, and the amount of rainfall and humidity drop accordingly, while the temperature rises. The part of Zagros that starts from north and extends to the Karoon basin is called moist Zagros, and the southeastern part of it is called as dry Zagros.

1.10.3 *Central Mountains*

These mountain ranges stretch in the Central Iran in a NW–SE direction (from Ghaflan-kuh and intersection of Zanjan and Ghezel Ozan rivers to Makran and slopes of Bazman Mountain).

These mountain ranges can be divided into smaller parts such as Ghaflan-kuh, Karkas-kuh, Shir-kuh, Kuh-banan, Laleh-zar mountains, and Jebal-barez. Some of the characteristics of these mountain ranges are low rainfall, sudden weather change, gusty winds, scarcity of permanent rivers, and turbulent seasonal rivers during severe rainfall.

The important peaks within these ranges are Karkas (3,895 m) on the southeast of Kashan, Shir-kuh (4,055 m) on the east of Yazd, Kuh-banan (3,775 m) on the north of Kerman, Laleh-zar (4,351 m) on the north of Baft, and Hezar with 4,465 m on the southwest of Rayen.

The length of the central mountain ranges up to the slopes of the Bazman Mountain is 1,460 km, and it covers an area of 143,000 km^2 (Jafari 2005; Orohydrography Map of Iran 1994). Some rivers like Halil-rud, Ghara-chay, Ghoh-rud, and Shoor originate from these mountains.

From a geological standpoint, most mountains in Central Iran are considered to be part of the Zagros structure because of their vicinity and climate and rainfall regimes. These mountains include those in Kordestan, Hamedan, Boroujerd, and south of Arak.

1.10.4 Eastern Mountains

These mountains form scattered spots on the west of Torbat-jam and stretch southwardly as long as 1,000 km between the Namak and Lut deserts on one side and Iran's border with Afghanistan and Pakistan on the other. They finally connect to the mountains on the north of the Oman Sea and the south of Baluchistan. These mountains cover an area of 133,000 km^2 (Jafari 2005; Orohydrography Map of Iran 1994).

Eastern-southeastern mountain ranges can be divided into the following sections:

(a) Southern Khorasan mountains known as Ghahestan (including mountains in Gonabad, Ferdows, Tabas, and Birjand areas)
(b) Mountains in Sistan-Baluchestan including Taftan, Koohak, Bazman, Nosrat-abad, Chah-shahi mountains, and mountains in Makran (including mountains in Sarbaz, Iranshahr, Fanooj, and north of Chah-bahar areas)

1.11 Seas

The Caspian Sea is located on the north and the Persian Gulf and Oman Sea are on the south of Iran.

1.11.1 Caspian Sea

The Caspian Sea, with an area of about 436,000 km^2, is the largest lake on Earth, whose elevation is 25–29 m below the open sea level. The depth of the Caspian Sea is 10–12 m in its northern part, 770 m in the middle, and up to 1,000 m in the southern part (Shahrabi 1994).

The Caspian Sea was a part of Para-Tethys and departed from the Black and Mediterranean Seas after the final stage of Alpine Orogeny in the form of its current shape forming a lake (Shahrabi 1994).

The amount of salinity in the Caspian Sea is 1/3 of that in open seas and oceans.

This sea had more than 70 names during past centuries, and it was named after the tribes living around this sea at the time. It is known as the Caspian Sea worldwide, except in Iran where it is called Khazar, the name of a tribe that used to live around the Caspian Sea in the sixteenth century.

More than two thirds of the volume of water of the Caspian Sea is located in its southern part, and considering the area of Iran, it can be stated that a significant part of this volume is within the Iranian territory.

The water sources feeding the Caspian Sea are (Shahrabi 1994)

- Surface runoff, streams, and rivers, of which 95–97% originate from rivers in Russia such as the Volga River, and only 2–3% is supplied by the rivers from Iran
- Annual rainfall that could raise the water level in the Caspian Sea by 125–135 cm
- Underground water and springs, which supply 30% of the water fed into the Caspian Sea

The Caspian Sea spreads more than 10° in geographic latitude, which results in a considerable temperature difference between its north and south ends. The average annual temperature rises when moving from northern part to south. The average annual temperature is 8 °C in the northern part and 18 °C in the southern part (Shahrabi 1994). The northern part of the Caspian Sea is mostly frozen during winter as opposed to the southern part that never experience glaciation.

Taking into account the significant position of the Caspian Sea between Asia and Europe, this sea has always been important to the people of the region and has been used as a route for sea transportation. Many biological species including fishes, seals, and crabs live in the Caspian Sea. It also contains large reserves of oil and gas, and in fact it is the second most important oil and gas reserve after the Persian Gulf (Ghorbani 2009a).

1.11.2 Persian Gulf

The Persian Gulf is a marginal sea with a length of 900 km and width of 240 km, whose depth varies from 35 to 100 m (Shahrabi 1994). It is located on the south of Iran and is connected to the Oman Sea and Indian Ocean via a narrow strait with a width of 60 km, called Hormuz. This gulf formed 35 million years ago and was much larger than it is today.

The Persian Gulf features a dry and subtropical climate, and due to its position between areas with continental climate, it experiences many seasonal changes. The temperature in summer reaches up to 50 °C. The dominance of northerly winds on its coast exposes it to permanent desert winds.

The deepest point of the gulf is near Ras al-Musandam (182 m), and the shallowest point is 30 m. In general, the depth of 1/3 of the gulf does not exceed 36 m (Shahrabi 1994). The volume of water filling the gulf is about 13,000 billion cubic meters, which flows from the northwest toward the southwest.

Much of its water is salty and only a small amount of water coming from the Zagros Mountains is fresh, but generally its degree of salinity is less than that of the oceans (Shahrabi 1994).

Historical evidence shows that around 2,600 B.C. sailors were living along the coasts of the Persian Gulf. It houses many islands that once were in the territory and jurisdiction of Iran, but only ten of them are currently part of the country. They include Larak, Hormuz, Qeshm, Hengam, Kish, Lavan, Khark, Lesser Tonb, Greater Tonb, and Abu Musa.

The Persian Gulf is rich in different flora and fauna. One of Iran's non-oil exports is the fishery products from this gulf. The gulf contains the largest oil and gas reserves of the country that have been exploited till date.

1.11.3 Oman Sea (Makran Sea)

The Oman Sea is part of the Indian Ocean and the only open seaway of Iran. The Persian Gulf is connected to the open seas in the world via the Strait of Hormuz and Oman Sea. This sea is bordered by Iran in the north, and Pakistan and India in the east, Arabian Peninsula in the west, and Indian Ocean in the south. The length of the Oman Sea from the Strait of Hormuz to Deccan is about 600 km, and it covers an area of 900,000 km^2. It is relatively deep and its depth near Chabahar is 3,398 m. The length of the coastline of the Oman Sea in Iran is 784 km.

Tropic of Cancer passes from the north of this sea, which makes the Oman Sea one of the warmest seas in southwest Asia. The maximum temperature of the surface of the sea reaches 33 °C in summer, and the minimum temperature is 19.8 °C (Shahrabi 1994).

There are many small local bays along the coast of the Oman Sea but are not suitable for large ships to dock because they are shallow and sandy. The water of the Oman Sea is somehow salty, and the tidal system is as irregular as it is in the Persian Gulf.

Due to dry weather, extreme heat, lack of rainfall, and lack of vegetation within the basin of the Oman Sea, permanent rivers are limited, and most of the rivers are seasonal with low or no currents in most of the year.

1.12 Lakes

Most lakes in Iran are the remnants of once a large sea, parts of which turned into swamps and parts still exist. The most important existing lakes are Orumiyeh, Qom Namak, Bakhtegan, Gav-khouni, Jazmourian, Zarivar, Maharlou, Parishan, Tar, and Hamoun.

Lake Orumyeh: This lake, with an area of more than 5,000 km^2, is one of the shallowest lakes in the world, which is located in the Western Azerbaijan Province, on northwestern Iran. It is 140 km long and its width varies from 15 to 50 km. The water is supersaturated by salts (23% almost twice as salty as the oceans), and thus has no fish species. The pH of water in this lake is between 7.2 and 7.6 (Shahrabi 1994).

Many aquatic and bird species live on the mud flats around the lake including white pelicans and flamingos. The most significant creature living in this lake is a rare species of crustaceans called *Artemia salina* that feeds on green algae.

Three well-known rivers called Talkheh-rud (160 km long), Zarrineh-rud (240 km long), and Simineh-rud terminate into this lake.

Qom Namak Lake: This lake covers an area of more than 30,000 km^2 located 120 km south of Tehran. The area of the lake immensely varies due to the changes in water input and seasonal conditions. Two-thirds of the area of the lake is covered with irregular polygonal salt sheets, and in some cases with salt humps measuring 6–20 cm in thickness (Shahrabi 1994).

There are six islands within this lake, five of them are small and one (Kuh-e Sargardan) large with an area of 100 km^2. They are made up of volcanic rocks.

The main water sources feeding this lake are Gharah-chay, Jajroud, and a branch of Rud-e Shoor.

Lake Bakhtegan (Neyriz): This lake, with an area of approximately 800 km^2, is located 160 km east of Shiraz. The length of Lake Bakhtegan is 77 km, and its width is 10 km (Shahrabi 1994). It is shallow so that the deepest point of the lake is 1.10 m. Since the main part of the lake is covered with salt, the water of this lake is very salty.

There are a few islands and peninsulas within this lake whose area depends on the annual rainfall rate. The major water source of this lake is the Kor River, which originates from the Baraftab and Mousa-khani Mountains. The elevation of this lake is 1,558 above the mean sea level (Shahrabi 1994).

Lake Gav-khouni (Gav-khouni Swamp): This lake or swamp is shallow and covers an area of about 480 km^2. Its elevation is 1,470 m (Shahrabi 1994). Gav-khouni is located southeast of Isfahan in Central Iran and is fed by the Zayandeh-rud River. Because most rivers and streams running into this swamp are dry during much of the year, the size of the area of the swamp varies in different seasons. The swamp is covered with canebrake, and the water of the lake is salty.

Arid climate reigns over this area, causing the evaporation rate to surpass water input, and therefore large areas surrounding the swamp are covered with salt marshes and sand dunes with diverse plant species. Of these plants, tamarisks and soaproots are the most well known. The name Gav-khouni comes from two ancient words, Gav (meaning large) and khooni (meaning well), which can be translated to "large well."

Lake Jazmourian (Jazmourian Swamp): This lake or swamp covers an area of approximately 1,100 km^2, and its elevation is 350 m (Shahrabi 1994). Due to specific climatic condition and evaporation, large part of this lake (59%) turns into salt marsh and mud flat for much of the year. Its water is also supersaturated with salt in dry seasons.

This shallow lake is bounded between two mountains, and two rivers of Bampour and Halil-rud are the main water supply to this lake. It has been named after a tiny local thorn plant.

Lake Zarivar: This lake is located 8 km west of the city of Marivan in the Kordestan Province. Since the water resource flowing into this lake comes from melting of snow reserves, this lake has freshwater. It covers an area of 8.5 km^2 and is surrounded by relatively high mountains. The elevation of Lake Zarivar is 818 m above the sea level. Its deepest part is 50 m and the average depth is 10 m (Shahrabi 1994).

The Cham River, along with other streams and springs, supplies the water of this lake. Dating studies on samples from this lake have shown that Lake Zarivar was formed around 200 B.C. in a cold climate condition (Shahrabi 1994). The weather of the area is cold in winter, mild to cold in autumn and spring, and mild in summer. For its beautiful sceneries and green nature, many tourists visit this area.

Lake Maharlou: This lake is located about 18 km southwest of Shiraz. The elevation of the lake is 1,460 m above the sea level and its maximum depth is 3 m (Shahrabi 1994). The evaporation rate is high due to its shallow depth. A part of its bottom is covered with salt. Water exists only in the central and northern parts of the lake with shallow depth and high salinity. The main water resources supplying this lake are rivers and streams coming from the northwest and southeast. Its surface area varies in different seasons and is a function of rainfall. Besides the high evaporation rate, the gypsum deposits of Sachoon Formation and also two salt domes on the east of this lake have significant impact on the extreme salinity of this lake.

Lake Parishan: Lake Parishan or Lake Famour is located 15 km southwest of the city of Kazeroon. The name Famour has been derived from the Famour Mountain on the northeast of the lake. It is a freshwater lake, and most of its water comes from the streams running down the Famour Mountain, and the rest is supplied by groundwater resources. The shallow depth of the lake causes parts of it to be covered with evaporative deposits in dry season. Lake Parishan covers an area of 43 km^2 and it is positioned in a shallow depression whose elevation is 820 m above the sea level (Shahrabi 1994).

Lake Tar: This lake is located 30 km east of Damavand city, and it is one of the mountainous freshwater lakes. Its elevation is 2,900 m above the sea level, and it is accessible through the Damavand–Tar road.

The length of the lake at its longest stretch is 1.3 km and its width is 400 m (Shahrabi 1994). The origin of the water resources pouring into this lake are springs in Gharah-dagh, Siyah-chal, and Shah-neshin mountains in the north, and seasonal streams in the south, which partly pour into Late Tar and partly feed the Tar River.

Lake Hamoun: This lake, with an area of about 1,800 km^2, is the largest lake in east of Iran and consists of three main basins of Yozak, Saberi, and Hirmand, which are separated in low rainfall seasons and during the 120-day winds but regroup into one body in high rainfall seasons (Shahrabi 1994). Hirmand is in Iran, and a part of the Saberi basin is in Afghanistan. Many rivers and streams discharge water into Lake Hamoun, but among them only the Hirmand River has the highest flow rate.

1.13 Watersheds

The network of running water in Iran can be divided into four areas with respect to natural water flows, course condition, slope of water resources, estuaries, and the basins (The Balance Report on Water Resources of Iran, Office of Water Resource Research Organization 2005–2006):

- North intake basin: The water flow direction within this basin is usually toward the north pouring into the Caspian basin.
- Western and southern intake basin: The water flow direction within this basin is toward the west and south, and ultimately ends in the Persian Gulf and Oman Sea.
- Lateral intake basin: The water flow direction is toward depressions and confined basins adjacent to Iran on the east or north side.
- Central intake basin: This includes confined basins within the central Iranian Plateau. Most of the water supply of Iran belongs to this area, but due to its climatic conditions the area is considered as dry with low rainfall. The course of the water flow within this area is controlled by the mountain ranges surrounding the plateau. Alborz and Zagros are the water distributor inside and outside the Iranian Plateau.

Considering the characteristics and geographical position of each one of these intake basins, 12 watersheds have been defined that gather water discharge from the aforementioned intake basins. Two of them are major basins located in north and south, another two basins are lateral, and the remaining eight basins are of confined type. Below is a list of these basins:

Isfahan and Gav-khouni basin, Jazmourian basin, Persian Gulf and Oman Sea basin, Lake Orumiyeh basin, Lake Hamoun basin, Caspian Sea basin, Lut Kavir basin, Gharaghorom basin, Namak Kavir (Dashte Kavir) basin, Neyriz and Shiraz basin, and Yazd and Ardestan basin.

1.14 Rivers

Due to the mountainous characteristics in parts of Iran, many present-day large and small rivers have been the source of life in various parts of deserts and mountains in Iran for a very long time.

In general, the rivers in Iran are divided into two categories (Afshin 1994):

- Caspian Sea
- Persian Gulf

1.14.1 Caspian Sea Category

Most rivers in this category originate from the northern slopes of the Alborz Mountains. Because of the short distance from Alborz to the Caspian Sea, these rivers do not stretch very long.

Aras River: This river is 800 km long and originates from the Hezar Berkeh (thousand pounds) mountains in Turkey, which forms part of Iran's border with Azerbaijan and Armenia, and discharges into the Caspian Sea around a place called Salian. Some of its branches flowing into Iran are Maku, Khoy, and Marand rivers.

Sefid-rud: It is called Armdus in Greek, Ghezel Ozan in Azerbaijani, and Nahrol Abyaz in Arabic literatures (Afshin 1994). In the medieval era, the whole river was called Sefid-rud, but nowadays only the part of the river running between Manjil and the Caspian Sea is called as Sefid-rud. This river originates from Chehel-cheshmeh in Kordestan. Its important tributaries are Gharanghou, Miyaneh, Hasht-rud, Zanjan-rud, Shal, Gadiv (which come from the Khalkhal area), and Shahrood River (which comes from the Taleghan and Alamut area that joins Sefid-rud before reaching Manjil Gorge). The whole river system eventually is called Sefid-rud and discharges into the Caspian Sea at a point called Hassan-kiyadeh.

Seh-hezar River: This river originates from Salim-bar and Kandovan, and after feeding the Tonekabon area, it pours into the Caspian Sea.

Chaloos River: It is 80 km long and originates from the northern slope of the Taleghan and Kandovan mountains. After two tributaries of Zanoos and Mikhsaz join the Chaloos River, it pours into the Caspian Sea.

Haraz River: This river originates from the Lar Valley in north of Damavand. Its significant tributary is the Noor River, which joins Haraz at a point called Kiyalvand, and once it passes Amol, it pours into the Caspian Sea in the Babolsar area. The Haraz River is 150 km long.

Babol River: This river is almost 78 km long and originates from Savad-kuh. After it passes west of Babol city, it pours into the Caspian Sea in the Babolsar area.

Talar River: This river also originates from Savad-kuh, and once it passes the city of Ghaem-shahr, it pours into the Caspian Sea. The Talar River is 150 km long.

Tajan River: This river originates from around Rud-bar-e Hezar Jarib and stretches for 120 km long. Once it passes east of Sari, it pours into the Caspian Sea at a place called KhazarAbad (Farahabad).

Neka River: This river originates from Shah-kuh in the Gorgan area, and its well-known branch is called Shoor-ab. The Neka River is 100 km long and pours into the Gorgan Bay.

Gorgan River: This river originates from the Aladagh mountains, and it is 300 km long. Some of its significant tributaries are the Nardin River that originates from the Jajarm and Ab-e Garm River, which joins the Gorgan River once it passes Sangsar.

The Gorgan River splits into two branches once it passes Kooklan and old city of Jorjan. One branch pours into Khajeh Nafs, and the other pours into the Gorgan Bay.

Atrak River: This river originates from the Hezar-masjed mountains and pours into the Hossein-gholi Bay in southeast of the Caspian Sea. Different sources cite the length of Atrak River as long as 500–600 km.

1.14.2 Persian Gulf Category

The most important rivers in this category are as follows:

Gamasiyab River: This river originates from the Chehel-nabaleghan mountains in Nahavand, and its tributaries are the Mahidasht, Ghara-sou, and Diner rivers, which join Gamasiyab in Kermanshahan. Southwardly from Sirvan this river is called Seymareh, and at a point called Pay-e Pol its name changes to Karkheh. Its water discharge is collected behind the Karkheh Dam and a part of it feeds the western part of the Khouzestan Plain and the rest joins the Tigris River in Iraq via the marsh lands along the Iran-Iraq border. Gamasiyab is 200 km long and 20–50 m wide. Its depth varies from 0.5 to 2 m (Afshin 1994).

Karoon River: This river (including its tributaries) is the longest river of Iran with a total length of 890 km and is the only river in Iran suitable for shipping. Its width varies from 250 to 900 m in different locations (Afshin 1994). Its main origin is in the Zard-kuh Bakhtiyari mountains. The most important tributaries of this river are Dez or Dezful River, which joins Karoon at a place called Band-e Ghir. Once Karoon passes the city of Ahwaz, it splits into two branches: Bahman-shir, which pours in the Persian Gulf, and Karoon, which joins the Arvand-rud River. Karoon is a permanent river, and its annual volume of water running through this river has been measured as much as 24 billion m^3 (Afshin 1994).

Jarrahi River: This river is one of the most important rivers pouring in the Persian Gulf and Oman Sea basin, and most of it flows in the Khouzestan Province. The Jarrahi River originates from the mountains around Khouzestan and has two tributaries: Ab-e Zolal and Maroon, which join each other at a place called Kalat and form the Jarrahi River. This river is a permanent river with the highest flow rate in March. Its width varies from 100 to 200 m, and the average depth is 2–3 m. The length of the Jarrahi River from Maroon to its estuary is 520 km (Afshin 1994).

Hendijan River: It is also called as Zohreh River and originates from Kohkilouyeh. Its major tributaries are Ab-e Shirin or Kheyr-abad and Ab-e Shoor or Shoolestan. The latter pours in the Persian Gulf once it passes Hendijan.

Dalaki River: This river originates from the mountains surrounding Dasht-e Arjan and has two tributaries: Dalaki and Shapour, which join each other at a place called Koolan and pour in the Persian Gulf once they pass Rohaleh. The Dalaki River is 225 km long, and the size of the area of its watershed is 5,620 km^2 (Afshin 1994). This is a permanent river, and most of its water comes from the carbonaceous springs. Due to the presence of salty geological formations on the course of this river, the water at the end of the Dalaki River tastes salty and is undesirable.

Karkheh River: One of the most important rivers in the Persian Gulf and Oman Sea basin is the Karkheh River, but it does not directly pour into the Persian Gulf. This river has several major tributaries, among which Gamasiyab and Gharah-sou are the most important. The overall length of the Karkheh River is 875 km, and its width varies from 150 to 200 m. The average depth of this river is 4–6 m (Afshin 1994). It is a permanent river that feeds many towns and cities alongside its course.

Halil-rud River: This river is one of the most important rivers within the Jazmourian basin, and also one of the rivers with most amount of water in Kerman Province. The Halil-rud River collects water coming from the southern slopes of Hezar, Bid-khan, and Barez mountains, and after feeding the towns and villages alongside, it pours in the Jazmourian basin. The length of this river is 440 km, and the size of the area of its watershed is 8,300 km^2, which expands to 20,000 km^2 at the end of the course (Afshin 1994). The Halil-rud River is a permanent river due to the presence of snowy heights and mountain rainfall in the northern part of the basin.

Mehran River: This is one of the rivers within the Persian Gulf and Oman Sea basin, which originates from the mountains south of Lar and pours in the Persian Gulf at a point north of Qeshm Island. The Mehran River is a torrential river and lacks significant base flow as it dries out in summer. The water is salty and undrinkable. The length of the river is 380 km, and the area of its surface is 8,400 km^2 (Afshin 1994).

Shoor River: This river originates from the Darab mountains and pours in the Persian Gulf at a point north of Qeshm Island. It is 45 km long, and the size of the area of its watershed is 80 km^2 (Afshin 1994). This is a permanent river, and its water at the end of the course tastes salty because it flows through salt-bearing geological formation on its journey to the Persian Gulf.

Minab River: It is one of the most important and largest rivers in the Hormuzgan Province, which originates from the mountains in east of Darab. It pours in the Persian Gulf at a point north of Strait of Hormuz. The Minab River is 300 km long, and it is a permanent river (Afshin 1994). The water regime of this river is controlled by rainfall and flood water.

Zayanedeh-rud River: This river is one of the most important and largest rivers within the central basins of Iran with massive amounts of water. It drains Isfahan and Gav-khouni watershed basin and ends in the Gav-khouni swamp. Zayandeh-rud is formed by the joining of springs, streams, and rivers originating from the eastern slopes of Zagros and high mountains of Zard-kuh Bakhtiyari (one of the highest points is 4,221 m above the mean sea level). Its course is meandering and flows through high mountains and as it enters the Isfahan Plain, its gradient decreases. After it feeds the Isfahan Plain, it pours in the Gav-khouni swamp. The length of the Zayandeh-rud River is 405 km, and the size of the area of its watershed basin is 31,000 km^2 (Afshin 1994). The water regime of this river is rainy and torrential.

There are some other rivers that originate from sources outside Iran, and some of these are given as follows:

Hirmand River: This river originates from the Kuh-e Baba Mountain in Afghanistan, and it is 1,100 km long. One of its major tributaries is called Arghandab. Most of the Hirmand River flows in Afghanistan, and only the last part of the river flows in Iran. Once it feeds the Sistan Plain, it pours in Lake Hamoun. This river is a vital vein for the people of this area. Its depth is about 3–4 m, and its width varies from 200 to 900 m (Afshin 1994).

The rainfall is low in the Hirmand watershed, and most of its water is supplied by melting of snow on the Hindu Kush Mountains, and therefore its water regime is irregular but permanent.

Harir-rud River: This river also originates from the southern slopes of Kuh-e Baba Mountain in Afghanistan, and once it feeds the city of Herat, it flows toward the north. The Harir-rud River forms part of the border between Iran and Afghanistan.

There are some rivers originating from inside Iran but their estuaries are outside Iran, and some of these are given as follows:

Sirvan River: The name of the river is in fact Diyaleh, and only the part of the river flowing in Iran is called Sirvan. The Diyaleh River originates from Kordestan, and it joins the Tigris River at a point southeast of Baghdad in Iraq.

Zab Saghir River: This river is formed by joining of small rivers originating from heights south of Lake Orumiyeh. It is 400 km long, and flows toward the southwest of Iran, eventually entering Iraq and joining the Tigris River.

1.15 Deserts

The Central Iran basin has plains filled with Quaternary or Plio-Quaternary deposits and sediments. The average elevation of these plains is 1,200 m above the sea level. Most of these plains are seen on central and Eastern Iran. In the central parts of these internal basins, there are large barren lands with low amounts of water and high degree of evaporation mostly in the form of salt flats (playas), which are called "Kavir."

The total area of the deserts of Iran is 34 million hectares, which is almost 20% of the country's surface area. The largest deserts of Iran are Lut and Namak located in east and center of Iran.

There are two types of terranes seen in deserts (1) relatively large and regular lands and (2) unstable lands.

According to Gabriel's theory (Gabriel 1934), unstable lands (deserts) are mostly seen close or adjacent to mountains, but large deserts are located in the center of basins.

Sven Hedin (1910) has mentioned in a report that the surface of the deserts consists of 10 cm of clay and mud, underlain by a layer of salt with a thickness of 15 cm, and ultimately there is water at the depth of 1 m or below.

Fig. 1.2 Zagros area forests (Photographed by Alireza Amrikazemi)

Hedin compares the deserts in Iran to a sea that has been overturned, meaning as moving down in depth, the amount of water increases. Unfortunately the water resource in deserts is salty and bitter, and it is not potable (it is to be noted that these shallow water supplies have been trapped within compact and impermeable clay deposits).

The thin layer covering the surface of desert becomes sticky in the rain, which makes traveling on these deserts hard. The salts in deserts are mainly common salt (NaCl). But other types such as $CaSO_4$, $CaCl_2$, $CaSO_4.2H_2O$, $MgCl_2$, and a little KCl are also found in deserts. Some deserts are rich in salt mineral deposits like Khour-Biyabanak, Mighan in Arak, and desert south of Varamin.

There are more than 60 deserts in Iran, and the most important deserts are Lut, Khour-Biyabanak, Jandagh, Abar-kuh, and Damghan deserts.

It is assumed that the formation of deserts in Iran was the result of final stages of Alpine Orogeny that caused the Iranian Plateau to lift up (especially in the southern parts) and eustatic fall of sea level in Late Miocene and Early Pliocene.

Also, the retreat of glaciers in areas north of Iran 12,000–15,000 years ago resulted in increased evaporation and decreased rainfall in the central parts of Iran, which in turn led to the transformation of lakes to deserts. This matter indicates that most deserts in Iran were once lakes. At present, there are no evidences to verify whether Lut Desert was once a lake. In some parts, the Lut Desert was probably filled with water.

Figures 1.2, 1.3, 1.4, 1.5, 1.6, 1.7, 1.8, 1.9, 1.10, 1.11, 1.12, 1.13, 1.14, 1.15, 1.16, 1.17, 1.18, 1.19, 1.20, and 1.21 show some natural sceneries of Iran.

Fig. 1.3 Ancient rural dwellings in Meymand built in tuff and breccia and covered with hard agglomerate (northeast of Meydook)

Fig. 1.4 Naghshe Rostam, Near Shiraz, Far Province (Photographed by Alireza Amrikazemi)

Fig. 1.5 A view of Takht-e Soleyman Spring and dried spring of Zendan-e Soleyman and ancient city of Ganjeh near Takab (used to be called Shiz in Arabic, but it is known today as Takht-e Soleyman) (Photographed by Alireza Amrikazemi)

Fig. 1.6 A view of Dena Mountains, Yasouj area, within Zagros Mountain range (Photographed by Alireza Amrikazemi)

Fig. 1.7 Sand dunes around Mesr Village (Jandagh suburbs) (Photographed by Mohsen Iranmanesh)

Fig. 1.8 A view of forest cover and rice fields in Lahijan (Photographed by Negar Ghorbani)

Fig. 1.9 Ripple marks on sands around Mesr Village (Jandagh suburbs) (Photographed by Mohsen Iranmanesh)

Fig. 1.10 A view of Zayandeh-rud River in Isfahan (Photographed by Mohsen Iranmanesh)

Fig. 1.11 Pol-e Dokhtar historic bridge in Miyaneh (Photographed by Mohsen Iranmanesh)

Fig. 1.12 Fist-shaped island in Lake Orumiyeh

Fig. 1.13 A view of young geologic formations in the form of island in Lake Orumiyeh

Fig. 1.14 A view of young faults (along the roadside of Zanjan-Tabriz Freeway) (Photographed by Mohsen Iranmanesh)

Fig. 1.15 A view of young faults (along the roadside of Zanjan-Tabriz Freeway) (Photographed by Mohsen Iranmanesh)

Fig. 1.16 Mud volcano (Chabahar) (Photographed by Alireza Amrikazemi)

Fig. 1.17 Sandstone layers within Upper Red Formation with cranes' nests on top (Location) (Photographed by Mohsen Iranmanesh)

Fig. 1.18 Shivand Waterfall in the Zagros area

Fig. 1.19 Margoon Waterfall in the Zagros area

Fig. 1.20 A view of erosion effects on Mishan Formation in Qeshm Island

Fig. 1.21 A view of prairies and forests in the northern part of Iran

Chapter 2
A Summary of Geology of Iran

Abstract This chapter presents an abridged description of the geology of Iran covering the following areas:

- Stratigraphy of Iran from Late Precambrian to Quaternary along with a brief introduction to various lithostratigraphic formations
- Structural units of Iran, orogenic phases, and major faults that played a significant role in the geologic history of the country
- Magmatic activities of Iran, including igneous phases, intrusive and extrusive rocks, and classification of ophiolitic complexes

The chapter provides the reader with a good knowledge about general geological characteristics of Iran in an abridged form.

Keywords Geology of Iran • Faults of Iran • Stratigraphy of Iran• Magmatism of Iran • Structural units of Iran • Orogenic phases of Iran • Basement of Iran

From a global tectonic point of view, Iran is part of the Alpine–Himalayan orogenic belt that extends from the Atlantic Ocean to the Western Pacific. Most European and Asian geologists believe that this belt represents the great Tethys sea once located between two large continents, Gondwana and Laurasia, during the Paleozoic–Mesozoic eras.

For a more detailed information on the geology of Iran, refer to the geology of Iran (Ghorbani 2012a).

2.1 Structural Divisions of Iran

Iran is divided into several structural units, each characterized by a relatively unique record of stratigraphy, magmatic activities, metamorphism, orogenic events, tectonics, and overall geological style. The systematic geological studies in Iran started in the late 1960s with the establishment of the Geological Survey of Iran. The tectonic and

M. Ghorbani, *The Economic Geology of Iran: Mineral Deposits and Natural Resources*, Springer Geology, DOI 10.1007/978-94-007-5625-0_2,

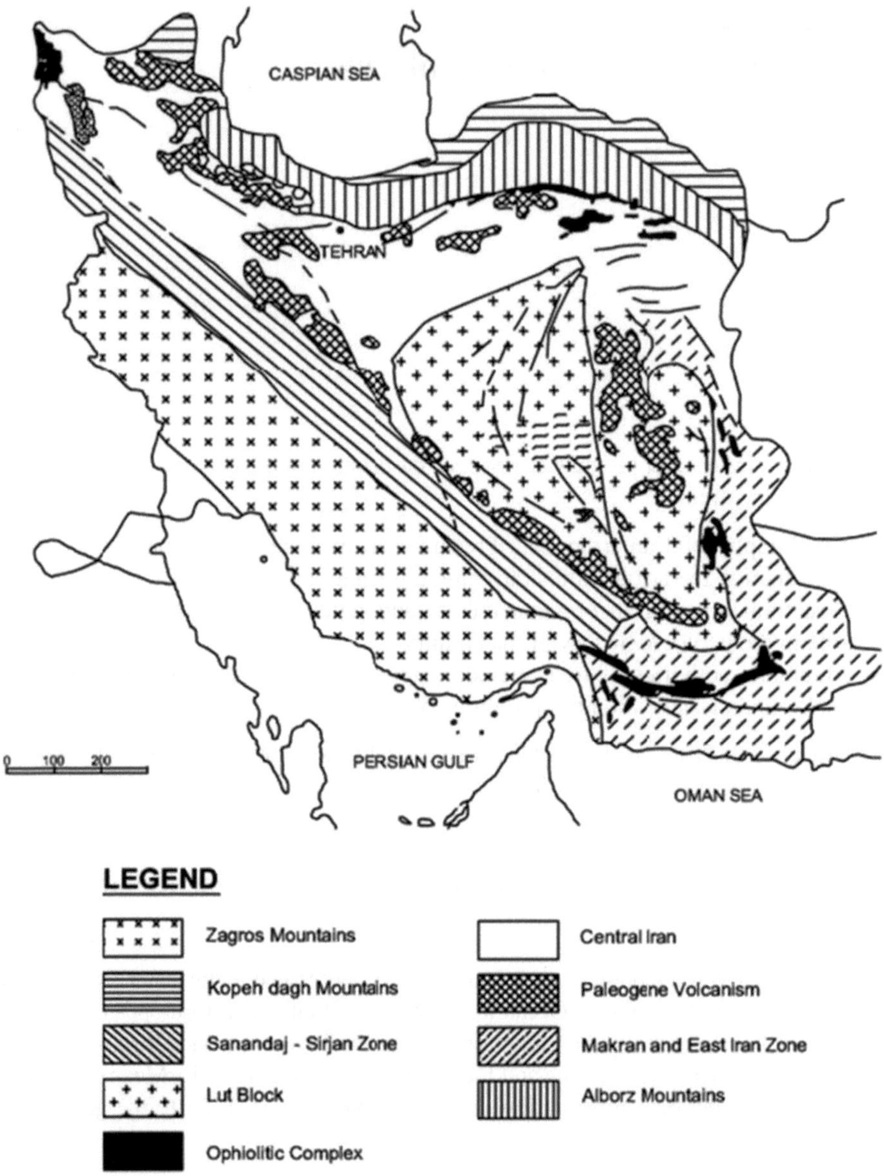

Fig. 2.1 Structural zones of Iran (Stöcklin 1968)

structural setting of Iran in the Alpine–Himalayan orogenic belt, and the structural evolution of Iran, has been the focus of many studies.

Using mostly the NIOC database, Stocklin and Nabavi (1973) were the first to publish the "Tectonic Map of Iran." They divided Iran into 10 structural zones (units) based on certain geological features (Fig. 2.1). This structural division remained a reference for Iranian geologists for almost three decades. However, the new observations and findings require a revision to this structural scheme. Following this structural division by Stocklin and Nabavi, some other structural divisions were presented, which are cited in the following section related to Central Iran. These newer structural schemes were mostly derived and inspired by the very first structural division presented by Stocklin.

In recent years, new interpretations and models have been offered regarding the geological setting of Iran (Nabavi 1976; Eftekharnejhad 1980; Nogol-e-Sadat 1993; Alavi 1993; Houshmand-Zadeh 1998; Aghanabati 2004). The following is a combined summary of the available data on various structural zones of Iran.

2.1.1 Central Iran

Located as a triangle in the middle of Iran, Central Iran is one of the most important and complicated structural zones in Iran. Here, rocks of all ages, from the Precambrian to the Quaternary, and several episodes of orogeny, metamorphism, and magmatism can be recognized. There is not a consensus regarding the boundaries of Central Iran.

According to Stocklin (1968), Central Iran is bordered by the Alborz Mountains in the north, Lut Block in the east, and Sanandaj–Sirjan in the south-southwest, whereas Nabavi (1976) considers the northern part of the Lut Block as a part of Central Iran. Nogol-e-Sadat (1993) extends the frontiers to the northeast as well as Eastern Iran and presents new subzones in his classification. Based on tectono-sedimentary features, Aghanabati (2004) believes that Central Iran and Sanandaj–Sirjan are parts of the central domain (Figs. 2.2, 2.3, 2.4, and 2.5).

2.1.2 Sanandaj–Sirjan

This zone is located to the south-southwest of Central Iran and the northeastern edge of Zagros range. In north and northeast, this zone is separated from Central Iran by depressions like Lake Orumiyeh, Tuzlu Gol, and Gavkhouni and faults like Shahr-e-Babak and Abadeh, and to the south-southwest by the main thrust fault of Zagros. A striking feature of this zone is the presence of immense volumes of magmatic and metamorphic rocks of Paleozoic and Mesozoic eras.

As far as the trends, and particularly the folding style, are concerned, some researchers consider the Sanandaj–Sirjan Zone to be similar to Zagros; however, considerable differences exist in rock types, magmatism, metamorphism, and orogenic events. There are some similarities between Sanandaj–Sirjan and Central Iran.

2.1.3 Zagros

This zone extends from Bandar Abbas in the south to Kermanshah in the northwest and continues through to Iraq. It is in fact the northeastern edge of the Arabian plate.

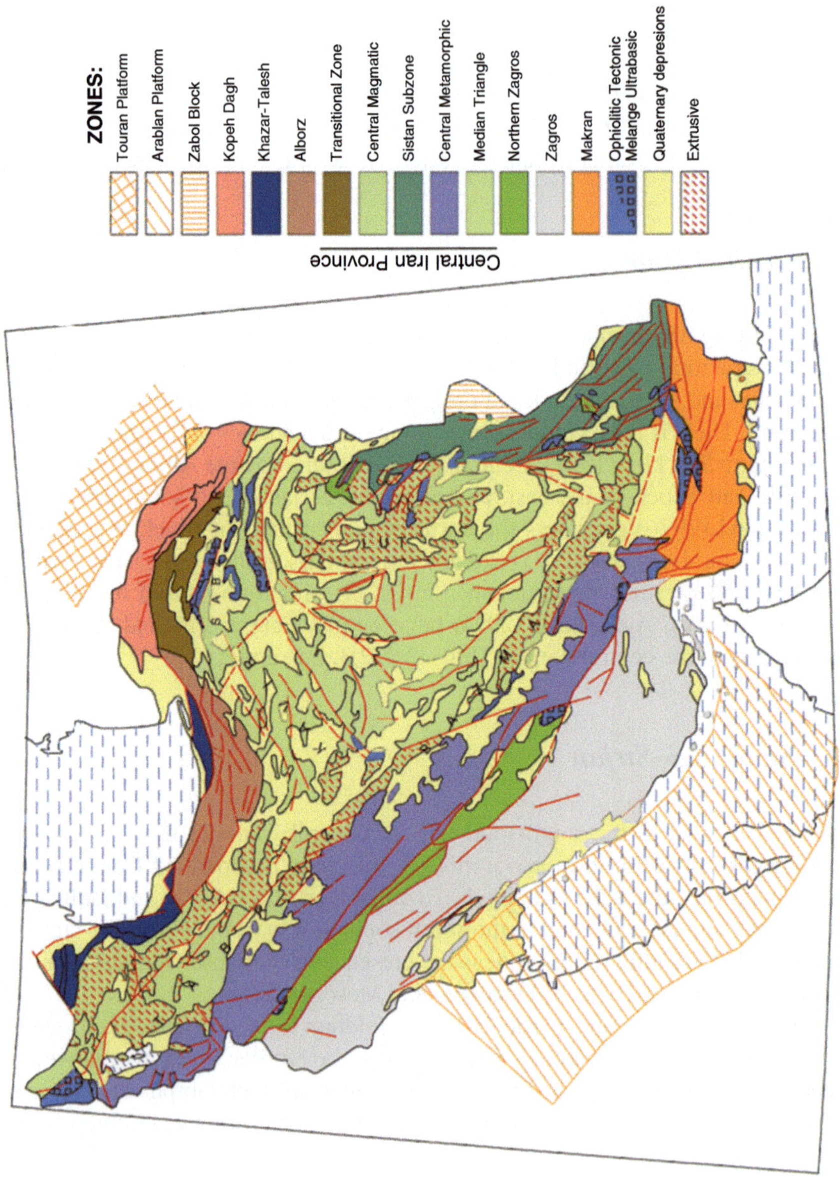

Fig. 2.2 Layout of main structural zones (Nogole-Sadat 1993)

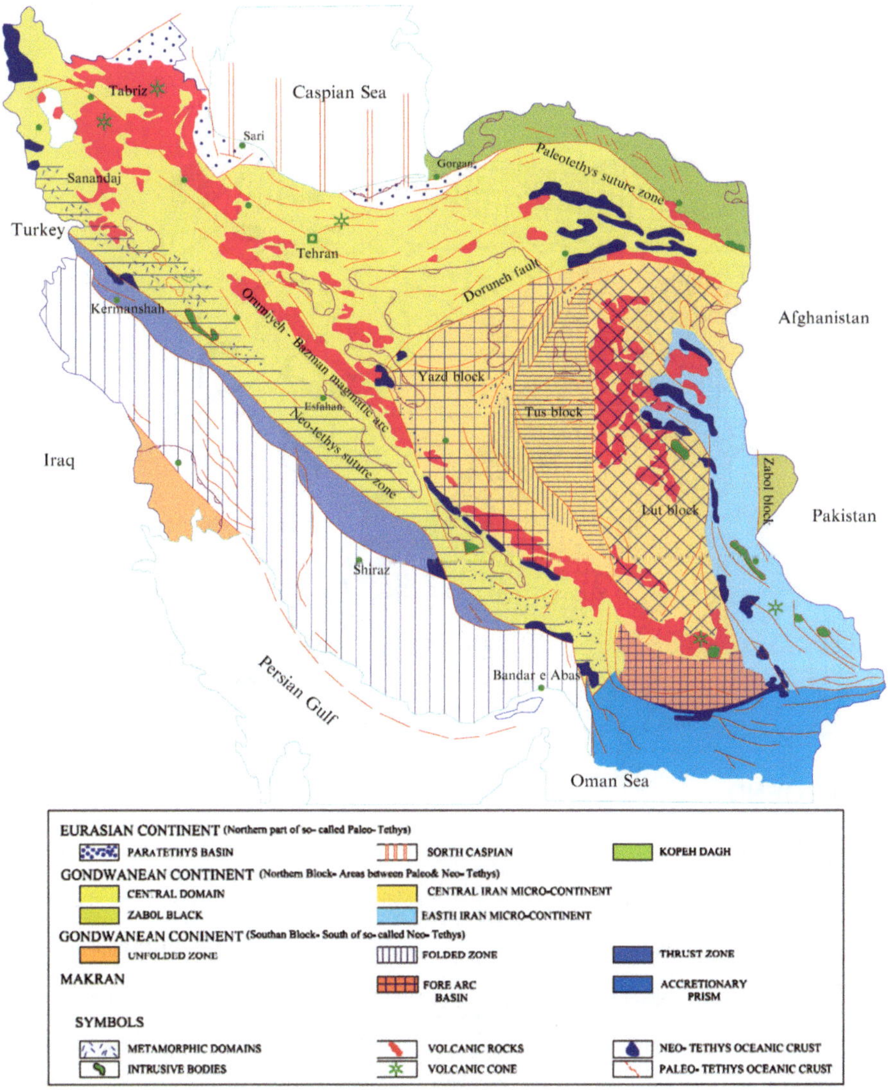

Fig. 2.3 Important tectono-sedimentary areas of Iran (Aghanabati 2004)

Some important features of Zagros include

- Absence of magmatic and metamorphic events after Triassic
- Low abundance of the outcrops of Paleozoic rocks
- Structurally consisting of large anticlines and small synclines
- Continuous sedimentation from Triassic to Miocene with negligible hiatuses

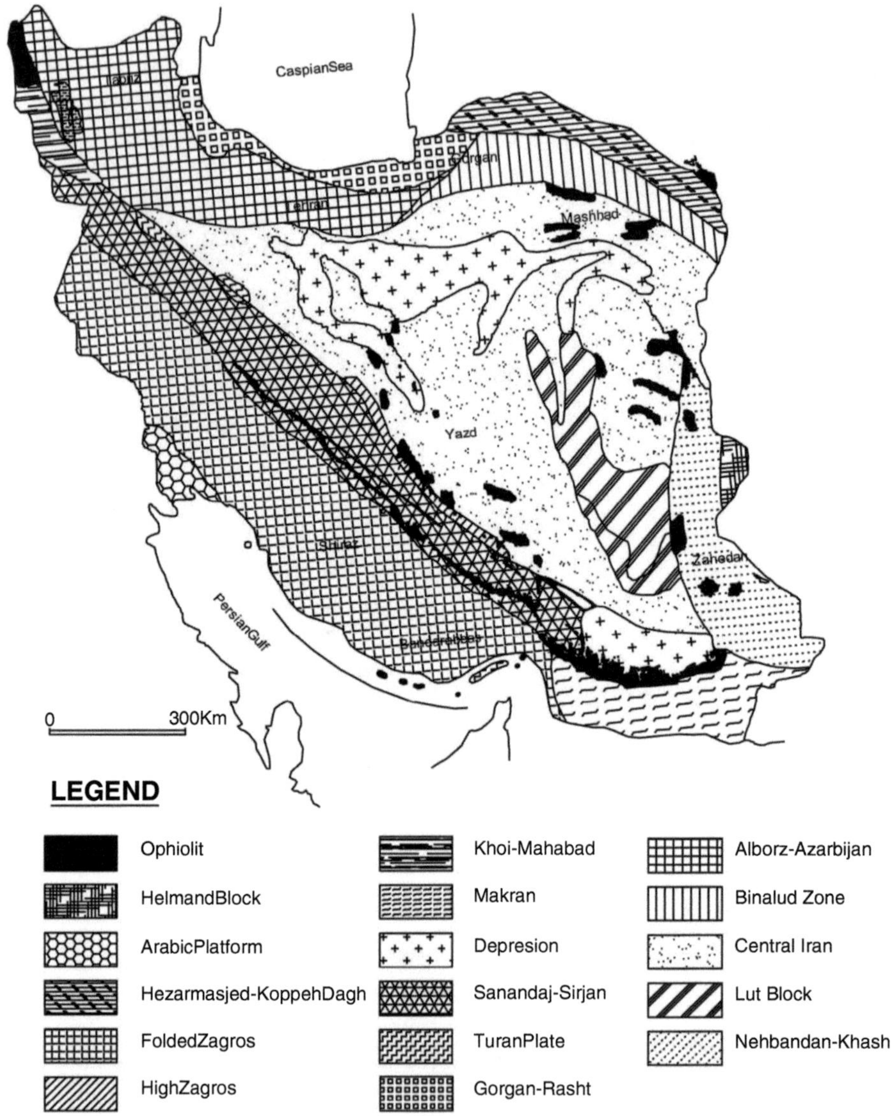

Fig. 2.4 Structural zones of Iran (Nabavi 1976)

2.1.4 Alborz

Alborz is located in northern Iran, parallel to the southern margin of the Caspian Sea. It is characterized by the dominance of platform-type sediments, including limestone, dolostone, and clastic rocks. Rock units from the Precambrian to the Quaternary have been identified, with some hiatuses and unconformities in Paleozoic and Mesozoic. Unlike its northern and southern boundaries (Caspian Sea and Central Iran, respectively), there is not a consensus regarding the eastern and the

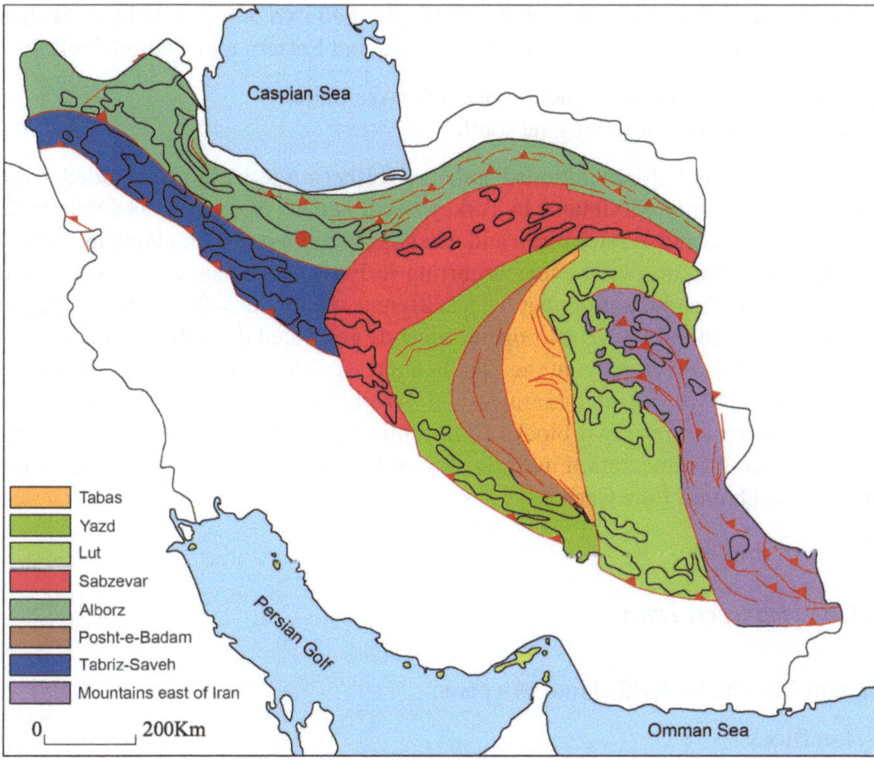

Fig. 2.5 Subzones in Central Iran presented by Alavi (1991)

western limits of Alborz. The Binaluod range in the east, although a continuation of Alborz, bears features comparable to those of Central Iran.

Nabavi (1976) considers Azerbaijan in the west as a part of Alborz, but Stocklin (1968), taking into account the structural features, considers this zone to be an anticlinorium in the northern margin of Central Iran with comparable stratigraphic and structural features. It is evident that the geological evolution of Alborz was different from that of Azerbaijan during the Cenozoic Era.

2.1.5 Azerbaijan

There is no agreement regarding the geological setting of Azerbaijan. Stocklin (1968) considers most of Azerbaijan as belonging to Central Iran. According to the author, the northeastern corner could be included in Alborz and the southeastern part in Sanandaj–Sirjan. Nabavi (1976) believes that most of Azerbaijan lies in a zone called Azerbaijan–Alborz, and as he indicates, this zone is bounded in the north by the Alborz fault, in the west by the Tabriz–Urumiyeh fault, and in the south by the Semnan fault. The eastern boundary with the Binaluod Zone is still controversial.

In a broader view, Innocenti et al. (1982) pictured two orogenic belts to explain the structural units of Western Iran, Azerbaijan, and Eastern and Central Turkey:

- Pantus, minor Caucasus, and Alborz Belt (Azerbaijan)
- Taurus–Central Iran Belt to the south

According to the aforementioned authors, Northern Azerbaijan is included in the Caucasus and Pontus Mountains in Turkey and Southern Azerbaijan is comparable with Central Iran and Western Iran and extends to the Taurus Mountains in Turkey.

The significant structural event occurring in Early Devonian was accompanied by faulting and fragmentation that led to different sedimentary facies in Azerbaijan (Eftekhar Nezhad 1975b). This orogenic episode generated the Tabriz fault, extending in a NW–SE direction from the Zanjan depression to the northern mountains of Tabriz (Mishu, Morou) and northwest of Azerbaijan and Caucasus. This event divided Azerbaijan into two blocks (Innocenti et al. 1976): one in northeast with subsidence and sedimentation in Early Devonian and the other in southwest, which remained high until Late Carboniferous.

2.1.6 Eastern Iran

Eastern Iran can be divided into two parts:

- Lut Block
- Flysch or colored mélange of the Zabol–Baluch Zone

Lut Block: Located to the west of the Zabol–Baluch Zone, the Lut Block is the main body of Eastern Iran. It extends for about 900 km in the north–south direction. It is bounded in the north by the Dorooneh fault and in the south by the Jazmourian depression. In the east, it is separated from Flysch Zone by the Nehbandan fault, whereas the western boundary with Central Iran is the Nayband fault and Shotori Mountains. The oldest units include upper Precambrian–Lower Cambrian schists overlain by Permian limestone and other Paleozoic sedimentary rocks.

Flysch Zone (Zabol–Baluch): This zone is located between the Lut Block to the west and Helmand (in Afghanistan) to the east. In contrast to the Lut Block, the Flysch Zone is highly deformed and tectonized and consists of thick deep-sea sediments like argillaceous and silicic shales, radiolarite, and pelagic limestone and volcanic rocks such as basalt, spilitic basalt, diabase, andesite, dacite, rhyolite, and subordinate serpentinized ultramafic rocks. The basement is likely composed of an oceanic crust. Most rock units in this zone fall into three main groups:

- Flyschoid sediments
- Volcanic, volcanosedimentary, and intrusive rocks
- Ophiolitic series

2.1.7 Southeastern Iran (Makran)

This zone is located to the south of the Jazmourian depression. Its western boundary is Minab fault; to the south, it is restricted by the Oman Sea, and to the east, it extends into Pakistan. The northern part is characterized by dominance of the east–west trending faults, with the Bashagard fault being the most important. Along these faults lie large sections of ophiolite series. The oldest rocks in this zone are the ophiolites of late Cretaceous–Paleocene overlain by a thick sequence (about 5,000 m) of sandstone, shale, and marl. The whole sequence is deformed prior to Early Miocene. Thick sequence of Neogene rock units, in excess of 5,000 m, covers the older series (Nabavi 1976).

2.1.8 Kopet Dagh

This basin is located in northeastern Iran. From Middle Jurassic, it was covered with a vast continental shelf sea (Berberian and King 1981). In this period of time and due to transgression as well as rapid subsidence basin, the western part became deeper. In this basin, a thick sequence of continuous marine and continental sediments was deposited (about 10 km). No major sedimentary gap or volcanic activities during Jurassic to Oligocene have ever been reported. This sedimentary complex provides suitable conditions for accumulation of hydrocarbons. Kopet Dagh sedimentary rocks were placed in their current position due to uplifting at the end of Miocene.

2.2 Ophiolite Series and Ultramafic Rocks of Iran

Ophiolite series and ultramafic rocks have a widespread occurrence in Iran and can be grouped as follows:

- Ultramafic and mafic units of Late Precambrian–Early Cambrian: Although comparable to modern ophiolites, these rocks do not display all typical features of an oceanic crust. The term "old ophiolite" might be a misnomer. These rocks are widespread in Takab and Anarak Regions. They might be representing a protorift.
- Ultramafic and mafic rocks of Upper Paleozoic: These rocks occur as metamorphosed as well as non-metamorphosed bodies in some areas like Fariman, Shanderman, and Asalam. These rocks display many typical features of modern ophiolites.

Ophiolite series of Early Cretaceous–Paleogene age: These rocks show typical features of ophiolitic sequences and are thought to be associated with the closure of

Neotethys. These ophiolite series are widespread in Iran. Some of the more important locations include

- Kermanshah–Neyriz–Oman Belt
- Makran (south of Jazmourian)
- Ultramafic–mafic rocks related to the Flysch Zone in the Khash–Nosrat Abad–Birjand Belt
- Ultramafic and mafic rocks north of the Dorooneh fault, Torbat Jam–Torbat Heidarieh–Sabzevar–Fariman Regions
- Central Iran–Naeen–Baft–Shahr Babak
- Khoy–Maku

Ultramafic and mafic rocks also occur in association with large gabbroic intrusions. This type probably resulted from differentiation in a large mafic magma chamber, comparable to those of the layered mafic intrusions. Examples occur in Sero, Urumiyeh, and in Masooleh, which are Late Cretaceous to Lower Oligocene in age.

2.3 Basement Rocks

Taking into account the available data on the geology of Iran and the Middle East, and comparing it with the Arabian basements, it seems that stabilization of the basement in Iran occurred in the Late Proterozoic to Early Cambrian. This is supported by similar Gondwanic features in both the Iranian and the Arabian basements.

It is believed that the Arabian shield was the continuation of Mozambique in Eastern Africa prior to the development of the Red Sea (BRGM and USGS and some German teams) (Geodynamic Report, GSI 1983).

The cratonization happened transitionally from Africa toward Iran as evident from the age of the basements; Central Africa has a basement of Late Archean–Early Proterozoic while in Northern Africa, there is no Archean basement; besides, the extent of Precambrian domains reduces from Central Africa toward north. Some authors (Hushmandzadeh 1998) believe that cratonization of Iran has been due to Baikalian, Asynitic, or Pan-African orogenies.

The oldest known sedimentary unit in Iran is the Kahar Formation, which is well exposed in Alborz and Azerbaijan. Kahar consists of shale, dolomitic sandstone, and tuff metamorphosed to slate and phyllite (B. Hamdi 1995). The uppermost layers are of about 650 million years old based on the paleontology and stratigraphic evidence; the basal layers are thought to be as old as 800 million years (Ghorbani 2012a, b). Based on field observations, an older age has been suggested for some metamorphic units in Central Iran and Takab Regions (e.g., Robat Poshtebadam, Saghand in Central Iran, Mahneshan in Azerbaijan; Haghipour 1974; Alavi Naeini et al. 1976); however, this is not supported by new data. The oldest rocks in Iran, based on the radiometric ages, are as old as 900 million years (Hushmandzadeh, unpublished). This was the time when Doran-type granites and Gharehdash series formed.

Comparing the geology of Iran and the Arabian Craton and concerning the intrusion of granitoid rocks as an essential component of the continental crust, it can be argued that cratonization of Iranian platform occurred in the Late Precambrian–Early Cambrian, prior to sedimentation of the Lalun sandstone.

2.4 Summary of Stratigraphy of Iran

2.4.1 Precambrian

It was believed that Precambrian rocks older than 1.5 billion years had extensive outcrops in Iran. However, recent data suggest that Precambrian domains were smaller in exposure and that they were all younger than 900 million years. In North and Central Iran, the Kahar and Gharehdash Formations and the lower half of the Soltanieh Formation are of Precambrian ages.

According to Hamdi (1985), the oldest rocks in Iran belonged to the Kushk Series consisting of clastic sediments, acidic volcanic, tuff, and carbonates (mainly dolomite). Other formations of the Late Precambrian–Early Cambrian ages include Rizu volcanic-sedimentary Formation, Dezu and Tashk Formations, Aghda limestone, Kalmard Series, Shorm Beds, and Anarak metamorphic units. The sedimentary facies of Precambrian–Lower Cambrian rocks in Northern Iran is different from that of Central Iran.

2.4.2 Paleozoic

Following the Pan-African orogenic episode, shallow marine sediments formed in Late Vendian. The influence of the orogenic episode is evident at the base of the Vendian sediments. Deposition of shallow marine sediments covered large areas in Iran during Paleozoic (e.g., Alborz, East of Iran, Zagros; Alavi-naini 2009).

There is strong stratigraphic evidence that transition from Vendian to Lower Cambrian was a progressive one, without hiatuses; there is no evidence of any orogenic or epeirogenic movements in Iran at this time (e.g., south of Zanjan, Valiabad Chalus, Shahin Dezh; Hamdi 1995; Alavi-naini 1993).

Early Cambrian started with an alternation of shale, phosphate-bearing limestone, and dolomite sitting conformably and transitionally over the Vendian dolomites. Transition from the Soltanieh Formation to Barut, Zaigoon, and Laloon Formations is very difficult to recognize in the field (Alavi-naini 1993).

Middle Cambrian is characterized by uplift and regression; however, a renewed progression at this time led to the deposition of the Mila and Kuhbanan Formations, consisting of limestone, dolomite, and shale, over older units. These formations bear trilobites and brachiopods of Middle and Late Cambrian (Alavi-naini 1993). In some areas, the Late Cambrian carbonate facies turns transitionally into Ordovician graptolite shales, known as Lashkarak Formation in Alborz, Shirgasht

Formation in Central Iran and Ilbeyk and Zardkuh Formations in Zagros. In the Kalmard area, Ordovician sediments are found on the top of the Vendian sediments through an angular unconformity (Aghanabati 2004).

In Late Ordovician, most parts of Iran were affected by epeirogenic movements (e.g. Alborz; Alavi-naini 2009); this coincides with Caledonian orogeny in Europe and some other parts of the earth. The epeirogeny caused a distinct hiatus at the Ordovician–Silurian boundary. Where present, the Silurian rocks in Iran consist mainly of limestone, sandstone, shale and volcanic materials, known as Niur Formation in Central Iran.

The Lower Devonian rocks have been reported from several localities in Central Iran (e.g., Tabas, Sourian, Kerman, Zagros; Ghorbani 2012a, b); however, they seem to be missing in Alborz and parts of Zagros. Upper Devonian is characterized by marine transgression, particularly in Alborz, that extends into Lower Carboniferous. With the exception of Tabas area, no record of Middle Carboniferous marine deposits has yet been discovered in Iran (Ghorbani 2012a, b). Upper Carboniferous deposits are not significantly present in Iran and have only been identified in several localities from index goniatites (Alavi-naini 1993). After a general regression and a distinct hiatus in Upper Carboniferous, Permian marine transgression deposits cover most parts of Iran (e.g., Alborz, Zagros, Central Iran; Aghanabati 2004). The Permian sediments are represented by Dorood sandstones, Ruteh and Nesen limestones in Alborz.

2.4.3 Mesozoic in Iran

The Lower Triassic sediments in Iran are mainly of shallow marine or continental shelf nature (e.g., Doroud sandstones and Elika dolomites in Alborz, Sorkh shales and Shotori dolomites in Central Iran; Aghanabati 2004; Alavi-naini 2009). A continuous Permian–Triassic sequence has been reported from several areas in Iran, including Jolfa (northwest of Iran), Abadeh (Southern Central Iran), and Southern Urumiyeh (the continuation of Taurus in Turkey), north of Kandevan and Southern Amol.

The transition from Middle to Upper Triassic coincides with Early Cimmerian orogenic episode, which led to the segmentation of the sedimentary basin into three sub-basins: Zagros in south and southwest, Alborz in north, and Central Iran (Aghanabati 2004).

The Lower Jurassic rocks conformably overlie the Upper Triassic units; so are the Early Cretaceous deposits over the Upper Jurassic strata (e.g., Zagros; Motiei 1993).

In North and Central Iran, the Upper Triassic and Lower–Middle Jurassic sediments have a detrital nature, consisting mainly of shale and sandstone with thicknesses varying from a few meters to more than 3,000 m. The presence of plant remains and coal beds suggests a continental or lagoon environment for the deposits (Aghanabati 1998). The Cretaceous deposits, characterized by diverse sedimentary facies, are widespread all over Iran. In Late Cretaceous, tectonic movements related to the Laramide orogeny affected most parts of Iran, leading to uplift, folding, and faulting (Ghorbani 2012a, b). This is a prelude to the significant developments in the geological evolution of Iran.

2.4.4 *Tertiary*

The Cretaceous–Paleocene boundary is characterized by striking changes in sedimentary environments (e.g., Alborz, Central Iran; Hajian 1996). An unconformity has been reported from many locations in Iran. Both continuous and discontinuous transitions have been discovered between Paleocene and Eocene strata; as is the case with Eocene and Oligocene (e.g., Central Iran; Hajian 1996). The Oligocene and Miocene stages are characterized by rapid subsidence, deposition, and facies changes in both marine and continental sedimentary basins (e.g., Mahneshan and Halab south of Zanjan; Rahimzadeh 1994). Oligocene sediments in most parts of Iran are of shallow marine character, turning into marine facies in Upper Oligocene through Lower Miocene (e.g., Qom; Rahimzadeh 1994). The Middle–Upper Miocene sediments are mostly of continental nature.

2.5 Stratigraphic Succession in Various Structural Divisions

A comparison of stratigraphic records in various structural divisions of Iran is shown in Fig. 2.6.

2.6 Overview of Distribution and Nature of Magmatic Rocks in Iran

Magmatic rocks of all ages, from the Precambrian to the Quaternary, are widespread in Iran (e.g., Doran granite, Zarigan–Narigan granite, Torghabeh granite, Ghaen granite, Chaghand gabbro, Alvand granite, Natanz granite; Ghorbani 2012a, b). A correlation exists between the distribution of magmatic rocks and certain types of ore deposits (e.g., iron deposits in Bafgh related to Zarigan–Narigan-type granites, Mazraeh copper deposit related to Sheyvar–Daghi granite, Sarcheshmeh porphyry deposit related to Sarcheshmeh porphyry body; Ghorbani 2002b). Several episodes of magmatic activities have been identified in Iran, which will be discussed next.

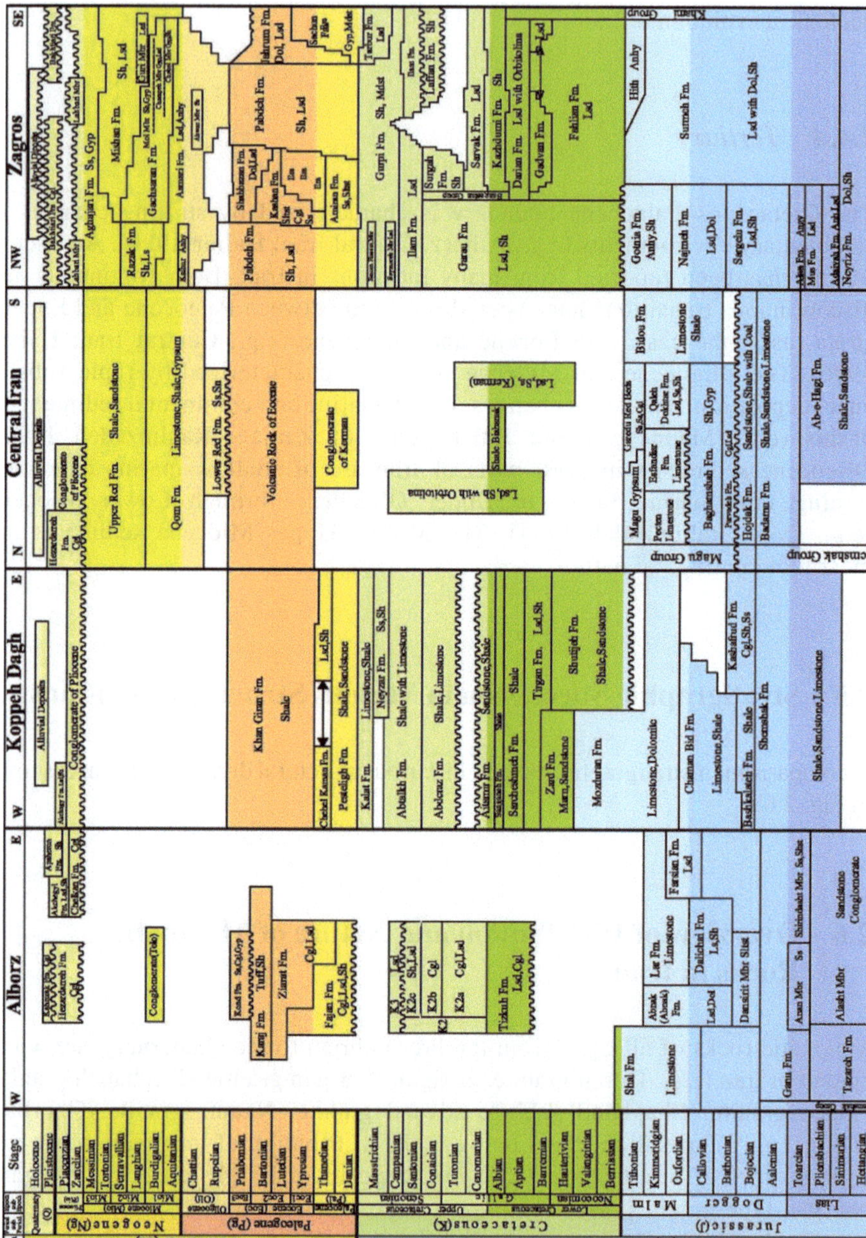
Alborz
Koppeh Dagh
Central Iran
Zagros
Stage
Cenozoic (Cz)
Neogene (Ng)
Paleogene (Pg)
Mezozoic (Mz)
Cretaceous (K)
Jurassic (J)
Holocene
Pleistocene
Piacenzian
Zanclian
Messinian
Tortonian
Serravallian
Langhian
Burdigalian
Aquitanian
Chattian
Rupelian
Priabonian
Bartonian
Lutetian
Ypresian
Thanetian
Danian
Maastrichtian
Campanian
Santonian
Coniacian
Turonian
Cenomanian
Albian
Aptian
Barremian
Hauterivian
Valanginian
Berriasian
Tithonian
Kimmeridgian
Oxfordian
Callovian
Bathonian
Bojocian
Aalenian
Toarcian
Pliensbachian
Sinimurian
Hettangian
Volcanic Rock of Eocene
Khan Giran Fm.
Alluvial Deposits
Conglomerate of Pliocene

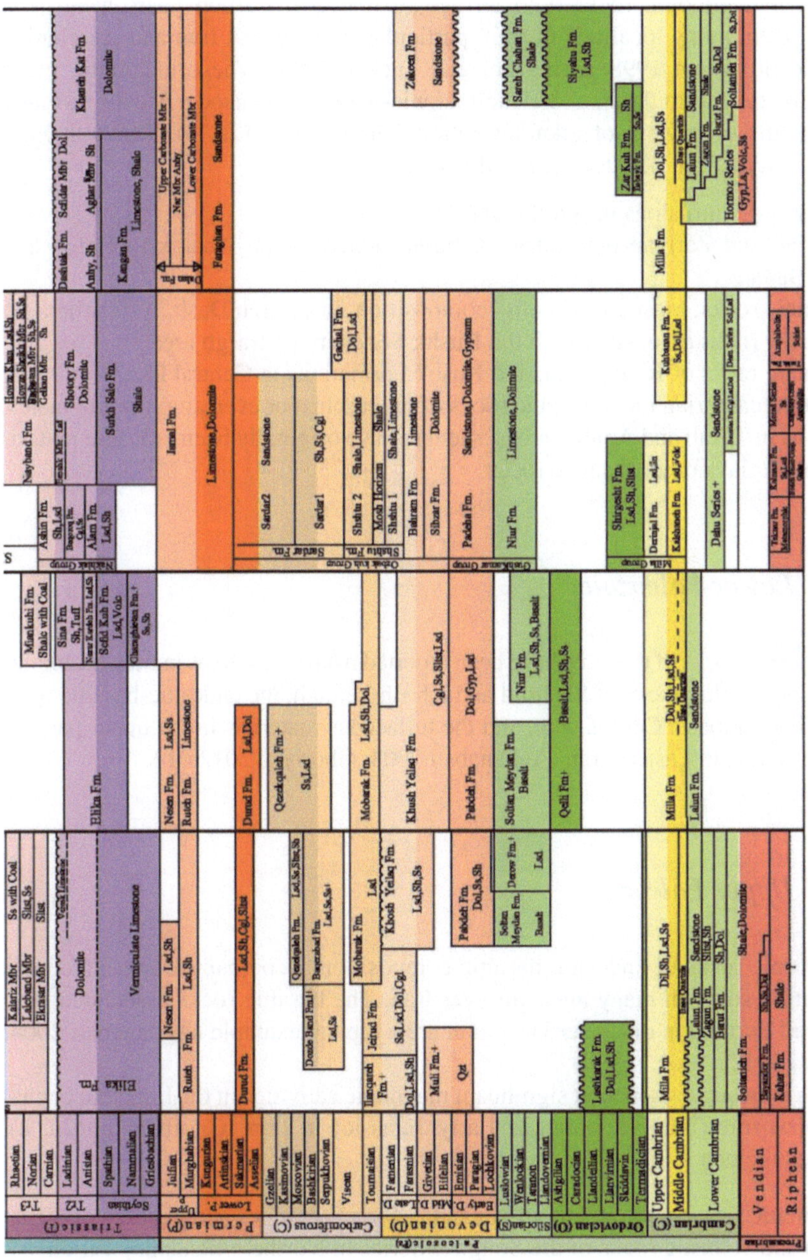

Fig. 2.6 Lithostratigraphic succession in various Structural Divisions of Iran (by Geological Society of Iran)

2.6.1 *Magmatic Rocks of Upper Precambrian–Lower Cambrian*

Volcanic and plutonic rocks with an age of 630–530 million years have been reported from many localities in Iran, particularly in Central Iran and Azerbaijan (Houshmand-Zadeh 1998; Ramezani and Tucker 2003). These magmatic rocks seem to be related to the Pan-African tectonic-magmatic episode. Most magmatic rocks of this time were of alkaline nature (Ghorbani 2012a, b). The following magmatic series can be attributed to this phase:

- Doran-type intrusions in Azerbaijan
- Narigan and Zarigan-type intrusive bodies extend from Anarak to Bafgh and Kuh Banan
- Volcanic rocks, mostly of rhyolite composition, in Ghareh Dash, Azerbaijan
- Volcanic rocks associated with the Kushk Series in the Bafgh area
- Volcanic rocks of Rizu, Dezu, and Kushk Formations in Central Iran
- Most Precambrian metamorphic rocks of greenschist or even amphibolite facies, such as in Takab and Anarak areas, seem to have originally been volcanic materials, either lava or pyroclastic rocks

2.6.2 *Lower Paleozoic*

The magmatic rocks of this time have been reported from many areas in Iran. Examples include the basaltic rocks of Shahrud and Khosh Yeilagh, the andesitic-basaltic units of Niur Formation in Central Iran, and the tuffaceous materials in the upper parts of Mila Formation in Eastern Iran (Aghanabati 2004; Ghorbani 2012a, b).

2.6.3 *Upper Paleozoic*

The volcanic rocks of andesitic-basaltic composition accompany Upper Paleozoic sedimentary strata in many areas all over Iran. The basaltic rocks associated with the Jeyrud Formation of Upper Devonian are a typical example (Aghanabati 2004; Ghorbani 2012a, b).

There is strong evidence of significant magmatic activities in the Late Paleozoic–Early Mesozoic (Early Permian to Early Jurassic) in Iran (Ghorbani 2012a, b). Examples include

- Magmatic rocks in the Southern Sanandaj–Sirjan (granites and gabbros of Sirjan area) and volcanic rocks of the Songhor Series in Northern Sanandaj–Sirjan
- Ultramafic and mafic rocks and their metamorphosed equivalents in Eastern Iran (Fariman area), Taknar Series, Gorgan schists, Shanderman mafic/ultramafic metamorphic series

2.6.4 Mesozoic

The Mesozoic magmatic rocks are associated with Cimmerian and Laramide orogenic events that caused continental and oceanic rifting, followed by closures and collisions in vast areas of Iran (e.g., Sanandaj–Sirjan; Omrani 2008). The Mesozoic magmatic rocks can be divided into three groups:

Volcanic rocks: These rocks occurred mainly as a result of extension or tension related to the continental rifting, or subduction of the developed oceanic lithosphere under the continental lithosphere (e.g., Central Alborz for continental rifting; Ghorbani 2012a, b; Saghez-Sanandaj axis for subduction, Tarkhani 2010).

Intrusive rocks: Many intrusive bodies of mafic to granitic composition, with ages varying from early Triassic to Late Cretaceous, have been identified in Iran (e.g., Boroujerd–Shamsabad axis; Masoudi 1997; Ahmadi-khalaji 2007).

In Triassic–Jurassic, the volcanic rocks predominated the plutonic rocks. They are mainly alkaline in nature and are more abundant in Sanandaj–Sirjan. In Jurassic–Cretaceous, the intrusive rocks exceeded the volcanic rocks; a significant number of batholiths in Iran occurred at this time (e.g., Alvand, Shirkuh, Kolah Ghazi, Shahkuh: Ghorbani 2012a, b).

2.6.5 Tertiary

Tertiary is of great concern in Iran because of the great volumes and highly diverse types of igneous rocks and associated mineral deposits. The magmatic rocks of this age are widespread all over Iran, except in Zagros and Kopet Dagh (Metalogenic Map of The Middle East 2011). Data from various structural zones indicate that the volcanic and plutonic activities started in Late Cretaceous, peaked at Eocene, and continued, with short stops, into Quaternary. The Quaternary volcanism produced very high peaks such as Damavand, Sahand, Sabalan, and many others. Some of the more important regions in terms of the Tertiary magmatic activities include

- Urumiyeh Dokhtar volcanic-plutonic belt
- Azerbaijan
- Tarom–Taleghan
- Central Alborz and its southern margins
- Kavir–Sabzevar
- Kashmar–Torbat-e Jam
- Lut and Kavir, Central Iran
- Sistan
- Bam, Bazman, and Taftan
- East Iran
- Southern Jazmourian–Sabzevaran

The origin and nature of the Tertiary magmatic rocks are controversial. Two different views exist:

1. Intracontinental rifts and aulacogens (Ghorbani 2003a)
2. Subduction of Neotethys oceanic crust under Iranian plate and collision of Arabian and Iranian plates in Late Cretaceous–Paleocene (Ghorbani 2003a, b, c, d, e, f)

2.7 Structural and Orogenic Events in Iran

Several orogenic events have been recognized in Iran, the most important of which are discussed next.

2.7.1 Pan-African

This is equivalent to Asynitic in the other parts of the earth. This event was associated with metamorphism, magmatism, folding, and faulting during Late Precambrian–Early Cambrian in Iran (e.g., south of Zanjan–Mahabad in Azerbaijan, Bafgh in Central; Nabavi 1976; Ghorbani 1999a). This tectonic phase started with tension or extension leading to the formation of rifts and generation of oceanic crust (e.g., in Takab and Anarak) and ended with folding, closure, metamorphism, growth of the continental crust, and development of regional faults.

2.7.2 Caledonian

There was no considerable folding or faulting related to this event in Iran. Caledonian in Iran is characterized by facies change in sedimentary basins, hiatuses, and epeirogenic movements (e.g., parts of Alborz, Zagros, and Central Iran; Nabavi 1976). This phase, starting from Late Cambrian, caused the marine facies of the Barut and Zaigoon Formations to change into the continental facies of the Lalun Formation, and continued on to Late Devonian.

2.7.3 Hercynian

The effects of this orogenic episode in Iran can be traced from Late Devonian to Middle Triassic. Due to the scarcity of magmatism, metamorphism, and folding related to this episode, the role of Hercynian in Iran is controversial; Hercynian in Iran is largely represented by extensional rather than compressional tectonics (e.g., Sanandaj–Sirjan; Hosseini 2011).

2.7.4 *Early Cimmerian*

Early Cimmerian tectonic event is one of the most important events in the geological history of earth. Many diverse features are associated with this phase, including metamorphism, magmatism, folding, faulting, creation of new basins, and facies change (Ghorbani 2012a, b). This event was associated with compressional tectonics in northern Iran and tensional tectonics in the south. There is evidence that the compressional phase was preceded by tension and rift development. The compressional phase, starting in Middle Triassic, finally led to the closure of the paleotethys (e.g., southeast to southwest of the Caspian Sea; Ghorbani 2002a, b, c).

2.7.5 *Middle Cimmerian*

The operation of this phase in Iran has been controversial. Aghanabti (1998) has recently presented rather strong evidence in support of this tectonic episode in Iran (e.g., Sanandaj–Sirjan and Central Iran).

2.7.6 *Late Cimmerian*

A significant tectonic event happened in Iran in Late Jurassic–Early Cretaceous times, about 140 m.a. This event is represented by folding, facies changes in sedimentary environments, angular unconformity, magmatism, and metamorphism (e.g., Alborz, Sanandaj–Sirjan, and Central Iran; Aghanabati 2004; Ghorbani 2012a, b).

2.7.7 *Laramide*

This event, occurring in Late Cretaceous–Eocene, played a great role in the geological evolution of Iran. This event started under a compressional regime, followed by an extension one (Ghorbani 2012a, b; Sadeghi 1999). The compressional regime, which was associated with significant intrusive magmatic activities, led to the closure of the oceanic basins and Neothetyan rifts. In some areas, slices of the oceanic crust have obducted onto the continental margins producing what we now call ophiolite assemblages or colored mélanges (e.g., mostly seen suture zone between Sanandaj–Sirjan and Zagros and alongside the Naybandan fault in east of Iran; Nogole-Sadat and Almasian 1993; Nogole-Sadat 1993).

2.7.8 *Pyrenean*

With regards to the geological evidence, this event was of compressional nature. This tectonic phase is represented by significant changes in the sedimentary environments, plutonism, and metamorphism (e.g., west of Central Iran, south of Central Alborz, Lut; Nabavi 1976).

2.7.9 Pasadenian

Pasadenian is the most important phase in forging the current shape of Iran. There are some orogenic events that occurred before the Pasadenian event, which were most likely the continuation of this phase (e.g., Alborz–Azerbaijan axis, Zagros, Central Iran; Tectonic and Seismotectonic Map of Iran 1993; Rahimzadeh 1994).

Chapter 3
History of Mining

Abstract Iran has been the pioneer in mining exploration and techniques for several millennia. The knowledge and talent of primeval Iranians, on the one hand, and high potential of mineral deposits, their diversity, and absence of dense vegetation cover, on the other hand, as well as the country's position at the crossing of various cultures and civilizations, have led to prosperous mining activities in ancient times and even after the renaissance, which still amaze the mining engineers about the Iranian expertise in mining. Interestingly, there are a few known metallic deposits in Iran that do not have the footprint of the ancient miners.

Hundreds of primordial mining sites of gold, copper, iron, lead–zinc, and silver have been discovered, which indicate centuries of mineral excavation. Numerous mining localities, tools, smelters, and miners' residence have existed since the fourth millennium B.C.

This chapter provides information about the mining activities in ancient Iran along with the mining techniques utilized.

Keywords History of mining in Iran • Ancient mining in Iran • Mining technology in ancient Iran • Gems in ancient Iran

3.1 Introduction

According to the results of archeological studies, the first metal utilized by man in its native form was gold and the first metal extracted through melting was copper. However, there are conflicting views regarding the use of copper. While some experts believe that the usage dates back to 12,000 B.C. (in Egypt) and others believe it to be earlier (as early as 20,000 B.C.), there are definite indications of copper smelting occurring around 6,000 years back on the route of the migration of the Arians (Zavosh 1968). The evidence of copper utilization in Iran (at Tal Eblis) dates back to around 9,000 B.C., and accordingly, Khoyei et al. (1999) argued that the Iranians were the pioneers in copper-smelting activities in the world. Claims of being

M. Ghorbani, *The Economic Geology of Iran: Mineral Deposits and Natural Resources*, Springer Geology, DOI 10.1007/978-94-007-5625-0_3,

the oldest civilization to extract and use metals are not an exaggeration when one considers the historical background and the natural riches of Iran.

The time of passage from the Stone Age (Neolithic) into the Bronze Age in Iran is considered to be the end of the seventh millennium B.C, whereas in Europe, the Neolithic continued till the fourth millennium B.C. (Alipour 1993).

The oldest artifact of copper in Iran is a copper necklace dated around the end of the seventh millennium B.C. (Alipour 1993). In 1966, in an exhibition titled "Seven Thousand Years of Iranian Art," a turquoise statue was exhibited. This in fact indicates that the extraction and utilization of turquoise goes back to that time. The remains of ancient smelting furnaces and their resulting slag at Tal Eblis, Siyalk (Kashan), and Ahangaran (Ghaen) are seen throughout the country, from Zagros and the Alborz mountain ranges up to the deserts of Yazd, Kerman, Qom, Kashan, Khorasan, and various locations in Baluchestan (Chehel Kure copper mine, abandoned lead mines of Kharestan, and Bidaster around Taftan). These are all indications of the advanced metallurgical knowledge and expertise of ancient Iranians in extracting metals from ores and minerals (Ghorbani 2002b).

The onset of civilization and migration of Arians in the third millennium B.C. resulted in a colossal revolution in Iran's mining industry. In spite of not having access to much metal resources, the Arians were aware of the value of different metals (Alipour 1993). It is therefore believed that the Indo–European civilization had advanced beyond the Metal Age before its emigration from Central Asia. Ghorbani (2002a, b, c) states that the term "aios" is a general term referring to metals in ancient Indo–European language: "aes" in Latin is used for copper and gunmetal, the Sanskrit "ayas" stands for iron, and Persian "ahan," Kurdish "asen," German "eisen," and English "iron" are probably variations of the original Indo–European term for iron. It is written in Old Persian texts that "Houshang Pishdadi," who extracted iron from rocks using fire, established the "Sade" festival. "Sade" was one of the festivals celebrated in ancient Iran. For more information about the "Sade" festival, please refer to "Wikipedia."

According to Straboun, the Greek word "cassiteris," meaning tin, and mineral cassiterite, a hard heavy dark mineral that is the chief source of tin, get their names from a place called "Cassian," which was situated on the southern coasts of the Caspian Sea (Alipour 1993).

Dorant, in his book *History of Civilization*, describes the Arians' knowledge of metals as follows: "this immigrant race (Arians) were familiar with metals; thus on arriving into the Iranian Plateau, they settled down, in order to avail themselves of the mineral riches such as copper, iron, lead, silver, gold, marble and precious stones." He continues further that "the Iranian civilization was much more advanced than the Egyptian."

From the early third millennium B.C., the writings of Sumerians, Babylonians, and Eilamians all point toward the technological and scientific achievements of ancient Iranians (Alipour 1993). Among the regions acclaimed for their metal industry and merchandizing, "Shahdad" stands out. Metallurgy, ceramic, serpentinite, and marble industry were highly developed, and various pieces of artifacts made up of semiprecious stones such as agate, lapis lazuli, and turquoise were manufactured. By the middle of the third millennium B.C., extraction of serpentinite was extensive and many objects were made from this stone; moreover, gold was widely utilized.

It has been proved that the passage from the Stone to the Copper Age was gradual. The beginning of the Bronze Age in Iran took place around 7000 B.C. at a time when this alloy was considered to be more valuable than gold and silver. The discovery of bronze led to higher utilization and manufacturing of this metal, thus raising the demand for raw material. At that time, Iran was considered to be one of the most important producers of copper, lead, and tin (though there are no known tin deposits presently in Iran, but Girshman (1954) describes such a deposit from Deh Hosein near Shazand), along with other rocks that were dominantly extracted from the area between Shazand and Nezam Abad, which was then known as the most famous mining locality in Iran. Moreover, the investigations carried out at Siyalk indicate that the trade of "sadaf" (chalcedony or white agate), lapis lazuli, turquoise, and "yashm" (jadeite) was very common in this period.

The utilization of iron in Iran dates back to a time around the second millennium B.C. The ever-increasing production of iron at the said time revolutionized the socioeconomic state of affairs. Countries possessing rich and diverse deposits of this metal, not reputed until then, found a prominent position in the political equations of the time. One such locality is Northern Iran. Excavations at Siyalk indicate a relative increase in the number of iron tools as compared to the bronze tools (Girshman 1934; Wulff 1966). It is believed that Shah Bolagh (south of Zanjan), Masoule in Northern Iran, and Golgohar (near Neyriz) were discovered during this time. The common utilization of iron in weaponry and farm implements drastically changed the state of affairs in trade at the beginning of the first millennium B.C.; prior to that, copper and its alloys dominated the markets.

3.2 History of Mining

3.2.1 Hakhamaneshian Period (550 B.C.–330 B.C.)

In the second half of the first millennium B.C., with the establishment of the mighty Hakhamaneshian (Achaemenian) Empire by Cyrus the Great, Iranian civilization attained its most glorious position. The knowledge of the earth and its potential resources improved, and exploitation of copper, gold, silver, lead–zinc, and other metals reached their peak. The copious reserve of gold in the Hakhamaneshian treasury is a witness to this claim. It is also believed that the largest deposits of gold in the Pre-Islamic Era were discovered at this time, which also witnessed an increase in gold production and making of the first gold coins of the world called "darik." Gold coinage became trivial in other countries, for example, the Roman Empire, after the reign of Darius the Great. Gold was also used in making different ornaments and utensils and as façade of Shoush and Takht-e Jamshid palaces according to Shoush Memorial.

At this period of time, the Iranians used iron in the production of steel (called "poulad"). Greek writers have described the steel weaponry and bridges made by the Iranians; the steel used in the construction of bridges was covered with tar and has not yet rusted. Iron and lead were utilized in the fortification of Takht-e Jamshid and Pasargad buildings. Dandamayev, the Russian explorer, writes that the opening paragraphs of the Bistoon Stone, which bears the names of Darius's ancestors, were covered with lead to extend its life span. Conquering of different countries by Cyrus the Great and Darius the Great led to advancement of the mining and metal industries. Girshman (1954) writes that among the objects found in Shoush, the weaponry were mostly made up of gunmetal, while in a slum area right next to Shoush, iron ornaments prevailed. During the reign of Cyrus the Great, silver coins were also made, which were paid as a part of wages to the laborers and soldiers (especially in the navy) (Alipour 1993). Silver was additionally used in making ornaments and utensils. According to the writings on the Claystone No. 52 found in Takht-e Jamshid, the major production center of iron weaponry was Neyriz in Fars.

Berberian and Manugian (1997) believe that the Iranians came to know about "simab" (mercury) during Hakhamaneshian's time, and realgar, borax, and alum were extensively used in various industries. Lines 37–40 of Darius states that the "blue crystals" (kaputaka or lapis lazuli) and "shangarf" (cinnabar) used here were from Sugudd while the dark blue crystals were from Kharazm. A writing in the Palace of Darius the Great at Shoush talks of "akhshaiye" (turquoise) (Zavosh 1968).

The extreme ingenuity and cleverness of the Iranians resulted in the most magnificent period of mining in the history of the country. The most important mines of this period include Zarshouran (Takab), Zarrin (Ardakan), Kuh Zar (Damghan), and Qal'e Zar (Birjand) gold deposits; Kuh Some (Firooz Abad) lead–zinc ore; Neyriz iron mine; and Neyshabour turquoise deposit.

3.2.2 *Seleucid Period (320 B.C.–63 B.C.)*

There is not much known about the technical aspects of mining in the Seleucid period. However, it is obvious that during the rule of Alexander the Great and his successors, the mining activity was on a decline. In fact, the overall situation of the country was exasperating due to internal insurgencies. Nonetheless, mining activity, as Girshman (1954) puts it, was carried out on a controlled extent under the supervision of government inspectors. Moein (1976) states that "at the time of Seleucid, iron, copper and lead were being extracted, concentrated and exported under the supervision of kings' administrators; the mining activities were under the monopoly of the kings and the mines were considered as kings' assets."

3.2.3 *Arsacid Period (247 B.C.–228 A.D.)*

The Parthian times witnessed a positive change in mining and related industries brought about by the need for weaponry and deployment of Prisoner of War in mining activities. The copper deposits of Eastern Iran (Ahangaran, Ghaen, Gazoo, etc.) were discovered and exploited at this time period. The mining evidence at Shadadi mines to the south of Damghan, such as Kuhzar and Roba'i, point to their activity at the time of Ashkanian. The proximity of these areas with the capital of Ashkanian (Sad Darvaze) is a further confirmation of their prosperity. Pelin, in his *Book XXXIV* (Chap. 14), states that the highest-quality steel was made in China with the second best in Pars.

3.2.4 Sassanid Period (224 A.D.–670 A.D.)

Organized mining activities were initiated at the time of Sassanian. Most of the lead–zinc, copper, gold, and iron deposits that were extracted then are still in production (e.g., Golgohar and Zarshouran).

Most of the metallic artifacts unearthed during excavations belonged to the Sassanian Period, which is a further proof of the advancement of gold, silver, lead–zinc, iron, and copper mining activities during this time. Most of the gold mines of the Hakhamaneshian (Achaemenian) Period once again became operative. The abundance of silver utensils at this time is an indication of widespread mine and metal industry; the lead–silver mines at Nakhlak, Kharestan, and Ahangaran and the lead mines of Kuhbanan, to name a few, were among the most important centers. The Chinese missionaries who visited Iran during the Sassanian times have written that "at Tisfoon, the capital of Sassanian, natural material such as gold, silver, corals, pearls, crystals, iron, copper, cinnabar, mercury etc. were common."

3.2.5 Islamic Era (651 A.D.–1230 A.D.)

Following the Arab invasion of Iran and subjugation of Iran, scientific and industrial depression dominated the country for over 200 years. The mining industry was no exception and, as a result, the activities waned. There are no indications of any mining work during this period, and most of the mines that were operative during the Sassanian Period were abandoned and never functioned again. Many such locations hold the clues to present-day exploration work.

With the inception of the uprisings in different parts of Iran, the antagonism of the "Abbasian Government" toward the Iranians subsided, and many key governmental positions were offered to them. Iranian art and culture flourished once again and embodied in achievements of the new religion, however. This period, which Adam Metz calls the "Islamic Renaissance" (Metz 1985), witnessed the establishment of various Iranian dynasties by Samanian, Ghaznavian, Al-e Bouye, Saljouqian, and Kharazmshahian. The advances in metal and ceramic industries during this time in turn resulted in the progress of mining techniques and activities. The famous "Saljuqi" tile enamel was procured from the copper–cobalt mine at Qamsar (Kashan).

Ahmad ebn-e Omar ebn-e Rasteh, Al-aalagh Al-nafiseh (translated by Hossein Qare-Chanlou 1986) writes, "there are occurrences of gold, silver, mercury and turquoise in Khorasan; gold, silver, copper, iron and sulfur in Kerman; gold and silver in Fars; coal in Bokhara, Yazd and Lorestan; lead in Yazd; gold, mercury and silver in Azarbayjan; and, silver in Esfahan, Bokhara, Sari and Balkh…. Silver mining was common in Samanian times up until the invasion of Moghols with Zar Afshan mines (situated in present day Tajikistan) produced most of the silver…. Iranian miners are skilled in excavating tunnels and shafts."

A number of works on the mining products and techniques were written by Ya'qoub ebn Es'haq Kandi, namely, Resale fi Anva' Al-Jawaher, Resale fi Alne'at Al-Jawaher va Ma'aden, and Resale fi Anva' Al-Jawaher va Samine.

3.2.6 Mughal Era (1230 A.D.–1357 A.D.)

The inefficiency and imprudence of the last king of Kharazmshahiyan and the instigations of Abbasian led to the invasion of the country by the Mongols. This led to a 400-year rule of terror by Teymour's successors, Atabakan, Jalayerian, Choupanian, Aq-Qoyounlou, Al-e Mozaffar, and quarrels between sovereigns or quelling of the rebellions. There is no evidence of any industrial progress (including mining) at this time, though because of growing reclusiveness among Iranian scholars, solitary arts such as poetry improved astoundingly.

At a time when every move of the people (wedding ceremonies, for instance) was taxed by the rulers, there is no documentation of taxing mines, thus pointing to the decline of mining activities.

During this era, many thinkers existed who talked theoretically about the natural resources in general; for instance, Khaje Nasir-o-din Toosi (1200 A.D.–1273 A.D.), who describes 71 ore minerals in his book *Tansookh Name-ye Ilkhani*. Mohammad ben Habib Bekran in *Jahan Name* (1208 A.D.) defines many precious stones and metals. Zakariya Qazvini (1205 A.D.) names around 120 minerals in his book *Ajayeb-al-Makhlouqat va Garayeb-al-Mojoudat*. *Arayes-al-Jawaher va Atayeb-al-Nafayes*, written by Abolqasem Abdollah Kashani (1300 A.D.), is about the mineral resources (this work is in fact an excerpt from *Tansookh Name*). Hamdollah Mostofi in *Nezhat-ol-Qoloub* has described the various minerals that were known to exist in Iran. Gems are also described in the book *Javaher Name-ye Soltani* by Mohammad ben Mansour Deshnaki (1478 A.D.).

3.2.7 Safavid Period (1501 A.D.–1786 A.D.)

The onset of the Safaviyan dynasty coincides with the many important changes in the global settings. On the one hand, the Ottoman Empire had reached its maximum extent and had conquered Constantinople. Terrified by the Ottomans' achievements, European countries had no choice but to support the newly constituted Safaviyan

Government on the east. The Austrian ambassador to Iran has been quoted by Abolqasem Taher (*Political and Social History of Iran*) as "between us and obliteration, Iranians stand." George Weston, in his book *History of Iran and the World*, describes Safaviyan as a "reign" that slowed down the Turks and "reduced Christians detriments." On the other hand, for the Iranian public, who had not experienced a strong central government, enthronement of Safaviyan brought pride and satisfaction.

All these issues made Iran the center of attention, and the European countries began to take efforts to establish political and economic ties with Safaviyan. The economic trade with Europe led to a revolution in exports and imports of the country. Many of the abandoned mines resumed works at the time of Shah Abbas. However, employment of skilled laborers (instead of slaves or prisoners of war) increased the cost of production, thus hindering the development of the mining industry. Tavernier (translated by Nouri 1990) writes, "minerals are extracted from mountainous regions of Iran. Extraction and consumption of copper has increased enormously. Tin is not known to occur in Iran. Iron and steel comes from Khorasan, which also supplies gold and silver.... Shah Abbas had tried to revive the gold and silver mines but not succeeded because of lack of revenues.... It is definite that in older times, gold and silver mines exited in Iran, and remnants of such activities are seen as deep excavations. But ever since the increase in imports from Ethiopia, Sumatra and China, Iranians have stopped exploration for these two metals."

By the end of the Safaviyan Period, the economic situation of Iran deteriorated again and, except the short reign of Karim Khan, the country was the scene of internal conflicts.

Zavosh (1968) is of the opinion that the Safaviyan Period was the period when the extraction of the known mineral riches of the country reached its highest levels. The Veshnave copper deposit, which was in existence since the Bronze Age, was actively exploited during this time (Gruppe 1971).

Mining evidences reveal that the lead and zinc oxide mines of the Kuh Banan–Behabad Belt were active at the time of Safaviyan, and *Tootia* produced at this region was exported to other countries.

3.2.8 Qajar Period (1794 A.D.–1925 A.D.)

With the establishment of the Qajar dynasty and relative stability of the country (especially at the time of Fathali Shah), the number of political and business envoys traveling to Iran increased.

Alexander Khudzco, the Russian Consul General to Mohammad Shah and Fathali Shah courts, writes (Sahami 1354), "Ore and mineral deposits of Iran, especially those of copper, are hardly matched in the world. However, Iranians are way behind the world technologically and cannot use the tools purchased from Europeans. The rich copper mines of Khorasan produce very little, which do not escape the raids by Turkmen. The iron mines of the country too are neglected. Arsenic in form of yellow orpiment and red realgar is found in Kordestan and Qazvin, and is exported to Osmani (Turkey). Sulfur is abundant in various regions and so is rock salt. Coal is readily found on the surface in considerable amounts."

Coleman Howar, the French orientologist, writes (Coleman 1984), "Mining operations have scaled down as compared to the ancient times. Oil probably occurs from Caucasus to Persian Gulf (but is not explored); rock salt and oil in Qeshm Island; ochre clay in Hormoz and Abu Moosa Islands; sulfur on the east and west of Bandar Lenge; and turquoise in Neyshabour. Badakhshan is famous for its lapis lazuli; Yazd for its shiny yellow marble; and Azarbayjan for its iron, lead and copper. Moreover, copper is extracted from Sabzevar and Qal'ezari regions in voluminous amounts." These writings convinced the statesmen to consider inducting modern and scientific techniques of mining.

The signs of advanced Western culture and industry infatuated the heads of state, in particular Crown Prince Abbas Mirza. The travelogue of *Lord Curzon* (translated by Vahid Mazandarani, 1362), which talks about the works of Abbas Mirza, describes his interest toward mining and industry as "early in the nineteenth century." Abbas Mirza, the courageous crown prince who dreamt of developing his country, supported the efforts of the Englishmen in the mines of Azerbaijan and, as a result, Williamson discovered the extensive copper deposits of Sheykh Darand (Torkamanchay) in 1810 A.D. and Sergeant Monight set up a forging factory in 1815. In the same year, Richard Wilbraham reported, "our guide Mr. Robertson, who is in charge of the miners said that he had seen no such deposit anywhere in the world; in the extensive area of the mine that covers several miles, there are iron-, tin- and copper-rich veins whose reserves seem to be never-ending."

At the time of Amir Kabir's premiership, major steps were taken toward exploitation of gold reserves of the country. The Moote gold deposit in the vicinity of Delijan, that is still productive, was among those that were explored.

In 1896 when the thought of development of the country through utilization of foreign investments initiated, various exploration and exploitation contracts were signed, for example, oil concession to D'Arcy and mining concession to Reuter.

After the victory of the constitutionalists (*Mashroote Khahan*) in 1906, concession for mining gold at Astane (near Arak) was given to private parties with no success. A scientific report by De Morgan on gold deposits of the country states, "Gold is naturally rare in Iran, found only in riverine deposits of Kordestan, Khorasan and Qare Dagh, but nothing is known on their mode of occurrence and methods employed for their extraction."

The thought of launching modern mining enterprises in Iran started at the end of the Qajar Period. At the time of Ahmad Shah, the last king of the Qajar dynasty, a census of all the active mines of the country was conducted in order to check the feasibility of the operations.

3.2.9 Pahlavi Period (1925 A.D.–1979 A.D.)

The political stability that prevailed in the early times of the Pahlavi rule opened the way for numerous development programs aimed at the industrialization of the country. This included the construction of railway network and the establishment of steel plants, thus increasing the demand for raw material and energy resources and triggering the need for exploration and exploitation of mineral deposits. The setting up of the Department of Industry and Mining within the framework of the Ministry of Industry and Agriculture in the early years of the Pahlavi dynasty was the first step toward achieving these goals. The main duty of this department was planning for industrial and mining activities based on modern techniques, and, as a result, contracts were made between the government and foreign industrial and mining concerns for the transfer of know-how.

The first government undertaking in mining was initiated in 1934 in Anarak and Shemshak. The Shemshak coal mines supplied coke to various industrial establishments in and around Tehran, such as the Rey Cement Factory.

In the same year, during the construction of the north–south railway, the coal beds of Zirab were discovered and the exploration work was started in the region. The Galandroud coal mine (south of Alamde, Mazandaran) was inaugurated in 1937.

The Lead–zinc mines in Anarak, Esfahan, and Yazd; copper mines in Abbas Abad and Zanjan; and iron reserves of Shams Abad (Arak) and Semnan were discovered at the same time through the efforts of Reza Shah to set up modern metal industries in the country. All of the exploration work was conducted under the guidance of the Directorate General of Mining within the Commerce, Labor, and Arts Ministry (Vezarat Bazargani, Pishe va Honar).

Though the early years were accompanied by technical supervision of foreign experts, with the return of the Iranian graduates from Europe (from 1934 onward) and the establishment of the Technical Faculty at the Tehran University, the supervisory and operation works were gradually transferred to the Iranians.

The onset of World War II delayed the establishment of the Geological Survey of Iran, which was originally planned for the year 1939, the same year that witnessed the passing of Mining Law of the country by the parliament (Majles) (Alipour 1993).

The climax of mining activities reached at the time of Reza Shah was not repeated in the years following World War II. The political instability resulting from the occupation of the country by Allied Forces in the early 1940s followed by political struggles between opposing parties resulted in the closing down of almost all mining activities including exploration. The nationalization of the oil industry by Mossadeq's cabinet was about to show its consequences in the mining sector, but the military coup terminated all plans.

Foreign technicians were employed in most of the key mining projects from 1950 onward. With oil being the main target of the exploration works, all investigations were centered in and around the sedimentary basins of the country by the oil companies (especially the National Iranian Oil Company – NIOC). However, in the following decade (1960s), the increased oil revenue of the country and political stability once again prepared the ground for industrial growth and thus the increased demand for raw material.

The Geological Survey of Iran (GSI) was finally set up in the year 1959, and since then, most of the mining exploration has been carried out under the supervision of this government institution.

Following the skyrocketing of oil prices in the 1970s, the mining sector was once again neglected, though the fundamental exploration works were continued by GSI and NIOC. With no market to offer their produce, the few private companies that operated in those years were run out of the interest and motivations of their owners.

The 5-year development plans of the government after the victory of the Islamic Revolution put the mining sector back into the progressive track, and now there are more than 2,000 active mines all over the country. Today, Iran stands among the world leaders of mining products in terms of quality and quantity (Ghorbani 2002a, b, c).

3.3 Mining Techniques in Ancient Iran

According to archeological surveys, Iran hosts the most ancient mining activities in the world (Alipour 1993). Ancient civilization, richness of natural resources, and records of ancient mining and metalworking are testimonies to such a claim (Zavosh 1976). The techniques involved in mining activities can be grouped under four categories, namely, exploration, exploitation, processing, and metallurgy.

3.3.1 Exploration

The extent and methods of exploration procedures carried out in ancient Iran are not known. However, it can be inferred from the old locations of such activities that the shafts and tunnels have been excavated meticulously based on scientific principles. In the early twentieth century when the modern exploration works began, the ancient excavations formed the basis of all the reconnaissance studies. The old works on the mineral deposits of the country are called "Shadadi" (Ancient), in the geological literature, a term referring to ancient abandoned mines where the modern Western mining techniques and/or equipment such as earthmoving machinery and explosives were not deployed (Ghorbani 2002a, b, c). Interestingly, no major ore deposit in the country has been discovered till date where no traces of ancient mining activities exist. For instance, all the large gold deposits of Iran that are exploited today were known to our ancestors for hundreds or even thousands of years; the Zarshouran deposit had its name before being known for its gold deposit, so are Kuh-e Zar (Damghan) and Zarin (Ardakan) – where the detailed explorations are in their final phases. Table 3.1 lists some of the ancient mining localities that have been named after a mining product.

3.3.2 Exploitation

Since the mining and drilling equipment in ancient years were not as efficient and modern as those found today, exploitation would face quite limitations. The miners exploited the rich and thick parts of the deposits, and, therefore, a winding network

Table 3.1 The relationship between the names of mining products and that of geographical locations

Part of the name	Localities of mines
Zar (gold)	Zarin, Zarshouran, Kuh-e Zar, Zargaran, Zarvand
Mes (copper)	Talmesi, Meskani, Mejdar (Mesdar) Kuh-e Mes, Chahmesi, Dare Mes, Sang-e Mes, Meskan
Zangar (copper)	Zangarlou, Zangalou
Sorme (lead–zinc)	Kuh Sorme, Khane Sorme
Noqre (silver)	Dare Noqre (near Golpayegan), Noqre Kamar, Kuh-e Zardan Noqrei (Baluchestan),
Ahan or Asen (iron)	Ahangaran (Malayer), Kuh-e Ahangaran (Shams Abad), Ahangaran (Qa'en) Asen Abad (Marivan)
Zaj or Zaglik (alunite)	Zajkan (Tarom), Dare Zaglik (Ahar)
Boraq (borax)	Boraq (Taft)
Gel (clay)	Gelkan, Gelou
Naft or Tashi (oil or gas)	Naftoon (Masjed Soleyman), Tape Tashi (Ramhormoz)

would be created in the ground wherein ancient work sites can be seen. The miners could only work down their way to the level right above the underground water table. They would dig through the deposit with narrow tunnels and holes in order to avoid the risk of mine collapse.

The evidences of ancient mining indicate that two common methods were used in Iran: underground and open-pit.

3.3.3 Processing and Metallurgy

Many researchers consider the ancient inhabitants of Egypt, Babylon, and Sindh as the pioneers in melting metals. However, it is evident from the excavations that the metallurgical activities in Iran predate all those of the other regions. Old furnaces and the remaining slags at Yazd, Kerman, Qom, Kashan, Arak, Khorasan, and Baluchestan all and all are indications of the expertise of our ancestors in metallurgical procedures and their advanced knowledge of mineral processing.

3.3.3.1 Processing of Zinc and Tootia Production

Chemically, *Tootia* is composed of zinc oxide. Most of the zinc extraction in ancient Iran was for the production of *Tootia*, whose solution, a strong disinfecting agent, was used in curing eye diseases. This compound was produced in large quantities during Sassanian times and was exported to various countries. Though zinc oxide can rarely be found in the nature, it can be readily formed by heating zinc-bearing rocks in open furnaces.

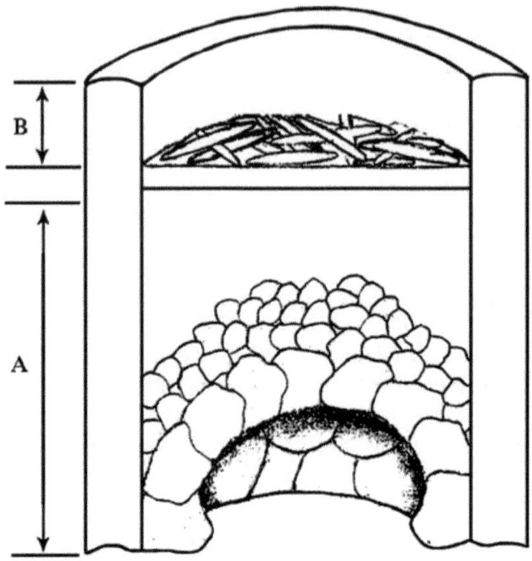

Fig. 3.1 *Tootia*-producing furnaces (Al-e Taha 1996)

The term *Tootia* (Moein 1976, A Persian Dictionary) is derived from the Persian word *Doodia*, a substance extracted from smoke *Dood* (Al-e Taha 1996), thus indicating the mode of production of the substance.

Tootia was produced in smelters, whose remnants are still present in Central Iran (Behabad and Kuhbanan), from the oxide ore of zinc. The furnace (Fig. 3.1) consisted of two sections: the lower section (section A), where the temperature reached around 900°C, held the zinc-bearing rocks, and in the upper section (section B), cones made up of porcelain were placed on a grid. Zinc fumes (in the form of smoke) expelled from the heated ores in the lower section were, on cooling, deposited on the surface of the porcelain cones in the upper section as soot, which was later scraped and powdered to form *Tootia* (Al-e Taha 1996).

3.3.3.2 Cupellation Technique and Gold Extraction

Cupellation and panning constitute the oldest methods used for the extraction of gold and other precious metals in ancient Iran. During the renaissance, this technique was widely used in Europe, especially for the extraction of silver. This method is used today in experimental “fire assay” of silver and gold.

This technique is based on the chemical and physical characteristics of lead, which in molten state displays a strong affinity toward heavy precious metals such as gold. It involves two distinct stages:

(a) Melting stage: Preparation of the molten ore that precedes the cupellation comprises of pounding the ore material with lead oxide and melting catalysts (ash and borax) to get a uniform mix. The resulting mixture is covered with a layer of common salt to avoid oxidation.

Next, the mixture is heated up to around 1,000°C. At this temperature, due to the chemical reaction between the various components, lead oxide is reduced to lead, which in turn segregates out at the bottom of the vessel. On its way out, lead carries all the heavy precious metals with itself and, thus, a layer of lead containing the precious metal is formed upon cooling. This layer can be separated by the use of a hammer.

(b) Cupellation stage: The precious metal that was extracted along with lead in the previous stage is separated here by oxidation and withdrawal of lead. The cooled lead extract of the previous stage is heated up to around 900°C in a cupel, which is a small porous bowl made of bone ash. The cupel is capable of absorbing a quantity of lead oxide equal to its own weight. After the absorption of lead oxide is complete, heavy metals remain in the form of minute nodules on the surface of the cupel.

3.4 Mining Provinces in Ancient Iran

The results of archeological investigations carried out in Iran indicate that many areas hosted ancient mining activities. Table 3.2 presents a summary of all the regions believed to have been mined for various metals (Fig. 3.2).

Table 3.2 Ming provinces in ancient Iran

Locality	Type of mine	Locality	Type of mine
Ahangaran	Copper, iron, lead, and mercury	Abbas Abad	Copper
Ahar	Gold, copper, and iron	Anarak	Copper, lead, mercury, and gold
Ardebil	Copper	Bafq–Kuh Banan	Copper, gold, and iron
Damghan	Gold, copper, and turquoise	Daran–Najaf Abad	Lead–zinc
Esfahan	Lead–zinc	Kharestan	Lead, mercury, and gold
Kerman	Copper, gold, and turquoise	Kuh Sorme	Lead–zinc
Lar-Asaji	Copper and gold	Masoule	Iron
Neiriez	Iron	Neishabour	Turquoise and gold
Qal'e Zari	Copper	Qom–Kashan	Copper, gold, and iron
Shams Abad (Lorestan)	Lead, gold, silver, iron, and tin	Tarom	Copper, gold, and lead
Takab	Gold, arsenic, and mercury	Torbat	Gold and arsenic
Zanjan	Iron and gold	Zarrin	Gold

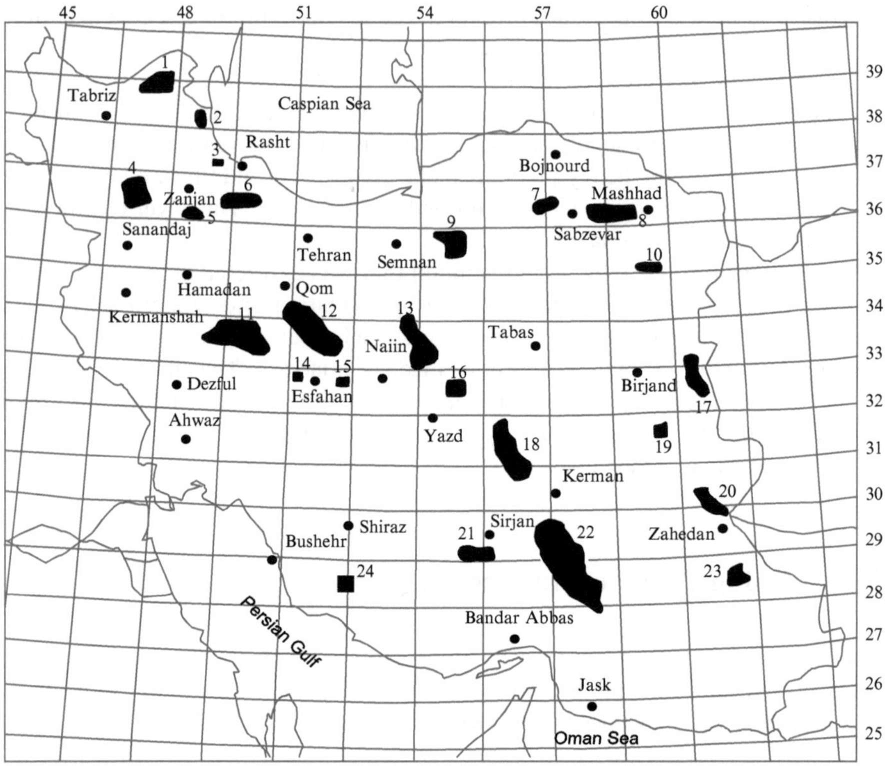

Fig. 3.2 Map of Iran showing the ancient mining localities discovered in archeological explorations [*1* Ahar area (Cu, Au, Fe), *2* Ardabil area (Cu), *3* Masooleh area (Fe), *4* Takab area (Shiz) (Au, As, Ag), *5* Zanjan area (Fe, Au), *6* Tarom area (Cu, Au, Pb), *7* Abbas Abad area (Cu), *8* Neyshabour area (Tr, Au), *9* West of Damghan (Au, Cu, Tr), *10* Torbat areas (Au, As), *11* Ahangaran–Shamsabad (north area of Lorestan) (Pb, Au, Ag, Fe, Sn), *12* Qom–Kashan area (Cu, Au, Fe), *13* Anarak area (Cu, Pb, Ag, Au), *14* Daran–Najaf Abad (Pb, Zn), *15* Esfahan area (Zn, Pb), *16* Zarin area (Ardakan) (Au), *17* Ahangaran area (Cu, Fe, Pb, Ag), *18* Bafq–Kuh Banan (Zn, Pb), *19* Ghaleh Zari area (Cu), *20* Lar–Asagi area (Cu, Au), *21* Neyriz area (Fe), *22* Kerman area (Cu, Au, Tr), *23* Kharestan area (Pb, Ag, Au), *24* Kuh Sormeh area (Pb, Zn)]

Chapter 4
Metallogenic and Mineralization Phases of Iran

Abstract This chapter describes the mineralization phases of all the minerals reported from Iran. Since there is a good association between the mineralization phases and orogenic and magmatic stages, it has been tried to briefly explain the latter as well.

This chapter depicts the type of mineral deposits formed in each geological period and orogenic phase, and also explains the location of these mineral deposits within various geological and mining zones. The temporal and spatial distribution of Iran's mineral deposits, ranging from Late Proterozoic to the present time, are clearly described.

Keywords Mineralization phases in Iran • Metallogenic phases in Iran • Temporal and spatial metallogeny in Iran

4.1 Introduction

The pioneering work for the classification of mineral deposits in Iran through time (from Precambrian to recent) was conducted by Momenzadeh (1981), followed by Momenzadeh and Walters (1984), Ghorbani and Momenzadeh (1995), and finally by Ghorbani (1995b to present). A good correlation exists between the mineralization phases and tectono-magmatic events, including Pan-African, Hercynian, Early Cimmerian, Late Cimmerian, Laramide, and Late Alpine, which played an important role in the formation of mineral resources. Considering the available data on magmatism, stratigraphy, structural geology, and basically the geological evolution of Iran, it is possible to categorize the mineralization phases of Iran as follows:

Late Proterozoic–Early Cambrian (coinciding with the Pan-African tectono-magmatic event)
Lower Paleozoic (corresponding to the Caledonian tectono-magmatic event)
Upper Paleozoic–Triassic (corresponding to Hercynian and Early Cimmerian)
Jurassic–Early Cretaceous (Late Cimmerian tectono-magmatic event)
Late Cretaceous–Lower Paleozoic (Laramide tectono-magmatic event)
Tertiary–Quaternary (Late Alpine tectono-magmatic event)

M. Ghorbani, *The Economic Geology of Iran: Mineral Deposits and Natural Resources*, Springer Geology, DOI 10.1007/978-94-007-5625-0_4,

4.1.1 Late Proterozoic–Early Cambrian

This phase spans a period of ~270 million years from 800 to 530 million years ago. During this period, some formations such as Kahar, Gharedash, Soltanieh, and Barut in Alborz and Kushk, Rizu, and Dezu Series were formed. This phase concluded before the sedimentation of Lalun. The phase ended around 530–650 million years ago.

During this phase, the volcano-sedimentary units of Kushk, Rizu, and Dezu series consisting of basic to acid volcanic and pyroclastic rocks formed in Central Iran. At the same time, rhyolite and quartz porphyry of Gharedash Formation in Azerbaijan and intrusion of Doran- and Zarigan-type granite in a vast region of Iran occurred.

The Doran-type granites mainly occur in northwest of Iran and the Zarigan-type in Central Iran. It seems that Doran granite is older than Zarigan and Narigan. All these granites were emplaced in the period 570–650 million years ago (Ghorbani 2012b). The magmatic episode locally continued to Early Cambrian; manifestations of these magmatic activities can be traced in the Hormoz volcanic series of the Persian Gulf, the Barut Formation of Alborz, and the submarine volcanics in the Dezu Formation of Central Iran.

Mineralization of this phase includes sedimentary and volcano-sedimentary iron deposits in the northwestern region: There are several iron reserves in the volcanic member of Gharedash and lower parts of Soltanieh Formation, which seem to be co-genetic, with their host rocks. Iron in the Gharedash volcanic member occurs in the matrix and is not economically important. But in the dolomitic lower part of Soltanieh, some important reserves occur that can be traced in the northwestern region of Iran extending from Arjin in Soltanieh up to Oshnavieh. Some iron deposits and occurrences of this type include Arjin, Shahbolagh, Mirjan, Trepaghlu, Ghalicheh Bolagh, Chahartagh, Kuh e Tekkeh, Ghiassi, Balestan (in the east of the Urumiyeh Lake), and Aghbolagh (Oshnavieh) deposits.

Volcano-sedimentary Iron deposits: Iron deposits occurring in the evaporitic Hormoz Formation in the Bandar Abbas Region and in the lakes of the eastern Persian Gulf seem to be co-genetic with Hormoz Formation. Typical examples include Tang e Zagh, Hormoz, Larak, and Gheshm deposits.

Volcano-sedimentary iron in Central Iran: There are two types of iron deposits in the Bafgh area in Central Iran: one is of volcano-sedimentary origin and the other is of magmatic origin (Ghorbani 2002a). The first type, which is manganese bearing, is older, less important, and approximately contemporaneous with the lead–zinc deposits of the Bafgh Region. Representative deposit of this type is Narigan iron–manganese deposit and part of Mishdowan iron deposits.

The second type that is of magmatic origin is younger and extends into Cimmerian. This type is characterized by the abundance of REE as compared to the other iron deposits of Central Iran.

Lead–zinc: Large deposits of lead–zinc occur in association with submarine volcanic activities in the Late Proterozoic (Ghorbani 2002a). This phase is represented by low/high-grade ores in the Takab Region: in western Iran (e.g., Anguran, Alamkandi, Poshtkuh, and Chichaklu) and Bafgh area in Central Iran (e.g., Kushk, Chah Mir, Nahrak, Chah Gaz, Zarigan, and Cheshmeh Firuz).

Data from Anguran and Kushk point to their same age (Late Proterozoic) and massive sulfide and Sedex type. Considering the lithology of the host rocks in Kushk and Anguran, mostly shale and carbonate, and their geodynamic setting, they seem to belong to a rift setting and can be called Kushk or Anguran massive sulfides. These deposits may represent a different type of massive sulfide as compared to the more common types (e.g., Koroko, Abitibi, and Besshi) (Ghorbani 2002a).

Gold: Muteh gold deposit, Golpayegan, occurs in the metamorphic rocks of Late Precambrian–Early Cambrian times. The Muteh gold mine bears features typical of mesothermal gold deposits (Ghorbani and Momenzadeh 1994).

Sedimentary phosphate: In Alborz and Azerbaijan, transgression of the Cambrian Sea on Proterozoic units started at Manicayan, and through the Tomanian and Atabanian, a sequence of shale, phosphate-bearing limestone and dolomite were deposited on the Vendian dolomite (Alavi 1993). Some examples are the phosphate deposits of Atabanian occurring in Dalir, Valiabad, and Firuz Abad in central Alborz as well as in the Soltanieh Mountains.

Salt: Huge reserves of salt occur in Hormoz Series that consists of volcanic-sedimentary materials.

Magmatic iron: In the Bafgh area overlying the lead–zinc and volcanic-sedimentary iron ores, high-grade magmatic iron ores associated with acidic to intermediate intrusions exist. Large deposits such as Chadormalu, Choghart, Lakkeh Siah, and Mishdowan belong to this phase. These deposits are generally rich in phosphate and poor in manganese; they are locally rich in REEs. Apatites are mined at Esfordi. Age dating on accessory apatite (Jafarzadeh et al. 1996) of these deposits revealed a Lower Cambrian to Ordovician time of formation.

4.1.2 Early Paleozoic

In contrast to the older beliefs of a very calm environment through Paleozoic, new data indicate occurrences of tectono-magmatic events and related mineralization in different regions in Iran during the course of Early Paleozoic (e.g., Lalun Silica deposit in Alborz, Soltanieh phosphate deposit in Alborz–Azerbaijan, Valiabad phosphate deposit in Alborz, Kan-e Madan copper deposit in Zagros; Ghorbani and Momenzadeh 1994; Ghorbani 2008g).

Presence of phosphate has been reported from the Ordovician sedimentary rocks in Tabas (Shirgasht Formation in the Kalmard and Rahdar), Kerman (Dahuieh,

Zarand Region), (Halalat and Bolourchi 1994), and Zagros (Zardkuh and Chalesheh areas) in the Zardkuh and Il Beyk Formations (Motiei 1993).

4.1.3 Late Paleozoic–Triassic

Metallic and nonmetallic ores occur in Devonian, Carboniferous, Permian, and Early–Middle Triassic in different regions of Iran (e.g., silica deposits in Kerman, Esteghlal fireclay deposit in Abadeh, iron deposits within northern Sanandaj–Sirjan Zone; Ghorbani and Momenzadeh 1994). There are not enough data on the geological evolution of Iran during Late Paleozoic–Early Mesozoic to precisely recognize the various mineralization phases. Some examples of Late Paleozoic–Early Mesozoic ore deposits include the following:

Copper–lead–zinc massive sulfide deposits: Largely occur in Upper Paleozoic rocks of Eastern Iran (Taknar deposit) and Sanandaj–Sirjan Zone (Chah Gaz deposit). Exploration in these regions, especially in the areas close to the two deposits, might prove to be very promising.

Iron and iron–manganese: Examples include

- Heneshk iron–manganese ore in the Dehbid area (Ghorbani 1990; Taraz 1974; Hushmandzadeh personal communication)
- Iron ore of Kalateh Naser in Ahangaran (Qayen), (Ghorbani 1990), Zafarabad in northwest of Divandarreh, and in the Hamedan area (Momenzadeh et al. 1995)
- Iron ore of Masuleh (Ghorbani et al. 1995)
- Hameh Kassi, Galali, and Baba Ali in northeast of Sanandaj–Sirjan in the Hamedan area (Ghorbani and Ojaghi 2003)
- Apooneh in northeast of Sanandaj–Sirjan Daran in the Esfahan area (Hosseini 2011)

Lead–zinc–silver: Based on new evidence, considerable lead–zinc–silver or zinc–lead and silver ores occurring in vast areas of Alborz (e.g., Duna, Elika), Central Iran (Kuh Banan, Bahabad and Ravar), and Zagros (Kuh e Sormeh deposit) during Triassic Period (Ghorbani et al. 2000). These deposits are often accompanied by fluorite and barite. The copper–lead–zinc deposits of Eastern Iran, earlier attributed to Devonian, are probably Triassic. In Late Paleozoic–Triassic, tungsten–gold ores occur in some areas in Binaloud, Southern Khorasan (Chah Calb and Chah Charnoy), Sanandaj–Sirjan Zone (Broujerd-Azna Belt, Eqlid–Dehbid Belt, Bvanat and Zartorosh), and likely in northern Alborz (Gorgan to Asalam) (Ghorbani and Momenzadeh 1994).

Nonmetallic reserves: The interval of Late Paleozoic to Early–Middle Triassic is significant in the generation of nonmetallic deposits, as large deposits of clay, bauxite, barite and fluorite, and as well as phosphates formed.

Fireclay: In Iran, most fireclay deposits occur in Late Paleozoic to Early–Middle Triassic. In the Abadeh area, there are several horizons of fireclay of Late Devonian to

Triassic age, for example, Esteghlal deposit. Some horizons of Kavir deposits are of Late Devonian–Carboniferous age, and most of them belong to Permian–Triassic.

In the Tabas area, the Cheshmeh Shotoran and Chah Kular deposits in Robatkhan are of Permo–Triassic age (Aghanabati 2004).

Bauxite: Most economic bauxite deposits in Iran occur during Permian–Triassic. Examples include Jajarm, Bukan, and Saqqez (Soheili 2004).

Sedimentary phosphate: The Devonian Jeyrood Formation is the main host to sedimentary phosphates in Iran. Phosphate deposits and occurrences have been reported from the Jeyrood, Lalun, Firuzkuh, Damavand, Shahrud, and Damghan Regions. It is noteworthy that large sedimentary phosphate deposits of Devonian age in the global scale have been discovered in Iran and Armenia (Halalat and Bolourchi 1994).

Barite and fluorite: In the latest stages of the Paleozoic–Triassic mineralization period, barite ore bodies, locally accompanied by fluorite, occur in many areas of Iran. Fluorite and barite deposits occur in Central Iran in Ardakan, Kamshajeh, Faskhood, Kashan, and Delijan as well as in central Alborz in the Triassic Shotori Formation and in eastern Central Iran in the Tabas area, which are accompanied by lead–zinc. All these deposits are of Early–Middle Triassic age (Ghorbani and Momenzadeh 1994; Ghorbani et al. 2000). The Ghahrabad fluorite deposit of Saqqez in Sanandaj–Sirjan Zone likely belongs to this phase.

Sedimentary silica: Silicic sandstones and conglomerates of Kereshk Group of Late Devonian–Carboniferous age in the Kerman Region and quartzite horizons in Doroud Formation of Early Permian in parts of Alborz (Masuleh) are noteworthy (Aghanabati 2004).

4.1.4 Jurassic–Early Cretaceous

This phase coincides with late Cimmerian orogenic episode that is represented by extensive magmatism and metamorphism in vast areas of Central and Eastern Iran, particularly in the Sanandaj–Sirjan Zone. This episode was associated with the formation of lead–zinc, barite, iron, iron–manganese, tungsten, gold, tin in east and northwest of Iran and Sanandaj–Sirjan Zone.

Lead–zinc and barite: There are many deposits of lead–zinc and barite in the Lower Cretaceous carbonate rocks. Some 300 deposits and occurrences of Pb–Zn–Ba of Lower Cretaceous age have been reported in the literature (Ghorbani et al. 2000) Examples include lead–zinc deposits in Malayer–Esfahan Belt in northern Sanandaj–Sirjan Zone (e.g., Irankuh, Robat, Emarat, Lakan, Ahangaran) and deposits in Yazd area, such as Mehdi Abad, Mansour Abad, Farah Abad, and Darreh Anjir.

Volcano-sedimentary Mn-bearing iron deposits: Several Mn-bearing iron deposits occur in Malayer–Esfahan Belt; typical examples include Shams Abad (Arak), Ahangaran (Malayer), Chahbasheh, Morcheh Khort, and Meimeh (Jafarzadeh et al. 1996).

Tungsten, gold, and tin: Deposits of tungsten and gold, locally containing tin, occur in Sanandaj–Sirjan (in Broujerd–Azna Belt) as well as in Abarkuh (tungsten and possibly gold), in Anarak in Central Iran, in Eastern Iran, Shirkuh (Yazd), Zarrin (Ardakan), and southern Birjand (Jahangiri 1999). There are not enough data to precisely identify the periods of tungsten–gold formation in Iran.

Copper, gold, silver, and manganese: These deposits have been discovered in metamorphosed volcanic-sedimentary units of Upper Jurassic–Cretaceous ages. The Copper deposits of possibly massive sulfide type have been identified in southeast of Kerman (e.g., Rameshk and Sargaz copper; Metalogenic Map of The Middle East 2011).

Kaolin and fireclay: Fireclay at the base of Liassic (base of coal horizons) in Alborz (Sangrood area) and kaolin in Khorasan (Kaftar Kuh, Gonabad) are examples of this type of mineralization.

4.1.5 Late Cretaceous–Paleogene (Laramide), Chromite

Associated with Laramide orogeny in Late Cretaceous–Paleogene times, significant ore formation occurred in many parts of Iran (Ghorbani and Momenzadeh 1994). There are several Late Mesozoic chromite deposits in northwest (Gheshlagh, Khoy), northeast (Sabzavar Region), Neyriz, Fanouj area in southeast, Khash area, Karvandar Belt, and Sefidabeh, which all belong to the Late Cretaceous–Paleogene age. Chromite deposits of Esfandagheh and Frayab in southern Iran are related to this phase (Ghorbani and Momenzadeh 1994; Metalogenic Map of The Middle East 2011).

Magnesite: Magnesite deposits associated with ophiolitic complexes in south of Birjand and east of Qayen in east of Iran and deposits in Sisitan–Baluchestan Province belong to this phase (Ghorbani and Momenzadeh 1994; Metalogenic Map of The Middle East 2011).

Manganese: Manganese deposits in Iran are mainly associated with ophiolitic complexes. Examples include Ab Band in Fars Province, Gounich in Khash, and some other manganese deposits in Baft (Kerman area) (Samani 1995).

Copper, gold, and silver: Massive sulfide deposits of copper, gold, and silver associated with Cretaceous ophiolites have been reported in the Haji-abad Bandar Abbas area (Sheikh Hadi) and the Fanuj–Esfandagheh Belt in southeast of Iran (Khoyi et al. 1999).

Phosphate: Phosphate deposits in Gurpi–Pabdeh Formations of Late Cretaceous–Paleocene age as well as economic deposits in Kuh e Lar, Sheikh Habil, Kuh e Rish, Kuh e Koomeh, and Kuh e Sefid occur in Zagros (Kohkiluyeh–Boyerahmad Province) (Halalat and Bolourchi 1994).

Bauxite: Occurrences of bauxite of late Cretaceous–Paleogene have been reported from Zagros.

4.1.6 *Tertiary–Quaternary (Middle–Late Alpine)*

This phase created the largest and most varied deposits of all ages in Iran. The deposits occur mainly in the volcanic, volcano-sedimentary, and intrusive rocks of Early Tertiary (Paleogene), Late Tertiary (Neogene), and Quaternary (Ghorbani and Momenzadeh 1994; Metalogenic Map of The Middle East 2011).

Tertiary–Quaternary magmatism mainly occurred in Urumiyeh–Dokhtar Zone extending along a belt from west of Urumiyeh Lake to Taftan in a northwest–southeast direction; moreover, it is also seen in the southern margin of central Alborz and in Azerbaijan on the northeastern Iran. It continues to Afghanistan on the east of Iran and scarcely in Kavir Desert (Ghorbani 2003a). Deposits related to these magmatic rocks are as follows

Copper: More than 95% known copper deposits of Iran occur in Tertiary; examples include deposits in the Urumiyeh–Dokhtar Zone, deposits in the Tarom–Hashtjin area, and Ahar deposits and copper deposits of Eastern Iran (Ghorbani 2008b).

Manganese: Significant manganese deposits occur in pyroclastic and volcanic rocks of Tertiary (like those of Qom–Naien Belt) and also in young Tertiary–Quaternary volcanics (like manganese deposits of Bostan Abad in Maragheh, Azerbaijan) (Samani 1995).

Lead–zinc: There are many lead–zinc, silver, and copper deposits in the Tertiary volcanic rocks, like Zeh Abad, Baycheh Bagh, Barik Ab, Ay Ghaleh Si, and many others (Ghorbani et al. 2000).

Iron: Several iron deposits are hosted in both volcanic-pyroclastic rocks and in plutonic rocks of Eocene to Pliocene ages (Jafarzadeh et al. 1996). Some of the deposits include

- Iron deposits of Urumiyeh–Dokhtar Zone like Niasar in Kashahn, Daran in Esfahan, Shahrak in Takab, and Kuh e Baba in takab
- Iron deposits of Tarom Region
- Iron deposits of Ahar area
- Iron deposits of Eastern Iran, like Sangan and Tanoorcheh

Antimony, arsenic, mercury, and gold: Many epithermal gold deposits accompanied by arsenic, antimony, and mercury occur in the younger Tertiary and Lower Quaternary volcanic rocks of Iran. The more important deposits are present mainly in northwest of Iran (like Ghorveh–Takab Belt), Anarak area in Central Iran, and in northeast of Iran (Kashmar-Torbat-e Heydarieh and Ferdows) (Ghorbani 1995a).

Nonmetallic deposits: All bentonite deposits, most kaolin deposits and all perlite, zeolite, alunite, and diatomite deposits belong to this phase (Ghorbani 1999d; Ghorbani 1994a; Ghorbani 2002a).

Celestite: These deposits occur at the northern margin of Central Kavir as well as in Zagros (Behbahan deposits) associated with evaporative sequences consisting of carbonates, calcium sulfates, and halite deposits (Ghorbani 2002a).

Barite: Barite of vein and lens types as well as stratiform type occur in the volcanic-pyroclastic rocks of Eocene and Oligocene in Qom, Saveh, Kashan, Delijan, and Ghazvin areas (Ghorbani 2010a, b, c).

Chapter 5
Metallogeny and Distribution of Minerals

Abstract This chapter explains the metallogeny, mineralization, and paragenesis of all mineral deposits. It is also tried herein to put forth the distribution of each type of mineral within the metallogenic and mining zones and provinces.

Mineral deposits are categorized based on their distribution and paragenesis to allow prediction of their reserves and dispersal in different geological epochs. The reader would acquire a good temporal and spatial knowledge of all types of mineralization in Iran.

Although the concept of metallogeny is mostly believed to be applicable to metallic mineral deposits, it has been tried to extend the application of metallogenic principles to nonmetallic mineral deposits as well.

Keywords Metallogeny of metalic minerals in Iran • Metallogeny of nonmetallic minerals in Iran • Distribution of metallic minerals in Iran • Metallogeny of nonmetalic minerals in Iran

5.1 Iron

5.1.1 Metallogenic Phases of Iron

Iron deposits are not restricted to any particular episode of the earth's history and are found throughout the geological timescale. Though most of the known iron reserves belong to Precambrian, there are known occurrences in Paleozoic, Mesozoic, and Cenozoic reported from the world over.

Iran is no exception with iron deposits distributed throughout its geological history. With origin and formation of the crustal rocks of Iran in Late Proterozoic and continued tectono-magmatic evolution, iron mineralization has occurred, and traces of such activity can be seen from the said time till Mio-Pliocene (Jafarzadeh et al. 1996; Ghorbani 2007a).

M. Ghorbani, *The Economic Geology of Iran: Mineral Deposits and Natural Resources*, Springer Geology, DOI 10.1007/978-94-007-5625-0_5,

It must be noted that the deposits of Late Proterozoic–Early Cambrian are more abundant than those of other time periods.

About four billion tons of iron ore of igneous (Sangan, Bafgh, Zanjan, Morvariyeh, Sorkhe Dizaj), sedimentary volcanic (Bandar Abbas, Shams Abad), and volcano-sedimentary (Soltanieh-Mahabad Belt, Hamekasi Complex origins) have been discovered till date (Fig. 5.1).

Late Proterozoic–Early Cambrian Iron Mineralization: In parallel with the Late Proterozoic igneous phase that continued till Early Cambrian (620–530 Ma), a number of iron deposits have been formed that are directly or indirectly associated with the igneous and sedimentary volcanic rocks of this period. These deposits can be grouped into two categories based on their time and mode of origin (Ghorbani 2007a):

1. Deposits of the volcanic and volcanic-sedimentary origin of Late Proterozoic that extend up to Early Cambrian. These uneconomical deposits include
   (a) Bafgh volcanic-sedimentary deposits (first phase of mineralization) such as the iron–manganese deposit of Narigan, the horizontal section of Mishdavan deposit. These deposits contain manganese, REEs, and uranium.
   (b) Sedimentary iron deposits of Soltanieh-Mahabad Belt that have volcano-sedimentary origin associated with the lower parts of Soltanieh Formation and sometimes Qareh Dash Series. These include Arjin iron deposits (Soltanieh Region), Shah Bolagh and Kavand (southwest of Zanjan), Mirjan–Ghalicheh Bolagh (Mahneshan Region), Alamkandi and Chahartagh (Takab area), Tekeh-Ghiyasi Mount (Soltanieh Region), Bastan (Mahabad area), and Agh Bolagh (Oshnaviyeh area).
   The iron deposits that are located within the Late Precambrian–Early Cambrian rocks of this belt show sedimentary structures, and their paragenesis includes goethite, hematite, magnetite, and siderite along with calcite, dolomite, and barite as gangue minerals.
   (c) Volcanosedimentary iron deposits are occurring within Hormuz Salt Series and their associated volcanics in Bandar Abbas and islands to its southeast such as Tange Zagh, Hormuz, Larak, and Qeshm, which are mostly of Early Cambrian age. Sometimes, these deposits occur within younger formations, which are due to erosion of iron ores from older rocks and their redeposition in younger formations.
   (d) Volcanic iron deposits mostly mixed with Qareh Dash Volcanics include Bardeh Rash Mount, Bichaghchi, and Hamam (Shahin Dezh area). Due to their association with rhyolitic rocks, they lack any economic value.
2. Deposits of igneous (orthomagmatic) origin along with their host rocks have been metasomatically altered under the influence of chemically active solutions emanating from the original source of magma (due to this phenomenon, these deposits are referred to as metasomatic in literature) (e.g., Choghart, Chadormaloo; Jafarzadeh et al. 1996). The rocks of Bafgh and Sirjan area that constitute the largest iron deposits of the country are of this type. They have been formed due to the plutonic phase following the volcanic activity that gave rise to the Rizoo and Dezoo Formations, for example, Zarigan and Narigan granites.

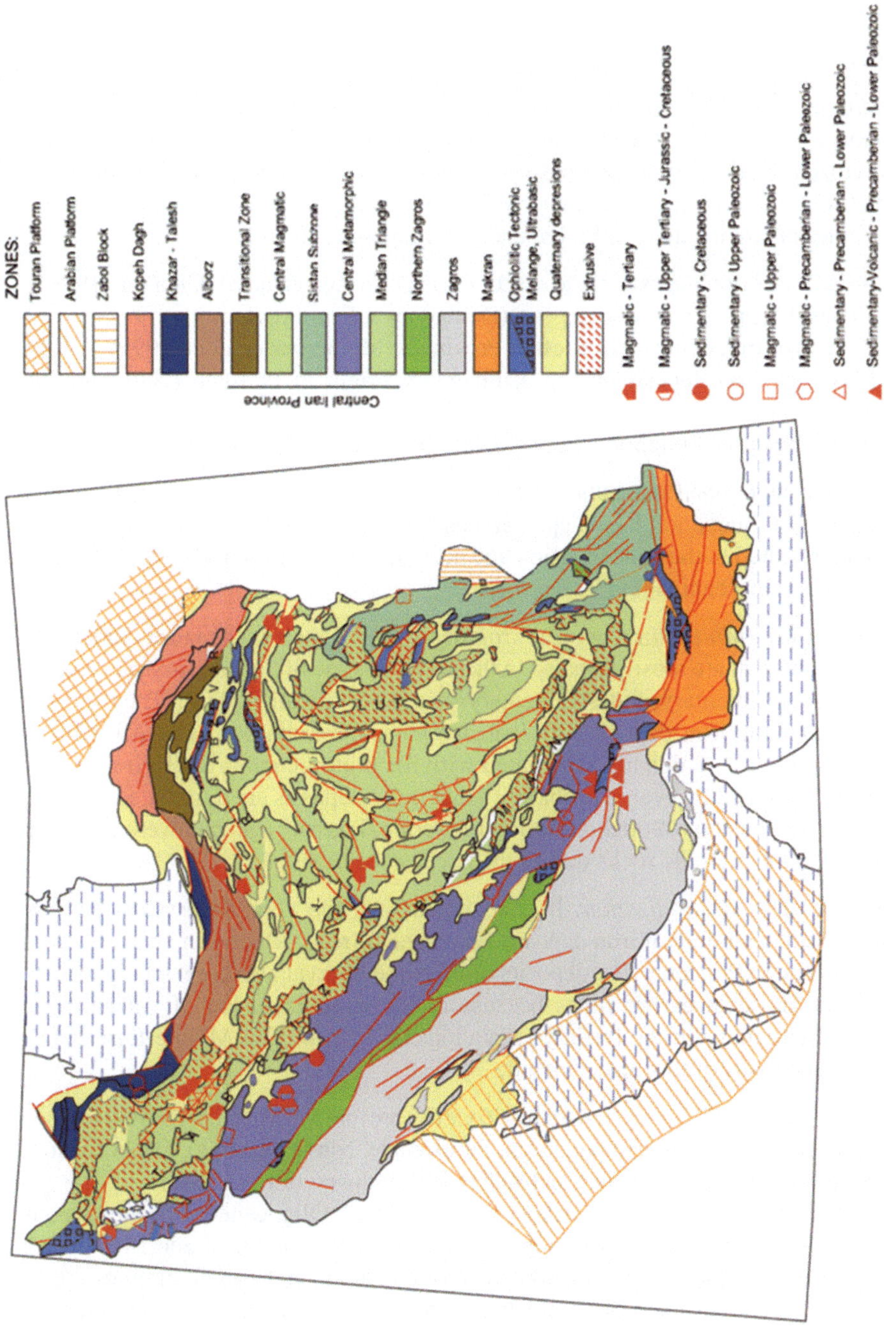

Fig. 5.1 Major iron-bearing areas of Iran

The igneous activity has involved an alkaline phase rich in iron and its derivatives whose REE content was very high. The iron deposits of this phase include Chadormalu, Choghart, Sechahoon, and Chahgaz (Zarand area), second phase of Mishdavan and Gol Gohar.

Paleozoic–Early Triassic Iron Mineralization: This phase of iron mineralization is comparatively poorer in iron content and deposit size as compared with the Late Proterozoic–Early Cambrian phase just giving rise to a few small iron deposits, which are as follows (e.g., Abpooneh in Esfahan area, Galali in Hamedan, Zafarabad in Kordestan, all located within Sanandaj–Sirjan zone; Hosseini 2011):

- Kalat Naser iron deposit on the east of Ghaen (Ahangaran area) (Ghorbani 1993b)
- Honeshk iron–manganese deposit (Dehbid area, Fars Province)
- Zafarabad iron deposit (Kordestan Province) (Momenzadeh et al. 1995)
- Masooleh iron deposit having sedimentary origin in Permian (Ghorbani and Mostafavi 1995)
- Iron deposits of Songhor area (Kanishire, Khosro Abad, and Deh Khazal)

All the mentioned deposits are hosted by iron-rich volcanic and sedimentary volcanic rocks metamorphosed up to greenschist facies. Further metasomatic alteration has resulted in the concentration and accumulation of iron in the form of layers and lenses (Ghorbani 2008g).

Mesozoic Iron Mineralization: The iron deposits (all with manganese content) of this time interval are mostly located in the northern zone of Sanandaj–Sirjan and are of volcanosedimentary and sedimentary origin (Ghorbani 2007d). They include skarn-magmatic iron deposits of Hamedan area, for example, Baba Ali, Chenar Bala, and Golali, which are associated with the plutonic gabbro–diorite–syenite bodies. There is, however, a probability that these deposits are associated with the volcanic-sedimentary Songhor Series of Permo-Triassic age but were enriched along with the plutonic rocks during the Cretaceous times.

Cenozoic Iron Mineralization: In association with the very widespread Tertiary magmatism, a number of iron deposits have been formed, which can be traced from Late Eocene to Pliocene. The deposits of this time interval are more widely distributed than those of earlier times because of the wide extension of volcanic and plutonic activities during Tertiary (Metallogenic Map of The Middle East 2011). The iron deposits of this phase include

(A) Iron deposits of Urumiyeh–Dokhtar Zone that are formed in parallel with Oligocene–Miocene volcanism and plutonism, for example, Niasar (Kashan), Daran (west of Esfahan), Shahrak (Zanjan), and Kuh Baba (southwest of Hashtrood).

(B) Magmatic iron deposit of Alborz associated with Oligocene–Miocene igneous activity in Western Alborz zone, for example, Sorkhe Dizaj and Morvariyeh (formed in association with Abhar, Khoram Dareh, and Zanjan granites).

It should be noted that a number of iron deposits in northern Azerbaijan are formed by the alteration of sulfide minerals (Ghorbani 2011). These deposits are generally in the form of gossan capping, for example, Damirchi (Tarom area) and other deposits of Khalkhal and Ahar areas. The Eskandarian (Khoi area) and Mazra'e (Ahar area) iron deposits of Azerbaijan are of skarn-magmatic type.

(C) Eastern Iran iron deposits including Sangan (one of the largest of the deposits) believed to have been formed during Late Eocene to Oligocene–Miocene. There are a number of views on the mode of formation of Sangan iron deposit:

- Ghasemipour (1976), in a report of Iran Barite Co., related the formation of Sangan ore body to metasomatic alteration to intrusion of Sarnosar Granite into the Mesozoic limestones.
- The report of Sangan Iron Mining Co. by Boroumandi (1982) ascribes the ore deposit to volcanogenic processes comparing it to those that gave rise to Taknar Rhyolites of Central Iran (Infra-Cambrian age); however, the author believes that Taknar Formation is probably of Paleozoic age (Ghorbani 2012a). The author also believes that Sangan iron mineralization occurred in Late Proterozoic due to the volcanic activities, but it became more concentrated during Tertiary as a result of younger volcanic activities that gave this deposit skarnic feature.
- Kermani and Forster (1991) considered the Eocene volcanism of Eastern Iran and western Afghanistan responsible for the formation of Sangan deposit.
- Bomeri (1992) stated a skarn origin for the Sangan ores.

Taking into account the extensive studies carried out on Sangan iron deposit and its morphology, it can be inferred that the ore body originated due to volcanogenic processes and later on was metasomatically altered by hydrothermal solutions emanating due to the igneous activity of Late Eocene–Oligocene (e.g., Sarnosar Granite) turning it into a skarn that is very well developed, especially in the excavated mining areas (Karimpour 1991).

Clarification of the association of Sangan deposit with any of Tertiary (Late Eocene–Oligocene) volcanics, Late Proterozoic (geological report of Khavaf 1:100,000 Quadrangle), or volcanic rocks equivalent to Taknar Series requires further detailed investigations. However, by reckoning the claims of Boroumandi (1982) and other experts of the Geological Survey of Iran, Sangan deposit may be considered as an equivalent of Bafgh and Esfordi deposits having been formed in the interval between Late Proterozoic and Early Cambrian (Table 5.1).

5.1.2 Distribution of Iron Deposits

The overall distribution of the iron deposits and indications of Iran can be classified as under (Fig. 5.1):

1. Central Iran mostly concentrated around Bafgh (approximately 2 billion tons) (Ghorbani 2007a)

Table 5.1 Temporal and spatial distribution of iron deposits

Age		Paragenesis		Location	Deposits and indication
Tertiary Late Eocene–Quaternary		Magmatic	Concurrent with Oligocene volcanic and plutonic activity	Urumiyeh–Dokhtar zone	Niasar (Kashan), Daran, Shahrak, and Kuh Baba
			Associated with Oligo-Miocene igneous activity	Alborz zone	Semnan, Sorkhe Dizaj, Morvariyeh (north of Khoram Dareh), Damirchi
			Metasomatic and volcanic; orthomagmatic	East of Iran	Sangan
Mesozoic	Cretaceous	Magmatic skarn, volcanogenic	Magmatic skarn	Hamedan region	Baba Ali, Chenar Bala, Golali
			Magmatic skarn	Northeast of Khoy	Eskandian (Late Cretaceous–Early Paleocene)
			Volcano sedimentary	Southwest of Arak	Shams Abad (lower Cretaceous)
	Triassic–Early Jurassic	Associated with plutonic bodies		Northern regions of Sanandaj–Sirjan zone	Hezar Khani, Khosro Abad, Charmale (Songhor), Zafarabad, Hamekasi Complex
		Sedimentary		Central Iran	Kharanagh Oligist, Rebat, Posht Badam
Late Paleozoic–Early Triassic		Volcanosedimentary		East of Iran	Kalat Naser (east of Ghaen), Ahangaran region
				Southern Sanandaj–Sirjan	Honeshk Iron and Manganese (Dehbid, Fars)
				Hamedan–Kordestan Region	Zafarabad (Kordestan)
				Alborz zone	Masooleh (Gilan)

Late Proterozoic–Early Cambrian	Magmatic	Zarigan–Narigan type granitic magmas		Central Iran	Choghart, Chadormalu, Sechahoon, Mishdavan, Esfordi, and most of the deposits at Bafgh area
		Mafic and Ultramafic Magmas		South of Sanandaj–Sirjan zone	Gol Gohar
	Directly or indirectly associated with plutonic or volcano sedimentary rocks	Volcanosedimentary, intermingled with Rizoo and Dezoo Formations		Central Iran	Manganiferous iron deposit at Mishdavan and Ghar Dareh Dahoo
		Volcanosedimentary, associated with Qareh Dash Volcanics and lower parts of Soltanieh Formation	Intermingled with Qareh Dash Volcanics	Azerbaijan	Bordeh Rash Kuh, Bichaghchi and Hamam (Shahin Dezh)
			Stratiform with Upper Kahar and Lower Soltanieh as host	Azerbaijan	Arjin, Shah Bolagh, Mirjan, Ghaliche Bolagh, Char Tagh, Balestan
		Volcanosedimentary, along with Hormoz Series and its associated volcanic rocks		Bandar Abbas and islands towards the south	Tange Zagh, Hormoz, Larak and Qeshm

2. Sanandaj–Sirjan Zone, including Sirjan (Gol Gohar with probable reserves of 1.2 billion tons), Arak (Shams Abad with a reserve of 100 million tons), and Hamedan (Hamekasi with a reserve of 50 million tons), (Jafarzadeh et al. 1996)
3. Eastern Iran (reserves of up to 700 million tons), including Sangan in Taibad area (620 million tons), Deh Zaman in Kashmar (approximately 20 million tons), and smaller deposits such as Tanourche (Jafarzadeh et al. 1996)
4. Kordestan region, including Shahrak iron deposits (40 million tons), Zafarabad, and Asen Abad (Marivan) (Jafarzadeh et al. 1996)

Other scattered reserves also exist that are not economically viable, for example, Soltanieh, Shahin Dezh, Shah Bolagh, Mirjan, Semnan, and Bandar Abbas (Larak, Hormuz, and Tange Zagh). Figure 5.1 illustrates the main iron-bearing regions of Iran.

Iron Deposits of Bafgh: Since the most important iron deposit of the country is located in the Bafgh region, their origin is considered in detail in the following.

Almost all the experts who studied the Bafgh iron deposit have unanimously related it with the volcanic activities responsible for the formation of the Rizoo rocks (Daliran 1990; Jafarzadeh 1981; National Iranian Steel Co., 1983; Code 8 Archive 19 Report of National Iranian Steel Co.).

The main controversies about the iron deposits of the Bafgh area arise due to the fact that in spite of their close association with granites, they are always observed along with Rizoo volcanics and that is because there is a small age difference between these granites and Rizoo volcanics and both belong to the very same magmatic series (Ghorbani 2012a). The author is of the opinion that the following observations could be used in resolving this controversy:

- All the deposits are associated with the volcanic rocks of the Rizoo Series. These rocks are well exposed in the area between Narigan, Esfordi, and Zarigan.
- The volcanic rocks of the Rizoo Series that were thought to be rhyolitic are in fact basic to intermediate in composition (Houshmandzadeh 1988).
- The deposits of Bafgh are rich in apatite so that in some regions (e.g., Zarigan and Esfordi) the ore is classified as apatite-bearing magnetite.
- The REE content of the apatites is high in all the deposits of the area (Halalat and Bolourchi (1994) on Choghart area; Russian experts; Code 6, Archive 3 NISCO).
- Some apatite grains contain iron, which points toward the simultaneous formation of apatite and iron ore.
- The host rocks of all iron deposits have undergone metasomatic alteration.
- The radiometric dating of the iron-bearing apatite minerals of various deposits in the Bafgh area reveals a Late Proterozoic–Early Cambrian age (Ghorbani 2012a).

Thus, it can be deduced that all the iron deposits of Bafgh have alkaline magmatic–volcanic origin formed in a rift setting (Ghorbani 2007c). Some of these deposits have undergone metasomatic alteration along with their enclosing rocks in later phases. The time of emplacement of iron ores cannot be simultaneous with the Rizoo Series; the larger ore bodies of Bafgh are associated with plutonic igneous activity that intruded the volcanosedimentary Rizoo and Dezoo Formations (or even older rocks). The iron deposits have been formed in two successive but separate phases (Ghorbani 2007c): the first phase is associated with the volcanosedimentary rocks of Rizoo and Dezoo Formations and the second phase is associated with the plutonic bodies. Since the volcanic rocks of Rizoo and Dezoo have high iron content, they have helped in further enrichment of the iron-bearing plutonic bodies of second phase. Conversely, the plutonics have played an important role in the concentration of iron in the Rizoo Series of rocks.

5.2 Manganese

5.2.1 Mineralization Phases of Manganese

The manganese ores of Iran date from Late Proterozoic to Pliocene, but most of the deposits belong to Tertiary (Samani 1995). The important mineralization phases of manganese in the order of decreasing age are as follows:

Late Precambrian–Early Cambrian: During this interval of time, volcanosedimentary and volcanic manganese deposits have been formed in Central Iran and Azerbaijan, for example, Narigan deposit in the Bafgh area and Amir Abad manganese indication on the west of Angooran deposit. The manganese-bearing horizon of this time is well correlated with the lead–zinc horizons of Kooshk and Angooran, the sedimentary iron deposits of Mahabad–Soltanieh Belt, and Soltanieh Dolomites (Ghorbani 2002a). Most of the manganese deposits of this horizon have been discovered, but the hope for new finds is high.

Late Paleozoic: The manganese deposits of sedimentary and/or volcanosedimentary type are found within the Late Paleozoic rocks of Eastern Iran, Central Iran, and Sanandaj–Sirjan, which are mostly manganese-bearing iron deposits, for example, Honeshk manganiferous iron deposits in Dehbid area and Kalat Naser manganiferous iron in Ahangaran, Ghaen (Ghorbani and Mozafarzadeh 2008).

Cretaceous: In Early Cretaceous rocks of Sanandaj–Sirjan, there are a number of manganese and manganese-bearing iron deposits that are volcanosedimentary in nature, for example, manganiferous iron deposit to the north of Shams Abad and Robat in the Arak area, manganese deposit of Moorcheh Khort, and Ahmad Roghani deposit of Malayer. In addition, all the manganese-bearing iron deposits of the Malayer–Esfahan Belt are placed within the Cretaceous rocks (Ghorbani and Mozafarzadeh 2008).

Late Cretaceous–Paleogene: The manganese deposits of this time interval are associated with ophiolitic suites within the oceanic crustal rocks (Samani 1995), for example, Ab Band deposits of Estahban area, Goonij manganiferous iron deposit of Khash, Chah Basheh in Esfahan, Benviyeh, Gol Kangoo, and Kooh Dom deposits in Anarak, Sanboort, and Asad deposits in Sabzevar, Baft manganese deposit of Kerman, and Baghereh manganese deposit of Torbat area.

Eocene–Oligocene: Most of the economically important manganese deposits of Iran belong to this interval of time and are of volcanic, volcanosedimentary, and hydrothermal origin within the Paleogene pyroclastic and volcanic rocks, for example, the Venarch deposit of Qom, Robat Karim in Saveh area, the Bozni deposit in Qom-Naein Belt, and the manganese deposits in south of Sabzevar (Ghorbani and Mozafarzadeh 2008; Samani 1995). The geological indicators point to the possibility of finding more economic manganese deposits in these areas, if further explored.

Late Miocene–Pliocene: A number of hydrothermal manganese deposits are associated with the andesitic and andesitic–basaltic rocks of the Mio-Pliocene age, mostly in Azerbaijan. In spite of their small sizes, these deposits are usually very high grade (Samani 1995). Examples of these deposits are Debakloo nearby Mianeh, Manamin in the Ardabil area, and Bostanabad east of Tabriz.

5.2.2 *Distribution of Manganese Deposits*

The manganese deposits of Iran are classified into three types (NISCO 1977):

(a) Hydrothermal vein deposits associated with volcano-plutonic processes
(b) Volcanosedimentary deposits
(c) Polygenetic deposits associated with limestone formations

The formation of manganiferous iron deposits of Central Iran is attributed to the Pan-African metallogenic phase (Samani 1995). The rift-type pyroclastic–magmatic series of Posht Badam-Bafgh region, subduction-related manganese deposits within metamorphic rocks having ophiolitic affinities such as Petyar and Torkamani, and the Late Precambrian aulacogen-rift type of Anarak are examples of this type.

The Paleozoic volcanic activity of Sanandaj–Sirjan has led to mineralization of manganiferous iron in the metamorphic complexes (Hosseini 2011, e.g., in Honeshk deposit).

Among the manganese deposits that are related with iron mineralization of Cretaceous times is the Shams Abad deposit in Arak region (Samani 1995). Moreover, the Ab Band deposit (Neyriz), the Goonij deposit (Khash), and Bensport and Asad deposits (Sabzevar) are also associated with the ophiolitic suites of Cretaceous–Paleogene occurring within the lowermost sedimentary horizons of such suites (Samani 1995).

The Volcanogenic manganese deposits are located in Qom-Naein and Sabzevar manganiferous provinces (Ghorbani and Mozafarzadeh 2008). These are either volcanosedimentary or hydrothermal in nature. Hydrothermal manganese deposits are the result of the passage of chemically active solutions through Paleogene pyroclastic and volcanic rocks, for example, Venarch, Robat Karim, and Bozni.

In addition to the above-mentioned deposits, younger volcanic activities have resulted in the formation of manganese deposits in some parts of Iran such as Mianeh (Debakloo) and Bostanabad east of Tabriz .

Tables 5.2 and 5.3 depict the geographic distribution of manganese deposits and indications of Iran while Table 5.4 lists the general characteristics of some of these deposits. The geographic distribution of manganese deposits can be viewed in Fig. 5.2.

Table 5.2 Provincial distribution of manganese deposits and indications

Sr. no.	Name of deposit or indication	Province	Sr. no.	Name of deposit or indication	Province
1	Debalkoo	East Azerbaijan	13	Bozni	Esfahan
2	Eideh Koochoo Ghani	East Azerbaijan	14	Sorkh Shad	Esfahan
3	Ghoopooz	East Azerbaijan	15	Chah Sefid	Esfahan
4	Vila Dareh	East Azerbaijan	16	Robat Karim	Tehran
5	Zarshloo	East Azerbaijan	17	Venarch	Tehran
6	Khalifeh Kamal	East Azerbaijan	18	Bensport	Khorasan
7	Manamin Khalkhal	East Azerbaijan	19	Asad	Khorasan
8	Koloocheh	East Azerbaijan	20	Nogh	Khorasan
9	Chai Talvar	East Azerbaijan	21	Bagh Qareh	Khorasan
10	Janbehan	East Azerbaijan	22	Salam Rud	Khorasan
11	Chahbashi	Esfahan	23	Honeshk	Fars
12	Benvid	Esfahan	24	Ab Band	Fars

Ghorbani and Mozafarzadeh (2008)

Table 5.3 Provincial distribution of manganiferous iron deposits and indications

Sr. no.	Name of deposit or indication	Province	Sr. no.	Name of deposit or indication	Province
1	Qezeljeh Maragheh	East Azerbaijan	11	Arad Kuh	Tehran
2	Kuh Zard	Esfahan	12	Mohammad Alikhan	Tehran
3	Khaloo Heydar	Esfahan	13	Bardashkhaneh	Khorasan
4	Torkamani	Esfahan	14	Abdol Abad	Zanjan
5	Moorchehkhort	Esfahan	15	Chah Gabri Se Kavir	Semnan
6	Hassan Rabat	Esfahan	16	Gonij	Sistan and Baluchestan
7	Mooteh	Esfahan	17	Sangaz (Sargaz)	Kerman
8	Chah Palang	Esfahan	18	Shams Abad	Markazi
9	Qal'e Gong	Esfahan	19	Malayer	Hamedan
10	Niyak	Tehran	20	Kuh Darbid	Yazd

Ghorbani and Mozafarzadeh (2008)

5.3 Chromite

5.3.1 Metallogenic Phases of Chromite

All economic chromite deposits of Iran have been formed in the Late Cretaceous–Paleocene time interval (Ghorbani et al. 2009). During Jurassic–Cretaceous (sometimes up to Paleocene), Alpine orogeny resulted in the closure of Neotethys Ocean forming numerous chromite deposits extending from Spain to Philippines. The Chromite deposits of Iran correspond to the middle to eastern parts of this ophiolitic belt constituting valuable chromium sources. However, some geologists (e.g., c, Sabze'ei 1985) attribute the formation of Iranian chromites to intra-cratonic

Table 5.4 General characteristics of some manganese deposits of Iran

Age	Name of deposit or mineral indication	General characteristic
Neogene and Quaternary	Debakloo, Eideh Koochoo Ghani, Ghoopooz, Vila Dareh, Manamin, Zarshloo, Khalifeh Kamal, Galoojeh, Chai Talvar	Hot spring deposits (terrestrial)
Tertiary (volcanic-plutonic)	Chah Sefid, Sargaz, Robat Karim, Bozni, Sorkhshad, Abdol Abad, Noogh	Hydrothermal–vein (terrestrial)
Late Cretaceous–Paleogene	Venarch, Asad, Bensport, Bagh Qareh, Ab Band, Gonij, Benvid, Salam Rud, Zaboli	Sedimentary–hydrothermal (marine) Volcanosedimentary (marine)
Early Cretaceous	Shams Abad, Chah Basheh	Sedimentary associated with iron (marine)
Paleozoic	Kalat Naser, Heneshk	Associated with iron
Late Precambrian	Narigan, Amir Abad (east of Angooran)	Volcanosedimentary (marine)

Ghorbani and Mozafarzadeh (2008)

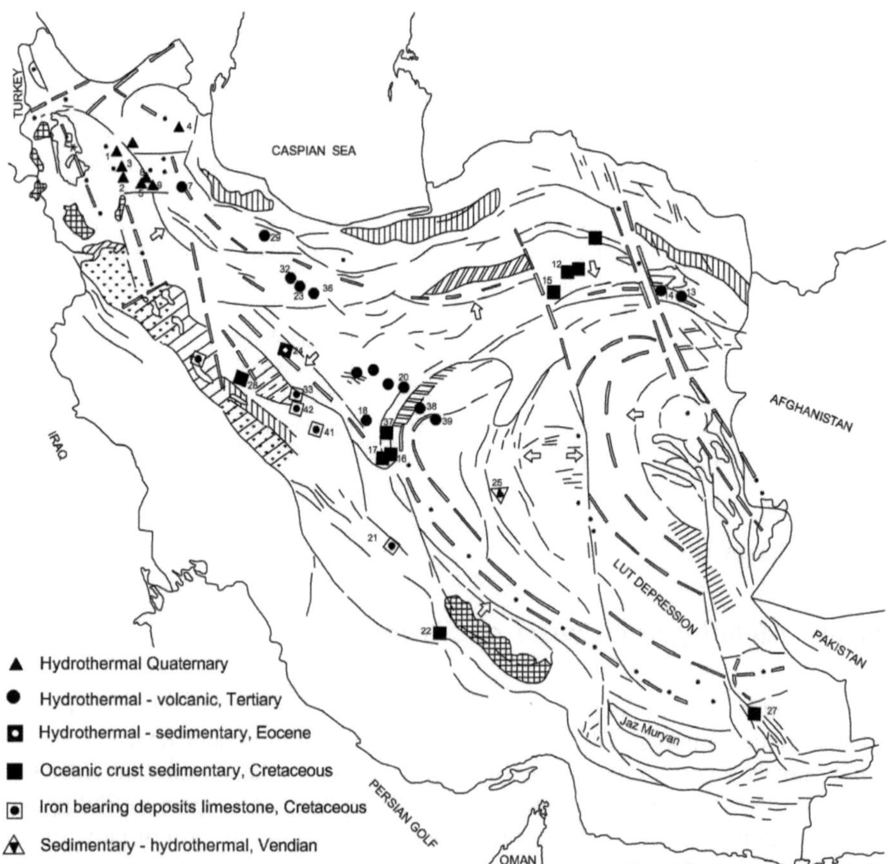

Fig. 5.2 Distribution map of manganese deposits of Iran (Samani 1995; with some modification by M. Ghorbani)

rifts. Sabze'ei (1985) is of the opinion that chromites have been formed in an intracratonic rift at Sanandaj–Sirjan during Late Ordovician–Early Silurian from komatiitic magma. This opinion is not acceptable to the author of this book. Sabze'ei attributes chromite mineralization of Esfandagheh and Faryab areas to this phase of magmatism. However, later on, Sabze'ei attributes Esfandagheh and Faryab to Precambrian. But the author believes that the age of ophiolitic complex (Jurassic–Paleocene) indicates that the chromites have been formed during Laramide orogeny in Neotethys (Metallogenic Map of The Middle East 2011; Ghorbani 2012a).

Chromite mineralization of Iran is of Alpine type, in which the ultramafic masses are mostly composed of harzburgite and lherzolite (Ghorbani et al. 2009). Examples of Alpine type include deposits of Greece, Cyprus, Turkey, Iran, Oman, and Pakistan, most of which are cumulative deposits of the Mesozoic and Cenozoic age. Deposits of Ural, Albania, Philippines, Turkey, and Iran constitute the largest Alpine-type chromite reserves of the world (Metallogenic Map of The Middle East 2011).

A number of chromite mineralizations have been reported from northwest (Gheshlagh Khoi area), northeast (Sabzevar area), and southeast (Fonooj area) of Iran, as well as Neyriz, Khash, Karevandar, and Sefidabeh that belong to Late Cretaceous (Metallogenic Map of The Middle East 2011). The chromites of the Esfandagheh–Faryab region most probably belong to the Late Cretaceous–Paleogene (Laramide) phase of mineralization.

5.3.2 Distribution of Chromite Deposits

The first report of chromite occurrence in Iran dates back to 1940 when a body of chromite was discovered at a distance of 14 km from Kahak village, 70 km from Sabzevar city, now known as Foroomad Chromite (Anaraki 1969). Taking into account the irregular to lensoid shape of chromite masses and their association with serpentinites as well as the overall characteristics of mafic and ultramafic rocks and serpentinized peridotites, most chromite bodies of Iran are believed to be of Alpine type (no Bushveld-type deposit has been reported till date) (Ghorbani 2002b; Ghorbani et al. 2009). The chromite deposits of Iran are concentrated in the following regions:

Esfandagheh–Faryab: Chromites of this area can be divided into two distinct belts, one in the north and the other in the south. The northern belt is located at a distance of 60 km to the southeast of Baft while the southern belt occurs on the northeast of Minab at a distance of 30 km.

The northern belt contains Lar chromite measuring 60 km in length and 5–10 km in width. A number of deposits have been discovered in this area since 1955, the most important of which are Abdasht, Soghan, Sheikh Aali, and Seikhoran that are located at an altitude of 2,150–2,300 m from mean sea level. According to Sabze'ei (1974), the chromites of this area belong to four distinct stratigraphic horizons:

1. Lower dunite
2. Harzburgite and pyroxinite layer

3. Upper dunite and harzburgite
4. Lower gabbro within pyroxinites and dunites

In the Abdasht area, chromites have been found in Sheikh Aali, Soghan, Abdasht, Bagh Borj, and Kamal Abad. Abdasht ultramafic massif measures 8 km in length and 5 km in width and is composed of a more or less regular alternation of dunite, harzburgite, lherzolite, and highly metamorphosed verlite. The probable reserve of Esfandagheh is estimated at eight million tons (Etemadi 1997).

The length of the southern belt (Faryab area) is almost 8 km while its width varies between 2 and 3 km. There are 23 mining localites in the area, the most important being Shahryar, Amir, Ebrahim, Nazafarin, Nader, Majid 1 and 2, Yasmin, Sarhang, Baharsang, Reza, Dastgerd, and Kuh-e Sorkh (Ghorbani 2008g). Faryab mines are located in an area where the major geological divisions of Iran change their trends from northwest–southeast to north–south. The area is at the conjunction of two main fault zones trending in north–south and northwest–southeast directions; they play a pivotal role in the formation of structural features of the area. The rock units of the area are as follows (Sabzehei 1974):

1. Paleozoic (Devonian) metamorphics belonging to Sanandaj–Sirjan zone
2. Ultramafic rocks of Sorkh Band massif; this unit is in fact the lower part of unit 1 metamorphics, separated due to tectonic upheavals
3. Glaucophane schist of colored mélanges
4. Colored mélanges

Folding and metamorphic activity within the area is not very widespread, but if present, the grade of ore as well as the thickness of the chromite horizon increases significantly. The enclosing rocks are structurally undisturbed, and, as a result, mineral indications are located easily. Recently, a new two million ton reserve has been discovered in the vicinity of Amir Mine at Faryab. The probable reserve of Faryab is around 30 million tons (Nasrollazadeh 1997).

Sabzevar–Torbat-e Jam: This region is located between Fariman and Shahrood. Chromite mineralization is in the form of lens-shaped bodies of varying sizes measuring from 10 to 350 m in length and 2–6 m in thickness. A total of 30 mines exist in the area, 23 are around Sabzevar and 7 in Torbat-e Heidariyeh, some of which are active. The amount of chromium oxide (Cr_2O_3) varies from 38 to 50%. Chromite occurs in east–west trending serpentinized dunites and harzburgite bodies. The shallower parts of these deposits have been excavated, and further work is suspended (Vatanpour 1998; Ghorbani et al. 2009).

The Sabzevar ophiolitic zone is located to the north of the Dorooneh (Great Kavir) Fault and south of the Shahrood Fault covering an estimated 2,000 sq. km. The exposed ophiolites in the north and south of zone differ from each other and have the following characteristics (Geological Map of Iran, 1:2,500,000, 1989; Alavi-tehrani 1979; Ghani et al. 2010).

The northern outcrops have general east–west trend and extend from Abbasabad to east of Soltan Abad (west of Neishabour). The ophiolites to the south of Sabzevar outcrop to the north of Bardeskan, west of Kashmar, and south of Dolat Abad (Darvishzadeh 1991). The ultrabasic masses of Sabzevar–Torbat-e Heydariyeh are in the form of scattered shallow (rootless bodies extending about 500 km in length and 10–50 km in width). The Joghtai Mountains are situated in the north of the area, while the western and southern parts are covered by Mesozoic sediments and recent conglomerates, respectively. The chromites of this area have leopard skin and stripped texture and are of low density. The most important deposit of the area is the one located at Foroomad, 40 km north of Abbasabad.

On the south of the Gaft area, a number of lens bodies of chromite are seen that run parallel to each other, the most important being the Great Gaft Lens measuring 65 m in length. The chromites of this area are dense having leopard-skin texture. In the Soroor village, 9 km to the north of Foroomad, there are two chromite lenses that have been mined out. Younger intrusive bodies of gabbro, diorite, and even granite occur in the north of Soroor and Gaft villages (Ghorbani 2008g).

The Mir Mahmood deposit is located at a distance of 8 km to the northwest of the Foroomad village, on the southern slope of the Joghtai Mountain. It consists of four lenses of chromite that yielded 20,000 tons annually (Vatanpour 1998; Ghorbani 2008g). Another deposit situated at about 6 km from Foroomad is the Ghandaviz deposit, which contains two large lenses of chromites and a few deserted excavation sites exist in Kuh Siyah to the north of Sabzevar. In all the said localities, the shallower parts of the ores have been removed, but it seems that the lenses extend far below the surface where they are still intact.

Two deserted chromite mines occur at Torbat-e Heydariyeh that contain nickel indications (Ghorbani et al. 2009). The chromite bodies of this area are more or less concordant, with mineralization being concentrated in the uppermost ultramafic horizons of ophiolite suites. However, smaller lenses of chromite are scattered through the deeper parts of ultramafics. Chromite occurs in the form of massive bodies of economic potentials in the north of Abdar and Sabzevar as well as in the Ziarat area. These cumulate bodies have undergone various types of deformation and fracturing due to tectonic forces, thus complicating the exploration work.

Numerous deserted mines are scattered at a distance of 45–55 km to the north of Torbat-e Heydariyeh that housed cumulate-type of chromite in association with dunites. Chromite grains measuring less than 3 mm, oriented parallel with foliation surfaces, show cataclastic texture whose fractures are filled with pentlandite mineral. In some cases, nickel is also present as traces in noritic rocks (Razmara 1990). A number of copper veins measuring a few centimeters in thickness and tens of meters in length cut across diabases and other rocks on the north of Sabzevar.

Neyriz: Distinct outcrops of Neyriz–Oman ophiolitic zone extend with a north–south trend from Dehbid to Neyriz. This ophiolitic band is cut across by Zagros Thrust further southward on the east of Neyriz, where it disappears. Nevertheless, it appears once again in the Oman mountains, and hence the name Neyriz–Oman has been given to it (Geological Map of Iran, 1:2,500,000, 1989; Metallogenic Map of The Middle East 2011).

The Neyriz ophiolite suite conformably overlies clayey limestones of Sarvak Formation (Cenomanian–Turonian age). Petrographically, the Neyriz ophiolites consist of the following rock units in the upward direction (Sheikhi-kahrizaki 2006):

1. Tectonized harzburgite with lherzolitic unit
2. Ultramafics with intercalations of dunite, chromite, wehrlite, websterite, and clinopyroxinite
3. Cumulate units including lower and upper gabbros
4. Layered dikes consisting of dolerite and basalt
5. Minor plagiogranites located between dunites and upper gabbros
6. Pillow lavas and radiolarites

The most important metamorphic rocks of Neyriz ophiolites include skarn and marble, which are in contact with peridotites and seem to be the result of metamorphism of peridotites and limestones. Among the chromite mines of the area, Khajeh Jamali, a deserted mine, is more important. The enclosing rocks of chromites consist of serpentinite, pyroxenite, and harzburgite.

Khoi: On the Iran–Turkey border, to the west of Lake Urumiyeh, a number of ophiolite–radiolarite suites exit that trend in the north–south direction. The area falls between 44°44′ and 44°47′ eastern longitudes and 38°39′ and 38°40 northern latitudes. The average altitude of the area is 1,950 m above mean sea level. Stocklin (1974) and Darvishzadeh (1991) are of the opinion that the ophiolites of northwestern Iran (Makoo–Khoi–Urumiyeh) are similar to the colored mélanges of Central Iran. The ophiolitic rocks of the area consist of the following units:

1. Ultramafics
2. Diabases
3. Leucocrates
4. Basalts and andesites
5. Pelagic sedimentary rocks

Sistan and Baluchestan: More than 50 deposits and indications of chromite occur in Sistan and Baluchestan Province, which can be divided into two categories according to their geological setting (Science and Technology Research Institute 1999; Ghorbani 2002b, 2008g):

(a) Makran chromites, which are in association with ophiolites of south Jazmourian and have an east–west trend extending from Kahnouj to south of Iranshahr, for example, Goutij, Sabze–Ghalman, Mokhtar Abad, and Serze in the Fonouj area.
(b) Iranshahr–Nehbandan chromite belt, which trend in the north–south direction. A number of chromite deposits are found in this belt (e.g., Kuh Zarij, Chehzar, Nosrat Abad, Sefidabeh).

The chromite deposits of Sistan and Baluchestan form small reserves that are dominantly aluminum-rich chromites with low-chromium content. Platinum has also been reported from a few chromite bodies of the area, for example, Chehzar (Khash).

Figure 5.3 illustrates the chromite-bearing regions of Iran.

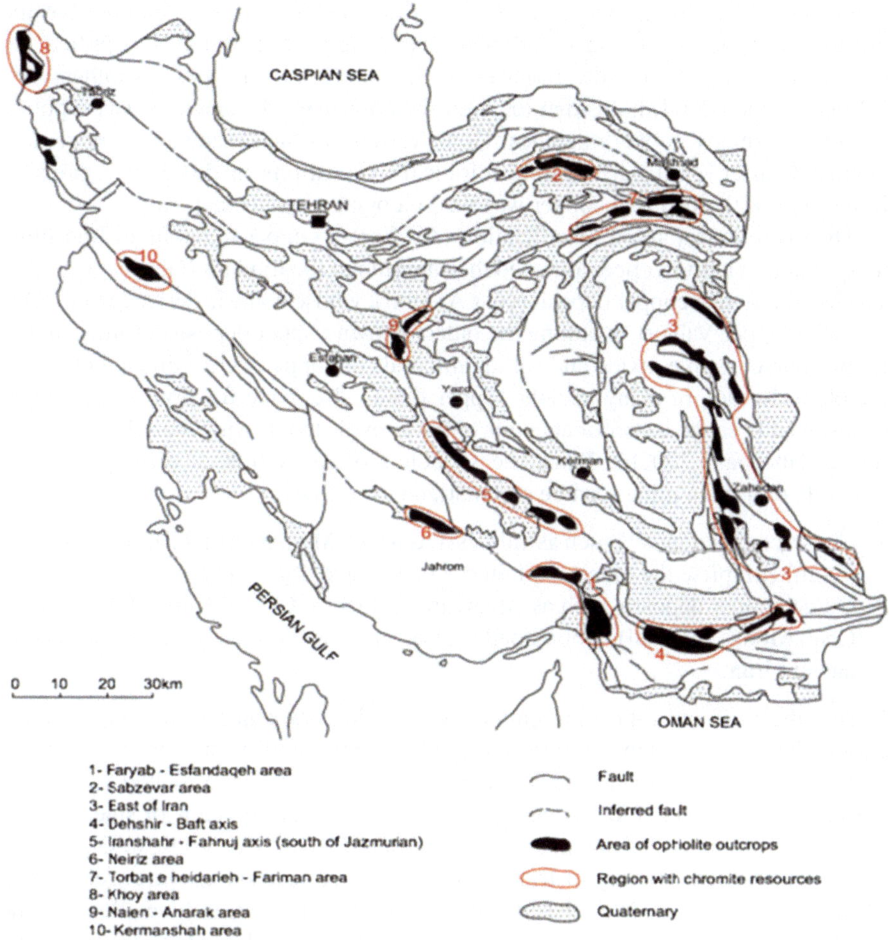

Fig. 5.3 Prioritized chromite-bearing areas of Iran

5.4 Copper

5.4.1 Introduction

The Chemical and physical characteristics of copper have made it one of the most useful metals. Properties such as electrical and thermal conductivity along with resistance to corrosion have widened the applications of copper in modern society.

Copper was the first metal discovered and utilized by man (Khoyi et al. 1999). There are many indications that the discovery of copper took place in Iran. The places that have yielded the most ancient evidences on copper smelting are the first and second levels of excavation at Siyalk (near Kashan), which is definitely older than Tal Eblis (Zawvosh 1996). Wertime (1973) is of the opinion that the oldest copper mine of Iran is Talmesi near Anarak that probably supplied the raw material needed for the furnaces at Siyalk (Zawvosh 1996). Another view (Momen-Zadeh 2004) holds that the required copper for the smelters was supplied from the Veshnave copper mine near Qom. Various artifacts made up of copper and gunmetal, discovered during archaeological excavations in different regions of Iran, point to the fact that copper was well recognized in ancient Iran.

The evolutionary trend of utilization of different copper minerals in Iran illustrates that first native copper attracted the attention of the Iranians (Ghorbani 2002a). The Mejdar native copper deposit (near Ardebil), whose name is derived from "Mes Darre" (Copper Valley), is among the oldest known copper deposits of the country. In the course of time, oxide and sulfide minerals of copper were extracted one after the other. In fact, in many ancient copper mines, the oxide fraction of the copper deposit had been the center of attention and was extracted, leaving the sulfide fraction intact (Ghorbani 2002b). Thus, the sequence of extraction of different copper minerals in Iranian mines can be summarized as follows:

1. Native copper deposits such as Talmesi (Anarak), Mejdar (Ardebil), and Veshnave (Qom) comprise the oldest exploited mines (since Shadadi Era).
2. Oxide copper deposits such as Ahangaran (Ghaen), Chehel Kure (Baluchestan), Chah Moosa (Toroud), Abbasabad (Sabzevar), and Kerman area were known in ancient Iran.

During the fifth and fourth millennia B.C., Iranian craftsmen were able to create enough heat to reach temperatures required for melting of most of the then-known raw materials and extract metals. Cuprite and malachite were melted using coal extracted from mines in the Anarak area, where there are remains of smelters that consumed these types of raw materials (Ghorbani 2002a). According to Diyakonov (1967) in "Ancient History of Iran," copper-melting procedures were well known in various parts of Iran. With the advancement of the knowledge of metallurgy in the Achaemenian period, finely crafted copper and gunmetal objects were created. Copper coins were also produced in this period of Iranian history. Enormous amounts of copper were used in the construction of "Ya'jouj-o-Ma'jouj" dam (also known as Kourosh Kabir dam) in the Gorgan area (Zawvosh 1996).

In the Islamic Period, the well-known Arab traveler Abudolaf (Minoreski 1963) has written about the greatness of the Neishabour copper mine. But it was not until the Safavid times as the foreign travelers described the mineral riches of Iran, where the vast deposits of the country are exposed to the world.

Chardon (1956) writes, "copper is found in Sari, Khorasan and Ghazvin; however, Iranian copper is not malleable. It has to be mixed with copper from Sweden and Japan to make it soft."

During the seventeenth century, Keary (from Keary Log Travel to Iran, 1969) states, "three miles from Tabriz, there exists a gold mine which is not operational because the extracted gold is not sufficient to cover the running cost of the mine. A copper mine is situated four miles from there whose produce is enormous and has huge revenues for the Royal Treasury."

Conte Shoushuar, a French ambassador of Qajariye times, writes (Zawvosh 1996), "Copper consumption is very high in Iran with no equivalents around the world. Every domestic device is made up of copper. Most of the copper utensils are produced at Kashan."

5.4.2 Copper Alloys

The alloys of copper known to ancient Iranians include

1. Brass (called "berenj" in Persian), made up of copper and zinc, was used in the making of tools and utensils. Due to the difference in the proportion of copper and zinc, the density of brass was estimated as 7.5 and 8.6 g/cm^3 by Biruni and Khazeni, respectively (Zawvosh 1996).
2. Homogenous mixture of copper and lead, called "petroy," was used in the manufacturing of pestles and mortars (Zawvosh 1996). This alloy is also named "Tal" in Tansuqname and Harayes Al-Javaher. Biruni writes in Al-Jamaher, "it is a type of shabah (a type of rock) called shabaha mefragha." In other works such as Rasa'el-e Ekhvan-ol-Safa, this alloy is referred to as "mefragh" (gunmetal) (Zawvosh 1996).

Zawvosh (1996) is of the opinion that in the Post-Islamic Period (i.e., eighth century A.D.), the name "mefragh" was given to an alloy of copper and lead. Nowadays, this term (mefragh or gunmetal) is used for an alloy of copper and tin, which was known in ancient Iran as "espidruy." Biruni and Khazeni have both talked about espidruy and have estimated its density as 9 and 8.9 g/cm^3, respectively.

It is now definite that there are no tin reserves in Iran (Ghorbani 2001). The localities famous for their tin mines, for example, Lorestan and Sistan, in fact host lead and copper mines. Thus, the author believes that the term mefragh in the old works actually indicates an alloy of copper and lead, since both these metals occur together in the ancient mines exploited in Iran.

5.4.3 Copper Metallogeny in Iran

More than 2,000 km of the global copper belt passes through Iran (extending diagonally from Azerbaijan on the northwest to Sistan and Baluchestan in southeast and from Semnan to Sabzevar in Khorasan) resulting in the occurrence of about 100 copper deposits that have been explored and surveyed (Fig. 5.4). Many of the world-famous economic geologists who have worked on the country's copper deposits believe that Iran is one of the places with very high tendencies of discovering new porphyry copper deposits. However, no detailed investigation has been carried out on the mineralogy and metallogeny of copper in Iran. The results of chemical analysis of magmatic or metamorphosed igneous rocks reveal the existence of nominal

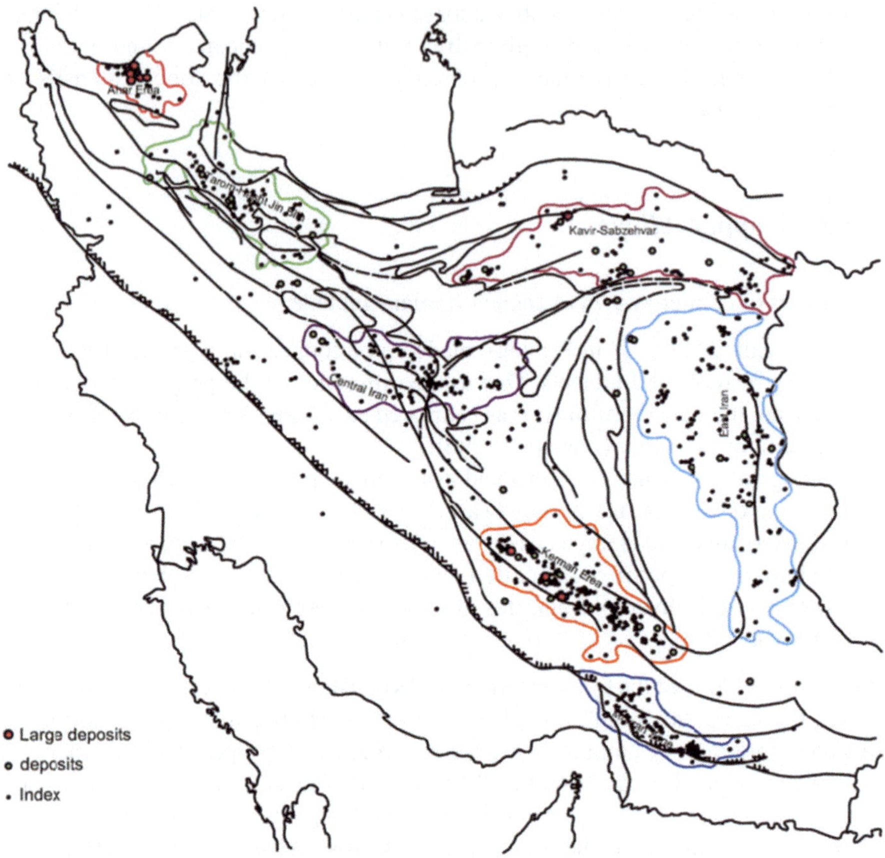

Fig. 5.4 Distribution map of copper deposits and indications in Iran

copper concentration in the pre-Tertiary intrusive igneous rocks of Iran. Only a few intrusive bodies of pre-Tertiary times are found that contain higher-than-average copper content, for example, Taknar Formation. Nevertheless, most of the volcanics, andesites, andesitic basalts, trachyandesites, and basalts of Middle Eocene have significant copper. At some of the metallogenic zones such as Kerman, Tarom, and Ahar, high anomalies of copper have been recorded.

The following generalizations can be enumerated for the Iranian copper deposits:

1. Most of the porphyry copper deposits of Iran are associated with the Tertiary intrusive and extrusive igneous rocks.
2. Older copper reserves are not comparable with the Tertiary deposits in terms of quality and quantity.
3. Wherever Tertiary copper is found, a definite relation exists between them and Eocene extrusive and shallow intrusive rocks.

From all the above-mentioned facts, it can be deduced that in areas such as Kerman, Ahar, Tarom, and Hashtjin, where the Eocene extrusive activity is voluminous, the volcanics have acted as the host rock, having been invaded by the copper-rich magmas of Late Eocene–Pliocene. Leaching and recycling of copper from the preexisting rocks and the chemically active solutions thus produced have resulted in the formation and enrichment of copper deposits.

5.4.4 Phases of Copper Mineralization

While there are indications of copper mineralization in Iran from Late Proterozoic to Pliocene, most of the copper deposits of the country belong to the Tertiary (especially Oligocene and Miocene) time (Ghorbani 2008b). The overall phases of mineralization of copper can be categorized into the following.

5.4.4.1 Late Precambrian–Early Paleozoic

No major copper deposit is known to have been formed at this interval of time. However, a number of indications (such as Yasouj and Shahrekord) occur within the rocks of these periods which are not so important economically (Kar Ma'dan indication at the foothills of Dena). Though the volcanic, sedimentary-volcanic, as well as metamorphosed igneous types of rocks (such as those exposed in Takab, Anarak, and Bafq) have low copper contents, their lead–zinc content is abnormally high espidruy.

5.4.4.2 Late Paleozoic

Traces of copper have been reported espidruy in association with some of the metamorphic rocks of Late Paleozoic, which are of igneous or pyroclastic origin (e.g., Taknar massive sulfide deposit associated with the Taknar Series). The lead, zinc, and copper ore body of Chah Gaz also belongs to this time interval. There is a high probability of discovering new deposits if the equivalents of these rocks (such as Gorgan Schists and Shandorman Metamorphics) are further investigated, particularly in localities where a follow-up magmatic activity has taken place.

5.4.4.3 Cretaceous–Paleocene

There are indications of massive sulfide reserves of copper associated with the Cretaceous and Paleocene ophiolite suites, for instance, copper indications at Haji Abad (Bandar Abbas), Sargaz (Jiroft area), and Rameshk (southwest Jazmourian) (Ghorbani 2006b).

5.4.4.4 Tertiary

Most of Iran's copper reserves have been formed during the Tertiary phase of mineralization. These deposits are closely related with the Tertiary magmatism that can be further divided into the following phases (Khoyi et al. 1999):

(a) Eocene: Being dominantly volcanic in nature, the rocks of this phase have high-background copper content. A number of deposits and indications are recognized within this phase.
(b) Late Eocene–Early Oligocene: Most of the intrusive igneous rocks, including granites, granodiorites, tonalities, and diorites of this interval of time, are associated with vein-type copper deposits, for example, Mazra'e and Ghal'e Zari.
(c) Late Oligocene–Early Miocene: During this interval, which is the second largest phase of Tertiary volcanism and magmatism, a number of porphyry- and vein-type copper deposits are formed in association with the igneous rocks.
(d) Late Miocene–Pliocene: Mostly includes small shallow intrusive bodies that are believed to contain most of the porphyry coppers of Iran. The igneous activity subsides in this interval, but there are a number of deposits and indications that have been formed during or just after this time; however, no information is available on them.

5.4.5 Distribution of Copper Deposits in Iran

Comparing the distribution of copper deposits of Iran with the maps showing the dispersion of magmatic rocks on the scales of 1:2500000 (Aghanabati 1991) and 1:1000000 (Emami et al. 1993), the following zones of copper mineralization can be recognized:

1. Urumiyeh–Dokhtar Zone, which is further divided into three areas, namely, Southern (Kerman), Central (Anarak-Kashan), and Northern
2. Taleghan–Tarom–Hashtjin Zone (western Alborz)
3. Sabalan Zone (Arasbaran area, west of Ardebil)
4. Kavir–Sabzevar Zone (Binaloud and Taknar)
5. Lut Zone
6. Makran Zone

5.4.5.1 Urumiyeh–Dokhtar Zone

The rocks of Urumiyeh–Dokhtar Zone characteristically consist of Eocene to Miocene diorites, granodiorites, and granites. There is no agreement among some experts on the mode of origin of these rocks (M. Sabze'ei, M.H. Emami, J. Omrani, H. Moein-vaziri); some experts (J. Omrani, H. Moein-vaziri) consider them as products of a magmatic arc (Ghorbani 2012a); some (M.H. Emami, ?. Amidi) believe that they are the result of rifting (Ghorbani 2004a). While the northern and central parts are comparatively much poorer in copper content, the southern parts of this zone are rich in terms of copper minerals. In fact, copper has been replaced by other metals such as iron, lead, and zinc in the north direction (Ghorbani 2006b).

The Kerman region, situated in the southern parts of the Urumiyeh–Dokhtar Zone, hosts 600 km of the copper belt with a varying width of 40–70 km (Ghorbani 2006b). Geographically, this zone trends northwest–southeast, extending from Shahr-e Babak to Bazman. More than 300 copper deposits and indications exist within this belt, 20 of which are thought to be of porphyry type (Ghorbani 2002a).

However, as mentioned earlier, the copper content of the zone reduces in the northward direction. In the central parts of Urumiyeh–Dokhtar, only a few porphyry copper deposits are known, for example, Khonj-Tal-e Siah and Kale Kafi (near Anarak) and Dare-Zereshk (Taft), whereas no such deposits exist in the northern parts.

5.4.5.2 Taleghan–Tarom–Hashtjin Belt

There are extensive deposits of copper in the Tarom region, all of which are associated with Eocene extrusive (mostly andesites) and intrusive (mostly tonalites and granites) rock bodies and show high copper anomalies.

The intrusive granite bodies of the Tarom area are of I-Type (Cordilleran) accompanied by extensive iron and copper mineralization. In the Taleghan area, a number of copper deposits and indications occur in association with monzonites and alkaline volcanic rocks (especially trachytes); the copper minerals of these rocks are of carbonate type (Ghorbani 2002b).

5.4.5.3 Arasbaran (Ahar Area, West of Ardebil)

The Ahar area is the only location that structurally and mineralogically satisfies the conditions to be termed Sabalan copper-rich zone. The area is very important from the point of view of copper mineralization and hosts a number of deposits. The northward extension of this zone comprises the copper–molybdenum deposits of Armenia.

The copper mineralization within the Sabalan Zone is of magmatic origin (Ghorbani 2011). Taking into account the habit and type of mineralization, the copper deposits of this zone can be divided as

(a) Porphyry copper deposits, which are molybdenum-rich, for example, Songoon and Kighal deposits (Lotfi et al. 1993)
(b) Vein and skarn copper deposits, which sometimes contain gold, for example, Mazra'e and Barmalek

5.4.5.4 Kavir–Sabzevar Zone (Binaloud and Taknar)

This zone is situated on the north of the Daroune fault to the south of the Miyamey fault and Binaloud Mountains. A number of copper ore deposits and indications are associated with the Tertiary volcanics of this zone. Most of the ore bodies are in the form of veins formed in conjunction with andesitic and basaltic volcanic rocks (Ghorbani 2002a).

5.4.5.5 Lut Zone

In the Lut Zone, numerous deposits and indications of copper occur along with the Tertiary volcanics, mostly in association with andesites, for example, Ghal'e Zari copper deposit (Khoyi et al. 1999).

5.4.5.6 Makran Zone

A belt of copper mineralization occurs to the north of Makran and south and southwest of Jazmourian. This zone trends northwest–southeast in the Faryab area but upon reaching Mokhtar Abad changes its trend to east–west (Ghorbani 2006b). Most of the 50 copper indications of this zone have been mineralized in association with flysch and volcanic–andesitic complexes. No detailed investigation has been carried out in the area, but the geological evidence is in support of massive sulfide origin of all the ore bodies.

The most important copper deposits of Iran along with details are presented in Tables 5.5 and 5.6.

The relative advantages of Iran's copper deposits over the rest of the world are

1. The average grade of known porphyry copper deposits with reserves exceeding 1.5 billion tons is above 0.65% (PRGC-Arian Zamin 2004).
2. Gold, molybdenum, and occasionally silver are associated with copper in all known porphyry copper deposits. These elements make Iran's copper reserves more valued. For example, the Sarcheshmeh copper deposit with an estimated reserve of one billion tons of 0.70% grade ore contains 270 ppb gold, 300 ppm molybdenum, and 3.09 ppm silver while the Songoon copper deposit has one billion tons of copper ore with an average of 0.70% copper and 50 ppm molybdenum (PRGC-Arian Zamin 2004).
3. Though the reserves of vein-type copper deposits of Iran are not as great as the rest of the world, presence of a few veins in almost all the mineralized zones can make the feed of a large smelting plant.
4. The vein-type deposits are usually rich in gold content (around 10 ppm), for example, Khalife Lou (Tarom), Mazra'e, and Anjard (both in the Ahar area) (Maghsoudi et al. 2005).

From the exploration point of view, the copper deposits of the country have the following advantageous characteristics (PRGC-Arian Zamin 2004):

1. Excepting a few, most of the deposits are either hosted by volcanic rocks (ranging in composition from dacite to andesite) or their intrusive equivalents of Miocene–Pliocene.

Table 5.5 Most important copper deposits in Iran

Name of deposit	Location	Reserves and grade	Details
Sarcheshmeh	Kerman	1.2 billion tons with an average grade of 0.68% of copper	Molybdenum (300 ppm), gold (270 ppm), and silver (309 ppm) are associated with this deposit
Meydouk	Kerman	Exploitable deposit is estimated at around 140 million tons with an average grade of 0.85%; comprising of 70 million tons of hypogene reserves having 0.70% grade and 70 million tons of supergene reserves having 1% grade	The results of drilling in the area reveal that the probable reserve of Meydouk is around 500 million tons and the copper grade increases with depth
Darre Zar	Kerman	1.1 million tons oxide reserve with 0.95% grade. 41 million tons of 0.70% proved reserves and 100 million tons of probable reserves	In spite of the absence of supergene deposit, the stripping volume is not great. Molybdenum grade reaches around 500 ppm
Songoon	Ahar	Proved and probable reserves estimated at about one and two billion tons, respectively, with an approximate grade of 0.70%	The supergene reserve is limited, and most copper is situated in hypogene level. Molybdenum grade is around 200 ppm
Mazra'e	Ahar	Total reserved measures at 431,062 tons with 1.7% copper grade. Exploitable reserve is around 387,900 while probable reserves is 860,000 tons	–
Anjard	Ahar	–	–
Darre Zereshk	Taft	Proved reserve is 29 million tons with 0.68% copper	The reserves are estimated for 3,079 m level using 21 boreholes
Kale Kafi	Anarak	Proved reserve is 13.5 million tons with 0.90% copper	Mineralization is in supergene region with very poor hypogene. Total drilling in the area 15 boreholes measuring 2,200 m
Aliabad	Taft	–	–
Taknar	Kashmar	–	–
Abbasabad	Shahrood	–	–
Ghal'e Zari	Eastern Iran	Proved reserve is 360 thousand tons with 0.30–0.50% copper	The deposit is of vein-type with high gold content of about 10 ppm
Lar	Zahedan	40 million tons of probable reserves with a grade of around 0.40–0.45%	–

Ghorbani (2002b)

Table 5.6 Details of important copper mines in Iran

Name of the mine	Geographic coordinates	Production 2000	2001	2002	Details Status	Main product	Type of mining	Proved reserve (million tons)	Probable reserve (million tons)	Annual production	Average grade
Sarcheshmeh	55°50′ 29°58′	0.1–0.12	0.1–0.12	0.1–0.12	Active	Cu	Open pit	826.5	1,200	0.14	Cu=0.68
Meydouk	55°10′ 30°10′				Being equipped	Cu	Open pit	180	500	0.05	Cu=0.85
Ghal'e Zari	58°55′ 31°49′	0.06	0.06	0.06	Active	Cu	Underground	0.36	1.315	0.1	Cu=0.5–3.0 Au=6 ppb
Mazra'e	47°04′ 38°39′	0.03 27–30% concentrate	0.03 27–30% concentrate	0.03 27–30% concentrate	Active	Cu	Underground	0.2	0.431	0.04 concentrate	Cu=1.70
Songoon	46°43 38°42				Being equipped	Cu	Open pit	1,000	2,000	0.6 copper	Cu=0.70
Taknar	57°46′ 35°22′				Being equipped	Cu, Pb, Zn	Underground	–	79	0.125 ore	Cu=1.50

Ghorbani (2002b)

2. The deposits are located in definite geological zones (Kerman in southeast, Arasbaran in northeast of Iran).
3. Due to similar paragenesis of most of the deposits, copper-bearing zones with specific mineralization characteristics can be defined.
4. Study of ore paragenesis and fluid inclusions reveals a forming temperature of 150–400°C (Bazin and Hübner 1969; Dimitrijevic and Djokovic 1973; Ghorbani 2008b).
5. Most of the copper-bearing zones (Kerman Belt especially in the Sarcheshmeh area, Arasbaran Subzone especially in Songun, and Masjed Daghi) show characteristic metasomatism similar to one another. This is even visible in the aerial photographs and satellite images (PRGC-Arian Zamin 2004; Ghorbani 2011).

Moreover, availability of the technology and know-how for copper exploration, processing, extraction, etc., along with inexpensive electric supply compared to the rest of copper-rich countries add to the above-mentioned facts to make Iran a promising country for future investments.

5.5 Lead and Zinc

5.5.1 Introduction

Lead: Due to the ease of extraction from its ore, lead was discovered very early in the course of human civilization (Alipour 1993). Lead has been exploited by man ever since 4000 B.C., and indications of its utilization in Iran during the third millennium B.C. are widespread (Zavosh 1976). Extraction of carbonaceous ores of lead started prior to other lead compounds since the melting process of carbonates of lead is simpler. It is interesting to note that in places where both sulfide and carbonate of lead occurred together, the extraction procedure was limited to the carbonates. In the pre-Islamic era, lead was made use of as slurry mix in the construction of heavy structures such as dams and bridges (Alipour 1993). Mo'ien (1976) states that "at the time of Selokids, iron, copper and lead were being extracted, concentrated and exported under the supervision of kings' administrators; the mining activities were under the monopoly of the kings and the mines were considered as kings' assets."

Lead sulfide, known as "Sorme" in ancient Iran, has been used as an eye remedy. Names such as Kuh-e Sorme (Firooz Abad, Fars), Khane-ye Sorme (Arak), and Dare-ye Sorme (Esfahan) are given to the localities hosting lead–zinc deposits.

During the Sassanid period and post-Islamic era, the exploitation of lead mines was mostly carried out with the intension of silver extraction. In many books, the lead deposits, then considered to be silver mines for their high silver content, are referred to. The abundant use of silver utensils in the Sassanid period is an indication of the extent of silver extraction from lead ores, since independent silver deposits are not known to exist in Iran even now. The amount of lead production was so great that Biruni in Al-Jamaher talks of the export of lead metal (Zawvosh 1969).

Zinc: The discovery of zinc and its utilization is a recent event in the history of Iran. Zinc was relatively unknown in ancient Iran. However, utilization of zinc compounds as medical remedies was well acknowledged. Most of the zinc extraction was for the production of "Tootia," which is the natural impure zinc oxide (ZnO) produced in lead–zinc smelting furnaces. Its solution, a strong disinfecting agent, was used in curing eye diseases. This compound was produced and exported to various countries (Zawvosh 1969).

The German term "Tuthi" (Moein 1976, A Persian Dictionary) is derived from the Persian word "Tootia," which in turn has been derived from the word "Doodia," a substance extracted from smoke "Dood" (Al-e Taha 1996).

Tootia was produced in smelters whose remnants are still present in Central Iran (Behabad and Kuhbanan) from the oxide ore of zinc. The furnace consisted of two sections: the lower section (section A), where the temperature was very high, held the zinc-bearing rocks, and in the upper section (section B), cones made up of porcelain were placed. The Zinc fumes (in form of smoke) expelled from the heated ores in the lower section were, on cooling, deposited on the surface of the porcelain cones in the upper section as soot, which was later scraped and powdered to form Tootia (Fig. 3.1).

5.5.2 Metallogenic Phases of Lead and Zinc

The most important phases of mineralization of lead–zinc deposits of Iran can be grouped as follows.

Late Proterozoic–Early Cambrian Phase: In this interval of time, following the magmatic activities associated with intracontinental rifts such as Takab-Anar-Bafq Rift (Ghorbani 1999f), many valuable reserves of massive sulfide, volcanosedimentary, and sedimentary type deposits of lead–zinc have been formed, for example, Koushk (dominantly zinc deposit), Chahmir, Zarigan (Bafq area), Angooran, and Alamkandi (Takab region).

In most of the deposits formed during Late Proterozoic–Early Cambrian, the ratio of zinc to lead is much greater than one. For instance, the Zn–Pb for Angooran is around 6 while that of Koushk is 4. (In fact, this characteristically high ratio of zinc has resulted in the proposal of a new type of paragenesis for these two ore bodies by Ghorbani 1999f).

Late Paleozoic: The Late Paleozoic rocks of Iran host a number of lead–zinc ore deposits and indications, for example, Anjir ore body (Yazd) in the Permian rocks of central Iran, Dona and Shakin deposits in the Permian rocks of the Alborz Mountains, and Bahram and Bive Zan deposits (southwest of Mashhad) in the Devonian rocks of Eastern Iran. Interestingly, there are valuable lead–zinc ore bodies of similar age in the Zagros region, for example, Kuh Sorme and Dasht-e Bahram deposits (Firooz Abad, Fars) in Dalan Formation (Ghorbani et al. 2000).

Triassic: Important lead–zinc deposits have been formed at Triassic times within dolomites and dolomitic limestones (Shotori Formation in Eastern and Central Iran, Elika Formation in Alborz and Azerbaijan, and their equivalents). It is worth mentioning that most of the fluorite reserves of Iran are the associated gangue minerals of this phase of lead–zinc mineralization (Ghorbani et al. 2000; Ghorbani and Momenzadeh 1994). Some of the ore bodies that were previously considered as Late Paleozoic are now known to belong to this phase. Some of these reserves are

- Lead–zinc reserves of Central Iran, for example, deposits of Behabad–Kuhbanan–Ravar triangle
- Lead–zinc reserves of Alborz, for example, Elika ore body

Cretaceous: Most of the known lead–zinc reserves of Iran are found in the Cretaceous rocks. The ore deposits and indications of lead and zinc with Cretaceous rocks as their hosts are recognized from all over Iran, for example, Sanandaj–Sirjan Zone (ore deposits of Malayer–Esfahan Belt) and Central Iran (Mehdiabad, Mansourabad, Farahabad, Dare Zanjir, and Nakhlak). In most of the ore bodies formed during this time period, especially those of the Malayer–Esfahan Belt, the Zn–Pb ratio is greater than one; this ratio reaches four in Irankuh. Most of the lead–zinc deposits of Cretaceous genetically are of Mississippi Valley or Sedimentary Exhalative (SEDEX) origin (Ghorbani et al. 2000; Ghorbani 2002b).

Tertiary: A number of lead–zinc deposits and indications are known from the Eocene and Oligo-Miocene volcanic, volcanosedimentary, and sedimentary rocks of Iran. If the ore formed during this time interval is hosted by volcanic rocks, ore bodies will be rich in copper minerals and be vein-type. Such lead–zinc reserves are mostly located within the Urumiyeh–Dokhtar volcanic strip, Azerbaijan, and Eastern Iran (Ghorbani 2007b). Generally, the Zn–Pb ratios of Tertiary times are less than one.

5.5.3 Spatial (Geographical) Distribution of Lead–Zinc Deposits and Indications

According to their geological characteristics, the areas of lead–zinc mineralization in Iran can be classified into the following categories (Ghorbani et al. 2000):

1. Malayer–Esfahan Belt
2. Azerbaijan
3. Central Iran (Bafq, Behabad–Kuhbanan, surrounding areas of Yazd and Sirjan)
4. Eastern Iran
5. Alborz

Figure 5.5 illustrates the distribution of lead–zinc deposits and indications in Iran.

To understand the evolutionary history of the lead–zinc metallogeny in Iran, the geological and mineralogical characteristics of each of the above-mentioned categories are briefly discussed in the following sections.

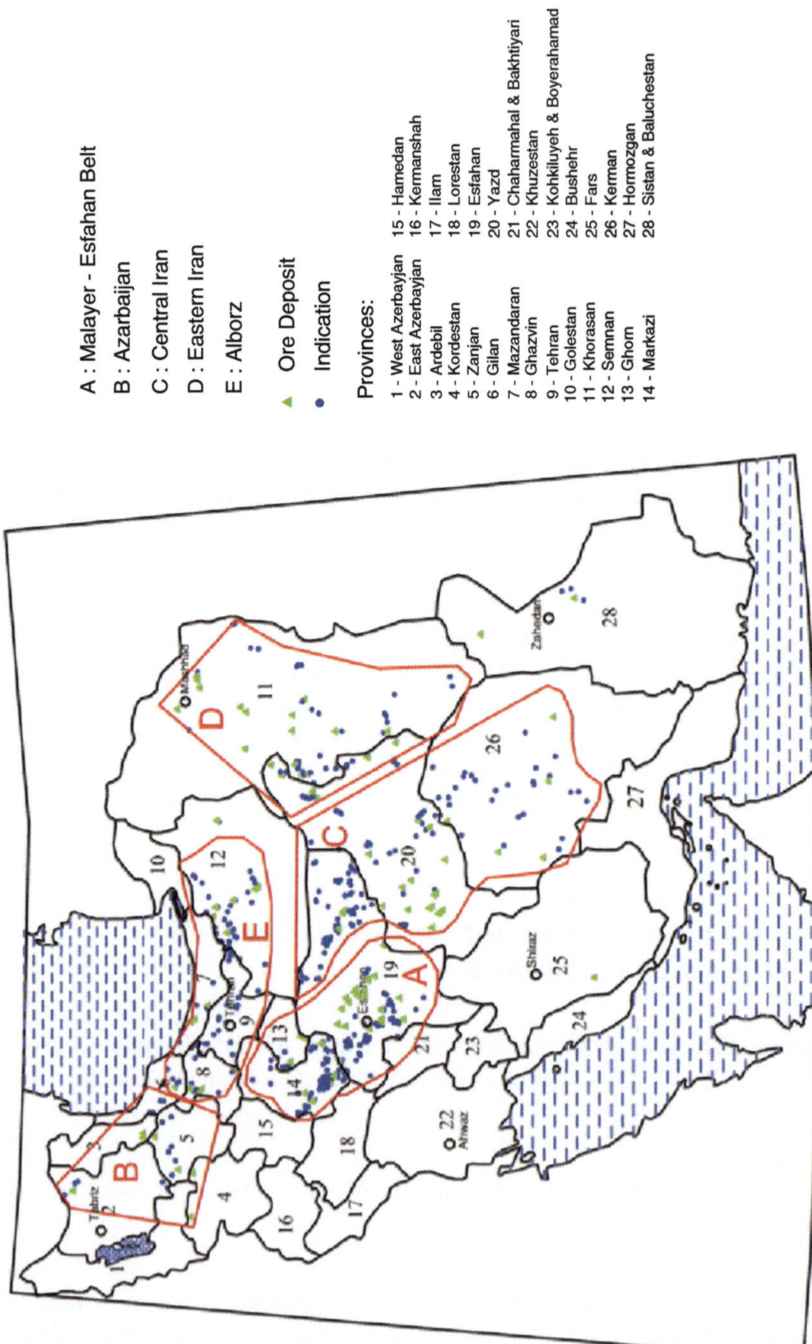

Fig. 5.5 Distribution map of lead–zinc ore deposits and indications of Iran (Ghorbani 2002b)

5.5.3.1 Malayer–Esfahan Belt

From the geological point of view, all the deposits of this belt are younger than Jurassic. In fact, the rocks hosting these ore deposits and indications are Cretaceous carbonates that overlie Jurassic shales; in fact, there are no Tertiary volcanogenic ore deposits in this belt. Nevertheless, in the northwest part of this belt, evidence of submarine volcanic activity in form of tuffs and lavas is visible in the surrounding areas of some ore bodies such as Shamsabad, Ashtiyan, and Khomein (Ghorbani 2002b; Ghorbani et al. 2000).

The host rock of all the ore deposits is limestone. The mineral composition of the ore bodies of this belt is of sulfide type; carbonaceous ores of lead–zinc are also present in shallower parts of the ore bodies. Even though lead and zinc are both present in all the deposits, the lead content of smaller ore bodies is greater as compared to the larger deposits where the situation is vice versa. The associated gangue minerals are dominantly composed of silica or lime; toward Esfahan, barite and fluorite also appear in the gangue reaching their maximum amount in the north and northeast of Esfahan City. Though the silver content of the ores is relatively high in this belt, it gradually decreases from Ahangaran (Malayer) toward Esfahan. This trend is accompanied by gradual increase in the zinc–lead ratio (Ghorbani 2002b).

The ore minerals of the smaller deposits are dominantly formed along fault plains, joint surfaces, and fractures. In Khomein, Shamsabad, and Ashtiyan, iron mineralization occurs side-by-side lead–zinc.

5.5.3.2 Azerbaijan Region

The lead–zinc ore deposits of Azerbaijan region can be categorized into two main classes on the bases of their time of origin, paragenesis, and host rock composition, which are enumerated below (Ghorbani 2002b; Ghorbani et al. 2000):

1. Pre-Tertiary deposits, dominantly Late Proterozoic–Early Cambrian, found in southern areas of Azerbaijan (especially in southern Zanjan, Takab, and Shahin Dez, e.g., Angooran, Alamkandi, and Poshtkuh).
2. Tertiary deposits, associated with Tertiary magmatism, found in north Azerbaijan–Tarom Belt, whose host rocks are mainly vein-type volcanics that form small bodies rich in copper, lead, zinc, and, occasionally, gold. Here the zinc content is usually greater as compared to lead.

5.5.3.3 Central Iran

According to the time of origination, there are three categories of lead–zinc deposits in Central Iran (Ghorbani 2002b, 2007b; Ghorbani et al. 2000):

1. Late Proterozoic deposits, for example, ore bodies of Bafq area (e.g., Koushk, Chahmir, Zarigan, God-e Vafadari, Cheshme Firoozi, Farak, Gonde Dar), which ore bodies are either oxide or sulfide type with sulfide type ores dominating. The host rocks are shales or carbonates, the former being in greater proportion. The ore bodies are mainly of stratiform type, and their origin is either massive sulfide or sedimentary exhalative.
2. Triassic ore deposits and indications in and around Behabad, Kuhbanan, Ravar, and Tabas (specially Ozbak Kuh) areas (e.g., Goujar, Deh Asgar, Gour, Tarzeh, and Taj Kuh deposits) that are mainly oxide ores formed along faulting surfaces of dolomitic hosts. The zinc content of most of these deposits is higher than lead occasionally reaching 15–20%.
3. Cretaceous ore deposits that are widely distributed in Central Iran and include Mehdiabad (near Yazd), Taft, and Anarak regions. All the ore deposits are identical in their mineralogical composition, host rocks, and stratigraphical horizon. They are dominantly concentrated in Yazd Province with Taft Formation as their host. In the mining localities, the mineralization has taken place at the contact between Sangestan Clastic and Taft Formations.

The enclosing rocks of the lead–zinc deposits of Central Iran are usually sedimentary in nature. The lead minerals are mostly of carbonate type; cerussite is more abundant as compared to galena. The zinc minerals are also of carbonate (smithsonite) and silicate (calamine) type; however, this is not the case for the larger deposits of Mehdiabad and Koushk. In most localities, the associated gangues are calcite (dominating in Behabad–Kuhbanan area) along with large quantities of iron oxide. However, calcite is associated with quartz in some place, and occasionally quartz is the only gangue present. Barite seldom occurs as the gangue mineral (except in Mehdiabad). Fluorite occurs as rare mineral in the deposits and indications of this zone.

5.5.3.4 Eastern and Southeastern Iran

The important point regarding the lead–zinc mineral deposits in eastern and southeastern parts of Iran is that their age variation is less compared to the rest of the country (mostly belong to Tertiary). There is no controversy regarding the hydrothermal origin of these deposits (Science and Technology Research Institute 1999; Ghorbani 2002b). Lead and zinc minerals occur in almost equal amounts in all the ore bodies of this region. Chemically, the ores are composed of sulfide. In southeastern regions of the country, for example, in north and northeast of Taftan, the silver content of the ore bodies is extraordinarily high. The host rocks are volcanics varying from andesites to dacites in composition. Pyrite, considered as the gangue in these deposits, is very abundant while copper-bearing minerals form minor mineral of these ores. Most of these deposits are of vein type. Till date, no major lead–zinc deposits have been discovered in this region.

5.5.3.5 Alborz Region

The lead–zinc deposits of the Alborz region are mineralogically dominated by the lead ores while the zinc minerals are rare. The host rocks vary from Permian to Tertiary, but most enclosing rocks belong to the Elika and Route Formations that are carbonaceous in composition. Nevertheless, some deposits have formed within Jurassic (Dalichai Formation) and Eocene Limestones (Ghorbani 2005j). In many localities, volcanic rocks, tuffs, and intrusive bodies occur in the vicinity of the mineral deposits. The lead minerals are mostly of sulfide type with carbonate compositions forming minor portion of the deposits. Barite is the chief gangue present in all of the ore bodies. Copper minerals are rare in the lead–zinc deposits of this region. Fluorite is so abundant in some of the deposits that they are more famous for their fluorite content, for example, Pachi-Miana deposit. The Silver content of the most of the ore bodies is high, and sometimes tonardite is also present in them, for example, Dona and Elika deposits (Ghorbani et al. 2000).

5.5.4 Economically Important Lead–Zinc Deposits

There are about 700 known lead–zinc ore deposits and indications in Iran, of which around 150 have been extracted so far. At present, mining activities within nine localities are in progress (Table 5.7).

Although the Geological Survey of Iran along with other government and private institutions has explored all the lead–zinc deposits of the country under the auspices of the Lead–Zinc Prospecting Project, detailed investigations of the metallogenic characterization and reserve estimation of the deposits are yet to be carried out.

Considering the results of geological studies and laboratory examinations extracted from the archives of the mining authorities (Ghorbani 2002b), along with the field observations of more than 200 localities in various parts of Iran, we have tried here to classify the reserves based on their geological characteristics.

The total amount of the known lead and zinc reserves of the country is estimated at over 200 million tons of ore amounting to about 18 million tons of zinc and five million tons of lead, taking into account the ore grades. Table 5.8 shows the specifications of some of the more important lead–zinc deposits of Iran.

5.6 Aluminum

5.6.1 Introduction

A large volume of bauxites are contained within stable continental settings such as shields and platforms in lateritic type and paraplatforms in tichon type. More than 92% of karstic bauxites are found within the orogenic belts (Shahriyari 1986).

Table 5.7 Production in most important active lead and zinc mines in Iran (2000–2002)

Name		Angooran	Koushk	Emarat	Irankuh	Nakhlak	Dona–Elika	Dare Anjir	Kuh-e Sorme	Mehdiabad
Geographic coordinates		47°24′ 36°37′	55°44′ 31°45′	49°36′ 33°51′	51°26′ 32°43′	53°50′ 33°33′	51°27′ 36°09′	54°13′ 31°34′	53°30′ 28°03′	55°01′ 31°29′
Location		120 km off Annan on Zanjan–Dandi Road	42 km from Bafq on Bafq–Behabad Road	46 km off Arak on Arak–Shazand Road	51 km off Esfahan on Esfahan–Shiraz Road	Nakhlak Lead Mine, on Naein–Anarak Road	95 km off Karaj on the road to Chalous	10 km west of Taft on the road to Deh Bala	25 km off Dalan refinery on road to Firoozabad	80 km off Yazd on the road to Kerman
Production (mt.)	2008	0.34	0.016–0.017 (con.)	0.0055–0.0735 (con.)	0.415 (sulfur) 0.002 (carbonate)	0.02	0.006	0.0015	–	0.001
	2009	0.4	0.016–0.017 (con.)	0.0055–0.0735 (con.)	0.296 (sulfur) 0.0067 (carbonate)	0.024	0.006	0.0015	0.003	0.001
	2010	0.45	0.012 (con.)	0.0055–0.0735 (con.)	0.14 (sulfur) 0.058 (carbonate)	0.019	0.006	0.001	0.0035	0.001

Details	Project status	Active	Active	Active	Active	Active	Active	Active	Active	Installation of equipment
	Main Metal (s)	Zn, Pb	Zn, Pb	Zn, Pb	Zn, Pb	Zn, Pb	Pb, Zn, Ag	Zn, Pb	Zn, Pb	Zn, Pb
	Mine type	OP	OP, UG	OP	OP	UG	UG	UG	UG	OP, UG
	Proved reserves (mt.)	21	6	2	9.92	0.549	0.5	–	0.4	45
	Probable reserves (mt.)	21.7	8.5	10.5	30–40	7	1	0.02–0.03	0.84	200
	Remaining reserves	12	2.5	2.5			<0.4			
	Average grade (%)	Zn = 28; Pb = 3–6	Zn = 12; Pb = 3	Zn = 6; Pb = 2.25	Sufide: Zn = 3–6; Pb = 1.8 Oxide: Zn = 2; Pb = 2	Pb = 8.33; Zn = 0.38; Ag > 70 ppm	Pb + Zn 6.7; Ag = 1,400 ppm	Pb + Zn 12–16	Pb + Zn 10–15	Pb = 2.7; Zn = 7.7

OP open pit, *UG* underground

Table 5.8 Characteristics of the most important lead and zinc ore deposits of Iran

Sr. no.	Name of the deposit	Geographic location	Reserves (million tons of ore)	Average grade	Details
1	Angooran	Southwest of Zanjan, Dandi Area	21 million tons of oxide ores remaining	Zn 28% Pb 3–6%	The said reserve is mainly oxide ore enclosed in carbonaceous host. However, the sulfide ore reserves are not yet explored. Probability of new finds is high
2	Mehdiabad	Yazd	45 million tons of higher grade sulfide reserves 104 million tons of lower grade sulfide reserves	Pb 2.7% Zn 7.7% Zn 5.5% Pb 2%	The northern limit of the reserve is not known, and thus, probability of an increase in the reserves up to 100 million tons exists. The oxide ore has a zinc grade of 15–20% and amounts to two million tons
3	Koushk	Bafq, Yazd	Remaining high-grade ores: 2.5 million tons; Low-grade ores in pyritized zone: 4–6 million tons	Average 15% Zn 12% and Pb 3% Zn 4–6% within pyritized zone	The major part of the ore body has been extracted (around 3.5 million tons). However, the low-grade pyritized reserves equal the whole deposit.
4	Irankuh Complex	Esfahan	Total remaining reserves: 8–9 million tons; An estimated equal amount has been so far extracted; Oxide zone reserves amount to one million tons	Zn 5% and Pb 1.8% in sulfide zone Zn 2% and Pb 2% in oxide zone	Goushfil (three million tons, Pb 2%, Zn 3%), Tape Sorkh (five million tons, Pb 1.5%, Zn 3%), Kolah Darvaze (exhausted), and God-e-Zendan deposits are included in this complex
5	Emarat	Shamsabad, Arak	Remaining reserves: ten million tons	Zn 6% Pb 2.26%	The probable reserves of the deposit had been estimated at ten million tons which is proved recently
6	Chah Gaz	Sirjan	Proved reserves: 0.4 million tons Probable reserves: two million tons	The average grade of lead–zinc is about 20%	The deposit includes both oxide and sulfide ores whose explorations are not yet complete

7	Chah Mir	Shitour, near Bafq, Yazd	Proved reserves unknown; probable reserves: two million tons	Zn 10% Pb 2%	Although the exploration work is not yet over, but all the evidence indicates that the mineralization processes are similar to those of Koushk deposit (especially the low-grade parts of Koushk)
8	Goucher	Behabad–Kuhbanan region	Proved reserves: 0.4 million tons Probable reserves: two million tons	Zn 15% or more Pb 5%	Mineralization is of oxide type. Zinc occurs as silicate and carbonate while lead is in the form of cerrusite
9	Kuh Sorme	Firooz Abad, Fars	Proved reserves: 0.4 million tons Probable reserves: one million tons	Zn 17% Pb 5–15%	The deposit has been explored and excavated simultaneously, and more than 500,000 tons of ores have been extracted till date
10	Elika and Dona	Chalous	Proved reserves: 0.5 million tons Probable reserves: one million tons	Pb 6.7% Ag 14 ppm	Reserves largely mined
11	Nakhlak	Anarak	Seven million tons	Pb 8.33% Zn 0.38% Ag $\geq$70 ppm	Reserves largely mined

Ghorbani (2002b) with some modification

Bauxite deposits are known from all over the world, except Antarctica. Vast reserves of bauxite exist in the tropical and subtropical regions of Australia, Africa, Asia, and North and South America (Caribbean region).

The Karstic bauxite deposits of Europe are better known as Mediterranean Belt deposits extending over Turkey, Hungary, France, and Yugoslavia (Bradossy 1993; Shaffer 1975). The Characteristic Asian bauxite deposits are found in Russia (Ural and Siberia), Irano-Himalayan (Iran, Afghanistan, Pakistan, Kashmir), and East Asia (China, Vietnam, Indonesia) (Metallogenic Map of The Middle East 2011; Ghorbani 2010b).

5.6.2 Temporal Settings of Bauxite Deposits

The Bauxite deposits are known from all geological epochs, and the oldest belongs to Cambrian. The Karstic bauxite deposits have been reported in all continents, except Africa, from all geological ages. Lateritic deposits are mostly confined to tropical and subtropical Cenozoic and younger horizons (Smirnov 1983). The statistics show that the bauxites of Paleozoic age constitute 25.1% of the world's total reserves while that of Mesozoic and Cenozoic constitute 35.3 and 39.6% (Smirnov 1983; Mineral Facts and Problems 1985; Shaffer 1975). The distributions of karstic bauxites reached several maxima during Devonian, Carboniferous, Upper Cretaceous, Eocene, Paleocene, Miocene, and Pliocene (Smirnov 1983; Mineral Facts and Problems 1985; Shaffer 1975).

The Bauxite deposits and indications of Iran have been classified according to their lithostratigraphic and chronostratigraphic characteristics with the largest temporal occurrence within Permian (47.61%). The contribution of other geological epochs is as follows: Triassic–Jurassic 21.8%, Permo-Triassic 19.04%, Triassic 9.52%, and Cretaceous 4.76% (Table 5.9) (Soheili 2004; Ghorbani 2010b).

Permian and Permo-Triassic: The most important bauxite deposits of Iran as well as other aluminum-rich deposits such as kaolinitic clays belong to this interval of time (Ghorbani 2002b, 2010b; Soheili 2004). The Permian and Permo-Triassic bauxite-bearing horizon is known from various regions of Iran due to the presence of a stratigraphic hiatus in the form of layers overlying the Rooteh Formation and overlain by the Elika Formation. Some of these areas are Jajarm, Boukan, Mahabad, Miandoab, Abgarm, Ghazvin, Firooz Kuh, and Ardakan. In these areas, the bauxite horizon occurs between the Late Permian rocks and Early Jurassic (Shemshak) sediments.

Triassic–Jurassic: These occur above the Elika and Shemshak Formations in and around Gano, Shahmirzad, etc. A few horizons of aluminum-rich fireclays and occasional boehmite and diaspora are found scattered as intercalations in the Jurassic rocks, for example, Sangrood\(Alborz) and Dopolan (Zagros).

Cretaceous: The Cretaceous of Iran is very important from the viewpoint of bauxite exploration. The Cretaceous deposits are only reported from Zagros, where the

Table 5.9 Temporal distribution of bauxite deposits and indications of Iran

Geologic period	Mineral deposit or indication
Permian	Aghajari, Gol Charmoo – Sari Qamish, Southeast Hossein Abad, Kaleijeh, Sarchaveh, Jajarm A Zone, Abgarm 1st and 2nd Horizons, Cha Azloo, Aliabad, Karaftoo, Naveh Qoran, Gaveshleh, Soleiman Kandi, North Yazd, Chekchek
Permo-Triassic	Ali Balta 1st Horizon, Chapoo, Qopi Baba Ali,Norooz Abad, Kavir Kuh, Abadeh, Kiasar, Robat Khan, Darvar, Ayenehvarzan, Sangrud, Dovilan, Dorak, Kuh Dena, Bazargan, Darsinoeid, Khampesian, Doolab, Gheshlagh
Triassic	Sadr Abad, Godar Zard, Nasr Abad, Sorkheh Hesar, Bolbolouyeh, Tabas
Triassic–Jurassic	Shahmirzad, Kal Jafar Agha, Abgarm 3rd Horizon, Siah Rudbar, Alamoot, Jajarm B Zone, Ganoo, Ali Balta 2nd Horizon, Shah Bolaghi
Cretaceous	Sarfaryab, Sardaryoosh, Semirom Ab Malakh

Ghorbani (2010b)

bauxite horizon is sandwiched between the Sarvak and Eilam Formations of Bangestan Group, extending at times even up to the Gurpi Formation (Ghorbani 2002a, b, c; Soheili 2004). It should be, however, kept in mind that this layer is the same as the famous refractory clay horizon. Examples of Cretaceous bauxite deposits include Sarfaryab, Sardariush, Semirom, and Ab Malakh.

5.6.3 *Distribution of Bauxite Deposits*

Following the first report of bauxite in 1958 from Bolbolouye in Kerman Province (Soheili 2004), geological surveys and exploration work led to the discovery of many karstic bauxite deposits. Apart from Jajarm, all the bauxite deposits of Iran are of diaspor/bohemite composition, whose origin is Mediterranean karstic bauxite with a lensoid carbonaceous footwall block (Soheili 2004). The intensity of karstification of the footwall block is medium to weak. Geographically, the bauxite deposits of Iran are found in the northeastern, northern, northwestern, central, and southwestern parts of the country. They are situated in eastern (Binaloud), central and western Alborz, Central Iran, and Zagros tectono-sedimentary divisions of the country varying in age from Permian to Late Cretaceous.

The bauxite deposits of Iran are not very extensive and, considering the conditions of their formation and the geological settings of Iran, it is not expected to discover any large (on the global scale) such deposits. Nevertheless, a number of economically important bauxite deposits have been reported from Jajarm (Shahrood), Boukan, Saghez, Firooz Kuh, and Abgarm (Ghazvin). As can be seen in Fig. 5.6 (Soheili 2004), the major part of the bauxites of Iran is found in East and West Azerbaijan, Chaharmahal and Bakhtiari, Khorasan, Semnan, Fars, Ghazvin, Kerman, Kohkilouye and Boir Ahmad, Kordestan, Mazandaran, Golestan, and Yazd.

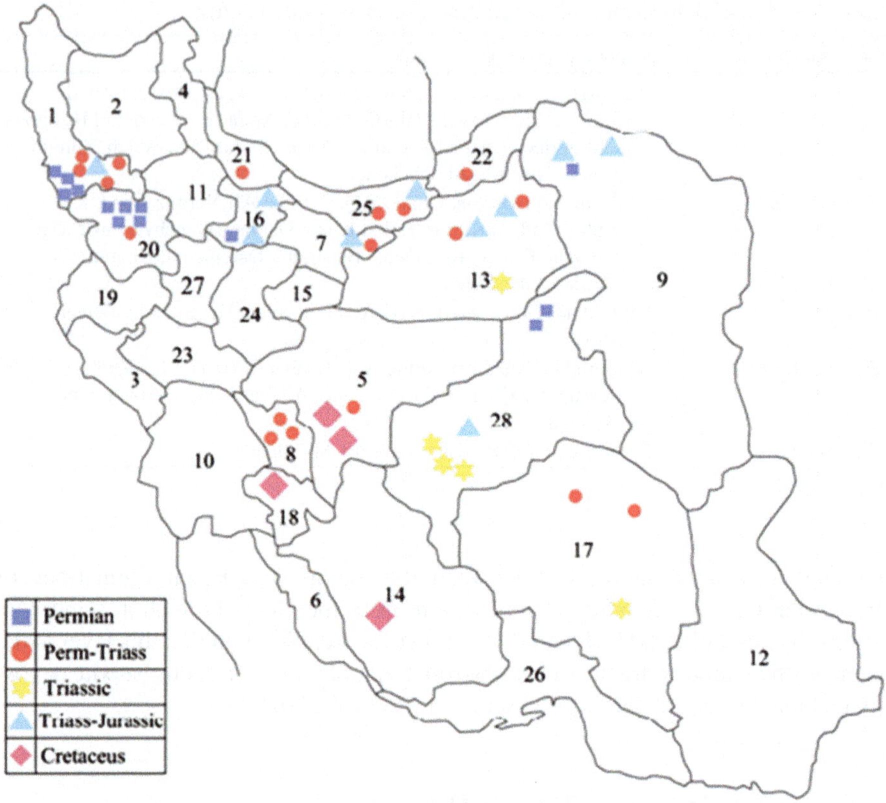

Fig. 5.6 Distribution map of bauxite deposits and indications of Iran (Soheili 2004)

In general, the bauxite deposits of Iran can be classified into the following categories based on their location:

- Kopet Dagh and eastern Alborz (including Jajarm deposit)
- Central Alborz
- Western Alborz and Azerbaijan including Alamout, Takab, Boukan, and Ajabshir
- Khuzestan plain
- Folded Zagros
- High Zagros
- Yazd

As mentioned earlier, in addition to bauxite, a number of other aluminum-rich minerals such as alunite and nepheline also exist in nature, which are excavated extensively in Russia for aluminum extraction. Fortunately, the alunite deposits of Tarom, Ghazvin, and Manjil, as well as the nepheline deposits of Bozghoush, Razgah (north of Mianeh), and Kalibar (Ahar area), are considerable in amount containing tens of millions of tons of such ores (Ghorbani 2002b, 2010b). If aluminum is extracted from such deposits, Iran will not only attain self-sufficiency but will become an exporter of alumina and aluminum products.

The geological settings of a number of bauxite deposits of Iran are described in the following paragraphs (Soheili 2004).

Siyah–Roudbar, Gorgan:
Footwall rocks: Triassic dolomites (Elika Formation)
Hanging wall rocks: Jurassic sandstones and shales (Shemshak Formation)

Shah Bolaghi, Damavand:
Footwall rocks: Triassic dolomites (Elika Formation)
Hanging wall rocks: Jurassic sandstones and shales (Shemshak Formation)

Abgarm, Ghazvin:
Two horizons of bauxite mineralization exist:

- Between Early Permian limestones (Rooteh Formation) and Late Permian rocks

Between Triassic limestones (Elika Formation) and Jurassic sandstones and shales (Shemshak Formation)

Sarchaveh, Boukan:
Two horizons of bauxite mineralization exist, the upper one being richer:

- Within Permian limestones (Rooteh Formation) in lower section
- Overlying Permian limestones

Sadr Abad, Yazd:
Footwall rocks: Middle Triassic dolomites (Shotori Formation)
Hanging wall rocks: Late Triassic–Early Jurassic detrital rocks (Nayband and Shemshak Formations)

Northern Yazd:
Footwall rocks: Permian limestone and dolomites
Hanging wall rocks: Early Jurassic sandstones and shales (equivalent to Shemshak Formation)

Chekchek, Yazd:
Footwall rocks: Permian dolomites
Hanging wall rocks: Jurassic sandstones and shales

Sarfaryab, Dehdasht:
Footwall rocks: Cretaceous limestones and marly limestones (Sarvak Formation)
Hanging wall rocks: Late Cretaceous limestones and marly limestones (Eilam Formation)

Table 5.10 lists the various specifications of the known bauxite deposits of Iran while Fig. 5.6 depicts the geographical distributions of these deposits.

Table 5.10 Mineralogical composition of representative samples of bauxite deposits of Iran

No.	Name of deposit	Primary mineral	Chemical formula	Accessory mineral	Chemical formula	Associated minerals
1	Ali Balta	Kaolinite	$Al_2Si_2O_5(OH)_4$	Quartz	SiO_2	–
				Anatase	TiO_2	
2	Aghajari	Kaolinite	$Al_2Si_2O_5(OH)_4$	Bohemite	AlO(OH)	–
				Anatase	TiO_2	
				Hematite	Fe_2O_3	
3	Chapoo	Diaspor	AlO(OH)	Pyrophyllite	$Al_2Si_4O_5(OH)_2$	–
		Chlorite	$(Mg,Fe)_6(Si,Al)_4O_{10}(OH)_8$			
4	Gol Charmoo–Sari Qamish	Hematite	Fe_2O_3			
		Diaspor	AlO(OH)	Chlorite	$(Mg,Fe)_6(Si,Al)_4O_{10}(OH)_8$	Quartz
		Anatase	TiO_2	Goetite	FeO(OH)	SiO_2
		Kaolinite	$Al_2Si_2O_5(OH)_4$			
5	Hossein Abad–Mahabad	Hematite	Fe_2O_3			Muscovite
		Diaspor	AlO(OH)	Chlorite	$(Mg,Fe)_6(Si,Al)_4O_{10}(OH)_8$	$Al_2Si_3AlO_{10}(OH)_2$
		Kaolinite	$Al_2Si_2O_5(OH)_4$	Pyrophyllite	$Al_2Si_4O_5(OH)_2$	Quartz
		Anatase	TiO_2			SiO_2
6	Kaleijeh Mahabad	Hematite	Fe_2O_3			
		Kaolinite	$Al_2Si_2O_5(OH)_4$	Halite	NaCl	Quartz
		Diaspor	AlO(OH)	Anatase	TiO_2	SiO_2
		Pyrophyllite	$Al_2Si_4O_5(OH)_2$			
6–1	Kaleijeh Mahabad	Hematite	Fe_2O_3	Anatase	TiO_2	–
		Diaspor	AlO(OH)	Kaolinite	$Al_2Si_2O_5(OH)_4$	
		Pyrophyllite	$Al_2Si_4O_5(OH)_2$			
7	Shah Bolaghi	Goetite	FeO(OH)	–	–	Calcite
		Kaolinite	$Al_2Si_2O_5(OH)_4$			$CaCO_3$
		Anatase	TiO_2			
8	Veresk Bridge–Firooz Kuh	Bohemite	AlO(OH)	Kaolinite	$Al_2Si_2O_5(OH)_4$	–
		Hematite	Fe_2O_3			

9	Dopolan	Diaspor	$AlO(OH)$	Anatase	TiO_2	–
		Kaolinite	$Al_2Si_2O_5(OH)_4$	Bohemite	$AlO(OH)$	
10	Jafar Agha	Diaspor	$AlO(OH)$	Anatase	TiO_2	–
		Hematite	Fe_2O_3			
		Kaolinite	$Al_2Si_2O_5(OH)_4$			
11	Jafar Agha	Hematite	Fe_2O_3	Diaspor	$AlO(OH)$	Calcite
		Kaolinite	$Al_2Si_2O_5(OH)_4$	Anatase	TiO_2	$CaCO_3$
12	Permo-Triassic Horizon	Diaspor	$AlO(OH)$			
		Kaolinite	$Al_2Si_2O_5(OH)_4$	Chlorite	$(Mg,Fe)_6(Si,Al)_4O_{10}(OH)_8$	Quartz
		Hematite	Fe_2O_3	Calcite	$CaCO_3$	SiO_2
		Anatase	TiO_2			
13	Shemshak Permo-Triassic Horizon					Calcite
		Diaspor	$AlO(OH)$	Anatase	TiO_2	$CaCO_3$
		Hematite	Fe_2O_3	Kaolinite	$Al_2Si_2O_5(OH)_4$	Quartz
						SiO_2
14	Southwest Jajarm	Dolomite	$CaMg(CO_3)_2$	Hematite	Fe_2O_3	Calcite
				Goetite	$FeO(OH)$	$CaCO_3$
15	Lateritic Horizon between Carboniferous and Triassic–Rey Abad	Quartz	SiO_2	Calcite	$CaCO_3$	Goetite
		Hematite	Fe_2O_3			$FeO(OH)$
		Kaolinite	$Al_2Si_2O_5(OH)_4$			
16	Zoo District	Diaspor	$AlO(OH)$	Hematite	Fe_2O_3	–
		Kaolinite	$Al_2Si_2O_5(OH)_4$			
		Anatase	TiO_2			
		Goetite	$FeO(OH)$			
17	Tagooie District	Diaspor	$AlO(OH)$	Anatase	TiO_2	–
		Kaolinite	$Al_2Si_2O_5(OH)_4$			
18	Gol Bini District	Diaspor	$AlO(OH)$	Anatase	TiO_2	–
		Hematite	Fe_2O_3	Chlorite	$(Mg,Fe)_6(Si,Al)_4O_{10}(OH)_8$	
		Kaolinite	$Al_2Si_2O_5(OH)_4$	Calcite	$CaCO_3$	

(continued)

Table 5.10 (continued)

No.	Name of deposit	Primary mineral	Chemical formula	Accessory mineral	Chemical formula	Associated minerals
19	Red Mud Factory			Anatase	TiO_2	
		Hematite	Fe_2O_3	Chlorite	$(Mg,Fe)_6(Si,Al)_4O_{10}(OH)_8$	–
		Gibsite	$AlO(OH)$	Bohemite	$AlO(OH)$	
				Nierrite	$Na_2Ca(CO_3)_2$	
20	Shahmirzad	Kaolinite	$Al_2Si_2O_5(OH)_4$	Hematite	Fe_2O_3	–
				Goetite	$FeO(OH)$	
21	Bolbolouyeh	Diaspor	$AlO(OH)$			
		Hematite	Fe_2O_3	Kaolinite	$Al_2Si_2O_5(OH)_4$	–
		Muscovite	$KAl_2Si_3AlO_{10}(OH)_{12}$	Anatase	TiO_2	
		Quartz	SiO_2			
22	Sarfaryab	Bohemite	$AlO(OH)$	Calcite	$CaCO_3$	–
				Anatase	TiO_2	
				Hematite	Fe_2O_3	
22–1	Sarfaryab	Bohemite	$AlO(OH)$	Diaspor	$AlO(OH)$	–
				Anatase	TiO_2	
		Hematite	Fe_2O_3	Kaolinite	$Al_2Si_2O_5(OH)_4$	
23	Chagharloo	Chlorite	$(Mg,Fe)_6(Si,Al)_4O_{10}(OH)_8$	Bohemite	$AlO(OH)$	–
		Ilite	$(K, H_3O)Al_2Si_3AlO_{10}(OH)_2$			
		Hydrated iron Oxide	$FeO(OH)_x$			
24	Aliabad – Karaftoo	Quartz	SiO_2	Hematite	Fe_2O_3	Natrolite
						$NaAl_3(SO_4)_2(OH)_6$
25	Tave Qoran	Quartz	SiO_2	Goetite	$FeO(OH)$	Titan
		Kaolinite	$Al_2Si_2O_5(OH)_4$			TiO_2
		Calcite	$CaCO_3$			Muscovite
						$KAl_2Si_3AlO_{10}(OH)_{12}$

26	Gaveshleh	Calcite	$CaCO_3$	–	–	Muscovite
		Quartz	SiO_2			$KAl_2Si_3AlO_{10}(OH)_{12}$
						Goetite
						FeO(OH)
27	Soleyman Kandi	Hematite	Fe_2O_3	Chlorite	$(Mg,Fe)_6(Si,Al)_4O_{10}(OH)_8$	Muscovite
		Quartz	SiO_2			$KAl_2Si_3AlO_{10}(OH)_{12}$
		Diaspor	AlO(OH)			
		Barite	$BaSO_4$			
28	Gheshlagh	Kaolinite	$Al_2Si_2O_5(OH)_4$	Bohemite	AlO(OH)	–
				Anatase	TiO_2	
				Rutile	TiO_2	
28–1	Gheshlagh	Bohemite	AlO(OH)	Anatase	TiO_2	–
		Kaolinite	$Al_2Si_2O_5(OH)_4$			
28–2	Gheshlagh	Bohemite	AlO(OH)	Anatase	TiO_2	–
		Kaolinite	$Al_2Si_2O_5(OH)_4$	Rutile	TiO_2	
29	Sadr Abad – Yazd	Diaspor	AlO(OH)	–	–	–
		Chlorite	$(Mg,Fe)_6(Si,Al)_4O_{10}(OH)_8$			
		Hematite	Fe_2O_3			

5.7 Gold

5.7.1 *Introduction*

Generally, the formation of gold deposits is not restricted to a particular geological period, but their richness varies largely in different intervals of time. Foster (1993) states that there were two major episodes of gold mineralization in the history of the Earth, namely, Late Archaean–Early Proterozoic (consisting of mesothermal quartz-vein deposits of Late Archaean and Late Archaean–Early Proterozoic conglomerates) and Mesozoic–Quaternary (epithermal and mesothermal deposits associated with the porphyries of the same age). Since there are no Archaean–Proterozoic terrains in Iran, presence of any gold deposits of this category is ruled out.

Nevertheless, the gold deposits of Iran have been formed throughout Late Proterozoic to Quaternary times by different paragenesis such as magmatogenic, volcanogenic, hydrothermal, metamorphic, ophiolitic, mélange-listvenite, and placer-type. (Ghorbani 2008a). Overall, the gold deposits of Iran can be genetically categorized into the following.

Gold Associated with Ophiolitic and Greenschist Basement of Late Proterozoic–Early Cambrian: Large parts of the Late Proterozoic–Early Cambrian rocks that appear to be of greenschist type were of igneous origin (e.g., basalt, ultramafics, tuffs) before being metamorphosed. These rocks are stratigraphically equivalent to the Kahar Formation, but some of them might be correlated with the Chopoghloo Shales or Baroot Formation.

It is interesting to note that the Late Proterozoic–Early Cambrian greenschists of basic igneous origin (especially ultrabasic igneous) have high gold content. As will be described in the following paragraphs, these rocks form the main source of the gold in Iran, and, thus, their origin should be investigated in detail, especially when the rocks have been affected by tectonic–magmatic phenomena. The reasons for the high content of gold in the Late Proterozoic–Early Cambrian basement are (Ghorbani 2008a)

1. The Clarke of concentration of gold in mafic–ultramafic rocks is higher than that of intermediate and acidic rocks. Thus, the greenschists of this type of origin are more attractive from the point of view of gold mineralization.
2. The older the rocks of mantle origin, the richer their gold content because gold has been separated due to the magmatic differentiation of mantle and accumulated in crust.

In general, the gold reserves associated with the basement rocks can be differentiated into two types:

1. Ores whose hosts are Late Proterozoic–Early Cambrian basement rocks; accumulation of gold occurred due to younger geological phenomena and circulation of hydrothermal solutions and is controlled by the position of fractures and joints and/or metasomatic zones (Ghorbani 2008a). Examples include the gold deposits of Mooteh (Esfahan) and Kavand (Zanjan).

2. Ores and indication of gold whose principal source lies within ophiolitic and greenschist basement but their mineralization and concentration has taken place within younger rocks through igneous (especially volcanic activities of Tertiary–Quaternary) processes and their associated hydrothermal solutions (Ghorbani 2008a). In other words, the source of the gold lied within mafic–ultramafic rocks metamorphosed to greenschist facies, but gold mineralization is a result of recycling from basement to younger rocks. Examples include the gold deposits of the Takab area (Zarshouran, Agh Dareh, Arabshah) and Anarak area (Khooni, Booteh Alam, Kooh Dom, Chah Mesi).

The mineralization of gold associated with the ophiolitic and greenschist basement of Late Proterozoic–Early Cambrian has been attributed to (Ghorbani 2008h)

1. Penetration and circulation of igneous material within the basement
2. Faulting and jointing of basement and younger rocks
3. Production and circulation of hydrothermal solutions of magmatic or atmospheric origin
4. Metasomatism associated with mineralization and deposition of gold

Gold Associated with Late Paleozoic–Mesozoic and Mesozoic–Cenozoic Thonaleitic to Granitic Intrusive Rocks: These reserves can be grouped as follows (Ghorbani 2008h):

1. Deposits whose ore minerals are within the groundmass of granites and the entire rock body or part of it has been considered as gold deposit (e.g., Torghabeh Granite in Khorasan)
2. Place deposits whose gold is resulted from the erosion of the above-mentioned granite masses (e.g., Astaneh Gold Deposit in Arak)
3. A combination of the above-mentioned types (Zarin Granite of Ardakan in Yazd)

It is worth mentioning that most of the Late Paleozoic–Cenozoic granite bodies are devoid of gold.

The geological investigations of the author show that only those bodies of granites that have passed through mafic rocks or their metamorphic equivalents, especially Late Proterozoic–Early Cambrian greenschist basement, host gold mineralization; Golpayegan–Arak granites, Sanandaj–Sirjan Zone and Zarin Granite of Central Iran can be given as examples. The granites of the Arak area have passed through the same schists that are exposed within the Mooteh area, and as mentioned earlier, the Clarke of concentration of gold within these rocks is higher than normal. In the Zarin area, there are schists that are genetically related with Zarin Granite that intrudes them. The high gold concentration at the contact of schists and granite indicates the association of the gold with schists.

Young Epithermal Gold Associated with Arsenic and Mercury Deposits and Indications (Carlin Type): Such deposits, which have recently attracted attention of the researchers from all over the world (e.g., Zarshooran and Agh Dareh in the Takab area; refer to Fig. 6.6), are comparatively poorer in their gold content with 2–3 ppm concentration. Mineralization in most of these deposits is a result of the activity of hydrothermal solutions of 60–250°C temperature that are directly or indirectly associated with magmatism (Ghorbani 2008h). In fact, none of these widely distributed deposits have been originally described as gold concentrations. They are mostly deposits of antimony, arsenic, mercury, etc., and in case they are systematically explored, can yield economically important gold reserves. The general geological and metallogenic characteristics of these deposits are as follows (Ghorbani 2008a):

1. The areas within which deposits and indications of antimony, arsenic, and mercury are found have continental basement with younger volcanics (basaltic to intermediate-acidic) extrusions. The basaltic and andesitic volcanic rocks are comparatively older and more extensive than the acidic extrusions. In spite of this, the acidic rocks form the cover and contain most of the mineralized zones.
2. Most of the deposits and indications of antimony, arsenic, and mercury are associated with and controlled by deep faults.
3. Geological and mineralogical evidence within deposits and indications of antimony, arsenic, and mercury of Iran point to the formation of the ore in association with low-temperature (epithermal) hydrothermal solutions originating from acidic volcanics of lower crust (Ghorbani 2008h).

Gold Associated with Ophiolitic Suites and Listvenites: Ophiolitic melanges which are suites of tectonized and crushed ultramafic rocks such as harzburgite, dunite, serpentinite, diabase, pillow lava, gabbro, diorite, and radiolarite in addition to listvenites (rocks metasomatized due to the circulation of CO_2-rich solutions).

Among the above-mentioned rocks, listvenites that have extensive exposures in Eastern Iran are now considered as sources of gold. Preliminary analysis of these rocks (Ghorbani 2008h) indicates the presence of gold at economical grades. Examples of this type of deposits include

1. Makoo gold-bearing mineral indication within ophiolitic melanges, which can be categorized as serpentinized type
2. Chah Charnofoozi (Kalateh Dagh, Nehbandan) gold indication whose gold is genetically related with ophiolitic melanges and listvenites limestones

Gold Associated with Copper Mineral Deposits and Indications: Most of the copper deposits of Iran have considerable gold content that occurs in the form of minor constituents along with various ores of copper (Ghorbani 2008a). The gold in such deposits cannot be economically extracted separately, but during the extraction of copper, it is produced as a by-product.

The gold-bearing copper deposits of Iran can be categorized as follows (Ghorbani 2008h):

1. Porphyry-type gold-bearing copper deposits of the Kerman area, for example, Sarcheshmeh, Meydook, Darehzar, and Chahargonbad.
2. Vein-type gold-bearing copper deposits that are very extensive with a high gold content of about 4 ppm, for example, the Qaleh Zari copper deposit. These are

further subdivided areally on the basis of their mineralogical and geological characteristics:

(a) Gold-bearing copper deposits of the Ahar area, including Mazra'e, Barmalek, Agha Mirza, Gol Dargh, Astamal, Qareh Chamanloo, and Zarnekab
(b) Gold-bearing copper deposits of Tarom-Taleghan, including Dizeh Jin, Chargar, Khalifeloo, and Rashid Abad
(c) Gold-bearing copper deposits of Eastern Iran, including Qale Zari and Chehel Kooreh
(d) Gold-bearing copper deposits of Kavir–Sabzevar, including Kuh Zar (Damghan) and Anarestan
(e) Gold-bearing copper deposits of Urumiyeh–Dokhtar, including Bayche Bagh and Avroos Morghi

5.7.2 Geological Distribution of Gold Reserves

The gold deposits and indication of Iran have been formed along with the formation of the crustal rocks of Iran during the Late Proterozoic–Phanerozoic times due to the tectonic and igneous activities, especially the Pyrenean (Eocene–Oligocene) and Post Pyrenean (Miocene–Pliocene), which also gave rise to most diverse mineral deposits of Iran (Ghorbani 2002b; Maghsoudi et al. 2005). Since all the major controlling factors of gold mineralization in the said period of time are faults, shear zones, and brecciation zones (except the Quaternary placer deposits), the mineralization of gold is of epigenetic type formed due to younger phases of deformation whose exact age is not yet known. Thus, the intervals of gold mineralization have to be ascertained through the estimation of the ages of the host rocks and are as follows.

Late Proterozoic–Early Cambrian: Gold deposits of this age have been reported from Takab, Anarak, and parts of the Sanandaj–Sirjan Zone. The host rocks for the gold mineralization include schists and intercalations of limestone, shale, dolomite, marble, and schists (Ghorbani 2002a; Maghsoudi et al. 2005). Gold mineralization in the Anarak area is dominantly within the metamorphic rocks consisting of schists, marbles, or contact of the two. The main structures controlling the mineralization of gold have been faults that caused the formation of polymetal deposits in the form of veins and veinlets (Ghorbani 2002a; Maghsoudi et al. 2005). At some locations, younger (Mesozoic and Cenozoic) intrusive masses are seen near the host units that point to the possible formation of the deposits due to the activity of hydrothermal solutions associated with them (e.g., Kuh-e Dom in the Anarak area, Fig. 6.4). Thus, considering the age of these masses and the tectonic processes that resulted in mineralization, the age of deposits should be much younger than their hosts. Moreover, the high-background gold content of the basement rocks hosting the deposits is also responsible for the increase in the gold content of the deposits. The most important gold deposit within the rocks of this time interval includes the Zarshouran deposit. Others include Khooni, God, God Morad, Anarak, Chah Mileh, Gora, Tavazoi and Torkamani (in Anarak area), and Gorgab III and IV (northeast of Kashan).

There are no doubts that mineralization of some of these deposits has taken place in more recent periods, but the hosts are all of Late Proterozoic–Early Cambrian age. It seems that the Late Proterozoic–Early Cambrian rock has a generally high-background gold content in some areas such as Anarak (Ghorbani 2002a; Maghsoudi et al. 2005).

Late Paleozoic–Early Triassic: Some of the gold-bearing rocks previously assigned to the Precambrian times were found to be of younger age (Late Paleozoic–Early Triassic) based on more recent studies (Ghorbani 2002a; Maghsoudi et al. 2005). These include the rocks of the Taknar Series that host a number of massive sulfide and hydrothermal (vein-type) gold and copper deposits and indications. Other examples of the deposits belonging to this interval of time are the gold-bearing polymetallic lead–zinc–copper deposit of Chah Gaz (Sirjan area), the Mooteh gold deposit near Delijan, and the Zartosht deposit of Kerman.

Triassic–Jurassic: Gold mineralization of this time interval has generally occurred in sandstones, shales, and limestones of Triassic and Jurassic (Shemshak Formation) which are lowly metamorphosed or highly deformed; in addition, in some intrusive masses of granitic and dioritic composition, deposition of gold has taken place near the contacts of the igneous body (Ghorbani 2002b; Maghsoudi et al. 2005). Mineralization of gold in these units is usually in the form of veins and veinlets within faulted, jointed, and fractured zones along with tungsten and copper and in a limited localities is associated with silver, arsenic, lead–zinc, bismuth, and antimony (Ghorbani 2002b; Maghsoudi et al. 2005).

Torghabeh, Tarik Dareh, Booteh Alam, Chah Palang, Chah Kalap, and Shend Mahmood are the most prominent examples of the deposits and indications of this time interval. It is, however, possible that some of the deposits of this interval are younger in age (Ghorbani 2002b; Maghsoudi et al. 2005), for example, Shend Mahmood is thought to be as young as Neogene.

Some of the gold-bearing rocks that were earlier assigned to Cambrian are found to be younger, and the geological evidence supports a (Late) Paleozoic to Early Triassic age for them (Ghorbani 2002b; Maghsoudi et al. 2005). Among these rocks are

1. Taknar Series, which contains a number of deposits and indications of gold along with massive sulfide and vein hydrothermal copper
2. Rocks of Chah Gaz area, which is a polymetal Pb–Zn–Cu deposit with considerable amount of gold
3. Deposits and indications of Mooteh gold deposit near Delijan
4. Zartosht deposits, Kerman

Late Cretaceous–Paleogene: The deposition of gold within the Cretaceous rocks is in the form of silicic and aplitic veins within the skarns at the contact of intrusive igneous rocks (granitic and dioritic in composition) with Cretaceous limestones (Ghorbani 2002a; Maghsoudi et al. 2005). These deposits are seen in Astaneh (Arak), Arasbaran, northeast of Kashan, and Yazd. In the Arasbaran area, copper mineralization is dominant and is sometimes associated with gold and silver (Ghorbani 2002a; Maghsoudi et al. 2005); examples include Mazra'e, Mardanel,

Agha Mira, and Abbasabad copper deposits. The only exception is Nabi Jan indication, where gold is associated with lead–zinc and iron skarns. The area to the northeast of Kashan hosts skarn mineralization of iron and iron–manganese type that includes gold, copper, and silver as accessories (Ghorbani 2002a; Maghsoudi et al. 2005); Gorgab I and II are examples. In Yazd, gold mineralization has occurred along with copper skarns in the Dareh Zereshk area (Ghorbani 2002a; Maghsoudi et al. 2005).

Associated with Laramide igneous-tectonic upheaval, gold mineralization within the short interval of Late Cretaceous–Paleogene has taken place in the form of listvenite-type mineral indications along with the ophiolites (Ghorbani 2002a; Maghsoudi et al. 2005). This type of mineralization occurs within Makoo–Oshnavieh and East Iran ophiolitic belts and includes Soladokal, Todan, Khangol, and Chaldoran (all in Makoo–Oshnavieh Belt) along with Hangaran, and Sahl Abad (East Iran Belt) indications are among those associated with the ophiolites of this time interval. In addition, gold mineralization in Saghez–Naghadeh region occurs in sheared metavolcanic rocks of Late Cretaceous (Ghorbani 2002a; Maghsoudi et al. 2005), for example, Karooyan and Barika gold deposits.

Tertiary: This interval can be said to be the most active magmatic phase in the geological history of Iran whose effects are seen throughout the country, except Kopet Dagh, so that the term "metallogenic era" is assigned to the Pyrenean tectonic event (Eocene–Oligocene) (Ghorbani 2002a; Maghsoudi et al. 2005). The most extensive and most diverse mineral deposits (including gold) have been formed in Iran during Tertiary.

Gold mineralization in the Tertiary times is generally within the extrusive igneous, pyroclastic, and sedimentary rocks or calc-alkaline intrusive igneous rocks (sub-volcanics) and their associated andesites and dacites of Eocene–Oligocene age and younger (especially Miocene–Pliocene). The solutions expelled from the igneous intrusives invading the Eocene and Neogene rocks bring about extensive silicic and argillic metasomatism, which result in the deposition of gold and other elements within faults and other fractures (Ghorbani 2002b; Maghsoudi et al. 2005).

The most important deposits and indications of gold attributed to this interval time include all the gold-bearing localities of Kerman, Tarom–Hashtjin, Moaleman–Torbat-e Heydariyeh (except Taknar), Central Alborz, Qorveh, and Mianeh along with most of the deposits and indications of Saveh–Kashan–Naein Belt, Takab, Arasbaran, and Lut.

Quaternary: The most important gold deposits of the Quaternary age include placer deposits of Kuh Zar (Damghan), Zar Mehr (Torbat-e Heydariyeh), and Angooran Chah (Mah Neshan).

5.7.3 Geographic Distribution of Gold Deposits

The gold deposits of Iran can be classified according to their structural–metallogenic zones and type of mineralization process as follows (Fig. 5.7) (Ghorbani 2002a, 2008a; Maghsoudi et al. 2005).

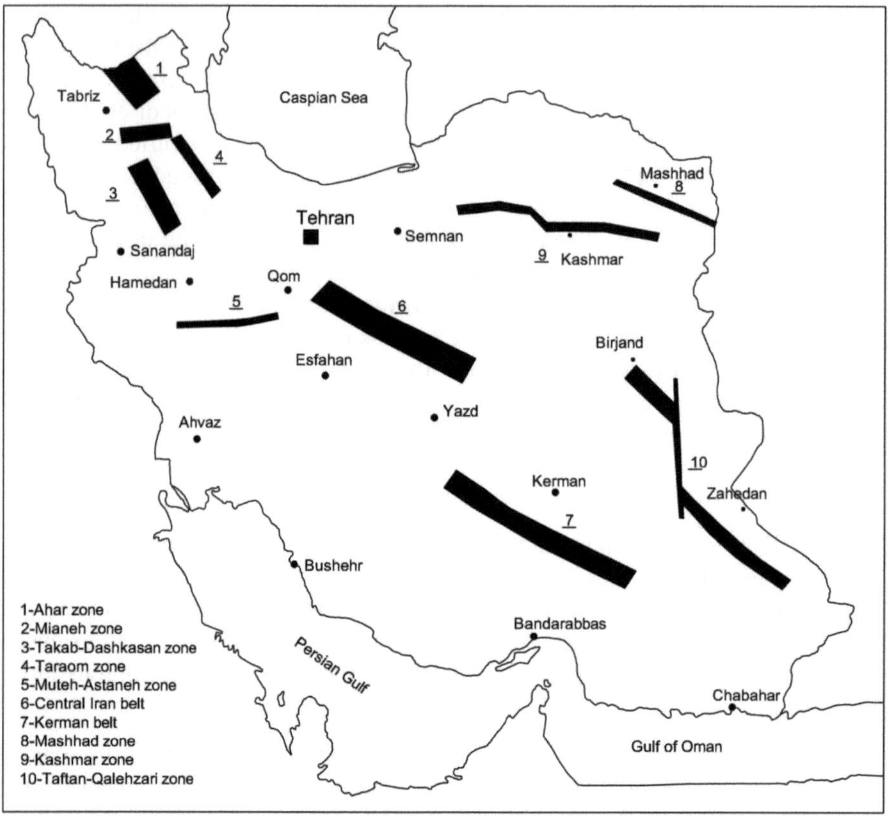

Fig. 5.7 The gold-bearing provinces and areas of Iran (Maghsoudi et al. 2005)

Sanandaj–Sirjan Gold-Bearing Province: Measuring about 1,500 km in length and 150–200 km in breadth, this province extends from the Urumiyeh Lake on the northwest to Minab fault on the southeast and includes many important gold deposits and indications that are related with shear zones. The genesis of gold which occurs in the form of veins, veinlets, and occasionally placers is attributed to magmatic and metamorphic processes (Ghorbani 2002a, 2008h; Maghsoudi et al. 2005). Based on the distribution of gold deposits and indications, the zone can be divided into northern (Golpayegan–Naghadeh–Makoo) and southern (Golpayegan–Sirjan) parts, with the northern part being richer in terms of gold. The northern part hosts Mooteh, Dashkesan, Karooyan, Barika, and Astaneh deposits along with Makoo–Oshnavieh, Alvand (Hamedan), Kondar (Aligudarz), Bakhtar Jaldian, and Kharabe (Piranshahr) indications. The southern part includes Zartosht and Chah Gaz deposits.

The most important known deposit of Sanandaj–Sirjan Province occurs at Dashkesan in the Qorveh area, which is related with the Quaternary volcanism.

Takab Gold-Bearing Zone: The most effective and diverse type of gold mineralization has occurred in the Takab area, resulting in the formation of Carlin-, vein-, polymetallic vein-, and placer-type deposits within the area (Ghorbani 2002a, 2008h; Maghsoudi et al. 2005). Except the placers, the mineralization of gold in the Takab area is generally related with hydrothermal solutions associated with the deep-seated and intermediate igneous bodies of Tertiary times (Ghorbani 2002a, 2008h; Maghsoudi et al. 2005). The most promising rocks of the area from the point of view of gold mineralization are young acidic domes intruded within the schists of the Precambrian age. The important gold deposits and indications of Takab include Zarshouran, Agh Dareh, Ay Qal'e Si, Toozlar, Bayche Bagh, Arabshah, and Angooran Chai Placer.

Arasbaran Gold-Bearing Zone: This area, which includes the highlands of Arasbaran and Qareh Dagh, is one of the most important and well-known metallogenic regions of Iran and hosts extensive copper–molybdenum mineralization containing gold. Gold occurs in three forms, namely, along with porphyry copper–molybdenum, vein-type, and skarn-type, the first two associated with the Tertiary intrusive and extrusive igneous rocks and the last one along the contact of the Tertiary intrusives and Cretaceous limestones (Ghorbani 2002a, 2008h; Maghsoudi et al. 2005). The gold deposits and indications of the area include Songoon, Masjed Daghi, Aharaf Abad-Hizeh Jan, Safi Khanloo-Naghdooz, Nabi Jan, Mazra'e, and Kharvanah.

Gold mineralization with small amount of copper, or without it, is related with vein-type silicic metasomatism confined to ancient calderas, for example, Masjed Daghi, Safi Khanloo-Naghdooz, and Sharaf Abad.

Tarom–Hashtjin Gold-Bearing Belt: Extending from the north to the southeast of Zanjan and covering parts of western Alborz structural zone, the Tarom–Hashtjin Belt contains gold mineralization along with lead, zinc, copper, and occasionally silver deposits in the form of veins and veinlets in Tertiary hosts (especially Eocene volcanics). Haj Aliloo (2000) has reported porphyry gold mineralization in Hashtjin area. No independent mineralization of gold has been reported from the area till date, but most of the copper and lead–zinc deposits of the area contain considerable amounts of gold. In other words, gold occurs as a by-product of copper and lead–zinc mineralization (Ghorbani 2002a, 2008h; Maghsoudi et al. 2005). Gold placer indication of Kalooch-Gilvan on the banks of Qezel Ozan River is also reported from the area. Other gold indications include Khalifeloo, Zeh Abad, Koohiyan, Dizeh Jin, Chal, Aliabad-e Moosavi, Lak, Somagh, Bashgol, Ab Torsh, Gav Kamar, Senjedeh, Zajkan, Marshoon, and Lehne.

Mianeh Gold-Bearing Zone: Trending in the east–west direction, the Mianeh Zone extends the area between Mianeh and Bostanabad covering the highlands of Bozgoosh. This zone is one of the promising areas of the country which has not yet been explored in detail. The indications of gold in the zone are located at Armoodagh and Aghoran (Ghorbani 2002a, 2008h; Maghsoudi et al. 2005).

Moaleman–Torbat-e Heydariyeh Gold-Bearing Belt: This highly promising belt is in the form of a stripe trending almost east–west and is situated on the northern margins of central desert of Iran. The Moaleman–Torbat-e Heydariyeh Belt has good potential for gold, copper, lead–zinc, silver, antimony, etc. (Ghorbani 2002a, 2008h; Maghsoudi et al. 2005). Gold mineralization occurs within the Precambrian–Paleozoic (polymetallic massive oxide and sulfide), Tertiary (in the form of veins of base metals and specularite-rich copper), and Quaternary rocks (in the form of placers). Examples include Arghash, Gandi, Kuh Zar (Torbat-e Heydariyeh), Kuh Zar (Damghan), Qal'e Joogh, Zar Mehr, Darestan, Kalateh Teymoor, Chah Moosa, Kakiyeh, Bid-e Mohammad Hassan, and Taknar.

Mashhad Gold-Bearing Zone: This zone extends from Torghabeh on the west of Mashhad to Janat Abad on the border with Afghanistan and Turkmenistan and contains Tarik Dareh and Torghabeh deposits (Ghorbani 2002a, 2008h; Maghsoudi et al. 2005).

Kerman Gold-Bearing Belt: The Kerman gold belt occurs in the southern parts of Urumiyeh–Dokhtar Metallogenic Province and includes 19 gold deposits and indication that are either associated with porphyry-type or vein-type copper–molybdenum deposits within the Tertiary intrusive rock (Ghorbani 2002a, 2008h; Maghsoudi et al. 2005). The porphyry deposits of Sarcheshmeh and Dareh Hamze (Jiroft) and vein deposits of Bazman, Abdar, Chah Mesi, Oroos Morghi, Chahar Gonbad, Baghdeh, and Dolat Abad fall within this belt.

All vein-type deposits, except Bazman, where silver is the major paragenesis for gold and copper and lead–zinc, are minor, and copper forms the most important paragenesis (Ghorbani 2002a, 2008h; Maghsoudi et al. 2005).

Saveh–Kashan–Naein Gold-Bearing Belt: Trending in the northwest–southeast direction within the Urumiyeh–Dokhtar volcanic zone, the Saveh–Kashan–Naein Belt is known to host skarn- and vein-type gold mineralization. Skarn mineralization occurs at the contact of Cretaceous limestones with Tertiary granodioritic masses, whereas vein and veinlet mineralization are in the Eocene volcanics (Ghorbani 2002a, 2008h; Maghsoudi et al. 2005). Examples include Gorgab I and II (both skarn-type), Qal'e Sardar, Maranjab, Ghasem Abad, Serajiyeh, Cheshme Talhe, and Kuh Dom (all of which are vein-type).

Yazd Block: Being situated on the western parts of Central Iran micro-continent, the block is bounded by the Dorooneh Fault on the north and the Naein-Baft Ophiolitic Belt on the west. Gold mineralization in this block has occurred in Anarak Metamorphic Complex (mostly in polymetallic form) along with porphyry copper, skarns and listvenite-type (Ghorbani 2002a, 2008h; Maghsoudi et al. 2005). The gold deposits and indications of Yazd Block include Khooni, East Khooni, God Morad, Gorva, Anarak, Chah Mileh, Torkamani, Tal Siyah, Ashin, Booteh Alam, Southern Chah Palang, Nadooshan, and Dareh Zereshk.

Lut Block: Measuring about 900 km in length, the Lut Block is bounded between the Nehbandan Fault (on the east) and the Nayband Fault (on the west). Gold mineralization in the block is associated with that of copper, silver, lead–zinc, antimony, and tungsten and occurs in the Upper Triassic–Jurassic volcanic-sedimentary rock successions and Tertiary volcanics (Ghorbani 2002a, 2008h; Maghsoudi et al. 2005). Qal'e Zari, Hirood, Chah Kalap, Shourab, Chah Zaghoo, and Khoonic comprise the gold deposits and indications of the block.

East Iran Gold-Bearing Zone: The gold-bearing region of East Iran measures 800 km in length and 200 km in breadth and falls in between the Lut and Damghan Blocks. Gold mineralization is either in the form of listvenite-type within the ophiolitic complexes or as veins, veinlets, and skarns related with hydrothermal solutions in association with the Tertiary granitoid bodies (Ghorbani 2002a, 2008h; Maghsoudi et al. 2005). The deposits and indications of gold in this zone include Hangaran, Sahl Abad, Dargiyaban, Siyah Jangal, Siyadetorgi, Shoor Chah, Almasaki, Toozagi, and Kuh Lar. Gold is dominantly associated with copper, arsenic, antimony, mercury, and lead–zinc.

Central Alborz Gold-Bearing Zone: Central Alborz trends in the east–west direction from the east of Talesh–Tarom Mountains to Gorgan. Pas Qal'e polymetallic deposit within the volcanosedimentary rocks of Karaj Formation is the only gold-bearing mineralization of the zone (Ghorbani 2002a, 2008h; Maghsoudi et al. 2005).

5.8 Antimony, Arsenic, and Mercury

5.8.1 Mineralization Phases of Antimony, Arsenic, and Mercury

Such deposits have not yet been exploited in Iran, and yet no detailed exploration and prospecting work has been done for antimony, arsenic, and mercury deposits, so the geology and metallogeny of such resources in Iran are almost unidentified.

To understand the metallogeny of these elements, some facts and characteristics are cited below for such deposits so that they will be usefully applied in future exploration works:

1. Antimony, arsenic, and mercury deposits in Iran can be identified in two completely separate time spans (Ghorbani 1995a, 2009c):

   - Late Proterozoic–Early Cambrian
   - Oligocene and Mio-Pliocene

   The Late Proterozoic–Early Cambrian rocks form the host and the surrounding rocks of some antimony, arsenic, and mercury deposits and indications like those in the Anarak and Takab areas.

   It has to be mentioned that although antimony and arsenic minerals are found in these rocks, younger Tertiary volcanic rocks mingle with the older ones. Geological evidence indicates that the formation of these minerals occurred in more recent time periods (after Eocene) and no concentration of these minerals has been indentified in rocks older than Eocene. Since the origin of these minerals is attributed to the Late Proterozoic–Early Cambrian rocks, it can be stated that mineralization of antimony and arsenic and mercury anomaly are the result of a recycling process that extracted these minerals from the Proterozoic basement and concentrated them in young Tertiary rocks (Ghorbani 1995a, 2009c).

Most antimony, arsenic, and mercury deposits and indications in Iran were formed in association with volcanic rocks younger than Eocene. Since the mineralization of antimony, arsenic, and mercury deposits and indications are seen in Precambrian rocks and they are also the product of recycling process by volcanism younger than Eocene, it can be concluded that the concentration of antimony, arsenic, and mercury minerals was associated with the volcanism and plutonism events younger than Eocene.

2. All areas wherein antimony, arsenic, and mercury deposits and indications are seen have continental basement, and these deposits are mostly found within young volcanic rocks with composition of basaltic, intermediate, or acid volcanics. Basaltic and andesitic volcanic rocks are older than acid volcanic rocks and are more widespread. Although acid volcanic rocks show less expansion, mineralization areas are mostly in connection with acid volcanic rocks, and basic and intermediate volcanic rocks within the mineralization areas are mostly covered by acid volcanic rocks (Ghorbani 1995a, 2009c).
3. Most antimony, arsenic, and mercury deposits and indications of Iran are in connection with deep faults (Ghorbani 1995a, 2009c).

   With regard to what has already been cited, it can be concluded that geological and mineralogical evidences of antimony, arsenic, and mercury deposits and indications of Iran point toward the concentration of these mineral deposits by low-temperature hydrothermal solutions (epithermal) that were in connection with the acid volcanic rock complex underneath the crust. Based on aforementioned geological evidences, for exploration and prospecting antimony, arsenic, and mercury deposits, it is recommended to consider those continental areas where young Tertiary volcanism (especially Eocene) occurred and deep faults are present. The volcanic rocks in such areas vary from basalt to rhyolite, but shallow small intrusive bodies (porphyry granite and porphyry tonalite) are seen within these volcanic rocks (Ghorbani 1995a, 2009c).

Within the young ophiolitic areas (such as colored mélange on the east of Ghaen and ophiolitic mélange in Khoy area) where Tertiary magmatism was active (Oligo-Miocene), the formation of antimony, arsenic, and mercury deposits was possible as well.

Antimony deposits are relatively young. It can be undoubtedly stated that other than few deposits belonging to Hercynian mineralization cycle almost no antimony deposit in Precambrian and Paleozoic rocks has yet been identified. Antimony deposits are mostly formed during Alpine orogeny (young Alpine in particular) (Ghorbani 1995a, 2009c).

Metallogeny of mercury is completely similar to that of antimony. No mercury deposit associated with Caledonian orogeny or older has yet been reported (Ghorbani 1995a, 2009c). Therefore, metallogeny of mercury took place in Late

Paleozoic or later. Some mercury deposits related to Hercynian orogeny (Late Paleozoic) have been reported, but main mercury deposits are formed during the Cimmerian and Alpine orogenic phases. Mercury deposits are mainly specific to folded and continental areas. They are found within deep faults and volcanic belts and are in connection with basaltic magmatism and the resulted anatexis phenomenon underneath the crust (Ghorbani 1995a, 2009c).

Large epithermal gold deposits accompanied with arsenic, and sometimes with antimony and mercury, formed in numerous and vast areas in Iran during the young Tertiary and Early Quaternary and in connection with the volcanic activities in these periods of time (Ghorbani 1995a, 2009c). Such deposits are mostly seen in areas northwest of Iran such as Ghorveh–Takab axis, northeast of Kashmar–Torbat-e Heydariyeh, and Ferdows.

5.8.2 Distribution of Antimony, Arsenic, and Mercury Deposits

Arsenic Deposits in Iran: So far, 17 arsenic deposits and indications have been identified in Iran, and currently only two of them (i.e., Zarshooran and Valiloo) are being exploited. The Zarshooran deposit is unique because of the degree of purity and crystallized and cauliflower realgar. Besides the degree of purity, it contains over 5,000 tons of gold with an average grade of 30 ppm (Ghorbani 1995a, 2009c).

The Most important arsenic deposits of Iran are

- Zarshooran in Takab area
- Chelpo in Kuh-e Sorkh in Kashmar area
- Dashkasan in Ghorveh area
- Valiloo in Ahar area
- Dastjerd realgar deposit Tabriz area
- Tikmeh-dash realgar deposit in Bostanabad area
- Aliabad realgar deposit in Kashmar area
- Gold-bearing arsenopyrite veins north of Torbat-e Jam
- Copper arsenate-bearing veins in Anarak

Antimony Deposits in Iran: More than 15 antimony deposits and indications have been identified in Iran, but only 3 of them are actively operating while the country imports a large amount of antimony to fulfill its needs.

Antimony deposits of Iran are concentrated in three areas:

- Ghorveh, Bijar, Takab area (Dashkasan, Moghanloo, Agh Dareh, … deposits)
- Central Iran area (Patyar, Torkamani, … deposits)
- Ferdows, Kashmar, Torbat-e Jam area (Chelpo in Kuh-e Sorkh, Shoorab, Sarghaleh, Shed-mahmoud, Kalehneginan, Torbat-e Jam, Ghasoon, … deposits)

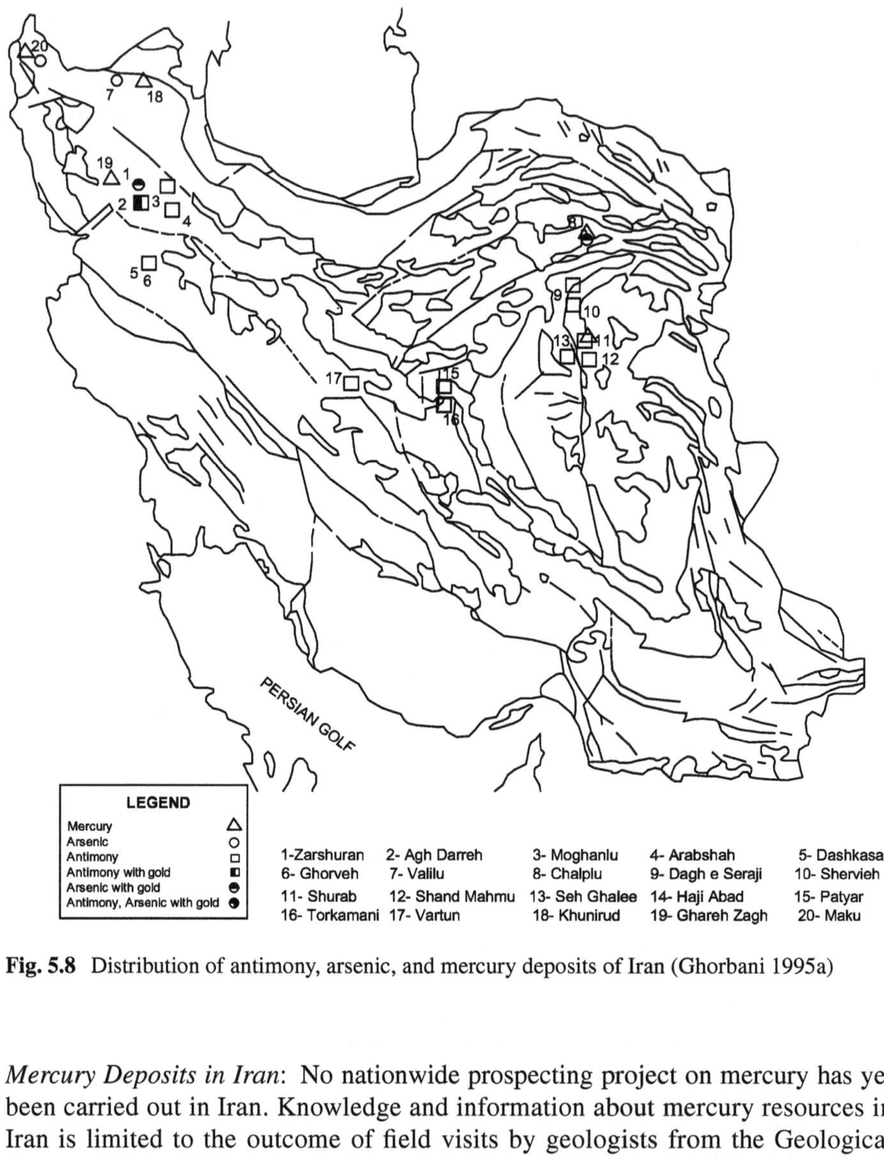

Fig. 5.8 Distribution of antimony, arsenic, and mercury deposits of Iran (Ghorbani 1995a)

Mercury Deposits in Iran: No nationwide prospecting project on mercury has yet been carried out in Iran. Knowledge and information about mercury resources in Iran is limited to the outcome of field visits by geologists from the Geological Survey of Iran (GSI) (e.g., B. Borna, M. Momenzadeh, N. Rashidnejad-omran), some information given by the local people and miners, and materials presented in some preliminary prospecting projects by the Bureau of Mines in Iran (Borna 1991; Momen-Zadeh et al. 1987).

Figure 5.8 illustrates the distribution of antimony, arsenic, and mercury deposits and indications of Iran.

Table 5.11 presents some geological/genetic characteristics of a number of antimony, arsenic, and mercury deposits and indications as well as their associated host rocks.

Table 5.11 Geological characteristics of some antimony, arsenic, and mercury deposits of Iran

General characteristics and type of deposit	Associated elements	Host rock	Explored areas and name of deposits
Hydrothermal deposits having old mafic and ultramafic basement	Sb + As ± Hg along with gold	Old carbonates and metamorphic rocks	Takab area (Zarshooran, Arabshah, Agh Dareh), Anarak area (Patyar, Torkamani)
Hydrothermal deposits having old volcanosedimentary basement	Sb + Hg ± As along with gold	Young carbonates and old metamorphic rocks	Takab area (Moghanloo-Agh Dareh, Qizghapan, Qareh Dagh)
Hydrothermal deposits having Mesozoic continental basement	Sb + Hg ± As ± (Pb ± Zn)	Mesozoic clastic rocks and/or younger igneous rocks	Ferdows area (Shoorab, Hesamieh, Se Qal'e, etc.), Qorveh area (Dashkesan)
Hydrothermal deposits having Cretaceous ophiolitic basement	Sb + Hg + As ± Au	Rocks of ophiolitic suites and/or their associated igneous rocks	Khoy area (Khan Geli), East of Qa'en (Haji Abad)
Hydrothermal deposits having volcanic basement	As	Sedimentary rocks	Ahar area (Valiloo)

Ghorbani (2009c)

Regarding the geochemistry, metallogeny, and geology of mercury deposits, their paragenetic relationship with antimony and arsenic deposits, and their coincidence with the geological events in Neogene–Quaternary of Iran, the existence of economically valuable mercury resources in Iran is undeniable. A few indications and signs of mercury resources have been reported in the following areas (Borna 1986; Ghorbani 1995a):

Shoorab: Considering the geological evidences and chemical analysis results, this mining area is very promising for mercury deposits

Makoo: Khan Goli and Kelisa Kandi areas

Ghara-dagh: Agh Dareh and Shakh-Shakh in the Takab area

Ghaen: Aliabad locality in the Haji Abad area

Ahar: Khoyna-rud locality

5.9 Nickel and Cobalt

5.9.1 Mineralization Phases of Nickel and Cobalt

Nickel and cobalt are usually found in ultrabasic rocks (such as peridotite and serpentine) along with copper and minerals of platinum group. Cobalt is also found in the form of high-temperature hydrothermal along with copper minerals in basaltic to andesitic rocks. Even with the significant distribution of ultrabasic rocks and relatively large number of chromite deposits in Iran, no noteworthy nickel and cobalt deposit has yet been identified along them. Nickel and cobalt indications identified in Iran are mostly of polymetal-type deposits.

5.9.2 Distribution of Nickel and Cobalt Deposits in Iran

The Most important known nickel and cobalt indications in Iran are given next:

Anarak Area: Nickel and cobalt minerals have been reported in ancient mines in Anarak such as Mes-kani and Talmesi and also within other mineral deposits such as She-barz, Chah-shoor, Gowd-morad, Cheshmeh Ab-e Shoorab, and Cheshmeh-karim. Talmesi and Mes-kani are the most important among the above-mentioned deposits.

Talmesi and Mes-kani Mines: Mineralization in these mines occurs in Eocene volcanic andesites. Based on the geophysical, geological, and geochemical studies carried out in the area, the basement consists of ultrabasic rocks. About 50 minerals have been identified in these mines that contain copper, iron, lead–zinc, nickel, cobalt bismuth, uranium, arsenic, etc.

Mahneshan, Zanjan (*Baycheh Bagh Deposit*): Volcanic rocks such as andesite, andesite–basalt, agglomerate, and andesitic breccia along with acid volcanic rocks are seen in this area. The basement of this area consists of ultramafic and mafic rocks. Mineralization is seen within the fractures due the function of high-temperature hydrothermal solutions (Ghorbani 2000h). More than ten mineral veins have been identified in this area that contain copper, lead–zinc, cobalt, nickel, bismuth, molybdenum, gold, uranium, etc. The most important cobalt minerals in these deposits are smaltite ($CoAs_3S_2$) and cobaltite (CoAsS).

Ghamsar Deposit: Sedimentary rocks such as limestone and sandstone outcrop in north and south of Ghamsar. In some places, basic dikes have penetrated into these sedimentary rocks, resulting in mineralization within fractures. Ancient mining works can be seen 1 km south and 6.5 km north of Ghamsar, wherein some signs and evidences of chalcopyrite, malachite, azurite, pyrite, erythrite, and limonite minerals exist within piles of gangue consisting of quartzite and calcite.

There are some other areas where nickel and cobalt have been reported as accessory minerals. For example, in the Eghlid area, mineralization took place in Paleozoic rocks, and ancient slags containing copper and cobalt spread out over an area of 1 km^2. Also, mineralization of cobalt has been reported in Early Cambrian rocks within the Zagros Zone (Shahrekord area, and Karoon-rud deposit northeast of Behbahan).

5.10 Titanium

5.10.1 Mineralization Phases of Titanium in Iran

The origins of titanium deposits are magmatic, placer, and metamorphic, with the latter two comprising the most important and economically valuable deposits:

1. Titanium placer deposits: Since titanium minerals (ilmenite and rutile) are resistant to weathering, they are mostly found as heavy minerals along with magnetite, monazite, zircon, sillimanite, garnet, kyanite, andalusite, corundum, tourmaline, spinel, staurolite, and epidote within river alluvium and beach sand. One of the important concentration places for titanium-bearing minerals is titanium-bearing black sands that are of specific importance when it comes to exploration and prospecting. The source rocks of placer-type titanium deposits are titanium-bearing rocks exposed to weathering and erosion. The titanium-bearing sands Cretaceous are more abundant (Ghorbani 2002a).
2. Titanium magmatic deposits: These deposits are in fact titanium-bearing magnetite and hematite that contain significant amounts of ilmenite and are mostly seen with layered ultramafic and mafic rocks (gabbro–pyroxenite–norite and anorthosite).[1]

[1] Most large titanium deposits of the world are associated with Proterozoic anorthosite rocks. Such rocks have not been found in Iran as yet.

The titanium-bearing magnetite deposits are divided into two groups (Ghorbani 2002a):

(a) Titanomagnetite deposits that are formed in anorthosite and gabbro-anorthosite rocks and contain ilmenite, magnetite, ilmenite-bearing magnetite, ilmenite-bearing hematite, and rutile.
(b) Titanomagnetite deposits within gabbro and norite rocks. Titanium deposits of Iran are of this group, which are the most important source of titanium.

5.10.2 Distribution of Titanium Deposits in Iran

Titanium deposits and resources in Iran have not completely been identified, and those that have been investigated are not in production yet. Currently, all required supply of titanium is imported (in the form of titanium dioxide). Thus, several exploration projects have been carried out in recent years in order to provide titanium resources in Iran, which are briefly described next.

Due to the high demand and price of titanium in Iran, exploration projects started in 1983 in various areas such as the Caspian Sea coast, part of the Oman Sea coast, Saghand, and Zanjan. However, low tonnage and grade of titanium resource in the aforementioned prospective areas as well as the problems in mineral processing and production had downplayed the promising ROI (return on investment) for these areas until the Kahnouj area (wherein ilmenite was reported back in 1975) was restudied in 1986. Primary studies and evaluations during prospecting and detailed exploration works resulted in the discovery of a placer deposit of ilmenite and one anomaly with an estimated reserve of 2.2 million tons and grade of 6% (Ghorbani and Iranmanesh 2012).

In this area, the N-S trending rock units of the Band-e Ziarat area are the source of ilmenite deposits found in river sands in east, west, and south of the area, which encompass Darreh Gaz, Dasht-e Manoujan, Darreh Bagh, and Tiyab River. Aside from Darreh Gaz, which contains the highest tonnage and grade, the rest of these areas feature tonnage of several tens of million tons and grade of 2–4%, which could be considered as promising titanium potentials in the future.

Exploration studies (in the framework of the Kahnouj Ilmenite Project; Kowsari et al. 1986) on titanium source rock (meso-gabbros) in the Kahnouj area point to the existence of 20 million tons of titanium-bearing minerals with a TiO_2 grade of 5.5%. Currently, the technological feasibility studies for this area within the framework of a project called "Equipment and development of titanium processing plant in Kahnouj" are underway.

In 1996, similar exploration studies were carried out by the Kavoshgaran Engineering Co. in northwest of Urumiyeh that led to the discovery of Ghara Aghaj titanium and phosphate deposit, and more exploration and further development were recommended for this area (Maghsoudi 2000).

Based on the petrographic and geological studies, the intrusive complex of Ghara Aghaj mainly consists of ultrabasic rocks with the composition of verlite, altered diorite, and gabbro to gabbro–diorite. The degree of mineralization in the ultrabasic rocks is higher compared to that in the altered basic rocks, and the ultrabasic unit is considered to be the host rock of mineralization. The minerals within the ore include ilmenite, titanomagnetite, magnetite, ulvospinel, hercynite, and apatite with a small amount of sulfide minerals (pyrite, chalcopyrite) (Alipour-asl 1998).

Estimation and evaluation of this deposit point to approximately 208 million tons of titanium minerals (probable reserve) with an average grade of 8.25% and 80 million tons of phosphate (probable reserve) with a grade of 4% of P_2O_5. Considering these grades, this is the first deposit of ilmenite–titanomagnetite and apatite type in the Middle East, which can be placed under the "exceptional" category from the standpoint of titanium reserves (Alipour-asl 1998).

Geochemical and heavy mineral studies in northwest of Urumiyeh resulted in the discovery and presentation of Kanik-Mangol titanium mineralization in the year 2000. Mineralization in this area is economically viable from two standpoints: Titanium mineralization occurs both in the source rock and the placer deposit (Maghsoudi 2000). Mineralization in the source rock is significantly extensive and is of the following three forms:

1. Massive
2. Layered within gabbros
3. Disseminated within ferro-gabbros

Based on chemical analysis on some samples from these rock units, the percentage of TiO_2 was shown to vary between 2 and 14%. The minerals within the ore include ilmenite, titanomagnetite, magnetite, ulvospinel, hercynite, and, in some localities, apatite.

The Kanik-Mangol placer, which has originated and formed the from the aforementioned rock units, displays highly variable grade in such a fashion that the amount of TiO_2 in the geochemical samples taken from alluviums range between 1 and 13.5%, and the amount of ilmenite in the heavy minerals samples is found to be as low as 1% to a maximum of 69% in the enriched areas of the placer (Maghsoudi 2000).

5.11 Tungsten

5.11.1 Mineralization Phases of Tungsten

Similar to tin and molybdenum, the reserves of tungsten increased temporally when moving from the old to the new eras. No tungsten deposit has been identified in Archean. A small number of tungsten deposits formed in Proterozoic, of which small tungsten-bearing pegmatites in the United States (Silver Hill) and tungsten-bearing skarns in Brazil (Brezhu) and Sweden (Igohellen) are the best known (Smirnov 1983).

During Paleozoic and due to Caledonian orogeny, larger tungsten deposits such as Boguty with hydrothermal origin formed in Kazakhstan (Smirnov 1983).

During Late Paleozoic and due to Hercynian orogeny, large tungsten deposits of skarn, greisen, hydrothermal, and porphyry types formed in many parts of the world, including Akchatav in Kazakhstan, Ingichke in Central Asia, Panasguelva in Portugal, Kings Island in Australia, and a tungsten porphyry deposit in southwest of England (with age of Devonian to Permian) (Smirnov 1983). Most of the economically valuable tungsten deposits formed in Mesozoic and Cenozoic due to Alpine orogeny, including tungsten deposits in southeast of China, which are the largest tungsten deposit in the world and belong to Mesozoic. The tungsten deposits in South Korea and Thailand also belong to this time period (Smirnov 1983).

The abundance of tungsten deposits in Phanerozoic as compared to Precambrian shows the connection of tungsten with continental rocks, especially the granitic ones.

5.11.2 Distribution of Tungsten Deposits in Iran

Tungsten mineralization in Mesozoic and Tertiary rocks of Iran has been reported (Jahangiri 1999). Much of the tungsten mineralization has been found in association of Jurassic–Cretaceous acid intrusive bodies and Tertiary magmatic rocks and also along with the mineralization of copper and molybdenum. In general, tungsten mineralization horizons in Iran can be categorized into the following (Jahangiri 1999):

1. Mesozoic mineralization (Jurassic–Cretaceous in particular) in relation with intrusive bodies: This type of mineralization is mostly seen in the Sanandaj–Sirjan Zone wherein tungsten mineralization is accompanied with tin, gold, and silver.

   Mineralization of tungsten and gold along with signs of tin within the Sanandaj–Sirjan Zone, along the Boroujerd–Azna axis, in Abar-kuh in Central Iran (iron, tungsten, and possibly gold), and east of Iran such as Anarak, Shirkuh, Zarrin-e Ardekan, and south of Birjand are among the noteworthy examples.

   Mineralization of copper, gold, silver, and manganese in the volcanosedimentary rocks of Mesozoic is seen within the metamorphosed areas. Some copper deposits of massive sulfide type have been identified in southeast of Iran.
2. Tertiary mineralization: This type of mineralization takes place in association with Eocene intrusive magmatism and is more widespread in Iran. It is accompanied with copper, molybdenum, and polymetal mineralization of Tertiary age (e.g., copper, molybdenum, and tungsten mineralization in Ahar and Tarom areas). Tungsten mineralization was anomalous during this time period, and no important tungsten deposit has yet been identified.

Mineralization of copper, gold, silver, and manganese is in association with Late Jurassic–Cretaceous volcanosedimentary rocks and can be seen in the metamorphosed rock units. In northwest and southeast of Iran, copper deposits, possibly of massive sulfide type, can be found. If the metallogenic characteristics of tungsten and geological features of Sanandaj–Sirjan Zone in areas like Dehbid, Eghlid, Abar-kuh, Aligudarz, Shamsabad, and Nezam Abad are compared with those in the east of Iran in areas like Deh Salm, Bazman, and Chah Kalab south of Birjan, there could be hope for discovery of economically valuable tungsten deposits (Table 5.12).

Table 5.12 Important tungsten deposits and indications of Iran

Name of deposit or indication	Location	Geological setting	Economic characteristics
Nezam Abad deposit	Southwest of Arak	Mineralization occurs within Jurassic sedimentary rocks and granitic massifs	Extraction of the deposit has been stopped due to its low grade
Chah Kalab deposit	200 km to the south of Birjan	Mineralization occurs within Jurassic rocks consisting of schists, sandstones, and marbles, within which Shah-kuh granite has intruded	Scheelite and ferberite constitute the major ore minerals that have close paragenesis with copper, zinc, and iron sulfides. The estimated reserve of this deposit is 78,000 tons of WO_3 by Akrami (1974)
Southern Chah Palang deposit	55 km to the southeast of Anarak	Mineralization is associated with the granite body that has intruded into schistose Jurassic rocks	This is a polymetallic deposit of copper in which wolframite occurs within quartz veins
Deh Salm indication	Southeast of Nehbandan, Qavar, Chahdashi, and Darvish regions	Mineralization is associated with Deh Salm intrusive body. Heavy mineral analysis of the rocks of the area reveals presence of wolframite, scheelite, and cassiterite	A strong tungsten and tin anomaly has been reported from the area
Tarik Dareh indication	30 km to the northwest of Torbat-e Jam and a distance of 10 km from Cheshme Gol Coal Mine	Mineralization is associated with the granite body that has intruded into schistose Jurassic rocks	Mineralization is in the form of scheelite along with arsenopyrite

5.12 Uranium

5.12.1 Mineralization Phases of Uranium in Iran

Generally, all uranium deposits were formed in continental platform environment, and most (about 70%) were in clastic rocks such as conglomerate and sandstone and few of them in igneous rocks.

Uranium is among those metals whose mineralization is in close connection with the genesis of granitic crust, its development, and formation of continental platforms. The uranium deposits of Archean (excluding Late Archean) have not been identified for the time periods where granitic crust did not exist. However, with the genesis of granitic crust that is at the border of Late Archean and Early Proterozoic, uranium deposits also came into existence, which form 20% of the uranium deposits of the world (Ghorbani 2007a).

During the passage from Early to Late Proterozoic, large uranium deposits of unconformity type were formed.

In general, those uranium deposits associated with granitic crust are accounted for most of the world's uranium deposits in Late Archean–Early Proterozoic and Early Proterozoic–Middle Proterozoic. More than 75% of exploitable uranium deposits of the world belong to this time period (Dohlkamp 1989; Ghorbani 2007a). The reducing condition in Early Proterozoic environments, which was very suitable for the formation of uranium deposits, can help explain and understand this episode.

A few uranium deposits of sandstone and magmatic types have been identified that were formed in Late Paleozoic and Early Mesozoic.

5.12.2 Distribution of Uranium Deposits in Iran

Considering the above-mentioned points, the important uranium deposits of the world belong to Late Archean–Early Proterozoic and the passage from Early to Middle Proterozoic; the uranium deposits attributed to other geologic times are insignificant (Dohlkamp 1989).

Late Archean, Early, and Middle Proterozoic terranes have no outcrop in Iran. In fact, the oldest rocks in Iran belong to Late Proterozoic, and therefore large uranium deposits in Iran are unlikely to exist. Some uranium deposits within Late Proterozoic and Early Cambrian have been reported in Iran (Saghand, Narigan, Anarak, and Gachin salt dome) (Ghorbani 2007a). Also, some evidence of uranium in association with the Tertiary magmatic rocks has been obtained in some parts of Iran such as Alamout and Ahar (Ghorbani 2007a), wherein the Atomic Energy Organization of Iran has been carrying out exploration projects. The author does not have much information about such uranium deposits, but considering the metallogeny of uranium and Iran's geological characteristics, the author does not expect valuable uranium deposits to occur in Iran.

5.13 Phosphate

5.13.1 Introduction

The phosphate deposits can be categorized into three types: sedimentary, igneous, and organic (guano).

The Iranian phosphate deposits are known from three geological time periods and are dominated by sedimentary type (though igneous deposits are also recognized) (Ghorbani 2008d).

Late Proterozoic–Early Cambrian phosphates are either igneous or sedimentary in nature. Sedimentary phosphates are intermingled in the rocks of Soltanieh Formation (especially its shale and middle dolomitic members) in Alborz and Central Iran while igneous phosphates are found in the rocks of the Rizoo Series and their associated intrusives in and around the Bafgh area. Transgression of Early Cambrian sea in Alborz and Azerbaijan over the Proterozoic rocks has resulted in the formation of phosphate-bearing shales, limestones, and dolomites during Tomanian and Atabanian conformably overlying the Vendian dolomites (Alavi Naeini 1993). The phosphate deposits of Dalir, Vali Abad, and Firooz Abad in Central Alborz (Atabanian Stage) are of this type. Similar phosphate deposits are also found in Mount Soltanieh.

Late Paleozoic–Triassic phosphates include deposits of Jeiroud, Laloon, Firooz Kuh, Shahrood, and Damghan areas. It is worth mentioning that the most important Devonian phosphate deposits of the world are found in Iran and Armenia (Halalat and Bolourchi 1994).

Laramide (Late Cretaceous–Tertiary) mineralizations are found in most parts of the world including the coastal regions of North Africa, West Africa, Middle East, southern Urals, Greece, Tunisia, Brazil, Columbia, and Venezuela as well as interior Paris Basin (Halalat and Bolourchi 1994). In the Zagros Zone of Iran, phosphate deposits are reported from Gurpi exposures around Behbahan. Other indications include Lar, Sheikh Habil, Kuh-e Rish, Kuh-e Kumeh, and Kuh-e Sefid (Ghorbani 2008d).

The various origins of phosphate deposits are described next.

Sedimentary Origin: Seawater saturated with P_2O_5 at high pressure and low temperature is the main source of sedimentary phosphate. Cold water – belonging to deep sea and/or polar mediums – causes more soluble phosphate compared to warm and shallow waters. When such cold waters migrate – by upwelling currents – to warm and shallow environments, they become fully, or partly, saturated with phosphate, so they acquire the ability to form sedimentary phosphate deposits. Generally, phosphate-bearing deposits are different in texture and composition, but their main controlling factors are always the same. Such parameters are used in interpreting and modeling sedimentation patterns. Generally, the sedimentary deposits of phosphate are widespread (Halalat and Bolourchi 1994).

Igneous Origin: Economic igneous resources of phosphate are observed in the form of intrusive bodies, hydrothermal veins, or deposits that result from the differentiation of alkaline magmas that formed ijolite, nepheline syenite, and carbonatite. Apatite resources related to carbonatites are commonly found in continental rift zones. Igneous phosphates usually contain hydroxyl fluorine apatite. Rarely, in some deposits such as Eppawala, high concentrations of chlorine are observed. Usually, minerals associated with apatite – such as vermiculite, anatase, pyrochlor, and monazite – contain rare earth elements (RRE), so they are valuable by-products. In fact, only 18% of phosphate production comes from igneous resources (Halalat and Bolourchi 1994).

Organic Origin (*Guano and Its Derivatives*): Guano deposits are formed by the feces of seabirds or bats, which are in small quantities. The most famous guano deposit is located in Morocco's Sahara, which is extracted at a few thousand tons annually (Halalat and Bolourchi 1994). There is no sign of such deposits in Iran.

5.13.2 Mineralization Phases of Phosphate Resources in Iran

Iran phosphate minerals and resources were formed in three time periods with igneous and sedimentary origins. These deposits are located in Central Iran and central Alborz geologic zones (Fig. 5.9).

Late Proterozoic–Early Cambrian: Late Proterozoic–Early Cambrian phosphates are either igneous or sedimentary in nature. Sedimentary phosphates are intermingled in the rocks of Soltanieh Formation (especially its shale and middle dolomitic members) in Alborz and Central Iran, while igneous phosphates are found in the rocks of the Rizoo Series and their associated intrusive in and around the Bafgh area. Transgression of Early Cambrian sea in Alborz and Azerbaijan over the Proterozoic rocks has resulted in the formation of phosphate-bearing shales, limestones, and dolomites during Tomanian and Atabanian which conformably overly the Vendian dolomites (Alavi Naeini 1993). The phosphate deposits of Dalir, Vali Abad, and Firooz Abad in central Alborz (Atabanian Stage) are of this type. Similar phosphate deposits are also found in Mount Soltanieh (Fig. 5.10a, b).

Ordovician mineralization phase is observed in east of central Iran at Tabas and Kerman regions; small phosphate deposits are also found in Mount Zardkouh in the Zagros Zone associated with sandstones and limey sandstones (Ghorbani 2008d).

Late Paleozoic–Triassic: There is a phosphate horizon in central Alborz within Jeiroud Formation. The phosphate deposits and indications belonging to this period include the deposits of Jeiroud, Laloon, Firooz Kuh, Shahrood, and Damghan areas in the central Alborz zone. It is worth mentioning that the most important Devonian phosphate deposits of the world are found in Iran and Armenia (Halalat and Bolourchi 1994), (Fig. 5.11).

Laramide Mineralization Phase: Laramide (Late Cretaceous–Tertiary) mineralization is found in most parts of the world, including the coastal regions of North Africa, West Africa, Middle East, southern Urals, Greece, Tunisia, Brazil, Columbia, and Venezuela as well as interior Paris Basin (Halalat and Bolourchi 1994). In the Zagros Zone of Iran, phosphate deposits are reported from Gurpi Formation

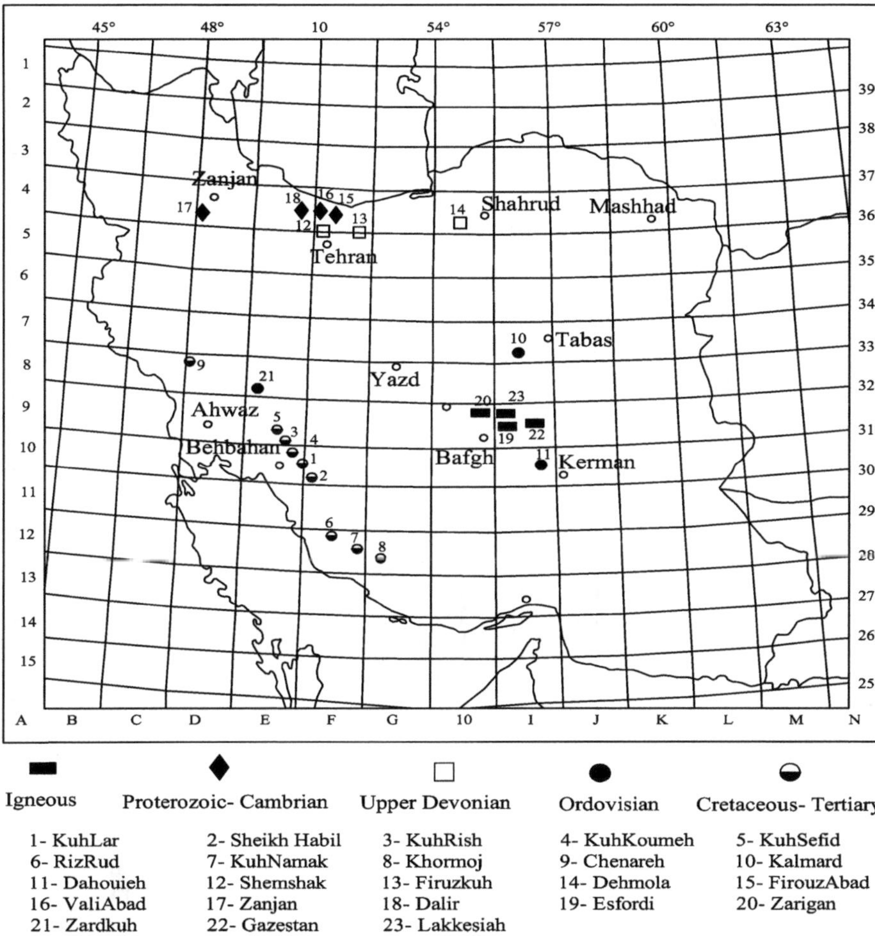

Fig. 5.9 Distribution of Iran phosphate deposits through different geologic periods (Ghorbani 2008d)

exposures around Behbahan with Late Cretaceous–Paleocene age (Fig. 5.12). Other indications include Lar, Sheikh Habil, Kuh-e Rish, Kuh-e Kumeh, and Kuh-e Sefid (Ghorbani 2008d).

5.13.3 Distribution of Phosphate Deposits in Iran

Two modes of origin, sedimentary and igneous, are considered for the phosphate deposits of Iran. Igneous phosphates are usually found in alkaline complexes such as syenite, ijolite, carbonatite, pyroxenite, and glymerite. Iranian igneous phosphates located at Esfordi, Zarigan, and Lakeh Siyah in Bafgh-Posht Badam Block are associated with alkaline igneous rocks (Halalat and Bolourchi 1994). Taking into

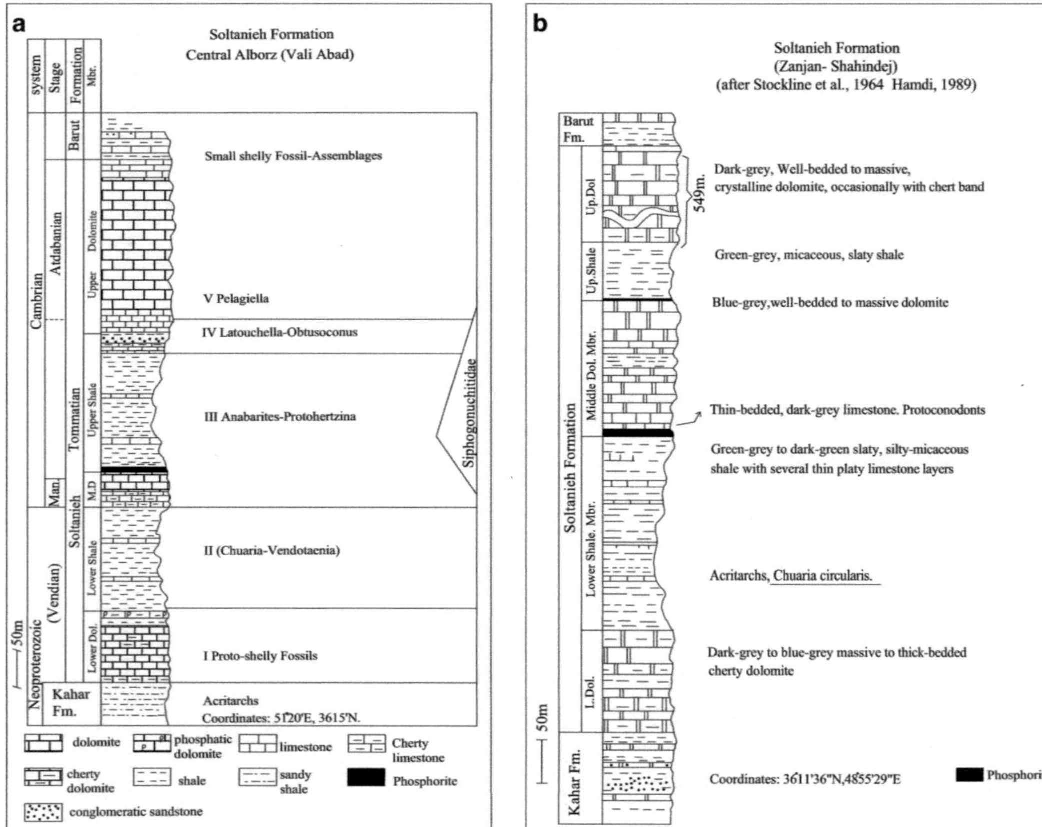

Fig. 5.10 Lithostratigraphic column sections of (**a**) Soltanieh Formation in central Alborz and (**b**) Soltanieh Formation in Zanjan–Shahin Dehz

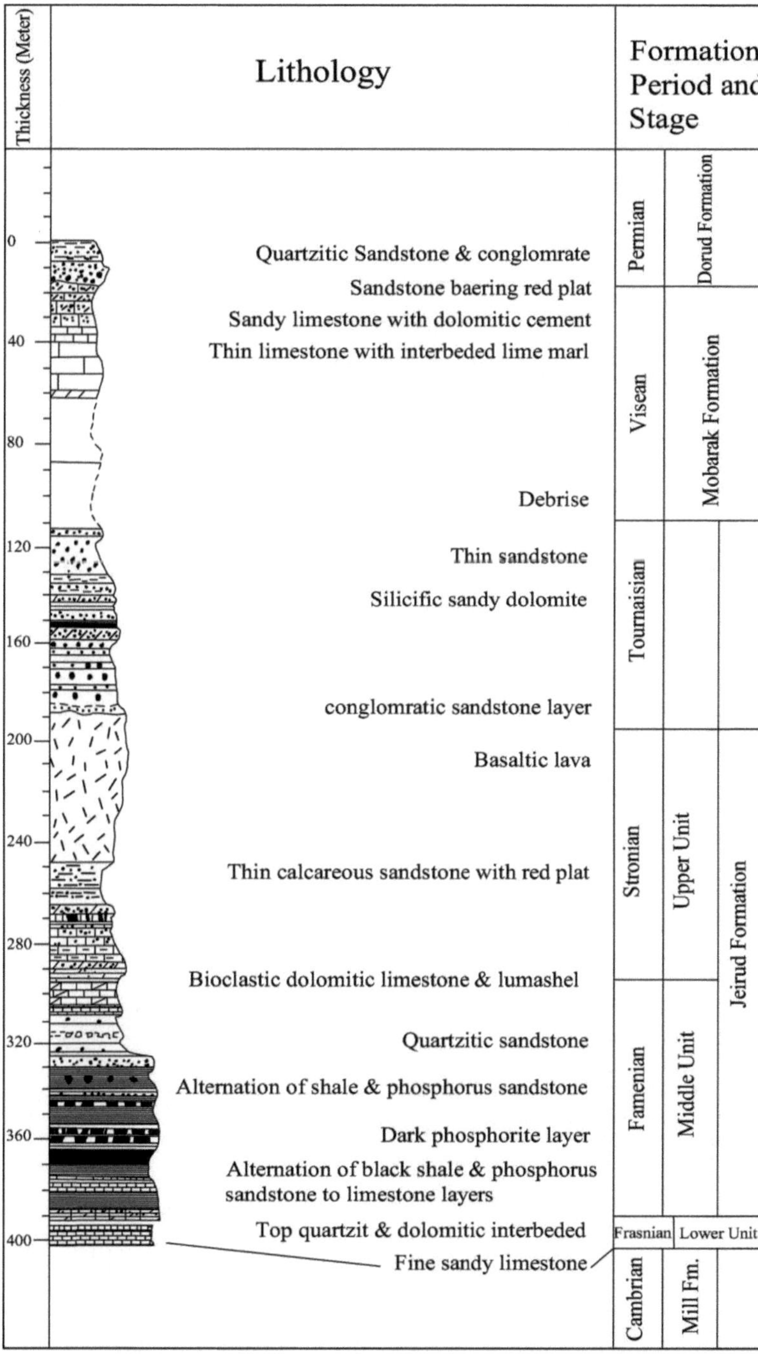

Fig. 5.11 Phosphorus lithostratigraphic column section of Jeiroud Formation in central Alborz (After Halalat and Bolourchi 1994)

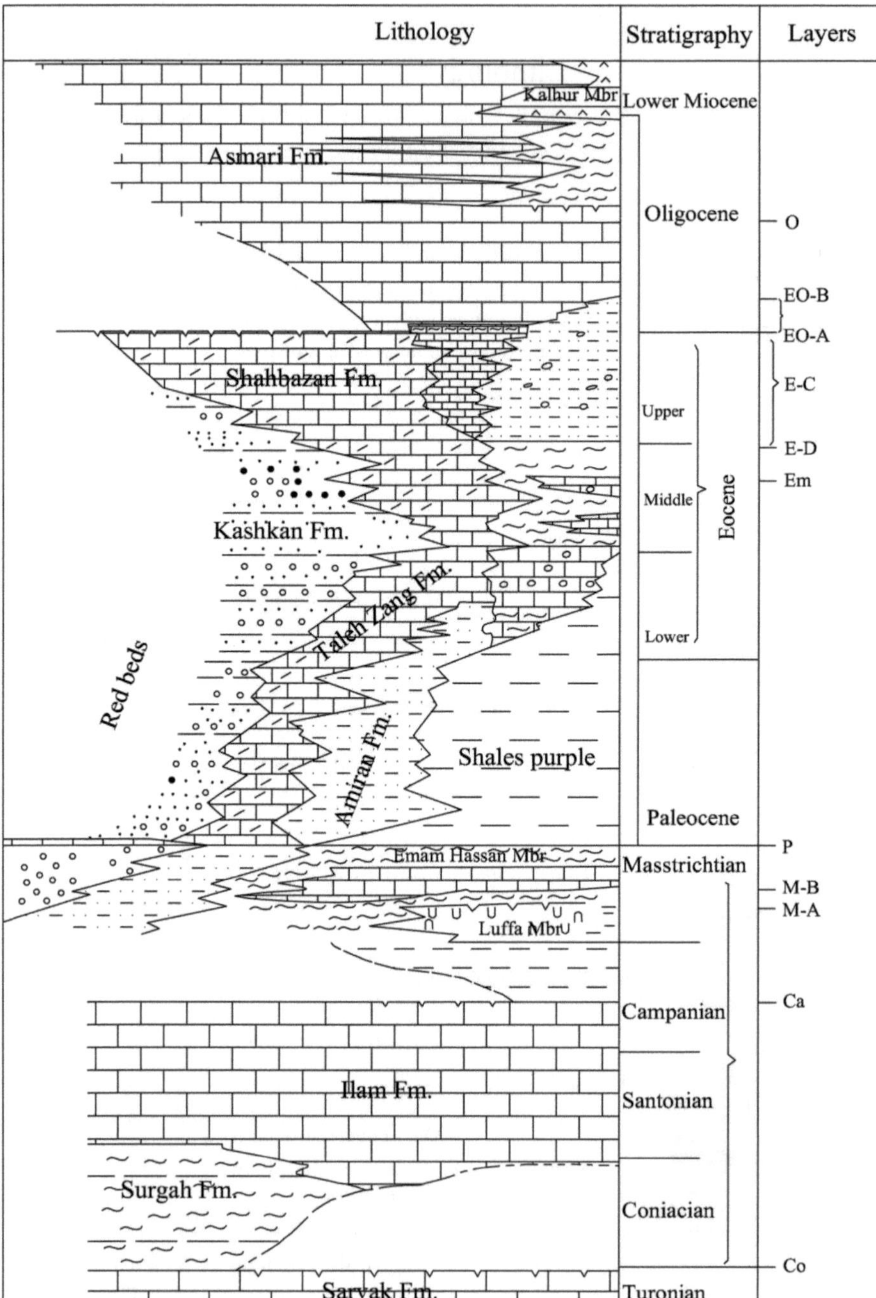

Fig. 5.12 Phosphate formations in Zagros (After Halalat and Bolourchi 1994)

account the extension of such rocks in various parts of Iran, the possibility of existence of igneous apatite reserves in alkaline rocks cannot be ruled out. A complex from volcanic–sedimentary rocks with gabbro to granite intrusive bodies with Early to Late Cambrian age occurs at the Bafgh region (Central Iran zone). There is a vast mineralization in the form of magnetite, apatite–magnetite, and apatite-bearing magnetite in the area (to see the related map, please refer to Chap. 6, Fig. 6.3).

All known igneous phosphate deposits and indications in Iran are limited to Bafgh, for example, Esfordi, Zarigan, Lakeh Siyah, and Mishdavan and Sechahoon. Also, there are considerable phosphate resources in association with Bafgh iron deposits like Chadormalu. High content of rare earth elements (REE) is observed within igneous phosphates in the form of apatite and monazite minerals (Halalat and Bolourchi 1994). If the whole iron-bearing rocks are extracted, they can provide a great quantity of phosphate and REE.

Sedimentary phosphates constitute the majority of produce. They are dominantly found in the rocks of Proterozoic–Cambrian, Ordovician–Silurian, Upper Devonian, and Cretaceous–Tertiary (Halalat and Bolourchi 1994).

The results of the investigations in various parts of the country have thrown light on a number of deposits and indications of phosphates in the outcrops of Soltanieh Formation including those on Tehran–Chalus route and Taleghan area of central Alborz; indications at Mount Soltanieh, deposits of Shahin Dezh, and indications on Ayeneh Varzan anticline are the other localities of this type (Halalat and Bolourchi 1994).

Proterozoic–Cambrian phosphate resources belong to shale member of Soltanieh Formation. The most important areas with known deposits are in central Alborz, Taleghan mountain range, Soltanieh mountain chain, and highlands of Takab–Shahin Dezh (Halalat and Bolourchi 1994). Such resources consist of low-grade phosphate with high impurities, but their quantities are such that they form considerable deposits. Some deposits of this type are Dalir, Valiabad, Zanjan, and Firooz Abad (Fig. 5.10a, b).

The most important Ordovician phosphate deposit occurs within the shale–sandstone sequences of Shirgesht Formation in Central Iran especially at Kalmard anticline. The sedimentary phosphates of this time interval can be observed in Central Iran (e.g., Kalmard and Rahdad deposits) and Zarand (Dahoueiyeh area). Overall, the phosphates of this interval have no economic value (Halalat and Bolourchi 1994).

Devonian phosphates constitute the most important sedimentary deposits in Iran. The most important phosphate-producing event in the world during the Devonian times occurred in Iran and Armenia (Halalat and Bolourchi 1994). In Iran, the phosphates of this time interval are found within the sandstones of Jeiroud Formation (central Alborz) or its equivalents. This phosphate deposit has a special importance. The phosphatic zone has a considerable thickness in excess of 70 m beginning with a limestone layer rich in brachiopods. This mining zone contains individual layers with higher than 18% P_2O_5 and phosphatic sandstone and limestone with black shale intercalations of lower than 18% P_2O_5. These are separated from each other by sandstones. In the Jeiroud area, this layer is associated with several layers of phosphate each measuring 1–2 m in thickness (the phosphate-bearing horizon is located at the lowermost parts of the section). Phosphate mineral in this formation is a type of cryptocrystalline Cl-bearing apatite (frankolite) (Halalat and Bolourchi 1994).

Among the sedimentary phosphates of Alborz, deposits such as Jeiroud, Kasil, Laloon, Gadook, and Dogol along with indications such as Tar Lake, Ordineh, Kuh-e Shourab, Kuh-e Zangi, Hamla, and Margdar are more important than the others.

Cretaceous–Tertiary phosphate deposits are dominant in Zagros. There are seven phosphogenic events during Coniacian to Oligocene in these regions, the most important being (both in terms of quality and quantity) Paleocene (occurring at the base of Pabdeh Formation) at Rizrud, Khourmoj, and Kuh-e Namak anticlines and Late Eocene–Oligocene at the Dehdasht–Behbahan region. The phosphate-bearing rocks of this period are important resources or hosts for petroleum (Ghorbani 2008d). Usually, source rocks such as Pabdeh Formation constitute the most important phosphate horizon of Zagros. This formation is the most significant source rock for the Asmari petroleum reservoir, which is the most important oil field in southern Iran. Chenareh, Rita, and Tal-e Zang anticlines in Late Cretaceous; Khourmoj, Kuh-e Namak, and Rizrud anticlines in Paleocene; and Kuh-e Lar, Kuh-e Siyah, Kuh-e Sepid, Kuh-e Kumeh, Kuh-e Rish, Kuh-e Nil, Seikh Habil, and southern tip of Bangestan anticline in Eocene–Oligocene are a few examples of phosphate mineralization in the above-mentioned regions (Halalat and Bolourchi 1994). Table 5.13 and Fig. 5.9 show the various types of phosphate deposits and their geological and geographical distribution in Iran.

5.14 Talc

5.14.1 Mineralization Phases of Talc in Iran

Since talc is a secondary mineral, it is not logical to define specific time periods for the formation of talc deposits. The origin and source rocks of talc in Iran are ultramafic and dolomitic rocks that show various ages from Late Precambrian to the Early Paleogene. These rocks transform to talc as they are exposed to metamorphism and metasomatic solutions rich in CO_2 and SiO_2. Most ophiolitic complexes in Iran belong to Late Mesozoic, and talc mineralization occurred in some of them due to metamorphism and metasomatism during Laramide orogeny (Ghorbani 2010a).

5.14.2 Distribution of Talc Deposits in Iran

The important talc deposits of Iran belong to the following three paragenesis:

1. *Talc deposits of dolomitic type*: These deposits are the product of metasomatism during metamorphism within the faulted zones in dolomites or dolomitic limestones and can be considered as talcification process of dolomites (Ghorbani 2010a). This type of talc deposits is seen in northern Sanandaj–Sirjan Zone (e.g., Masoud-abad talc deposit, Azna talc deposit…).
2. *Talc deposits of serpentine type*: These deposits are associated with metamorphosed ultramafic rocks. If solutions rich in CO_2 and SiO_2 are present in the

Table 5.13 Important phosphate deposits and indications of Iran

Name of deposit or indication	Geographic location	Probable reserve (million tons)	Average grade P_2O_5 (percentage)
Kuh-e Lar	Charam, Kohkilooyeh	350	8
Kuh-e Kume	Southwest of Dehdasht	22	9.3
Kuh-e Rish	North of Behbahan	10.6	11.2
Sheikh Habil	North of Dehdasht	1	22
Kuh Sefid	South of Izeh	17	12.5
Riz Rud	East of Bushehr	160	8.25
Khur Moj	North of Kangan	56	8.5
Kuh-e Namak	Northeast of Andimeshk	24	8.2
Chenar	North of Tehran	5	3.2
Shemshak-Jeirud	Northeast of Firooz Kuh	73	9.13
Firooz Kuh-Gadook	Northeast of Firooz Kuh	56	12
Deh Mola	Southwest of Shahrood	9	10
Kalmard anticline	Southwest of Tabas	6.2	5
Dahooeiye, Zarand	East of Zarand	1.13	7.5
Dalir	South of Chalous	23	11.5
Vali Abad	South of Chalous	3	9
Zanjan	Southwest of Zanjan	12	–
Firooz Abad	South of Chalous	40	8
Esfordi	Northeast of Bafgh	15	12
Zarigan	Northeast of Bafgh	0.5	3

Modified after Halalat and Bolourchi (1994)

environment during the metamorphism of ultramafic rocks, talc would form (the author is of the opinion that solutions poor in CO_2 and SiO_2 will give rise to serpentine). This type of deposits is found within ophiolitic complexes, especially in Sistan–Baluchestan where the basement is ophiolitic; also, it occurs in areas where basic volcanic rocks cut across ultramafic basement such as Kharestan, northeast of Taftan, and vicinity of the Hamont Mountain along the Iranshahr–Magasi (Zaboli) road.

3. *Talc deposits of talc-schist type*: These deposits form during metamorphism in a closed system with clay-bearing dolomitic rocks or dolomite-bearing clays where silica solutions are also present (Ghorbani 2010a, b, c). These deposits are called talc-schist (talc is one of the components but does not have high quality) and are noteworthy within the Flysch Zone in east of Iran.

5.15 Mica

5.15.1 Mineralization Phases of Mica in Iran

More than 50 mica deposits and indications have been discovered in Iran so far (Fig. 5.13).

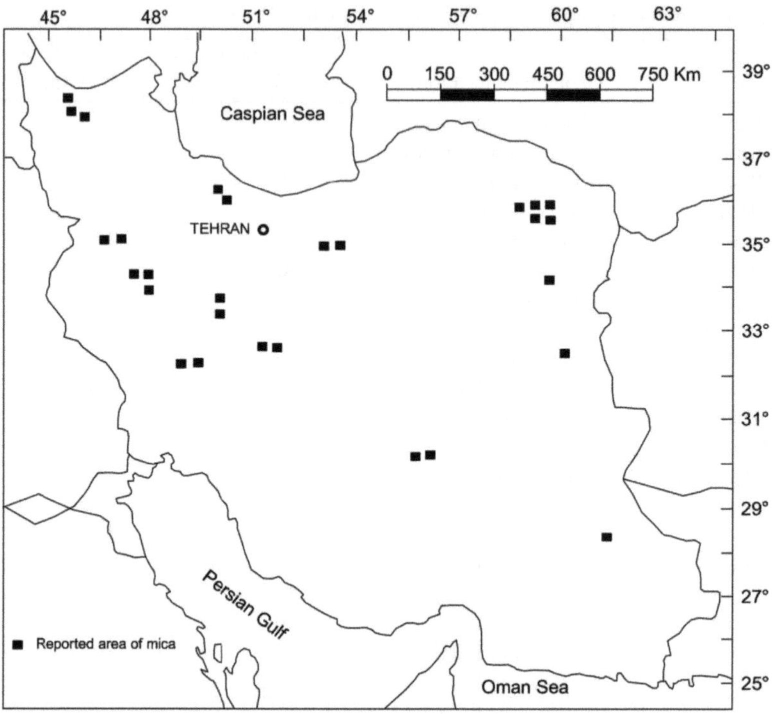

Fig. 5.13 Distribution map of mica deposits of Iran (Ghorbani 2011)

All mica deposits of Iran are in association with pegmatitic parts of granitoid batholiths and sometimes in metasomatized zones in lamprophyre dikes. In other words, mica deposits are mostly found in places where igneous and metamorphic complexes are seen together (Ghorbani 2011).

5.15.2 *Distribution of Mica Deposits in Iran*

Some of these igneous–metamorphic complexes suitable for formation of mica deposits are

- Hamedan igneous–metamorphic complex
- Aligudarz–Boroujerd igneous–metamorphic complex
- Igneous–metamorphic complex north–northwest of Late Urumiyeh
- Deh Salm igneous–metamorphic complex
- Masouleh–Amlash igneous–metamorphic complex

Table 5.14 shows the important specifications of mica deposits in Iran.

Table 5.14 Geological characteristics of mica deposits of Iran

Sr. no.	Geographic location	Occurrence	Enclosing rock	Type of mica	Associated minerals
1	Qareh Bagh, west Azerbaijan	Metamorphic zone	Metamorphic	Phlogopite	Green apatite, amphibole
2	Yarim Ghiyeh, west Azerbaijan	Metamorphic zone	Metamorphic	Phlogopite	–
3	Zarean, west Azerbaijan	Metamorphic zone	Metamorphic	Phlogopite	–
4	Ali Gavaber, Amlash, Gilan	?Pegmatite	Metamorphic	–	–
5	Masouleh, Gilan	?Pegmatite	Schist	Muscovite and phlogopite	Quartz and sodic feldspar
6	Jandagh, Esfahan	Pegmatite	Schist	Muscovite	Quartz, sodic and potassic feldspar
7	Zanjanbar, Kashan, Esfahan	Metasomatic zone	Limestone	Phlogopite	–
8	Gheshlagh, Khorasan	Pegmatite	Granite	Muscovite and minor amount of phlogopite	Quartz, potassic feldspar
9	Ching Kalagh, Khorasan	Pegmatite	Granite	Muscovite and minor amount of phlogopite	Sodic and potassic feldspar
10	Deh Gheibi, Khorasan	Pegmatite	Granite	Muscovite and minor amount of phlogopite	Sodic and potassic feldspar
11	Khajeh Morad, Khorasan	Pegmatite	Granite	Muscovite and minor amount of phlogopite	Sodic and potassic feldspar
12	Sangan Khorasan	–	–	–	–
13	East Birjand, Khorasan	–	–	–	–
14	Nehbandan, Khorasan	Pegmatite	Metamorphic	Muscovite	Sodic feldspar and quartz
15	Shotor Kuh Khaftari, Semnan	Metamorphosed sedimentary layers	Metamorphic	Amphibole and phlogopite	–
16	Cheshmeh Gorgab, Semnan	Metamorphosed sedimentary layers	Metamorphic	?	?
17	Kuh Gabri, Kerman	Granite	Metamorphic	?	?
18	Sirjan, Kerman	Pegmatite	–	–	–
19	Qorveh, Kordestan	Pegmatite	Metasomatic	Muscovite and phlogopite	Sodic feldspar and quartz
20	Garmab, Kordestan	Pegmatite	Metasomatic	Muscovite and phlogopite	Sodic feldspar and quartz

(continued)

Table 5.14 (continued)

Sr. no.	Geographic location	Occurrence	Enclosing rock	Type of mica	Associated minerals
21	Cheshmeh Chalan Choolan, Lorestan	Pegmatite	Granite	Muscovite	Feldspar and quartz
22	Mola Taleb Lorestan	Pegmatite	Granodiorite	Muscovite	Feldspar and quartz
23	Dehno, Markazi	Pegmatite	Granodiorite	Muscovite	Feldspar and quartz
24	Shazand, Markazi	Pegmatite	Granodiorite	Muscovite	Feldspar and quartz
25	Dareh Kashkin, Sistan and Baluchestan	Granite	Granite	Phlogopite	–
26	Mangavi, Hamedan	Pegmatite	Metamorphic	Muscovite	Quartz and sodic feldspar
27	Arzanfood, Hamedan	Pegmatite	Metamorphic	Muscovite	Quartz and sodic feldspar
28	Dehno Asadolah Khan	Pegmatite	Metamorphic	Muscovite	Quartz and sodic feldspar

Ghorbani (2011)

5.16 Boron and Borax

5.16.1 Mineralization Phases of Borax in Iran

The age of all borax deposits of Iran ranges between beginning of Late Miocene to Early Quaternary, but it seems they were mostly formed during a time span at the Miocene–Pliocene boundary.

5.16.2 Distribution of Borax Deposits in Iran

When the formation settings (i.e., such as stratigraphy, lithology, tectonics, source rocks, climate, physiography, and geomorphology) of borax deposits around the world are compared, Iran will pose as a very promising place for discovering valuable borax deposits. With the exception of the Gharagol deposit in Zanjan, which is now being actively exploited and a lot of information is available about its geology and mineralogy, some haphazard exploration and prospecting works were carried out in different parts of the country, and none came up with satisfactory results because of inadequate knowledge about the mineral properties, mineralization processes, and mechanism of deposition of borax in specific geological settings. A list of the most important borax mineral indications identified in Iran so far is presented next:

- Khatoon-abad-e Sirjan borax mineral indication (Deh-shotoran borax)
- Mohammad-abad Orion borax mineral indication (south of Sabzevar)
- Nadooshan borax mineral indication (Taft, Yazd)
- Borax mineral indication with magnesium sulfate (Miynaj, Mahneshan)
- Borax mineral indication in Kafeh Noogh and Kashkooyeh (Rafsanjan)
- Ashin borax mineral indication (north of Anarak)
- Isisoo borax mineral indication (north of Late Urumiyeh)

5.17 Magnesite

Magnesium is naturally found in the form of magnesite $MgCO_3$, brucite $Mg(OH)_2$, dolomite $MgCa(CO_3)$, and magnesium-bearing silicates. Magnesite is the most economically important mineral of this group. Rocks rich in magnesium such as ultramafic and mafic rocks and dolomitic rocks are excellent sources for the formation of magnesite. Dolomitic magnesite deposits are good source of magnesium. The Magnesite deposits in the aforementioned rocks form in various paragenesis, including (Ghorbani 2007g)

- Hydrothermal magnesite deposits
- Replacement hydrothermal magnesite deposits
- Seepage magnesite deposits
- Sedimentary magnesite deposits

The conditions and settings for the above-mentioned types are available in most parts of Iran, but no plan has yet been presented for the exploration of sedimentary magnesite.

The magnesite deposits of Iran were formed in a time span ranging from Late Cretaceous to Quaternary, and because of the characteristics of magnesite mineralization, no exact time can be defined for the formation of magnesite deposits in Iran.

5.17.1 *Mineralization Phases of Magnesite*

The magnesite deposits in east of Iran, which are within the ophiolitic complexes south of Birjand and east of Ghaen in Sistan–Baluchestan, are associated with Late Cretaceous–Paleogene mineralization phase (concurrent with Laramide orogeny). Although the possibility of formation of magnesite deposits beyond this time period through the end of Tertiary cannot be ruled out, most of the magnesite deposits and all of their source rocks formed during Late Cretaceous–Paleogene (Ghorbani 2007g).

5.17.2 *Distribution of Magnesite Deposits*

Although rocks rich in magnesium are found in many parts of Iran (ophiolitic zones), most of the magnesite deposits of Iran are located in east of Iran (south of Khorasan and Sistan–Baluchestan provinces). Table 5.15 presents the most important magnesite deposits of Iran.

5.18 Celestine (Strontium Sulfate)

5.18.1 *Mineralization Phases of Celestine Deposits in Iran*

During young Tertiary and in association with the Tertiary–Quaternary phase, celestine deposits in the form of intercalations with carbonate, calcium sulfate, and barium sulfate were formed concurrently along the northern edge of the Central Iran desert and in Zagros (e.g., north of Behbahan).

5.18.2 *Distribution of Celestine Deposits in Iran*

Strontium usually concentrates in celestine mineral, and strontium deposits consist of this mineral. There are significant celestine deposits in Iran including those in Namak Kavir, Behbahan, Bushehr, salt domes in Zagros, and Qom

Table 5.15 Important magnesite deposits of Iran

Sr. no.	Name of mine	Proved reserve	Probable reserve
1	Chahkhoo	254,000	–
2	North Afzal Abad	170,000	252,000
3	Gomanj, Ghaen	170,000	2,700,000
4	South Afzal Abad	99,000	–
5	Hoz Sefid Arabkhaneh	212,000	–
6	East Hoz Sefid Arabkhaneh	190,000	–
7	Torshak Mohamadi	1,546,000	–
8	Tabas Mesina	–	–
9	Sarlord	162,000	–
10	Shir Koohak	1,450,000	–
11	Kasrab	23,000	36,000
12	Tak Siyah	72,000	105,000
13	Rezgh Soleiman	64,000	80,000
14	Kalateh Ali Mohamad	17,500	–
15	Shoor-o-Shirin Av	104,000	–
16	Cheshmeh Lagoori	–	131,000
17	Farmaj	36,000	–
18	Espiki	142,000	238,000
19	Hasan Abad Korin	220,000	547,000

(Ghorbani 2002a). In general, celestine mineralization occurred in two zones of Central Iran and Zagros. The Malek-abad celestine deposit in northeast of Namak Kavir is the most important celestine deposit in Central Iran and has been exploited for many years.

Celestine Mineralization in Alborz: Celestine can directly deposit from seawater or form during the early stages of diagenesis process. Celestine mineralization has been reported from a few horizons within Zagros, two of which are economically valuable and one has been exploited.

Celestine Deposit in Salt Domes in Zagros: The origin of strontium in Hormuz Formation is syngenetic and is accompanied by salt, gypsum, and marl.

Celestine Mineralization in Razak–Gachsaran Formation: Razak Formation is correlated with Gachsaran Formation and is separable from it in the Fars area due to the lateral lithologic changes. The rocks within this formation include marl, marl–limestone, gypsum, and anhydrite. Deposition of celestine in this formation occurred during cementation process as most celestine minerals are surrounded by calcite (Ghorbani 2002a).

Celestine with post-diagenetic origin is identifiable through studying inclusions and remnants of tiny crystals of calcite floating within celestine minerals. Strontium is normally crystallized as celestine where enough sulfate ions are available in the environment. These ions could be supplied from the seawater or groundwater that once was flowing through the gypsum rock units (Ghorbani 2002a, b, c). The possibility of such scenario to have occurred in Gachsaran Formation is higher due to the high content of gypsum and anhydrite in this formation, and this is the reason for deposition of economically valuable celestine deposits (e.g., Lilak-e Behbahan celestine deposit) within Gachsaran Formation, which consists of limestone, anhydrite, gypsum, and marl belonging to Early Miocene (Ghorbani 2002a). Asmari Formation overlies Gachsaran Formation conformably.

In Lilak celestine deposit, there are two layers with thicknesses of 1.5 and 2 m stretching for about 1,000 m; they form the main body of the deposit. The host rocks include gypsum, anhydrite, marl, and foraminiferal limestone. The replacement of limestone, gypsum, and anhydrite with celestine can be observed in places where the layers outcrop and points to the fact that the formation of celestine veins was syngenetic and metasomatic in origin (Romanko et al. 1985).

Celestine deposits are in the form of small and large crystals with compact massive texture containing celestine-enriched masses having small pores and cavities filled with needle-shaped crystals. Celestine minerals in this area are mostly seen along with remnants of carbonate materials from host rocks and show great resemblance to celestine minerals found in Central Iran (Romanko et al. 1985).

Celestine Deposits in Asmari Formation: Asmari Formation has an evaporite member that consists of gypsum, anhydrite, marl, and thin-layered limestone, which belongs to Early Miocene. This segment is called Kalhor member and is mostly seen in Lorestan, but it can be traced in Khuzestan Province as well. Sometimes, evidence and sign of celestine is found in this member (Ghorbani 2002a).

Celestine Deposits in Central Iran: Celestine mineralization in Central Iran can be studied and traced in evaporite member of Qom Formation (which contains acid volcanic rocks) in areas such as Anarak, Qom, Varamin, and Garmsar (Ghorbani 2002a).

The most important and the only active celestine deposit in Central Iran is the Malek-abad celestine deposit located in northeast of Namak Kavir, which is about 200 km southeast of Tehran. The celestine minerals are seen along with intercalations of marine series of fossil-bearing limestone, marl, shale, breccia, salt-bearing clay, and chert of Early Miocene, which are known as Qom Formation (Ghorbani 2002a). To describe the celestine horizon in Central Iran, Malek-abad deposit is briefly described here.

The economically valuable concentration of celestine within the three horizons of marl–gypsum member (Member C_2) of Qom Formation may be seen in the form of a layer with or without gypsum. The position of major celestine horizons depends on the overall thickness of Qom Formation as well as the distance of Member C_2 from the base of Qom Formation in any given area (Ghorbani 2002a). However, small amounts of celestine and some nodules are also seen locally in marls within Members C_1 and C_3, or limestone–marl and marl–limestone in the upper part of Member D, and also along the contact between Qom Formation and Upper Red Formation (Ghorbani 2002a).

The tonnage of this deposit with regard to average thickness of 2.5 m for celestine-bearing layer has been estimated at around 470,000 tons (Ghorbani 2002a). The deposit is being exploited in an open-pit fashion.

5.19 Barite

5.19.1 Mineralization Phases of Barite Deposits in Iran

Barite deposits have various origins including sedimentary, volcanosedimentary, and hydrothermal (Ghorbani 2002a). Barite mineralization in Iran can be traced from Late Precambrian through Pliocene, but major barite mineralization events occurred during the following time periods.

Barite Mineralization in Late Precambrian–Early Cambrian: Mineralization during this period took place as syngenetic with deposition of dolomites of Soltanieh Formation in Alborz and Azerbaijan or equivalent rocks in Central Iran (Ghorbani 2002a). Although barite mineralization in the aforementioned rocks is seen in many locations, few valuable deposits have been identified during this period.

Barite Mineralization in Permian–Triassic: Extensive barite mineralization is seen in Permian and especially in Triassic dolomitic rocks in Central Iran, Alborz, and Sanandaj–Sirjan. Within this period mineralization of barite along with fluorite occurred in large areas of Iran. Barite along with fluorite deposits in Central Iran (Ardekan, Kamshecheh, Faskhood, Atashkooh, Kashan, and Delijan areas), in Alborz (Elika Formation), and in eastern part of Central Iran (Ravar–Kuhebanan) accompanied by lead and zinc, belong to Early–Middle Triassic (Ghorbani 2002a). The Ghahrabad fluorite deposit in Sanandaj–Sirjan probably belongs to this time period. The fluorite deposits in Alborz, which are mainly located in Elika Formation, belong to this time period as well. In general, barite mineralization in Late Paleozoic–Triassic can be categorized into the following groups (Ghorbani 2002a):

(a) Alborz: Barite mineralization accompanied by lead and zinc occurred in the Elika and Duna areas, and barite mineralization along with fluorite took place in Pachi Miyana.
(b) Central Iran: Especially in Ardekan area, extensive barite mineralization hosted by dolomites occurred in the form of veins and lenses. Kamshecheh deposit (which is in fact a barite and fluorite deposit) is an example of such scenario.

Barite mineralization in Central Iran and Alborz during Triassic was mostly accompanied by fluorite mineralization; in other words, fluorite deposits of this time period are present in areas with more barite mineralization.

Barite Mineralization in Cretaceous: Barite deposits belonging to Cretaceous (mostly Early Cretaceous) are mainly seen in Central Iran and Sanandaj–Sirjan zones. The barite deposits of this time period are mostly seen in vicinity of lead and zinc deposits (e.g., Mehdiabad) and close to some lead and zinc deposits along the Malayer–Esfahan axis (Ghorbani 2002a).

Barite Mineralization in Tertiary: In Tertiary, an extensive barite mineralization in connection with volcanic and volcanosedimentary rocks took place in south of central Alborz, Azerbaijan, and Urumiyeh–Dokhtar volcanic belt (especially in its central parts). The barite deposits in form of veins, lenses, and layers are abundantly seen within Eocene and Oligocene volcanic and pyroclastic rocks in some areas such as Qom, Saveh, Kashan, Delijan, and Qazvin (Ghorbani 2002a).

5.19.2 Distribution of Barite Deposits in Iran

Important barite deposits of Iran are distributed in the following areas (Khoshjou 1999; Ghorbani 2002a):

Central Alborz – two types of mineralizations are seen in this area:

1. Barite mineralization syngenetic with sedimentary carbonates in the northern part of central Alborz
2. Barite mineralization within volcanosedimentary rocks in the southern part of central Alborz, especially along the Karaj–Takestan axis.

Central Iran – two types of mineralizations are seen in this area:

1. Barite mineralization within sedimentary carbonates (dolomite) in the Ardekan area
2. Barite mineralization within volcanic and volcanosedimentary rocks along the Qom–Kashan axis

Several barite deposits have been discovered in Kordestan and Western Azerbaijan provinces in recent years, and therefore these areas have turned into new prospects for barite exploration.

As mentioned earlier, barite deposits are geographically distributed in many parts of Iran, but the most important and active barite deposits and mines are found in the following areas (Fig. 5.14):

- Kamshecheh in Ardestan, Esfahan Province (10)
- Haji Abad in Zarin, Yazd Province (7)
- Ardekan, Yazd Province (3)
- Darreh Kashan, Esfahan Province (2)
- Jasb in Delijan, Esfahan Province (23)
- Darreh Lar in Tehran, Tehran Province (14)
- Haft-har in Aghda, Yazd Province (1)
- Elite (13), Alborz Province

Chari in Ab Torsh, Kerman Province (17)

5.20 Feldspar Mineralization

5.20.1 Mineralization Phases of Feldspar Deposits

The chemical composition of feldspar group is aluminosilicates of sodium, potassium, and/or calcium; rarely, barium is seen in their chemical composition.

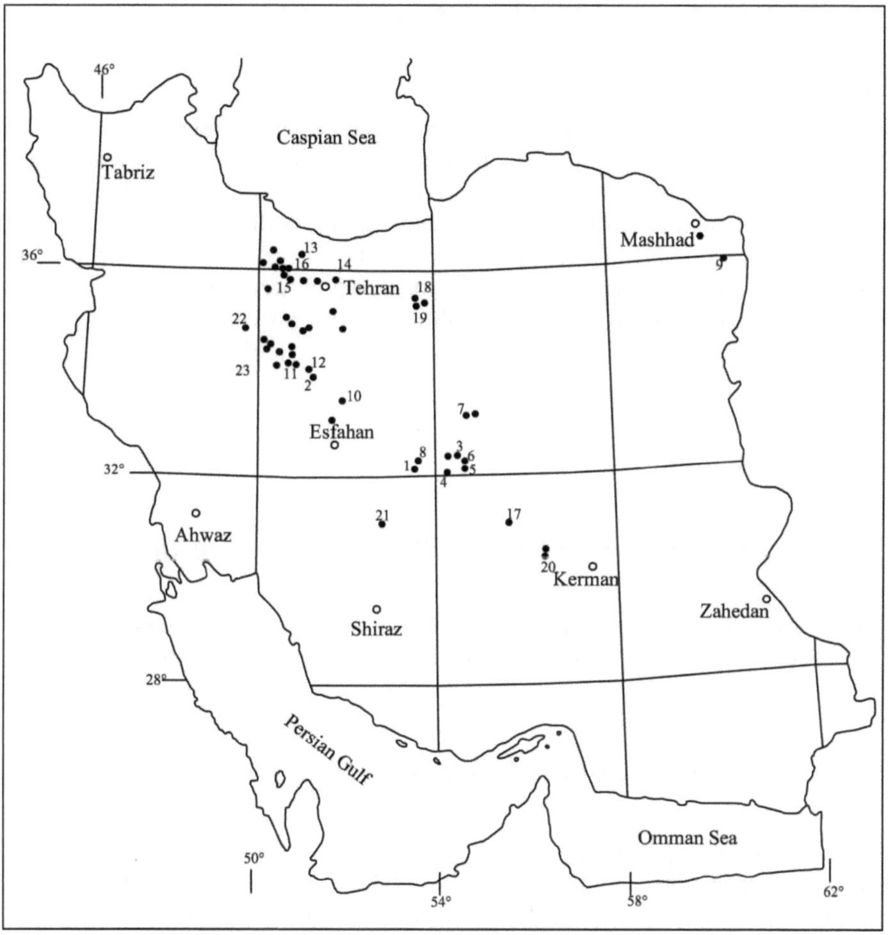

Fig. 5.14 Distribution map of barite deposits of Iran (Khoshjou 1999) (*1* Haft-har, *2* Darreh Kashan, *3* Ardekan, *4* Dorbid, *5* Niyook, *6* Dasht-e Deh, *7* Haji Abad, *8* Hoodeh, *9* Ghara Gheytan, *10* Kamshacheh, *11* Tappeh-sorkh Bichegan, *12* Vavan, *13* Elite, *14* Lar, *15* Ahoorak Taleghen, *16* Seplark, *17* Chari-Abtorsh, *18* Garmab, *19* Chah Shirin, *20* Baghin, *21* Khaneh Hozeh-abad, *22* Tang-abad, *23* Jasb)

There are plenty of feldspar resources in Iran, and it is probable that with more exploration works, many new deposits will be found. In order to get acquainted with feldspar reserves in Iran, the various types of such reserves are briefly described here. Generally, from a genetic standpoint, feldspar deposits in Iran can be divided into two groups (Ghorbani 2002a):

1. Feldspar deposits associated with acid intrusive bodies
2. Feldspar deposits associated with acid volcanic rocks and tuffs

5.20.2 Distribution of Feldspar Deposits of Iran

There are more than 300 granitoid intrusive bodies in Iran, and 15% of them contain pegmatitic zones rich in feldspar (sodic and potassic). On the other hand, there are leucocratic intrusive bodies that lack ferro-magnesium minerals and instead consist of feldspar and quartz minerals. In parts of these bodies, the content of quartz is low and feldspar is high (Ghorbani 2002a). Such intrusive bodies can be considered as multi thousand-ton potential feldspar reserves. The feldspar deposits in leucocratic bodies occur within the Tertiary volcanic rocks, which went through fine differentiation process (in magma chamber) resulting in the formation of horizons (layers) of lava and white tuff rich in feldspar, which form large feldspar deposits.

5.21 Rock Salt

5.21.1 Mineralization Phases of Rock Salt Deposits in Iran

The largest salt deposits of Iran belong to the Hormuz Series that consists of volcanosedimentary rocks and belongs to Late Precambrian–Early Cambrian.

5.21.2 Distribution of Rock Salt Deposits in Iran

Iran is rich in salt deposits, and they are found in various forms as salt domes, closed lakes, and open seas. Evaporite series from Precambrian to present have been identified in many parts of Iran, but there are two major salt horizons as described next (Helmi 2000):

One is a salt horizon within the rock units equivalent to the Hormuz Series, which mostly spread in south of Iran.

The second one is within the Tertiary rocks found in the Lower Red, Qom, and Upper Red Formations in Central Iran, Azerbaijan, east of Iran, and Gachsaran Formation in Zagros.

Salt deposits and indications of Iran can be categorized into the following groups (Helmi 2000):

1. *Salt domes*: These domes contain evaporative rocks including salt rock and domes in the core of anticlines and along the fault lines. The most important salt domes of Iran are

- Qeshm salt dome, 110 km southwest of Qeshm Island
- Kangan salt dome, 55 km from Kangan
- Siyalou salt dome, at kilometer 19 of the Bandar Abbas–Sirjan road

- Gachin salt dome, 50 km west of Bandar Abbas
- Aliabad salt dome, on the northwest of Ravar
- Ismail-abad salt dome, 15 km northwest of Ravar
- Qom salt dome, 20 km from Qom

It is worth mentioning that most of the salt deposits in the Persian Gulf salt domes stand above the sea level, for example, those in Hormuz, Qeshm, Abu-musa, and Greater and Lesser Tunbs.

There are 115 salt domes in south of Iran, of which 101 are located between Bandar Abbas and Sarvestan and 14 in south of Kazerun (Helmi 2000). Besides, within the salt-bearing basin of Ardekan, salt domes outcrop in various shapes. As many as 80 salt domes have been counted in north of desert near Ardekan–Yazd road including Namak-e Kalout, Haji Abad, and Rastagh in Yazd (Helmi 2000).

Such deposits are also seen in other parts of Iran. There are many salt domes in Azerbaijan, and some of them are being exploited. Some examples of these salt domes are Douzkan, Alachigh, Khajeh, Shekarbolaghi, and Khoy-dou. There are also many salt domes along the Garmsar–Semnan axis, and some of these are of high purity (Helmi 2000).

2. *Brines*: In many parts of Iran, there are water bodies that run through salt-bearing formations and dissolve salt, and some of them are economically worthy of exploitation. Some of these brines are

- Rayen brine: It is located along the Rayen–Sardouiyeh axis. This brine is the product of evaporation of saturated salt water in the area.
- Brine in Lut–Shahdad desert: This is located in northeast of Kerman at the edge of Lut Kavir.
- Noogh brine: This is located 45 km northwest of Rafsanjan.

3. *Salt springs*: In some parts of Iran, there are brine springs whose outlet is near salt domes or within salt-bearing formations. Such brine springs are also seen in the Zagros Zone. For example, there are many salt springs seen at 30 km east of Shooshtar to Sardasht in north of Dezful, all of which flow through Gachsaran Formation. Four of these springs are worth mentioning:

- Brine spring in Aghili great plain
- Bardaleh bring spring
- Sardasht brine spring
- Gotvand brine spring

4. *Salt-bearing evaporite deposits within closed basins of Iran*:

- Hoz-e Soltan salt lake in Qom
- Lake Maharlou
- Lake Urumiyeh, the largest and saltiest perennial lake in Iran

5.22 Potash Mineralization in Iran

5.22.1 Mineralization Phases of Potash Deposits in Iran

Potash deposits as a result of evaporation of water within intracontinental epeirogenic basins as well as intracontinental basins formed due to rifting. Epeirogenic basins mostly belong to Paleozoic, whereas rift basins belong to Mesozoic and Cenozoic (Helmi 2000).

5.22.2 Distribution of Potash Deposits in Iran

Potash deposits in Iran are found in two forms: rock potash deposits, which are the product of sedimentation during geologic times, and saltwater potash deposits, which are currently being formed within places like playas. Of course, some alkaline rocks in Iran such as syenites rich in lucitite can be considered as a source of potassium. In general, potash-bearing areas in Iran include the following basins (Helmi 2000):

1. Garmsar–Qom–Semnan basin: Various amounts of potassium have been reported from different salt mines within this basin, the highest from the Rahrahak salt mine (as high as 32% of KCl).
2. Azerbaijan–Zanjan–Hamedan basin: The evaporites in this basin belong to Miocene. The salt-bearing layers are undisturbed and have not been affected by severe diapirism (the micro-folds of pre-doming phase can still be observed within domes), and in some cases, no diapirism has occurred (e.g., Maman salt mine at 25 km northeast of Mianeh where the potassium-bearing layers are very thin and the percentage of potassium content in some samples reaches up to 50%).
3. Great Kavir of Iran basin: Khoor playa is one of the largest playas within Great Kavir with an area of 1,800 km^2. It is a suitable locality for exploitation of potash from potassium-bearing salt waters.
4. Gavkhooni swamp basin: This basin has an area of 550 km^2, with a potash content (KCl) of about 5.7 g/L.

Exploration and prospecting for potash resources have also been carried out in Khorasan, Hormuzgan, and salt domes in Zagros.

5.23 Silica Mineralization in Iran

5.23.1 Mineralization Phases of Silica Deposits in Iran

Throughout the geologic times, silica deposits with various origins of sedimentary, metamorphic, pegmatitic, metasomatic, hydrothermal, and placer were formed in Iran (Ghorbani 1994d).

Temporally, economically valuable silica deposits of Iran mostly belong to Cambrian, with those belonging to Mesozoic ranking next (Ghorbani 1994d).

5.23.2 Distribution of Silica Deposits in Iran

The Kahar Formation in Azerbaijan and Alborz contains layers of quartzite that, considering the tonnage and grade of SiO_2, can yield productive deposits. For example, in the Alamkandi area, there are valuable silica deposits within some metamorphic rocks equivalent to the Kahar Formation (Ghorbani 1994d).

5.23.2.1 Mineralization of Silica in Paleozoic

Silica deposits in Paleozoic rocks are divided into three groups based on their time of formation, quality, and quantity (Ghorbani 1994d):

Group One – Cambrian Sedimentary Silica Deposits in Association with Lalun Formation: The silica deposits of this period account for most of the silica deposits of Iran; they are widely seen in Alborz and Central Iran. The deposits of Alborz (especially those between Firouzkuh and Soltanieh) are related to the upper part of Lalun sandstone (Top Quartzite). In Kerman and Yazd provinces in Central Iran, silica deposits such as Shajareh, Dahouiyeh, Darkaj, and Hashish are related to Top Quartzite of Lalun Formation or its equivalents. The silica deposits within this horizon (especially in central Alborz and Soltanieh area) are sufficient in quantity and quality and are utilized in the glass, molding, and cement industries; however, they are not of much use in the ferrosilicate industry.

Group Two – Late Paleozoic Sedimentary Silica Deposits in Association with Siliceous Conglomerate and Sandstone: With the exception of few cases, these deposits are not noteworthy in terms of number, quantity, and quality. Among the economically valuable deposits within this group, the Rudkard and Chenariyeh deposits in Kerman Province are the most important and belong to Silurian–Devonian (probably up to Carboniferous). They are associated with the phosphate member of the Kereshk group. The next most important are the siliceous conglomerate and sandstone layers in Sardar Formation (especially in its upper part) that belong to Visean Stage. In many places in the center, west, and southwest as well as Alborz, there is a white to red-spotted white siliceous layer with a thickness of 40–60 m. An example of this siliceous layer is a horizon with reasonable silica grade in the Talesh Mountains north of Masouleh that stratigraphically is laid at the base of Permian.

Group Three – Silica Deposits with Igneous–Metamorphic Origin: These deposits belong to Paleozoic and can be seen within the Sanandaj–Sirjan zone. The Koli-kosh deposit is an example of such silica.

5.23.2.2 Mineralization of Silica in Mesozoic

Silica deposits in Mesozoic rocks are divided into five groups based on their genesis (Ghorbani 1994d):

(A) *Silica deposits of igneous–metamorphic origin*: The silica deposits of this origin have the most abundance within the Sanandaj–Sirjan Zone. The silica deposits within Hamedan phyllites are of this type. Since the origin of some deposits in this zone is metamorphic, sometimes igneous, and sometimes both, they are here named as igneous–metamorphic deposits.

Most of these deposits are located in the northern part of the Sanandaj–Sirjan Zone. Most pre-Cretaceous formations in this zone have been metamorphosed. In addition to extensive metamorphism in this zone, numerous basic, ultrabasic, and granite-granitoid intrusive bodies were injected into this zone that resulted in the silicification phenomenon. Metamorphism caused several veins and lenses of pure quartz to form within schists and gneiss and also close to pegmatitic bodies particularly in the northwestern part of this zone, which are the Golpayegan, Doroud, and Hamedan areas. The silica deposits of this zone are not large in terms of tonnage, but they are of good quality. The following factors played a role in the formation of these deposits:

1. Few of these deposits are directly associated to magmatic activities accompanying pegmatitic phase.
2. Many of these deposits are syngenetic with the metamorphic rocks, and their host rocks are of metamorphic schists. The origin of silica within these schists and metamorphic rocks can be attributed to the following:

   - Silica deposits are the end result of injection of silica solutions released from acid magmatic bodies (e.g., Alvand granite) and emplaced within the surrounding rocks. Such phenomenon cannot be applicable for all silica deposits because many of these in the eastern part of Hamedan (Malayer) are quartzites that are older than the igneous bodies (schist rocks surrounding quartzites belong to Jurassic whereas Alvand acid bodies are attributed to Cretaceous).
   - The silica deposits are syngenetic with schists, were metamorphosed along with schists, and turned into quartzite. The contradictory behavior compared to the host rocks is due to their different reaction to folding during metamorphism, leading to their lens-shaped masses.
   - Lenses, veins, and thin layers of silica were formed concurrent with metamorphism so that during the increase of pressure and temperature in sedimentary rocks, silica solutions moved through the rock layers. These solutions were stopped as they hit intercalated carbonate rocks where silica solutions started to etch the thin-layered carbonates. As pH unbalanced in the environment, soluble silica began to precipitate and formed the deposits we see today.

(B) *Sedimentary silica deposits*: The Mesozoic sedimentary silica deposits in Iran are more or less deformed. None of these deposits are suitable for the glass industry, but they can be consumed in other industries such as casting and cement. Some of these deposits are given next:

- Early Triassic silica horizon with a thickness of 5 m, which has been reported in Alborz and some parts in Central Iran
- Silica layer within quartzitic conglomerate and sandstone in Semnan area in Alborz (Chashm and Larestan localities), which is associated with Early Jurassic rocks
- Early Cretaceous silica deposits in Yazd Province, which are associated with siliceous sandstone in Darreh-zanjir Formation

(C) *Silica deposits of placer origin*: These deposits consist of quartz or quartzitic pebbles and sands (which are the product of erosion of granites) that deposit within water channels and river grooves, sea coasts, sand dunes, or the foothill of granitic mountains. The size of the large pebbles is about 10 cm, which are used as silica balls. Notwithstanding the numerous exploration works carried out so far, no noteworthy deposit of this type has yet been found in Iran. The silica deposit in Shirkuh granite in Yazd and the Rig Sefid silica deposit in Shahr-e Babak are good examples of this type of deposit. The purity of silica in these deposits is not desirable, and they have specific granulation. These silica deposits are seen in different grain sizes, which suit other applications.

(D) *Silica deposits of hydrothermal origin*: These deposits are rare and are small in size. The Mizooj deposit in Qazvin is an example of such deposits.

(E) *Alteration silica deposits*: There are extensive alteration zones along the Takestan–Jolfa axis, especially in northwest of the Takestan and Meshkinshahr–Ahar area, wherein valuable silica deposits are found. These alteration zones can be differentiated as argillic, siliceous, and hematitic. The argillic zone is more extensive than the other two, and the hematitic zone is the least extensive one. The argillic zone is at the bottom and the siliceous zone at the top, with the hematitic zone sometimes seen between these two zones. These silica deposits are not suitable for the glass industry due to the low percentage of SiO_2 but can be consumed in the casting and cement industries.

5.24 Kaolin and Fireclay

5.24.1 Mineralization Phases of Kaolin and Fireclay Deposits in Iran

The kaolin and refractory deposits of Iran are found in association with two time periods of Late Paleozoic–Triassic and Jurassic–Cretaceous more than any other geologic time periods (Ghorbani 1994e).

Mineralization of Kaolin and Refractories in Late Paleozoic–Triassic: Most of the kaolin and refractory deposits of Iran were formed during Late Paleozoic–

Triassic. These deposits are mainly located within the following areas (Ghorbani 1994e):

- Numerous horizons of refractory clay belonging to Late Devonian–Triassic are seen in the Abadeh area. Among these deposits are Esteghlal and some horizons of Kavir deposits belonging to Late Devonian and some horizons of Kavir deposits with Carboniferous–Permian age and Triassic kaolin-bearing horizon within Shoorjestan Formation.
- Cheshmeh Shotoran and Chah-koolar deposits of Permo-Triassic in Robat Khan area in Tabas.
- Norooz Abad, Ali-baltoo, and Chalpoo deposits of Permo-Triassic in Shahin Dezh area.
- Amin-abad deposit in east of Tehran within central Alborz.

Mineralization of Kaolin and Refractories in Jurassic–Cretaceous: The refractory deposits at the base of Lias (at the base of coal horizons) in Alborz (Sangrood area) and the alteration kaolin deposit in Khorasan (Kaftar-kuh in Gonabad) are examples of such deposits formed during the Jurassic–Early Cretaceous mineralization phase (Ghorbani 1994e).

5.24.2 Distribution of Kaolin and Fireclay Deposits in Iran

The kaolin deposits of Iran can be divided into two types based on mineralogy, chemical composition, formation setting, and even application (Ghorbani 1994e):

1. Sedimentary-type deposits, which are the product of erosion and sedimentation of volcanic rocks. These deposits mostly belong to Late Paleozoic–Early Mesozoic, but some of them were formed in Cretaceous.

The geographic and temporal distribution of this type of deposits is described next (Fig. 5.15).

Abadeh area: This area houses most kaolin deposits and potentials. Kaolin mineralization in this area occurs in three horizons:

- Upper Devonian horizon: Among the deposits in this horizon, Kavir 5 and Esteghlal deposits are the best and richest ones.
- Carboniferous–Triassic horizon including Kavir deposits.
- Triassic horizon associated with Shoorjestan Formation.

Central Iran and Alborz: This includes the kaolin and refractory deposits formed within the Permo-Triassic strata in east of Iran (Robat Khan, Tabas, and Gonabad), northwest of Iran (Shahi Dezh), and central Alborz (south of Amin-abad east of Tehran). Most refractory deposits of Iran are seen in these areas.

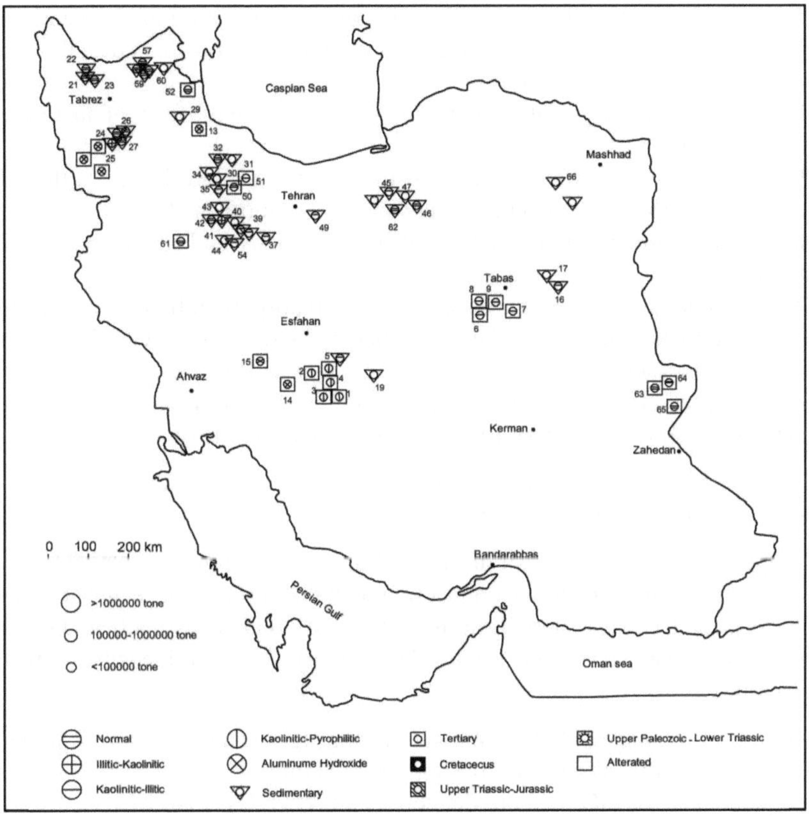

1- Esteghlal	12- Norooz-abad	23- Zonooz green clay	34- Nikooiyeh	45- Ghooshe kaolin-alunite	56- Issti-soo yellow clay
2- Kavir 5	13- Sang-rud	24- Issti-soo	35- Ghazan-daghi	46- Gandi	57- Ghalandari ziling clay
3- Kavir 16	14- Poshteh-Samirom	25- Maneshakeh	36- Shabolagh	47- Komboloo refractory	58- Ahar mixed clay
4- Kavir 17	15- Dooplan refractory	26- Azar-goyooni kaolin	37- Kooshk-e Nosrat	48- Darreh-jazin	59- Ahar white clay
5- Vijeh	16- Baghe-siyah Gonabad	27- Abak	38- Parandak	49- Kritoneh clay	60- Ghalandari-Ahar ziling clay
6- Cheshmeh Shotoran	17- Kabootar-kuh kaolin	28- Gavazn	39- Tavakol-abad kaolin	50- Soorojin clay	61- Lalehjin clay
7- Robat-khan	18- Makki Kashmar	29- Zajekan Qazvin	40- Hajib kaolin	51- Niyagh clay	62- Dasht-e Kalat
8- Tabas ball clay	19- Abdullah Shahbaz	30- Qazvin	41- Kooshk clay	52- Bibi-janloo clay	63- Brick clay Shileh
9- Chah-bid clay-like deposit	20- Robat Shah-abbasi	31- Abdol-abad	42- Shirin-goo	53- Ab-garm Mahalt clay	64- Shahr-e Sookhteh clay
10- Chelpo Shahin-dezh	21- Zonooz	32- Ab-torsh kaolin	43- Shoorjehbolagh-bidloo	54- Kondaj kaolin-clay	65- Varmal clay
11- Shahin-dezh	22- Blootlook Zonooz	33- Kaolin deposit near Alan Ghaya	44- Deposit near ghezeljeh	55- Gray clay	66- Ghasem-abad

Fig. 5.15 Distribution map of kaolin and fireclay deposits of Iran (Ghorbani 1994e)

Zagros: The kaolin and refractory deposits in Zagros are found in Upper Cretaceous and in association with Sarvak and Ilam Formations (Shahreza area).

In all these deposits, kaolin is the dominant mineral with a little quartz content. Minerals rich in aluminum such as diaspor, boehmite, and, sometimes, pyrophyllite are also seen. Iron minerals such as limonite and hematite are mostly seen in their paragenesis, and sometimes, the grade is so high that no sorting is usually needed after extraction (Ghorbani 1994e).

These deposits usually contain titanium, which exists in the form of anatase mineral (Ghorbani 1994e).

The percentage of SiO_2 in these deposits does not exceed 55%, but the Al_2O_3 content is above 25% and sometimes reaching 49% (Ghorbani 1994e).

Most of these mineral materials are of high plasticity and relatively good fire-resistance.

These sedimentary deposits are usually in alternation with limestone, dolomite, shale, and bitumen-bearing shale layers of shallow marine origin (Ghorbani 1994e). They are the product of erosion and alteration imposed on rocks rich in feldspar. Some of the characteristics of these deposits are their association with epeirogenic carbonate sediments and lack of fossil flora (Ghorbani 1994e).

2. Hydrothermal or alteration-type deposits, which are the result of Tertiary volcanic activities in Alborz, Azerbaijan, and east of Iran.

Except the Kaftar-kuh deposit in eastern Iran, which belongs to Jurassic, the rest of the deposits of this type are in connection with the Tertiary rocks and are located in the north and northwest of Iran as well as in other areas such as Gonabad, Kashmar, and Saveh (Ghorbani 1994e). The characteristics of deposits of this type are given next:

- Absolute dominance of kaolin mineral, having significant amounts of quartz and sometimes alunite, absence of diaspor and boehmite.
- The percentage of SiO_2 is more than 60% and Al_2O_3 less than 24%.
- Associated with these deposits, volcanic rocks such as tuff, andesite, and dacite as well as intermediate to acid intrusive bodies at shallow depths are abundant (especially along the Takestan–Jolfa axis), and their age ranges from Eocene to Oligocene.

Late Eocene and Oligocene magmatism caused an increase in geothermal gradient in many places in Iran and emplaced numerous intrusive bodies at shallow depths (Ghorbani 1994e). As a result of this magmatism, hydrothermal solutions were mobilized (these hydrothermal solutions either originated from magma or the product of the heating up of groundwater).

Based on available facts, these hydrothermal solutions contained great amounts of sulfate ions and caused an extensive alteration along the Takestan–Jolfa axis. If the hydrothermal solutions contained more sulfate ions (high acidity or low pH), the alteration process would lead to the formation of alunite, whereas the lower degree of acidity of the hydrothermal solutions (high pH) would result in kaolinization (Ghorbani 1994e).

5.25 Bentonite

5.25.1 Introduction

Iranians were aware of bentonite back in 1,000 B.C. and differentiated its various types. They used it as detergent and medicine. However, the mass production of bentonite began when oil explorations in the southern part of Iran brought up new applications for this mineral. Currently, 50% of bentonite reserves of Iran are consumed by the oil industry, and the rest is used in other industries such as ceramics, pelletization, paint, and various oil refining applications (Ghorbani 1991a). So far, 105 bentonite deposits and indications have been identified and explored, but only a few are being exploited to fulfill the needs of the country. The data presented herein are the result of a nationwide research on the bentonite deposits of Iran targeted at understanding their physicochemical properties and genesis (Ghorbani 1994a).

5.25.2 Distribution of Bentonite Deposits in Iran

In general, 105 bentonite deposits and indications have been identified in Iran, most of which are located in central Iran, eastern Iran, Alborz, and Azerbaijan (Fig. 5.16).

These deposits are mostly found in the structural units with significant Tertiary volcanic activities, which are clearly visible by comparing Fig. 5.16 with Fig. 2.2 in Chap. 2.

5.25.3 Specifications of Bentonite Deposits in Iran and Their Genesis

The primary mineral of all bentonites in Iran is montmorillonite. In most of these, bentonites minerals like cristobalite, quartz, and calcite are found as secondary minerals. It is interesting to note that in the bentonite deposits of Iran, montmorillonite and kaolinite minerals are not paragenetic. Evidences from mineralogical studies and field observations indicate that wherever these two minerals are seen together, the minerals and the host rocks show stratiform structure, and the minerals are the result of erosion and sedimentation (they are not in situ) and do not form high-quality deposits.

There are a few cases of bentonite deposits in Iran where pieces of albite, feldspar, and some minerals from the source rock are found and therefore do not produce

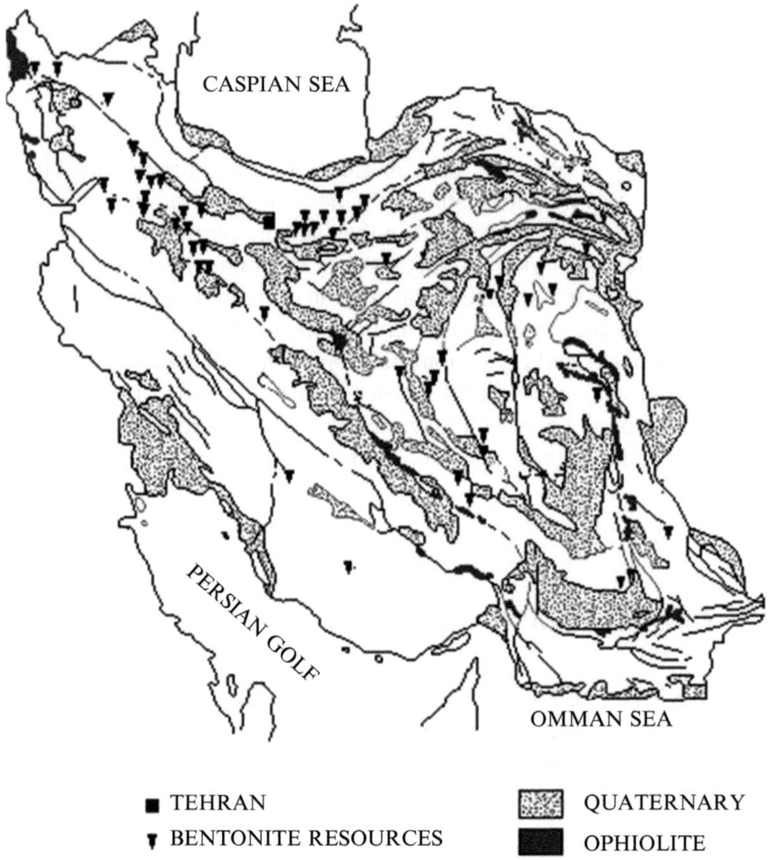

Fig. 5.16 Distribution map of bentonite deposits of Iran (Hejazi and Ghorbani 1994)

high-quality deposits. In other words, such bentonite deposits are the product of surface alteration (Ghorbani 1991a).

5.25.4 Chemical Composition of Bentonite Deposits in Iran

Studies on the chemical composition of bentonite deposits of Iran reveal the following six regimes (Table 5.16) (Ghorbani 1991a):

1. The silica percentage for most bentonite deposits within the same zone is alike, for example, the percentage of SiO_2 in all bentonite deposits with age of Eocene around Central Iran desert is 58–63%, and those with age of Oligo-Miocene located in east of central Iran is about 55%. This fact seems to be due to the composition of the source rock, pH and Eh conditions, and the depth of the basin at the time of bentonite formation.

Table 5.16 Chemical composition of bentonite deposits of Iran

Deposit name	Total	H_2O-	H_2O+	MgO	CaO	Al_2O_3	SiO_2	FeO^+ Fe_2O_3	K_2O	Na_2O
Goosh-mir	100.51	6.06	6.52	2.02	6.8	12.43	63	0.59	0.44	2.65
Hessami-ye Ferdows	101.54	4.3	14.37	2.52	9.81	12.66	50	1.88	3.67	2.33
Jahrom	100.77	9.38	8.02	5.24	1.61	21.04	53.06	1.13	1.29	0
Kheyr-abad-e Kerman	98.43	8.03	11.3	2.9	1.94	14.4	56.49	1.24	0.62	1.51
Khoshab-e Kashmar	100.6	12.81	11.6	2.67	10.57	14.34	41.6	4.95	0.28	1.78
Kilan	99.67	7.68	5.56	4.36	2.66	17.11	54.38	4.3	1.3	2.32
Mehrjan	99.97	5.19	4.41	2.67	1.54	15.09	66.79	1.69	0.26	2.33
Rashm	99.46	5.49	4.99	1.92	0.77	16.65	62.07	2.16	1.14	4.27
Siah Kuh	99.21	4.91	4.03	1.92	1.33	13.37	68.6	1.37	0.91	2.77
Soosan-var	89.67	3.1	5	1.11	2.03	10.04	62.53	1.2	1.13	3.53
Tafresh	98.87	9.46	6.31	3.75	1.47	13.85	53.09	3.2	1.3	1.44
Zarrin	99.45	3.76	8	1.81	6.66	14.24	55.9	3.74	1.31	4.03

Ghorbani (1991a)

2. The percentages of Al_2O_3 in all bentonite deposits of Iran are very close to one another, and different bentonite-bearing areas do not seem to be in conflict regarding this matter. The fact is that Al_2O_3 is an oxide with low mobility, and its amount does not change much during the bentonization process. This is especially true for bentonites of Iran as the source rocks for all of them are acid volcanic rocks, whose amount of Al_2O_3 does not fluctuate significantly.
3. The amount of iron oxide in the bentonite deposits of Iran is low. The iron content in bentonite is reflected in its color. When the bentonite is light in color, the iron content is low and under 2%, while those with red, yellow, or green color point to high percentage of iron oxide. However, it is to be noted that the color of the bentonite deposits in adjacent areas does not differ much and with the same token, the amount of iron oxide in such deposits does not vary much either. This finding points to the fact that the chemical composition of bentonite-forming volcanic rocks in adjacent areas is approximately similar. Field observations made by the author confirm this matter.
4. The percentages of sodium and potassium oxides are low in all bentonite deposits of Iran while the amount of sodium is usually more than potassium. In those deposits where the NiO_2 percentage is significantly high, halite is found as the secondary mineral. Halite is seen in desert areas of Iran and is associated with severe evaporative condition in deserts that occurs after bentonization process.
5. There is an inverse relation between the amount of H_2O and SiO_2 in bentonite deposits; the percentage of H_2O in bentonite with high percentage of SiO_2 is low and vice versa.
6. The percentages of MgO and CaO are variable because these two components appear in bentonite in two forms:

   (a) In the structure of montmorillonite mineral
   (b) In calcite or dolomite minerals that exist in bentonite as secondary minerals

In general, the number of calcium bentonite deposits is far less than sodium bentonites.

5.25.5 Age of Bentonite Deposits in Iran

All bentonite deposits of the world have almost been formed during Jurassic to Pleistocene. There are many bentonite deposits in the United States, Europe, and Asia with age of Cretaceous. However, most bentonite deposits belong to Tertiary. No bentonite deposits with age of Cretaceous or older have yet been identified in Iran. All bentonite deposits of Iran are younger than Cretaceous, and almost all of them have been formed in Tertiary. In general, no bentonite deposits with age of older than Eocene and younger than Late Miocene have been found in Iran. This age range exactly matches the time during which volcanic activities, especially submarine volcanism, occurred in Iran. The volcanic activities during Tertiary are traceable in Iran, but there were three periods of time where such activities took place most extensively (Ghorbani 1991a):

- Middle Eocene
- Late Oligocene–Early Miocene
- Middle–Late Miocene

It must be noted that all bentonite deposits in Iran formed during these three time intervals.

5.25.6 Environment of Bentonite Deposition

Before proceeding to describe the setting for such deposits, some facts and features about bentonite deposits of Iran need to be cited:

1. Except for a few cases, the bentonite deposits of Iran were formed in situ, meaning no transportation has taken place.
2. All bentonite deposits of Iran conform to their surrounding rocks (host rocks) and show stratiform shapes.
3. The thickness of these deposits varies even within a short distance.
4. In some cases, the bentonite deposits are seen in the form of two colored layers (e.g., the bentonite deposit in the Tafresh area, where the green layer is at the bottom and the yellow layer on the top; the bentonite deposit in Oushar Alamdar in the Jolfa area, wherein the lower layer is whiter with better quality and the top layer is brownish with lower quality; the bentonite deposit in Panjeh Sofla in the Bijar area, with the white layer at the bottom and the red layer on the top). The assumption is that the bentonite color does not change diagenetically because bentonite is impermeable, and even under extreme oxidizing conditions, its color remains unchanged.
5. Minerals like muscovite, biotite, and rounded quartz are not seen in bentonite deposits of Iran and, if present, are anhedral and angular. In fact, no clastic minerals are found in the bentonite deposits of Iran.
6. Although bentonite can preserve fossils in good condition, no fossils have yet been reported in the bentonite deposits in Iran.
7. The chemical properties of the bentonite deposits of Iran point to alkaline pH ($pH=8$), and field observations indicate that such properties have not changed after they were formed.
8. Organic bituminous materials that give dark color to rocks are not found in the bentonite deposits in Iran. The white color of these deposits is a proof.
9. Taking into account the aforementioned facts and the field studies on all bentonite-bearing areas of Iran, it can be stated that the environmental setting for the formation of these deposits was of shallow marine type, either coastal or lagoonal in connection with open sea. Such environments were oxidizing with a pH of 8–9 and salinities higher than the ocean.

There are other chemical evidences that point to oxidizing conditions of the formation of these deposits:

(A) The amount of SiO_2 in these bentonite deposits is less than the source rocks (the source rocks for most of the bentonite deposits in Iran are acid volcanic rocks). Since SiO_2 is an acidic oxide, it is very soluble in neutral or acid environment and needs alkaline environment to deposit.

(B) The relative percentage of Al_2O_3 in bentonite deposits is higher than the source rock. This indicates that not only Al_2O_3 did not dissolve but was stable during the formation of bentonite.

5.25.7 Origin and Genesis of Bentonite Deposits in Iran

Before discussing the origin of these deposits, some facts need to be mentioned:

1. Cristobalite is seen as a secondary or accessory mineral in many bentonite deposits around the world, and it has been found in 30 bentonite deposits in Iran as well. Cristobalite forms at higher temperature and lower pressure than that required for quartz, and it exists in volcanic rocks such as rhyolite to dacite. It can be clearly stated that no cristobalite exists in plutonic bodies.
2. All bentonite deposits in Iran are younger than Eocene and formed during the Eocene and Oligo-Miocene periods.
3. No bentonite deposit has yet been identified around the plutonic bodies. This statement is based on the observations and studies done over the surrounding rocks of more than 300 intrusive bodies, and it seems there is no connection between the igneous intrusive bodies (even those of acid type) and bentonite deposits. However, it is to be noted that the reference is made to those bentonite deposits with more than 80% of montmorillonite content.
4. As mentioned earlier, strong volcanism occurred in most parts of Iran in Tertiary. In Alborz and Central Iran areas during some period of time, especially in Middle Eocene, volcanic and pyroclastic formations are dominant and more visible than other rock units. Such activities began from Lower Eocene and reached their peak in Middle Eocene. They were replaced by plutonic activities in Late Eocene–Early Oligocene, and from Late Oligocene through Pliocene, they soared again, finally slowing down from the beginning of Plio-Quaternary until settling down as of now.
5. In all areas, the bentonite deposits are accompanied by the volcanic rocks, especially the acid ones, and tuffs, and there is a very close connection between them.
6. In those parts of Iran where the volcanic activities were weak or absent, signs of bentonite are either rare or wanting.
7. In some bentonite deposits in Iran, pieces of volcanic glass are found (e.g., bentonite deposits in east of Iran, Qazvin area, and Angouran area around Zanjan).

Based on the facts and observations mentioned earlier, the lithologic characteristics of rocks surrounding the bentonite areas, as well as the mineralogical and chemical studies on the bentonite deposits of Iran, the origin of bentonite deposits and the their formation settings can be depicted as follows.

Acid pyroclastic rocks are the source rock for bentonite deposits, which are of rhyolitic and dacitic types, and they are always seen in contact with bentonite deposits. Field observations on 100 bentonite deposits and indications as well as the mineralogical studies on minerals, host, and surrounding rocks undoubtedly indicate a very close and direct connection between volcanic rocks (especially acid tuffs) and bentonite deposits, and the deposits are the product of transformation and alteration of acid tuffs. Besides, some minerals left in the bentonite bodies have acidic composition such as cristobalite, quartz, and alkali-feldspar. The chemical composition of bentonite indicates that the very first source rock of bentonite is of acidic type. Even the studies on rare elements in bentonite deposits and acid tuffs in a few areas indicate a great resemblance between the rare elements in bentonite deposits and those in acidic tuffs (Ghorbani 1991a; Ghorbani and Tajbakhsh 2010).

5.25.8 Formation of Bentonite Deposits

Before proceeding to the formation of bentonite deposits, some facts need to be cited (Ghorbani 1991a; Ghorbani and Tajbakhsh 2010):

1. Out of 105 bentonite deposits in Iran, only 3 contain montmorillonite and kaolinite together, where montmorillonite is the primary mineral and kaolinite the secondary. All three deposits have been transported to their current positions due to sedimentary processes. No kaolinite mineral exists in any of in situ bentonite deposits.
2. No bentonite deposit has been identified around the granite bodies in Iran.
3. The source rocks of bentonite deposits are acidic tuff, rhyolitic and dacitic pyroclastic rocks.
4. The environment of deposition of bentonite is coastal or lagoonal constantly affected by tidal waves. This environment has alkaline pH and high Eh.

It is to be noted that the general belief is that during the alteration and weathering processes, feldspar transforms to clay minerals and eventually bauxite, as described next:

$$\text{Alkaline feldspar} \xrightarrow{\text{Weathering}} \text{montmorillonite} + \text{alkaline cations} + \text{SiO}_2$$

$$\text{Montmorillonite} \xrightarrow{\text{Weathering}} \text{kaolinite} + \text{alkaline cations} + \text{SiO}_2$$

$$\text{Kaolinite} \xrightarrow{\text{Weathering}} \text{bauxite}$$

The common thought is that bentonite and kaolinite deposits are the product of alteration and weathering of feldspar-bearing rocks in suitable climate conditions, but contrarily, surface alteration and weathering processes could not produce bentonite and kaolinite deposits out of igneous and metamorphic rock, and the following facts confirm this view (Ghorbani 1991a; Ghorbani and Tajbakhsh 2010):

(a) Minerals that are the product of alteration of igneous or metamorphic rocks are not found in bentonite.

(b) Quartz with clastic origin is rarely seen in bentonite deposits.
(c) Not even one bentonite deposit has been reported around the large plutonic bodies in Iran.

Considering the aforementioned facts, it should not be assumed that valuable montmorillonite deposits are formed from the alteration of igneous and metamorphic rocks. In fact, common clay deposits used in bricks and ceramic form around the plutonic bodies.

Considering the aforementioned facts, the formation of bentonite can be explained by ejection of tuffs with rhyolitic and dacitic compositions into a suitable environment (shallow basin with alkaline pH and high Eh), thus forming economical and valuable deposits before entering diagenesis (Ghorbani 1991a; Ghorbani and Tajbakhsh 2010). Bentonitization is a physiochemical process that could take place in different forms, including (Ghorbani 1991a) the following:

(a) Explosive or fairly calm volcanic eruptions spray out volcanic particles of ash size into the sedimentary basin. Provided the suitable condition is in place, these particles stay suspended for a period of time while descending toward the bottom of the basin (under Stoke's law), and the bentonization process could take place while these particles shower down onto the bottom of the basin.
(b) While these particles settle down at the bottom (like snow covers the ground), before undergoing diagenesis, and because of the porous nature of the mineral body, some SiO_2 and alkaline elements are washed out in the form of solutions; this is how volcanic materials with feldspatic composition transform to montmorillonite (bentonite).
(c) The transformation of volcanic materials to bentonite after diagenesis is much harder compared to the same process before it.
(d) Volcanic tuffs may transform to bentonite deposits when exposed to precipitation and weathering, but these deposits are not desirable in terms of quality, and no such deposit has ever been exploited in Iran.

5.26 Diatomite

5.26.1 Distribution of Diatomite Deposits

Diatomite deposits in Iran have been studied in the Azerbaijan area in three localities.

(A) Some diatomite deposits are seen within the Tabriz area:

- Deposits in Agh-yoghoosh Pass and Khalat-pooshan Valley in east of Tabriz, where the thickness of diatomite deposits reaches up to 30 m in some places
- Deposits in Eeli-goli and Emamiyeh

(B) Diatomite deposits have been identified in east of Sahand Mountain in Khosroshahr, Oskou and Azar-shahr, and lake deposits including tuff, pumice, diatomite, and conglomerate have been found in the vicinity of the Mamagah Village. The content of diatomite in these deposits varies from 2 to 36.1%.

(C) The physical, chemical, mechanical, and microscopic studies and analysis on diatomite samples taken from south of Ardebil and east of Sareiyn have proved satisfactory and promising results from grade and tonnage standpoints.

5.27 Perlite

5.27.1 Mineralization Phases of Perlite Deposits in Iran

The perlite is a vitreous acidic to intermediate volcanic rock with 2–5% water content in its composition. The texture of perlite is vitreous (perlitic), with conchoidal fracturing looking like centric facets or onion layers (Ghorbani 1999d). Similar to most volcanic glasses, perlite is unstable too. Since perlite is a volcanic glass, it starts to crystallize as time goes by and loses its properties, which is the reason for high-quality perlites to be found in Tertiary and Quaternary only.

5.27.2 Distribution of Perlite Deposits in Iran

The perlite deposits in Iran are found along with the Tertiary volcanic rocks.

There are volcanic rocks within the perlite-bearing areas of Iran, which are sometimes mixed with shallow marine sedimentary rocks. These volcanic rocks include andesite, trachyte, and rhyolite and associated tuff, which belong to Eocene, Oligo-Miocene, and Mio-Pliocene (Ghorbani 1999d). Since these volcanic rocks have significant distribution in Iran, it is very likely to identify more perlite deposits.

Large perlite deposits are seen east of Mianeh–Tabriz road in the Sefid Khaneh area (46 km northeast of Mianeh) and also in the vicinity of the Tarom Village in west of Mianeh.

Other promising areas in terms of perlite exploration are Birjand, Ferdows, Tabas, Naein, Kashan, and Sistan–Baluchestan.

5.28 Dolomite

5.28.1 Mineralization Phases of Dolomite Deposits in Iran

Dolomite is a member of carbonate rocks and is considered as a common marine sediment. When it comes to exploring the dolomite deposits, the general fallacious assumption is that they would form wherever and whenever carbonate deposits form. The studies of dolomitization of carbonate rocks during various geologic times indicate that the formation of dolomite rocks/deposits similar to other mineral deposits in the world follows specific rules, which leads to the formation of such deposits within specific time and places. For example, in most parts of the world, carbonate rocks display a more significant presence in Triassic. The dolomite deposits in Iran can be traced from Late Precambrian to Miocene. However, most of the dolomite deposits of Iran belong to the following time periods (Ghorbani 2002a).

5.28.2 Distribution of Dolomite Deposits in Iran

Mineralization of Dolomite in Late Precambrian–Early Cambrian: Dolomite rocks and carbonate sequences are remarkably present during this period of time, and generally it can be stated that in most parts of Iran where Late Precambrian–Early Cambrian rocks outcrop, dolomite rocks are also seen. Of dolomite rocks associated with this time period, those in Soltanieh Formation in Alborz and Azerbaijan, dolomite layers in Barout Formation, and dolomite layers within the Kooshk, Rizoo, and Dezoo Series in Central Iran are worth mentioning (Ghorbani 2002a).

Mineralization of Dolomite in Paleozoic: Within some carbonate sequences in Paleozoic of Iran, dolomitic units are present, which are cited next (Ghorbani 2002a):

- Parts of Mila Formation in Alborz belonging to Late Cambrian–Ordovician
- At the base of Nivar Formation in Ozbak Kuh belonging to Silurian
- Parts of Padeha and Sibzar Formations in east of Iran belonging to Devonian
- Parts of Mooli Formation in Azerbaijan (Makoo) belonging to Devonian
- Jamal Formation in Central Iran and east of Iran belonging to Permian
- Locally, in upper part of Ruteh Formation in Sanandaj–Sirjan belonging to Permian
- Parts of Dalan Formation in Alborz belonging to Permian

Mineralization of Dolomite in Triassic: Most dolomite deposits of Iran belong to Triassic (Middle Triassic in particular) as dolomite deposits are likely present wherever Triassic carbonate rocks are found. Some of the dolomite-bearing Triassic formations are cited next (Ghorbani 2002a):

- Elika Formation in Alborz and Azerbaijan: This formation consists of limestone, limestone–dolomite, and dolomite. Its upper parts, with age of Middle Triassic, are more dolomitized as compared to the other parts of the formation.
- Shotori Formation in Central Iran and east of Iran.
- Abadeh area hosts Permian–Triassic Type Section, and those parts of the section with age of Middle Triassic are dolomitized.
- Khaneh-kat Formation in Zagros with age of Middle Triassic, wherein abundant dolomite deposits have been reported.

Mineralization of Dolomite in Jurassic: Dolomite horizons in the Jurassic rocks are less abundant as compared to the Triassic rocks, but dolomitization is seen in parts of Jurassic formations (mostly Upper Jurassic), some of which are briefly cited next (Ghorbani 2002a):

- Mozdouran Formation in Kopet Dagh: Dolomite horizons are found at the base and middle parts of this formation.
- Lar Formation in Sanandaj–Sirjan Zone: Parts of this formation are dolomitized in the Kaboudar-ahang area.
- Sourmeh Formation in Zagros: Parts of this formation are dolomitized in the Fars area.

Mineralization of Dolomite in Cretaceous: Although Cretaceous sedimentary sequences in Iran are mostly of carbonate type, most of these consist of limestone with rare dolomitic limestone (Ghorbani 2002a).

Mineralization of Dolomite in Tertiary: Dolomitization scarcely took place in Tertiary carbonate sequences. However, within some carbonate formations, some dolomitized areas are seen, which are briefly cited next (Ghorbani 2002a):

- Chehel-kaman Formation in Kopet Dagh: Some parts of this formation are dolomitized.
- Asmari Formation in Zagros: Sometimes, dolomite horizons are locally seen within the Asmari Formation (Chenareh area near Andimeshk).
- Jahrom Formation in Zagros: Similar to the Asmari Formation, sometimes dolomite horizons are locally seen within the Jahrom Formation.
- Qom Formation: In Naein and Shemshed-zardireh areas, the upper parts of the Qom Formation are dolomitized.

5.29 Tar and Natural Asphalt

Application of tar and petroleum seepages goes back to ancient times, when people in southwest of Iran used these for fuel and medication purposes. In ancient Egypt, these materials were also used in mummification.

Natural asphalt is a product of crude oil. It forms as it migrates toward the ground surface (or close to the surface) emplaces in a suitable place and dries up.

As opposed to the old days when natural asphalt was only considered as a fuel source, nowadays this material is acknowledged as an industrial material. It is only found in a few countries, and Iran is one of those countries that has numerous tar and natural asphalt deposits.

5.29.1 Origin of Tar and Natural Asphalt

Almost in all oil fields in the world, the presence and existence of tar and asphalt has been identified (Ghazvini 2002). Masjed Soleyman is one of these oil fields.

Based on the studies and surveys carried out on various petroliferous basins, the characteristics of tar and natural asphalt outcrops in Iran are somehow similar to those studied in Oklahoma in the United States and Maracaibo Lake in Venezuela. In southwest of Iran (in Naft Shahr in Iran and Naft-khaneh in Iraq), the oil-bearing layers (which formed oil reserves) underwent structural deformation and folded, and their oil content seeped to surface. This scenario is in accordance with that of Venezuela. Nowadays, it is believed that the rupture and tearing up of caprock of oil reserves and consequently the seepage of hydrocarbon materials led to the formation of tar and natural asphalt along the edge of continental plates. Such a situation and condition is traceable in southwest of Iran within some areas such as Ghasr-e Shirin, Guilan-e Gharb, Soomar, and Dehloran (Ghazvini 2002).

5.29.2 Distribution of Tar and Natural Asphalt Deposits in Iran

These materials are found in Kermanshah Province (Naft Shahr, Guilan Bakhtar, and Ghasr-e Shirin), Ilam Province (Soomar, Mehran, Dehloran), and Khuzestan within Gachsaran, Taleh Zang, and Asmari Formations. The seepages of tar and natural asphalt originated from the reservoir rocks of Asmari and Taleh Zang. Table 5.17 shows some tar and natural asphalt deposits of Iran.

The source rocks of the Asmari reservoir in Naft Shahr and anticlines in Guilan-e-gharb and Soomar underlie TZ2 Member of Taleh Zang Formation, which formed in Middle Eocene and contains great amounts of organic materials. Shale, marl, and marl rich in bitumen are exposed on the surface near Emam Hassan Anticline and stretch out in the southeastern direction toward the Anjir-Kalak area (Ghazvini 2002).

5.30 Dimension and Ornamental (Decorative) Stones

5.30.1 Standards of Dimension Stones

For dimension stones to remain intact during cutting and finishing, they need to possess certain properties and specifications. These physical and mechanical properties and specifications have to meet international standards of valid and credible institutions such as ASTM International. The standard requirements for dimension stones defined by ASTM International are presented in Tables 5.18, 5.19, and 5.20 (Kassiani-Avval 2004).

Table 5.17 Geological characteristics of natural tar deposits of Iran

Sr. no.	Name of deposit	Geographic location	Host rock	Probable reserves (tons)	Quality of tar
1	Sadeghi natural tar	9 km to the northwest of Gilan Gharb	Gachsaran Formation	91,700	80% Carbon, 0.7% sulfur, 9% ash
2	Goor Sefid (Marjani)	1.1 km to northwest of Gilan Gharb	Gachsaran Formation	180,000	83% Carbon, 1.3% sulfur, 4% ash
3	Soomar	8 km to the east of Naft Shahr	Gachsaran Formation	Unknown	76% Carbon, 1.5% sulfur, 13% ash
4	Palayesh and Yavari	20 km to the northwest of Gilan Gharb	Gachsaran Formation	Unknown	70% Carbon, 10.5% sulfur, 18% ash
5	Ahangaran	30 km to the northwest of Behbahan (Likak)	Gachsaran Formation	Unknown	Saltwater springs containing tar flow during summer season
6	Mamatin	40 km from Ramhorhoz	Gachsaran Formation	Unknown	Around 8–10 tar springs occur

After Ghazvini (2002)

Table 5.18 ASTM required specifications of construction granite

Sr. no.	Physico-mechanical specifications	Amount	Method of experiment
1	Maximum water absorption (weight percentage)	0.4	ASTMC 97-96
2	Minimum density (kg/m^3)	2,560	ASTMC 97-96
3	Minimum compressional strength (MPa)	131	ASTMC (reapproved, 1994) 87
4	Minimum rupture module (MPa)	10.34	ASTMC (reapproved, 1994) 87
5	Minimum erosional strength (hardness)	25	ASTMC (reapproved, 1994) 241-90
6	Minimum bending strength (MPa)	8.27	ASTMC 880-96

Table 5.19 ASTM required specifications of construction limestone

Physico-mechanical specification	Amount	Classified group	Standard of experiment
Maximum water absorption (weight percentage)	12	Low density	ASTMC 97
	7.5	Medium density	
	3	High density	
Maximum density (kg/m^3)	1,760	Low density	ASTMC 97
	2,160	Medium density	
	2,560	High density	
Minimum compressional strength (MPa)	12	Low density	ASTMC 170
	28	Medium density	
	55	High density	
Minimum rupture module (MPa)	2.9	Low density	ASTMC 99
	3.4	Medium density	
	6.9	High density	
Minimum erosional strength (MPa)	10	All three groups	ASTMC 241

Table 5.20 ASTM required specifications of marble

Physico-mechanical specification	Amount	Classified group	Standard of experiment
Maximum water absorption (weight percentage)	0.2	All four groups	ASTMC 97
Maximum density (kg/m^3)	2,595	Calcite	ASTMC 97
	2,800	Dolomite	
	2,690	Serpentine	
	2,305	Travertine	
Minimum compressional strength (MPa)	52	All four groups	ASTMC 170
Minimum rupture module (MPa)	7	All four groups	ASTMC 99
Minimum erosional strength (MPa)	10	All four groups	ASTMC 241
Minimum bending strength (MPa)	7	All four groups	ASTMC 880

5.30.2 Distribution of Dimension Stones

The most common dimension stones used as façade in buildings are granite, marble, onyx, travertine, metamorphosed limestone, sandstone, and slate (metamorphosed shale which is foliated and gives large flat slabs). Iran is among the countries that possess great potentials and capabilities in the production of dimension stones. In spite of the relatively large volume of extraction of dimension stones, these potentials and capabilities did not receive the attention they deserved in the past decade (Ghorbani 2009b).

There is a wide range of sedimentary, igneous, and metamorphic dimension stones in Iran. The chemical composition and properties, color, and age range of dimension stones in Iran are very variable, and these characteristics are briefly described next (Ghorbani 2009b):

1. *Igneous rocks* – The following types of igneous rocks are found in Iran:

- Intrusive igneous rocks: granitoids (granite, granodiorite, monzonite, and syenite), gabbro, gabbro–diorites, and ultramafic rocks
- Extrusive and pyroclastic igneous rocks: dacitoids, basalts, andesites, and tuffs

So far, more than 500 bodies of igneous rocks have been identified in Iran. Most of the intrusive bodies that give away good size and uniform blocks with pleasant colors belong to Mesozoic and Tertiary, of which some examples are

- Intrusive bodies along Takestan–Tarom–Hashtjin axis, Ahar, and Siyah-rud
- Intrusive bodies within Urumiyeh–Dokhtar axis (in Qom, Natanz, and Anar)
- Granitoids along Zahedan–Saravan axis
- Granitoids and gabbros within Sanandaj–Sirjan (northern Sanandaj–Sirjan in particular)
- Granitoid and gabbroid intrusive bodies in Masouleh–Astara
- Granitoid intrusive bodies along Karaj–Taleghan–Kelardasht axis
- Small number of intrusive bodies northeast of Iran (Taknar–Binaloud and Kavir–Sabzevar Zones)

2. *Metamorphic rocks*: The dimension stones of metamorphic type such as marble, onyx, schist, hornfels, and gneiss are found in metamorphic areas in west of Iran like Toyserkan, Mahabad, and Naghadeh. Also, some beautiful schists and amphibolites have been identified in east of Iran within Birjand and east of Ghaen areas, which will provide important resources of dimension stones for Iran if the small-size blocks of these rocks become demandable in the future.
3. *Sedimentary rocks* – Good-size sedimentary rocks with beautiful colors and textures have been identified in different stratigraphic horizons, and the most important ones are the following:

Crystallized Limestones in Ruteh Formation: These rocks produce very good size and white blocks in many parts of Iran, especially within the Sanandaj–Sirjan Zone.

Lar and Delijan Limestone Formations: These rocks produce good-size blocks in parts of Alborz, Central Iran, and Sanandaj–Sirjan in particular.

Hojedk Sandstone Formation: This formation is in east of Central Iran (west of Tabas along the Khoor–Tabas road), which produces good-size black blocks.

Cretaceous Limestones: Some Cretaceous limestones in Central Iran, Bafgh, Kerman, east of Iran, Sanandaj–Sirjan Zone (Esfahan, Neyriz, Piranshahr, and Kermanshah), and Zagros Zone in Chaharmahal and Bakhtiari (Sarvak Formation) produce good-size blocks in white, gray, and off-white colors.

Fajan Conglomerate: Fajan and Kerman conglomerates in Alborz and Central Iran produce good-size blocks in beautiful colors.

Qom Formation: Sometimes, Member F of Qom Formation in Central Iran (Abadeh, Ab-e Garm Qazvin, south of Abhar) produces good-size blocks in yellowish white to gray color.

Asmari Formation: In some parts of Zagros (Kermanshah, Chaharmahal, Bakhtiari, Kohkilouyeh, and Boyer Ahmad), Amari limestone produces reasonable good-size blocks that are suitable for use as façade stones in buildings.

Late Miocene to Quaternary Travertines: During this period, travertine deposits were formed in some parts of Iran, and some of them produce good-size blocks and are being exploited.

Dimension stones are considered among the high mining potentials and capabilities of Iran. Due to the high quality, beautiful color, and texture of dimension stones of Iran, these resources are distinguished and superior as compared to their counterparts from other parts of the world, and in some cases, they are one of a kind in the market. Therefore, the dimension stones industry can be accounted as a strong point of mining sector from different aspects such as exploitation, domestic production, export, and foreign exchange revenue.

Based on studies, reports, and statistics, the known exploitable dimension stone deposits of Iran can be summarized as travertine (59 million tons), onyx (3.6 million tons), marble (500 million tons), crystallized limestone (53 million tons), granite (60 million tons), and ornamental limestone (100 million tons).

The quantities presented above are a small portion of the proven reserves of dimension stones in Iran because the actual reserves are believed to be more than several billion tons. It can be stated that the dimension stones in Iran are considered as an endless source of income.

5.31 Rare Earth Elements (REEs) Resources in Iran

No fundamental and integrated exploration work has yet been carried out on REEs in Iran, and at present, available information about REE is limited to a few reports and theses (Ghorbani 2011). Some studies on various parts of Iran including Central

Iran, alkaline rocks in Eslami Peninsula, iron and apatite in Hormuz Island, Kahnouj titanium deposit, granitoid bodies in Yazd, Azerbaijan, and Mashhad and associated dikes, and finally placers related to Shemshak Formation in Marvast, Kharanagh, and Ardekan indicate high concentration of REE in magmatogenic iron–apatite deposits in Central Iran (Yazd area) and placers in Marvast area in Yazd.

Within Central Iran, significant amounts of REEs are found in phosphate, iron–phosphate deposits, and apatite-bearing metasomatic rocks in the Esfordi, Lakeh Siyah, and Gazestan localities, and also in Choghart and Chadormalu apatite-bearing magnetite deposits (Ghorbani 2011).

Based on the studies carried out on the Esfordi and Lakeh Siyah areas, high amount of REEs exists in connection with apatite veins, apatite-bearing magnetite masses, and apatite-bearing actinolite–tremolite deposits. Litho-geochemical studies on the above-mentioned areas point to a direct relationship between light rare earth element (LREE) and heavy rare earth element (HREE) themselves and also between LREE and HREE and the content of phosphorus and calcium. In other words, with an increase in phosphorus and calcium content, the amount of LREE and HREE rises as well, which indicates the direct connection between REE mineralization and apatites in the area in such a way that sometimes the total amount of REE in apatite crystals reaches up to 2.5% (Ghorbani 2011).

During the course of these studies, the elements thorium and uranium showed direct relationship with REE and phosphorus, which could mean the enrichment of these elements is associated with the metasomatism event in the area (Maghsoudi et al. 2005). Considering the high concentration of REEs (about 0.5%) in gangue produced in the Esfordi mine, the exploitation of REE as a by-product from the Esfordi phosphate mine can be taken into consideration (Ghorbani 2011).

Based on the reports by the National Iranian Steel Company (NISCO 1975), REE is seen in apatites within Choghart iron deposit. Apatite minerals are of fluorapatite type and contain monazite inclusions. The amount of REE is 0.5% in apatites while apatite minerals occupy 5.3% of the ore. In Choghart deposit, LREE has more concentration than HREE.

Monazite-bearing alluvium in the Marvast area was first studied and identified within the framework of geochemical exploration of heavy minerals along the Yazd–Sabzevar axis and within the Marvast 1:100,000 geological map. The result of these studies showed that the average grade of monazite in Marvast placers was about 150 g/ton. Europium (Eu) showed notable percentage within Marvast monazite concentrates (Alipour-asl 2003).

Based on the studies by Alipour-e Asl, not much hope exists for the presence of economically valuable deposits of monazite and REEs within the Marvast area and other parts of the Sanandaj–Sirjan Zone. However, due to the significant expansion of Late Triassic–Jurassic sedimentary sequences within the Sanandaj–Sirjan Zone as well as the high physical and chemical stability of monazite, the existence of placer-type deposits of monazite is not far-fetched.

Chapter 6
Metallogenic and Mining Provinces, Belts and Zones of Iran

Abstract Tectonically, Iran is a part of the Alpine–Himalayan belt. The consequences of most of the orogenic phases recorded in this tectonic belt, especially the Alpine orogeny, are traceable in Iran. All the Mesozoic and Cenozoic structural divisions of the country have, in effect, formed or evolved in response to the Alpine–Himalayan orogeny.

Combining the tectonic and economic evidences available, it can be deduced that all the areas of the country have been developed under the influence of the Alpine orogeny, and therefore it is possible to classify various rocks and their associated mineral deposits into a number of metallogenic divisions based on the developmental stage of this orogeny. Taking into account the concepts of metallogeny, geological features, and mineralization characteristics of the deposits, three categories of metallogenic divisions can be considered for Iran as mentioned below:

1. Metallogenic provinces: Central Iran, Sanadaj-Sirjan, Orumieh-Dokhtar, Northeastern, Eastern and Southeastern
2. Metallogenic belts: Malayer-Esfahan Lead-Zinc Belt, Kerman Copper Belt, Esfandaghe-Faryab Chromium-Bearing Ophiolite Belt, Hashtjin-Tarom Metallogenic Belt, Zagros Oil and Gas Belt, Taknar Belt, etc.
3. Metallogenic zones: Takab, Anarak, Bafq, Tabas, Abadeh, Behabad etc.

6.1 Introduction

Metallogenic investigations are amalgamation of studies in various earth science disciplines that try to find the natural sequence of processes governing the mode of formation as well as the spatial and temporal distribution of ore deposits. In view of metallogeny, it is possible to study orderly distribution of the metallic ores and non-metallic mineral deposits as well as fossil fuels. Simply put, metallogeny is a science that discusses what (kind of mineral deposit), where (in which structural division), when (in which geological period), and how (under the influence of which factors) mineral deposits form.

M. Ghorbani, *The Economic Geology of Iran: Mineral Deposits and Natural Resources*, Springer Geology, DOI 10.1007/978-94-007-5625-0_6,

During the first three decades of the twentieth century, metallogeny was used to combine the mining and geological information in order to direct the exploration operations. Today, after almost a century of practice, metallogeny attempts to unify geological insights with tectonic and magmatic views creating various theories of mineralization.

In order to study and classify the metallogeny of a locality, one has to gather information on the stratigraphy, magmatic and metamorphic activities, and tectonic processes and recognize all the geological phenomena that have influenced the area.

Metallogenic investigations are in their infancy in Iran and the number of researchers who have worked on this topic is few. Though preliminary metallogenic maps presenting limited information on some economically important regions of the country exist (e.g., Kerman area, Anarak area, Malayer–Esfahan axis), a complete metallogenic map of Iran is yet to be produced.

This chapter is an attempt to introduce the concept and terminology involved in metallogeny and lay out the metallogenic subdivisions of the country within the Alpine–Himalayan orogenic belt.

6.2 Categorization of Metallogenic and Mining Areas in Iran

Metallogenic study in Iran is at the onset of its course, and no metallogenic province, belt, or area has been defined in Iran to this date. Considering tectono-magmatic, stratigraphic, and metamorphic events, sedimentary facies, and economic geological studies, the following metallogenic regions are introduced. This categorization was presented for the first time in Ghorbani (1999f).

Metallogenic provinces of Iran are as follows (Fig. 6.1):

- Central Iran
- Urumiyeh–Dokhtar metallogenic province
- Sanandaj–Sirjan metallogenic province
- Northeast metallogenic province (Taknar, Kavir, Sabzevar Belts)
- Alborz metallogenic province
- Southeast and east of Iran metallogenic province
- Zagros oil and gas province
- Kopet–Dagh oil and gas province

Metallogenic (mining) belts of Iran are as follows:

- Malayer–Isfahan lead and zinc belt
- Kerman copper belt
- Esfandagheh–Faryab chromite ophiolitic belt
- Khash–Nehbandan Belt (with chromium, copper, and magnesium deposits)
- Qom–Naein Belt (manganese, barite, copper deposits)
- Kavir–Sabzevar Belt (copper, chromium, gold, iron deposits)
- Taknar Belt (copper, gold, arsenic deposits)

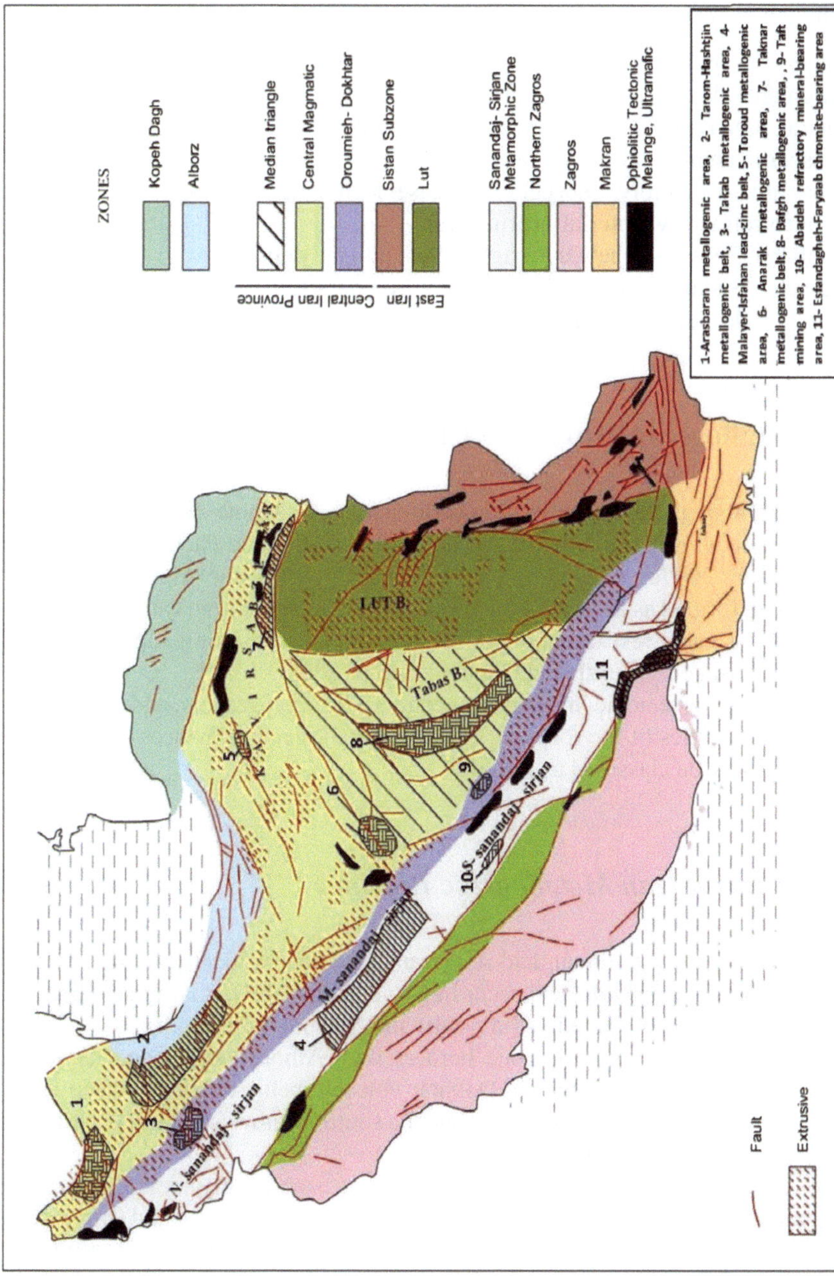

Fig. 6.1 Structural map, metallogenic and mining provinces

- Tarom–Hashtjin Belt (copper, iron, lead, zinc, gold deposits)
- Maku–Khoy–Urumiyeh Belt (gold, mercury, copper, chromium, iron deposits)

Metallogenic areas are as follows:

- Takab area (gold, arsenic, antimony, mercury, lead, zinc, poly-metal deposits)
- Bafgh area (iron, lead, zinc, apatite, REE deposits)
- Anarak area (copper, lead, zinc, gold, iron, antimony, arsenic deposits)
- Arasbaran area
- Tabas area (refractory, fluorite, manganese, lead, zinc deposits)
- Abadeh refractory material-bearing area
- Kuhbanan–Ravar–Behabad triangle (lead and zinc deposits)
- Qorveh–Asadabad area (antimony, gold, iron deposits)
- Taft area (lead, zinc, copper deposits)

6.3 Metallogenic Provinces of Iran

From the point of global tectonics, Iran is situated in the Alpine–Himalayan tectonic belt that extends from the Atlantic Ocean on the west to the Pacific Ocean on the east (Ghorbani 2007a). According to the prevailing views, the Alpine–Himalayan belt has been formed as a result of the closure of an ancient ocean called Tethys that trended in the east–west direction in between the then-existing supercontinents, namely, Laurasia on its north and Gondwana to its south.

Considering the tectonic, magmatic, metamorphic, stratigraphic, and sedimentary facies characteristics of the rocks of Iran, the metallogenic provinces and zones of the country may be classified as follows.

6.3.1 Central Iran Metallogenic Province

Central Iran is one of the main and important units that is located in the center of Iran in the form of a triangle, and according to Boulin (1991), it is situated between two orogenic belts of Paleotethys (in north) and Neotethys (in south).

The north boundary of this zone borders with Alborz heights and its south–southwest border meets Urumiyeh–Dokhtar volcanic belt. A series of steep faults, which were active until Mesozoic, separate the south–southwest part of Central Iran from Sanandaj–Sirjan (Sengör 1991).

The eastern boundary of this province has not been clearly identified because some geologists considered the Lut block as a part of Central Iran and some others recognize it as a separate entity. This zone is the largest and most complicated geological unit in Iran as it went through numerous incidents and experienced numerous phases of metamorphic, magmatic, orogenic, and folding activities.

Nogol–Sadat (1978, 1993) has divided Iran into nine major areas, and two lateral stable cratonic areas (Abadan and Sarakhs) based on evidences such as fortification of basement of structural stages, stratigraphic sequences, modes of deformation, magmatic–metamorphic regions, and related mineralization (see Chap. 2 and Fig. 2.2). According to such divisions, Central Iran covers a large area where deposits from Precambrian (large thickness) through recent occur.

Central Iran is bounded by the Nayband fault in the east, the Great Kavir (Dorouneh) fault in the north and northeast, and the Dehshir fault in the west and southwest. Ophiolites outcrop along these faults (Nogole-Sadat 1993).

According to current information, the oldest rocks within Central Iran belong to the Kooshk, Rizu, and Dezu Series, which are of volcanosedimentary type. The volcanic rocks are mostly acidic to intermediate and their nature tends to be alkaline to calc–alkaline (with age of Late Precambrian) (Hamdi 1995; Aghanabati 2004; Ghorbani 2012a). In the Bafgh region, a large-scale magmatism with composition of granite to quartz–porphyry to basic bodies took place, which is equivalent to Dowran granite (Ghorbani 2002b). The magmatic activities in this period of time coincide with the beginning of rifting, during which the Kooshk, Rizu, and Dezu Series came into existence (Ghorbani 2012a). The rocks of the Dezu Series are more basic than those that belong to the Rizu Series, and they tend to be more alkaline in nature.

Once the volcanic activities slowed down and ceased, the intrusive magmatism started and granite bodies such as Zarigan with age of 520–540 Ma were formed (Ramezani et al. 2003). Such granitization led to the fortification of basement of Iran, and also caused metamorphism.

Widely distributed outcrops of high-grade metamorphic rocks of Precambrian affinity have not been observed in Central Iran, and the higher grade metamorphism in Central Iran belongs to younger ages (Houshmand-Zadeh 1997). Concurrent to the formation of the above-mentioned rocks, mineralization of iron, lead–zinc, manganese, phosphate, and REE took place at large scale (Ghorbani 2002a).

The Paleozoic deposits in Iran and Central Iran, particularly in the middle triangle, began with continental facies sediments like red sandstone, which are covered by a white layer/horizon of quartzite with thickness of 50 m (e.g., Lalun Formation, Mila Formation, Niur Formation) (Hamdi 1995).

Significant stratigraphic hiatus occurred within Paleozoic in the Late Ordovician–Silurian and Devonian, which is an evidence of structural movements of Early Paleozoic (Caledonian), followed by erosion (Haghipour 1977).

To the belief of Haghipour et al. (1979), the movements in these areas were merely epeirogenic and should not be considered as an orogenic phase of Hercynian.

Moreover, other movements took place at the end of Carboniferous, which led to a pause in sedimentation in Late Carboniferous (Alavi-naini 2009).

Triassic deposits formed in eastern, central, and western parts of middle triangle, but they are much more extensive in the eastern part (especially Tabas) as opposed to the other parts of the triangle, because of the activity of Kalmard, Kuh–Banan, and Anar faults. The deposits of Middle and Upper Triassic were laid over the other deposits with a disconformity. The magmatic activities were the result of Early Cimmerian orogeny, and Esmailabad and Dehbid granite bodies are two examples related to such magmatic activities (Ghorbani 2002a).

Additional signs and evidences of metamorphism associated with this orogeny are seen in central part (Saghand in particular) (Aghanabati 2004). Extensive lead–zinc mineralization occurred in Triassic rocks within Bahabad–Ravar–Kuhbanan triangle and formed numerous deposits such as Gowjar, Gour, Deh-askar, Taj-kuh, Bash-koun, Boneh-anar, Gijar-kuh, and Tarz (Ghorbani 2002a).

In most parts of the middle triangle, facies such as sandstone, shale, and marl were formed during Jurassic. The Late Cimmerian orogenic phase caused facies transformation in these areas during Middle Jurassic. This phase was a compressional phase accompanied with metamorphism and magmatism, and Shir–Kuh granite is a product of such activities. This granite body cut through the Jurassic deposits, while Early Cretaceous deposits overlie it (Aghanabati 1998).

In Early Cretaceous, marine transgression took place within the middle triangle in a large scale and gave rise to rock facies such as conglomerate, sandstone, and clastic limestone. At the end of Cretaceous (Maastrichtian–Paleocene), most part of deposits in the middle triangle underwent severe folding accompanied with metamorphism and led to the formation of a disconformity between deposits of Paleocene and Late Cretaceous (Shemirani 1988).

In parts of the middle triangle, sedimentary facies belonging to Cenozoic (Paleocene) began to form with basal conglomerate and sandstone. They overlie the older rocks with disconformity (Hajian 1996). Magmatism in the middle triangle during Cenozoic took place both internally and externally. Following the compression phase of Late Cretaceous (Laramid), which was accompanied with metamorphism, folding and uplifting, and emplacement of ophiolites within the perimeter of the middle triangle, intrusive bodies with granodiorite composition and calc–alkaline nature were born (Ghorbani 2002a).

Meanwhile, during Eocene, volcanic activities produced large volumes of volcanic rocks with composition of andesite–dacite along the faults. Such activities are extensively seen on east of the Dehshir–Baft fault, and is still going on in the form of travertine formation (Ghorbani 2003a). Due to younger igneous activities, vast metallic mineralization such as iron, gold, copper, antimony, lead–zinc occurred in Anarak area with ultramafic–mafic basement (Ghorbani 2002a).

During Quaternary and concurrent to final shape-up of the highlands, many sedimentary basins lost their connections with seas and turned into vast plains where evaporative sediments such as gypsum and salt along with clay and marl with desert characteristics were formed (Ghorbani 2002a). Among the most important deserts in the middle triangle, Ardekan, Abarghou, Bafgh, and Biyabanak are worth mentioning.

In the following sections, the important areas within this metallogenic province comprising of Anarak, Takab, and Bafgh are briefly described.

6.3.1.1 Bafgh Metallogenic Area

This metallogenic zone lies in the eastern part of the Central Iran metallogenic province. The Bafq zone is bounded on the south by the Kuhbanan fault while it is separated from the Tabas block by the Naein Basement and Kalmard faults on the northeast. The northern boundary of the zones is marked by the Chapedony and Saghand–Posht Badam fault, whereas the Kuhbanan fault forms the western limits (Nogole-Sadat 1993).

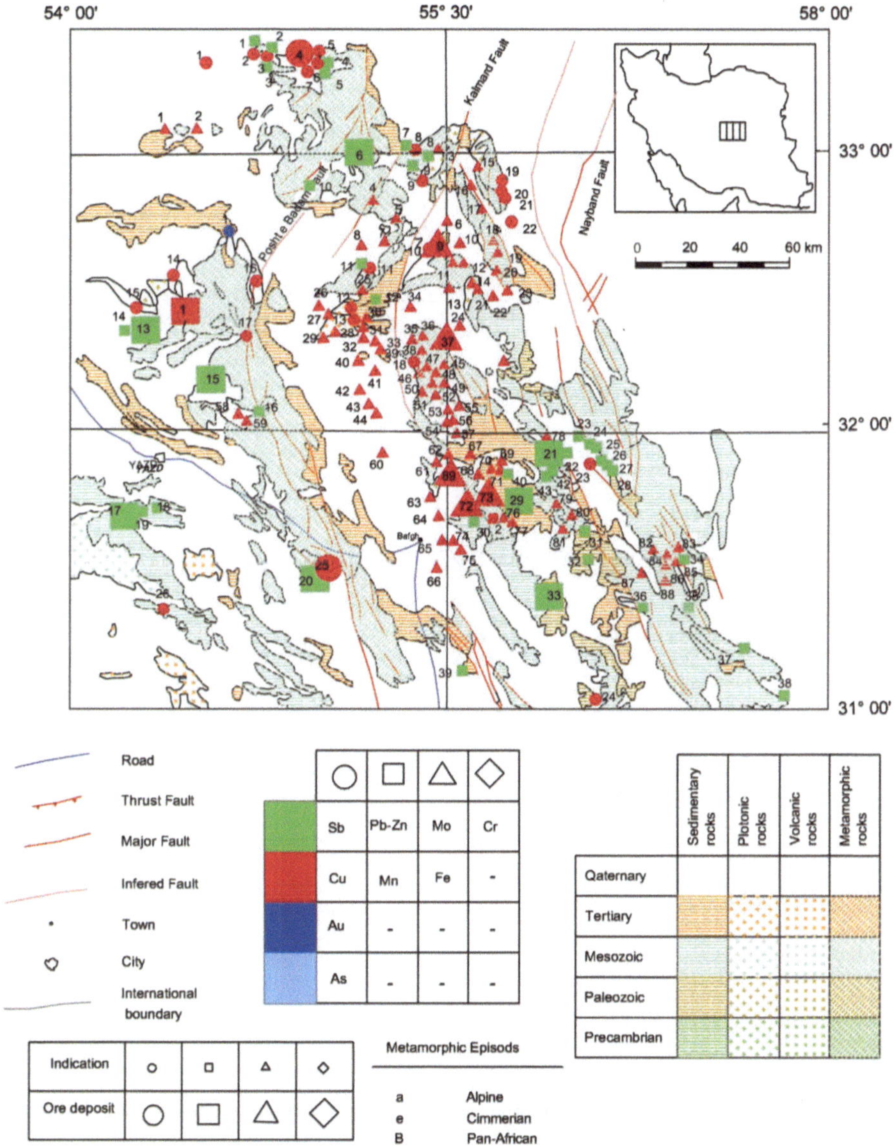

Fig. 6.2 Metallogenic map of Bafgh area (Ghorbani 2002a)

The Bafq metallogenic zone is the oldest metallogenic zone of Iran. It belongs to the Late Precambrian–Early Cambrian times and hosts vast reserves of iron, lead–zinc, and phosphate (Fig. 6.2).

Tectonically, the Bafgh area is a section of the central part of the Central Iran tectonic zone. The oldest rocks, which outcrop in Central Iran, are of Late Proterozoic–Early Cambrian age (Fig. 6.3) and are composed of volcanics, volcanoclastics,

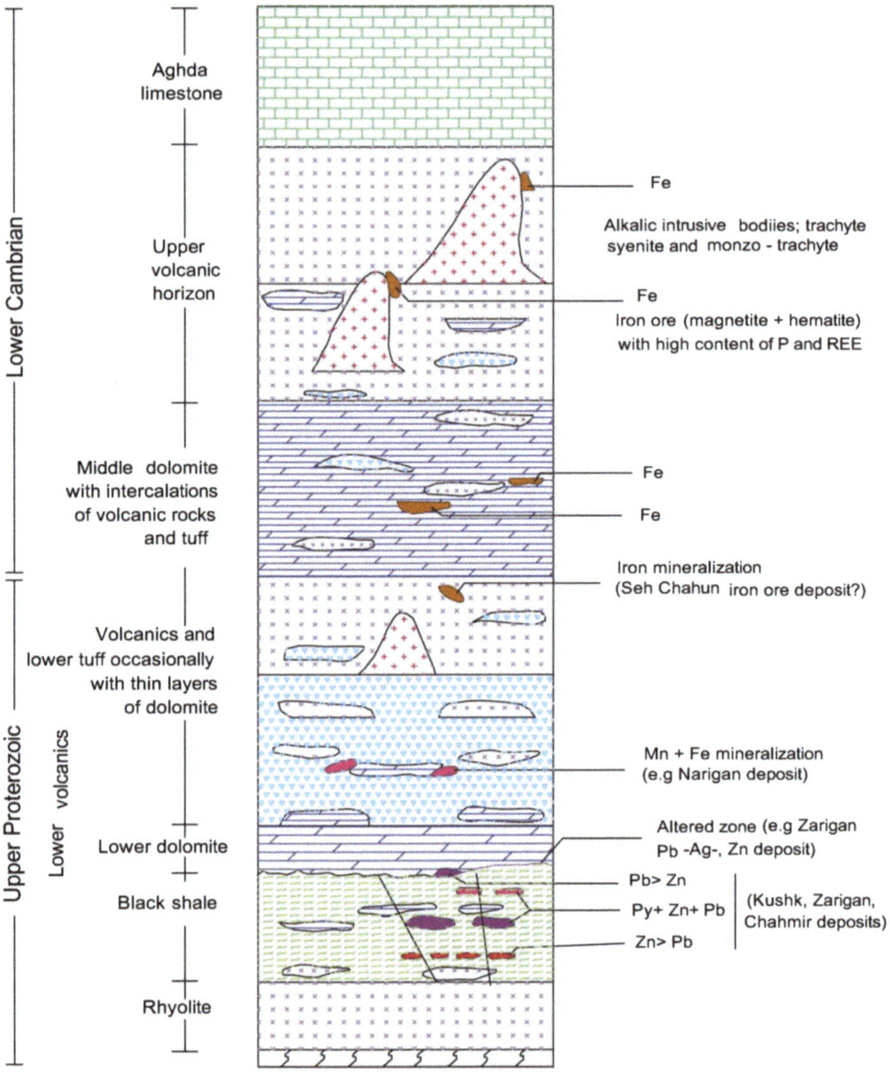

Fig. 6.3 Lithostratigraphy of Precambrian and Cambrian rocks and mineralization associated with them (Ghorbani 2002a)

detrital, and carbonate (mostly dolomitic) compositions. These rocks are considered as equivalents of the Morad Series and Kahar Formation by various experts (including F. Daliran 1990; Momen-Zadeh and Heidari 1995; Heidari 1996; Ghorbani 1999f). However, the petrological studies do not confirm the alkaline nature of these rocks in some cases (Heidari 1996). The composition of rocks varies from rhyolite to basalt. But the majority of the volume of these rocks is andesite and trachyte. Among the oldest outcrops in the Bafgh area are the Rizu and Dezu Series, which are composed mainly of volcanics, volcanosedimentaries, and dolomites. These rocks in the Kushk area overlie "Kushk Series" (Hamdi 1995). Some geologists consider that these rocks are formed in the rift environment (Daliran 1990; Samimi 1991; Momen-Zadeh and Heidari 1995; Ghorbani 1999a). Similar rocks are also found in the north of the Arabian Plate in a rift system (Ghorbani 2002a). Some of the authors consider an island arc or inter arc basin geological environment for the formation of these rocks (Heidari 1996).

The intrusive rocks that occur within these rocks are composed of a spectrum of compositions from granite to gabbro. These intrusive rocks are usually alkaline. The age of intrusive rocks is 540–600 Ma. The mineralization of Kushk deposit is considered 580 ± 20 Ma (Romanko quoted by Houshmanzadeh, personal communications). Based on field observations and laboratory analyses, some of the ore deposits are genetically related to these old rocks. Pb, Zn, Fe, and Mn are the elements that make up mineralizations in these rocks. This stage can be considered as the first mineralization phase. The sediment hosted lead–zinc deposits of the Bafgh area belong to this phase. The Mishdovan iron and Narigan iron–manganese mineralizations are also belong to this phase. The iron ore contains some apatite and is mainly hematite with high Mn-content so that in Narigan it makes up a manganese deposit. The Ti-content is high and the V content is moderate. The mineralization of lead–zinc, as well as the iron–manganese, is stratiform to stratabound within the shales, carbonates, rhyolitic pyroclastics, and lavas and andesitic lavas. The iron ore is occasionally interbedded with jaspilite (rarely magnetite flow). An intercalation of evaporates, carbonates and pyroclastic, and detritic rocks covers the above-mentioned Morad Series (or Kushk Series in the Kushk area) (Ghorbani 2002a). The micropaleontological studies (Hamdi 1989) indicated an Early Cambrian age for these rocks. The volcanic activity is continued from Late Precambrian to Early Cambrian without serious changes in composition.

The alkaline magmatism, expressed mainly by the granite intrusions of Early Cambrian (540–530 Ma.), is extended in Central Iran, like many parts of the Gondwana Land. These granites are well exposed in an area from Anarak to Kuhbanan, as well as in the Bafgh area. The sodic granites of Narigan and Zarigan are the outstanding examples of these granites. The intrusion of granites is associated by metamorphism in some parts (Houshmand-Zadeh 1997). This alkaline magmatism is expressed by intermediate rocks in some areas as well. It seems that this magmatism, which are considered as the second phase of magmatism and mineralization (Fig. 6.3), has caused extended iron mineralizations in the Bafgh area, which occurs after the first phase (Ghorbani 2002a). (The first phase we explained as the main Zn, Pb mineralization phase.)

The formation of the main iron ore deposits in the Bafgh area is considered to be related to this second phase. The mineralization in this phase is rich in P and REE so that the Esfordi phosphate deposit, with very high REE content, has been formed in this phase. The Chador Malu and Choghart iron ore deposits are also formed in this phase. Metamorphism and metasomatism are extended in some areas like Esfordi and Choghart. The metasomatism, which is caused by the second magmatism phase, may have caused recycling of the iron ore from the first phase and caused an enrichment of iron in the second phase. Therefore, the iron ores, which are formed in the second phase, may be the product of both phases (Ghorbani 2003b).

Some riebeckite granites are generated in the second phase of magmatism. Apatite and REE are associated the granite intrusives, Rizu and Dezu Series, but not with the rock sequence older than the Rizu Series, that is, the Morad Series and Kushk Series. The basement in the Bafgh area, on the other hand, seems to be of ultramafic alkaline nature (Samani 1988; Ghorbani 2002a). These rocks underlie the above-mentioned old rocks in the Bafgh area. Therefore, it can be said that the ultramafic alkaline rocks have had an influence in the formation of iron ores in the Chador Malu, Choghart, Seh Chahun, Esfordi, and Mishdovan deposits.

Tectono-Magmatic Phases in Central Iran: The Bafgh area is characterized by its iron ore and lead–zinc deposits. Both these deposits are generated in the Pan-African tectono-magmatic phases, which have occurred between 540 and 640 Ma (Ghorbani 2007c).

The other mineralizations phases in Central Iran are Middle Triassic and Lower Cretaceous. In Middle Triassic, lead–zinc deposits have formed in Behabad–Kuhbanan block (east, southeast of the Bafgh area), and lead–zinc and manganese deposits have been formed in the Yazd block (west of the Bafgh area) (Ghorbani 2002a).

Three tectono-magmatic phases can be seen in the Bafgh metallogenic area as described next:

(A) *Pan-African Tectono-Magmatic Phase*: The metamorphosed volcanosedimentary rocks of Central Iran, which were mentioned earlier, are not well studied from the tectono-magmatic point of view. Although most experts (e.g., F. Daliran, A. Houshmandzadeh) consider them as rifting products, they may be also the products of simple pull-apart basin tectonics due to the movement of transform faults of the Precambrian basement (Ghorbani 2002a). Due to these transform faults, the Central Iran block is considered to be broken so that Bafgh and its neighboring blocks, like Tabas and Behabad–Kuhbanan–Ravar blocks, went through different episodes in the Paleozoic times (Alavi 1993; Nogol 1993). For example, the Paleozoic rock sequence in the west of the Torkamani fault in the Anarak area is identical to the Alborz and East Iran rocks. Whereas the rocks between the Chapedoni and Poshtebadam faults lack, likely, the lower Paleozoic rocks (Ghorbani 2012a).

(B) *Early and Late Cimmerian Orogeny* (*Tectono-Magmatic*) *Phases*: The Early Cimmerian orogeny phase is mainly identified by regional metamorphism. The magmatic activity is not very intensive in this orogeny phase. This phase, in the Bafgh area, is expressed mainly by a regional metamorphic episode. It is well recognized in the area east of the Poshtebadam fault. The same orogeny phase has affected the pink granites of the Pan-African phase (which was mentioned before) and transformed them to gneiss, like the Zamanabad gneiss in the Poshtebadam area (Aghanabati 2004; Ghorbani 2012a). These granite–gneiss rocks show an age of 220 Ma. by K/Ar dating (Emami 2000). The Early Cimmerian metamorphism does not occur with the same intensity in the whole Central Iran. It is manifested by a weak metamorphism in some parts. The more evident effect of this phase is a horst–graben and rough morphology, which has caused different sedimentary regimes in different parts of Central Iran (Houshmand-Zadeh and Posht-kouhi 1996). The rapid variation of thickness and lithology of Biabanak detrital formation (Biabanak shale) is an example. The thickness of these detrial rocks in west of the Poshtebadam fault is much more than that in its east side, and the lithology is also different (gray shale in the west and red sandstone and conglomerate in the east).

As several Zn, Pb occurrences are located in the neighboring area of Bafgh toward the east, it is of interest to give a general description about their geotectonic position. These deposits are all located in a tectonically discriminated area, (i.e., the Behabad–Kuhbanan block) from the eastern Bafgh–Saghand block. The Bafgh–Saghand block, being the tectonic unit which hosts the Kushk-type Zn, Pb deposits and the iron ore deposits of the Bafgh area, is confined between the Yazd block from the west, the Tabas block from the east and the Bahabad–Kuhbanan block from the south east. In the Bahabad–Kuhbanan block, as opposed to the Bafgh–Saghand block, the uppermost Proterozoic–Lower Cambrian rocks are absent. The Upper Paleozoic–Triassic rocks, on the other hand, are extensive. They are composed of shales and sandstones of Devonian to Permian, the carbonates of Permian (Jamal Formation), the red shales and sandstones of Permian (Sorkh shale Formation), and the dolomites of Triassic (Shotori dolomites). Some 40 Zn, Pb occurrences are known, and most of them are hosted by the Triassic carbonates (Shotori dolomites). The ores are mainly oxidic and most of them are indeed zinc mineralizations. Some of these occurrences are Tajkuh, Ahmadabad, Bajgun, Abbid, Tappeh Sorkh, Deh Asgar, Goejar, Tarzeh, Gowr, Magasu, Gharu, Karvangah, Khorand; Gijar Kuh, Rig, and Kalaghi. These Lower Triassic hosted mineralizations are indeed stratabound, occasionally stratiform (Ghorbani 2000e).

(C) *Laramide Tectono-Magmatic Phase*: The Laramide tectono-magmatic phase affects Central Iran including the Bafgh area at the end of Cretaceous and beginning of Tertiary. Extended granite–granodiorite bodies have intruded older rocks during this event. The Chapedoni gneisses are of this type. The dating of Chapedoni gneisses showed 57 Ma (J., Zhion, China expert, oral talking). This age is quite different from the previous assumption of the Precambrian age for these rocks (Haghipour 1974). The Laramide phase has caused extended mineralizations in the Anarak area (not in the Bafgh area). The Kal e Kafi Cu, Mo, Au, Khuni Au, Nakhlak Pb, Ag, Talmesi, Cu, and Meskandi Cu mineralizations are among them.

Conclusions on the Geological Potential of the Bafgh Area: During the Late Proterozoic–Early Cambrian time, volcanic rocks (mainly felsic and alkalic), magmatic and sedimentary iron oxide ore (magnetite and hematite), massive sulfides or sedex Pb–Zn deposits and magmatic phosphate, uranium and REE formed in Central Iran, especially in the Bafgh–Poshtebadam area (Ghorbani 2002a). During the same time, evaporite deposits, sedimentary phosphate, hydrocarbons, and some other ore mineralizations formed in south, southwest, northeast, and northwest of Iran. The volcanic rocks, which are associated with evaporite deposits and the evaporites in the Kerman area, became more abundant and more extensive in the Bafgh–Saghand area in Central Iran (Hamdi 1995). Iron ores show a good coincidence with the magmatic rocks. Magmatic phosphorus (apatite), uranium, and REE have also close coincidence with the magmatic rocks. The exhalative Zn, Pb massive sulfides or sedex deposits (in Kushk, Chahmir, and Zarigan are classic examples) show a closer coincidence with the volcanoclastic and sedimentary rocks rather than the intrusives (Ghorbani 2000c, 2002a). These Zn, Pb do not show direct association with the magmatic rocks as the iron and phosphorus mineralizations do.

The Zn, Pb mineralizations in the Bafgh area can be considered as a distal product of the Pan-African magmatism, with a center where the iron mineralizations, as magnetite and hematite formed in direct association with the magmatic rocks, as seen along the Bafgh–Poshtebadam axis (Ghorbani 2000e, 2002a).

6.3.1.2 Anarak Metallogenic Area

This metallogenic zone is located at the center of Central Iran metallogenic province. The zone is bounded on the north by the extension of the Dorouneh fault, on the southwest by the Nain–Zevar ophiolite zone, and on the south by the Nain–Anarak depression. The two basement faults (Dorouneh fault and Naeen–Shah–kuh fault, which is to the south of Doroneh and parallel to it) probably pass through this metallogenic zone (Nogole-Sadat 1993).

The Anarak zone, with its characteristic metallogenic, magmatic, and metamorphic features along with widespread Late Proterozoic rock exposures, forms an interesting metallogenic zone that holds a number of economically important ore mineral deposits (Fig. 6.4).

Situated in the middle of Central Iran geological division, the Anarak Metallogenic Province is bounded on its north by extension of the Darooneh fault, on its southwest by the Naein–Zavar ophiolitic zone, and on its south by the Naein–Anarak embayment. The Darooneh fault and the one parallel to it on the south pass through this metallogenic province (Nogole-Sadat 1993).

Anarak is one of the most interesting metallogenic provinces of Iran from the point of view of metallogeny, igneous activity, metamorphism, and distribution of Upper Proterozoic rocks. Detailed geological and metallogenic investigations have been carried out in the area by Russian geologists under the supervision of the Geological Survey of Iran, the results of which have been published in the Anarak Metallogenic Map (Metallogenic Map of Anarak 1984) and briefly described hereunder.

The Anarak metallogenic area has been divided into six blocks by Russian geologists, namely:

1. Anarak–Khor block
2. Posht Badam block
3. Chah Palang–Bayazeh block
4. Naein–Zavareh Ophiolites block
5. Kuh Dom–Qale Sardar block
6. Late Alpine depression and intramountain basins block

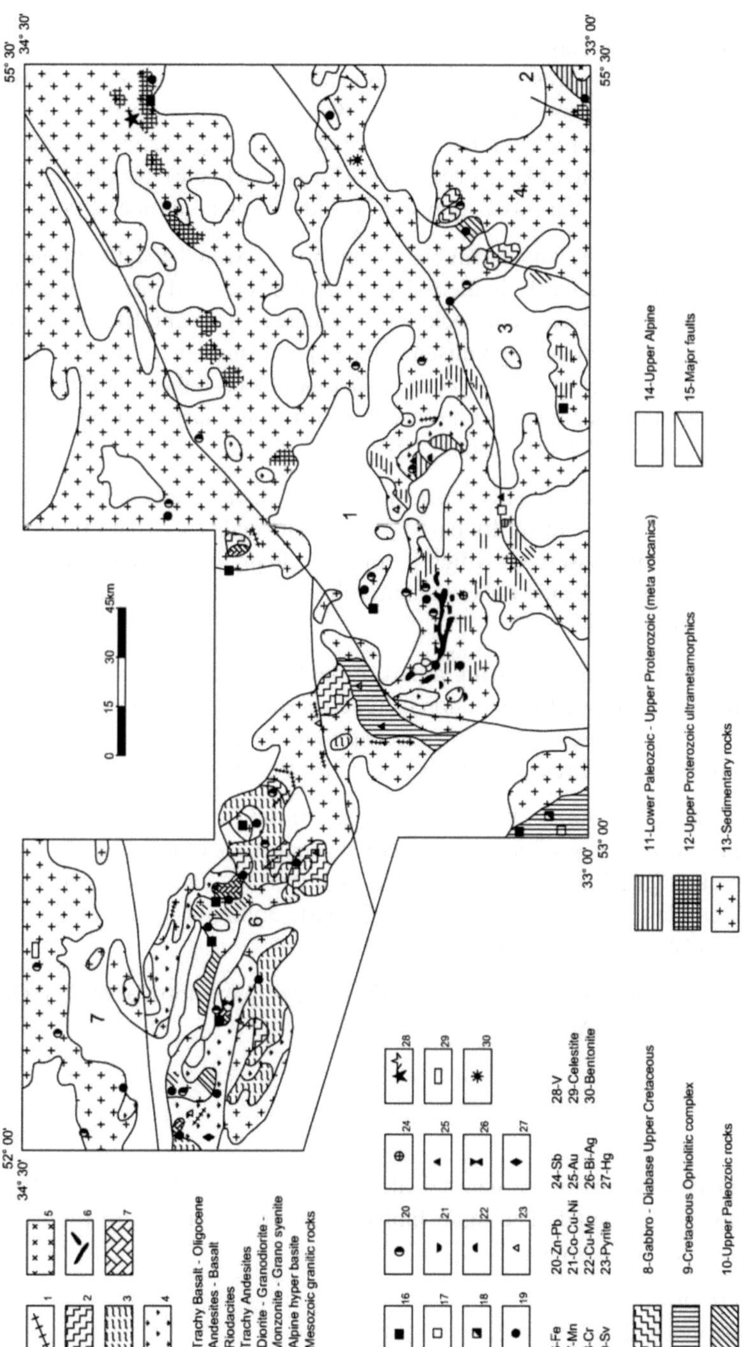

Fig. 6.4 Schematic view of magmatic and mineralization events in Anarak area (Metallogenic Map of Anarak 1984)

Anarak–Khor Block: This oval-shaped block is extended from the northeast to the southwest measuring 250 km in length and 60 km in breadth. The Biyabanak folded zone and Chah Palang–Bayazeh block with Cimmerian and Laramide age form its southern limit. On the west the block is bounded by the Naein–Zavar ophiolite band, while it terminates into the Oligocene–Miocene molasses of the Lut Desert northward. The basement rocks of this block are the greenschist facies metamorphic rocks (epidote amphibolite of Late Proterozoic–Early Paleozoic), which were originally graywacke, arkose, sandstone, claystones, and basic and ultrabasic volcanics.

Posht Badam Block: This unit is highly metamorphosed and locally turned into migmatized containing smaller bodies of granitic gneisses and anatactic granites. The lower parts of the basement rocks are truncated by Posht Badam complex (which is metamorphosed up to greenschist facies) and small metagabbroic bodies. The original rocks of the area have been sandstone, meta-diabase, schist, marble, dolomite, and holnfels in contact with Mesozoic intrusions (varying in age from 140 to 40 Ma) (Aghanabati 2004; Ghorbani 2012a).

Chah Palang–Bayazeh Block: This is a very complex block bounded on its south by the Anarak–Khor block and on its east by the Biyabanak zone. The block has a 200-m thick Baikalian basement which is folded in many phases and is composed of marble, schist, and volcanic rocks metamorphosed in greenschist facies (epidote amphibolite) of Late Proterozoic.

Naein–Zavar Ophiolite Block: Being a part of the Sabzevar–Naein ophiolite zone, this block has been formed by extrusion of igneous rocks and deposition of marine sediments during the Late Cretaceous–Eocene times and later on deformed due to the closure of the basin. Presently, the block exhibits a colored mélange structure.

Kuh Dom–Qale Sardar Block: Kuh Dom is a part of the Central Iran volcanic belt, and more than 80% of its surface is covered by volcanic rock of Eocene with varying basement rocks. Qale Sardar overlies folded metamorphic basement, which is not exposed. The oldest outcrops of the area are Triassic dolomites and Jurassic clayey shales and sandstones (Shemshak Formation).

Late Alpine Embayment Block: This block covers areas to the north of the Lut Desert and southwest of Qom–Ardakan.

Faults

The importance of faults in the geological framework of the Anarak metallogenic zones cannot be neglected. The main faults of the area that control the trend of mineralization are as follows:

1. Lut desert fault
2. Baba Bozorgi fault
3. Torkamani fault
4. Bayazeh fault
5. Chapdoni fault

Geology of Anarak

The rocks of the Anarak area, also known as Anarak metamorphics, are a part of Iran basement complex dating back to Late Proterozoic–Early Cambrian. The term Anarak Metamorphics was introduced by Davoudzadeh (1969) for these rocks and later on a Russian group used it extensively for the metamorphic rock exposures of Anarak and Khor regions. The results of investigations carried out by Russian geologists in Anarak 1:250,000 geological quadrangle show the presence of five different complexes, namely (Table 6.1),

1. Chah Gorbeh complex
2. Morghab complex
3. Patyar complex
4. Mohammad Abad complex
5. Doshakh complex

Moreover, within the Khor region, the metamorphic equivalents of Anarak are also divisible into five complexes, namely (Table 6.2):

1. Chah Gorbeh complex
2. Patyar complex
3. Kaboodan complex
4. Doshakh complex
5. Posht Badam complex

The Kaboodan complex is itself divisible into four units, which in ascending order are schist (equivalent of upper parts of Chah Gorbeh), marble (similar to Lakh Marble of Anarak), schist–marble, and quartzite (both equivalent to Patyar complex). Moreover, the Chah Gorbeh complex is found to be equivalent to Mohammad Abad, whereas Doshakh is part equivalent to Chah Gorbeh and Patyar. Thus, Chah Gorbeh forms the lowermost section of Anarak metamorphics and Kaboodan can be correlated with the upper parts of Chah Gorbeh, Lakh marble, and Patyar complex. Further, Doshakh is part equivalent to all of the above-mentioned complexes. Altogether, therefore, only seven complexes exist in both Anarak and Khor regions (Table 6.3).

Chah Gorbeh Complex: This complex comprises of five members, which in ascending order are as follows:

1. Sebarz schist
2. Lower marble
3. Lower schist
4. Middle marble
5. Upper schist

The Chah Gorbeh complex is equivalent to parts of Soltanieh Formation in northern Iran and Rizoo Series of Kerman Province.

Sebarz Schist: This member forms the lowermost parts of Chah Gorbeh complex and is dominantly composed of pale green muscovite schists. These are of Late Proterozoic age and have an average thickness of 1,000 m.

Table 6.1 Late Proterozoic–Early Paleozoic stratigraphic column of Anarak Region

Early Cambrian	Marble; dolomite	1,000 m	600 m		Doshakh complex	Lakh marble		Anarak metamorphics
Late Proterozoic	Chlorite–Muscovite and Chlorite–Epidote–Actinolite schist; quartzite; marble; dolomite		4,100 m	1,800 2,000		Mohammad Abad complex	Patyar complex Morghab complex	
	Muscovite; Muscovite–Chlorite and Muscovite–Epidote–Chlorite schist; quartzite; marble lenses			3,100 3,800				
	Muscovite–Shlorite, Epidote–Chlorite, Epidote–Actinolite, and Muscovite–Carbonate schist; marble			1,600 2,100		Chah Gorbeh complex		
	Muscovite Granite; Granite Gneiss; Amphibole Micaschist and Micagneiss	950 m			Pol Khavand			

Almasian (1998)

Table 6.2 Late Proterozoic–Early Paleozoic stratigraphic column of Khor Region

Ordovician	Metamorphosed sandstone; phyllite; dolomite; schist; metamorphosed volcanics	Posht Badam complex	Doshakh complex	Kaboodan complex	Patyar complex	Anarak metamorphics
Cambrian	Metamorphosed dolomite, limestone, and sandstone					
Late Proterozoic	Schist; marble			Chah Gorbeh complex		
				Jondagh complex		
	Gneiss; anatactic migmatite; amphibolite	Chapdony complex				

Aistov et al. (1984)

Table 6.3 Stratigraphic units of Anarak Region

Morghab unit		Muscovite, muscovite–chlorite, epidote–chlorite, chlorite–epidote–amphibolite schists
Patyar unit		Chlorite–muscovite, chlorite–epidote, actinolite, epidote–albite–amphibole, chlorite–epidote–emphibole schist; dark brown dolomitic marble; black quartzite; quartz schist; chert intercalations
Lakh marble unit		Thinly-bedded white marble; carbonate schist; schist; boudinage dolomite; dolomitized limestone; solution, collapse, and fault breccias
Chah Gorbeh complex	Upper schist member	Epidote, tourmaline, albite, muscovite–epidote–glaucophane, chlorite, muscovite–epidote–albite schist
	Middle marble member	Metasomatic schists with dolomite and marble intercalations
	Lower schist member	Muscovite–albite–epidote, sericite–chlorite, carbonate schists with small lenses of marble and quartzite
	Lower marble member	Metasomatic schist with marble intercalations
	Sebarz schist member	Muscovite, muscovite–chlorite, muscovite–carbonate schist with small lenses of marble, quartzite, and ultramafics
Anarak ophiolites		Serpentinite; serpentinized peridotite; gabbro; diabase; small lenses of listhonite and rodnegite; pillow lavas; schist intercalations in the uppermost parts

Almasian (1998)

Lower Marble: Having been formed during Late Proterozoic, it consists of alternating layers of gray marbles and micaceous schists reaching a total thickness of 700 m. The marbles show slumping in the form of recumbent concordant folds.

Lower Schist: This member is dominantly composed of muscovite, chlorite, biotite, epidote, and carbonate schists with intercalations of quartzite and marble. The thickness of the lower schist member varies between 500 and 600 m and is of the Late Proterozoic age.

Middle Marble: Having been formed in the Early Cambrian times, the middle marble is made up of gray-layered marbles and schists that measure around 800 m in thickness.

Upper Schist: Epidote, chlorite, biotite, and albite schists with intercalations of marble and quartzite constitute the upper schists of the Chah Gorbeh complex. The thickness of this member at the core of Lakh Mount is about 1,500 m and is equivalent to the upper shale member of the Soltanieh Formation.

Lakh Marble Unit: Overlying the Chah Gorbeh complex is the most prominent lithological unit of Central Iran, that is, Lakh marble. This unit is equivalent to the upper dolomite member of Soltanieh in Alborz (Hamdi 1989), the keybed of the Saghand Region (Haghipour 1974), and the uppermost marble of the Rizoo Series (Hokride et al. 1964).

Patyar Unit: The most complete section of this unit is located on the east of the abandoned antimony mine near Patyar. This unit is composed of chlorite, epidote, mica, and glaucophane schists along with dolomite, quartzite, and cherty dolomites. Stratigraphically, it is of Early Cambrian age and equivalent to the Baroot Formation of northern Iran and parts of the Dezoo Series of Kerman (Hokride et al. 1964)

Morghab Unit: The upper unit of Anarak Metamorphics is the Morghab Unit described from the outcrops on the north of Moala village. The rocks constituting this unit are dominantly chlorite, epidote, and mica-schists with numerous quartz veins. It overlies the topmost carbonaceous rocks of Patyar unit and is of the Early Cambrian age.

Paleozoic Rocks

The rocks of the Paleozoic Era are widely exposed on the southeast of the Torkamani–Ordib fault. A complete succession of Paleozoic rocks ranging in age from the Ordovician to the Permian outcrop around Abdolhossein Mount, which are comparable with the Tabas area. It should be mentioned that the rocks of the Shirgesht Formation lie over the metamorphics of Anarak with an angular unconformity.

Shirgesht Formation (Ordovican): Greenish grey sandstone, conglomerate, tuff, and ignimberite with limestone intercalations in its lower parts.

Niyoor Formation (Silurian): White and red sandstones, diabase, quartzitic lenses, and dolomitized limestone.

Padoha Formation (Lower Devonian): Diabase, quartzite, and dolomitic limestones.

Sibzar Dolomite (Middle Devonian): Dolomite and dolomitic limestone.

Bahram Limestone (Upper Devonian): Dominantly composed of limestone and dolomitized limestone with marly intercalations.

Shishtoo Formation (Carboniferous): Mainly consists of limestone, sandstone, siltstone, and shale.

Sardar Formation (Carboniferous): Dark limestone and shale with limey intercalations.

Jamal Formation (Permian): Limestones.

Mesozoic Rocks

The Permo–Triassic rocks outcropping to the south of Abdolhossein Mount correlate well with the Shotori Formation of Tabas. The Chah Palang Formation

(Jurassic–Cretaceous) has limited exposures in and around the abandoned mines at Torkamani and is dominantly composed of sandstones and fine-grained conglomerates.

The Cretaceous rocks of the Anarak–Khor region are the most prominent Mesozoic rocks with considerable exposure in the Khor area. Overall, seven formations of Cretaceous age have been recognized that can be grouped into two categories, namely, Lower and Upper Cretaceous. The Lower Cretaceous succession consists of orbitoline limestone, sandstone, marl, and conglomerate while the Upper Cretaceous sequence is made up of limestone, sandstone, and conglomerate.

Cenozoic Rocks

Having considerable exposures in the Anarak area, the Cenozoic succession is mainly clastic in nature and has substantial exposures. It can be divided into the following:

1. Kerman conglomerate (Paleocene): Conglomerates, sandstones, and occasional limestone layers.
2. E_1a unit (Lower Eocene): Trachyandesites, trachytes, shaushonite, tuffaceous andesite, tuffite, and sandstone.
3. E_2a unit (Lower Eocene): Andesites, tuffs, and andesite–basalts.
4. Sahlab Formation (Middle and Upper Eocene): Conglomerate, marl, and sandstone.
5. E_0 Unit (Upper Eocene–Lower Oligocene): Sandstone, marl, and conglomerate.
6. Lower Red Formation (Oligocene): divided into three sections, namely:

   (a) O_1c made up of conglomerate and sandstone.
   (b) O_1sm made up of sandstone, marl, and conglomerate.
   (c) O_1m made up of sandstone, marl, and limestone.

Quaternary Rocks

The rocks of this time interval can be grouped into three main categories as follows:

1. Ancient Quaternary: Alluvial and fluvial sediments consisting of gravel, sand (sometimes accompanied with chalk), coarse sand, and conglomerate.
2. Young Quaternary: Alluvial and fluvial sediments consisting of gravel, sand, clayey sand, conglomerate, lake, and Aeolian sand.
3. Recent Sediments: Fluvial sediments, gravel, clayey sand, salty clayey sand, aeolian sand, and conglomerate.

Igneous Activity

Widespread igneous activity accompanied by large-scale mineralization is seen throughout the Anarak area (Table 6.4).

Table 6.4 Major phases of igneous activity and mineralization in Anarak

Igneous phase		Rock types	Age	Main elements and minerals
Middle Alpine		A series of volcanic rocks along with their plutonic associates (including andesite–basalts, rhyolite–trachyte, acidic, and basic intrusives)	Paleogene	Cu, Au, Zn, Fe, Ag, Bi, Ni, FeS_2, Barite, Alunite, Bentonite, Te, Mg
Laramide	Early Alpine	Ophiolitic suite	Early Cretaceous	FeS_2, Cu, Mn
		Acidic to intermediate intrusives	Triassic–Jurassic	Zn, Pb, Cu
Caledonian–Hercynian		Volcanosedimentary rocks	Paleozoic	No mineralization
Pan-African		Metamorphosed mafic and ultramafic rocks along with migmatites and anatactic granites	Early Cambrian–Late Proterozoic	Pb, Zn, Fe; high concentration of Sb, Ni, Co, Mn, Ag, Au, and Bi

Though there are traces of igneous activity throughout Late Precambrian–Neogene, the major phases can be categorized as follows.

Late Proterozoic–Early Cambrian Phase: This phase resulted in the formation of mafics, ultramafics, basic tuffs, and anatactic granites and is most probably associated with the continental rifting during Late Precambrian–Early Cambrian.

Mesozoic Phase: The Mesozoic igneous rocks of Anarak can be grouped into two categories:

1. Acidic to intermediate intrusive rock (Triassic–Jurassic) with a compositional range of granodiorite–granite, syno–granite, quartz diorite, and plagiogranite.
2. Igneous rocks belonging to ophiolite suites (Cretaceous) including dunite, herzburgite, plagiogranite, spillite, cratophyre, basalt, diabase, rhodengite. These are emplaced in the Naein–Zavar block and are associated with chromite, manganese, and copper mineralization.

Cenozoic Phase: The rocks of this phase have been dominantly formed in Eocene times and constitute the majority of the igneous rocks of the Anarak area. They include extrusive (andesite, basalt, trachyte, trachyandesite, and rhyolite) and intrusive (monzonite, grano–synite, diorite, and granodiorite) igneous rocks. Extensive mineralization of copper, molybdenum, nickel cobalt, bismuth, lead–zinc, gold, etc., has taken place along with the Cenozoic phase of igneous activity.

Mineralization in Anarak Metallogenic Province

Anarak is one of the richest metallogenic provinces of Iran from the point of view of the diversity of mineral species and shows well-defined metallogenic zones (Fig. 6.5). Moreover, it is one of the oldest mining localities of Iran, and modern mining operations began operating from this province.

Mineralization in the Anarak area from the point of view of mineral types and their origins, habits, and paragenesis is much varied, so that a large number of iron, copper, molybdenum, antimony, gold, lead–zinc, arsenic, nickel, cobalt, manganese, chromite, and strontium mineral bodies have been observed till date, which all have different modes of origin such as hydrothermal, sedimentary, magmatic, metamorphic, etc.

Copper Mineralization

It is a widespread mineralization event in the area. Apart from the ore deposits that are specific copper minerals, many other polymetallic and complex deposits (e.g., Cu–Mo, Cu–Ni–Co) have also been reported that can be categorized under hydrothermal deposits into plutogenic, volcanogenic, metasomatic, telethermal, and skarn as follows.

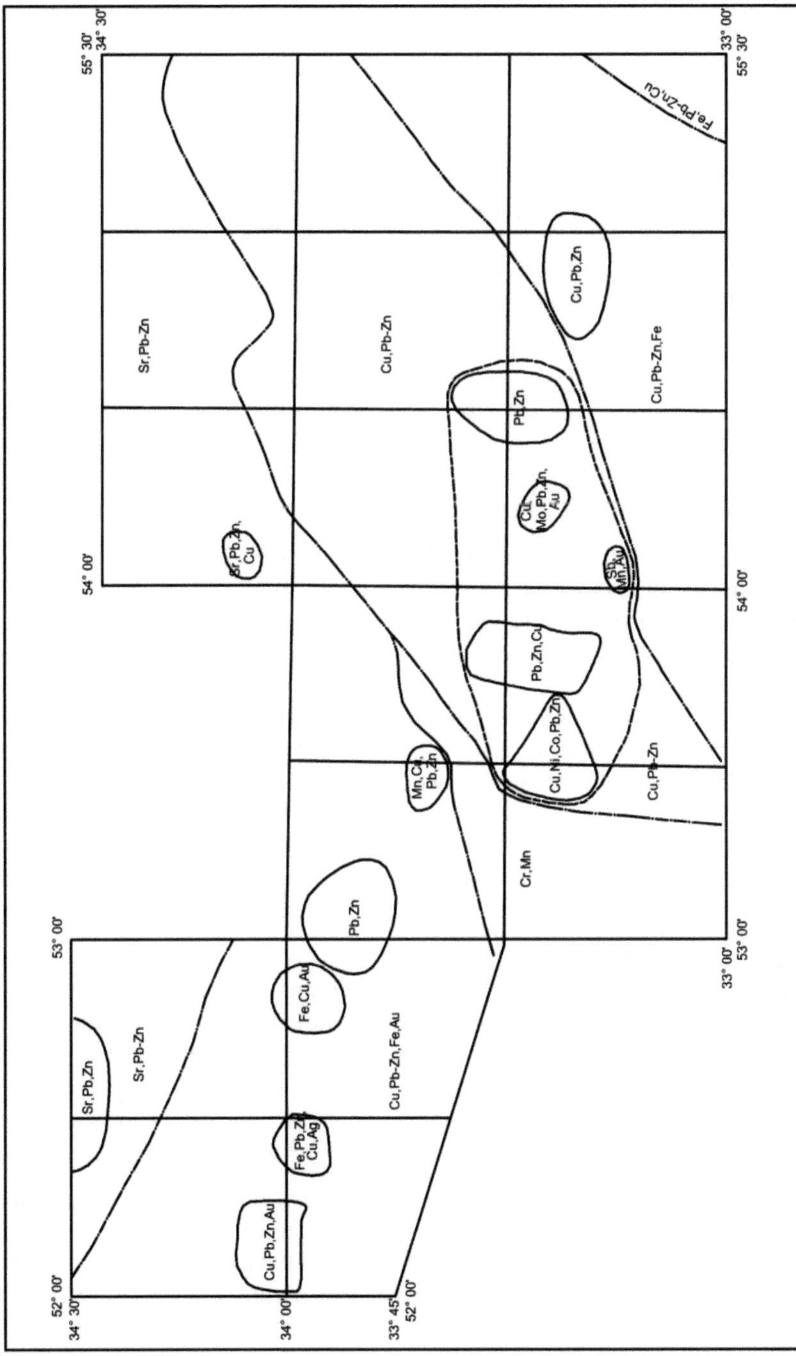

Fig. 6.5 Metallogenic zoning of Anarak area (Technoexport 1984)

Plutogenic Deposits: These deposits are usually associated with Tertiary magmatic activity, for example, Kuh Dom, Kal Kafi, Khonj, and Tale Siyah intrusives. Copper mineralization of this type occurs in granitoids, metamorphic aureoles, and rocks situated at a definite distance from the plutons in the form of mineral-bearing metasomatic zones and fault-associated veins.

Metasomatic Deposits: The type of copper deposition usually occurs in parallel with metasomatism of the host rock. The veins are covered with a thin layer of metasomatic material on their margins and show scattered mineralization, while the veinlets have their ores in the form of layers or masses. The most common type of ore mineral is chalcopyrite while bornite, chalcocite, covellite, native copper, galena, sphalerite, molybdenite, and gold occur on its side.

Volcanogenic Deposits: Some of the volcanic areas of Eocene time are important from the point of view of copper mineralization. Most of these areas have their ore deposits limited to volcanic rocks and their associated fault system. The ore deposits are in the form of stockworks and veins, the former being accompanied by host rock metasomatism, thus producing a compositionally uncomplicated ore with chalcocite, chalcopyrite, covellite, bornite, and native copper along with some oxide minerals.

Telethermal Deposits: This type of mineralization mostly occurs within the Anarak area proper. The major telethermal copper deposits are formed in the Anarak–Khor block in the basement or its overlying Oligocene–Eocene strata. Nevertheless, the telethermal copper have also been reported from the fault zones of Kuh Dom volcanic rocks.

The distinction between telethermal and volcanogenic mineralization is not an easy task, but the former are usually vein-type deposits showing stratiform and linear mineral zone characteristics, whose host rocks have not been metasomatized much.

The ore accumulations mostly contain pyrite, chalcopyrite, and chalcocite. The cobalt content of these deposits varies from 0.08 to 10 while silver and lead show nominal concentration and zinc, arsenic, manganese, and strontium are locally present.

Copper–Molybdenum Mineralization: Like copper mineralization, deposition of copper–molybdenum is associated with Eocene granitic intrusions, but is limited to the Kal Kafi area. The host rocks of Kal Kafi are schists and marbles of Late Proterozoic. Veins of this type have been reported from the junction of two fault system occurring within aplitic–granitic dikes and porphyritic granodiorites where the surrounding rocks have undergone hydrothermal metasomatism.

Lead–Zinc Mineralization: The mineralization of lead and zinc was very widespread in the Anarak area where the deposits of these two metals are morphologically and mineralogically very diverse. Genetically speaking, all the processes that have led to lead–zinc mineralization were hydrothermal in nature and can be grouped into three different groups, namely, plutogenic, volcanogenic, and telethermal. Nevertheless, minor skarn processes of lead–zinc mineralization have also been observed.

All the ore deposits possess varying amounts of lead–zinc and silver with occasional copper. Based on this characteristic, the ore bodies have been classified into the following:

1. Group A: with high lead content (Pb:Zn is around 1:0.1–0.2), for example, Rizab Maryam, Aroosan, Zeh, Namak, Gorgab III, Bandkal II, etc.
2. Group B: with equal amount of lead and zinc, for example, Chah Mileh, Negin, Goorche, Berenj II, etc.
3. Group C: with subordinate lead content (Pb:Zn is around 1:3–1:12), for example, God, Pis Kuh, and Chah Gorbeh.

In all the above-mentioned groups, the silver content varies from 1 to 100 g per ton of lead–zinc ore.

Iron Mineralization: The ore deposits and indication of iron within the Anarak Metallogenic Province have been attributed to hydrothermal, volcanogenic, and skarn process and have been the center of attention in the past due to the high manganese content.

The hydrothermal iron occurs in small deposits in the form of short veins of quartz, hematite and occasionally siderite. On the other hand, the volcanic rocks of the Eocene times host the volcanogenic irons in their crushed zones. The skarn-type iron is seen along the outer contact of Eocene granitoids and is of contact metasomatic magnetite type.

The largest skarn iron is North Chah Palangi, which occurs within Upper Proterozoic–Lower Paleozoic rock successions and is metamorphosed.

Chromium Mineralization: Small scattered chromite bodies have been reported from the Naein–Zavar block within the dunite herzburgite rocks of this metallogenic province.

Manganese Mineralization: Occurring in the form of small ore deposits and indications, the manganese deposits of Anarak are dominantly of hydrothermal type. Mineralization (both carbonaceous and oxide type) has resulted in the formation of carbonate ore deposits of iron and manganese, which include manganese-bearing ankerite and siderite and metasomatically altered dolomites. The manganese content of these rocks varies between 2 and 4 while that of iron ranges from 15 to 20.

Oxide ores with high manganese content occur in the form of veinlets with their largest measuring 50 m in length and 0.4–3.5 m in thickness. They consist of pyrolusite (up to 35), psilomelane (45), other manganese minerals (10), and quartz (10) with an average manganese content of 23.1. The total amount of oxide ore of manganese is estimated at 15,000 tons and that of carbonaceous iron–manganese at ten million tons.

Sedimentary manganese deposits are reported from the western parts of the Anarak Province in the base of the Oligocene–Miocene limestones of Qom Formation, where two manganese-bearing strata have been observed. In the lower stratum, manganese mineralization is in the form of thin lenses of pyrolusite within the conglomerates of Qom Formation. The thickness of ore layer varies from 1.5 to 3 m while its length is about 1.4 km. The upper horizon that measures 1–1.5 m in thickness and extends for 1.8 km occurs in the top 20 m of the organic limestone layer and is in the form of mixed pyrolusite–pyrite nodules. Locally, veins of pyrolusite are also seen in the limestones. The manganese content of the upper manganese stratum is estimated between 0.7 and 1.2.

Polymetallic Mineralization: Another characteristic of Anarak Metallogenic Province is the existence of polymetallic mineralization of the following types:

- *Pyrite Mineralization along with Ag, Au, Mo, Zn, Pb, and Cu*: Being of volcanogenic-hydrothermal origin, this type of mineralization is reported from trachyandesites, andesite basalts, and occasionally volcanosedimentary dacites. The only exception is that of Chah Sefid, which is associated with Upper Cretaceous cratophyres.
- *Polymetallic Stratiform Mineralization along with Cu, Pb, Zn, Au, Ag, and Sr*: This type of polymetallic mineralization occurs in Lower to Middle Eocene and Eocene–Oligocene volcanosedimentary rocks located on the southern extension of rupture on North Anarak Mountains. Mineralization is limited within thinly-bedded sandstone, limestone, and sometimes conglomerate.

Molybdenum Mineralization: Occasional molybdenum mineralization has been reported from the outer contacts of Kal Kafi intrusive and Eocene volcanics on the western parts of Anarak Province. The metal occurs in high concentrations within silicified fault zones that bear minor sulfides.

Tungsten Mineralization: Chah Palang is the only ore deposit that bears tungsten. It is in fact a copper–tungsten ore body placed within highly-deformed clayey–sandy rocks of Triassic–Jurassic (Shemshak Formation). Minor quantities of scheelite also occurs in the eastern contact of Kal Kafi intrusive (skarn zones) and Kuh Dom (quartz–scheelite veins).

Bismuth Mineralization: Various amounts of bismuth minerals are reported from a number of polymetallic ore deposits such as Tal Siyah, Kuh Dom, Khooni, Kal Kafi, Gorgab IV, Chah Palang, etc.

At the outer margin of Kal Kafi intrusive, weak bismuth mineralization is seen along with molybdenum, gold, tungsten, and copper. Bismuth occurs in the form of high-grade aureoles at the center and the western parts of the Kuh Dom volcanic zone near the Gorgab intrusive body. In the pyrite mineralization zone of Eocene volcanic rocks of Kuh Dom, Gorgab and Kuh Yakhab, scattered anomalies of bismuth, copper, lead–zinc, molybdenum, and silver exist while at Chah Ali Khan gold–quartz veins, a copper–bismuth anomaly has been reported. High concentration of bismuth (along with silver, antimony, lead, and zinc) also occurs near the granitic gneiss intrusives of the Torkamani fault zone.

Mercury Mineralization: Minor mercury mineralization has taken place along large fault structures; silicification along the intersection of north–south and east–west trending faults is accompanied by 0.001 concentration of mercury in bands measuring 2–200 m in width and 0.5 km in length.

Antimony Mineralization: Occurring on the east of Anarak, Patyar, and Torkamani deposits that have been mined in the past show trends of antimony mineralization. Both these deposits possess antimony–quartz of telethermal origin.

Strontium Mineralization: Celestine occurs in some of the ore deposits and indications of the northern parts of Anarak Metallogenic Province. Celestine mineralization is limited to the evaporitic limestones in the lower parts of Qom Formation (Oligocene–Miocene). In spite of lithological and stratigraphical limitation, celestine seems to be of epigenetic (telethermal) origin, formed due to pore filling and metasomatic replacement of carbonate rock on the basis of habit, texture, and structure.

Gold Mineralization: Extensive gold mineralization is believed to have occurred in Anarak with the most prominent cases occurring within quartz veins, granitoid bodies (Kuh Dom, Eastern Khooni), Eocene intrusives and extrusives (Chah Alikhan), and serpentinized hyperbasalts (Sebarz fault zone). Apart from these, gold also occurs in the form of characteristic indications (gold content varying between 1 and 4.2 ppm, occasionally reaching 26 ppm) within polymetallic Khooni, Gorgab II, and Kal Kafi deposits. Moreover, small amount of gold (around 1 ppm) is reported from Cu–Ni–Co polymetallic deposits (God Morad, Sebarz), copper deposits (Douma, Tal Siyah), pyrite (Gorgab, Mazra'e, Zavar), lead–zinc deposits (Gourche, Berenj I and II, Zeh, Goud), antimony deposits (Torkamani), tungsten (Chah Palang), and other minerals.

Silver Mineralization: Silver is found in varying amounts (from a few grams/ton to more than a kilogram/ton) in the polymetallic and lead–zinc deposits. There is a definite order in the occurrence of silver within various ore deposits so that the maximum amount of it (1 kg/ton) is reported from the Band Gol II, Aroosan, Nakhlak, and Zeh ore deposits.

Mineralization of Nonmetals: Small deposits and occurrences of nonmetallic minerals have been reported from Anarak Metallogenic Province. To name a few, muscovite, hydromagnesite, montmorillonite, gypsum, anhydrite, halite, and turquoise. Most of these have been excavated in the past and at present limited mining activity is undergoing in these areas.

6.3.1.3 Takab Metallogenic Area*

In two separate time intervals, one during the Late Precambrian–Early Cambrian and the other at Tertiary (Neogene), the prevailing geological conditions, namely, tectonic, magmatic, metamorphic, stratigraphic, and mineralogic characteristics, have made Takab quadrangle one of the most important metallogenic provinces of the country with no compare.

General Geology of Takab Region

Lithostratigraphy: Igneous, sedimentary, and metamorphic rocks ranging in age from Late Precambrian to Quaternary outcrop in the Takab Region. The spatial

*The whole section is from Ghorbani 1999f

and temporal distribution of these rocks is not uniform. Stratigraphically, rocks belonging to Late Precambrian–Early Cambrian and Oligocene–Miocene intervals of time are more extensive in terms of areal distribution and lithological variations.

Alavi and Bolourchi (1982) have divided Precambrian rocks of Takab quadrangle into three categories:

1. Metamorphic rocks including amphibolite, marble, mica-schist migmatite of greenschist and amphibolite facies underlying lowly metamorphosed Kahar Formation, forming the basement complex and being exposed in Azerbaijan, are considered as the oldest rocks of Iran.
2. Low-grade metamorphics of clastic and tufaceous origin named Kahar Formation are widespread on the east and west of Takab quadrangle.
3. Doran granites.

The results of investigations reveal that the ages assigned to rocks considered older than Kahar Formation are just based on their degree of metamorphism and not on the fossil content or radiometric dating. The paleontological evidence presented by Hamdi (1995) and the field observations by the present author (1999) indicate that these rock are not older than Kahar Formation and in places are equivalent to topmost beds of Kahar. Moreover, rocks formerly thought to belong to Late Precambrian–Early Cambrian Doran Granites are not everywhere the same; only some outcrops resemble Doran type lithologically but are much younger in age. Therefore, rocks belonging to Kahar Formation along with those of Qareh Dash Series, Doran Granite, and Bayandor Formation are all of Precambrian age.

Additionally, there are problematic rock exposures in the central Takab Region of Late Precambrian–Early Cambrian age that cannot be differentiated readily and hence are referred to as Upper Precambrian–Lower Cambrian rock.

Soltanieh Formation: This formation is very widespread in Takab quadrangle and conformably overlies Kahar Formation. In the central parts of the quadrangle, Soltanieh Formation is traceable but with a different facies. In fact, all the marbles marked on the 1:250,000 map of Takab belong to Soltanieh Formation. Based on their fossil content, an Early Cambrian age is assigned to these rocks.

Baroot Formation: The exposures of Baroot Formation in Takab quadrangle are limited to the northeast of Shahin Dezh only.

Takab Complex Series: The rocks belonging to this series are confined to the central part of the quadrangle map. Alavi and Bolourchi (1982) have described the present series as comprising of four complexes:

1. Sursat Mount complex with mica-schist, gneiss, migmatite, and quartz-schist rocks, being exposed on the northwest of Takab and east of Shahin Dezh around Zikandi, Khangholi, and Khazaei villages.
2. Mahneshan complex comprising of Poshtkuh marble and Aghkand gneiss on the east of Mahneshan.

3. Kheirabad complex comprising of Angooran gneisses, schists, and marbles occuring near Kheiabad village.
4. Amirabad complex with gneiss, migmatite, and mica-schist rocks around Amirabad village.

The last three complexes are classified by Ghorbani (1999f) within one Great Precambrian–Cambrian Takab complex. He further considered the whole structure as being confined to the central part of the quadrangle calling it the central great complex of Takab quadrangle. As stated earlier, these rocks (marked "PCM" in Fig. 6.6) have been regarded as Precambrian rocks older than Kahar in age but recent work by Ghorbani (1999f) revealed that they are equivalents of Upper Kahar, Soltanieh, and Baroot Formations.

It must be noted that the rocks of Takab complex are only age-equivalents of Upper Kahar, Soltanieh, and Baroot, but differ greatly with the said formations in Alborz and Azerbaijan, taking into account facies, lithology, and degree of metamorphism. However, there exist certain similarities between them and the exposures in Anarak Region.

The rocks of Takab complex play an important role in interpretation of metallogenic characteristics of the area as well as elucidating the tectonic and magmatic conditions prevailed during Late Precambrian–Early Cambrian times in Iran and can be classified as follows:

1. Rocks with ultramafic, mafic, and volcanoclastic (basic tuffs) origin that are seen today as schists, serpentinite, amphibole–epidote schist, chlorite schist, epidote–amphibole schist, epidote–actinolite schist, amphibolite, actinolite–tremolite amphibolite, garnet amphibolite, and orthogneiss. Wherever they occur, these rocks are all important from the metallogenic point of view.
2. Rocks of sedimentary origin, mostly of pelitic and carbonaceous type, that appear within carbonate–pelitic and graywacke sequences.

Most of the rocks of the above-mentioned groups have metamorphosed within greenschist and amphibolite facies. Rocks of the first group mostly underlie rocks of the second group with a gradational contact so that at places the topmost layers of the first group have carbonaceous intercalations.

Zaigoon and Laloon Formations: These rocks occur mostly to the north of Shahin Dezh. Wherever the rocks of Zaigoon and Laloon Formations (especially the latter) crop out, they are similar to the exposures of Alborz and Azerbaijan.

Upper Paleozoic

The only Upper Paleozoic rocks that are exposed within the quadrangle map belong to Permian having two different facies, namely, clastic (Dorood Sandstone Formation) and carbonaceous (Rooteh Limestone Formation). A suite of igneous rocks are associated with these rocks everywhere. The most important characteristic of the Upper Paleozoic rocks of the area is the presence of a fireclay horizon (Permo–Triassic in age).

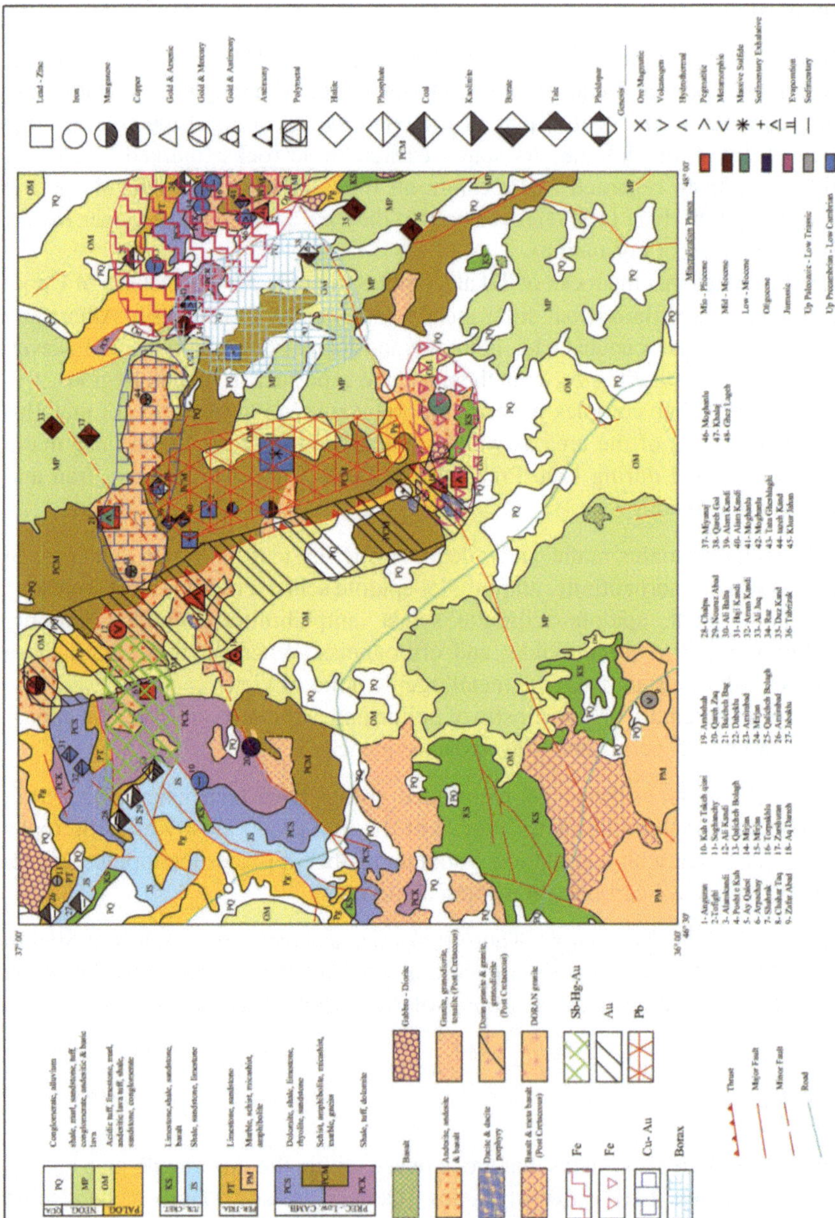

Fig. 6.6 Metallogenic map of Takab area

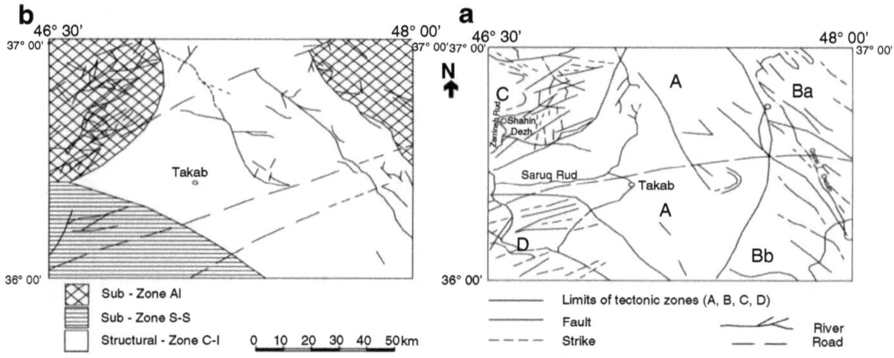

Fig. 6.7 Structural units in Takab region (**a**) by Alavi and Bolourchi (1982); (**b**) by Ghorbani (1999f)

Mesozoic

Similar to the above-mentioned category, the Mesozoic rocks of Takab quadrangle area crop out dominantly on the east, that is, to the north and south of Shahin Dezh. Mesozoic rocks also occur on the east of Mahnehsan where they overlie the Permians. However, they are absent in the central parts of the quadrangle.

Tertiary

The overall effects of stratigraphic phenomena, facies conditions, and tectonic forces during Tertiary could be grouped into three different intervals.

1. Paleocene–Eocene rocks of Shahin Dezh area that resemble to those of south central Alborz lithologically.
2. Oligocene–Miocene rocks whose outcrops come next to Precambrian–Cambrian rocks area-wise. In the central parts of the quadrangle, they are more extensive than Precambrian–Cambrian rocks and overlie the latter. Lithologically, lavas, tuffs, and tuffaceous rocks dominate over sedimentary rock successions.
3. Pliocene–Quaternary rocks that have considerable extension over the area of the quadrangle overlie the older rocks with an angular unconformity.

Structural Geology

Takab quadrangle is very interesting from the structural point of view. Geologically and structurally, parts of the quadrangle map are closely associated with Central Iran, while others closely resemble Alborz and Sanandaj–Sirjan. Major structural features such as horsts, grabens, unconformities, thrusts, and orogenic phases (especially younger Alpine phases) are visible in the area.

Alavi and Bolourchi (1982) tectonically divided Takab quadrangle into five zones denoted by A, C, Ba, Bb, and D as seen in Fig. 6.7a. The description of each zone is discussed in detail in the report accompanying the map of the area. The present author believes that there are no structural and geological differences between Ba and C zones, or A and Bb zones. Thus, on the basis of stratigraphic, magmatic, sedimentary, structural, and overall tectonics, Takab quadrangle can only be divided into two different subzones and one zones as illustrated in Fig. 6.7b.

1. A_1 subzone, which is geologically very similar to Alborz and resembles it. This zone covers the northeastern and northwestern marginal parts of Takab quadrangle.
2. C-I zone, which shows all the geological characteristics of Central Iran. This zone covers the central parts of Takab quadrangle and forms the main metallogenic region.
3. S-S subzone, which possesses all the prominent aspects of Sanandaj–Sirjan zone and can be considered to be a part of it. It covers the southeastern part of the quadrangle.

Of the above-mentioned zones, C-I zone and S-S subzone, though basically different, have certain characters in common, but the A_1 subzone is totally different from the other two zones.

The faults within the quadrangle area can be divided into two groups, namely, those trending northwest–southeast, having formed in Precambrian times, which can be traced in an en echelon pattern outside Takab quadrangle; and those trending northeast–southwest perpendicular to the first group, having formed in Mesozoic times. The Tertiary metamorphism in the Takab area is closely associated with these two groups of faults especially with their points of intersection

Orogenic Phases

Most of the orogenic phases and tectonics upheavals that are reported from Central Iran can be traced in Takab quadrangle especially in the C-I zone. The most important deformation phases in the area are as follows.

Late Precambrian–Cambrian Orogenic Phase: An extensional phase of deformation has taken place in Takab during Late Precambrian whose effects can be traced up to Central Iran. This has resulted in the formation of a rift in Central Iran that was active in Late Precambrian–Cambrian and extended well into the C-I zone of Takab quadrangle. Subsequently, a compressional phase of deformation occurs, which results in northwest–southeast trending en echelon faults, thus reducing the width of the C-I zone. The compressional phase has risen the C-I zone in the form of a horst emerging it from under the sea. It has also produced metamorphism at regional scale within the area.

The Late Precambrian–Cambrian orogeny culminated before the deposition of Laloon sandstone. It is equivalent to the Pan-African orogeny that took place between 630 and 530 Ma worldwide.

Mesozoic (Cimmerian and Laramide) Orogenic Phase: This phase of deformation has dominantly affected the S-S and A_1 zones and includes the following processes:

1. Faulting and jointing with a northeast–southwest trend.
2. Unconformities and hiatuses within Mesozoic and Tertiary rocks.
3. Slate and greenschist facies metamorphism within S-S zone.

Tertiary Orogenic Phase: Contrary to the Mesozoic orogeny, the Tertiary phase has put its signature all over the Takab area, but is more evident in the C-I zone. For the sake of convenience, the Tertiary orogenic phase is divided into the following stages:

1. Paleocene orogeny, which is a continuation of the Laramide orogeny of Mesozoic Era, has resulted in the emergence of land areas within the S-S and A_1 zones.
2. Eocene–Early Oligocene orogeny, whose signature can be traced all over Takab quadrangle, is an extensional stage and corresponds to the extensional phase of deformation within Central Iran.
3. Late Oligocene–Early Miocene orogeny, associated with increasing depth of the sea due to extension and subsidence, thus replacing the red continental sediments with limey–marly successions. The sequence is composed of limestones intermingled with volcanics such as tuffs and tuffites extending up to the end of Burdigalian.
4. Therefore, the rocks corresponding to Qom Formation are mostly volcanic in origin. Tuffs and tuffites are dominantly seen within the C-I zone. However, the lavas outcrop along the northwest–southeast trending faults, especially at the junction of two sets of fractures.
5. Middle Miocene orogeny, which is reflected in the change of limestone facies to silty and clayey conglomeratic sandstones.
6. Late Miocene–Early Pliocene orogeny, being the most effectual tectonic phase within Takab quadrangle, is most evident in the C-I zone. It includes deformation and folding of older sediments (especially the Oligocene–Miocene rocks) producing the present morphological features. The angular unconformity between Oligocene–Miocene rocks and the overlying Pliocene conglomerates, extensive magmatic activities, formation of various ore deposits (gold, arsenic, antimony, iron, lead and zinc), a change from extension to compression mode of deformation resulting in faulting and overthrusting with northwest–southeast trend, and doming up of salt deposits of Middle Miocene age are all consequences of this stage of orogeny.

Igneous Activities

The Takab area has experienced widespread magmatic activity from Late Precambrian to Quaternary. During some intervals of time, for example, Late Precambrian–Early Cambrian and Oligocene–Miocene, the magmatic processes have dominated the geological phenomena acting in the area. The magmatic activities of the Takab area can be classified into the following.

Late Precambrian–Early Cambrian: The magmatic activity that took place at this time interval can be divided into two types: acidic magmatism with intrusives dominating over extrusive rock types, for example, Doran-type granites and Qare Dash Formation, and mafic–ultramafic magmatic activity with volcanism as the prevailing process. All the igneous activities of this time period within Takab quadrangle can be related to a rift on the basis of lithology, geochemistry, metallogeny, and field evidences. The rifting started in the Late Precambrian and ended by Early Cambrian also affecting Central Iran.

Magmatism and associated metamorphism has produced small bodies of granitoids (tonalites and alkaline granites) similar to Doran Granites. According to Ghorbani (1999f), the formation of such igneous bodies seems to be associated with metamorphic activities occurring at two time intervals: Late Precambrian–Early Cambrian corresponding to the closure of the rift, and Eocene–Oligocene related to a core complex.

Cretaceous: Scattered volcanic rocks belonging to Cretaceous times are seen throughout the Takab area.

Tertiary–Quaternary: With the extrusives prevailing over magmatics, the rocks range from olivine basalts to rhyolites. The volcanosedimentary rocks of Eocene times similar to Alborz Region are only seen around Shahindezh. However, extensive volcanic rocks including rhyolites, dacites, acidic tuffs, andesites, and basalts along with minor intrusions of gabbro and alkaline granites occur throughout Upper Oligocene to Lower Quaternary (Fig. 6.8) with two climaxes, one in Late Oligocene–Early Miocene (acid volcanics more than andesites) and other in Late Miocene–Pliocene. The latter can further be divided into basaltic–andesitic volcanics that start with dacitic tuffs and culminate into basic dikes, and acid plutonics in the form of small granitic to quartz porphyritic intrusives.

Metamorphic Activities

Extensive metamorphism occurs in Takab quadrangle. Excepting the Tertiaries, most of the rocks in the central parts of the region (especially Upper Cambrian–Lower Cambrian rocks) have experienced some form of metamorphism ranging from low to high grade.

The metamorphic rocks of Takab have been divided into five categories by Pelissier and Bolourchi (1967) and Alavi and Bolourchi (1982) which are as under:

1. Sursat Mount complex to the east of Shahindezh composed of mica-schist, gneiss, and migmatitic gneiss. Small granitic bodies are also seen.
2. Mahneshan complex on the northeast of Takab consisting of Aghkand gneiss, Poshtkuh schists, and Almaloo white marbles.
3. Kheirabad complex, which is further divided into the following:
   (a) Dava Yataghi gneiss made up of almandine-bearing layers with schist intercalations.
   (b) Alamkandi amphilolite composed of amphibolites, gneisses, and marbles.
   (c) Angooran schist including all types of schists, marbles, gneisses, and volcanics.

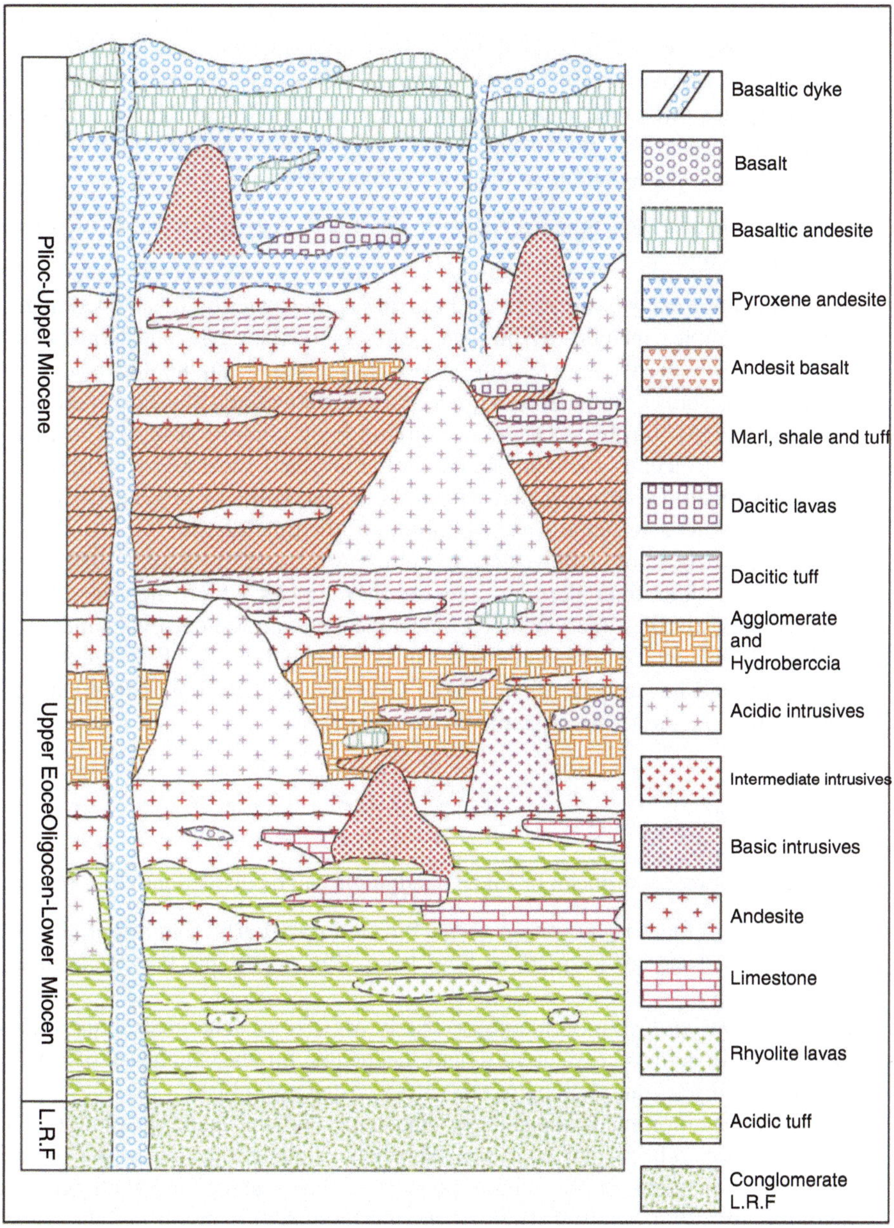

Fig. 6.8 A schematic view of Tertiary magmatism in the Takab area

4. Amirabad complex on the south of Dandi and south–southeast of Angooran mine consisting of mica-schists, amphibolite schists, marbles, and migmatitic gneisses. Pelissier and Bolourchi (1967) has assigned these rocks to greenschist facies.
5. Metamorphics occurring at the southwest terminus of Takab Region including amphibolites, gneisses, mica-schists, and marbles. Alavi and Bolourchi (1982) believe these rocks to belong to the pre-Permian times.

All the metamorphic and igneous rocks of the Takab region can be divided into three groups as follows:

1. Takab metamorphic super-complex: Occurring at the center of Takab quadrangle, the super-complex includes Mahneshan, Amirabad, Kheirabad, and Sursat complexes all of which have been metamorphosed in schist and amphibolite facies. The rocks include schists, amphibolites, gneisses, migmatites, and anataxic granites.
2. Low-grade metamorphics: Originally having been composed of clastics, carbonaceous–tuffaceous rocks of Kahar, Soltanieh, and Baroot Formations, the rocks have been metamorphosed within the lower limits of greenschist facies and outcrop on eastern and western borders of the quadrangle.
3. Metamorphic rocks of southwestern terminus: Alavi and Bolourchi (1982) have assigned an older than Permian age for these rocks which include schists, mica-schists, marbles, recrystallized limestones, metamorphosed volcanics, and spilites. However, Ghorbani (1999f) has correlated these rocks with Rooteh Formation and hence believes they are of Permian age having been metamorphosed due to Cimmerian orogenic phase.

Metamorphic Phases

The processes of metamorphism in Takab quadrangle can be divided into four phases.

1. Pan-African metamorphic phase (Late Precambrian–Early Cambrian): Kahar, Soltanieh, and Baroot Formations have been affected by this low-grade metamorphic phase.
2. Early Cimmerian metamorphic phase (Early Mesozoic): The greenschist facies marbles, calc-schists, and greenschists occurring on the southeast of Takab (Sanandaj–Sirjan zone) belong to this phase of metamorphism.
3. Laramide metamorphic phase (Late Cretaceous): Parts of Takab quadrangle that fall within Sanandaj–Sirjan Zone have experienced metamorphism of this phase.
4. Core complex metamorphic phase (Tertiary): Being associated with igneous activities resulting from intrusion of mantle diapirs, this phase has metamorphosed most of the rocks of central parts of the quadrangle.

Metallogeny

The overall structural, magmatic, metamorphic, and stratigraphic conditions prevailing through Late Precambrian–Early Cambrian and Tertiary (especially in Neogene times) have resulted in the consideration of Takab quadrangle (dominantly its central parts) as a major metallogenic zone with no equivalents. The following mineral deposits and indications occur in Takab quadrangle (Fig. 6.6):

- Lead–zinc, for example, Angooran, Alamkandi, Poshtkuh, Molla, and Ayghal'e-si with more than 30 million tons of lead–zinc ores.
- Iron, for example, Shahrak, Mirjan, Ghaliche Bolagh, Chehar Tagh, Kuhbaba, and Zafarabad with an estimated reserve of more than 50 million tons.
- Manganese, for example, Dabaklou and Amirabad.
- Gold, for example, Zarshouran, Agh Dare, Zarinabad, Ghoozlou, and Arabshah gold indication. The investigations show that a minimum of 100 tons of gold must occur at Zarshouran and Agh Dare.
- Antimony, Arsenic, and Mercury, for example, Moghanlou, Agh Dare, and Ghizghapan antimony deposit; Zarshouran arsenic deposits and Arabshah arsenic indication; and, Shakh–Shakh mercury indication and Qare Dash mercury deposit.
- Copper, for example, Baiche Bagh multimetallic deposit that include copper, lead–zinc, cobalt, nickel, bismuth, etc., and Tataqeshlaqi and Tazekand indications with an estimated 100,000 tons of reserves.

The nonmetallic reserves of Takab Region are as follows:

- Phosphate, for example, Amankandi and Hajkandi.
- Feldspar, for example, Moghanlou, Khalaj, and Qezelche.
- Talc, for example, Alamkandi.
- Dolomite, for example, Bardlou.
- Rock salt, for example, Douzkank, Raz, and Qareh Aghaj.
- Boron, for example, Qareh Gol and Mianj.
- Fireclays, for example, Chaplou, Norouzabad, Alibaltou, and Aghajir.
- Coal, for example, Mirjan and Ghaliche Bolaghi.

The paragenesis of metallic and nonmetallic deposits of the Takab area has taken place in the following phases of mineralization:

Late Precambrian–Early Cambrian Mineralization Phase: Various metallic and nonmetallic indications and deposits such as lead–zinc, iron, manganese, copper, barite, feldspar, talc, and phosphate exist in the Takab area that have been formed during time phase of mineralization. In addition to these, gold, arsenic, antimony, mercury, lead, and zinc deposits of Tertiary have been derived from the rocks formed during the Late Precambrian–Early Cambrian times.

In general, the mineral deposits and indications of this phase can be grouped in the following on the basis of their texture, structure, and mode of origin:

1. Massive sulfide and volcanosedimentary (sedex type) occur in the central part of Takab quadrangle. The stratabound examples include Angooran, Alamkandi and Poshtkuh lead–zinc deposits, Amirabad manganese deposits, Chehar Tagh iron deposits, and Khoor Jahan copper deposit, while the stratiform instances occur at

Alamkandi and Poshtkuh lead–zinc deposits and Chehar Tagh iron deposit (Fig. 6.9).

The geological evidence shows that the minerals occurring in these deposits are mostly related with basic and acidic volcanic activities and their associated sedimentary rocks. Thus, the massive sulfide deposit of Angooran is of Besshi type while its equivalents (e.g., Chehar Tagh iron, Alamkandi lead–zinc, etc.) are of sedex type due to the association of their hosts with sedimentary rocks.

One of the paragenetic characteristics of these deposits is the great abundance of zinc in some of them, for example, Angooran, or high percentage of lead, for example, Alamkandi. However, copper is always in minor quantities.

Based on the geological features of the central parts of Takab quadrangle, characteristics of massive sulfide and sedex deposits, comparison with other areas of Iran (such as Anarak), it can be deduced that the massive sulfide (and sede) deposits of Takab have been formed in a rift zone similar to the Red Sea though some aspects of these deposits differ from each other.

2. Stratiform, stratabound deposits related to Soltanieh Formation such as iron associated with barite on the east of Mahneshan, iron and phosphate ores of Shahindezh. The shale layers of Soltanieh (specifically the lower and upper shales) in and around Amankandi and Hjikandi villages on the east of Shahindezh show traces of phosphates, but the main concentration occurs in the middle dolomite sequences of this formation. The phosphate deposits of Shahindezh are very similar to those of the Zanjan and Alborz regions.
3. Nonmetallic deposits associated with plutonism and metamorphism (such as feldspar, talc, and silica), though being formed after this time, have originated from the rocks of the Late Precambrian–Early Cambrian times.
   Occurrence of lead, zinc, iron, manganese, barium, copper, arsenic, silver, antimony, and gold in younger rocks can be a result of recycling of these elements from the Late Precambrian–Early Cambrian rocks.
4. Sedimentary and volcanosedimentary deposits dominantly iron- and phosphate-bearing are not only stratabound but stratiform. These are not directly related to volcanic activity and vague relation exists between some of them and Qare Dash Series. All of the sedimentary and volcanosedimentary deposits occur in Soltanieh Formation, the iron deposits associated with the lower dolomite succession at the base of Soltanieh, while the phosphates occur within the middle dolomites and the upper parts of the lower shale and the lower parts of the upper shale succession of Soltanieh Formation.

The mineralogical composition of iron deposits consists of hematite, magnetite, siderite, and their gangues such as carbonate and barite. Wherever Qare Dash rock successions lie next to these deposits, the mineralogy of iron deposits changes to hematite–magnetite. But far from them, the mineralogical composition changes to hematite–siderite with abundant barite.

At times, anomalous occurrence of hematite is observed within the Qare Dash rock series. At Bardarash Mount where the highest thickness of the Qare Dash Series outcrops, an indication of hematite exists.

Similar deposits are also present outside Takab quadrangle. The author has surveyed many iron deposits intermingled with dolomites of Soltanieh (especially the lower dolomite succession) from Arjin in Soltanieh area to Oshnavieh. All these deposits compare from the point of view of reserves, mineralogy, and paragenesis.

Late Paleozoic–Early Triassic Mineralization Phase: Minor amount of mineralization occurs due to this phase, which is as follows:

1. Fireclays of Shahindezh corresponding to other fireclay horizons of southern Alborz and Abadeh areas.
2. Iron deposits related to Permian volcanism. These have been formed under marine conditions along with Rooteh limestone. The most important among them are Zafarabad volcanic iron deposit and Soghanchi sedimentary iron deposit. The former occurs within the Sanandaj–Sirjan zone and corresponds to the iron deposits of Songhor while the latter exists at Shahindezh and corresponds to the Masooleh iron of the Alborz Region.

Jurassic (Mesozoic) Mineralization Phase: No characteristic mineralization phase has occurred in Takab Metallogenic Province with the Mesozoic with the exception of coal indications within Shemshak Formation in Shahin Dezh and Mahneshan areas. These coal-bearing strata reach considerable thickness in Mirjan and Nasir Abad mines (east of Mahneshan).

Tertiary Mineralization Phase: The most effective and most diverse phases of mineralization in Takab Metallogenic Province have taken place during Tertiary times, whose results can be seen in the central parts (C-I zone) of the quadrangle only.

Tertiary mineralization phase has dominantly taken place during Neogene and, therefore, these younger phases of mineralization are described hereunder; however, since Neogene igneous activity played a major role in producing various mineral deposits, we shall consider that first and then describe the metallogeny of this time interval.

The neogene phases of mineralization are closely associated to the igneous activity of the Takab Region, which can be classified into the following:

1. Eocene volcanic phase in Shahin Dezh and Mahneshan with no significant mineralization.
2. Eocene–Oligocene plutonic phase (in the form of diapirism in C-I zone); no information is available on the exact age of this phase, but it can be considered as an equivalent to the metamorphism in the region.
3. Late Oligocene–Early Miocene volcanic phase including rhyolitic–ignimbritic rock types and their associated tuffs, andesites, and small plutonic bodies. This phase reached its climax during Aquitanian and diminished by the end of Early Miocene. Diverse mineralization including Shahrak Iron and some antimony and mercury are associated with the acidic rocks of this phase.
4. Early–Middle Miocene igneous phase, which followed the acid volcanic and pyroclastic phase and produced andesite–basalt suites with no mineralization.

5. Late Miocene–Pliocene phase that started explosively producing volcanic breccia and andesitic bombs and ended with basaltic–andesitic flows; small acidic plutonism also accompanied this phase, which in the view of the author played a major role in mineralization (Fig. 6.9).

Oligocene Phase

The acid plutonic phase of mineralization has taken place during Eocene–Oligocene, resulting in the formation of mercury mineral index at Qareh Zagh.

Early Miocene Phase

This phase of mineralization that was accompanied by deepening of the sea and deposition of Qom Formation in the Takab basin produced acidic extrusives intermingled with calcareous sediments. The volcanism that began in Late Oligocene continued well into Miocene so that the lower parts of Qom Formation are mostly tuffaceous in nature. The limestones of Qom Formation are interlayered with acidic tuffs, ignimbrite and rhyolite underlying the limestones, and andesite having interfingering relation or seldom overlying it. This has resulted in extensive mineralization in the region, which is as follows:

- Iron mineralization: Shahrak iron, which is the largest deposit of the northwest Iran, is the result of this phase. The field observations, geological, mineralogical and geochemical evidence all point to the volcanic origin of this iron deposit.
- Antimony, mercury, lead, and zinc mineralization: Having taken place in and around Zarshuran (Balderghani), Ghizghapan, and Agh Dareh, this phase is probably associated with the Early Miocene acidic sub volcanic that was laid down along with Qom Formation because the host rock and the mineralized zone have been deformed together. Baycheh Bagh mineralization (especially its nickel and cobalt indications) are results of this phase.

Middle Miocene (Evaporitic) Mineralization: After the subsidence of volcanic activities at the end of Early Miocene, the environmental conditions of the Takab basin of deposition (forming Qom Formation) changed gradually from marine to nonmarine (continental). However, the marine condition persisted up to Middle Miocene in Mahneshan and Dandy due to the graben structure, but with a shift toward evaporitic facies. As a result of the composition of the volcanics, extensive salt deposits have been formed. Most of the springs of Mahneshan that pass through the lower and middle part of the Lower Red Formation are rich in potassium, sodium, and magnesium.

Late Miocene–Pliocene Mineralization: As stated before, the most intensive phase of igneous activity (volcanism and intrusion of small acidic bodies of granite and porphyry granite type) took place at this interval of time in the center of Takab

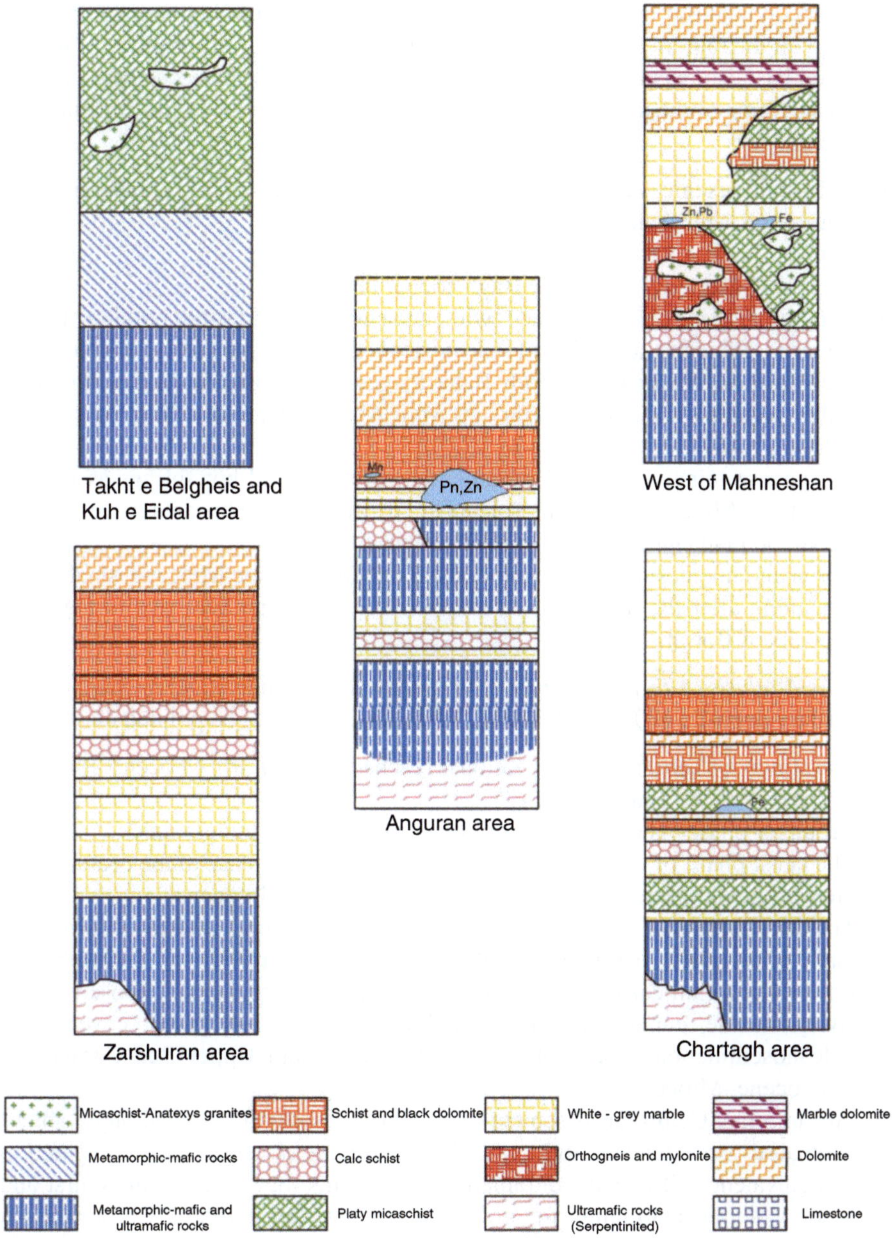

Fig. 6.9 Precambrian-Early Cambrian rocks and mineralization in Takab area

Province, indications of which are still existing in the form of hot springs scattered throughout the region. Simultaneously, the most diverse and valued phase of mineralization has also occurred reaching its climax at the end of Miocene/beginning of Pliocene. Some of the more important mineralization belonging to this time interval are as follows:

- Gold and arsenic mineralization: Gold–arsenic deposits of Zarshuran, Agh Dareh, and Arabshah.
- Copper, lead, and zinc mineralization: Lead–zinc deposits of Ay Qal'e Si, Arpachay, and parts of Bayche Bagh copper deposits.
- Iron mineralization: Associated with andesite and andesitic basalts of Late Miocene–Pliocene, a number of volcanogenic iron deposits and indications have been formed, for example, Kuh Baba magnetite deposit situated within Kuh Baba volcanic rocks; Baba Nazar and Amir Abad indications within Amir Abad volcanic rocks.
- Manganese mineralization: Hydrothermal solutions and hot springs associated with Late Miocene–Pliocene volcanism in the Kuh Baba area have formed Debakloo manganese deposit.
- Boron and magnesium sulfate mineralization: During Late Miocene, small lakes remaining after the drying up of Early–Middle Miocene Sea have formed rich deposits of Boron.

Genetic Analysis of Tertiary Deposits: In order to better understand the paragenesis of Tertiary deposits of Takab Province, the following should be pointed out:

1. The climax of Tertiary mineralization is reached during Early Miocene–Pliocene, referred here as "younger phase."
2. "Younger phase" is recorded at the center of Takab quadrangle only.
3. The latest phase of igneous activity has similarly taken place at the center of the quadrangle only, reaching its height at Early Miocene–Pliocene.
4. The central parts of Takab quadrangle where the "younger phase" occurred has mafic–ultra mafic basement which are associated with Upper Precambrian–Lower Cambrian carbonaceous–pellitic rocks. These rocks are exposed extensively at the center of Takab quadrangle. They are partly covered by volcanosedimentary rocks of Lower Red, Qom, and Upper Red Formations (Oligocene–Miocene).
5. Precambrian–Cambrian basement of middle parts of Takab quadrangle shows anomalous occurrence of gold, arsenic, mercury, lead, and zinc as compared to its equivalents. The mafic–ultramafic rocks of this time interval show first and second degree anomalies for gold and arsenic, respectively. Carbonaceous rocks have anomalous lead–zinc concentrations while volcanics, acidic tuffs, and pellites show antimony and mercury anomalies.
6. The basement rocks of the younger deposits are as follows:

   (a) The Agh Dareh gold deposit overlies mixed mafic to ultramafic basement while Ghiz Ghapan and Moghanloo antimony indications, Shakh–Shakh

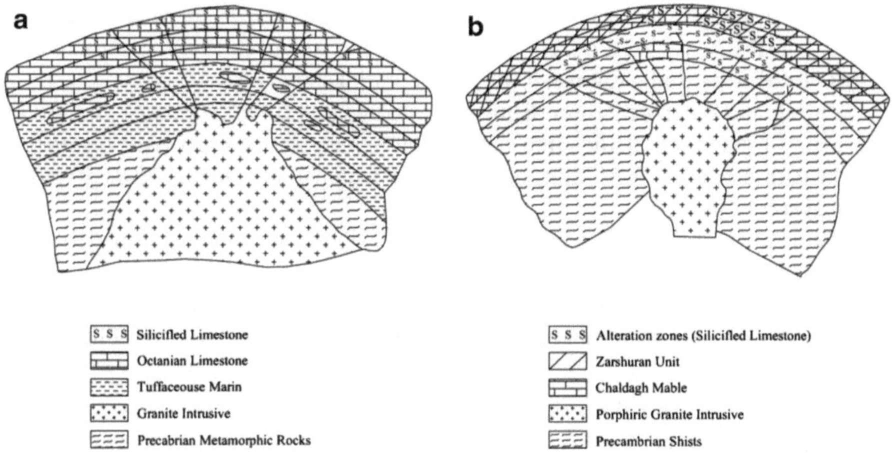

Fig. 6.10 (**a**) Schematic model of gold mineralization in Agh-Darreh; (**b**) schematic model of gold mineralization in Zarshooran

and Qare Zagh mercury indications overlie metamorphosed acid extrusive, pellitic, and granitic basement.

The Bayche Bagh polymetallic deposit has a mafic–ultramafic basement which is interrupted by Bayche Bagh fault through which Miocene volcanics erupted (Fig. 6.10a).

(b) The Zarshuran gold deposit and Arabshah indications are of pyroxinite and peridotite type overlain by carbonaceous and pellitic successions (Fig. 6.10b)

(c) The Ay Qal'e Si lead–zinc deposit overlies metamorphosed mafic, carbonaceous and pellitic basement.

7. Sulfur isotope measurements ($\partial S^{34}/S^{32}$) indicate significant resemblance between Late Precambrian–Early Cambrian lead–zinc deposits (Angooran and Alamkandi) and younger deposits, for example, Ay Qal'e Si.

Based on mineral paragenesis, fluid inclusion studies, and the control of mineralization processes by faults in Zarshuran, Ay Qal'e Si, Agh Dareh, Moghanloo, Arabshah, and Bayche Bagh, the importance of hydrothermal processes in formation of these deposits is revealed.

Based on the above-mentioned studies and considering the temperature of hydrothermal solutions in these deposits, the following sequence from left to right is applicable:

$$\text{Agh Dareh} \rightarrow \text{Zarshuran} \rightarrow \text{Ay Qal'e Si} \rightarrow \text{Bayche Bagh}$$

The results of fluid inclusion studies indicate that the temperature of hydrothermal solutions at Zarshuran must have been between 180 and 220°C (Moritz et al. 1993).

The same investigation has determined the temperature of hydrothermal solutions at lower than 180°C for Agh Dareh, more than 220°C for Ay Qal'e Si, and more than 300°C for Bayche Bagh (above 400°C in the initial stages).

Folding and other structural disturbances of younger Alpine phases along with intrusion of magmatic bodies into the shallower parts of the crust have resulted in the formation of a series of faults and fractures which conducted the flow of hydrothermal solutions.

Chemical analysis of younger rocks shows no significant anomaly of ore-forming elements.

Field association of small acidic intrusive bodies, epithermal mineralization, and the intensity of hydrothermal activity was distinct during Late Miocene–Pliocene so that all the travertine layers and the argillitic and silicic alteration zones of the area are associated with the intrusives, whereas at places where the extrusives predominate, alteration is not significant.

Genetic Model of Takab Mineralization

Considering the above-mentioned facts, the central parts of Takab quadrangle are believed to have been intensively impacted by igneous activities during Late Oligocene to Early Quaternary, the effects of which are seen in the form of various volcanic rocks such as rhyolites, basalts, and tuffs along with small intrusives of gabbro and alkaline granites. These igneous activities were followed by intensive hydrothermal actions that are still continued today in the form of hot travertine-producing springs.

Melting of Precambrian–Cambrian basement rocks (crustal magma) during the later phases of igneous activity and its mingling with magma originating from the mantle resulted in a rise of the thermal gradient (thus weakening the bonds of rare elements in the older rocks) and circulation of hydrothermal solutions (produced from igneous rocks, atmospheric precipitation, and metamorphic activity). The mineralizing solutions thus formed were trapped within the fractured rocks which had the capacity of reacting with them and producing ore deposits.

In line with this hypothesis, the younger deposits of Takab region can be grouped into two categories:

1. Ay Qal'e Si, Moghanloo, Shakh–Shakh, Arpachai deposits; parts of Bayche Bagh mineralization are of epithermal to mesothermal type whose hydrothermal solutions were derived from small intrusive bodies. The age of these ore deposits and indications and the role of the basement rocks are of importance.
2. Bayche Bagh polymetallic deposit, especially its nickel and cobalt, and Tatagheshlaghi and Tazekand copper indications are among the high-temperature hydrothermal deposits whose solutions originated from andesitic volcanic rocks. The formation temperature of these deposits was higher than that of epithermal deposits and the role of basement was more subdued.

6.3.2 *Urumiyeh–Dokhtar Metallogenic Province*

This belt lies on the west–southwest of Central Iran and north of Sanandaj–Sirjan provinces in the form of a volcanic arc. The rocks of Urumiyeh–Dokhtar metallogenic province consist of thick piles of volcanics and pyroclastics in spread from Sahand on the northwest of Iran to Bazman in the southeast. The igneous activity that has produced these rocks can be traced all along the Alpine–Himalayan orogenic belt (Ghorbani 2004a). On the basis of magmatic and mineralogical characteristics (especially that of copper), the province can be divided into three districts as southern (Kerman), central (Taft, Anar, and Kashan–Qom), and northern (Tafresh–Takab). The following are the major specifications of Urumiyeh–Dokhtar metallogenic province:

1. The general trend of the province is northwest–southeast.
2. The sequences consist of volcanic and pyroclastic rocks.
3. The volcanic sequences comprise of various types of lavas and clastics ranging in composition from basalt to rhyolite.
4. The pyroclastic successions alternate with the volcanic rocks and are composed of various clastics.
5. Intrusive rocks having granitic to gabbroic compositions of Paleocene to Pliocene age cut across the older rocks.
6. Older (Precambrian–Paleozoic) rocks form the basement of the northern parts while they are rare in the south.
7. Volcanic activity was not uniform throughout Tertiary reaching maxima at Middle Eocene, Late Eocene–Early Oligocene, Late Oligocene–Early Miocene, and Miocene–Pliocene.

The intrusive igneous rocks of this province are characteristically composed of diorites and granites, whereas the extrusives are of basaltic–rhyolitic composition. Associated with them, there are large deposits of porphyry copper, magmatic iron, epithermal gold, volcanogenic manganese, and hydrothermal barite, lead and zinc (Ghorbani 2002a, 2004a).

The igneous rock successions (especially the Eocene volcanics) reduce in volume and diversity in the northward direction so that their outcrops are nearly absent in the north (Ghorbani 2012a). Moreover, mineralization (especially that of copper) show the same trend and there are no reports of any porphyry deposits in the north.

6.3.2.1 Kerman Copper Belt

The Kerman copper-bearing belt in the southern part of the Urumiyeh–Dokhtar magmatic zone is one of the high-potential copper-bearing areas in Iran wherein several porphyry copper deposits have been found so far. From structural and geological standpoints, the Kerman copper-bearing belt is part of the Urumiyeh–Dokhtar magmatic zone and because of specific characteristics it differs from the middle and northern parts of this magmatic zone (Fig. 6.11).

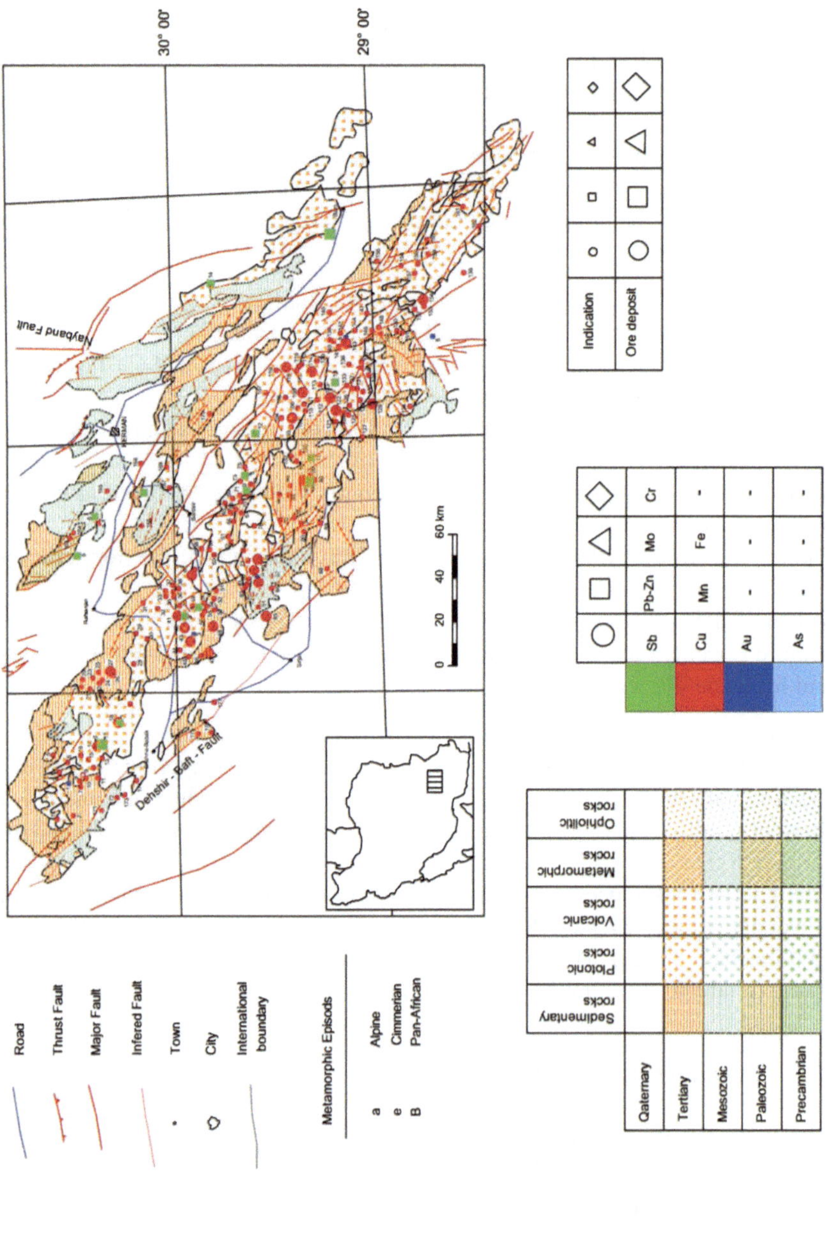

Fig. 6.11 Metallogenic map of Kerman copper belt (Ghorbani 2006b)

Based on the exploration works done by the Yugoslav team back in 1970s in the Kerman region (Dimitrijevic and Djokovic 1973), the geological and structural units of Kerman have been categorized as follows:

- Rafsanjan axis (includes Morad, Joopar, and Gook blocks)
- Dehaj–Sardouiyeh axis
- Color-melange axis
- Sirjan axis
- Esfandagheh–Sabzehvaran (Jiroft) axis

Out of aforementioned five axes, Dehaj–Sardouiyeh and the northern part of Esfandagheh–Sabzehvaran (Jebal–Barez) axes contain porphyry deposits. As a result of the exploration works by the Yugoslav team in the 1970s, Rio Tinto (2001–2009), and the recent exploration works by the National Iranian Copper Industries Company (NICICO), several subzones were identified and introduced, of which Abdar–Dehaj, Sirjan (or Sarcheshmeh–Sirjan), and Jebal–Barez subzones exhibit porphyry copper deposits with the first two subzones being more important.

Considering the features of magmatic phases, alteration, and porphyry mineralization, there is not much of a difference between the Abdar–Dehaj and Sarcheshmeh–Sirjan subzones, and therefore the author divides the whole Kerman copper-bearing belt into two subzones of Dehaj–Sardouiyeh and Jebal–Barez (Figs. 6.12 and 6.13).

The Kerman copper-bearing belt forms the southern part of Urumiyeh–Dokhtar metallogenic province, and it is the richest copper-bearing belt in Iran with more than 200 identified copper deposits and indications, of which some are of porphyry type. This belt forms the southern extension of the Urumiyeh–Dokhtar magmatic zone, and it is 450-km long and 80-km wide. This belt was cut through by many faults; some of these were in the N–S trend and some in the NE–SW direction.

Magmatic activities within Kerman copper-bearing belt had a significant and widespread distribution, and started from Early Eocene and continued till Quaternary. Most volcanic activities in this copper-bearing area belong to Eocene (Ghorbani 2004a). These rocks have been studied under the title of three complexes called Bahr–Asman, Razak, and Hezar (from older to younger).

In general, magmatic activities in Tertiary–Quaternary within Kerman belt took place in four temporal phases (Ghorbani 2012a):

- Eocene (Middle Eocene in particular): With large volumes of volcanic rocks especially andesite.
- Oligocene: Although volcanic activities were weaker than Eocene but intrusive activities showed up in the form of batholith especially in the Jebal–Barez area.
- Miocene: Main activities during this time period consist of diorite–tonalite porphyry intrusive bodies, in which most copper deposits formed.

Mio–Pliocene: Magmatic activities in Mio–Pliocene began with forming acid domes and in some parts ended with volcanic activities.

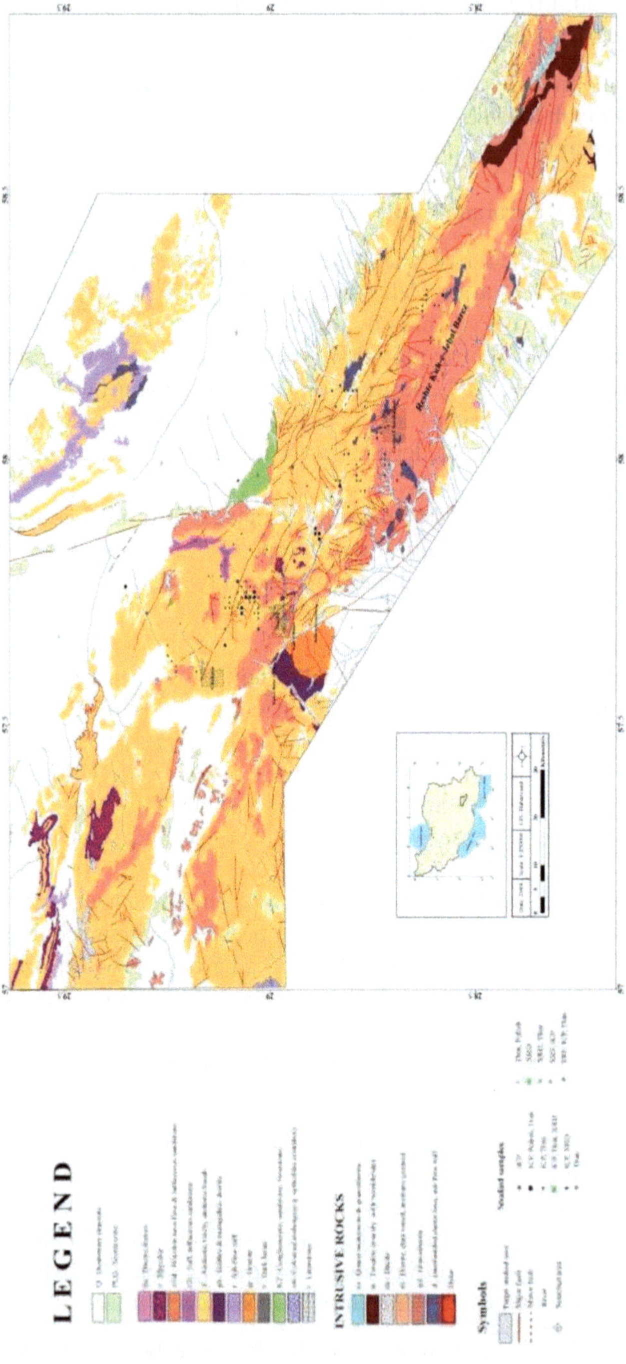

Fig. 6.12 Geological map of Jebal Barez (Arian Zamin Co. 2010b)

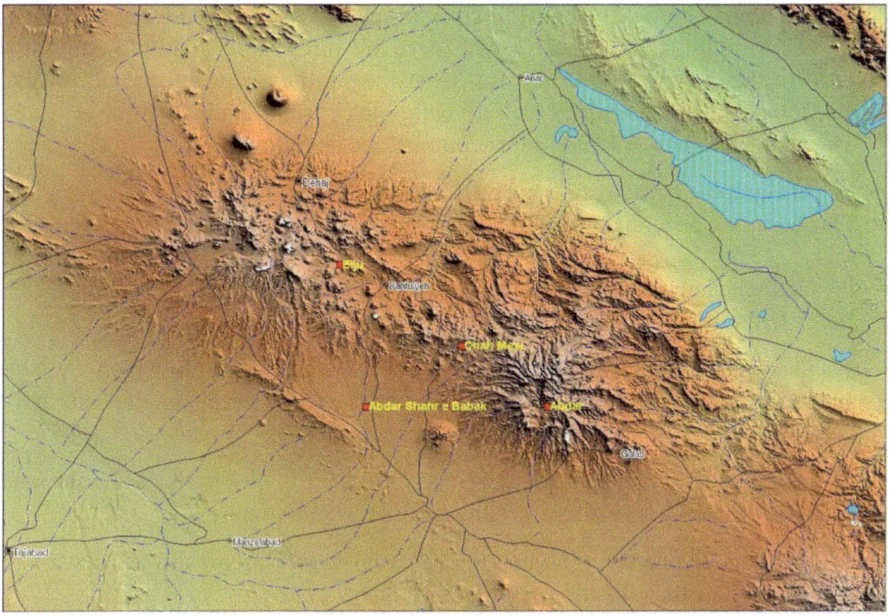

Fig. 6.13 Porphyry localities in Abdar–Dahaj area (Parvin-pour 2007)

Five magmatic phases that occurred from Eocene to Quaternary are traceable in this belt (Ghorbani 2012a):

- Early Eocene to Late Eocene volcanic phase with a variety of volcanic rocks such as andesite, andesite–basalt to trachyte with small volume of acid rocks, breccias, agglomerates, and tuffs.
- Eocene–Oligocene volcanic phase with acid and ultra-acid rock composition.
- Miocene porphyry intrusive phase resulted in bodies with composition of porphyry quartz–monzonite to porphyry tonalite.
- Mio–Pliocene volcanic and sub volcanic phase, which is the most important phase in this area. Most high rugged peaks are the result of this phase (Ayoub, Mardvar and Dehaj–Keder Mountains).
- Plio–Quaternary volcanic and sub volcanic phase, which is the final and youngest phase in the area, led to the formation of domes and craters (Aj Crater, Mozahem, Hezar, and Amiralmomenin Mountains).

The copper content in the background of Eocene andesite rocks is more compared to other andesitic rocks in the area. Various alteration zones such as propylitic, sericitic, and argilitic are common in most areas, and in some parts, potassic and silicic zones are seen as well. It is to be noted that in some known porphyry deposits such as Meydook, Sarcheshmeh, and Darreh-zar, all of the aforementioned alteration zones are detectable, and the propylitic zone has a significant expansion (Arian Zamin 2010b).

Magmatism during Early–Middle–Late Eocene in Jebal–Barez subzone is almost similar to that of in other parts of the Kerman copper-bearing belt; however, the acid magmatic products such as agglomerates, breccias, and tuffs are not significant.

The volcanic of Oligo–Miocene–Pliocene are not considerably present either, but Oligocene batholiths are widely seen. There are three large granitoid batholiths outcrop in this subzone and they are, from the southeast toward the northwest, Rigan, Jebal–Barez, and east of Delfard batholiths (Arian Zamin 2010b).

The characteristics of Jebal–Barez complex are as follows:

- Its petrographic composition varies from diorite to granite.
- This complex penetrated in a thick sequence of Eocene volcanic and pyroclastic rocks.
- Invaded volcanic rocks display metamorphic halo at the contact line with Jebal–Barez batholith.
- There were frequent tardy intrusions in Jebal–Barez batholith, which are considered as porphyry bodies wherein explorations for porphyry deposits have been done.
  - The composition of some of these tardy intrusive bodies is alkali–granite, and some of them are of alkali–granite to alkali–syno–granite.

Some of these porphyry bodies in the Jebal–Barez area show evidences of porphyry copper mineralization.

The role of fractures and linear structures in the formation of porphyry deposits within the Kerman copper-bearing belt is as follows:

- Most young volcanic rocks formed at the intersection of two fault systems.
- Based on field observations and studies, it has been determined that the thickness of volcanic rocks is much larger than it is in other parts of the area.
- Intrusive porphyry bodies, especially the fertile ones, are seen at the intersection of fault systems, particularly where the fault systems are of strike-slip type.
- In general, wherever the fractures are more congested, more mineralization signs are seen.

The differences between the Dehaj–Sardouiyeh and Jebal–Barez subzones are as follows:

- The volcanic rocks in the Dehaj–Sardouiyeh subzone show more variety than those in Jebal–Barez.
- The volcanic rocks in the Dehaj–Sardouiyeh subzone have extensive variety in petrography and geochemistry, and can be traced from Eocene to Quaternary.
- There are large batholiths in Jebal–Barez subzone, whereas no batholith is seen in the Dehaj–Sardouiyeh subzone.
- Intrusive bodies in the Dehaj–Sardouiyeh subzone are mostly small with fairly similar composition. Although some of these bodies are the product of frequent intrusions, no tardy intrusion is found there.
- Porphyry bodies within the Jebal–Barez subzone are in fact of tardy intrusion type, which seem to be associated with differentiation of Jebal–Barez batholith complex. Therefore, the level of acidity in rocks increases gradually as moving from very first phase (the large batholith) toward the final phase (tardy alkali–granite).

- Porphyry copper mineralization in the Jebal–Barez subzone is very different from porphyry mineralization in the Dehaj–Sardouiyeh subzone or copper mineralization in any other part of this area.
- In addition to porphyry copper mineralization, copper mineralization is found in the form of veins and veinlets in volcanic and intrusive rocks, which do not exist in the Jebal–Barez subzone.
- Copper contamination in volcanic and intrusive rocks in Dehaj–Sardouiyeh is higher than it is in the Jebal–Barez subzone.
- Copper grade in porphyry bodies in the Jebal–Barez subzone fluctuates significantly as in some parts the copper content reaches up to 0.1, whereas copper grade variation in porphyry bodies in the Dehaj–Sardouiyeh subzone follows much fair distribution order.

6.3.3 Sanandaj–Sirjan Metallogenic Province

This province is called Sanadaj–Sirjan (Stöcklin 1968), Urumiyeh–Esfandaghe (Takin 1971), Esfandaghe–Marivan (Nabavi 1976), and Manujan–Marivan (Houshmand-Zadeh and Berberian 1972). Sanandaj–Sirjan Province is in fact a division of Central Iran. However, some geological characteristics of its Mesozoic and Tertiary sequences contradict those of Central Iran.

Sanandaj–Sirjan Province extends in the form of a magmatic–metamorphic belt from Urumiyeh on the northwest to Esfandaghe on the southeast, in between Zagros and Central Iran. Urumiyeh–Dokhtar volcanic arc, and Sirjan, Marvdasht, Gavkhooni, and Kabudar–Ahang depressions separate this province from Central Iran while Zagros thrust fault (ZTF) marks its boundary with the Zagros Mountains.

Eftekharnejad (1980) has divided Sanandaj–Sirjan Province into northern and southern areas based on geological characteristics. But taking into account the metallogenic characteristics of the province, it may be divided into three districts, namely, southern, central, and northern districts. The southern district covers the area between Sirjan and Esfahan and is separated from the central district by the extension of the Dorouneh fault. In turn, the central district is set apart from the northern by the extension of the South Tehran fault. The continuation of South Tehran and Dorouneh faults forms the western and the eastern borders of central district and corresponds with the Malayer–Esfahan lead–zinc belt.

The southern district is characterized by mineralization of iron, iron–manganese (Gole–Gohar, Heneshk, Baft), and lead–zinc–copper (Chah–Gaz, Ghanat–Marvan) of Late Precambrian to Cretaceous. The discovery of large deposits of iron and iron–manganese deposits similar to Hanshak manganese, Baft manganese, and Chah–Gaz lead–zinc–copper is very probable within this district.

The central district, informally further divided into northern and southern stripes, bears a higher potential of ore deposits compared to the other two districts. The more important deposits of this district include the Malayer–Esfahan lead–zinc belt and Shams–Abad metallogenic zone (Astaneh gold, Shams–Abad iron, Deh–Hossein tungsten–tin, and Mooteh gold camp).

The northern district includes deposits of iron at Hamekasi (Hamedan), Songhor and Zafar–Abad (Divan Darreh), gold and gold–antimony at Dashkessan (Ghorveh), and Alout (Saghez). The most important characteristic of Sanandaj–Sirjan is the temporal correspondence of mineralization with various phases of Alpine orogeny.

6.3.3.1 Malayer–Esfahan Metallogenic Belt

Being a part of the Sanadaj–Sirjan Zone, this belt is one of the main lead–zinc mineralization zones of Iran. The belt is bounded on the south by the Zagros Mountains and on the north by the Orumieh–Dokhtar volcanic strip (Fig. 6.14). The Malayer–Esfahan Belt covers major parts of Markazi, Esfahan, and Hamedan Provinces and minor parts of Lorestan and Chahar–Mahal–o–Bakhtiyari, which extends up to Qom, Tafresh, and Qahavand on the north, Boroojerd, Aligoodarz, Shahrekord, and Boroojen on the west, Shahreza–Naein trunk road on the south, and Ardestan, Kashan, and Qom on the east; the area has a cold mountainous climate type.

The western limits of the area extend from northwest to southeast and pass through a hilly region to the south of Malayer, Sarband Mountains, Aligoodarz Mountains, Dalan Mount, and Shahrekord highlands while the eastern limits of the belt extend from northwest to southeast and include Saveh highlands, Qom–Kashan–Natanz Plain, Karkas hills, and Ardestan–Naein Mountains. The northern limits include, from north to south, Tafresh Mountains, Malayer Mountains, Sarband Mountains, and Shazand Mountains while the southern boundary passes through Naein highlands, Gavkhooni Swamp, and Shahreza highlands in the north–south direction.

General Geology

The area of the Malayer–Esfahan Belt is geologically, sedimentologically, and structurally similar to Central Iran. The belt demonstrates several unconformities during the Mesozoic and Cenozoic times that are also visible in Northern and Central Iran; however, it can be distinguished from the latter by the absence of Tertiary volcanic rocks, underdevelopment of Tertiary sediments and having a trend similar to that of Zagros. The orogenic phases that have affected the area are Pan-African, Early Cimmerian, Late Cimmerian, Laramide, Pirenne, and younger Pliocene phases (Aghanabati 2004; Ghorbani 2002a).

Overall, it can be considered that the geomorphological features of the area have been formed in relation with Alpine orogeny and the general trend of fold axes follows that of Zagros (Alavi 1993).

From a structural point of view, the belt is limited by two large basement faults (Ghorbani 2006b). The first is an extension of the Dorooneh fault that forms the northern limit of the belt. Cretaceous and younger rock successions cover this basement fault, but certain associated structures are visible on the surface. The second basement fault is the one that passes through southern central Alborz, and similarly, covered by Upper Mesozoic and Tertiary rocks, and can be distinguished by the associated structures.

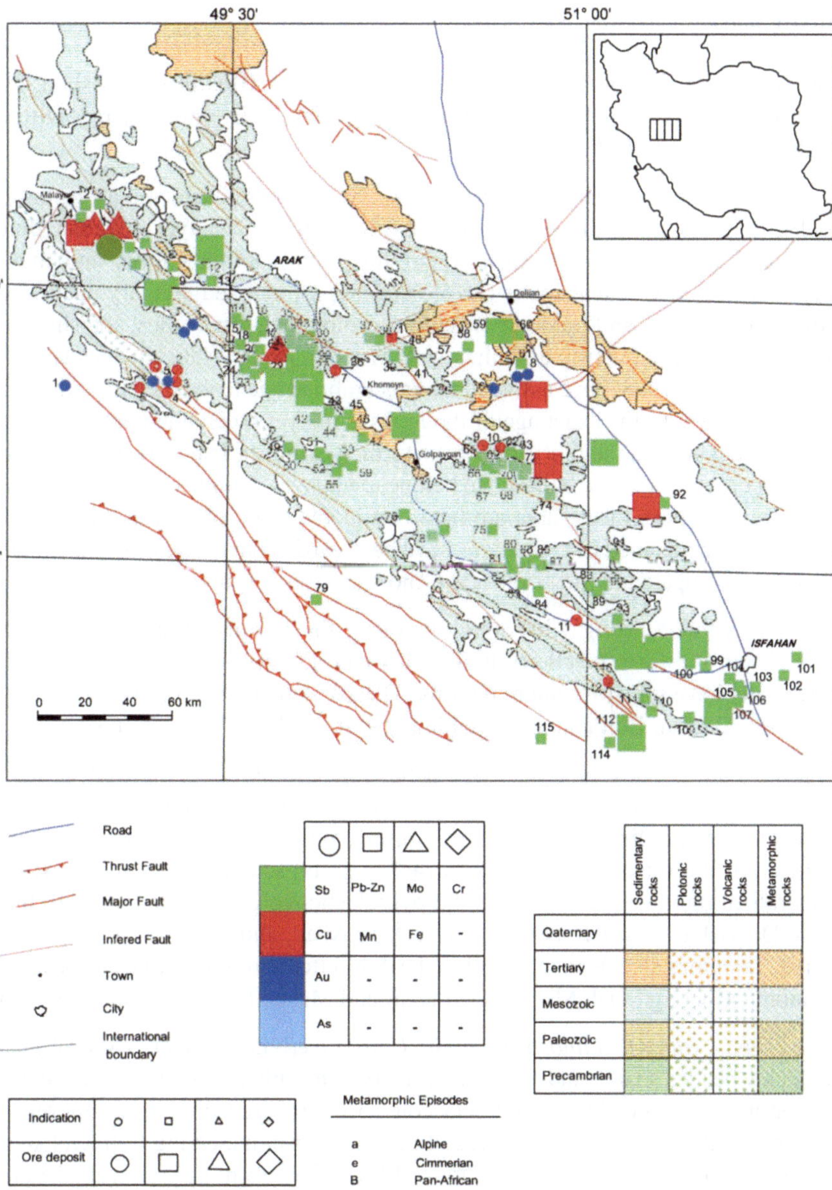

Fig. 6.14 Metallogenic map of Malayer Esfahan area (Ghorbani 2002a, 2006b)

Stratigraphy

Based on the geological maps (1:250,000 and 1:100,000) published by the Geological Survey of Iran, the following stratigraphic units outcrop within the Malayer–Esfahan Belt:

Upper Precambrian–Early Paleozoic Rocks: Sometimes considered as the basement rocks, they comprise the oldest rocks exposed in the Malayer–Esfahan Belt and consist of metamorphics (schists, quartzites, amphibolite, gneiss, and marbles), metamorphosed volcanic rocks (acidic and basic), and sedimentary succession of Soltanieh, Zagoon, and Laloon Formations (dolomites, sandstones, and shales). The sedimentary rocks outcrop at Golpaygan, Mooteh, Azna, Mahalat, and Aligoodarz.

Thiele et al. (1968) have assigned the Precambrian age to the metamorphic rocks of the area. However, the exact age of the rocks is not yet ascertained; in fact, the metamorphic rocks earlier thought to have been formed during the Pan-African (Precambrian) phase of orogeny are now believed to belong to younger (Cimmerian) phases. Thus, the metamorphic rocks of the area are here considered to be of Precambrian–Paleozoic age. Wherever detailed investigations have been carried out on these rocks, higher gold content as compared to Mesozoic and Cenozoic rocks is found to exist.

Permian Rock: Starting with conglomerates, the Permian rock units of the belt continue in the form of cherty limestones and dolomites which are crystalline in nature and have schistose intercalations. These rocks are exposed on the west of the Malayer–Esfahan Belt and are suitable to be utilized as construction stones.

Triassic Units: They include pyroclastic–volcanic suites composed of alternate acidic tuffs and acidic to intermediate lavaflows with carbonaceous intercalations (limestone, marly limestones, and dolomites); at times, the intercalations change to shales. All the rocks are metamorphosed to green schist facies. Vein-type copper mineralization is very common in the pyroclastic–volcanic suites (e.g., Kohandan, Dare Darvishan, Sivaleh, Nosrat Abad, Aziz Abad, and Kan Sorkh copper indications). Talcum is also deposited within the dolomites and schists that can be extracted economically (e.g., Deh Moosa, Deh Haji, and Masood Abad talcum deposits).

Late Triassic–Jurassic Lithostratigraphic Units: Overlying the older rocks, a sequence of thickly bedded shale, sandstone, slate, and phillite with quartz intercalation exists that is overlain unconformably by the Cretaceous rocks. This sequence is metamorphosed in the western direction, with the degree of metamorphism increasing from Golpaygan to Boroojerd and Malayer and further toward Hamedan where they are called Hamedan phyllites by Seyyed Emami et al. (1971). The author believes that these rocks are of Late Triassic–Early Jurassic age and are equivalent to Nayband and Shemshak Formations. The Late Triassic–Jurassic rocks are not important from the point of view of mineralization (whether metallic or nonmetallic), but the lead–zinc deposits of the belt have been attributed to them by some experts due to their serving as the host for such deposits. Moreover, at places where intrusive rocks have been emplaced, graphite deposits, and indications are found; economic silica lenses are also common at many localities.

Cretaceous Rock Formations: The prograding conglomerate that lies unconformably over the Jurassic rocks marks the beginning of the Cretaceous succession (Thiele et al.

1968; Momen-Zadeh 1976). Sandstones and orbitoline limestone (Aptian–Albian) follow in the sequence. The Cretaceous rocks of this belt are dominantly carbonaceous in nature. Occasionally, submarine volcanic rocks underlie the carbonates.

Thiele et al. (1968) have classified the Cretaceous rocks into the following units (from older to younger):

1. Sandy conglomerate horizons of upper Middle Cretaceous
2. Thinly layered limestones and marls
3. Massive orbitoline limestone
4. Thinly bedded limestone with shale and marl intercalations

The main host of the lead–zinc and iron–manganese deposits of Malayer–Esfahan Belt is Cretaceous rocks so that all the economic deposits of these metals are found in Cretaceous limestones. Momenzadeh (1976) is of the opinion that these deposits are confined to certain geologic horizons (Fig. 6.15).

Tertiary Untis: Being exposed in the central parts of the belt, the Tertiary rocks are not so important and are seen as peripheral volcanic and sedimentary rocks.

Igneous Activity

The igneous rocks of the Malayer–Esfahan Metallogenic Belt can be divided into two categories (Ghorbani 2002a):

Plutonic Rocks: Granitic intrusive bodies that have intruded into the metamorphosed Precambrian–Paleozoic rocks (Mooteh area) are granitic to granodioritic in composition. Thiele et al. (1968) have assigned these rocks to Precambrian. Whereas the intrusives that cut across Jurassic shales and sandstones are granitic to granodioritic in composition (e.g., granitic bodies of Boroojerd, Astaneh, and Kolah Ghazi), the plutonic bodies intruded within the marls and shales of Cretaceous are smaller and vary in composition from granite to diorite and gabbrodiorite.

There is a lot of controversy regarding the age of the intrusive bodies but recent works have revealed a Mesozoic (Jurassic–Cretaceous) age for them (e.g., Boroujerd granite, Astaneh granite, Alvand granite). The intrusive bodies play no role in the formation of lead–zinc, copper, and iron in this belt, but some of the gold deposits and indications are either directly or indirectly associated with them (e.g., Astaneh and Mooteh granite). The author believes that if the magma that formed the intrusives has passed the metamorphosed Precambrian–Paleozoic igneous rocks, the chances of gold mineralization increase.

Some of the intrusive bodies of the Malayer–Esfahan Belt are of S-type and thus the probability of tungsten, tin, and molybdenum mineralization around them is high; the Nezam Abad tungsten indication, Deh Hosein and Chal-e Homa tin deposits fall in this category. A few nonmetallic deposits such as silica and graphite are directly or indirectly associated with these intrusive bodies (Jahangiri 1999; Ghorbani 2002b).

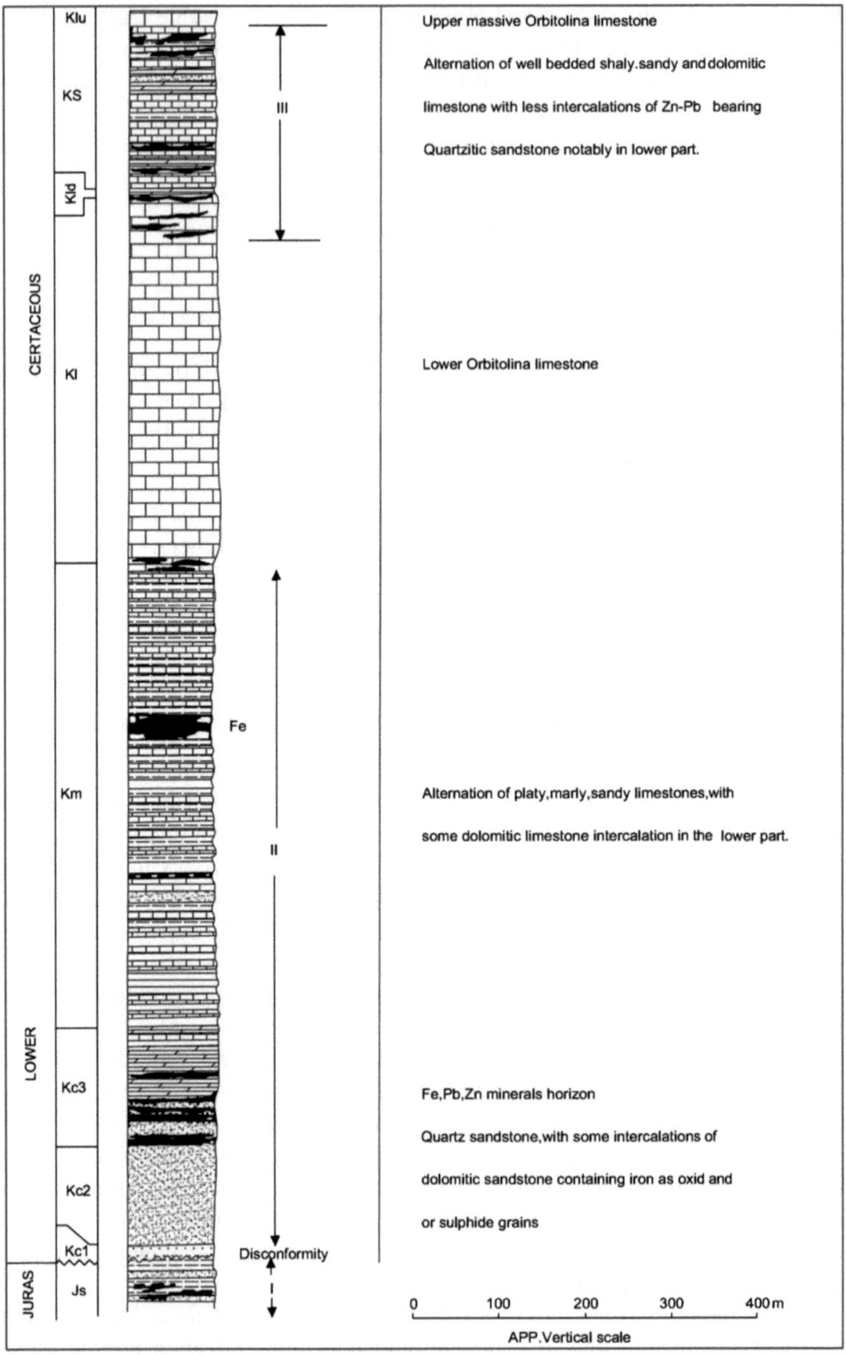

Fig. 6.15 Schematic view of lithostratigraphy of Malayer-Isfaha axis along with mineralized horizons (Momen-zadeh 1976)

Volcanic Rocks: The volcanic rocks of the Malayer–Esfahan Belt can temporally be divided into four groups:

Precambrian–Paleozoic: These rocks, which are completely metamorphosed, vary in composition from basalt to rhyolite and are exposed in and around Mooteh, Azna, and Golpaygan. The geochemical analysis of these rocks points toward the presence of high gold content.

Upper Paleozoic–Lower Triassic: The volcanic rocks of this time interval vary from andesite to rhyolite and are intermingled with sedimentary rocks. The underlie Upper Triassic–Jurassic sandstones, shales, phyllites, and slates with no outcrops except at a few locations such as Aligoodarz. Due to their limited exposures, the mineralogy and petrology of the volcanic rocks of Upper Paleozoic–Lower Triassic have not been studied in detail. However, they seem to be responsible for the copper mineralization within the area and might be responsible for the origin of lead–zinc deposits within the younger rocks.

Lower Cretaceous: At parts of Malayer–Esfahan Belt (Ashtiyan, Shams Abad, etc.), volcanic rocks of andesitic composition are exposed, which are mostly covered by Middle to Upper Cretaceous sediments. Based on field evidence, their extensive distribution at the Sanandaj–Sirjan Zone, and silicic metasomatism of Cretaceous rocks of Robat Makan (Ashtiyan), these rocks seem to be limited to the western limits of the belt only. The volcanic rocks of the Cretaceous age have been directly or indirectly responsible for lead–zinc, iron, and manganese mineralization within Cretaceous hosts (at least in the western limits).

Tertiary: These have very limited exposures within the Malayer–Esfahan Metallogenic Belt and only outcrop toward the eastern limits. No major mineralization has been reported from these rocks.

Metamorphism

Limited metamorphism has occurred around the intrusive bodies within the Malayer–Esfahan Belt that can be grouped into two main metamorphic facies: The area on the southwest to the northwest of the belt that hosts most of the mines and within which the change in facies is gradual from low temperature–low pressure conditions to low temperature–high pressure, and the area on the north which a part of Malayer–Esfahan metamorphic zone and within which the change of facies occur discontinuously from low temperature (above) to high temperature–medium pressure (Hosseini 2011).

Mineralization

Apart from extensive lead–zinc mineralization that will be covered in detail in the later parts of the book, the following mineralizing zones have been recognized in the Malayer–Esfahan Belt:

1. Shams Abad–Nezam Abad (Shazand) metallogenic zone with lead–zinc, gold, silver, tungsten, and tin mineral deposits

2. Ahangaran metallogenic zone with lead–zinc, silver, iron, and manganese mineral deposits
3. Mooteh gold-bearing zone

Some of these localities are presented in Fig. 6.13.

Lead–Zinc Mineralization: Over 120 deposits and indications of lead–zinc have been reported from the Malayer–Esfahan Belt with a proved reserve (extracted and in place) of more than 60 million tons (Ghorbani 2000b). Half of these fall within the Esfahan region, whereas the other half are located in Khomein, Ashtiyan, Malayer, and Shams Abad areas. The deposits and indications are of stratabound type and are limited to Lower Cretaceous limestones, limey-dolomitic shales, and sometimes sandstone, although some are hosted by older rocks (e.g., Sakiye-bala deposit). Lead–zinc mineralization is in the form of layered veins and lenses.

Silicic, dolomitic, and calcic metasomatism is seen in association with these deposits in the western parts of the belt around Rabat, Kalishe, and Emarat, while on the east (Anjire–Tiran, Esfahan), such metasomatic effects are absent (Ghorbani 2000b).

Various opinions about the mode of origin of the lead–zinc deposits of the Malayer–Esfahan Belt have been put forth that can be grouped into two categories:

1. Sedex-type origin dominantly supported by Momenzadeh and coworkers who have studied the entire area of the belt.
2. Mississippi Valley Type (MVT) origin supported by various works each having studied a part of the belt.

It seems that both types of mineralization processes have played a role in the formation of lead–zinc deposits with sedex type being dominant in the west belt and Mississippi type in the east. However, some of these deposits in the west are MVT. Considering the characteristics of the Cretaceous host rocks of these deposits, the author believes that the solutions produced by metamorphism affected Triassic–Jurassic slates and phyllites during Late Jurassic and Early Cretaceous that resulted in the deposition of lead–zinc minerals within the faults zone, which follows and exhibits mineral formation of Mississippi type.

Gold Mineralization: Gold mineralization occurs at two horizons in the Malayer–Esfahan Belt:

1. Within Precambrian–Paleozoic rocks along fault zones, for example, Mooteh and Azna gold deposits, though the deposition process seems to be of younger age.
2. In association with some Mesozoic intrusive bodies, for example, Astaneh Granite.

Iron and Manganese Mineralization: Iron and iron–manganese mineralization has taken place within Precambrian–Paleozoic and Cretaceous rocks of this metallogenic belt with the former (Precambrian–Paleozoic) being uneconomic.

Tin and Tungsten Mineralization: Along the hornfelses and granites or sometimes even within the metamorphosed Lower Cretaceous rocks, tin and tungsten (e.g., Nezam Abad–Arak) or only tin (e.g., Deh Hosein and Chaleh Homa – south of Astaneh; Ashna Khor and Khak Abad – east of Aligoodarz) have taken place.

Copper Mineralization: At places where the Upper Paleozoic–Lower Triassic basement rocks are exposed, indications of copper mineralization have been observed. No detailed investigations have been performed on such deposits.

Nonmetallic Minerals: The Malayer–Esfahan Metallogenic Belt includes many metallic and nonmetallic deposits and indications that were briefly discussed under the geology of the area. The nonmetallic deposits of the belt comprise talc, graphite, barite, and constructions stones which are economically very important.

6.3.3.2 Esfandaghe–Faryab Chromium Ophiolite Belt

This belt is located on the southern termination of Urumiyeh–Dokhtar, bounded on the east by the Jazmoorian depression, on the south–southwest by Makran and Zagros, and on the north by the volcanic rocks of Urumiyeh–Dokhtar (Fig. 6.16).

There are two chromite-bearing regions within this belt, namely, Esfandaghe and Faryab, which are the largest chromium reserves discovered in Iran. All the chromite deposits occurring in Esfandaghe–Faryab are of Alpine type.

6.3.4 Northeast Metallogenic Province (Taknar and Kavir–Sabzevar Metallogenic Belts)

This province is bounded to Miami and Neyshabour–Torbant–jam faults in north. In fact, it is a trapezoid block located between Binaloud, Alborz and Lut zones, which encompasses the cities of Torbat–jam, Kashmar, Bardskan, Neyshabour, Sabzevar, and Shahroud and also stretches toward south of Damghan and Semnan desert (Fig. 6.17).

From geological and metallogenical standpoints, this province can be divided into two belts of Kavir–Sabzevar and Taknar. Before proceeding to describe these two belts, the faults within this metallogenic province are cited first.

The Miami fault separates Sabzevar from eastern Alborz zone. This fault is parallel to the Dorouneh fault. It fault trends in the NE–SW direction and turns to the Neyshabour–Torbatjam fault with the NW–SE direction, which separates the Alborz zone from Binaloud zone.

In fact, two faults of Miami and Neyshabour–Torbatjam can be considered as the Miami–Torbatjam fault (Lindenberg et al. 1983).

6.3.4.1 Geology of Taknar–Torbatjam Metallogenic Belt

Taknar zone is bounded to two major faults: Dorouneh fault in the south and Riyoush (Taknar) fault in the north. Both faults trend in approximately E–W direction. The Riyoush fault merges with the Dorouneh fault at north of Dorouneh village and draws the western boundary of the Taknar zone.

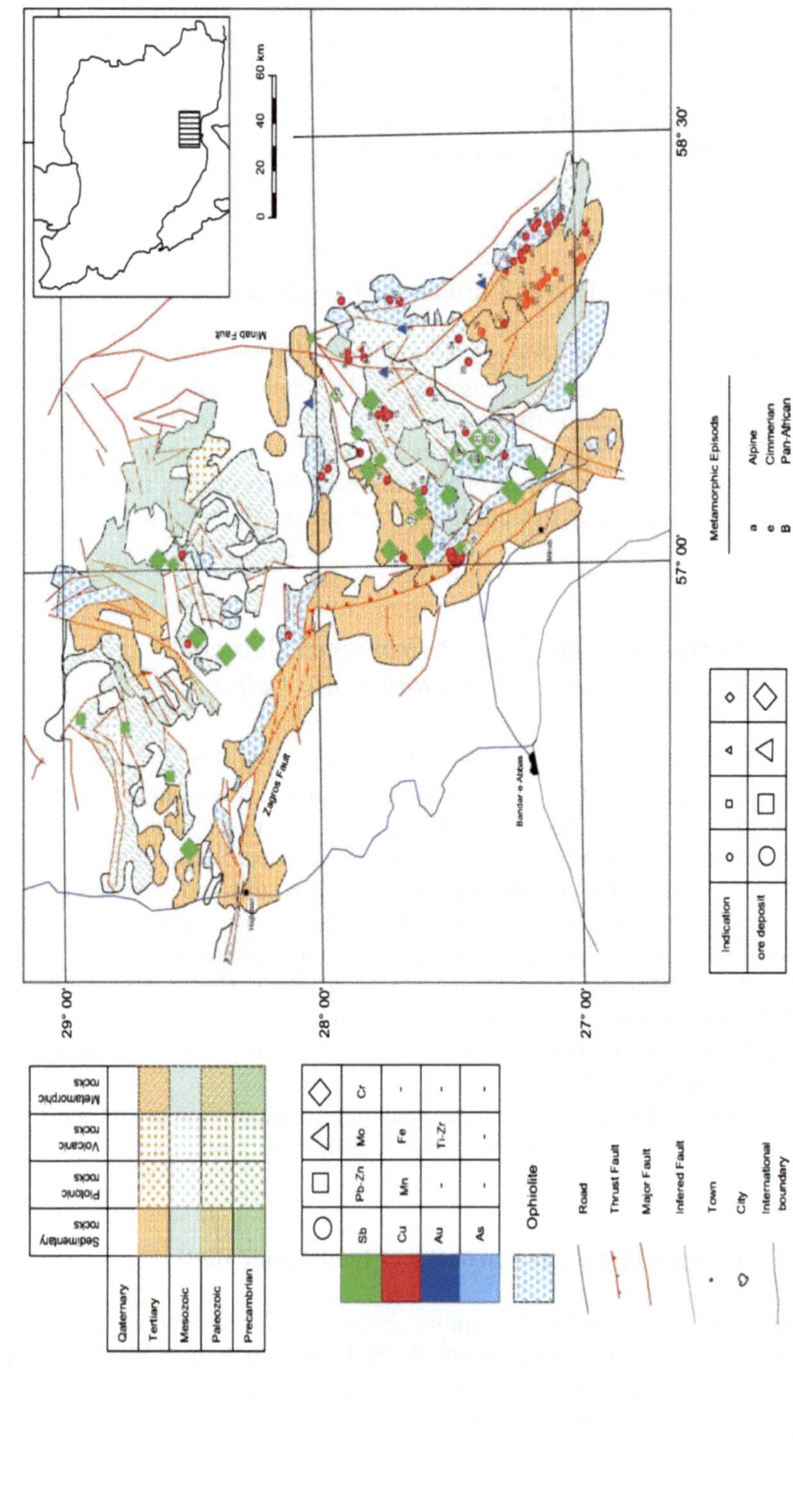

Fig. 6.16 Metallogenic map of Esfandagheh – Faryab area (Ghorbani 2002a, 2006b)

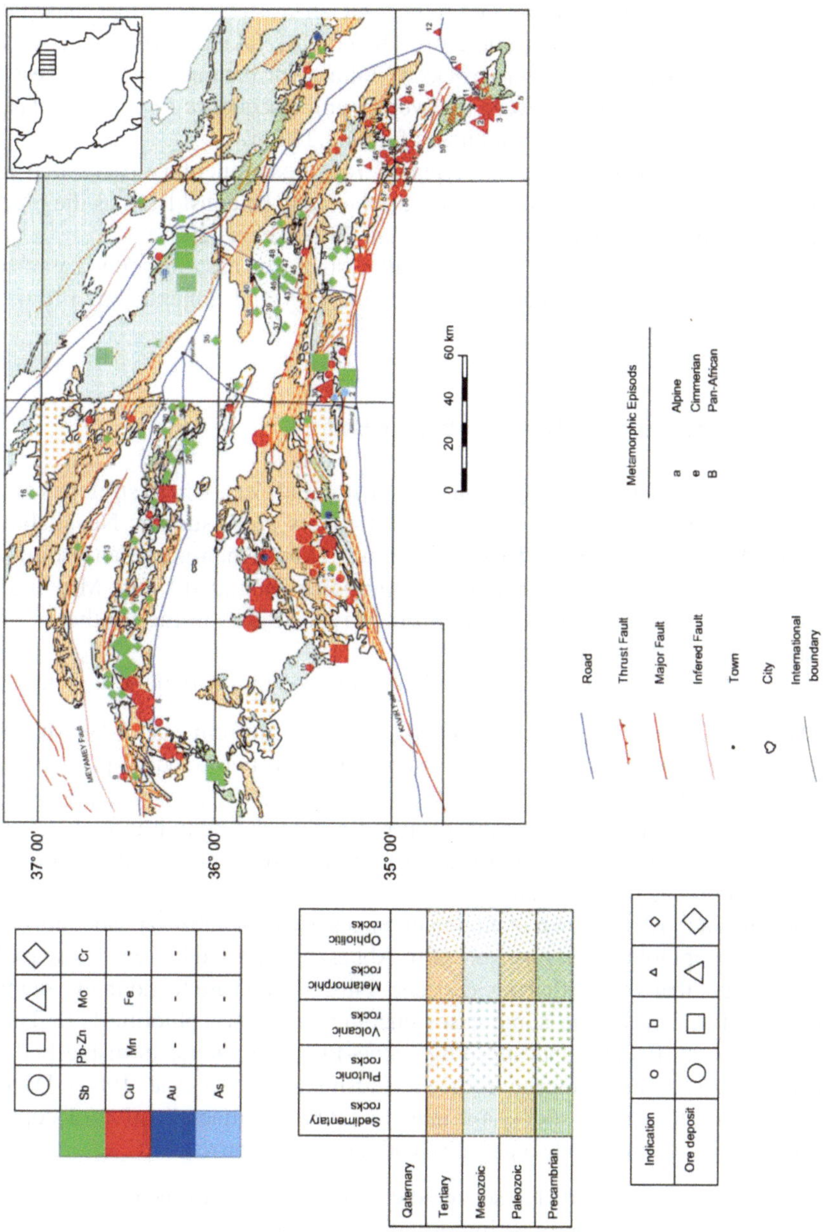

Fig. 6.17 Metallogenic map of Taknar and Kavir- Sabzevar belts (Ghorbani 2000b, 2002a)

The Taknar metallogenic zone in north of the Dorouneh fault uplifted as a wedge-shaped block and caused the Paleozoic, Mesozoic, and Cenozoic rocks to show up at the surface (Boroumandi 1982; Ghorbani 2012a). Due to the function of nearby faults, the western boundary of this belt turned into an erosion fenêtre that caused the old rocks of Taknar Formation to outcrop. The core of Taknar Formation consists of volcanic and volcanosedimentary rocks of Taknar Formation with a thickness of 1,650–1,850 m. The top of this core is covered with a sequence of younger clastic and carbonate sediments with a thickness of 200–2,250 m.

Taknar Formation is made of a series of volcanosedimentary rocks that are host rock to Taknar massive sulfide deposit, which can be considered as the oldest mineralization phase within this belt (Ghorbani 2002a; Feyzi 2005).

Due to disorderliness in the Riyoush fault system, the eastern boundary of the Taknar zone features significant diversity in lithology and mineralization (during Tertiary in particular).

6.3.4.2 Mineralization in Kavir–Sabzevar Belt

This belt is called as the Sabzevar zone by most geologists, which stretches from east of Neyshabour to Sarkavir Semnan and covers areas such as Neyshabour, Sabzevar, Abbasabad, Miami, Sarkavir Semnan (Moaleman area), and south of Damghan. The Kavir–Sabzevar metallogenic belt is bounded to the Miami and Neyshabour–Torbantjam faults in the north, Riyoush jault in the southeast, and Dorouneh fault in the southwest.

This metallogenic and mineralization belt is divided into two main sections:

(a) The ophiolitic section, which is a band stretching out from east of Neyshabour to west of Foroumad. This band is separated into two unequal parts by the Neyshabour Plain.

Wide chromite mineralization occurred within the ophiolitic body of this band whose chromite deposits are of high-grade type chromites in Iran (e.g., Foroumad deposit, Gaft deposit).

(b) The volcanic rocks are seen in mountain ranges between Sabzevar and Ghoochan within the ophiolitic body as well as in Abbasabad–Sarkavir area. These rocks are younger than ophiolitic body (Eocene–Pliocene) and are found in the vicinity of ophiolitic belt and in some cases encompassing the ophiolitic belt. These rocks include Eocene andesite, Oligocene–Pliocene dacitic intrusive bodies, alkaline basalt, and Miocene–Pliocene shoshonite (Spies et al. 1983). During the formation of these rocks, a relatively extensive mineralization of copper, gold-bearing copper, and sometimes lead and zinc occurred.

As mentioned earlier, wide but scattered chromite deposits formed, among which the Foroumad deposit is the most important deposit discovered so far. The mineralization evidences identified within the ultramafic part of ophiolites in Sabzevar bespeak of high likelihood of existence of high-grade chromite deposits. Therefore, more precise and detailed exploration work is required throughout this area.

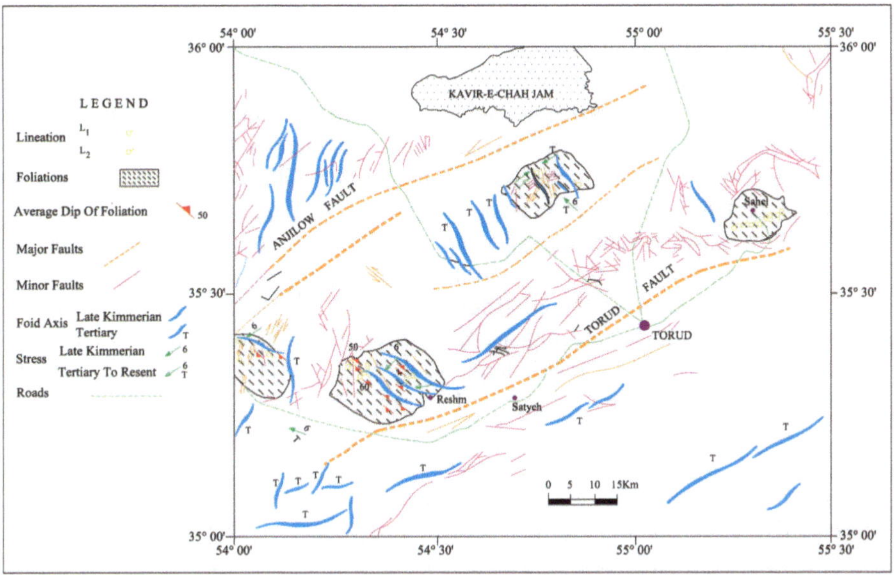

Fig. 6.18 A view of Toroud mining area between two Anjiloo and Toroud faults (After Houshmand-Zadeh et al. 1978)

Within post-ophiolitic rocks of Sabzevar mineralization of gold, gold-bearing copper, lead and zinc, and copper is seen, mostly in the form of vein, within the volcanic rocks of the Kavir–Sabzevar zone. No sign of porphyry copper mineralization has yet been identified along this belt (Ghorbani 2000f).

In general, the mineralization of copper, gold, lead, and zinc within the Kavir–Sabzevar metallogenic belt can be divided into two mining districts:

- Abbasabad mining district in southeast of Miami with copper and gold-bearing copper mineralization
- Sarkavir mining district with zinc, lead, copper, and copper–gold mineralization

There are many nonmetallic deposits such as bentonite and kaolinite; feldspar and silica have been discovered within the Kavir–Sabzevar Belt.

6.3.4.3 Toroud Metallogenic Area (Cu, Fe, Pb–Zn, Au)

The Toroud–Chah Shirin metallogenic area is located in the northern part of Central Iran zone, south of Damghan. The area is bounded between two faults: Toroud fault in the south and Anjilou fault in the north, both trending E–W. Other faults with N–S trend are also seen in the area (Fig. 6.18).

The Toroud–Chah Shirin area mainly consists of volcanic to volcanosedimentary rocks such as andesite, andesite–basalt, volcanic andesite–trachyte breccias, and acid tuffs, which are cut through by intermediate to acid intrusive bodies including diorite, monzonite, syenite, tonalite, and granodiorite. The magmatic rocks in the area generally fall in alkaline series bucket and feature an intracontinental rift. This volcanic-intrusive belt forms a volcano-plutonic arc in the form of an elongated ellipse whose major axis trends northeast–southwest. One of the significant features in this area is the presence of Eocene and Eo–Oligocene magmatism (Ghorbani 2012a).

There are many mineral deposits and indications with metallic variety seen in this area, which are distributed in a specific order (e.g., Chah-mousi copper deposit, Gandi gold deposit, Abolhasan gold deposit, Kuh-zard and Bagho gold deposits, Khan-jaroo lead–zinc deposit). Mineralization in some of these deposits follows exactly east–west direction (e.g., Abolhasani gold deposit), and in some others, north–south direction (Chah-mousi copper deposit). Generally, mineralization is bounded and controlled by major faults in the area. So far more than 20 metallic deposits such as Cu, Au, Pb, Zn, Fe, and turquoise have been identified in this area, and although they are small in size all of them have significant grade that makes them economically valuable in the current situation.

The Toroud area witnessed magmatic activities since very old geologic time. These activities began weakly in Paleozoic and Mesozoic due to tectonic events, and hit their peak in Tertiary, which left significant volumes of igneous rocks throughout during this time. In Tertiary, the magmatic activities began in the Lutetian stage and continued through Middle Eocene and came to a halt in Late Eocene, and then resumed their activity at the end of Late Eocene to Early Oligocene and the nature and mechanism of these activities shifted from volcanism toward plutonism (Ghorbani 2007a). However, the intensity of such activities was not as immense compared to the primary phase, and they shut down very soon.

The volcanic rocks in the area can be summarized as follows:

- The volcanic rocks in the southern part of the area consist of olivin–basalt, differentiated olivine–basalt, and pyroxene–andesite, which intercalate with sedimentary rocks (especially nummulite-bearing limestone and sandstone).
- The volcanic rocks in the northern part of the area mostly include rocks with more acidic composition leaning toward more andesite type and less basalt and dacite.
- The volcanic rocks in the southern part include lavas in the form of flows and columns with aphanitic texture.
- Toward the north, there are rocks of intermediate to more acidic nature, which are characterized with porphyritic texture and plagioclase phenocrysts such as amphibole, pyroxene, and biotite.
- The volcanoclastic rock complex is seen as intercalation with lavas in different parts of the area.

The plutonic rocks in the area can be summarized as follows.

- The volcanic rocks in Toroud–Chah Shirin area have been cut through by intermediate to acid intrusive bodies in form of dome, stock, or dike.
- These can be as large as a few km^2, with the largest being Chalou (25 km^2).
- Except for one intrusive body (Gandi in northeast Moaleman), the rest of these plutonic bodies feature microgranular or porphyritic texture and are considered as semi-deep type of body (Ghorbani 2012a).

- The petrographic composition of these bodies mainly includes monzodiorite–granodiorite–granite, and intermediate to basic dike cut through these bodies.
- The nature of the magmatic series of Eocene volcanic rocks in the Toroud area and associated intrusive bodies is mainly high-potassium calc–alkaline to alkaline (shoshonite).
- It is to be noted that the volcanic rocks in the Toroud area belong to one magmatic phase with age of Eocene, and the intrusive bodies that were injected through these rocks later on show genetic relation with these volcanic rocks (Ghorbani 2007a).

Alteration in Toroud Area

Since the intrusive bodies in the area are more acidic than volcanic rocks, the hydrothermal solutions produced by these intrusive bodies caused silicic alteration in most cases, and therefore silicic alteration is the dominant type in the Toroud area.

Four types of alterations are seen in the area as described next:

- Silicic alteration: Such alteration is mostly bounded and controlled by faults, and it is traceable wherever the intrusive bodies are widely spread. Silicic alteration is very important in regard to gold mineralization. As an example, silicic alteration is seen north of Gandi where Abolhasani gold deposit is located and is spread out for a few kilometers.
- Propylytic alteration: This alteration type is seen in many places within the Toroud metallogenic area although they are mostly local. There are some specific places where propylytic alteration is seen:
  - In places where basic dikes cut through surrounding rocks in the area (along the edges of dikes)
  - Along the contact line between volcanic rocks and intrusive bodies
  - In fractures in hard volcanic rocks
  - It is to be noted that propylytic alteration had no role in mineralization in the area, and it only helps to determine the mineralization borders and limits when it comes to iron mineralization studies.
- Argillic alteration: This type of alteration is widely seen in pyroclastic rocks especially in tuffs. Argillic alteration had a major role in kaolinite and bentonite mineralization along the Rashm-Gandi axis (north of Moaleman).
- Potassic and sericitic alteration: Although some weak signs of sericitic alteration are seen and traceable in some mineral veins, no sign of potassic alteration has been found in the area so far.

In general, two specific types of mineralization are seen in the area: metallic and nonmetallic. Metallic mineralization includes iron, copper, gold, lead, and zinc. For nonmetallic mineralization, we can refer to bentonite, kaolinite, feldspar, and semi-precious stones.

Metallic mineralization in the area can be summarized as follows:

- Veins and veinlets: Most metallic mineralization is seen in the form of veins and veinlets, which are mostly controlled by the faults and are of hydrothermal origin.

- Lens and massive: Such mineralization type is mostly specific to bentonite, kaolinite, and feldspar deposits, which generally originated from acid tuffs.
- Lens: Most iron deposits in the area are lentoid such as Baloujeh and Panj-kuh in north of the Toroud.
- Porphyry: The reality is that the geology, magmatism, and alteration of the area indicate that the Toroud area lacks porphyry copper mineralization for the following reasons:
  - Only two magmatic phases occurred in the area.
  - No sign of porphyry alteration is seen in the area.
  - No sign of porphyry copper found as studied by the aerial and satellite images.
  - From the tectonic and structural geology standpoints, Toroud is not suitable for the formation of porphyry copper deposits.

Metallogenic Districts in Toroud Area

Considering the characteristics of this metallogenic belt, five metallogenic districts can be defined within this belt:

1. Moaleman: There are three major rock units in this district:
   - Sedimentary rocks including carbonate, sandstone, and shale, which had almost no role in mineralization in the area.
   - Volcanoclastic rocks, especially acid tuffs, spread out discontinuously from Rashm to northeast of Gandi. These tuffs are the source for bentonite and kaolinite deposits in the area such as the Rashm bentonite, Soosanvar bentonite, and Gandi kaolinite deposits.
   - Volcanic and sub volcanic rocks that caused significant mineralization and include andesite, andesite-basalt, trachyte, rhyolite, and acidic sub volcanic rocks.

Mineralization in these rocks is summarized as follows:

- Gold: Gandi and Kakiyeh gold deposits
- Copper and gold: Abol-hassani copper and gold deposit
- Iron: Chalou iron deposit
- Lead and gold: Cheshmesh sefid lead and gold indication

2. Robaei: This district is located east of Rashm gorge and consists of three major rock units:
   - Paleozoic metamorphic rocks including marble-schist
   - Cretaceous carbonate sedimentary rocks
   - Intermediate to acid intrusive bodies that are not large

In this metallogenic district, the metamorphic and carbonate sedimentary rocks played the role of host rock.

Mineralization in these rocks is summarized as follows:

- Lead and zinc: Khanjar Rashm, Anarou, Robaei, Sang-kar, Tanoreh, Tangeh, Par-magasou, Akhori, Ghale-dokhtar, and Bozeh

- Iron: Heshdeh deposit and Kam-anjir iron indication
- Manganese: Chah-gabri deposit

3. Kuhzar–Baghou: There are also three rock units in this district that played a role in the mineralization and accumulation of mineral deposits, and can be divided as follows:
   - Volcanic rocks with composition of mostly andesite, andesite–basalt, and trachyte–andesite
   - Tonalite to granite intrusive bodies
   - Pliocene to Early Quaternary alluvium

Mineralization in these rocks is summarized as follows:

- Gold: Kuhzar gold deposit originated from intrusive bodies and associated solutions.
- Turquoise: Baghou turquoise deposit, which is associated with the andesite rocks in this district.
- Placer gold: This type of placer gold accumulation can be remarkable within the whole Pliocene and Early Quaternary alluvium in northern plains of this district.

4. Polymetal district north of Toroud: It is located north of Toroud (Fig. 6.19) and mainly consists of Eocene volcanic and pyroclastic rocks and Eocene–Oligocene intrusive bodies. These bodies penetrated into volcanic and pyroclastic rocks and released hydrothermal solutions into these rocks that led to the formation of polymetal deposits. Some of these polymetal deposits are as follows:
   - Gholeh Kaftaran
   - Chah Mousa
   - Chah Galeh
   - Cheshmeh Hafez
5. Chah Shirin–Kalout: This metallogenic district is seen on the west of the Toroud area and mainly consists of Eocene volcanic and pyroclastic rocks, and metamorphic rocks. Oligo–Miocene intrusive bodies cut through the aforementioned rock units, and the major factors in mineralization in this district were the very same hydrothermal solutions released from these intrusive bodies. Copper, lead, and zinc are the main mineralization in this district, and some of the significant deposits are as follows:
   - Chah Faragh
   - Chah Farsakh
   - Chah Farakh
   - Chah Bod
   - Sar-shirin
   - Kalout
   - Kalout-e Boland
   - Sar Kalateh Mehran 1 and 2

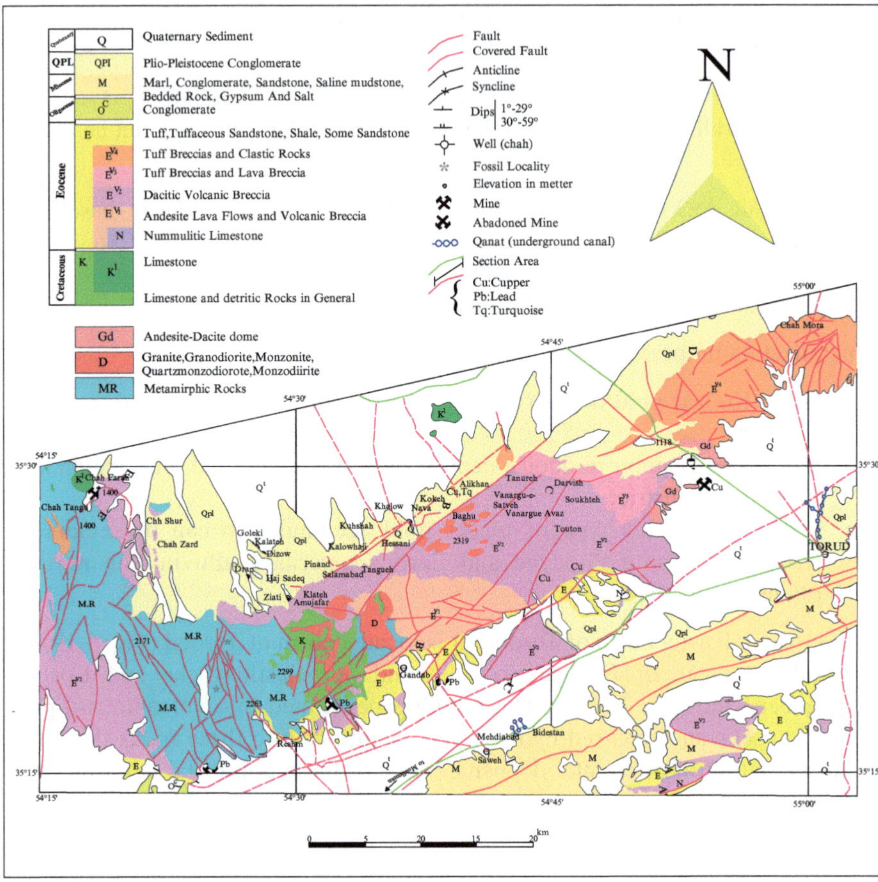

Fig. 6.19 Metallogenic map of Toroud area

6.3.5 *Alborz Metallogenic Province*

Alborz is a mountain range in north of Iran with a length of about 1,000 km and its width varies between 50 and 100 km. It covers an area of about 5,200 km^2. This mountain range starts from Gorgan stretching like an arc along the south side of the Caspian Sea and ends in Astara. From a geographic standpoint, this mountain range trends in the east–west direction and stretches from Azerbaijan to Khorasan. Eastwardly it enters into Afghanistan and westwardly sneaks into Caucasus (Fig. 6.20).

6.3.5.1 Tarom–Hashtjin Metallogenic Belt

From a structural point of view, this belt is confined from the north by the Sefidrood fault, from the south by the continuation of the Tabriz–Soltanieh and Soltanieh–Takestan faults, and from the west by the Astara–Marivan fault (Ghorbani 1999f).

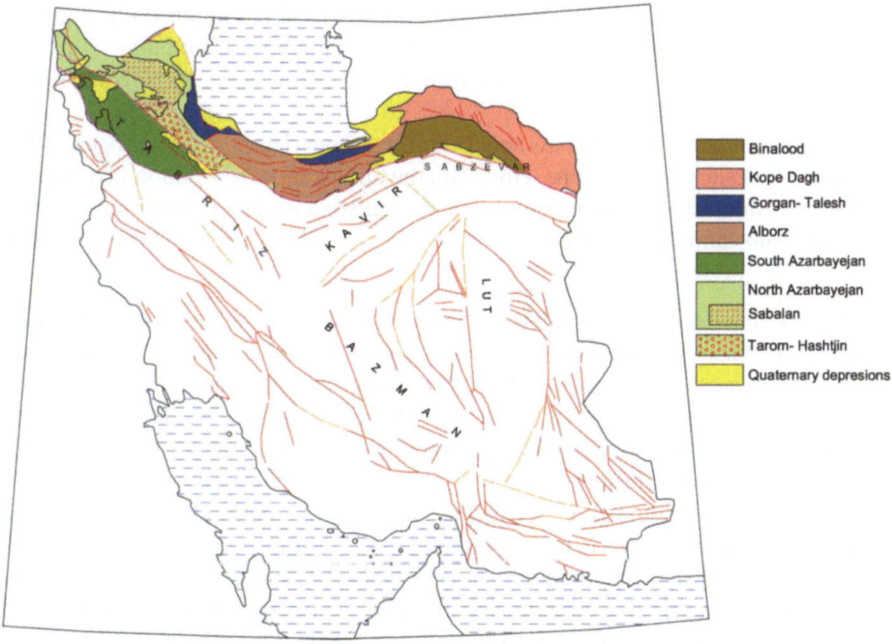

Fig. 6.20 Structural zoning of Alborz and adjacent structures in west and east (Ghorbani et al. 2005)

This belt covers the area between Qazvin (west of Taleghan) and north–northwest of Mianeh and is bounded on the north by the Manjil Embayment and Talesh Mountains and on the south by the Abhar–Znjan–Mianeh axis.

Structurally, the Tarom–Hashtjin Belt is limited by three major faults, namely, Sefidrud fault on the north, extension of Tabriz–Soltanieh and Soltanieh–Takestan faults on the south, and Astara–Marivan fault on the west (Ghorbani 2006b)

Volcanic rocks of Tarom vary in composition from rhyodacite and dacite to basalt, which are in the form of lavas, tuffs, and tuffites.

The intrusive bodies consist of granite, alkaline granite, granodiorite, monzogranite, monzonite, quartz monzodiorite, syenite, alkaline syeite, and quartz syenite.

The nature of magmatic melt of these rocks is shoshonite and potassium-rich calc–alkaline I-type granite.

Shallower igneous bodies in the form of porphyrite with a general composition inclined toward that of monzonite are seen along the main intrusive body, which played an important role in the formation of hydrothermal metasomatism and mineralization processes

Geology and Mineralization Processes

This belt covers the area between Qazvin (west of Taleghan) and north–northwest of Mianeh and is bounded on the north by the Manjil Embayment and Talesh Mountains and on the south by the Abhar–Znjan–Mianeh axis.

Structurally, the Tarom–Hashtjin Belt is limited by three major faults, namely, Sefidrud fault on the north, extension of Tabriz–Soltanieh and Soltanieh–Takestan faults on the south, and Astara–Marivan fault on the west (Ghorbani 1999f; Fig. 6.21).

Many experts (e.g., M. Nabavi) consider the Tarom–Hashtjin Metallogenic Belt as a part of western Alborz zone. However, the author is of the opinion that, at least in Tertiary times, due to the mode of metamorphism and mineralization processes, there are substantial differences between the two, which can be summarized as follows:

1. Rocks considered as equivalents of Karaj Formation in the Tarom–Hashtjin Belt are more basic in composition and as a result the proportion of volcanic lava flows increases as compared to pyroclastics contrary to the composition of Karaj Formation, where the latter are the preeminent rock successions. In fact, the central parts of Alborz are dominantly composed of acidic tuffs while the major volume of rocks in Tarom–Hashtjin is andesites and andesitic basalts.
2. In the Tarom–Hashtjin Belt, the ancient rocks are not exposed within the central parts and the basement is mostly composed of Tertiary intrusives. However, in central Alborz, the ancient basement is in excess of younger intrusives.
3. Copper, gold, lead–zinc, barium, and manganese mineralization is very common in the Tarom–Hashtjin Belt, whereas in central Alborz they are rare.
4. The volcanic rocks of Tarom vary in composition from rhyodacite and dacite to basalt, which are in the form of lavas, tuffs, and tuffites. Petrographical studies carried out on the rocks of Tarom–Hashtjin reveal the presence of basalt, andesite–basalt, andesite, trachyte, latite, trachyandesite, dacite, rhyodacite, ignimbrite, and acidic to intermediate tuffs with andesites forming the bulk of the rocks. Moein–Vaziri (1996) is of the opinion that most Eocene volcanics of Tarom (Tarom–Hashtjin) are potassic–alkaline, sodic–alkaline, or calc–alkaline in nature and andesites are of shoshonite composition. Feldspathoid are rarely formed in the Eocene volcanic rocks but contain considerable percentages of orthopyroxene.

Zareie (1992) showed the relationship between the acidic and basic intrusive bodies within the area, many of which, in the Tarom–Hashtjin Belt, cut across Eocene volcanics everywhere. They are in turn overlain by Qom Formation with a nonconformity and are thus Late Eocene–Oligocene in age. The intrusive bodies consist of granite, alkaline granite, granodiorite, monzogranite, monzonite, quartz monzodiorite, syenite, alkaline syenite, and quartz syenite. Studies by Peyrovan (1992), Torkaman (1997), and Moayyed (2001) have all indicated that the intrusive bodies are of I-type. The field, petrographic, and mineralogical observations by the author points to Cordillerian I-type nature for these rocks.

Haj Aliloo (2000) has determined the depth and temperature of formation of the igneous intrusives at 1,400–3,000 m and 700–880°C, respectively. He further concludes that due to high water content and sulfur fugacity, extensive hydrothermal metasomatism along with scattered vein mineralization has occurred within Eocene tuffites and volcanics. Petrological evidence indicates a post-collisional magmatic arc for the origin of the plutonic activity within the area (Moayyed 2001).

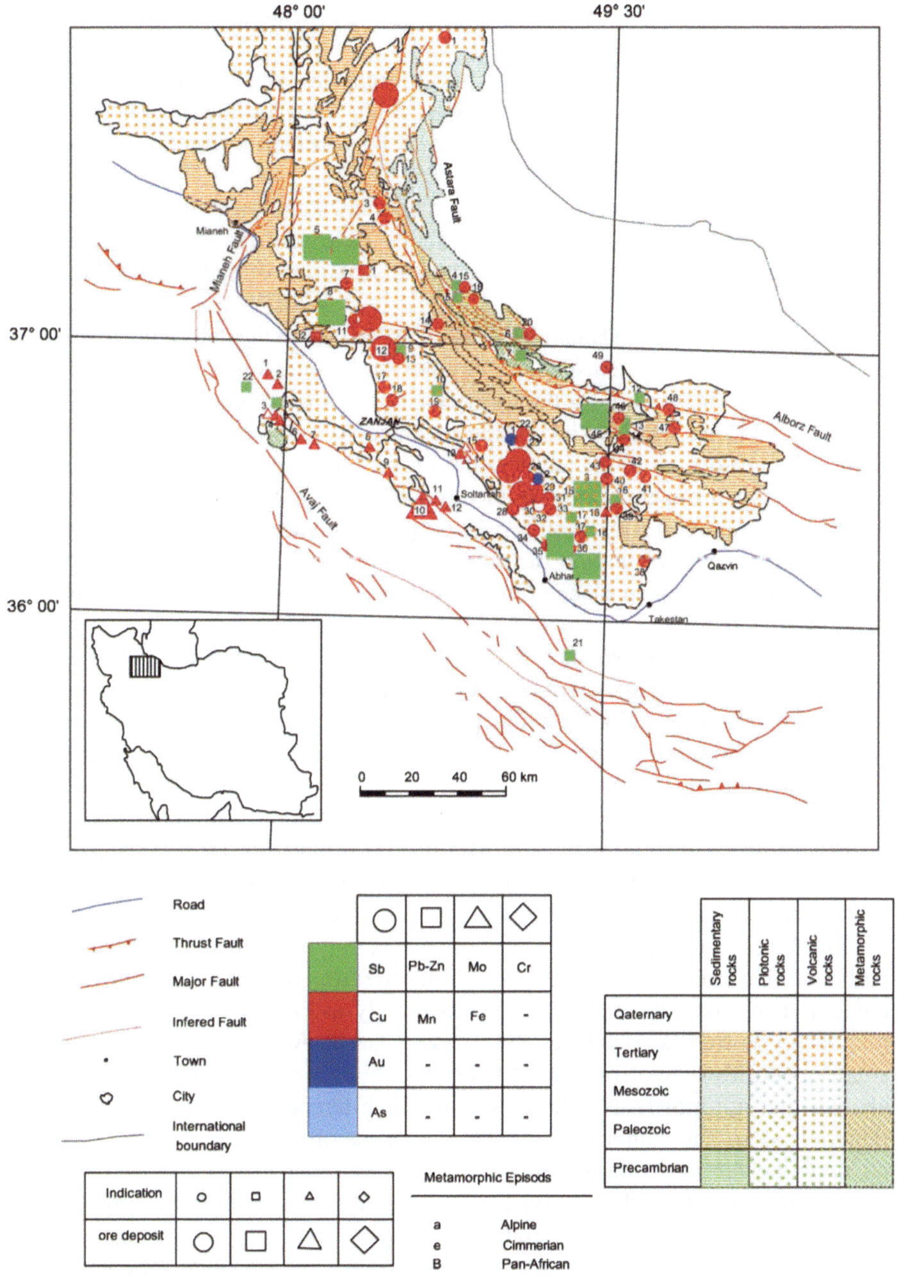

Fig. 6.21 Metallogenic map of Tarom–Hashtjin Belt (Ghorbani 2000b)

The nature of magmatic melt of these rocks has been shoshonite and potassium-rich calc–alkaline I-type granite.

Shallower igneous bodies in the form of porphyrite with a general composition inclined toward that of monzonite are seen along the main intrusive body, which played an important role in the formation of hydrothermal metasomatism and mineralization processes (Arian Zamin 2009).

Structural elements such as axes of the anticlines and synclines as well as the east–west northwest–southeast trending faults have all been responsible for the emplacement of the intrusive bodies and the distribution of hydrothermal metasomatic alteration zones.

Thermometric studies (Hajali-lou 2000) carried out on the vein-type mineralization zones around the metasomatic aureoles and silicic veins within them point toward a high temperature for the formation of the mineral bodies. Moreover, it is also observed that with the increasing distance from the intrusive body and the center of metasomatism, the temperature drops. However, there is no evidence of porphyry copper mineralization in the area. Chlorite thermometry reveals a 340–365°C temperature of formation for the propylitic metasomatic zone and medium temperature for the formation of argillites (Hajali-lou 2000). Further, lead–zinc mineralization has taken place at the same time, or shortly after the formation of the propylite zone.

Other metasomatic zones, such as serricitic, argillic (weak, medium, and advanced), silicic, chloritic, propylitic, zeolitic, and alonitic, have also been recognized in the area of the Tarom–Hashtjin Metallogenic Belt (Hajali-lou 2000).

Due to the presence of two distinct phases of metasomatism (local and regional), the relationship between metasomatism and mineralization requires to be accurately investigated to throw light on the local metasomatic activity within the area.

Mineralographic studies by Haj Aliloo (2000) have revealed the presence of bornite, chalcocite, covellite, chalcopyrite, malachite, azurite, dijenite, pyrite, arsenopyrite, galena, sphalerite, tetrahedrite, freibergite, sericite, magnetite, limonite, hematite, and lepidocrocite in the vein-type mineralized zones of the Aghkand–Hashtjin Region. Copper mineralization is mainly in the form of chalcopyrite, covellite, bornite, and occasionally as chalcopyrite. Electron probe microanalysis studies (Hajali-lou 2000) have also thrown light on the presence of small quantities of wolframite, shellite, enargite, tennantite, bismuthinite, and ekanite that are mostly associated with the high-temperature pegmatitic phase of mineralization. The author believes that this phase of mineralization can be traced all along the Tarom–Hashtjin Belt with minor variations.

The presence of shoshonite-type (potassium-rich) intermediate-depth plutonic bodies, high sulfur fugacity of hydrothermal solutions, extensive development of silicic and alunitic metasomatism, and arsenic- and antimony-bearing telluride composition in Hashtjin area along with small quantities of copper, silver, and molybdenum have lead Haj Aliloo (2000) to the conclusion that there are high chances of formation of mesothermal and porphyry-type gold deposits in the area. This is further supported by the existence of copper (vein-type or otherwise), lead–zinc, and iron deposits that show traces of gold.

The component areas of this metallogenic belt, namely, Tarom and Hashtjin, slightly differ from each other from the point of view of mineralogy and igneous petrology, and thus the belt can be divided into two parts based on the following distinctions:

1. The bulk of the host rocks in both the areas consist of Karaj Formation, but acidic and intermediate units are more in Tarom as compared to Hashtjin. Further, the composition of Karaj-equivalent volcanic rocks of Tarom is similar to porphyry-textured andesite, whereas at Hashtjin they are of andesitic basalt type. The distribution of acidic tuffs such as rhyolite and dacite in Tarom is more extensive; this explains the presence of widespread hydrothermal metasomatic products in this part of the belt.
2. Plutonic bodies of the Tarom area are comparatively more acetic and diverse than those of Hashtjin.
3. Though similar trend of mineralization is observed in both areas (dominated by lead–zinc, copper, iron and manganese), copper and iron are more widespread in Tarom. The average copper content of plutonic bodies of Tarom varies between 220 and 260 ppm, while lead, zinc, and silver are around 600, 800, and 8 ppm, respectively. Plutonic rocks of Hashtjin have copper content of 30 ppm, lead content of 40 ppm, and zinc content of 70 ppm, which is another reason for the scarcity of copper deposits in the area (Haj Aliloo 2000).
4. Extensive iron mineralization is seen along with intrusive bodies of Tarom only.
5. Alunite deposits are more in Tarom as compared to Hashtjin, for example, Yoozbash Chai, Kamar Rud, Nasr Abad, Sirdan, Zajkan, and Zajkandi. This has been attributed to the type of host (potassium- and aluminum-rich) rock and sulfur fugacity of hydrothermal solutions
6. The distribution of zone of serricitization in Hashtjin is more widespread as compared to the Tarom Region.
7. Greissen-type metasomatism has not been reported from Tarom, but the silicified zone (silica cap) is more extensive than Hashtjin.
8. Signs of gold mineralization are seen in both Hashtjin and Tarom areas. In Tarom, gold is associated with copper deposits and indications, for example, Khalifeloo, Abbasabad, and Chargar. Moreover, all the silicified and hematitized zones show gold mineralization, for example, Asadi and Ghez Ghal'e villages. Haj Aliloo (2000) has reported porphyry gold mineralization at Hashtjin where the percentage of molybdenum is extremely low.
9. Minor amounts of tungsten have been shown to exist in both Tarom (Kuhian) and Hashtjin (Kamar and Shah Alibeigloo villages).

6.3.5.2 Ahar–Jolfa Metallogenic Area (Arasbaran)

This metallogenic zone covers the areas to the north and west of Meshkin Shahr, Ahar, Kalibar, Varzeghan and Siyah Rud, and Arasbaran and Qareh Dagh Hills.

The Ahar–Jolfa metallogenic zone is limited on the south–southwest by the Tabriz–Soltanieh fault, on the east by the Ardebil–Mianeh fault, and on the

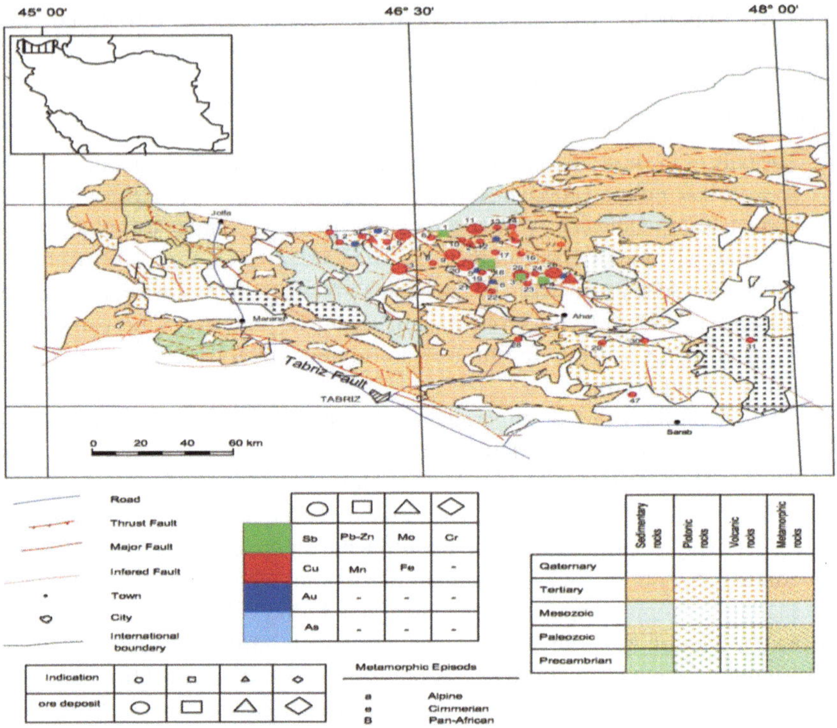

Fig. 6.22 Metallogenic map of Ahar- Jolfa area (Ghorbani 2011)

northeast by the embayment near the east–west trending Moghan fault. The extension of this metallogenic zone on the north reaches the Minor Caucasus metallogenic zone (Fig. 6.22).

Around Sefid Rud, the Alborz structural zone changes its east–west trend to northwest–southeast. This change is brought about by the movement along three faults, namely, Sefid Rud, Aras, and Talesh, the first two being left lateral while the last being right lateral, making the block on the southwest to move relatively down, while the block in between the Aras and Talesh faults on the northeast to move relatively upward, resulting in the formation of a tensional zone limited by the two faults (Talesh and Aras) leading to igneous activity during Quaternary (Sabalan Volcano).

Geology of Ahar Area

Structural Geology

The overall structure of Ahar–Jolfa Metallogenic Province has been formed as a result of Alpine activity, the most intense of which was the Laramide orogeny at the end of Cretaceous. Following this phase, the Pyrenean phase resulted in outpouring of extensive lavas and pyroclastics. During the Late Eocene–Early Oligocene, tectonic upheavals associated with emplacement of plutonic masses are dominant, while in Miocene widespread extrusion of lavas and formation of small intrusive bodies. The latter has continued up to the Quaternary (Ghorbani 2011).

According to Lescuyer and Riou (1976), following the granitization phase that acted upon the Azerbaijan Plateau during Oligocene and resulted in the strengthening of the plateau, Azerbaijan behaved like a microcontinent. This microcontinent was situated between the Talesh Mountains and Caucasus Minor around 20 million years ago (Aquitanian) connecting the Zagros thrust zone to the Caucasus suture zone by transform faults (Zonenshain et al. 1989). Thus, the intense tectonic movements that compressed the region in the north–south direction pushed the said microcontinent toward north, resulting in the emergence of Caucasus from under the sea. Simultaneously, the movement along the Soltanieh–Tabriz fault caused the microcontinent to rotate and shift toward the north.

Most probably this rotational movement has resulted in the structural displacement and reorientation of eastern parts of the Azerbaijan Plateau.

According to Zonenshein and Lopishoun (1986), the northward movement of Arabian Peninsula during Middle Pliocene (3.5 Ma) has resulted in the activation of Talesh and Caucasus transform faults, leading to the separation of Caspian sedimentary basin from the Black Sea. Therefore, the Azerbaijan Plateau is presently in a state of motion in the north–northeast direction and the Sabalan and Sahand Plio–Quaternary volcanisms and the earthquakes within the area are the results of the said motion.

The most fascinating tectonic aspect of Azerbaijan is the association of Mid-Alpine orogenic phase with an enormous folding phenomenon that led to the uplifting of the Qareh Dagh area and emplacement of intrusive bodies along the Sabaln–Songoon axis toward terminal Alpine (these are associated with volcanic activity and mineralization elsewhere). The present features of the area have all been developed through folding during the Alpine, Austrian, and Pasadenian orogenic phases (Ghorbani 2011).

The northern parts of the area have probably been uplifted following the Alpine tectonic activity. The presence of fragments of orbitoline limestone within the Upper Cretaceous conglomerates is the testimony to the said movements. There are no evidences of any stratigraphic hiatus between the Lower and Upper Cretaceous rocks as well as the Upper Cretaceous and Paleocene rocks in the Qareh Dagh area.

The tectonic activity of the Late Cretaceous time is only seen in the north of the area, which resulted in the southward displacement of the east–west trending folds. Fissure eruptions that followed the compressional movements of Late Cretaceous covered the pre-Paleogene topography under a thick column of volcanic lava. A subsiding basin formed immediately after Early Eocene as evidenced by the marl and limestone rocks covering the Eocene volcanics (Cheshmoghan Valley in Astamal area) (Ghorbani 2011).

Miocene sedimentary basin had a structural trend independent of all the prior basins. Contrary to the northern basin, an unconformity exists between Neogene rocks and those older. Structural deformation during Late Miocene has led to the folding of the Moghan Basin. While the diapiric structure of the Khaje Miocene basin was being formed, the southern Miocene basin was being uplifted through intense folding and faulting. The rise of the Arasbaran–Qousheh and Dagh–Sabalan range along with continuous Plio–Quaternary inter-mountain embayment separated Varzeghan, Ahar, and Meshkin Shahr from the southern basin. The fracturing resulting from this phenomenon caused extensive volcanic activity during the Pliocene and Quaternary times that formed Sabalan, after which the structural displacement was confined to Tabriz, Qousheh Dagh, and Moghan high-angle faults.

Lithostratigraphy

Though the majority of rock types in Ahar–Jolfa are Tertiary igneous, older rocks are also exposed in the area. The main rocks of Ahar–Jolfa Metallogenic Province are described in order from the oldest to youngest in the following paragraphs (Ghorbani 2011).

Pre-Late Cretaceous Rocks: The rock formations that have been formed prior to Late Cretaceous can be grouped into two broad categories:

- Metamorphic rocks exposed to the north of Kalibar, east of Allahyarloo and a few other localities whose composition is close to graywacke, phyllites, and marble.
- Jurassic–Cretaceous volcanosedimentary rocks that overlie the metamorphic are exposed on the northern parts of Ahar quadrangle. They include volcanic, pyroclastic (andesitic composition), and sedimentary rocks (limestone, marly limestone, and marl). These rocks are, in turn, covered by orbitolina-bearing limestones and marls of Aptian–Albian age. However, the Aptian–Albian rock formations are limited to the northwestern parts of Ahar quadrangle, that is, around the Aras River.

The schists and carbonates (Kalibar and Salavat Carbonates) of Albian to Turonian age along with acid volcanics and pyroclastics (and their intermediate submarine equivalents) are scattered in the northern parts of the quadrangle (Ghorbani 2011). Field evidence reveals that the submarine volcanic activity erupting acid to intermediate lavas extend their presence from end Albian times (with their height in northwest of Ahar quadrangle). From Late Turonian times they change in nature to basic types, while during Cenomanian once again the acid–intermediate–basic sequence is repeated. Paleocene times witness a totally different sequence of volcanic activities.

Paleocene–Eocene Volcanosedimentary Rocks: Though in continuation of Cretaceous volcanosedimentary activities, the Paleocene–Eocene submarine volcanism was more widespread both geologically and geographically in the Ahar–Jolfa zone (Ghorbani 2011). As a result, large areas of the northern half as well as the central parts of the zone have been covered by the products of this activity that are mostly andesitic in nature. Due to the extent and intensity of sub-marine volcanic activities, their extrusive and pyroclastic products dominate over the sedimentary rocks of this time which are limited in their exposures.

In Middle Eocene, volcanic activity was further aggravated in the form of alkaline basalts (olivine latite), basalt, cratophyric and spillitic lavas, trachytic lavas and pyroclastics, latite, dacite and trachyandesite along with tuff, volcanic ash, and colloidal material with ignimbrite texture are widely distributed in the northern and central parts of the quadrangle. This volcanic activity slowed down by the end of Eocene and the beginning of Oligocene, marking a quiet change from marine to semi-continental and continental environmental conditions. Simultaneously, continental lavas and pyroclastics replace sub-marine volcanic activity (Ghorbani 2011).

While the area was largely supplied through intensive volcanic activity during the time interval of Mid to Late Eocene, tuffs, tuffites, and nummulitic marly-limestones were deposited in limited thicknesses, whose outcrops are seen scattered on the west and southwest of Ahar and Kighal–Barmalek areas.

Oligocene–Miocene Rock Suites: As mentioned earlier, the volcanic activity that started in Cretaceous in the form of sub-marine lavas and pyroclastics changed its nature to continental by Oligocene, giving rise to small- to medium-sized dikes composed of andesite, alkaline syenites, and nepheline syenites. Their extrusive equivalents outcrop as ignimbrite, acidic domes, and dacitic volcanic breccias formed either on land or under very shallow marine conditions (Ghorbani 2011)

The associated sedimentary rocks include conglomerates, sandstones, reddish marls and conglomerates, gypsiferous marls with sand, and silt intercalations mainly exposed in the southern parts of Ahar–Jolfa Methallogenic Province.

Pliocene Volcanosedimentary Rocks: Shallow marine environmental conditions of Miocene were gradually replaced by fresh water ponds and lakes of Pliocene, while the igneous activity was once again dominated by extrusive phenomena; however, in the form of pyroclastic material only on land. The pyroclastics include trachyandesite domes, ignimbrites, and conglomerates that are largely scattered in the southern half of the Ahar–Jolfa zone and around Varzeghan. While the pyroclastic rocks are mostly exposed around Varzeghan and to the south of Sabalan, and the trachyandesites and ignimbrites are found only to the south of Sabalan, their sedimentary equivalents, that is, conglomerates and siltstones, are seen in the area to the east of Ahar and west of Varzeghan (Ghorbani 2011).

Quaternary Lavas, Pyroclastics, and Sediments: Quaternary lavas and pyroclastics cover parts of southern area of northeast Ahar quadrangle. Lavas having extruded from Sabalan Volcano are scattered on the south of Meshkin Shahr while lavas and pyroclastics suites were formed on the north of Varzeghan. The main craters of Neogene volcanoes that resulted in the formation of volcanic suites of north Varzeghan were apparently situated in the area between Kighal and Songoon in the north–south direction, and between Laleh Bejan and Sari Chaman in east–west direction (Ghorbani 2011).

Following the Quaternary volcanic activities or during the quiescence interval between different phases, sedimentary rocks were deposited that are presently exposed on the western parts of Ahar quadrangle (Tehran Padir 2002). Mid Quaternary onward, the volcanic activities of Ahar area were limited to two successive eruptions of Sabalan Volcano, the first resulting in the formation of trachyandesites and the second forming acidic lavas and breccias along with Pellean-type ignimbrite.

Igneous Classification of Ahar–Jolfa Zone (Ghorbani 2011)

Being one of the most active areas of Iran from the point of view of igneous activity, Ahar–Jolfa is comparable with the Kerman area only. Though most parts of the Orumieh–Dokhtar zone including Tarom–Hashtjin, Kavir–Sabzewar, and Anarak, have witnessed igneous activity during Tertiary, no other area has displayed volcanic and mineralization characteristics comparable with Ahar–Jolfa. Thus, it can be deduced that Ahar–Jolfa is not similar to any other parts of Azerbaijan and resembles the Kerman area in terms of Tertiary magmatic diversity.

While igneous activity started during the Late Cretaceous times in the Ahar–Jolfa zone and continued intermittently till Quaternary, in other parts of Iran it is not so persistent.

Taking into account the lithostratigraphic succession of Tertiary igneous rocks in Ahar–Jolfa, the activity can be classified into four dominant phases:

1. *Phase A*: Having started in the Late Cretaceous times, this phase reached its climax during Eocene (especially in Early to Middle Eocene). The beginning of this phase corresponds to the onset of the overall igneous activity within the area. It somewhat slowed down during Paleocene but once again heightened in Eocene, forming enormous volumes of andesitic–dacitic volcanic rocks and acidic tuffs that constitute the basement rocks of the area. This phase culminates with the formation of a large caldera and small intrusive bodies. The major igneous activity of this phase consists of volcanism.
2. *Phase B*: Following the intense volcanism of Early to Middle Eocene, the overall igneous activities came to a halt. However, a renewed phase of intrusive igneous activity started by the end of Late Eocene and Early Oligocene with no signs of any volcanism. A number of batholiths and stokes consisting of granite, granodiorite, syenite, nepheline syenite, monzo–syenite, gabbrodiorite, and diorite were emplaced during this phase, for example, Shiyor Dagh, Ordoobad and Yousefloo Granite–Granodiorite–Thonalite, and Kalibar Nepheline–Syenite.
3. *Phase C*: A time of quiescence followed the intense igneous phase of Late Eocene–Early Oligocene in Ahar–Jolfa (and Azerbaijan as a whole). However, this did not last for long and another phase of igneous activity began in Early Miocene (or according to some experts Late Oligocene) initially in the form of volcanism (but more diverse in its rock types) forming andesite, trachyandesite, trachyte, and occasionally acidic rhyolite. Along with these, a number of small intrusive bodies (mostly in the form of intermediate to acidic domes and dikes) were also emplaced. This phase reached its climax during Late Miocene–Pliocene when the small intrusives dominated over the volcanics. Most of the researchers (including those of Varzeghan 1:100,000 quadrangle) consider these as Plio–Quaternary in age. However, the author believes that this phase cannot be earlier than Late Miocene–Pliocene.
4. *Phase D*: Being the last phase of igneous activity in the Ahar–Jolfa zone, the rocks of this phase are basalts and andesitic basalts of Quaternary age. The Quaternary basalts of Ahar–Jolfa are quite widespread, almost found in every locality that experienced Phase C.

Petrology of Ahar–Jolfa Igneous Rocks (Ghorbani 2011)

Results of geochemical analyses of 60 samples collected from various phases of igneous rocks of the Ahar–Jolfa zone by X-ray fluorescence technique reveal that Phases A, B, and C are genetically related with one another while Phase D is different from the rest. Comparing the diagrams and curves drawn for Phases A, B, and C, it can be said that

1. Harker diagrams of all three phases are indicative of magmatic differentiation. This view is further supported by the curves of MnO vs. SiO_2 and MgO vs. SiO_2, or similar variation in TiO_2 and Al_2O_3. Thus, it can be inferred that a common parent magma was responsible for the production of Phases A, B, and C. However, the role of the crustal rocks in magmatic differentiation over a long period of time cannot be ruled out.

   To prove this, the variation in trace elements within the average samples of Phases A, B, and C were subjected to further investigations. The resemblance of average diagrams in Phases A and B is so great that the presence of one common parent magma cannot be ruled out. But Phase C, particularly when compared to Phase A, is somewhat different. The main difference between Phases A and C is that the volcanic rocks of the former are dominantly composed of andesites while that of the latter are dacitic in composition. As a result, the effect of crustal rocks in Phase C has been greater, so the magma giving rise to Phases A and B has somewhat changed over time under the influence of composition of the crustal rocks as well as the duration of differentiation, variation in melting percentage, etc., resulting in gradual enrichment of alkaline elements.
2. The relationship between Phases A, B, and C is further confirmed through the relationship of their nature of origin. According to the results of geochemical analyses, all the three phases contain calc–alkaline to sub-alkaline rocks (though some of the rocks of Phase B fall under the alkaline category, the reason for which is contamination of magma due to assimilation of crustal xenoliths).
3. Tectonic settings of all three phases are similarly based on the results of geochemical analyses. Phases A and C are basic to calc–alkaline rocks formed in island arc setting and are thus indicative of convergent plate margins. The diagrams drawn by Pierce et al. for the rocks of Phase B show that they have been formed on the border of island arc and continental environments; thus, crustal contamination, which is a characteristic of such settings, has led to an increase in yttrium and nobelium content of magma, inclining their chemical characteristics toward those of continental interiors.

Considering the petrological and petrographical characteristics of the Ahar–Jolfa metallogenic zone, as proposed herein, it can be amended to the status of metallogenic province. A research proposal for detailed investigations of this zone can be planned in the form of two doctoral dissertations (on intrusive and extrusive igneous rocks, respectively), a number of master dissertations on each of the intrusive bodies, enlightening the geodynamics of the body, isotopic composition, and making comparisons with Caucasus.

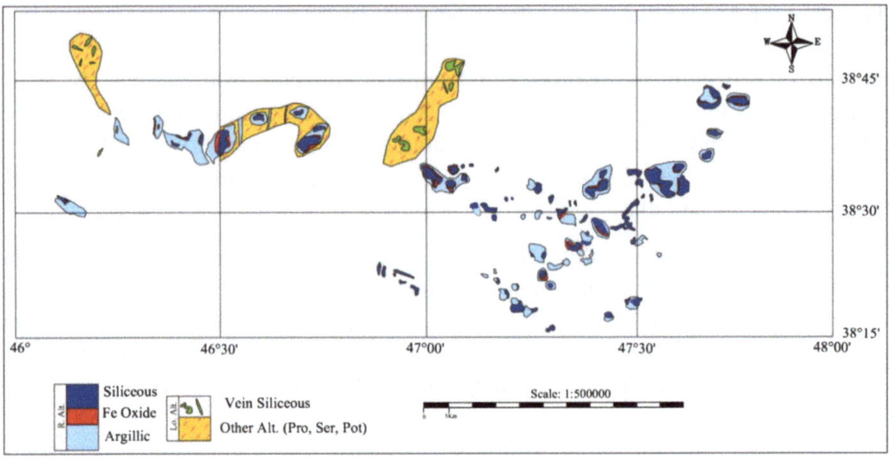

Fig. 6.23 Alteration along Ahar–Jolfa axis (Ghorbani 2004c)

Alteration

The Ahar–Jolfa metallogenic zone is incomparable in terms of alteration characteristics, and due to this quality, it has always been the center of attention of economic geologists (Fig. 6.23).

Alteration has not affected all the rocks within the Ahar–Jolfa zone with the same intensity and extent. The following need to be considered when discussing the alteration characteristics of these rocks:

1. The same alteration trend, as seen in the Ahar–Jolfa zone, is also witnessed in other areas of northwestern Iran, such as Hashtjin, parts of Tarom (on the north of Takestan), etc., so that an alteration zone can be recognized to cover the entire Takestan–Jolfa Belt. The zone of alteration is in the form of a regional alteration zone and will further be discussed elsewhere.
2. A specific relationship exists between lithology and alteration in areas affected by regional alteration (other areas that have experienced local alteration do not exhibit this type of relationship). Alteration is more extensive wherever tuffaceous rocks occur.
3. Detailed field investigations of batholithic bodies within the area, such as Shiyor Dagh, Kalibar, Ordubad, Arasbaran, and Yousefloo, indicate that they were not directly responsible for the alteration because alteration zones are seen even within them.
4. Large intrusive bodies of the area that are composed of gabbrodiorite to granite and range in age from Late Eocene to Early Oligocene have not undergone any alteration, while the rocks that host them show extensive alteration effects.

5. Apart from the regional alteration that is seen in the area, localized effects of alteration agents within the limits of Phase C of igneous activity were common. These alterations are usually productive and their characteristics differ from those of the regional alterations.
6. Late Cretaceous–Eocene andesitic volcanic rocks, shallow intrusive rocks (porphyritic bodies), and their associated tuffs or even their clastic sedimentary rocks that form the basement rocks of the region are generally rich in pyrite mineral. In some places, pyrite constitutes more than 5% of the volume of the rock and is primary in nature. The Astamal area exhibits a very interesting alteration condition with large amounts of pyrite whose Fe^{2+} ions have been released during the alteration. These irons have then been transformed into Fe^{3+}, resulting in the formation of iron hydroxide. The released sulfur irons are oxidized and form sulfate minerals, which in turn make the solutions more acidic, leading to further alteration of the surrounding rocks.

Regional Alteration

Types of Alteration

Various types of alteration processes that have affected the Ahar–Jolfa metallogenic zone can be grouped in the following, each having its own mineralization potentials (Ghorbani 2011):

1. Regional alteration
2. Local alteration
3. Siliceous alteration resulting in the formation of silica veins

The characteristics of this type of alteration are as follows:

- It is observed over a large area of the Ahar–Jolfa zone, extending from Takestan to Jolfa, that is, entire northwest of the country.
- Mineralogically, this type of alteration can be divided into three distinct zones, namely, lowermost argillaceous zone (which is the thickest of all, wherever present, and includes kaolinite, illite, feldspar, quartz, alunite, and gypsum); middle hematitic zone (or iron hydroxide zone, which contains a combination of iron oxides and hydroxides such as hematite, goethite, and limonite along with jasper, quartz, chalcedony, and gypsum, and sometimes argillaceous minerals such as montmorillonite and kaolinite, not more than 5 m in thickness); and upper siliceous zone (which usually forms the highest elevations of alteration zones with a thickness of up to 100 m, dominantly consisting of quartz and chalcedony and occasionally hematite, goethite, and jasper).

The results of geochemical analyses (ICP technique) on 20 samples collected from these three alteration zones reveals that the products of the regional alteration phase are barren from the point of view of gold, antimony, arsenic, lead–zinc, copper, and silver mineralization. This type of alteration is limited to the Eocene faults and calderas.

Local Alteration

Contrary to the regional alteration, local alteration is spatially scattered and mineralogically diverse, with no distinct zonation as described above. Some areas that have been affected by this type of alteration include Songoon, Zarinehcab, Kighal, Barmalek, Anjard, Mivehrud, and Nojeh Mehr. The specifications of this alteration are as follows:

- It is spread in the form of small limited areas.
- Though a number of diverse zones are present, they are not easily distinguished. These include propylitic, potassic, and phyllitic zones.
- The zoning, if present, includes propylitic, potassic (occasionally albitic), phyllitic, argillaceous, and siliceous (dominantly in form of veins) zones, the last two being very limited in extent.
- It is associated with Miocene–Pliocene volcanism and, hence, is absent where Phase C of igneous activity has not been effective.

Either distinct veins or anomalies of zinc, silver, molybdenum, gold, and copper are observed in all areas that have experienced this type of alteration. Interestingly, an agreement exists between the geological map, elemental anomaly map (produced by the National Iranian Copper Company and Geological survey of Iran), and the alteration map of the area (prepared by the present author); in fact, the areas of Phase C igneous activity, geochemical anomalies of gold, copper, lead–zinc, antimony, molybdenum and silver, and the local alteration zones overlap one another.

Siliceous Alteration

It is observed along with the previous two types of alteration, however, this type of alteration is distinguished from regional and local types by its vein form and limitation to the fault contacts of calderas formed by Phases A and C igneous activities. The results of geochemical analyses of 20 samples produced by this type of alteration point to the gold-bearing nature of veins.

Considering the facts about the nature of alteration in the Ahar–Jolfa zone, regional and local alteration areas must be considered for further exploration to allow resolving them. It is worth noting that regional and local alterations sometimes overlap. As a result, mineralogical and petrological studies must be carried out in order to distinguish the phase of igneous activity.

Mineralization (Ghorbani 2004c)

The Ahar–Jolfa zone is one of the richest Tertiary metallogenic zones of Iran dominantly associated with the Tertiary igneous rocks. Mineralization of copper, molybdenum, gold, silver, iron, lead–zinc, arsenic, antimony, and mercury has taken place in the form of porphyry, skarn, and vein deposits (Fig. 6.24), which are explained in the following sections.

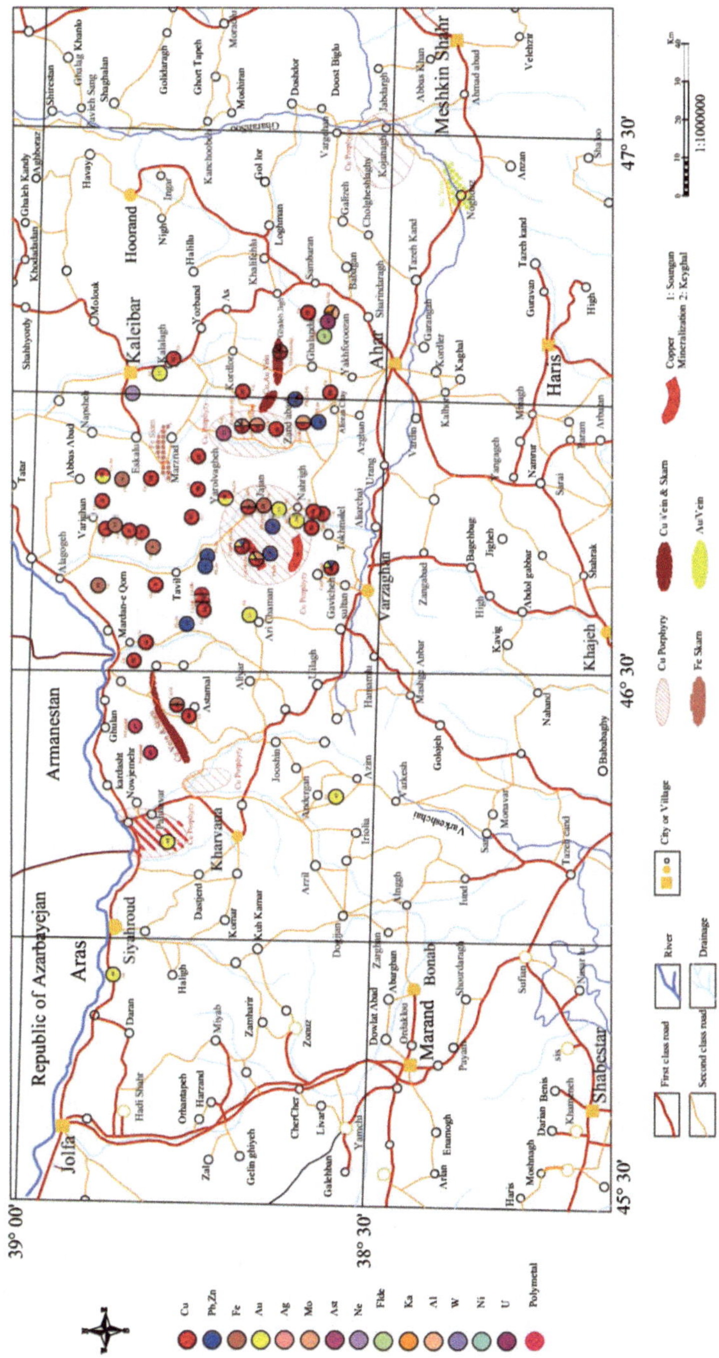

Fig. 6.24 Distribution of some mineral deposits and indications in Ahar area (Ghorbani 2004c)

Porphyry Mineralization

Coming next to the Kerman region, porphyry copper and molybdenum mineralization of the Ahar–Jolfa zone is the most prolific area of such elements in Iran. Occurrence of Songoon (with more than one billion tons of copper ore) and Agarat (to the north of Songoon, in Armenia) porphyry copper–molybdenum deposits are proofs of this claim. Other indications of porphyry copper deposits are seen in Khoveinard (south of Songoon), Lalehbajan, south of Nabijan, Kighal, Kharvanah (Siah Rud axis), west of Meshkin Shahr (Doostbeigloo); whether these localities contain economic porphyry copper deposits is a matter of speculation that needs further detailed investigation.

The evidence of the porphyry copper deposit in Iran (mentioned in Chap. 5) are all observed in the Ahar–Jolfa zone; these include large volumes of Eocene volcanic rocks with high copper content of the background (in fact, the Late Cretaceous and Paleocene rocks of the zone also exhibit such characteristics); successive metamorphic phases; and suitable structural conditions.

Skarn Mineralization

This type of mineralization is very widespread within the Ahar–Jolfa Metallogenic zone, giving rise to copper, gold, and iron deposits and indications.

Skarn-type deposits are of two types, namely, exogenic and endogenic. Both are seen at the contact of igneous and carbonate sedimentary rocks, or intrusive and extrusive igneous rocks. Skarn deposits of copper are seen in areas such as Songoon, south of Nabijan, west of Bandrigh, Mesgar area, Soot, south of Marzrud, and south of Sari Dareh, as well as in Anjard, Zandabad and Mazra'e deposits, whereas gold mineralization has been reported from the area surrounding Kalibar nepheline–syenite body in Marzrud, Mesgar and south of Ghajar Kuh (Ataloo 1998). There is also a possibility that the gold deposit at Kharvana is of this type (Rashid, personal communication).

Vein Mineralization

With the volcanic carbonates and intrusive igneous rocks acting as the host rock, vein mineralization is seen throughout the Ahar–Jolfa zone. Iron mineralization with Oligiste and magnetite minerals is seen around the copper deposits. Near the Songoon porphyry deposit, in an area around 250 km^2, another mineralization zone is seen (Khoveinard, Khonindiraj, Lalebejan, Baloojeh, Mahmood Abad, Honarikandi, Bandrigh, Arpaligh, and Barazin). Porphyry-, scattered-, and vein-type deposits of gold, silver, mercury, and antimony are seen in these localities. Lead–zinc and iron ores are seen in addition to copper deposits.

The vein-type deposits in the Ahar–Jolfa metallogenic zone are somewhat similar to the vein deposits of Tarom–Hashtjin Belt, but with fundamental differences.

Tarom–Hashtjin mineralization is typically different from Ahar–Jolfa in geological characteristics. These differences have led to dissimilar trend in mineralization process in the two regions. It can be said that the Ahar–Jolfa zone is in general richer than Tarom–Hashtjin. The major differences existing between the two regions are described hereunder:

1. The igneous activity of Ahar–Jolfa can be traced from Late Cretaceous to Quaternary and can be divided into four distinct magmatic phases while in Tarom–Hashtjin only two such phases are recognizable. The recurring magmatic activity along with structural characteristics and fracturing pattern in Ahar–Jolfa has provided ground for porphyry copper mineralization. Feeding from old porphyry systems, Hercynian, for example, can be responsible for this kind of mineralization (Moayed, personal communication).
2. Oligocene intrusive bodies within Tarom–Hashtjin are both lithologically and genetically uncomplicated, whereas those of Ahar are varied in the said two aspects, for example, Kalibar nepheline–syenite. In addition, the host of Ahar region is more diverse; carbonate rocks that form the majority of hosts in this area, and hence allowing skarn-type of mineralization possible, are totally absent from the Tarom–Hashtjin Belt.
3. Lead–zinc and iron–zinc mineralization is much more developed in the Tarom–Hashtjin Belt. Moreover, molybdenum mineralization in association with either porphyry or vein-type deposits is much more developed in Ahar–Jolfa.
4. Structurally, though the Tarom–Hashtjin Belt is surrounded by major faults, there are no disturbances in the central parts of the region. This trend is reverse in Ahar–Jolfa, where the central parts are affected more by the complex set faults and joints (Figs. 6.22 and 6.24).

The specifications of the most important mineral deposits in the Ahar–Jolfa area are presented in Table 6.5.

Interrelationship Between Mineralization, Igneous Activity, and Metasomatism (Ghorbani 2004c)

The relationship between the mineralization process, igneous activity, and metasomatic alteration of rocks in the Ahar–Jolfa Metallogenic zone can be described as under.

Although Phase A of magmatic activity has a high copper, iron, and gold in its background and is very rich in pyrite mineral, it shows no specific mineralization but has been mainly responsible for enriching the later phases of magmatic activity.

Phase B has intruded the sedimentary rocks (especially older limestones) and the volcanics resulting from Phase A leading to vein- and skarn-type mineralization of copper, iron and gold, for example, Mazra'e, Anjard, Mesgar, and Zandabad copper deposits.

Phase C of the igneous activity in the Ahar–Jolfa zone, which was in the form of volcanism and shallow intrusions (some of the intrusions have no outcrops on the surface) during Miocene and Pliocene, has resulted in the formation of localized

Table 6.5 Specifications of most important mineral deposits in Ahar–Jolfa area

Sr. no.	Name of deposit or indication	Paragenesis	Host rock	Geological characteristics
1	Songoon	Chalcopyrite, magnetite, pyrite, and molybdenite	Volcanic rocks, granodiorite, porphyry monzonite	Porphyry mineralization is also seen in the surrounding areas of Songoon
2	Mazra'e	Chalcopyrite, magnetite, hematite, chalcocite, covellite, pyrite, bismuth, wolframite	Cretaceous limestones, Shiyordagh Granite	Skarn-type of mineralization
3	Anjard	Chalcopyrite, pyrite, molybdenite, magnetite	Cretaceous limestones; volcanic rocks, Shiyordagh Granite	Skarn-type of mineralization
4	Godal	Chalcopyrite, magnetite, malachite	Cretaceous limestones; volcanic rocks, Shiyordagh Granite	Skarn-type of mineralization similar to Mazra'e
5	Zand Abad	Chalcopyrite, pyrite, magnetite, molybdenite, malachite, azurite, galena, sphalerite	Cretaceous limestones, Shiyordagh Granite	Skarn-type of mineralization
6	Saleh Dareh	Chalcopyrite, pyrite	Volcanic rocks; granitic intrusive bodies	Vein-type of mineralization
7	Jouiband	Galena, sphalerite, chalcopyrite, covellite, dignite, malachite, pyrite	Volcanic rocks, granophyre, limestones	Vein-type of mineralization; two indications are seen at a distance of 2 km from each other
8	Barmalek	Galena, sphalerite, pyrite, tetrahedrite, chalcopyrite, malachite, azurite, lucite	Volcanic rocks, porphyry intrusive bodies, monzonite, granodiorite	Vein-type of mineralization; a silicic zone is seen at a distance of about 1 km from it
9	Kighal	Chalcopyrite, magnetite, molybdenite, cassiterite, galena, sphalerite	Volcanic rocks, small intrusive bodies	Evidence of porphyry mineralization is seen
10	Goosh Abad (Oran), Siyah Lakan, Aghababa Sang, Zarineh Rekab	Copper ore (chalcopyrite), gold, lead	Volcanic rocks, intrusive bodies	The four indications mentioned here are similar to one another and are considered as copper-gold deposit

11	Astamal, Avan, Avansar, Cheshmeh Ghan	The main phase of mineralization is that of pyrite and copper minerals are little in amount	Volcanic rocks, acidic tuffs, small intrusive bodies	Scattered or vein-type pyrite mineralization (as well as copper) within small Eocene intrusive bodies
12	Qareh Chiler (Golan)	Chalcopyrite, pyrite, molybdenite, galena, tetrahedrite, chalcocite, covellite, bornite, limonite, malachite, azurite	Ordobad, intrusive bodies	Ore mineralization in the form of vein along with silica
13	Qareh Dareh	Molybdenite, pyrite, chalcopyrite, malachite	Mineralization in granitic rocks of silicic zones	Ore mineralization in the form of vein
14	Chamtal	Chalcopyrite, magnetite, covellite, bornite, gold	Granitic intrusive bodies, volcanic rocks, metamorphosed limestones	Skarn-type of mineralization
15	Daghar Dagh	Malachite, magnetite	Limestone bodies, granodioritic intrusive bodies	Skarn-type of mineralization
16	Mesgar	Malachite, pyrite, hematite	Pyroclastic and carbonaceous rocks	Vein-type of mineralization
17	Anbaloo	Chalcopyrite, magnetite, malachite	Limestones, schists, and intrusive bodies	Skarn-type of mineralization; a number of mineralized zones are seen in Anbaloo area including Abbasabad
18	Barghzar	Chalcocite, malachite, magnetite	Metamorphosed limestones, andesites	Mineralization seems to be of skarn-type

Ghorbani (2004a, b, c)

productive metasomatism giving rise to porphyry- and vein-type mineralization (most of the mineral deposits of Ahar–Jolfa are associated with this phase of igneous activity).

Thus, the mineralization processes of the Ahar–Jolfa Metallogenic Belt are classified into the following categories:

1. Vein-type copper deposits, gold-bearing coppers of hydrothermal, and skarn origin.
2. Porphyry copper deposits; only Songoon has so far been found to contain economic deposits but signs of other such bodies are also indicated in various localities and with further exploration, especially within local metasomatic areas, other such deposits are hoped to be discovered.
3. Gold deposits seen in vein zones are usually found along with siliceous zones that are associated with old calderas. It is suggested that all the siliceous veins of the area that are not related with the metasomatic siliceous zones should be further investigated.
4. Skarn- and metasomatic-type iron mineralization in the area, which is of two types:
   (a) Skarn- and metasomatic-type iron ore deposits, for example, Mahmoud Abad deposit.
   (b) Metasomatic iron ore deposits formed from alteration of pyrite, for example, Cheshmghal.

Of these, the former are economically important while the latter is not.

Last but not least, if it is intended to sum up all that has so far been described in one paragraph, it would be as follows.

A survey of igneous phases of Ahar–Jolfa indicates that Phase C (i.e., Miocene–Pliocene Phase) has had the central role in mineralization of gold, copper, mercury, etc. Therefore, all the future studies must target the rocks formed by Phase C. The most important metasomatic process affecting the rocks of the area is in local extent; however, the regional metasomatism seen in this zone has produced distinct argillitic, hematitic, and silicic zonations that are futile from the point of view of gold mineralization. Siliceous vein metasomatism should be further explored for gold, especially where such alterations are associated with older calderas. As a result, old calderas form important targets for future investigations.

6.3.6 Southeast and East of Iran Metallogenic Province (Science and Technology Research Institute 1999)

6.3.6.1 Geology of East and Southeast of Iran

From a geological standpoint, east and southeast of Iran can generally be divided into three sections:

- Lut block, which covers most part of east of Iran
- Flysch zone or colored-melange zone or Zabol–Baluch zone
- Makran zone

6.3.6.2 Lithostratigraphy of East and Southeast of Iran

Most rock units in east and southeast of Iran belong to Cretaceous and after. However, several rock units older than Cretaceous outcrop in some places such as Lut zone, northern Makran, and northeast of Flysch zone. Pre-Cretaceous sedimentary units (Paleozioc, Triassic, and Jurassic) in east and southeast of Iran feature continental facies, and for that matter this area is similar to Central Iran. As a matter of fact, in places where Cretaceous rocks form the basement (especially in the Lut zone and Bazman Region) not much of a difference is seen between east and northeast of Iran and Central Iran.

Cretaceous in east and southeast of Iran: Cretaceous played a key role in east and southeast of Iran in regard to the tectono-magmatic events, sedimentary basins, and formation of various rock units like ophiolitic complex, carbonate rocks and flysches, and later events that occurred in Tertiary were the continuation of developments in Cretaceous.

Lut block and Afghan block that once had formed one single continent separated from each other and an ocean formed between them. Formation, evolution, closure, and orogeny that resulted in closure of this ocean are considered as major developments in east and southeast of Iran. Cretaceous sediments are seen in relatively large areas within the Lut and Flysch zones, and north of Makran zone.

Early Cretaceous deposits in east and southeast of Iran including shale, marl, graywacke, and plagic and non plagic limestone are insignificant compared to the Late Cretaceous deposits. There are various volcanic rocks seen inside flysch sequence, whose age could reach to Paleocene. The formation of these volcanic rocks could be due to genesis of a rift and its expansion during Late Cretaceous. The expansion and development of such rifts within the Flysch and Makran zones during Cretaceous through Paleocene led to the formation of ophiolitic complexes in the form of areas and axes, and some of them are cited below:

- Ophiolitic axis south of Jazmourian, Kannouj–Iranshahr
- Taftan ophiolitic area
- Nosrat-abad ophiolitic area
- Lar-Asgi ophiolitic area
- Iranshahr–Zabol ophiolitic area
- Ophiolitic area around Hemont Mountain
- Birak Mountain ophiolitic area (Zaranj Mountain area)

The reason for diversity in these ophiolitic areas is structural activities along faults as well as intrusion of magmatic diapirs. However, as presented in Metallogeny in the Sistan–Baluchestan report, Nogol Sadat and Houshmandzadeh (1999) believe these various ophiolitic axes are in connection with several deep rift axes and do not consider a uniform setting for these ophiolitic axes.

Tertiary in east and southeast of Iran: In Late Cretaceous and Early Tertiary, the rift basins between the Lut and Afghan blocks and also northern Makran basin deformed and began to close out, which was concurrent to uplifting of the Lut and Bazman blocks. These activities accompanied with facies changes in the Flysch zone and Lut, and magmatic activities in such fashion that flysch facies (Paleocene–Eocene) were replaced by sandstone and conglomerate facies in the Lut and Bazman areas, and volcanic activities intensified and continued throughout Tertiary. In the southern part of Flysch zone, flysch deposits show significant changes in thickness and contain more volcanic materials, and in some places the age of flysch deposits reaches to Late Eocene.

In south of Flysch zone and Kahnouj–Iranshahr ophiolitic strip (i.e., Makran zone), the flysch facies become younger southwardly and temporally stretch to Miocene.

6.3.6.3 Structural Characteristics of Southeast and East of Iran

Similar structural features seen in other parts of Iran also control the general structure of east and southeast of Iran. These structural elements can be categorized into the following groups:

- Faults and fractures
- Folds
- Lineaments
- Ring structures

6.3.6.4 Mineralization in East and Southeast of Iran

So far vein-type mineralization of copper, lead, zinc, antimony, mercury, and gold has been reported in the Lut block. Vein-type deposits that have been studies are cited below:

1. Ghaleh–zari gold-bearing copper deposit
2. Shoorab–Kalehneginan antimony and mercury deposit

It seems the mineralization in Lut block is mostly in the form of vein, and evident examples of such mineralization are Ghaleh-zari and Shoorab. In general, geological and mineralization evidences in north and northwest of Lut zone (as moving toward Tabas block) point to mineralization of copper, lead, and zinc with supremacy of lead and zinc over copper, and copper mineralization accompanied with gold in south and southeast areas. Exploration studies with regard to geological characteristics in south and southeast of Lut zone can lead to useful results. The geological characteristics of the Zabol–Baluch and Makran zones are very similar to each other. These two zones both have ocean crust basement covered with younger sediments, and to some extent the younger magmatic activities affected them.

From a metallogenic standpoint, northern Makran which has an ultramafic basement, displays outcrops mostly of ultramafic rocks. The Northern Makran zone and southern parts of Flysch zone that contain ultramafic rocks are significant from a metallogenic standpoint. Such similar characteristics could be due to the similarity in basement, magmatism, and structure in these two zones. During the course of a study carried out on metallogeny of the Flysch and Makran zones, metallogenic zonation was presented for these two zones as they are briefly explained next (Fig. 6.25 and Table 6.6).

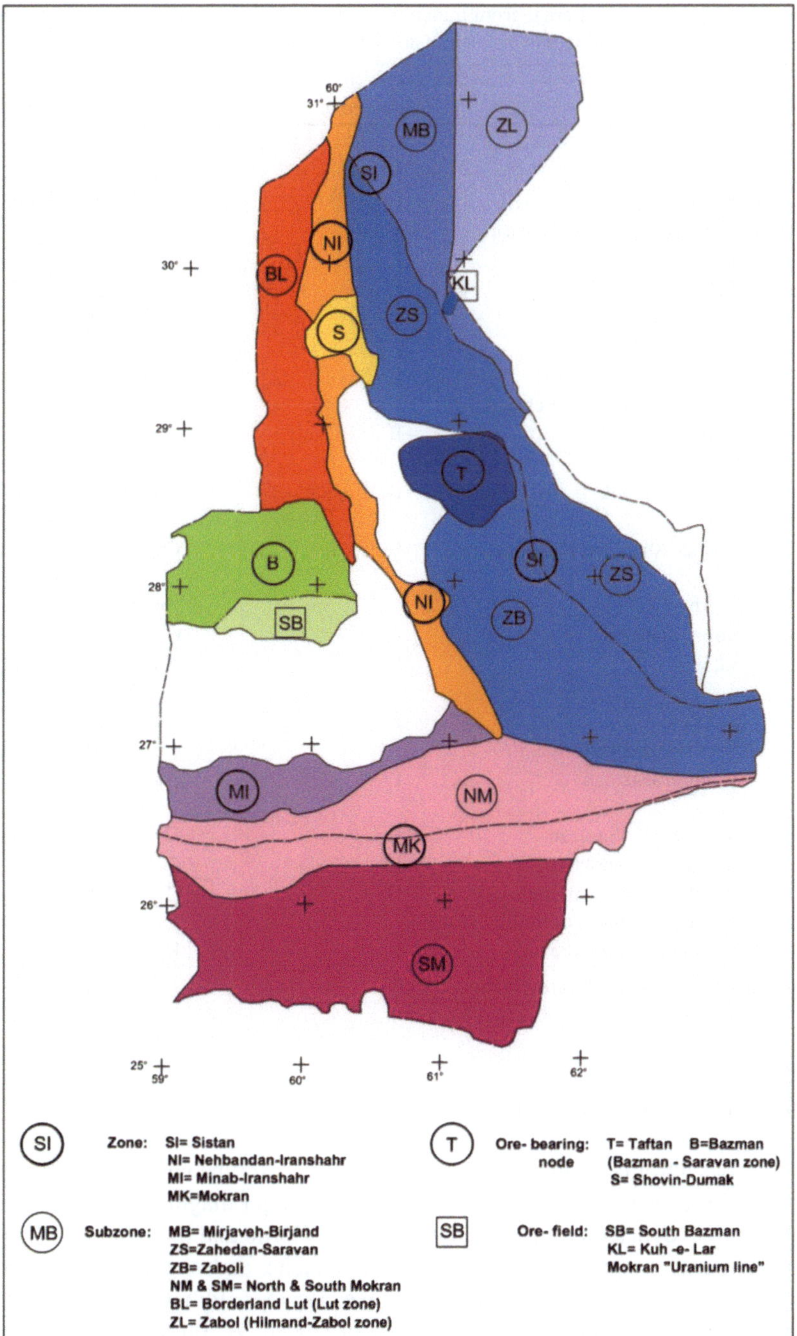

Fig. 6.25 Metallogenic zonation of Sistan-Baluchestan (Science and Technology Research Institute 1999)

Table 6.6 List of important areas from mineralization standpoint in south and southeast of Iran, for which detailed exploration and evaluation are recommended

Number	Name	Group type and degree of importance	Primary mineral (accessory mineral)
1	Bandan	III	Cr, Asb, Mn, AU-Cu
2	Kuh-e Janja	II	Au, Cu
3	Nehbandan–Iranshahr axis (northern ophiolitic strip)	IV	Cr, Cu (Au), Ni, Co, Mg
4	Kuh-e Siasetorgi	I	Au, Cu
5	Kuh-e Asagi	II	Au, Pb, Zn, Cu
6	Kuh-e Kalehgar and Dopar Kuh	III	Au?, Ag?
7	North of Nosrat–Abad	II	Cu (Au), Cr, Mg
8	Chehel–Kooreh, Pourchangi	IV	Cu, Pb, Zn
9	North of Zahedan	II	Au, Ag?
10	Kuh-e Lar	I	Au, Cu (Mo)
11	Shovin–Domak	II	Co, (Au), Cr, Ni, Co, Fe, Mn, Mg, Asb
12	Nosrat–Abad and Kahoorak	I	Cu, Ni, Co Nosrat-abad: Cu Kahoorak:
13	South of Zahedan	I	Au, Ag
14	Kuh-e Sefidak	I	Cu?, Au?
15	Taftan	I	Au, Mo, Cu, Ag, Pb, Zn, (Aln/Si)
16	Bazman	I	Cu, Au, Mo
17	North of Bazman	I	Au, As, Mg, (Mo,W)
18	South of Bazman	I	Au, Cu, (Mo), Pb, Zn, W?, V?
19	Kuh-e Sefid–Saravan	II	Cu, Au, Pb, Zn, Mo?, W?
20	Rootak–Maksookhteh	III	Cu?, Ni?, Co?, Cr?
21	Minab–Iranshahr metallogenic strip (southern ophiolitic strip)	III	Cr, Cu, Mg, Cu?
22	Makran uranium stretch	II	U

Science and Technology Research Institute (1999)

6.3.6.5 Taftan Mining Area

This area has ultramafic–mafic basement of ocean crust type with age of Cretaceous covered with Crateceous–Paleogene flysch and younger magmatic rocks.

The intrusive rocks include shallow bodies of diorite and monzonite–to–granite, and the extrusive rocks consist of rhyolite, rhyodacite, andesite, and associated tuff (Kharestan area) with age of Oligo–Miocene. Besides, Plio–Quaternary volcanic rocks with composition of andesite–dacite produced by Taftan volcano are seen around Taftan Volcano. The structural geology of this area is affected by doming in Taftan with spiral–radial system, and this doming event resulted in outcrop of ultramafic–mafic units. Associated hydrothermal solutions caused metasomatosis and metamorphosis of ultramafic and mafic rocks, as well as sulfide-type mineralization of lead, zinc, copper, molybdenum, silver, gold, antimony, and arsenic.

The studies carried out in the Kharestan area indicate richness of silver within vein-type lead and zinc deposits.

An extensive alteration of argillic, silicic, and propylitic type is seen in volcanic and flysch rocks of this area, especially the Kharestan area. Besides, alteration and metasomatism of lisvenitic, rodingitic, and talcified types are also found in ultramafic rocks. Within the Kharestan area there are a few mineralization zones such as Kuhe Zardan-noghrei silver-bearing lead and zinc deposit, Div-chah lead, zinc, copper and gold deposit (especially gold), and Tilouei lead, zinc, and antimony deposit.

6.3.6.6 Bazman Mining Area

The basement of this mining area is formed of Paleozoic–Mesozoic rocks that were cut through by Bazman granitoid intrusive complex in Late Cretaceous (with age of 70 Ma). The northern part of Bazman volcano is covered with Eocene–Miocene volcanic rocks (mostly andesite), over which Bazman volcanic rocks with composition of andesite, dacite and in some cases basalt sit.

Similar to the Taftan area, alteration of argillic, silicic, and propylitic types is also seen in Bazman area but with lower degree.

Mineralization types found in Bazman area include skarn (copper and gold), volcanic hydrothermal, and plutonic hydrothermal (gold, silver, mercury, tungsten and molybdenum–gold, and silver mineralization in Giyaban).

6.3.6.7 Iranshahr–Minab Mining Area

An ophiolitic strip is seen southeast of Jazmourian, which is the most complete ophiolitic series in Iran. Some of the rocks within this ophiolitic complex are peridotites including verlite, harzburgite, lherzolite, and dunite. These peridotites are seen in the form of intact massive and tectonized.

Serpentinization is developed so extensively in tectonized peridotites that in some cases the original rocks are hardly recognized. However, serpentinization in massive peridotites is seen in the form of veins dunites, harzburgites, lherzolites, and verlites.

Gabbro rocks are usually seen next to peridotites along a fault contact and include troctolite, plagioclasite, gabbro–norite, leucogabbro, and oralite.

The rocks in the area consist of olivine–gabbro, gabbro, plagiogranite, diabase, espilite, basaltic–espilite, sedimentary rocks with deep-sea facies including limestone and radiolarite-shale, and metamorphic rocks with blueschist and greenschist facies.

From a mineralization standpoint, there are two types of mineralization seen within this ophiolitic strip.

Chromite Mineralization: Numerous chromite deposits but in small size are found in ultramafic rocks, dunites, harzbugites, and serpentenites. Chromite mineralizations are seen in two forms:

- Spotted chromite in dunites
- Chromite in the form of dashed thin layers looking like beads in serpentine and serpentenized harzbugite

Some of the chromite deposits and indications found in this area are Kootij, Sarzeh, Gondtihan, Kooshook, Mokhtar-abad, Rameshk, and Mirab.

Copper Mineralization: Such mineralization is seen in the eastern part of this ophiolitic strip in an area between Mokhtarabad and east of Minab. It seems the mineralization of these deposits is of massive sulfide and hydrothermal type.

6.3.6.8 Nehbandan–Iranshahr Mining Area

This is a narrow strip which can be considered as a suture zone from a structural geology standpoint. This area has linear blocks with N–S fracturing systems. There are ophiolitic complexes, Cretaceous sedimentary complexes, and Tertiary magmatic rocks seen along this suture strip, in which alteration, metamorphism, and metasomatism are observed.

Mineralization of chromium, copper, iron, magnesium, and gold is seen along this strip, but the noticeable mineralization includes numerous chromites in the form of small to medium deposits and valuable magnesium deposits.

6.3.6.9 Mirjaveh–Birjan Mining Area

This area is part of the Flysch zone in east of Iran. The rocks in this area include Paleogene–Cretaceous flysch, ultramafic rocks that form the basement and outcrop through the erosion fenêtre. The Oligo–Miocene magmatic rocks consist of shallow intrusive bodies of diorite, monzodiorite and syenite, and associated volcanic rocks including trachyte, andesite, trachy-basalt, and Plio–Quaternary basalt and andesite. From a tectono-magmatic standpoint, the area features folds and magmatic rocks seen as dome-ring bulges. There are some places with evidences of skarn, hornfels, metasomatism, and argillic and propylitic alteration within this area. Mineralization in this area can be categorized into three following forms:

1. Magmatic mineralization such as chromium mineralization in ultramafic spread, and sometimes chromium–manganese–asbestos.
2. Skarn-hydrothermal mineralization including mineralization of gold–silver, zinc–copper, lead–gold, and copper–gold within ultramafic complexes and magmatic–flyschy rocks.
3. Hydrothermal mineralization.

6.3.6.10 Zahedan–Saravan Mining Area

This area in fact is a structural strip with granitoid intrusive bodies. Along this strip and within the folded flysch deposits and volcanic rocks belonged to pre-Oligocene, there is a series of granitoid bodies that were cut through by groups of parallel dolerite dikes. These intrusive bodies caused alteration, metasomatism, and contact metamorphism in the surrounding rocks. The hydrothermal solutions released by these intrusive bodies caused mineralization of metals such as copper, lead, zinc, gold, silver, molybdenum, and tungsten.

In addition to the mining areas already cited in the report on the metallogenic study of Sistan–Baluchestan, some other mining areas have been also mentioned and metallogenically defined in that report such as Zabol, Makran, and southern edge of Lut zone, which are not to be described here due to their low importance from a mineralization standpoint. However, there are some areas in southeast of Iran that are significant from a talc mineralization standpoint.

6.3.6.11 Kuh-e Hamont Mining Area (with Significant Talc Mineralization)

Kuh-e Hamont, the highest mountain in southeast of Iranshahr, is a noticeable case from a geological standpoint, and talc mineralization in this area is one of the outcomes of formation of this mountain. From a geological standpoint, Kuh-e Hamont is to be considered as Hamont complex, which is in fact a core complex. The most important rocks found in this complex are as follows:

- Ultramafic and mafic rocks that have been mostly metamorphosed at high-grade greenschist level including actinolite, schist, serpentinite, serpentinite schist, and talc-schist.
- Magmatic rocks in the form of two groups of granitic rocks and volcanic rocks.

Regarding the formation of the aforementioned rocks, Hamont complex, and its general geomorphology, it can be stated that considering the geological characteristics of the Makran zone and Hamont Mountain this area belongs to pre-Eocene with a simple structure meaning it is an ophiolitic basement overlaid with Eocene shale and sandstone deposits.

During Late Eocene and Oligocene, a magmatic diapir (probably with andesitic–basaltic composition) uplifted from mantle and intruded into the crust. The intrusion of this magmatic diapir caused partial melting of crust and therefore formation of intrusive bodies (granitoid) and volcanic rocks, which led to the formation of Kuh-e Hamont.

Intrusion of magma from mantle into the crust led to metamorphism in the form of core complex, which means the temperature and pressure the magmatic diapir had while ascending metamorphosed the host rocks and surrounding rocks. In addition to the aforementioned metamorphism by hydrothermal solutions produced due to magmatic and metamorphic activities, metasomatism phenomenon is also seen in these rocks especially in the fault zones. Formation of talc and serpentine around Kuh-e Hamont is the product of metasomatism in such way that in areas where hydrothermal solutions were rich in CO_2 and SiO_2 the ultramafic rocks transformed to talc, and in areas poor in CO_2 and SiO_2 serpentine was formed.

6.3.7 Zagros Oil and Gas Province (Motiei 1993)

The Zagros sedimentary basin is the most important oil and gas province of Iran situated between the Arabian shields and Sanandaj–Sirjan Zones. It is bounded on its north and northeast by Zagros main thrust, on its southeast by Oman Mountains, and on its west and southwest by Arabian shield.

A complete sequence of rock ranging from Lower Cambrian to Pliocene can be observed within the area of this province (Fig. 6.26). The basin possessed a platform environment of deposition between Early Cambrian to Middle Triassic, in line with other places of the country, occasionally experiencing epeirogenic movements. From then onward, till Miocene, sediments dominantly composed of carbonates along with minor amounts of sandstone, shale, and marl have been deposited in a continuously subsiding basin reaching a thickness of about 10,000 m. The presence of evaporite deposits as well as minor stratigraphic hiatuses within this sequence point toward epeirogenic movements within the basin of deposition.

Ever since the end of Miocene, the previously formed rock successions have been complexly folded and overthrust due to intense orogenic movements thus emerging from sea water. As a result, the younger sediments (Pleistocene) lie over them with an angular unconformity.

Zagros Province, which is basically a part of Alp–Himalaya orogenic belt, trends in northwest–southeast direction extending from the northwest of the country and continuing up to Strait of Hormoz. The deformations of the province have been attributed to the collision of Central Iran and Arabian Plates (traditionally, the upheavals connected with the final filling phases of the geosyncline between the two).

A number of classifications have been put forward for Zagros Province with every scheme having its own considerations. Nogole-Sadat and Almasian (1993) has delineated the latest scheme dividing this province into two major and minor zones, namely, Northern Zagros (the minor) and Zagros (the major).

6.3.8 Kopet–Dagh Natural Gas Province

This basin is located northeast of Iran. From Middle Jurassic on it was covered with a vast continental margin sea (Berberian and King 1981). In this period of time and due to sea progression as well as rapid subsiding of the basin, the depth of the water in western part was deeper. In this basin, a thick sequence of continuous marine and continental sediments was deposited (about 10 km). No major sedimentary gap and

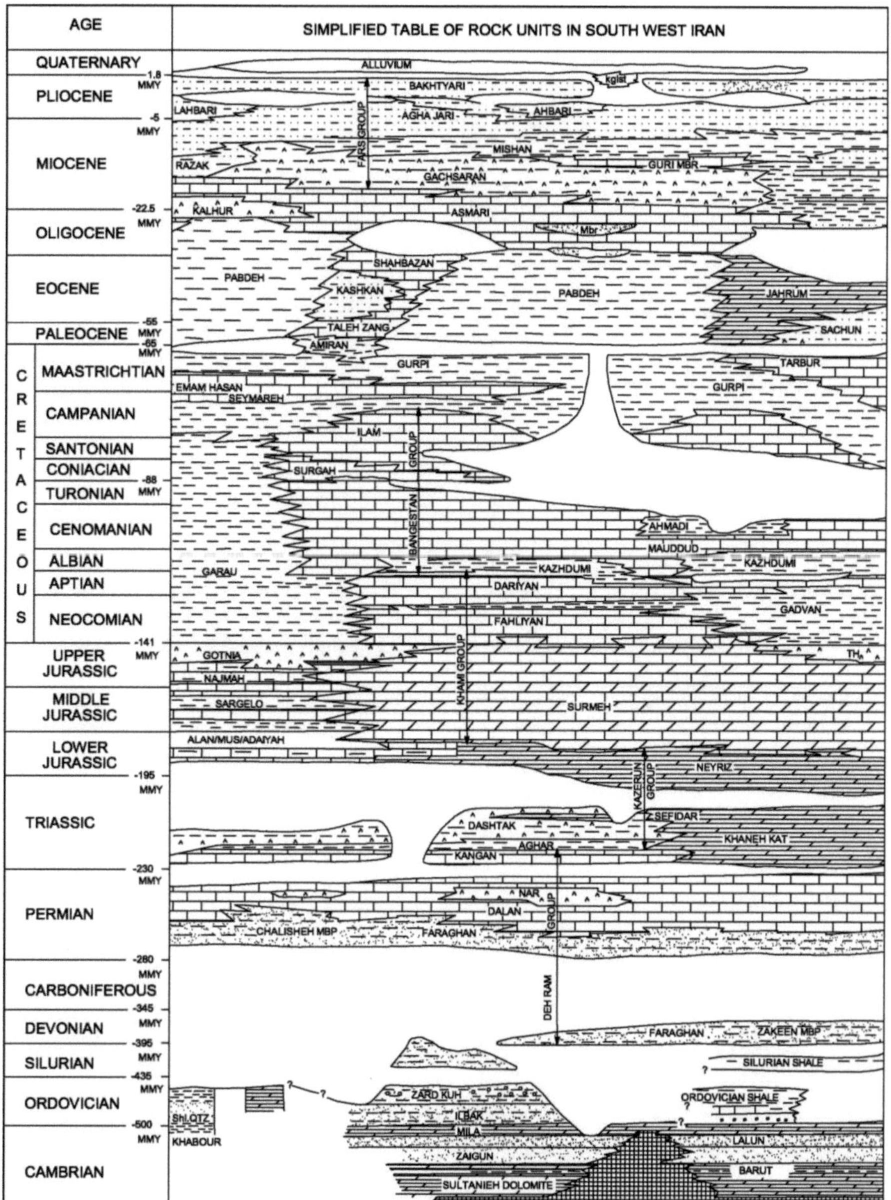

Fig. 6.26 Lithostratigraphy of Zagros zone (Motiei 1993)

volcanic activities during Jurassic to Oligocene have ever been reported. This sedimentary complex provided a very suitable condition for accumulation of hydrocarbons. Kopet–Dagh sedimentary rocks placed in their current position due to an uplifting at the end of Miocene.

Chapter 7
The Position of Iranian Mining Industry in the World

Abstract This chapter presents an economic, statistical, and comparative review on various mineral deposits (metallic and non metallic) as well as fossil fuel reserves in Iran as compared to those in the rest of the world.

This is a comprehensive collection of data about Iran's mineral deposits, production, and trade based on the available data sources, and gives the reader a good account of Iran's position in the world as far as its share in mineral deposits, production, and mineral trade is concerned.

The author has tried his best to provide a comprehensible outcome in qualitative and quantitative assessment of each mineral deposit based on available data, world class logical and comparative analytical methods, and Iran's geological features.

7.1 Introduction

Considering the geological condition of Iran and its position on the Alpine-Himalaya orogeny belt, Iran, from a mining point of view, is one of the gifted countries of the world. As recent statistics (2004) indicates, the proven and probable reserves of metallic and nonmetallic deposits of the country are proclaimed to be around 55 billion tons. There are 62 types of mineral deposits in Iran, which puts it in the category of countries having large number of mineral deposits with high variety. Table 7.1 shows various mineral deposit categories in Iran.

M. Ghorbani, *The Economic Geology of Iran: Mineral Deposits and Natural Resources*, Springer Geology, DOI 10.1007/978-94-007-5625-0_7,

Table 7.1 Types of mineral deposits of Iran

Metallic minerals	Nonmetallic minerals	Nonmetallic minerals	Building materials	Fossil fuel
Copper	Turquoise	Sepiolite	Building and decorative stone	Oil
Iron	Phosphate	Bitumen	Gypsum	Gas
Manganese	Salt	Orpiment	Rubble stone	Condensate
Chromite	Sulfur	Asbestos	Pumice	Coal
Gold	Sodium sulfate	Sylvite	Perlite	
Molybdenum	Kaolinite	Chalk	Pozzolan and pozzolanic tuff	
Lead	Bentonite	Vermiculite	Scoria	
Aluminum	Talc	Borax		
Antimony	Mica	Barite		
Arsenic	Feldspar	Magnesite		
Mercury	Silica	Celestite		
Silver	Fluorite	Garnet		
Polymetal	Alunite	Dolomite		
Pyrite	Nepheline syenite	Zeolite		
Uranium	Limestone	Diatomite		
Zinc	Ochre	Agate		
Tungsten	Fireclay	Bauxite		

7.2 Brief History

According to archeological surveys, Iran hosted the most ancient mining activities in the world (Alipour 1993). Ancient civilization, richness of natural resources, and records of ancient mining and metalworking are testimonies to such a claim (Zavosh 1976). The history of mining activities in Iran can be divided into two distinct periods:

1. Shadadi (ancient) mining period: The term "shadadi" refers to ancient abandoned mines where the modern western mining techniques and equipment such as earth-moving machinery and explosives were not deployed (Ghorbani 2002a). The mines exploited during this period can be grouped into three categories on basis of the type of the mined material:

    (a) Metallic ore mines of iron, copper, gold, lead, zinc, and silver
    (b) Nonmetallic mines and quarries of China clay (used in ceramic and tile making), "gel-e sarshouy" (a variety of bentonite used as soap in ancient Iran), and bentonite and serpentine (used in clayware)
    (c) Precious and semi-precious stones such as "firouze" (turquoise), "zebarjad" (olivine), "dor-e kuhi" (quartz), "gamest" (amethyst), "lal" (spinel), "yashm" (jadeite), "yaghout" (ruby), and "safir" (corundum)

2. Modern mining period: Mining activity utilizing modern western techniques that began at the time of Fathali Shah through the efforts of Abbas Mirza (his crown

prince) remained more or less dormant till Reza Shah's time. Then, with the onset of modernization of the country based on western principles and establishment of various metal industries, modern mining activities were initiated to supply an increased demand of raw materials. These efforts have continued till date, and as a result, more than 5,500 ore deposits and indications are known to exist in the country (Ghorbani 2002a) (Table 7.2).

7.3 Mining Activities in Iran

As stated in the previous section, there are more than 5,500 known ore deposits and indications in Iran (Ghorbani 2007a), a number of which are inactive, mined out, or not yet extracted. Table 7.3 shows the status of the countries in mining sector as per government statistics.

7.4 Share of Mining Sector in Iranian Economy

Considering the wide range and diversity of mining products of the country and the international position among the resource-rich countries, the mining sector is expected to play a pivotal role in Iran's economy. But owing to the dominance of the oil sector, the share of the mining sector has been reduced to a negligible figure (Table 7.4).

It can be deduced from Tables 7.5 and 7.6 that during the period 1996–2000, the highest share of the mining sector in the gross domestic product (GDP) of the country based on the current prices has been 0.72% (for the year 1998), while the same based on the fixed prices of 1990 has been 1.25% (for the year 2001). Interestingly, the share of the mining sector during the same time period based on current prices has always remained around 0.66%, whereas the figures based on the fixed prices of 1990 have increased from 0.97 to 1.25%. This can be attributed to the lower growth of the prices in the mining sector as compared to the general trend of inflation in the country. Moreover, due to the increased production of the mining sector and stable prices, the share of the sector in the GDP of the country based on the current prices has remained more or less unchanged. Taking into account the natural riches of Iran, the mere 1% share of mining in the GDP of the country cannot be considered as appropriate for the sector.

The employment figures of the mining sector (Table 7.1) once again reveal the insignificance of this sector from the point of job creation in the country. The total number of employees in the mining sector in the year 2001 was 55,560.

Though the number of active mines in Iran has grown fourfold during the period 1986–2001, the number of employees of the mining sector has increased by 10% only. Whereas in the year 1986, an average of 64 employees was engaged in every mine; in the year 2001, the figure reduced to 19. The trend is partly ascribed to increase in the number of small private mines that employ fewer workforces.

Table 7.2 Statistics of mines of Iran listed by the type of mineral deposit

Row	Mineral	Active	Inactive	In development	Total	Governmental	Cooperative	Private sector	Total	Staff	Tonnage (1,000 tons)
1	Limestone	280	69	25	374	10	39	325	374	5,446	11,660,635
2	Gypsum	281	135	21	437	1	103	333	437	3,647	1,336,892
3	Rubble stone	931	327	70	1,328	2	112	1,214	1,328	9,175	7,267,182
4	Crystalline limestone	293	148	19	460	0	73	387	460	6,840	665,940
5	Travertine	155	36	16	207	2	33	172	207	3,515	350,307
6	Crystallized white marble	156	25	5	186	0	45	141	186	3,829	249,148
7	Marble	105	26	2	134	0	18	115	134	869	6,276
8	Granite	232	125	18	375	0	21	354	375	4,928	476,691
9	Light weight aggregate	85	16	9	110	0	16	94	110	838	109,846
10	Silica	113	46	10	169	0	11	158	169	1,268	753,475
11	Kaolinite	20	14	4	38	0	4	34	38	382	48,175
12	Industrial soil	125	51	8	184	0	23	161	184	1,148	115,146
13	Feldspar	51	38	5	94	0	8	86	94	600	39,700
14	Talc	16	9	1	26	0	2	24	26	133	5,263
15	Mica	4	2	0	6	0	1	5	6	57	327
16	Dolomite	48	10	2	60	0	3	57	60	478	312,637
17	Borax	1	1	0	2	0	1	1	2	16	25
18	Rock salt	85	31	2	118	0	32	86	118	1,355	411,295
19	Orpiment	0	2	0	2	1	0	1	2	45	7
20	Celestite	4	6	0	10	0	1	9	10	42	1,115
21	Barite	95	35	6	136	1	9	126	136	1,246	25,239
22	Perlite	12	2	1	15	0	1	14	15	181	59,969
23	Diatomite	2	2	0	4	0	0	4	4	38	1,077
24	Bentonite	41	19	2	62	0	5	57	62	401	32,378
25	Sodium sulfate	17	26	0	43	0	15	28	43	339	10,004
26	Fireclay	47	8	6	61	2	7	52	61	1,188	47,407

27	Magnesite	13	12	0	25	0	2	23	25	248	3,398
28	Asbestose	0	0	0	0	0	0	0	0	0	
29	Sea shell	3	4	0	7	0	1	6	7	12	12,930
30	Ochre	5	3	1	9	0	0	9	9	142	1,064
31	Phosphate	2	0	0	2	2	0	0	2	108	16,500
32	Turquoise	1	1	0	2	0	1	1	2	94	49
33	Pegmatite	1	0	0	1	0	0	1	1	11	205
34	Chalk	3	3	0	6	0	2	4	6	20	37
35	Alum	0	3	0	3	0	1	2	3	9	87,704
36	Fluorine	21	3	5	29	0	2	27	29	403	1,536
37	Bauxite	10	2	0	12	2	3	7	12	573	22,841
38	Bitumen	8	5	0	13	0	0	13	13	92	149
39	Coal	111	45	4	160	27	15	118	160	25,285	520,401
40	Iron	73	16	10	99	8	9	82	99	6,621	3,706,793
41	Manganese	18	9	4	31	0	0	31	31	657	4,971
42	Molybdenum	0	0	0	0	0	0	0	0	0	0
43	Antimony	1	0	0	1	1	0	0	1	12	5
44	Titan	0	0	0	0	0	0	0	0	0	0
45	Lead and zinc	21	15	7	43	2	3	38	43	1,884	222,655
46	Chromite	36	13	3	52	1	4	47	52	3,638	20,655
47	Copper	17	14	3	34	7	1	26	34	8,099	1,067,900
48	Gold	3	1	2	6	0	0	6	6	453	18,747
49	Nepheline syenite	2	0	0	2	2	0	0	2	444	1,296,000
50	Zeolite	8	6	0	14	0	0	14	14	95	8,833
51	Cobalt	0	0	0	0	0	0	0	0	0	0
52	Pyroxene	1	0	0	1	0	0	1	1	16	675
53	Vermiculite	1	0	0	1	0	0	1	1	16	47
54	Graphite	1	1	0	2	0	0	2	2	7	112
55	Agate	3	0	0	3	0	0	3	3	13	1
56	Talc schist	2	0	0	2	0	0	2	2	8	346
57	Anhydrite	0	1	0	1	0	0	1	1	0	820

(continued)

Table 7.2 (continued)

Row	Mineral	Active	Inactive	In development	Total	Governmental	Cooperative	Private sector	Total	Staff	Tonnage (1,000 tons)
58	Garnet	0	2	0	2	0	0	2	2	26	4,090
59	Potassium nitrate	0	1	0	1	0	0	1	1	0	500
60	Andalusite schist	1	0	0	1	0	0	1	1	18	68,000
61	Oligist+	1	0	0	1	0	0	1	1	16	156
62	Chalcedony	0	1	0	1	0	0	1	1	16	175
63	Jasper	1	0	0	1	0	0	1	1	11	250
64	Magnesia	0	1	0	1	1	0	0	1	207	32,000
65	Sylvite	0	1	0	1	0	0	1	1	0	4,500
66	Iodine	1	0	0	1	0	0	1	1	20	350
Total		3,568	1,372	271	5,212	72	627	4,512	5,212	97,278	31,110,781

Source: Statistical Center of Iran, website (http://www.amar.org.ir/Default.aspx?tabid=133)

Table 7.3 Iran mining sector at a glance

Year	No. of mines			No. of employees				Salaries and service compensation (Rls. million)	Value of products (Rls. million)
		Type of ownership			Production				
	Total	Private	Public	Total	Technical	Nontechnical	Administration and services		
1986	718	436	282	46,295	19,324	17,888	9,085	46,171	74,273
1991	1,920	1,393	527	50,322	30,170	13,047	7,106	99,682	424,870
1996	2,704	2,137	567	53,046	23,259	16,372	13,415	480,155	1,941,600
1998	2,436	2,027	406	52,732	22,593	16,174	13,965	645,892	2,694,965
1999	2,507	2,057	450	55,575	27,601	14,511	13,463	700,226	3,380,525
2000	2,656	2,316	342	56,559	27,755	15,998	12,806	802,103	4,177,181
2001	2,886	2,543	343	55,560	28,013	14,939	12,608	899,382	5,675,787
2002	2,955	2,623	332	55,112	26,501	16,416	12,195	1,124,668	6,485,824
2003	3,125	2,747	378	56,553	28,046	16,722	11,786	1,299,285	7,854,232
2,004	3,224	2,872	352	55,703	21,669	16,773	11,349	1,526,023	12,312,116
2005	3,324	2,998	326	54,854	15,292	16,825	10,912	1,752,762	16,770,000
2006	3,582	3,303	269	60,062	18,020	16,146	12,810	2,307,840	22,642,407
2007	3,799	3,552	247	66,250	19,971	18,010	12,867	3,325,650	30,647,053

Source: Statistical Center of Iran website (http://www.amar.org.ir/Default.aspx?tabid=133)

Table 7.4 Production of various materials in operating mines (1,000 tons)

Product / Year	2001	2002	2003	2004	2005	2006	2007	2008
Coal	1,380	2,557	2,108	2,003	1,899	2,413	2,525	2,593
Iron	11,441	14,597	14,556	17,350	20,145	20,648	35,195	27,476
Lead–zinc	579	608	686	838	991	1,072	1,932	1,245
Gold[a]	352	120	418	494	571	742	856	829
Copper[a]	453	418	465	509	553	790	34,637	35,506
Chromite	174	222	150	192	234	269	221	233
Manganese	140	555	114	104	115	94	185	121
Bauxite	65	20	406	873	1,341	1,395	451	727
Kaolinite and fireclay	1,660	1,723	2,073	1,957	1,842	2,716	252	945
Dolomite	281	611	679	541	403	514	666	585
Bentonite	175	157	188	224	261	189	201	356
Magnesite	115	70	88	91	95	189	110	121
Sodium sulfate	369	362	318	523	728	280	–	–
Barite	238	193	197	214	231	226	296	343
Ochre	76	86	32	17	3	–	–	–
Fluorine	34	55	43	55	68	63	71	66
Borax	3	2	2	2	2	2	–	–
Phosphate	33	70	202	263	324	67	252	253
Rock salt	1,538	1,510	1,553	1,792	2,032	2,618	2,897	2,477
Silica	1,678	2,221	1,755	1,929	2,103	2,252	–	–
Perlite and pumice[b]	826	610	970	872	775	898	–	–
Feldspar	157	257	297	294	291	418	611	648
Talcum	4	60	87	79	71	69	74	50
Asbestos	4	2	1	1	1	0	0	0
Sea shells	96	48	154	136	118	99	92	31
Mica	5	2	35	2	1	2	7	5
Gravel and sand[b]	46,147	74,854	81,701	69,092	56,484	76,102		
Decorative stones	10,325	10,717	11,795	10,595	9,396	9,908	4,182	3,977
Rubble stone[c]	9,359	9,781	8,334	10,011	11,688	16,441	35,037	37,878
Ballast[b,c]	802	776	429	251	73	394	–	–
Limestone	43,498	43,452	46,664	48,412	50,160	62,959	60,710	79,019
Gypsum	8,347	8,338	11,877	11,511	11,145	11,878	17,351	17,691
Total	140,354	175,054	188,377	181,227	174,144	215,518	198,811	213,175

Source: Statistical Center of Iran website (http://www.amar.org.ir/Default.aspx?tabid=133)
[a]Data for gold production in 1991 are included with copper ore
[b]In 1,000 m^3
[c]Data for ballast production in 1991 are included with rubble stone

Moreover, improved technology and deployment of modern mining equipment is the major factor in the lower-than-expected number of employees. This can be better understood when the ratio of technical to nontechnical workforce is taken into account. During the 15-year time interval of 1986–2001, the nontechnical employees of the mines have been on decline, while the technical as well as the administrative staff have grown.

Table 7.5 National income and gross domestic product based on fixed prices of 1990 (Rls. billions)

	Year					Percent variation	
	1997	1998	1999	2000	2001	2000	2001
Mining sector	401.1	410.6	499.1	523.2	573.4	4.8	9.6
Overall country	41,308.5	40,337.6	42,129.8	44,144.7	46,039.2	4.8	4.3
Percentage	0.97	1.02	1.18	1.19	1.25	–	–

Source: Economic Trends of Central Bank of Islamic Republic of Iran (various issues)
Note: No data available for years after 2001

Table 7.6 National income and gross domestic product based on current prices (Rls. billions)

	Year					Percent variation	
	1997	1998	1999	2000	2001	2000	2001
Mining sector	1,522.2	1,914.0	2,397.7	2,943.1	3,667.1	22.7	24.6
Overall country	231,615.2	265,590.9	353,651.5	468,865.3	553,364.9	32.6	18.0
Percentage	0.66	0.72	0.68	0.63	0.66	–	–

Source: Economic Trends of Central Bank of Islamic Republic of Iran (various issues)
Note: No data available for years after 2001

The comparison of the growth rate of the mining products over the period 1986–2001 is an attestation to the technological advancement of the mining sector. Ever since 1996, the production growth rate has been higher than the rate of growth of the number of mines, despite of the reversed trend during the years 1986–1996. Examination of the growth rate of the value of the products as compared to their volume, over the period 1991–2001, along with the increased number of active mines demonstrates that up until 1996 investments in smaller mining firms were on the agenda, but afterward large mining operations were favored by the government. Moreover, the volume to value ratio indicates that the operations have tended toward exploitation of higher-valued mining materials. All the above-mentioned facts point to the deployment of a more comprehensible and orientated mining policy since 1996.

The value added of the mining sector for the year 2001 has amounted to Rls. 4,618,050 million while the per-employ value added stood at Rls. 83.12 million, showing a very high rate of return on the investment in this sector. Table 7.7 lists the amounts of investment in the mining sector of the country over the period 1986–2001; the fluctuation in the investment has resulted in economic instability of the sector.

Due to the improper structure of the exports, the mining sector has had a small share in the non oil exports of the country. The total exports of the mining sector rose from USD 67.7 million in 1997 to USD 181.6 million in 2001. Moreover, the share of the mining sector in the overall non oil exports of the country over the same interval of time increased from 2.35 to 4.30%. Table 7.8 enlists the performance of the mining sector in the non oil exports of the country.

Table 7.7 Amount of investment in the mining sector of Iran (Rls. billion)

	1986	1991	1995	1996	1997	1998	1999	2000	2001	2002	2003	2005	2006	2007
Total investment	6	106	434	294	523	255	403	382	595	711	857	255	173	258
Added value	–	364	1,034	1,421	1,594	2,003	2,495	3,068	4,618	5,060	5,772	13,503	16,666	23,733

Source: Economic Trends of Central Bank of Islamic Republic of Iran (various issues)

Table 7.8 Non oil exports of Iran (US$ millions)

Years		1997	1998	1999	2000	2001
Total non oil exports		2,876	3,013	3,362	3,763	4,224
Export of mining products	Total	67.7	49.5	95.7	133	181.6
	Mineral aggregates and metals	45.1	12.8	36.3	37.7	77.3
	Cement, stones, construction materials	22.6	36.7	59.4	95.3	104.3
Share of mining in total exports (percentage)		2.35	1.64	2.85	3.53	4.30

Source: Economic Trends of Central Bank of Islamic Republic of Iran (various issues)
Note: No data available for years after 2001

Table 7.9 Government facilities extended to the mining sector (Rls. million)

Development facilities of the government	1991	1998	1999	2000	2001
Mining exploration	40,784	39,802	55,840	110,133	106,867
Mining equipment	126,352	100,627	116,435	136,304	107,561
Technical assistance	0	0	0	64,200	19,000
Total	167,136	140,429	172,275	310,637	233,428

Source: Economic Trends of Central Bank of Islamic Republic of Iran (various issues)
Note: No data available for years after 2001

It is evident from the above-mentioned facts and figures that the contribution of the mining sector to the overall non oil exports of the country (both in terms of volume and value of products) is not in terms with the country's position among those rich in natural resources.

The various facilities extended by the government financial institutions to the mining sector are presented in Table 7.9. Considering the fact that the figures are not in accordance with the actual need of the sector, the need for deregulation and private investment, as well as foreign investment, cannot be denied. Nevertheless, the role of the government in exploration and research should be further strengthened.

The quantity and quality of the proved mineral resources of the country and the relative position of Iran within the global framework of countries rich in natural resources, on the one hand, and the small share of mining sector in the GDP, employment, and exports of the country, on the other hand, reveal the demand for further investment in the sector. In view of the limitations of the government funds and facilities, the need for foreign investment is obvious. However, the restrictions exerted by the constitution of the country on the investment by foreign financial institutions have formed the major obstacle on the way of disinvestments, leading to weakening of this important sector of the economy. Tables 7.10, 7.11, and 7.12 summarize the information available for the mining sector as per the latest government statistics.

7.5 Mineral Resources of Iran

The results of economic geology investigations and statistical reports on the known mineral deposits and indications, encompassing metals, nonmetals, and fossil fuels, of Iran are compared with the world mining statistics, hereunder, to illustrate the standing of the country in terms of resources, production, and trade of various mining products. Though the information on Iranian mines is collected from various published and unpublished reports presented at national and international levels, they are only the best estimates obtainable and not the exact numbers resulting from scientific research. It is tried to compile all the available

Table 7.10 Value of the products of various materials in operating mines (Rls. million)

Product \ Year	1991	1996	1998	1999	2000	2001	2002	2003	2004	2005	2006	2007
Coal	78,243	357,943	494,023	594,799	522,087	570,566	799,739	841,165	819,137	797,109	1,077,642	1,427,715
Iron	19,237	231,205	495,271	595,193	780,799	947,000	1,437,617	1,675,940	3,222,727	4,769,515	5,483,753	9,576,836
Lead–zinc	17,866	92,546	78,732	152,840	223,631	195,458	177,608	189,647	353,996	518,345	1,853,634	1,784,373
Gold[a]	0	3,042	3,329	3,909	4,133	9,415	4,002	28,819	68,397	107,975	113,156	195,740
Copper[a]	73,698	269,038	305,209	340,091	457,705	1,145,592	742,074	851,700	3,112,739	5,373,778	7,500,248	8,449,079
Chromite	8,460	80,024	88,156	84,495	62,543	70,388	88,513	85,547	122,625	159,703	297,223	244,018
Manganese	1,047	9,344	19,503	15,303	20,793	31,023	177,443	39,833	60,751	41,837	51,383	84,811
Bauxite	754	13,194	21,350	21,743	2,298	4,253	1,772	57,650	93,814	129,978	153,928	127,506
Kaolinite and fireclay	1,002	36,169	42,790	38,211	76,405	98,729	89,224	160,585	135,856	111,128	236,033	201,866
Dolomite	772	6,493	9,299	7,710	11,016	14,337	18,733	28,553	26,865	25,178	29,395	52,929
Bentonite	243	1,490	2,448	4,051	5,686	7,123	9,266	9,539	11,995	14,451	12,382	19,321
Magnesite	400	10,716	17,718	17,967	31,658	21,069	20,674	24,067	22,399	20,731	46,722	32,098
Sodium sulfate	438	7,729	4,214	7,482	18,110	7,622	39,174	51,876	35,223	18,570	14,306	46,499
Barite	1,497	8,351	11,345	14,895	20,366	24,215	20,230	24,566	37,094	49,623	39,997	66,128
Ochre	836	2,317	2,507	3,491	5,431	6,181	11,158	5,150	2,813	477	41	500
Fluorine	919	4,448	6,252	6,653	10,675	11,118	17,600	12,216	18,760	25,305	53,959	58,485
Borax	38	2,235	888	2,317	2,203	2,312	2,263	3,259	3,050	2,841	3,328	3,180
Orpiment	313	144	1,131	719	580	879	613	1,650	825	0	180	560
Phosphate	–	600	1,155	553	19,322	39,484	23,095	67,670	91,218	114,766	31,018	25,153
Rock salt	3,220	26,390	31,392	32,037	43,904	38,076	39,140	46,497	58,034	69,571	109,273	159,849
Silica	2,059	18,314	26,183	27,997	46,432	49,152	98,677	67,035	77,195	87,355	148,891	189,067

Perlite and pumice	970	3,672	4,058	5,115	7,925	10,147	16,742	26,818	24,643	22,468	24,926	37,928
Feldspar	393	3,158	5,047	5,614	12,175	9,474	14,390	17,460	18,354	19,248	32,814	35,761
Talcum	222	1,635	1,543	2,581	4,208	11,491	4,193	6,760	6,032	5,304	6,959	8,671
Asbestos	685	1,978	2,477	2,730	5,190	3,450	2,366	3,181	3,085	2,990	0	0
Sea shells	654	3,415	3,083	10,063	7,316	7,674	3,650	22,240	19,790	17,340	12,786	18,570
Mica	89	1,294	342	1,245	5,798	4,293	2,106	14,001	7,876	1,752	4,405	5,152
Turquoise	80	325	495	442	503	954	954	1,542	3,148	4,755	7,625	10,000
Gravel and sand	78,693	277,148	279,929	371,280	453,361	543,928	798,523	929,846	1,057,592	1,185,339	1,548,204	2,147,538
Decorative stones	76,266	238,796	437,254	604,752	790,178	1,013,763	1,214,525	1,499,976	1,509,672	1,519,369	1,916,393	3,050,306
Rubble stone[b]	5,258	57,910	56,782	58,178	154,940	309,179	144,171	191,449	301,283	411,118	487,803	737,096
Ballast[b]	0	8,457	11,660	13,984	22,557	17,147	16,770	13,158	7,428	1,698	19,448	45,283
Limestone	28,229	124,664	159,815	258,759	263,316	363,066	346,650	672,250	795,881	919,513	1,079,502	1,516,897
Gypsum	11,543	37,036	63,450	65,863	83,768	87,085	102,168	182,588	201,837	221,087	236,511	277,459
Total	414,124	1,941,220	2,688,827	3,373,442	4,177,181	5,675,643	6,484,723	7,854,233	12,332,134	16,670,217	22,633,868	30,641,624

Source: Statistical Center of Iran website (http://www.amar.org.ir/Default.aspx?tabid=133)

[a]Data for gold production in 1991 are included with copper ore

[b]Data for ballast production in 1991 are included with rubble stone

Table 7.11 Employment, services, salaries paid, and value of products of active mines of various provinces

	No. of mines			No. of employees					
		Type of ownership			Production			Salaries and service compensation	Value of products[a]
Province	Total	Private	Public	Total	Technical	Nontechnical	Administration and services	(Rls. million)	(Rls. million)
West Azerbaijan	145	125	20	1,344	843	237	264	20,848	89,967
East Azerbaijan	148	138	11	1,055	467	327	261	13,973	88,520
Ardebil	31	23	8	213	130	38	45	2,683	10,048
Esfahan	207	177	30	4,363	2,230	1,079	1,055	77,366	427,428
Ilam	30	28	2	326	143	71	112	3,317	18,424
Bushehr	74	68	6	1,243	582	211	450	31,618	261,677
Tehran	94	84	10	1,198	675	367	156	18,298	124,272
Chahar-Mahal and Bakhtiyary	54	46	8	428	233	48	146	7,528	64,763
Khorasan	261	253	8	4,421	2,101	1,529	791	50,584	280,415
Khuzestan	138	134	4	1,862	798	700	364	25,245	97,095
Zanjan	64	59	5	679	366	201	112	12,779	104,331
Semnan	139	123	16	3,113	1,732	739	642	43,863	151,006
Sistan and Baluchestan	40	29	11	383	187	103	93	5,266	27,852
Fars	146	125	21	2,976	1,778	553	644	37,697	368,415
Ghazvin	69	61	8	540	295	115	130	7,003	144,104
Qom	58	54	4	910	245	466	198	10,889	54,133
Kordestan	140	115	25	970	377	456	137	10,551	62,601
Kerman	101	94	7	10,001	5,259	1,548	3,196	219,332	1,483,731
Kermanshah	76	53	23	540	315	112	112	5,677	25,474
Kohkilooye and Boyr-Ahmad	64	59	5	374	120	171	83	3,698	11,243

Golestan	44	33	11	1,670	914	492	263	21,852	58,715
Gilan	35	30	5	759	269	366	124	8,109	66,108
Lorestan	90	76	13	2,354	1,174	454	726	38,421	253,928
Mazandaran	111	97	14	5,063	1,905	2,055	1,103	98,244	197,935
Markazi	151	129	22	1,542	853	415	274	19,069	95,282
Hormozgan	45	37	8	769	362	279	128	12,048	80,249
Hamedan	209	181	28	1,176	451	565	160	9,292	62,160
Yazd	123	112	11	5,290	3,210	1,241	839	84,135	965,910

Source: Statistical Center of Iran website (http://www.amar.org.ir/Default.aspx?tabid=133)

Table 7.12 Employment, services, salaries paid, and value of products of active mines of various materials for 2001–2002 fiscal

	No. of mines			No. of employees					
		Type of ownership			Production				
Activities	Total	Private	Public	Total	Technical	Nontechnical	Administration and services	Salaries and service compensation (Rls. million)	Value of products (Rls. million)
Coal	83	69	14	13,954	7,284	2,709	3,961	237,707	570,566
Iron	24	19	5	3,288	2,073	573	642	69,284	947,000
Lead–zinc	11	9	2	1,553	803	480	270	43,133	195,458
Gold	3	2	1	101	69	12	20	1,536	9,415
Copper	9	4	5	2,459	1,326	419	714	80,572	1,145,592
Chromite	23	23	0	2,121	772	1,235	114	35,680	70,388
Manganese	9	9	0	522	100	322	100	6,265	31,023
Antimony	0	0	0	0	0	0	0	0	0
Bauxite	6	4	2	120	62	49	9	1,420	4,253
Kaolinite and fireclay	105	100	5	1,370	482	561	327	20,126	98,729
Dolomite	16	14	2	144	87	34	23	1,786	14,337
Bentonite	17	16	1	112	67	26	19	1,300	7,123
Magnesite	24	22	2	327	109	181	37	4,183	21,069
Sodium sulfate	35	34	1	232	71	127	34	1,623	7,622
Barite	56	54	2	573	274	183	116	7,125	24,215
Ochre	6	5	1	56	29	14	13	961	6,181
Fluorine	9	9	0	239	89	105	45	3,364	11,118
Celestine	0	0	0	0	0	0	0	0	0
Borax	1	0	1	24	10	2	12	594	2,312
Orpiment	1	0	1	12	8	3	1	236	879
Alunite	1	1	0	8	3	3	2	44	144
Phosphate	2	0	2	155	119	28	8	2,264	39,484
Rock salt	61	58	3	742	346	278	117	7,229	38,076

Silica	92	86	6	694	283	303	108	8,746	49,152
Perlite and pumice	30	28	2	182	77	54	51	1,832	10,147
Feldspar	17	16	1	127	54	47	26	1,217	9,474
Talcum	7	6	1	98	54	31	13	1,477	11,491
Asbestos	1	1	0	131	45	61	25	1,872	3,450
Sea shells	6	6	0	267	64	165	38	1,471	7,674
Mica	7	7	0	67	27	30	10	720	4,293
Turquoise	1	0	1	24	7	10	7	303	954
Gravel and sand	1,028	843	185	9,507	4,569	2,693	2,245	116,335	543,928
Decorative stones	503	466	37	9,071	4,871	2,211	1,989	131,718	1,013,763
Rubble stone	305	286	19	2,827	1,325	795	707	46,506	309,179
Ballast	8	3	5	217	108	78	31	3,451	17,147
Limestone	180	149	31	2,796	1,656	668	472	42,622	363,066
Gypsum	199	194	5	1,441	690	449	302	14,678	87,085
Total	2,886	2,543	343	55,560	28,013	14,939	12,608	899,382	5,675,787

Source: Statistical Center of Iran website (http://www.amar.org.ir/Default.aspx?tabid=133)

Note: No data available for years after 2001

Table 7.13 Iron reserve provinces of Iran

Province	Mining area	Reserves
Khaf	Sangan mining complex	900 million tons
Bafq	Choghart, Chadormalou, Sechahoon, Chahgaz, Narigan	2 billion tons
Sirjan	Gol Gohar Mining Complex	1.2 billion tons
Total		4.1 billion tons

data on Iranian mines gathered from field works and visits on Iranian deposits by the author, Iran Statistics Center, USGS, and World Mineral Statistics in order to reach a better conclusion. However, more attention is paid to the qualitative aspects of some minerals.

7.5.1 *Iron*

Over 200 ore deposits, indications, or anomalies of iron are recognized in Iran whose reserves amount to almost 4.5 billion tons of iron ore. The average grade of iron in these reserves varies from 45 to 60%. More than 90% of Iran's iron ore deposits are located in the three provinces listed in Table 7.13, while Table 7.14 presents a comparison of Iran's iron ores with the rest of the world.

The share of Iran from the overall iron reserves stands at 2.7–3%, while the share in the world production varies from 0.9 to 1.0%.

7.5.2 *Manganese*

Over 45 ore deposit and indication of manganese are recognized in Iran, out of which 10 deposits are of medium tonnage and the rest are either small deposits or indications, revealing the fact that Iran is poorer in terms of manganese as compared to iron. Therefore, although the iron production has gone up in recent years, the manganese production has stayed unchanged. However, new exploration might result in the discovery of valuable manganese ore deposits. Tables 7.15, 7.16, and 7.17, respectively, illustrate the most important manganese ores, manganese-bearing iron ores, and the status of manganese production in Iran.

7.5.3 *Chromium (Chromite)*

The known chromite ore deposits of Iran are of Alpine type and therefore hard to estimate their reserves. However, considering the number of chromite ore deposits and indications (almost 200) and the vast distribution of rock types that are associated

Table 7.14 Reserves, production, and trade of iron in Iran as compared to the world[a, b, d]

World					
	Production				
Reserves	2010	2011	Resources	Export	Import
330 billion tons[c]	2.59 billion tons	2.8 billion tons	800 billion tons	1.041 billion tons	1.041 billion tons
Iran					
Reserves	Production		Resources	Export[e]	Import[e]
4.6 billion tons[b]	27 million tons (ore)[d]		8 billion tons	1.8 million tons	3.9 billion tons

[a]Mineral Commodity Summaries 2008
[b]Probable reserves
[c]Proved reserves
[d]In recent years, it has been fluctuating, but the overall growth rate has been around 6 percent; U.S. Geological Survey, Mineral Commodity Summaries, January 2012
[e]Australian commodity statistics, 2011; Statistics from Iranian Ministry of Mines and Metals as of 2009 and Islamic Republic of Iran's Custom Administration as of 2010

Table 7.15 Important manganese deposits of Iran (Samani 1995)

Name of the deposit	Proved reserve (tons)	Probable reserve (tons)
Venarch	6,000,000	–
Dabakloo	–	1,100,000
Benvid	–	70,000
Chah Sefid	40,000–50,000	–
Robat Karim	386,000	500,000–200,000,000
Bozni	110,000	300,000
Ab Band	50,000–100,000	–
Asad	50,000	–
Bensivert	20,000	
Chah Bashe	1,030,000	182,000,000
Baft	120,000	360,000
Zabol	5,000	–

Table 7.16 Manganese-bearing iron deposits of Iran

Name of the deposit	Proved reserve (tons)	Probable reserve (tons)
Shams Abad	100,000,000	140,000,000
Heneshk	300,000–500,000	75,000,000
Narigan	3,000,000	–

with chromite, the possibility of discovering small- and medium-sized reserves (or even large-sized bodies, on the basis of geological characteristics) cannot be ruled out. Most of the chromite deposits of Iran are situated in Faryab, Esfandaghe, Makran, Khash-Nehbandan Belt, Sabzevar, and Neyriz (Fars). The important chromite

Table 7.17 Reserves, production, and trade of manganese in Iran as compared to the world[a]

World					
	Production				
Reserves	2010	2011	Resources	Export	Import
630 million tons	13.9 million tons	14 million tons	5.2 billion tons	7–8 million tons	7.5 million tons
Iran					
Reserves	Production		Resources	Export	Import
10.19 million tons	121,450 tons		100 million tons[b]	3,290 tons	68,450 tons

U.S. Geological Survey, Mineral Commodity Summaries, January 2012
Statistics from Iranian Ministry of Mines and Metals as of 2009 and Islamic Republic of Iran's Custom Administration as of 2010
[a]Mineral Commodity Summaries 2008
[b]Manganese-bearing iron ore

Table 7.18 Important chromite deposits of Iran

Name of the mine	Location	Proved reserve (tons)	Probable reserve (tons)
Esfandaghe	Baft-Kerman	500,000	1,000,000–2,000,000
Faryab	147 km north of Bandar Abbas	–	2,000,000–3,000,000
Foroumad-Gaft	Sabzevar	–	200,000
Arzou	Hormozgan	700,000	>1,000,000
Dareh Bid, Dareh Parand	Sabzevar	–	200,000
Sefid Abe	Zabol	–	>1,000,000

deposits of Iran are listed in Table 7.18, and the status of Iran's deposits as compared to the world in Table 7.19.

7.5.4 *Molybdenum*

There are considerable amounts of molybdenum associated with the porphyry copper deposits of Sarcheshme, Meydook, Songoon, and Dareh Zar. In addition, a number of vein-type copper deposits of Iran show high contents of this metal, for example, Senj copper deposit (Karaj). Table 7.20 shows the amount of reserves, production, and trade of molybdenum in Iran as compared to the world figures.

7.5.5 *Copper*

Around 500 ore deposits and indications of copper are recognized in Iran, while only about 100 of them are surveyed and explored. Presently, there are

Table 7.19 Reserves, production, and trade of chromite in Iran as compared to the world[a]

World					
	Production				
Reserves	2010	2011	Resources	Export	Import
>480 million tons	23.7 million tons	24 million tons	12 billion tons	7.6 million tons	7.2 million tons
Iran					
Reserves	Production		Resources	Export	Import
10 million tons	233,380 tons		100 million tons	200,000 tons	547 tons

U.S. Geological Survey, Mineral Commodity Summaries, January 2012
Statistics from Iranian Ministry of Mines and Metals as of 2009 and Islamic Republic of Iran's Custom Administration as of 2010
[a]Mineral Commodity Summaries 2008

Table 7.20 Reserves, production, and trade of molybdenum in Iran as compared to the world[a]

World					
	Production				
Reserves	2010	2011	Resources	Export	Import
10 million tons	242,000 tons	250,000 tons	14 million tons	115,000–125,000 tons	145,000–170,000 tons
Iran					
Reserves	Production		Resources	Export	Import
50,000 tons	2,500 tons[b]		0.5 million tons	1,000 tons	1,000 tons

Statistics from Iranian Ministry of Mines and Metals as of 2009 and Islamic Republic of Iran's Custom Administration as of 2010
Statistics from Iranian Ministry of Mines and Metals as of 2009 and Islamic Republic of Iran's Custom Administration as of 2010
[a]Mineral Commodity Summaries 2008
[b]Average production of concentrated molybdenite over the past 5 years was 1,900 tons

10 active copper mines in Iran whose reserves amount to 3 billion tons of ore, containing 30 million tons of copper metal comprising 9% of the world's known copper reserves. The deposits can be geographically divided into six major provinces, namely, Kerman (which holds the largest reserves of the country), Azerbaijan (dominantly situated in the Ahar-Jolfa belt in Arasbaran area), South Khorasan–North Sistan (only reconnaissance surveys carried out), Minab-Kahnooj (only reconnaissance surveys carried out), Sabzevar-Semnan (all are of vein type with no porphyry deposits yet reported), and Tarom. The reserves, production, and trade of copper in Iran and the rest of the world are illustrated in Table 7.21.

Table 7.21 Reserves, production, and trade of copper in Iran as compared to the world[a, b]

World					
	Production				
Reserves	2010	2011	Resources	Export[b]	Import[b]
690 million tons	15.9 million tons	16.1 million tons	3700 million tons	5.2 million tons	5.03 million tons
Iran					
Reserves	Production		Resources	Export	Import
30 million tons (3 billion tons of ore)	0.2 million tons (Cu)		250 million tons	0.15 million tons (Cu)	–

U.S. Geological Survey, Mineral Commodity Summaries, January 2012
Statistics from Iranian Ministry of Mines and Metals as of 2009 and Islamic Republic of Iran's Custom Administration as of 2010
[a]Mineral Commodity Summaries 2008
[b]Australian commodity statistics, 2010

Table 7.22 Reserves, production, and trade of lead in Iran as compared to the world[a]

World					
	Production				
Reserves	2010	2011	Resources	Exports	Imports
85 million tons	4.14 million tons	4.5 million tons	1.5 billion tons	1.07 million tons	1.499 million tons
Iran					
Reserves	Production		Resources	Exports[b]	Imports
5 million tons (Pb)	1.245 million tons		1 billion tons (Ore)	360,940 tons	–

[a]Mineral Commodity Summaries 2008
[b]In the form of lead concentrate

7.5.6 *Lead–Zinc*

The lead–zinc ore deposits and indications in Iran number more than 600. Taking into account the average grade of the ores, the amount of lead and zinc metals is estimated at 18 and 5 million tons, respectively. There is a high probability of finding new large reserves as the exploration techniques advance. Tables 7.22 and 7.23 show, respectively, the status of Iran's lead and zinc deposits as compared to the world figures.

7.5.7 *Aluminum (Bauxite)*

Bauxite reserves of Iran are not significant in amount, and considering the conditions required for the formation of this mineral as well as the geological setting of the country,

Table 7.23 Reserves, production, and trade of zinc in Iran as compared to the world[a, d]

World					
	Production				
Reserves	2010	2011	Resources	Exports	Imports
250 million tons (Zn)	12 million tons	12.4 million tons	1.9 billion tons	4.052 tons	4.418 tons
Iran					
Reserves[b]	Production		Resources	Exports[c]	Imports
200 million tons (Ore)	1.245 million tons		1 billion tons (Ore)	47,875 tons	16,688 tons

U.S. Geological Survey, Mineral Commodity Summaries, January 2012
Statistics from Iranian Ministry of Mines and Metals as of 2009 and Islamic Republic of Iran's Custom Administration as of 2010
[a]Mineral Commodity Summaries 2008
[b]Total lead–zinc ores
[c]In form of zinc concentrate
[d]Australian commodity statistics, 2010

Table 7.24 Economic characteristics of bauxite deposits of Iran (Soheili 2004)

Name of the deposit	Reserve (million tons)	Percentage of Al_2O_3	Dominant mineral
Jajarm	10.6 (Proved)	47	Bauxite
Siyah Roudbar (Gorgan)	14 (Probable)	39 (Average)	Dominantly diaspore
Shahbodaghi (Damavand)	2 (Probable)	50 (Average)	Dominantly diaspore
Ab-e Garm-e Ghazvin	14 (Probable)	40 (Average)	Dominantly diaspore
Sarchaveh	3.5 (Probable)	42 (Average)	Dominantly diaspore
Sadr Abad (Yazd)	2 (Proved)	45	Diaspore
Shomal (Yazd)	2 (Proved)	45	Diaspore
Chekchek (Yazd)	1 (Proved)	39	Diaspore
Dehdasht	2 (Proved)	55	Boehmite and diaspore

the discovery of extensive bauxite deposits is not anticipated. However, there are a few sizable bauxite deposits that are economically exploited. These are listed in Table 7.24, while the status of the country from the point of view of aluminum resources is compared with that of the world in Table 7.25.

7.5.8 Arsenic

Although there are no estimates of the amount of arsenic reserves in Iran, the presence of 17 ore deposits and indications of this metal points toward the vast occurrence of arsenic in Iran. Two ore bodies at Zarshouran and Valilou are

Table 7.25 Reserves, production, and trade of aluminum, bauxite, and alumina in Iran as compared to the world[a]

World					
Reserves (bauxite)	Production (aluminum) 2010	Production (aluminum) 2011	Resources (bauxite)	Exports	Imports
29 billion tons	40.8 million tons	44.1 million tons	>55 billion tons	11.5 million tons	12.2 million tons
Iran					
Reserves (bauxite)	Production (aluminum)		Resources (bauxite)	Exports	Imports (alumina)
39 million tons	0.727 million tons		Not promising	–	0.152 million tons

U.S. Geological Survey, Mineral Commodity Summaries, January 2012
[a]Mineral Commodity Summaries 2008

Table 7.26 Reserves, production, and trade of arsenic in Iran as compared to the world[a, b]

World					
Reserves	Production 2010	Production 2011	Resources	Exports	Imports
1 million tons	52,800 tons	52,000 tons	11 million tons	27,000 tons	20,000 tons
Iran					
Reserves	Production		Resources	Exports	Imports
0.05 million tons	2,000 tons		0.5 million tons	–	–

U.S. Geological Survey, Mineral Commodity Summaries, January 2012
[a]Mineral Commodity Summaries 2008
[b]Australian commodity statistics, 2010

exploited presently. The Zarshouran ore deposit is unrivaled all over the world for purity, texture, and crystal form of its mineral (realgar). The most important arsenic ore deposits of Iran are Zarshouran, Chalpou (Kuh Sorkh, Kashmar), Dashkasan, Valilou, Dastjerd, Tikme Dash, Ali Abad, Torbat-e Jam (arsenopyrite), and Anarak (copper arsenate). Table 7.26 illustrates the status of arsenic reserves of Iran as compared to that of the world.

7.5.9 Antimony

Fifteen ore deposits and indications of antimony are known to occur in Iran, out of which only three are actively mined. The geographic distribution of antimony in Iran can be categorized into the following three regions:

1. Ghorve, Bijar, Takab (Dashkasan, Moghanlou and Agh Dare)
2. Anarak area in Central Iran (Patyar, Torkamani)

Table 7.27 Reserves, production, and trade of antimony in Iran as compared to the world[a, b]

World					
	Production				
Reserves	2010	2011	Resources	Exports	Imports
1.8 million tons	167,000 tons	169,000 tons	5.1 million tons	27,700 tons	62,600 tons
Iran					
Reserves	Production		Resources	Exports	Imports
0.3 million tons	106 tons		0.5 million tons	–	–

U.S. Geological Survey, Mineral Commodity Summaries, January 2012
[a]Mineral Commodity Summaries 2008
[b]Australian commodity statistics, 2010

Table 7.28 Reserves, production, and trade of rhenium in Iran as compared to the world[a]

World					
	Production				
Reserves	2010	2011	Resources	Exports	Imports
2,500 tons	47.2 tons	49 tons	6,000 tons		–
Iran					
Reserves	Production		Resources	Exports	Imports
–	1,000 tons		–	–	

U.S. Geological Survey, Mineral Commodity Summaries, January 2012
[a]Mineral Commodity Summaries 2008

3. Ferdos, Kashmar, and Torbat Jam area in Eastern Iran

The statistics regarding the reserves, production, and trade of antimony is presented in Table 7.27.

7.5.10 Rhenium

Some of the porphyry-type copper deposits of Iran contain significant quantities of trace elements such as rhenium. Table 7.28 illustrates the status of the country compared to the rest of the world for rhenium.

7.5.11 Gold

The number of deposits and indications of gold in Iran is 146, but no exact figure on the amount of reserves of this precious metal is at hand due to the unavailability of modern exploration techniques. Nonetheless, taking into account the geological evidence of gold mineralization, as well as the information on well-explored gold

Table 7.29 Recently surveyed gold deposits of Iran (Ghorbani 2008a, b, c, d, e, f, g, h)

Name of the deposit	Probable reserve (tons)	Average grade (ppm)
Zarshouran (Takab)	100	4–10
Agh Dare (Takab)	30	2.7
Dashkasan (Qorve)	63	–
Moote (Delijan)	14	2.5–4
Kuh-e Zar (Torbat Heydariye)	15	3.3
Kharvana (Tabriz)	30	4

Table 7.30 Reserves, production, and trade of gold in Iran as compared to the world[a, b]

World						
		Production				
Reserves (proved)	Basic reserves	2010	2011	Resources	Exports	Imports
51,000 tons	90,000 tons	2,560 tons	2,700 tons	–	–	
Iran						
Reserves (probable)	Basic reserves	Production		Resources	Exports	Imports
330 tons		3 tons		>1,000 tons	–	–

U.S. Geological Survey, Mineral Commodity Summaries, January 2012
[a]Mineral Commodity Summaries 2008
[b]Australian commodity statistics, 2010

deposits of the country, the gold reserves of Iran are optimistically estimated at 1,000 tons. The specifications of the major gold deposits of the country are enumerated in Table 7.29, while the status of Iran among the gold-producing countries of the world is shown in Table 7.30.

7.5.12 Phosphate

Over 80 phosphate occurrences and indications are known to exist in Iran. Based on the available exploration data, the proved reserves of the country can be genetically categorized into sedimentary and igneous types. Tables 7.30 and 7.31 show the most important phosphate deposits and the global position of Iran from the point of view of phosphate reserves, production, and trade, respectively (Table 7.32).

7.5.13 Fluorite

Known fluorite resources of Iran number more than 30 with reserves estimated at 500,000 tons with base reserve of around 1 million tons. Presently, four localities

Table 7.31 Major phosphate deposits of Iran (Halalat and Bolourchi 1994)

Name of the deposit	Geographic location	Probable reserve (million tons)	Average grade of phosphate (%)
Kuh-e Lar	Charam, Kohkilouye	350	8
Kuh-e Koome	Southeast of Dehdasht	22	9.3
Kuh-e Rish	North of Behbahan	10.6	11.2
Sheikh Habil	North of Dehdasht	1	22
Kuh-e Sefid	South of Eize	17	12.5
Rizroud	–	160	8.25
Khur Moj	East of Bushehr	56	8.5
Kuh-e Namak	North of Kangan	24	8.2
Chenar	Northeast of Andimeshk	5	3.2
Shemshak-Jiroud	North of Tehran	73	9.13
Firooz Kuh- Gadouk	Northeast of Firoozkuh	56	12
Deh Molla	Southwest of Shahroos	9	10
Kalmard Anticline	Southwest of T2abas	6.2	5
Dahooieye (Zarand)	East of Zarand	1.13	7.5
Dalir	South of Chalous	23	11.5
Vali Abad	South of Chalous	3	9
Zanjan	Southwest of Zanjan	12	–
Firooz Abad	South of Chalous	40	8
Esfordi	Northeast of Bafq	15	12
Zarigan	Northeast of Bafq	0.5	3
Total		883.43	8.78

Table 7.32 Reserves, production, and trade of phosphate in Iran as compared to the world[a]

World					
Reserves (proved)	Basic reserves	Production	Resources	Exports	Imports
12 billion tons	24 billion tons	158 million tons	Numerous	1.365 million tons	0.5 million tons
Iran					
Reserves (probable)	Basic reserves	Production	Resources	Exports	Imports
400 million tons	800 million tons	0.25 million tons	5 billion tons	18,470 tons	409 tons

U.S. Geological Survey, Mineral Commodity Summaries, January 2012
Statistics from Iranian Ministry of Mines and Metals as of 2009 and Islamic Republic of Iran's Custom Administration as of 2010
Australian Commodity Summaries, 2008
[a]Mineral Commodity Summaries 2008

are extracted. The most important fluorite reserves of Iran are situated in the following provinces:

- Mazandaran with 83,000 tons
- Khorasan with 326,000 tons

Table 7.33 Reserves, production, and trade of fluorite in Iran as compared to the world[a]

World					
	Production				
Reserves	2010	2011	Resources	Export	Import
240 million tons	6.01 million tons	6.2 million tons	500 million tons	–	–
Iran					
Reserves	Production		Resources	Export	Import
1.4 million tons	66,129 tons		>40 million tons	0.07 million tons	–

U.S. Geological Survey, Mineral Commodity Summaries, January 2012
[a]Mineral Commodity Summaries 2008

Table 7.34 Reserves, production, and trade of rock salt in Iran as compared to the world[a, b]

World					
	Production				
Reserves	2010	2011	Resources	Export	Import
Numerous	280 million tons	290 million tons	Numerous	25 million tons approx.	26 million tons
Iran					
Reserves	Production		Resources	Export	Import
> 10 billion tons	2.477 million tons		Numerous	0.126 million tons	426 tons

U.S. Geological Survey, Mineral Commodity Summaries, January 2012
Statistics from Iranian Ministry of Mines and Metals as of 2009 and Islamic Republic of Iran's Custom Administration as of 2010
[a]Mineral Commodity Summaries 2008
[b]Australian commodity Summaries, 2008

- Esfahan with 120,000 tons
- Kordestan with 5,000 tons

7.5.14 Rock Salt

Iran is famous for its numerous salt diapirs (salt domes) that are in fact mountains (made up of rock salt) which have pierced their overlying rock layers and become exposed at the surface. There are more than 100 such structures with very high salt content, each of which can be considered as an independent rock salt deposit (Table 7.33).

Additionally, there are many salt water springs with high salt content that can be taken as a source of dissolved (aqueous) salt. In some lakes such as Oroumiye and Maharlou, the salt content is so high that extraction by means of decantation method is possible. Table 7.34 illustrates the position of Iran from the point of view of rock salt reserves and production among other countries.

Table 7.35 Reserves, production, and trade of boron in Iran as compared to the world[a]

World					
	Production				
Reserves	2010	2011	Resources	Export	Import
210 million tons	4.08 million tons	4.3 million tons	N.A.	142,000 tons approx.	185,000 tons
Iran					
Reserves	Production		Resources	Export	Import
25,000 tons	1,000–1,500 tons		N.A.[b]	–	73.5 tons

U.S. Geological Survey, Mineral Commodity Summaries, January 2012
Statistics from Iranian Ministry of Mines and Metals as of 2009 and Islamic Republic of Iran's Custom Administration as of 2010
[a]Mineral Commodity Summaries 2008
[b]With new finds, it is expected to go beyond 1 million tons.

Table 7.36 Reserves, production, and trade of kaolin in Iran as compared to the world[a, b]

World					
	Production				
Reserves	2010	2011	Resources	Export	Import
N.A.	33.1 million tons	33.3 million tons	Numerous	28.5 million tons approx.	23.0 million tons
Iran					
Reserves	Production		Resources	Export	Import
382 million tons	945,758 tons		> 2 billion tons	38,179 tons	49,696 tons

U.S. Geological Survey, Mineral Commodity Summaries, January 2012
Statistics from Iranian Ministry of Mines and Metals as of 2009 and Islamic Republic of Iran's Custom Administration as of 2010
[a]Mineral Commodity Summaries 2008
[b]Australian Commodity Summaries, 2008

7.5.15 Boron (Borax)

There are seven known deposits and indications of boron in Iran. But taking into account the conditions favorable for formation of economic boron bodies, one can expect discovery of more such deposits in various parts of the country. The boron production and reserves of Iran as compared to the rest of the world are presented in Table 7.35.

7.5.15.1 Kaolin

Over 70 deposits and indications of kaolin occur in Iran, most of which are suitable to be utilized in the ceramic industry. Table 7.36 indicates the reserves and production of kaolin in Iran and the world.

Table 7.37 Reserves, production, and trade of bentonite in Iran as compared to the world[a]

World					
	Production				
Reserves	2010	2011	Resources	Export	Import
N.A.	10.6 million tons	11.3 million tons	Numerous	57 million tons approx.	40 million tons
Iran					
Reserves	Production		Resources	Export	Import
> 100 million tons	356,980 tons		> 1 billion tons	68,580 tons	1,478 tons

[a]Mineral Commodity Summaries 2008

7.5.16 Bentonite

Ever since the ancient times, bentonite has been utilized in Iran, and at present there are more than 100 deposits and indications of it in various parts of Iran. Table 7.37 compares the reserves, production, and trade of bentonite in Iran with the world.

7.5.17 Vermiculite

The available information on vermiculite deposits of the country is incoherent and limited. Until today, there has not been any systematic exploration or survey carried out on the vermiculite deposits of Iran. There are numerous indications of vermiculite in Kerman, Shahin Dezh (West Azerbaijan), Amlash (Gilan), and Kaleybar (East Azerbaijan), with the last two being actively extracted.

7.5.18 Mica

Various reserves and resources of mica are known to occur in Iran so that there are over 40 deposits and indications of this mineral having an overall reserve of more than 1 million tons. However, there are only two active mines where mica extraction is currently undergoing, namely, Masoule and Qare Bagh. Table 7.38 lists the status of mica reserves, production, and trade in Iran and the rest of the world.

7.5.19 Strontium

Considerable reserves of celestite (strontium sulfate) are known, and ten economic deposits are being extracted that are distributed in south of Varamin and Garmsar, Anarak, and North of Khuzestan. The statistics of strontium reserves and production are presented in Table 7.39.

Table 7.38 Reserves, production, and trade of mica in Iran as compared to the world[a]

World				
Reserves	Production	Resources	Export	Import
991 million tons	5.2 million tons	1.469 billion tons	–	–
Iran				
Reserves	Production	Resources	Export	Import
1 million tons	5,930 tons	5 million tons	–	1,499 tons

U.S. Geological Survey, Mineral Commodity Summaries, January 2012
Statistics from Iranian Ministry of Mines and Metals as of 2009 and Islamic Republic of Iran's Custom Administration as of 2010
[a]Mineral Commodity Summaries 2008

Table 7.39 Reserves, production, and trade of strontium in Iran as compared to the world[a]

World					
	Production				
Reserves	2010	2011	Resources	Export	Import
6.8 million tons	405,000 tons	380,000 tons	1.2 billion tons	–	–
Iran					
Reserves	Production		Resources	Export	Import
2 million tons	2,000 tons		5 million tons	–[b]	–

U.S. Geological Survey, Mineral Commodity Summaries, January 2012
[a]Mineral Commodity Summaries 2008
[b]In the year 1990 Iran was ranked fourth among strontium exporting countries

7.5.20 *Barite*

The number of barite deposits and indications in Iran exceeds 50 with overall reserves of 10 million tons. Nonetheless, there are high chances of discovering new reserves. The present status of production and trade of barite in Iran and the world is shown in Table 7.40.

7.5.21 *Diatomite*

A number of diatomite deposits have been studied in Azerbaijan Province, a few having considerable reserves. No data are, however, available on these deposits (Table 7.41).

7.5.22 *Feldspars*

Half of the 100 known feldspar deposits and indications of Iran are currently being extracted. Genetically, these can be divided into two categories, namely, those

Table 7.40 Reserves, production, and trade of barite in Iran as compared to the world[a]

World					
	Production				
Reserves	2010	2011	Resources	Export	Import
240 million tons	7.85 million tons	7.8 million tons	880 million tons	148,000 tons	148,000 tons
Iran					
Reserves	Production		Resources	Export	Import
5.3 million tons	343,700 tons		100 million tons	70,460 tons	76 tons

U.S. Geological Survey, Mineral Commodity Summaries, January 2012
Statistics from Iranian Ministry of Mines and Metals as of 2009 and Islamic Republic of Iran's Custom Administration as of 2010
[a]Mineral Commodity Summaries 2008

Table 7.41 Reserves, production, and trade of perlite in Iran as compared to the world[a]

World					
	Production				
Reserves	2010	2011	Resources	Export	Import
Numerous	1.82 million tons	1.8 million tons	–	–	–
Iran					
Reserves	Production		Resources	Export	Import
0.5 million tons	9,600 tons		–	–	–

U.S. Geological Survey, Mineral Commodity Summaries, January 2012
Statistics from Iranian Ministry of Mines and Metals as of 2009 and Islamic Republic of Iran's Custom Administration as of 2010
[a]Mineral Commodity Summaries 2008

associated with acidic intrusive bodies and those of volcanic and tufaceous origin (Table 7.42).

7.5.23 *Perlite*

The large reserves of perlite deposits of Iran are associated with shallow marine acid volcanic rocks. Table 7.43 shows the statistics on perlite in Iran as compared to the rest of the world.

7.5.24 *Dimension Stones*

Iran has very high potentials for the production and export of various dimension stones, but, in spite of growing production of mines during the past decade, in many instances, these potentials have been overlooked. Construction, decorative, and facing

Table 7.42 Reserves, production, and trade of perlite in Iran as compared to the world[a]

World					
	Production				
Reserves	2010	2011	Resources	Export	Import
Numerous	20.6 million tons	20.7 million tons	–	–	–
Iran					
Reserves	Production		Resources	Export	Import
21 million tons	648,350 tons		Numerous	104,981 tons	20,748 tons

U.S. Geological Survey, Mineral Commodity Summaries, January 2012
Statistics from Iranian Ministry of Mines and Metals as of 2009 and Islamic Republic of Iran's Custom Administration as of 2010
[a]Mineral Commodity Summaries 2008

Table 7.43 Reserves, production, and trade of perlite in Iran as compared to the world[a]

World					
	Production				
Reserves	2010	2011	Resources	Export	Import
700 million tons	1.67 million tons	1.9 million tons	7.7 billion tons	–	–
Iran					
Reserves	Production		Resources	Export	Import
250 million tons	40,307 tons		1 billion tons	227 tons	753 tons

U.S. Geological Survey, Mineral Commodity Summaries, January 2012
Statistics from Iranian Ministry of Mines and Metals as of 2009 and Islamic Republic of Iran's Custom Administration as of 2010
[a]Mineral Commodity Summaries 2008

stones of Iran are among the highest qualities available all over the world. In terms of variety of color, texture, quality, and economic value, some of these reserves are unique and can be extracted and exported, creating jobs and income for the country. In the year 1999, there were 1584 active mines producing more than 66 million tons of dimension stones (567 of these mines produced about 7 million tons of decorative and facing stones). The data regarding this category of mining products are presented in Table 7.44.

7.5.25 Magnesite

Most of the 100 plus magnesite deposit of Iran occur in Khorasan Province around Birjand. Tables 7.45 and 7.46 show the most important magnesite deposits and the global position of Iran from the point of view of magnesite reserves, production, and trade, respectively.

Table 7.44 Reserves, production, and trade of dimension stones in Iran as compared to the world[a]

World				
Reserves	Production	Resources	Export	Import
N.A.	85 million tons[b]	Numerous	11 million tons	15 million tons
Iran				
Reserves	Production	Resources	Export[c]	Import
> 100 million tons	10–11 million tons	N.A	880,000 tons as dimension stones[d]	–

[a]Mineral Commodity Summaries 2008
[b]The unprocessed dimension stones extracted amounted to 104.5 million tons
[c]The total exports of construction stones were around 393,000 tons
[d]In the year 1998, exports amounted to 129,000 tons

Table 7.45 Major magnesite deposits of Iran (Ghorbani 2007a, b, c, d, e, f, g, h)

Name of the deposit	Proved reserve (tons)	Probable reserve (tons)
Chah Khou	254,000	N.A.
North Afzal Abad	170,000	252,000
Gounij	170,000	2,700,000
South Afzal Abad	99,000	N.A.
Hoz Sefid	212,000	N.A.
East Hoz Sefid	190,000	N.A.
Torshak Mohammadi	1,546,000	N.A.
Tabas Mesina	N.A.	N.A.
Sar Lord	162,000	N.A.
Shir Kuhak	1,450,000	N.A.
Kasr Ab	23,000	36,000
Tok Siah	72,000	105,000
Rezq Soleyman	6,400	8,000
Kalate Ali Mohammad	17,500	N.A.
Shour-o-Shirin	104,000	N.A.
Cheshme Lagouri	N.A.	131,000
Farmaj	3,600	N.A.
Espiki	142,000	238,000
Hasan Abad-e Kourin	220,000	547,000

7.5.26 Fossil Fuels (Petroleum, Natural Gas, and Coal)

Coal reserves of Iran constitute a nominal 1% of the world reserves, but having around 11% of the total petroleum resources and 16% of the gas reserves, and still being a potential target for further exploration activities with many reserves still to be found, Iran has a distinctive position from the point of view of fossil fuels. The statistics on the fossil fuels of the country are presented in Tables 7.47, 7.48, and 7.49.

Table 7.46 Reserves, production, and trade of magnesite in Iran as compared to the world[a]

World					
	Production				
Reserves	2010	2011	Resources	Export	Import
2.2 billion tons	5.76 million tons	5.9 million tons	12 billion tons	4.5 million tons	4.3 million tons
Iran					
Reserves	Production		Resources	Export	Import
2.45 million tons	0.1 million tons		10 million tons	49 tons	–

U.S. Geological Survey, Mineral Commodity Summaries, January 2012
Statistics from Iranian Ministry of Mines and Metals as of 2009 and Islamic Republic of Iran's Custom Administration as of 2010
[a]Mineral Commodity Summaries 2008

Table 7.47 Reserves, production, and trade of crude oil in Iran as compared to the world[a]

World				
Proved reserves 2008	Production 2011	Resources	Export 2008	Import 2008
948 billion tons	835.796 million barrels per day	Almost twice the amount of reserves	14.54 billion barrels	14.54 billion barrels
Iran				
Proved reserves	Production 2011	Resources	Export 2012	Import 2008
1.3 billion tons	1.61 billion barrels	250 billion barrels	2,200 billion barrels per day	50.49 million barrels per day

U.S. Energy Information Administration
[a]BP Statistical World Review of Energy 2008

Table 7.48 Reserves, production, and trade of natural gas in Iran as compared to the world[a]

World				
Proved reserves 2009	Production 2010	Resources	Export 2008	Import 2008
188 trillion m^3	112,090 billion cubic feet	–	885 billion m^3	903.1 billion m^3
Iran				
Proved reserves 2011	Production 2010	Resources	Export 2010	Import 2010
29.6 trillion m^3	5,161 billion cubic feet	50 trillion m^3	5.4 billion m^3	5.2 billion m^3

U.S. Energy Information Administration
[a]BP Statistical World Review of Energy 2008

Table 7.49 Reserves, production, and trade of coal in Iran as compared to the world[a]

World				
Proved reserves	Production	Resources	Export	Import
8,261 billion tons	7.985 billion ton	–	1.21 billion tons	1.178 billion tons
Iran 2010				
Reserves	Production	Resources	Export	Import
1.71 billion tons	1.29 million tons	–	34,000 tons	708,000 tons

U.S. Energy Information Administration
[a]BP Statistical World Review of Energy 2008

Table 7.50 Water budget of Iran[a]

Average total annual precipitation	412 billion m^3
Total surface water including springs	99.7 billion m^3
Total surface water excluding springs	81 billion m^3
Total groundwater recharge	50.72 billion m^3
Total ground and surface water resources	131.72 billion m^3
Total groundwater discharge	54.72 billion m^3
Total water regulated through dams	35 billion m^3
Total runoff to the outside of the country	16.52 billion m^3
Total runoff into the country	8.08 billion m^3
Total amount of water utilized in the country	75 billion m^3

[a]The balance report on water resources of Iran, Office of Water Resource Research Organization, 2005–2006

7.5.27 Water Resources (Budget)

The annual precipitation in Iran amounts to 412 billion m^3 of water, which is described in the form of the water budget of the country as under (Table 7.50).

Chapter 8
Energy Resources, Production, and Consumption in Iran Compared with the World

Abstract This chapter covers the energy resources in Iran. These resources are divided into two groups:

- Non renewable energy resources, which mainly include oil, gas, and coal
- Renewable resources whose significance has been noticed recently

The reserves of non renewable resources or fossil fuels are cited in this chapter and compared to those in the Middle East and the rest of the world. The exploration potentials of such energy resource are also discussed.

This chapter also discusses the significant non renewable energy resources and their status quo. Among this type of energy resources geothermal, solar and running water resources are included.

8.1 Types of Energy Resources

Energy is one the most important requirements of human life and has been utilized since the ancient times till the present industrial age. The amount of consumption has been proportional to the growth rate of the society.

Availability of accurate information on the various energy resources and their consumption forms the foundations of any energy-related study or survey regarding the production and supply of energy because they constitute the primary capital of any trade or demand of energy; recognition of the sources and patterns of consumption can determine the extent of possible economic activities.

Energy resources are divided into two types, namely, nonrenewable and renewable. The former are not replenished in a foreseeable time and their consumption leads to their depletion, while the latter can be continuously used because of their replenishment over a comparatively shorter time.

M. Ghorbani, *The Economic Geology of Iran: Mineral Deposits and Natural Resources*, Springer Geology, DOI 10.1007/978-94-007-5625-0_8,

8.2 Nonrenewable Energy Resources

These primarily include fossil fuels, which are described in the following.

8.2.1 Crude Oil

Due to its high thermal quality, low cost of production, and ease of transportation, crude oil is presently the best source of energy and is predicted to retain this position till the year 2050.

While the proved reserves of crude oil constitute less than 14% of the global reserves of energy, they provide over 40% of the world's energy supply. According to official statistics (OPEC website), the proved crude oil reserves are estimated at 1.5 trillion barrels, and the consumption of crude oil in the year 2010 stood at 87 million barrels per day (Index mundi 2012).

8.2.2 Gas

The term natural gas refers to a combination of light gaseous hydrocarbons sometimes with acidic or neutral impurities that are contained in natural reservoirs (Ghorbani 2007a, b, c, d, e, f, g, h). Refined natural gas is produced from raw natural gas and contains over 80% methane and no acidic gases. The proved reserve of the global gas resources stands at 185 trillion m^3 (bp Website 2012).

8.2.3 Gas Condensates

Gases produced from various reserves may contain a considerable amount of light hydrocarbons, which are heavier than methane and are separated as liquid phase in different stages of processing (Ghorbani 2007a, b, c, d, e, f, g, h). The amount of gas condensates depends on the nature of the gas reserve.

8.2.4 Tar Sands

Heavy crude and tar (weighing over 900 kg/m^3), which are deposited as sand beds, are called tar sands (Ghorbani 2007a, b, c, d, e, f, g, h). Due to their high viscosity, these energy resources, dominantly found at the surface of the earth (occasionally up to a depth of 750 m), are not readily extracted and are usually produced synthetically.

Substantial reserves of tar sands are found in Canada and the United States, while minor quantities are reported from Russia, China, Nigeria, and Romania. The largest reserve of tar sand is located in Canada whose total amount of extractable oil is estimated at 630 billion barrels (Ghorbani 2007a, b, c, d, e, f, g, h); this constitutes about a quarter of the oil and tar present in tar sands globally.

The tar sand resources are presently extracted in Canada and China, dominantly from surface reserves. However, the existing opencast technologies can be economically applied to a depth of 100 m, and for deeper reserves, underground techniques need to be employed.

Advances have been made in the extraction and processing technologies of heavy crude and its hydrocarbon derivatives. The modern method of steam-assisted gravity drainage (SAGD) utilizes two horizontal shafts one above the other, the upper shaft conducting steam which causes reduction of viscosity in heavy crude and tar thus fluidizing it toward the lower shaft, where it is pumped up to the surface. The cost of operations was about USD 7.5 per barrel of tar in the early 2000s (Ghorbani 2007a, b, c, d, e, f, g, h). This method is presently exploited at Athabaska, Canada, where about 20,000 barrels of heavy crude are extracted. The overall production of this type of resource does not seem to reach 500,000 barrels globally.

The cost of conversion of heavy crude to synthetic oil in the early 2000s was estimated at USD 15–19 per barrel, thus making the total cost about USD 25 (Ghorbani 2007a, b, c, d, e, f, g, h). Considering the present day crude prices, extraction of tar sands is economically viable and growing demands for crude in the future make tar sands a precious resource of energy in the years to come.

8.2.5 Kerogens

Argillaceous rocks containing comparatively heavy hydrocarbons called kerogens were known to North American Indians since ancient times, and scientific investigations on methods for their extraction and conversion to crude have been carried out over the past century. Many advances have been made in the production of (Iran and Germany joint seminar 1998) liquid and gaseous fuels from kerogens in European countries such as Sweden, France, Spain, and Estonia, as well as South Africa and China.

There are many types of kerogens, the premium type yielding about 130 l of crude per ton of rock (Ghorbani 2007a, b, c, d, e, f, g, h). The extractable reserves of kerogens are estimated at over 2,000 billion barrels (almost twice the proved global crude oil reserves), most of which are located in the United States (Ghorbani 2007a, b, c, d, e, f, g, h). However, since kerogens have not been explored due to their economic infeasibility, their probable reserves must be much higher than the said estimates.

The usual mode of extraction is removal of argillaceous rocks from the mine and separation of crude from them at a temperature of about 500°C. The extract contains some amounts of nitrogen, oxygen, and sulfur, thus increasing the cost of production of crude taking into account environmental requirements. Moreover, the water required for the process varies between 120 and 300 m^3 per 100,000 barrels whose disposal imposes extensive environmental problems (Ghorbani 2007a, b, c, d, e, f, g, h). Recently, methods of extraction similar to that for sand tars have been utilized which remove crude oil by underground heating of kerogens. However, various issues such as high costs of extraction of kerogens as well as separation of contaminants have posed a barrier to development of their usage.

8.2.6 *Other Nonrenewable Resources*

These include coal and uranium resources. The estimated global reserves of coal stand at 948,453 million tons, which are dominantly located in the United States (23%), Russia and CIS countries (23%), and China (11%).

The proved world reserves of uranium are estimated at 3.1 million tons, which is equivalent to 1,367 billion barrels of crude if used in thermal reactors and equivalent to 8,200 billion barrels of crude if used in breeder reactors.

Table 8.1 shows the reserves of various fossil fuels in different countries.

Oil and gas reserves within countries in the Middle East and North America have significantly changed during 2003–2008 in such manner that the total oil reserve in the Middle East reached up to 755 billion barrels and 70.94 billion barrels in North America in 2008.

8.3 Fossil Fuel Reserves of Iran

8.3.1 *Crude Oil*

The first oil well of Iran hit oil in the year 1908 in the Naftoon oil field near Masjid Soleiman. This well, which was dug using steam-driven percussion drilling machinery, reached a depth of 360 m.

By the end of the year 2003, the total proved crude oil reserves of Iran was estimated at 130.7 billion barrels constituting 11.4% of the total world reserves. Accordingly, Iran ranks fifth in terms of proved reserves after Saudi Arabia, Iraq, UAE, and Kuwait.

The total extractable crude oil and gas condensate reserves of Iran in the year 2004 stood at 132.74 billion barrels. Considering the same amount of extraction as that of the year 2003, the reserves will last till 2030. According to official statistics of the Energy Ministry, 82.2% of Iranian reserves lie onshore, while the rest 17.8% are located offshore (Table 8.2).

Primary reserves: The amount of crude oil that can be produced using the reservoir potentials without applying any augmentation methods is called primary reserve.

Table 8.1 Proved reserves of crude oil, gas, and coal in different countries by the end of the year 2003

	Petroleum		Gas		Coal	
Country	Billion barrel	Percent	Trillion m^3	Percent	Million ton	Percent
America						
USA	30.7	2.7	5.23	3	24,994	25.4
Canada	16.9	1.5	1.66	0.9	6,578	0.7
Mexico	16	1.4	0.42	0.2	1,211	0.1
Venezuela	78	6.8	4.15	2.4	479	[a]
Argentina	3.2	0.3	0.66	0.4	–	–
Ecuador	4.6	0.4	–	–	–	–
Brazil	10.6	0.9	0.25	0.1	11,929	1.2
Peru	1	0.1	0.25	0.1	–	–
Columbia	1.5	0.1	0.11	0.1	6,648	0.7
Other countries	3.3	0.3	1.77	1	2,696	0.3
Europe and Eurasia						
England	4.5	0.4	0.63	0.4	1,500	0.2
Italy	0.7	0.1	0.22	0.1	–	–
Denmark	1.3	0.1	0.09	0.1	–	–
Romania	0.9	0.1	0.31	0.2	1,457	0.1
Azerbaijan	7	0.6	1.37	0.8	–	–
Uzbekistan	0.6	0.1	1.85	1.1	–	–
Turkmenistan	0.5	[a]	2.9	1.6	–	–
Russia	69.1	6	47	26.7	157,010	15.9
Kazakhstan	9	0.8	1.9	1.1	34,000	3.5
Norway	10.1	0.9	2.46	1.4	–	–
Germany	–	–	0.21	0.1	66,000	6.7
Other countries	2.1	0.1	3.37	0.86	95,403	9.6
Middle East						
UAE	97.8	8.5	6.06	3.4	–	–
Iran	130.7	11.4	26.69	15.2	197.4	–
Syria	2.3	0.2	0.3	0.2	–	–
Iraq	115	10	3.11	1.8	–	–
Saudi Arabia	262.7	22.9	6.68	3.8	–	–
Oman	5.6	0.5	0.95	0.5	–	–
Qatar	15.2	1.3	25.77	14.7	–	–
Kuwait	96.5	8.4	1.56	0.9	–	–
Yemen	0.7	0.1	0.48	0.3	–	–
Other countries	0.1	[a]	0.14	[a]	–	–
Africa						
Angola	8.9	0.8	–	–	–	–
Algeria	11.3	1	4.52	2.6	–	–
Libya	36	3.1	1.31	0.7	–	–
Egypt	3.6	0.3	1.76	1	–	–
Nigeria	34.3	3	5	2.8	–	–
Other countries	7.6	0.7	1.19	0.7	–	–

(continued)

Table 8.1 (continued)

	Petroleum		Gas		Coal	
Country	Billion barrel	Percent	Trillion m^3	Percent	Million ton	Percent
Asia and Oceania						
Australia	4.4	0.4	2.55	1.4	82,090	8.3
Indonesia	4.4	0.4	2.56	1.5	5,370	0.5
Brunei	1.1	0.1	0.35	0.2	–	–
Thailand	0.7	0.1	0.44	1	1,268	0.1
China	23.7	2.1	1.82	1.4	114,500	11.6
Malaysia	4	0.3	2.41	–	–	–
Vietnam	2.5	0.2	–	0.5	150	–
India	5.6	0.5	0.85	1.4	84,396	8.6
Other countries	1.3	0.1	2	100	4,697	0.4
All of the world	1,147.7	100	175.78		984,453	100

Source: Energy Report (2003)
[a]Reflects numbers less than 0.5
– Reflects numbers not available

Table 8.2 Liquid hydrocarbon reserves of Iran at the beginning of the year 2004

			A total store		
Location	Mayadeen	Extractable reservoir	Preliminary	Secondary	Total
Aridity	Petroleum	90.39	115.52	26.72	142.24
	Gassy condenses	13.41	15.69	2.13	17.82
	Total	103.8	131.21	28.85	160.06
Marine	Petroleum	13.42	12.45	6.26	18.71
	Gassy condenses	15.52	15.74	0.22	15.96
	Total	28.94	28.19	6.48	34.67
Total	Petroleum	103.81	127.97	32.98	160.59
	Gassy condenses	28.93	31.43	2.35	33.78
	Total	134.74	159.4	35.33	194.73

Source: Energy Report (2003)

Secondary reserves: Continued production from oil fields gradually decreases the reservoir pressure and consequently reducing the efficiency of the wells culminating in drying up of the wells. To continue production, the reservoir pressure can be augmented by various ways and the resultant crude is termed secondary reserve.

Iran has six onshore oil fields, which are located in Khuzestan, Bushehr, Eilam, Fars, and Kohkiluye and Boirahmad. The last discovered oil field is Azadegan, which is located in the Azadegan Plain, with an estimated reserve of about five billion barrels (including gas condensate trapped in the caps). The Azadegan oil field was discovered in the year 1999.

The offshore reserves of Iran lie in Khark, Lavan, Bahregan, and Siri Regions. In addition to the above-mentioned independent oil fields, there are a number of shared fields with the neighboring countries which are mostly offshore (Table 8.3).

Table 8.3 Shared (onshore and offshore) oil fields of Iran in the year 2003

Field name	Neighboring country
On-shore	
Dehloran	Iraq
Paydar Ghard	Iraq
Naft-shahr	Iraq
Off-shore	
Foroozan	Saudi Arabia
Esfandyar	Saudi Arabia
Salman	Dhabi – United Arab Emirates
Mobarak	Sharjah – United Arab Emirates
Nosrat	Dhabi – United Arab Emirates
Farzam	Dhabi – United Arab Emirates
Arash	Kuwait
Southern Pars Oil-bearing Bed	Qatar
Alborz Oil-bearing Block	Republic of Azerbaijan

Source: Energy Report (2003)

The oil resources of the Caspian Sea have been in the center of attention in the recent years. The Caspian Sea Region is strategically important because it will become one of the major oil- and gas-exporting regions of the world in the near future. The probable in-place reserves of the southern Caspian are estimated at around 32.85 billion barrels.

Extraction of these reserves has been hampered by various technical and political issues such as disagreement among the bordering countries of the Caspian Sea about their share of different resources. Another difficulty is the depth of operation; all the recognized structures, except Rudsar, are situated at a depth of 400–800 m below the water level.

Khazar Oil and Gas Co. has carried out detailed exploration studies on the Iranian part of Caspian to recognize probable productive structures. Table 8.4 lists the important structures on the Iranian part of Caspian. Map 8.1 illustrates the geographic location of various Iranian oil fields in the western and southern regions of the country.

8.3.2 Gas Condensates

In the lower parts of some independent gas fields where temperatures are higher (around 150°C), heavier liquid hydrocarbons are vaporized to form gas. This gas is immediately liquefied once it is extracted and cooled and is called gas condensate. The condensate composition is near that of pure gasoline and is usually called natural gasoline. If the condensate is accompanied by butane, propane, and ethane, it is called natural gas liquid (NGL).

Generally speaking, the gas condensates are lightweight hydrocarbons that are present in the expelled gases from the oil fields or independent gas fields. Expulsion

Table 8.4 Important geological structure on the southern Caspian Region

Square	Depth (meter under the sea surface)	Petroleum in probable place	Confine situation
1 (Alborz)	500	20,000	Shared with Azerbaijan Republic
29 (Chaloos)	800	3,000	17–22% of Caspian sea waters
6	750	3,000	14%
24 (Noorvardian)	700	2,550	17%
23(Ramsar)	600	1,500	11%
8	550	1,400	Near to Iran beach
7	750	900	17%
18 (Roodsar)	80	500	Near to the beach
Total	–	32,850	–

Source: Energy Report (2003)

from the reservoir, which causes a reduction in pressure and temperature of the gas, results in the formation of gas condensates proportionate to the volume of the extruded gas. Specifications of the gas condensates are very similar to that of crude oil.

The gas condensates from the oil fields are usually injected into the crude oil before export. The daily gas condensate production of the country in the year 2003 was 155,300 barrels, 29% higher than the previous year.

8.3.3 Gas

Iran possesses the second largest gas reserves of the world after Russia. The amount of gas reserves of the country at the end of the year 2003 stood at 26.69 trillion m^3 as compared to the global reserves of 175.78 trillion, constituting 15.2% of the world total.

The onshore gas reserves of Iran are about 12.3 trillion m^3 (45.9%), while the offshore extractable reserves stand at 14.39 trillion m^3 (54.1%). Considering the geological setting of the country, the discovery of new large reserves of oil and gas cannot be ruled out. Table 8.5 demonstrates the amount of reserves in various natural gas fields of the country.

8.3.4 Coal

Systematic prospecting for coal in Iran began in the mid-1920s by German and Iranian experts. However, the discovery of extensive reserves of oil and gas in the country pushed coal production aside till the time of inception of steel industry and construction of blast furnace which consumes large quantities of coal.

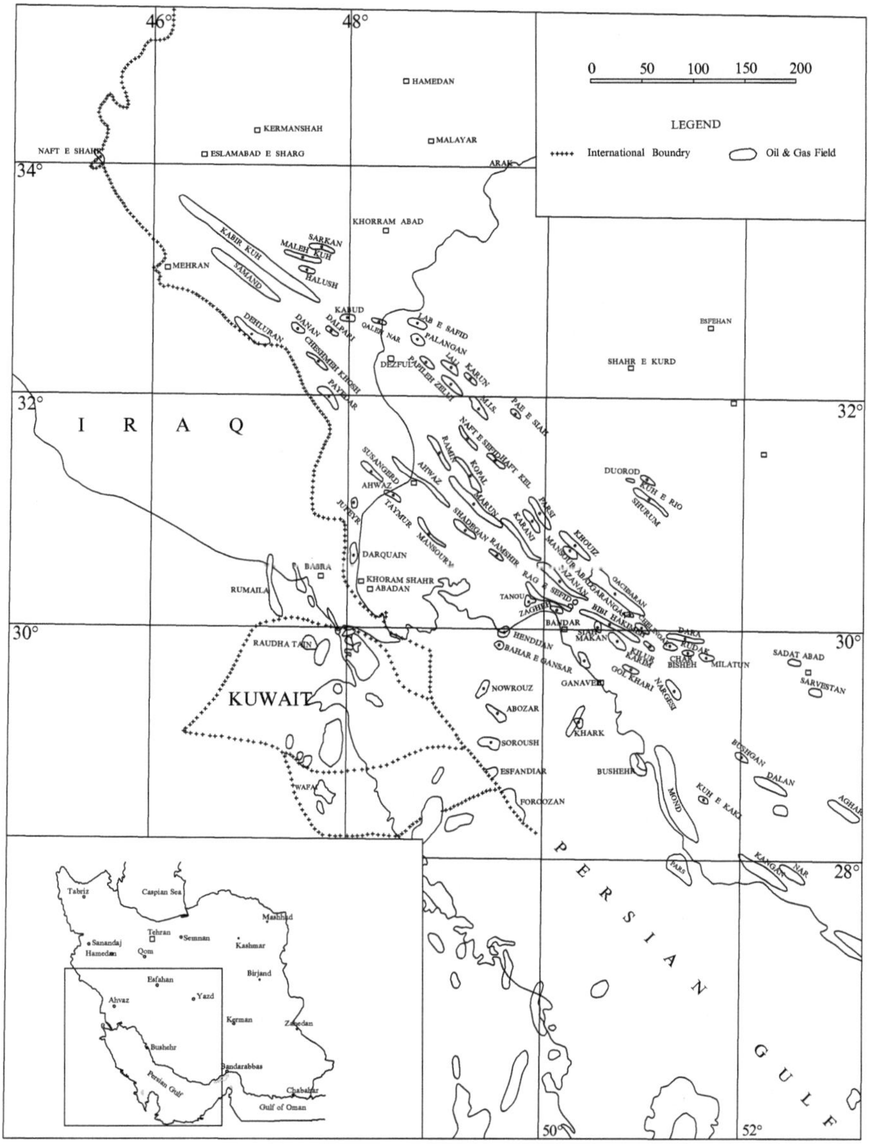

Map 8.1 Geographic location of Iranian oil fields on the south and the west of the country

Coal production in Iran began in the year 1969 by a number of government companies (subsidiaries of National Iranian Steel Co.). Considering the investigations carried out on coking coals and incomplete exploration of thermal coals of the country, the official statistics for the coal reserves of the country has always been subject to fluctuations. The latest figures (2003) put the overall reserves at 560 million tons 44.7% of which are coking and the remaining 55.3% thermal variety.

Table 8.5 Natural gas fields of Iran

Square	Location	Extractable reservoir (billion m^3)
Porch and inception	Fars Province	563.1
Nar	Bushehr Province	374.7
Kangan	Bushehr Province	670.5
Sarajeh	Ghom Province	6.96
Khangiran	Khorasan Province	679.2
Gonbadeli	Khorasan Province	1.8
Goorzin	Hormozgan Province	26.2
Sarkhoon	Hormozgan Province	242.2
Northern Pars	Hormozgan Province	1,505
Southern Pars	Hormozgan Province	7,929 (shared with Qatar)
Tabnak	Fars-Bushehr and Hormozgan border	545 (million barrels of gas)
Lavan	Hormozgan (marine)	176
Hengam	Hormozgan (marine)	Shared reservoir with Oman – doing descriptive researches
Salman	Hormozgan (marine)	Shared reservoir with Abu Dhabi
Mobarak	Hormozgan (marine)	Shared reservoir with Sharja – lack of capability and development possibility
Persian Gulf	Marine	Shared reservoir with Arabia, contract square exploration

Source: Energy Report (2006)

The overall coal reserves of Iran (including proved, probable, and prospective) amount to about 10,938.4 million tons, which is approximately 1.1% of the world total. Major Iranian coalfields are located in Kerman, Naiband, Alborz, northeastern Khorasan, Kashan, Isfahan, and Azerbaijan (Maragheh) (Table 8.6).

Table 8.7 shows the status of Iran from the point of view of fossilized fuel reserves in the world.

8.4 Fossilized Fuel Production of Iran and the World

8.4.1 Crude Oil

The crude oil production of the Middle East and North Africa (MENA) Region and of the world is presented in Table 8.8. The total crude oil production of the world in the year 2003 stood at about 3,697 million tons, while the OPEC countries supplied 1,466.9 million tons (39.7%). Saudi Arabia constituted the highest share of the global production (12.8%) followed by the United States (9.2%). Iran, UAE, and Kuwait ranked next to Saudi Arabia in the Middle East with 5.1, 3.2, and 3% shares of the global production, respectively.

Iran produced 190.1 million tons (5.1%) of crude oil in the year 2003, placing fourth in the world ranking of oil producers.

Table 8.6 Coal resources and reserves of Iran (million tons)

Title	Definite reservoirs	Probable reservoirs	Possible reservoirs	Remain reservoirs
Kerman	206.1	106	1,125	1,437.1
Alborz Central	138.5	497	1,947	2,582.5
Alborz Eastern	17.8	101	145	263.8
Alborz Western	2	3	_	5
Tabas	191	459	6,000	6,650
Total	555.4	1,166	9,217	10,938.4

Source: Energy Report (2003, 2006)

Table 8.7 Extractable fossilized fuel reserves of Iran as compared to the world (2006)

Type of fossil fuel	Unit	Reservoir		Iran scantling through the world
		Iran	The world	
Extractable petroleum	Billion barrels	137.6	1,261	12.7
Extractable natural gas	Trillion m^3	29	185.02	15.1
Extractable coal	Billion tons	10.9	826	1.1

Source: Energy Report (2008)

Table 8.8 Crude oil production of MENA region and the world

Country \ Year	2000	2001	2002	2003
Algeria	66.8	65.8	70.9	79
Egypt	38.8	37.7	37	36.8
Iran	189.4	184.6	168.8	190.1
Iraq	127.3	116.5	99.7	65.9
Kuwait	104	101.9	91.8	110.2
Libya	69.5	67	64.7	70
Oman	47.6	47.5	44.5	40.7
Qatar	38.7	38.4	35.1	41.2
Saudi Arabia	450.6	434.1	417.3	474.8
Syria	27.4	29	28.4	29.5
United Emirates	117.3	115.5	100.4	117.8
Yemen	21.3	22.3	21.8	21.4
The World	3,604.4	3,585.7	3,561.7	3,697

Source: Energy Report (2006)
Among the above-mentioned countries, Iraq has shown significant progress

About 58% of the growing demand of the world for crude oil over the next 20 years will be provided from the Persian Gulf Region. This is in spite of the fact that the share of this region in the past 20 years has been only 17%. To maintain its position as the second largest producer of crude among the OPEC countries, the Islamic Republic of Iran must be able to cover about 12% of this growing demand, and this requires special development strategies and investment for the oil sector of the country.

Table 8.9 Gas production of MENA region and the world in the years 2003 and 2008

	The rate of production (billion m^3)	
Country	2003	2008
Russia	578.6	–
United States of America	549.5	–
Canada	180.5	–
England	102.7	–
Algeria	82.8	–
Iran	79	116.3
United Arabia Emirates	44.4	55.24
Bahrain	9.6	13.45
Syria	6.3	–
Saudi Arabia	61	78.1
Oman	16.5	24.05
Qatar	30.8	76.62
Kuwait	8.3	12.8
All of the world	2,618.5	3,065.58

Source: Energy Report (2006)

8.4.2 *Gas*

In the year 2003, the total global production of natural gas stood at 2,618.5 billion m^3, the highest share of which was supplied by Russia and the United States with 22.1 and 21%, respectively. In Table 8.9, the major gas producers of the Middle East and some of the world's largest producers are listed.

With a production of 79 billion m^3 (3% of the world total), Iran ranks first among the Middle Eastern countries in terms of natural gas production. Globally, Iran ranks sixth following Russia, the United States, Canada, Britain, and Algeria. This is, however, not an optimum position taking into account the second rank of the country in terms of reserves.

8.4.3 *Coal*

The total global production of coal in the year 2003 amounted to 3,525.28 million tons with China (1,183.3 million tons) and the United States (767.9 million tons) occupying the first and second positions, respectively. According to official statistics, Iran produced about 0.84 million tons in the said year contributing 0.02% to the world production.

Table 8.10 demonstrates the status of Iran from the point of view of fossilized fuel production in the world in the year 2003.

Table 8.10 Fossilized fuel production in Iran as compared to the world in the year 2003

		Reservoir		
Type of fossil fuel	Unit	Iran	The world	Iran scantling in global production(%)
Simple oil	Million tons	190.1	3,697	5.1
Natural gas	Trillion m^3	79	2,618.5	3
Coal	Million tons	0.8	3,525.3	0.02

Source: Energy Report (2006)

Table 8.11 Share of natural gas in providing the energy for various sectors of Iran during the period 1996–2003

Illustration	1996	1997	1998	1999	2000	2001	2002	2003
Indoor, commercial, and public	39.7	42.7	43.5	48.3	51	52.3	55.3	58.5
Industry	45.4	42.8	39.4	48	47.1	44.6	45	47.4
Transportation	–	–	–	–	0.01	0.01	0.02	0.02
Powerhouses	55.1	60	70.7	70.3	70.8	69.8	71.5	74.9
Refineries	48.1	49.1	55	61.2	66.5	74.6	81.5	86.1

Source: Energy Report (2006)

8.5 Fossilized Fuel Consumption in Iran and the World

The published figures indicate that the total world consumption of crude oil in the year 2003 has been 3,636.6 million tons in which Iran had a share of 1.5% with 54 million tons. The annual consumption of petroleum products (liquefied gas, gasoline, kerosene, diesel fuel, and heating oil) in the country grew by 1.85% over the period 1996–2003. Gasoline with 8.81% and kerosene with –4.16% had the highest and lowest growth rates over the said period, respectively. The consumption of major petroleum products amounted to 72,561 million liters in the year 2003, showing a 0.6% growth over the previous year. The highest share of consumption belonged to gasoline and diesel fuel, while the lowest was that of liquefied gas.

Iran consumed 80.4 billion m^3 of natural gas during the year 2003, which is 3.1% of the world's consumption. Table 8.11 illustrates the share of natural gas in various sectors over the 1996–2003 period.

Global coal consumption in the year 2003 touched 3,560.16 million tons, 1.4 million tons (0.04%) of which was utilized in Iran. The largest consumers of coal in the world are steel industry and thermal power station. Similarly, the Isfahan steel plant was the largest consumer of coal in Iran in the year 2003 with 1,502,000 tons, most of which was used for coke production. Imported coal constituted 59% of the supply, while the remaining 41% was procured domestically.

Among other consumers of coking coal are ferroalloy, steel, forging, refractory, sugar, battery, automobile, and railway industries.

8.5.1 Consumption Pattern

With growing consumption of oil and gas, crude export of the country has decreased. While in the year 1918 over 69% of Iran's crude production was exported, this figure reduced to 57% in the year 1996, in spite of a 30-fold increase of production. In other words, population growth has caused the domestic consumption of various petroleum products to rise. Continuation of this trend will result in insufficiency of the production to supply the domestic demands. It must be noted that a great part of this demand is due to the per capita growth of consumption. The per capita consumption of energy grew twofold over the period 1976–1996 to 10.3 barrels of crude per year. The important point is that the growth has taken place mostly in the nonproductive sectors, causing a drop in the GDP per capita from Rls. 333,870 in 1976 to Rls. 247,320 in 1996. Household and business sectors constituted for 37.9% of the energy consumed.

Iran had a capacity to produce about 4 million barrels of crude, refine 1,347,000 barrels, and produce 311.78 million m^3 of gas per day. To achieve sustained growth, the said capacities must increase, and the present reserves must be developed.

Moreover, investment in the developing sectors such as oil- and gas-processing industry and a change in the present consumption pattern are a necessity, which can be achieved by strategic planning to divert energy consumption from nonproductive sectors toward productive sectors and optimal utilization of energy to promote industry.

8.6 Global Status of Iran in Terms of Energy Resources

Fortunately, Iran is among the richest countries of the world in terms of oil and gas reserves. Five Middle Eastern countries, namely, Saudi Arabia, Iran, Iraq, Kuwait, and UAE, together boast of a crude reserve of 702.7 billion barrels, constituting 61.2% of the world total.

The important point is that despite the fact that the crude reserves of Iran at the same level as Iraq, Kuwait, and UAE, the population of the country is many times more than their total populations and as a result the per capita reserve of the country is much less comparatively. Table 8.12 compares the status of Iran in terms of the energy resources with the world.

8.7 Nuclear Energy in Iran and the World

In today's industrial world where the scientific and technical advancement of every country is considered as an index of development and social welfare, nuclear technology has a significant importance. Utilization of nuclear science in industry, medicine, agriculture, and other fields indicates how it has affected the human society.

Table 8.12 Fossilized fuel energy resources of Iran as compared to the world in the year 2003

Simple oil							
Iran				World			
Reservoirs (billion barrel)	Production (billion ton)	Exports (billion ton)	Imports (million ton)	Reservoirs (billion barrel)	Production (million ton)	Exports (million ton)	Imports (million ton)
130.7	190.1	136.1	–	1,147.7	3,697	1,770	1,770
Natural gas							
Iran				World			
Reservoirs (trillion m^3)	Production (billion m^3)	Exports (billion m^3)	Imports (billion m^3)	Reservoirs (trillion m^3)	Production (billion m^3)	Exports (billion m^3)	Imports (billion m^3)
26.69	79	3.52	4.92	175.78	2,618.5	454.87	454.87
Coal							
Iran				World			
Reservoirs (million ton)	Production (million ton)	Exports (million ton)	Imports (million ton)	Reservoirs (million ton)	Production (million ton)	Exports (million ton)	Imports (million ton)
10,938	0.8	0.03	0.59	984,453	3,525.3	661.26	666.46

Source: Energy Report (2006)

Table 8.13 Uranium ores that can be produced at less than USD 130 per kilogram

The name of the country	Reservoirs
Canada	333,834
United States of America	345,000
Nigeria	102,227
Namibia	170,532
Australia	73,500
Iran	370?
South Africa	315,330
Russia	143,020
Kazakhstan	530,460
All of the world	3,169,238

Source: Energy Report (2006)

Nuclear energy is the most important source of energy after oil and gas and is utilized more in advanced countries to produce electricity.

The proved global reserves of uranium are estimated at 3.1 million tons, which equals 8,200 billion barrels of crude oil is used in breeder reactors.

Table 8.13 depicts the uranium reserves of some countries and the world that can be produced at less than USD 130 kg^{-1}.

The global share of nuclear power stations in producing electricity is about 17% (2,535 billion kWh), which is supplied by over 438 plants.

Since the inception of the Iran Atomic Energy Organization in the year 1974, all the activities have concentrated on the establishment of nuclear reactors. A number of contracts were signed before the victory of the Islamic revolution in the year 1979, which were subsequently canceled.

In order to save the nonrenewable energy resources of the country and diversify the power generation trend, the government approved plans to produce 10–20% of the energy by nonconventional (mainly nuclear) methods in the year 1982. Presently, phase I of Bushehr nuclear reactor with a capacity of 1,000 MW is being constructed.

The proved reserves of uranium in Iran are about 3,600 tons, while the probable reserves are estimated at about 20,000 tons. About 400 tons of enriched uranium can be produced from 2,500 tons of uranium ore, and this can supply Bushehr nuclear reactor with the capacity of 1,000 MW for 17 years.

8.8 Renewable Energy Resources of Iran

All the renewable energy resources of Iran are under the supervision of Renewable Energy Organization of Iran (REOI), a subsidiary of Energy Ministry. REOI closely works with other institutions like Iran Atomic Energy Organization and International Energy Institute. The renewable energy resources of the country are briefly described in the following paragraphs.

The Presence of the Caspian Sea to the north whose level is about 28 m below the global mean sea level, Alborz Mountains on its southern shore with heights of

Table 8.14 Nominal capacity and actual production of operational hydroelectric plant of Iran

Powerhouse name	Province	Year of inception	Capacity (MW)
Big powerhouses			
Amirkabir	Tehran	1961	91
Doz	Khoozestan	1962–1971	520
Sefidrood	Gilan	1964	87.5
Latian	Tehran	1969–1987	45
Zayanderood	Esfahan	1970	55.5
Aras	Eastern Azerbaijan	1971	22
Shahid Abbaspoor	Khoozestan	1977–2003	2,000
AbiKalan	Tehran	1988	115.5
Giroft	Kerman	1988	30
Soleiman Mosque	Khoozestan	2002–2003	1,000
Karkheh	Khoozestan	2002–2003	399
Moghan	Western Azerbaijan	2002	13
Subtotal			4,378.5
Small powerhouses			
Mahabad	Western Azerbaijan	1972	6
Doroodzan	Fars	1989	10
Forghan	Central	1996	10.4
Asiabak	Central	1997	5.2
Jannat Roodbar	Mazandaran	1996	1
Sar Rood	Khorasan	1987	0.065
Ardeh	Gilan	1991	0.125
Shahid Talebi	Fars	1994	2.25
Gamasyab	Hamedan	1999	2.8
Dare Takhte 2	Lorestan	2001	0.9
Karnogh	Ardebil	2002	0.05
Yasooj 7 (Karik1)	Kohkilooye va boyr Ahmad	1994	2.5
Pol Kaloo 1	Kohkilooye va boyr Ahmad	2004	4
Subtotal			45.29
Total			4,423.79

Source: Energy Report (2003)

over 5,000 m, and open sea on the southern border, there is considerable elevation difference within the country. Considering the average annual rainfall of the country, which is around 250 mm (1,500 mm at Bandar Anzali and 50 mm at Yazd), the potentials for utilization of hydroelectric power are substantial.

The first hydroelectric power plant of the country was established in Hamedan in the year 1929 with two power generators. Later on, the Shushtar hydel plant with a capacity of 1,000 kW became operational. The first major hydel power plant of the country was constructed at Karaj near Tehran in the year 1961, producing 150 MW. The Shahid Abbaspour plant with a capacity of 2,000 MW is the largest hydroelectric project of the country at present, but the Karoon III plant (3,000 MW) will come into production in the near future. Table 8.14 lists the various operational hydroelectric power plants of the country.

8.8.1 Hydroelectric Energy

The overall hydroelectric potential of the country is estimated at 50 billion kWh per annum. Karoon River alone can produce 30 billion kWh of electricity annually, while Dez River, Karkheh River, and others can produce 9, 6, and 5 billion kWh per annum, respectively. The hydroelectric projects of the country can be categorized into three groups:

1. *Under-Construction Projects*: Masjid Soleiman, Karkheh, Karoon III, Karoon IV, Seymareh, Gatvand, Roodbar (Lorestan), and Siyah Bisheh
2. *Investigation Projects*: Twenty-two projects of over 100 MW capacity were under investigation in the year 2003 and were to start in three phases.
3. *Small- and Medium-Sized Projects*: There are a number of projects which will come into operations in the next 10 years mostly situated in the southwest of the country.

8.8.2 Solar Energy

No doubt, sun is the largest energy resource for the earth. Solar energy in the equatorial region up to the 40° north and south latitudes is suitable for producing electricity. Considering the large areal extent of Iran and its geographic position (between 25° and 40° N), setting up of solar power plants in many regions is feasible.

The average solar energy reception of the country is 5 $kWh/m^2/day$ or 1,825 $kWh/m^2/year$. Thus, the total annual solar energy reception of the country amounts to 3×10^{15} kWh/year, which is about 100 times the total oil and gas reserves of the country. Results of investigations indicate that suitable provinces for solar energy installations are Kerman, Sistan, and Yazd.

The first solar power plant of the country came into operation at Shiraz in the year 1996. The plant, which utilizes 48 solar collectors, produces 250 kW of electricity. In addition, a number of experimental small household solar power projects have been incepted in Yazd Province.

8.8.3 Wind Energy

The geographic position of Iran, which is a low-pressure region bordering the high-pressure areas to the north and northwest, causes winds to blow during summer and winter. During summer, the moist winds of Atlantic that blow from the northwest and the monsoonal winds of the Indian Ocean from the southeast affect Iran. In winter, the lowering of the pressure over Iran causes the cold winds of Central Asia and the moist winds of Atlantic and Mediterranean pass over Iran. The well-known winds of Iran are the Siyah wind of Eastern Iran, northern wind on the shores of the Persian Gulf,

Table 8.15 Investigative, planned, and operational projects to utilize wind energy in Iran

Project name	Geographical location	Inception year	Capability of annual energy production (million kWh)
Ministry of Energy			
600 kW turbine	Gilan-Manjil	1376	2
10 kW turbine	Eastern Azerbaijan- Esko	1376	0.025
Taking information from wind statistics record stations	Gilan, Eastern, Western Azerbaijan Ardebil	1379	–
600 kW turbine optimization	Gilan and Eastern Azerbaijan	1382	–
Supply the Atlantic country wind	All over the country	1382	–
Binalood windy powerhouse	Khorasan	1380	102
Atomic Energy Organization			
Plan establishment of 1 MW windy electricity turbine	Gilan (Manjil and Roodbar)	1373	4
Plan establishment of 10 MW windy electricity turbine	Gilan (Manjil, Roodbar, and Harzvil)	1375	04-Mar
Plan establishment of 90 MW windy electricity turbine	Gilan	1378–1379	90–240
Calculating of country wind potential	–	1368	–

Source: Energy Report (2003)

the Khoshabad wind of Gogan Plain, the Diz wind between Mashhad and Neishabour, and the Sam wind of Khuzestan.

In order to develop, promote, plan, control, and manage various projects to utilize the potentials of wind energy in the country, the Energy Ministry has set up a number of recording sites to register the wind date and investigate the feasibility of wind energy farms. Accordingly, a wind farm was set up by the ministry in collaboration with the Iran Atomic Energy Organization at Roodbar and Manjil in the year 2001 with a capacity of 1 MW. There have been proposals to establish a number of wind farms totaling 100 MW capacity within the framework of the Second Five-Year Development Program (Table 8.15).

8.8.4 Geothermal Energy

Decay of radioactive material, molten core of the earth, orogeny, and overburden pressure of sedimentary basins are the sources of geothermal energy within the earth's crust. The resulting heat is transferred through convective currents to the surface where it dissipates through surface waters and atmosphere.

The location of geothermal reservoirs usually coincides with that of young active volcanoes. Taking into account the passage of parts of global volcanic belt through

Iran, there is a vast potential of utilization of this type of energy in the country. Quaternary volcanoes are present in the north, northwest, southeast, and central Iran, while active faults as well as hot springs are seen throughout the country. According to studies carried out by the Energy Ministry, Iran has a potential of 600 GJ of directly extractable energy.

The first geothermal study was carried out by an Italian company under the supervision of the Energy Ministry during the 1975–1979 period. After a period of inactivity, the work restarted in the year 1995 when a number of wells were drilled around Meshkin Shahr in Azerbaijan. According to a priority plan set out by the Energy Ministry, areas around Mount Sabalan, Mount Damavand, Makoo-Khoi, and Mount Sahand have been given 1–4 priorities for production of geothermal energy.

Mount Sabalan: Meshkin Shahr, Borjloo, and Sarein are located in this region with a thermal reserve of 19.2×10^{18} J. The first two have been chosen for producing electricity, while the third is proposed for direct use of geothermal energy. The progress made till the end of the year 2000 includes prospecting, topographic mapping, geothermal anomaly mapping, geophysical profiling, detailed geological survey and 3-D modeling of geothermal reservoir at Meshkin Shahr, drilling site planning, and international tender to choose the contractors.

Mount Damavand: Situated near Tehran, the thermal energy potential of this area is estimated at 5.11×10^{18} J. The area is proposed for electricity production as well as direct use of the geothermal energy.

Makoo-Khoi Region: Situated on the northwestern borders of Iran with Turkey, the area has a thermal potential of 40×10^{18} J. Since Makoo-Khoi is one of the cold agricultural areas of the country, direct utilization of the geothermal energy available can help heating up of the farm houses and agro-industries.

Mount Sahand: Covering lands of about 11,000 km^2 around the extinct quaternary volcano between Tabriz and Maragheh cities, the area has 23×10^{18} J geothermal energy potential that can be extracted both directly and indirectly.

Meshkin Shahr Geothermal Power Plant: The most important geothermal power project of the country is being worked out at Meshkin Shahr, where over 10,000 m of explorative drilling has been carried out to test the feasibility of the geothermal potential of the area.

The results obtained indicate the presence of a geothermal reservoir measuring 5 km^2 and having a temperature of 245°C. This reservoir is capable of supplying a 60 MW power plant at Meshkin Shahr. Surface surveys further prove the presence of additional reservoirs outside the drilled region that can further increase the power potentials by another 200 MW.

The exploration drilling at Sabalan is the first such experiment in the Middle East, putting Iran among the 22 countries that utilize geothermal power for the production of electricity.

8.8.5 *Tidal Energy*

The Renewable Energy Organization of Iran currently has two experimental projects to produce electricity from tidal waves at Chabahar (Oman Sea) and Kish Island (Persian Gulf).

The most suitable place for installing tidal energy plants is the Strait of Hormuz (the Persian Gulf outlet), where about 500 km^3 of water passes through the 50-km-wide strait at the time of tides. However, considering the volume of traffic and the high cost of construction, there is no scope of utilization of this energy in the foreseeable future.

Chapter 9
List of Mineral Deposits and Indications of Iran

Abstract This chapter provides a long list of all known mineral deposits and indications in Iran with geographic coordination, size, host rock, age, genetic and mineralization, associated orogenic phase, deposit shape or morphology and paragenesis.

9.1 Introduction

It is attempted herein to present a complete list of mineral deposits and indications identified in Iran till date along with their geologic and metallogenic specifications such as size, host rock, surrounding rock, shape, and orogenic phase in which mineralization occurred.

When reviewing information within these tables, two points need to be taken into consideration:

1. Geographic coordinates of some of the deposits and indications cited in these tables have been derived from geological maps with scales of 1:100, 000 and 1:250, 000, and therefore they might not exactly match the present day GPS readings of the very same mineral deposit or indication. The reason is very clear; back in the days when such information were gathered, use of GPS was not common.
2. The findings and facts on economic geology and metallogeny of these mineral deposits and indications (on the genesis of deposits in particular) presented in these tables are the results of studies and field visits by the author and therefore could not be specifically referenced to in these tables. The documentation of the literature, if any, is listed among the references at the end of the book.

M. Ghorbani, *The Economic Geology of Iran: Mineral Deposits and Natural Resources*, Springer Geology, DOI 10.1007/978-94-007-5625-0_9,

Complete list of iron mineral deposits and indications of Iran along with their detailed specifications

		Lat			Long								
No.	Name	D°	M	S″	D°	M	S″	Size	Host rock/age	Genetic/ mineralization	Orogenic phases	Morphology	Elem para
1	Farur	26	17	27	54	31	12	S[a]	Volcanic rocks and limestone/Upper Precambrian–Lower Cambrian	Volcanosedimentary	Pan-African	Stratiformbed	
2	Gheshm	26	56	44	56	3	9	S	Volcanic rocks and limestone/Upper Precambrian–Lower Cambrian	Volcanosedimentary	Pan-African	Stratiformbed	
3	Hengam	26	39	49	55	53	7	S	Volcanic rocks and limestone/Upper Precambrian–Lower Cambrian	Volcanosedimentary	Pan-African	Stratiformbed	
4	Lark	26	52	22	56	22	18	S	Volcanic rocks and limestone/Upper Precambrian–Lower Cambrian	Volcanosedimentary	Pan-African	Stratiformbed	
5	Ak Kahour	27	40	55	55	38	14	S	Marl, dolomite, gypsum, sandstone, conglomerate, shale, and carbonate and silicate layers bearing iron/ Cretaceous–Paleocene	Sedimentary	Laramide–Pyrenean	Stratiformbed to lens	
6	Ardan	27	40	0	55	26	0	S	Rhyolite, tuff, and dolomite/Upper Precambrian–Lower Cambrian	Volcanosedimentary	Pan-African	Stratiformbed	
7	East of Jiroft fault	27	15	32	57	59	46	I[a]	Ophiolite complex/upper Cretaceous	Unknown	Laramide	Unknown	
8	Hormoz	27	4	22	56	28	23	S	Volcanic rocks and limestone/Upper Precambrian–Lower Cambrian	Volcanosedimentary	Pan-African	Stratiformbed	
9	Soleymani	27	49	54	57	44	4	I	Diorite, monzonite, and syenite/ monzonite	Unknown	Unknown	Unknown	
10	Tange Zagh	28	3	52	55	58	37	S	Volcanic rocks, dolomite, and limestone/Upper Precambrian–Lower Cambrian	Volcanosedimentary	Pan-African	Stratiformbed	
11	Gol gohar	29	5	0	55	20	0	L[a]	Schist and gneiss/Upper Precambrian–Lower Cambrian	Metasomatic	Pan-African	Massive, lens	
12	Shateri	29	44	17	60	7	7	I	Diorite, monzonite, and syenite/ unknown	Unknown	Unknown	Unknown	
13	Charvak	30	24	0	56	32	0	I	Limestone and dolomite/Lower Cretaceous	Unknown	Unknown	Unknown	
14	Mohammad Abad	30	22	4	60	6	1	I	Unknown	Unknown	Unknown	Unknown	

15	Anomaly No.X	31	52	46	55	41	41	I	Sandstone and shale/Paleozoic	Unknown	Unknown	Unknown
16	Anomaly No.I A	31	46	6	55	26	12	I	Volcanosedimentary rocks/Upper Precambrian	Volcanic	Pan-African	Stratiformbed
17	Anomaly No.I B	31	43	36	55	27	38	S	Volcanics/Precambrian	Volcanic	Pan-African	Unknown
18	Anomaly No.I D	31	41	38	55	29	12	I	Volcanics/Precambrian	Volcanic	Pan-African	Lens
19	Anomaly No.II A	31	38	16	55	28	55	I	Volcanics/Precambrian	Volcanic	Pan-African	Unknown
20	Anomaly No.II B	31	38	0	55	30	41	I	Metamorphic and metasomatic rocks/ Precambrian	Magmatic–metamorphic	Pan-African	Unknown
21	Anomaly No.II C	31	36	43	55	32	40	I	Metamorphic and metasomatic rocks/ Precambrian	Magmatic–metamorphic	Pan-African	Unknown
22	Anomaly No.III	31	32	20	55	27	53	I	Metamorphic and metasomatic rocks/ Precambrian	Magmatic–metamorphic	Unknown	Unknown
23	Anomaly No.IV	31	55	58	55	36	40	I	Metamorphic and metasomatic rocks/ Precambrian	Magmatic–metamorphic	Pan-African	Unknown
24	Anomaly No.V B	31	45	21	55	55	44	I	Sandstone and shale/Paleozoic	Unknown	Unknown	Unknown
25	Anomaly No.V C	31	42	53	55	59	17	I	Sandstone and shale/Paleozoic	Unknown	Unknown	Unknown
26	Anomaly No.VIII	31	45	5	55	34	5	M	Granite, volcanics, and sandstone/ Upper Precambrian–Lower Cambrian	Magmatic	Pan-African	Unknown
27	Anomaly No.XI	31	54	19	55	43	12	I	Sandstone and shale/Paleozoic	Unknown	Unknown	Unknown
28	Anomaly XIIA	31	52	30	55	36	36	M	Volcano–sedimentary rocks/ Precambrian–Cambrian	Volcanosedimentary and metasomatic	Pan-African	Stratiformbed to lens
29	Anomaly No.XII B	31	53	1	55	29	5	I	Diorite and gabbro diorite/ Precambrian–Cambrian	Magmatic	Pan-African	Unknown
30	Anomaly No.XIII A	31	59	27	55	53	28	I	Porphyry granite/Upper Precambrian	Magmatic	Pan-African	Unknown
31	Anomaly No.XV	31	57	2	55	19	1	I	Gabbro, pyroxenite, marble, and diorite/Precambrian	Magmatic	Pan-African	Disseminated
32	Cheshmeh firuz	31	52	30	55	36	36	I	Acidic to intermediate volcanics and dolomite/Upper Precambrian–Lower Cambrian	Volcanosedimentary	Pan-African	Lens, vein

(continued)

(continued)

No.	Name	Lat D°	Lat M	Lat S″	Long D°	Long M	Long S″	Size	Host rock/age	Genetic/ mineralization	Orogenic phases	Morphology	Elem para
33	Choghart	31	45	0	55	30	0	L	Alkali granite, volcanics, sandstone, and schist/Upper Precambrian–Lower Cambrian	Magmatic	Pan-African	Massive	
34	East of Bafgh	31	51	9	55	54	47	I	Alkali granite, acidic volcanics, dolomite, and limestone/Upper Precambrian–Lower Cambrian	Magmatic	Pan-African	Massive	
35	Esfordi	31	47	19	55	38	19	M	Alkali granite, acidic volcanics, dolomite, and limestone/Upper Precambrian–Lower Cambrian	Magmatic	Pan-African	Massive	
36	Heneshg	31	51	0	53	17	0	S	Sandstone and shale/upper Paleozoic–Lower Triassic	Volcanosedimentary	Hercynian	Stratiformbed	Mn
37	Lakkeh siah	31	47	3	55	42	7	S	Alkali granite, acidic volcanics, dolomite, and limestone/Upper Precambrian–Lower Cambrian	Unknown	Unknown	Unknown	
38	Mishdovan	31	50	48	55	31	12	M	Alkali granite, volcanics, and sandstone/Upper Precambrian–Lower Cambrian	Magmatic	Pan-African	Massive	
39	Mobarakeh	31	37	13	55	32	9	I	Metamorphic rocks/Precambrian	Unknown	Pan-African	Unknown	
40	Nargun	31	43	18	55	43	34	I	Alkali granite, acidic volcanics, dolomite, and limestone/Upper Precambrian–Lower Cambrian	Magmatic	Pan-African	Massive	
41	Narigan	31	43	4	55	41	2	S	Volcanics and sandstone/Upper Precambrian–Lower Cambrian	Volcanosedimentary	Pan-African	Massive	Mn
42	NO.XL	31	35	33	56	25	42	I	Unknown	Unknown	Unknown	Unknown	
43	No.XXXV	31	31	23	56	24	36	I	Unknown	Unknown	Unknown	Unknown	
44	No.XXXVI	31	32	42	56	22	29	I	Unknown	Unknown	Unknown	Unknown	
45	No.XXXVII	31	27	8	56	22	53	I	Sandstone and shale/Jurassic	Unknown	Unknown	Unknown	
46	No.XXXVII I	31	28	36	56	16	34	I	Sandstone and shale/Jurassic	Unknown	Unknown	Unknown	
47	No.XXXVIX	31	34	43	56	19	36	I	Unknown	Unknown	Unknown	Unknown	

48	North of Sechangi 1	31	52	36	55	40	45	I	Sandstone and shale/Paleozoic	Unknown	Unknown	Unknown
49	North of Sechangi 2	31	52	26	55	43	4	I	Sandstone and shale/Paleozoic	Unknown	Unknown	Unknown
50	Sechahun	31	53	0	55	42	0	M	Diorite, volcanosedimentary/Upper Precambrian–Lower Cambrian	Volcanosedimentary	Pan-African	Massive
51	Sheytoor	31	39	6	57	57	5	I	Sandstone and shale/Upper Paleozoic	Unknown	Unknown	Unknown
52	Sorkh-e-Nadoushan	31	46	7	53	37	7	S	Tuff and andesite/Eocene; limestone and dolomite/Cretaceous	Hydrothermal	Pyrenean	Unknown
53	Zarand (Jalal Abad)	31	0	57	56	25	8	I	Volcanosedimentary/Upper Precambrian–Lower Cambrian	Volcanosedimentary	Pan-African	Lens
54	Anjireh	32	3	35	54	41	3	I	Phyllite and schist/ Jurassic–Cretaceous	Unknown	Unknown	Unknown
55	Anomaly No.XVI A	32	5	3	55	17	3	I	Covered with Tertiary and Quaternary alluvium	Magmatic	Pan-African	Disseminated
56	Anomaly No.XVI B	32	7	0	55	15	0	I	Unknown	Magmatic	Pan-African	Disseminated
57	Anomaly No.XVI C	32	10	5	55	13	5	I	Unknown	Magmatic	Pan-African	Disseminated
58	Anomaly No.XXIV A	32	24	58	55	1	55	I	Granitoid/Upper Precambrian–Lower Cambrian	Magmatic	Pan-African	Unknown
59	Anomaly No.XXIV B	32	21	59	55	3	43	I	Granitoid/Upper Precambrian–Lower Cambrian	Magmatic	Pan-African	Unknown
60	Anomaly No.VIII B	32	3	21	55	30	0	I	Granitoid/Upper Precambrian–Lower Cambrian	Magmatic	Pan-African	Unknown
61	Anomaly No.VIII C	32	1	58	55	32	0	I	Granitoid/Upper Precambrian–Lower Cambrian	Magmatic	Pan-African	Unknown
62	Anomaly No.XIII B	32	3	5	55	29	3	I	Porphyry granite/Upper Precambrian	Magmatic	Pan-African	Lens
63	Anomaly No.XIII C	32	2	2	55	31	4	I	Porphyry granite/Upper Precambrian	Magmatic	Pan-African	Disseminated
64	Anomaly No.XIV B	32	9	5	55	28	7	I	Granitoid/Proterozoic; shale and sandstone/Paleozoic	Magmatic	Pan-African	Unknown
65	Anomaly No.XIV C	32	9	3	55	25	2	I	Volcanic rocks and limestone/ Cretaceous	Magmatic	Pan-African	Disseminated
66	Anomaly No.XIV D	32	7	8	55	31	7	I	Gabbro and volcanosedimentary/ Proterozoic	Magmatic	Pan-African	Disseminated

(continued)

(continued)

No.	Name	Lat D°	Lat M	Lat S″	Long D°	Long M	Long S″	Size	Host rock/age	Genetic/ mineralization	Orogenic phases	Morphology	Elem para
67	Anomaly No.XIX A	32	20	17	55	13	42	I	Volcanosedimentary rocks/ Precambrian	Volcanosedimentary	Pan-African	Unknown	
68	Anomaly No.XIX B	32	18	23	55	13	39	I	Covered with shale, sandstone, and limestone/Paleozoic	Magmatic	Pan-African	Unknown	
69	Anomaly No.XIX C, D	32	14	8	55	16	0	I	Shale, sandstone, and limestone/ Paleozoic	Volcanosedimentary	Pan-African	Unknown	
70	Anomaly No.XVI A	32	5	3	55	17	3	1	Gabbro, diorite, schist and amphibolite/Precambrian	Magmatic	Pan-African	Disseminated	
71	Anomaly No.XVII A	32	13	0	55	24	0	I	Schist, amphibolite, and granite/ Precambrian	Skarn	Pan-African	Lens	
72	Anomaly No.XVII B	32	13	2	55	28	5	I	Granite, schist, and amphibolite/ Upper Precambrian–Lower Cambrian	Skarn	Pan-African	Lens	
73	Anomaly No.XVII C	32	11	3	55	27	8	I	Gnisse, schist, volcanosedimentary rocks, and granite/Upper Precambrian– Lower Cambrian	Magmatic	Pan-African	Lens	
74	Anomaly No.XVII D	32	11	1	55	30	8	I	Gnisse, schist, volcanosedimentary rocks, and granite/Upper Precambrian– Lower Cambrian	Magmatic	Pan-African	Disseminated	
75	Anomaly No.XVIII B	32	15	3	55	29	8	I	Granite, schist, and amphibolite/ Upper Precambrian–Lower Cambrian	Magmatic	Pan-African	Unknown	
76	Anomaly No.XVIII C	32	15	0	55	45	2	I	Limestone and sandstone/Jurassic; sandstone and limestone/Upper Cambrian	Unknown	Unknown	Unknown	
77	Anomaly No.XX (Saghand) A	32	31	13	55	36	4	I	Granitoid and schist/Upper Precambrian–Lower Cambrian	Magmatic	Pan-African	Unknown	
78	Anomaly No.XxB	32	30	8	55	37	37	I	Granitoid and schist/Upper Precambrian–Lower Cambrian	Volcanosedimentary	Pan-African	Unknown	
79	Anomaly No.XX C	32	30	8	55	45	5	I	Limestone and dolomite/Cretaceous	Unknown	Unknown	Unknown	

80	Anomaly No.XXI A	32	38	55	55	41	38	I	Unknown	Magmatic	Pan-African	Unknown
81	Anomaly No.XXI B	32	41	46	55	41	6	I	Unknown	Magmatic	Pan-African	Unknown
82	Anomaly No.XXI C	32	34	34	55	42	32	I	Unknown	Magmatic	Pan-African	Unknown
83	Anomaly No.XXII A	32	36	9	55	32	4	I	Schist and amphibolite/Upper Precambrian–Lower Cambrian	Volcanosedimentary	Pan-African	Unknown
84	Anomaly No.XXII B	32	36	12	55	33	39	I	Schist and Amphibolite/Upper Precambrian–Lower Cambrian	Volcanosedimentary	Pan-African	Unknown
85	Anomaly No. XXIII A	32	40	0	55	27	6	I	Schist and amphibolite/Upper Precambrian–Lower Cambrian	Volcanosedimentary	Pan-African	Unknown
86	Anomaly No.XXIII B	32	40	25	55	33	7	I	Unknown	Unknown	Unknown	Unknown
87	Anomaly No.XXIV C	32	26	1	55	1	1	I	Granitoid/Upper Precambrian–Lower Cambrian	Magmatic	Pan-African	Unknown
88	Anomaly No.XXIX	32	20	9	55	22	6	I	Granite, schist, amphibolite, and volcanosedimentary rocks/Upper Precambrian–Lower Cambrian	Magmatic	Pan-African	Disseminated
89	Anomaly No.XXVI	32	44	21	55	29	46	I	Schist and amphibolite/Upper Precambrian–Lower Cambrian	Volcanosedimentary	Pan-African	Unknown
90	Anomaly No.XXVI	32	56	44	55	37	26	I	Granite and syenite/Paleozoic, Schist and amphibolite/Upper Precambrian–Lower Cambrian		Pan-African	Disseminated
91	Anomaly No.XXX	32	27	53	55	23	14	I	Unknown	Unknown	Unknown	Unknown
92	Anomaly No.XXXII	32	41	7	55	15	27	I	Unknown	Unknown	Unknown	Unknown
93	Chador malu	32	17	0	55	32	0	L	Alkali granite, acidic to intermediate volcanics, and metamorphics (greenstone, mica-schist, marble)/ Upper Precambrian–Lower Cambrian	Magmatic	Pan-African	Massive
94	Chah Basheh	32	28	0	53	17	30	I	Tuff/Eocene	Hydrothermal	Pyrenean	Unknown
95	Chah gaz	32	8	0	55	29	0	L	Granite, andesite, and dacite/Upper Precambrian–Lower Cambrian	Magmatic	Pan-African	Massive
96	Chah Jamal (Iron and Alum)	32	20	43	55	12	2	I	Schist and gneiss/Mesozoic	Hydrothermal	Pyrenean	Unknown

(continued)

(continued)

No.	Name	Lat D°	Lat M	Lat S″	Long D°	Long M	Long S″	Size	Host rock/age	Genetic/ mineralization	Orogenic phases	Morphology	Elem para
97	Chah Komu	32	53	2	52	36	2	I	Unknown	Unknown	Unknown	Unknown	
98	Chapduni	32	50	40	55	13	24	I	Schist and amphibolite/Upper Precambrian–Lower Cambrian	Volcanosedimentary	Pan-African	Unknown	
99	Dig-e-dam Khomar	32	15	7	55	8	22	I	Unknown	Unknown	Unknown	Unknown	
100	Diposit No.XVII I	32	18	30	55	45	2	I	Unknown	Unknown	Unknown	Unknown	
101	Dureh 1	32	19	30	60	29	55	I	Diorite, syenite/Tertiary	Magmatic	Pyrenean	Unknown	
102	Dureh 2	32	18	45	60	30	50	I	Diorite, syenite/Tertiary	Magmatic	Pyrenean	Unknown	
103	Dureh(1)	32	18	45	60	30	5	I	Unknown	Unknown	Unknown	Unknown	
104	Dureh(2)	32	18	45	60	30	0	I	Unknown	Unknown	Unknown	Unknown	
105	Duzakh darreh	32	23	0	55	32	0	I	Granitoid, volcanic, dolomite/Upper Precambrian–Lower Cambrian	Magmatic	Pan-African	Unknown	
106	Fakhrabad (Red soil)	32	16	48	55	1	15	I	Granodiorite/Tertiary	Hydrothermal	Pyrenean	Unknown	
107	Gazik 1	32	17	55	60	30	33	I	Diorite, syenite/Tertiary gabbro and diorite/Mesozoic	Hydrothermal	Pyrenean	Unknown	
108	Golmandeh	32	40	11	55	27	38	I	Schist and amphibolite/Upper Precambrian–Lower Cambrian	Volcanosedimentary	Pan-African	Unknown	
109	Haft tal (1)	32	22	30	53	52	33	I	Unknown	Unknown	Unknown	Unknown	
110	Haft tal (2)	32	21	5	53	45	39	I	Unknown	Unknown	Unknown	Unknown	
111	Haft tal (3)	32	20	14	53	52	3	I	Unknown	Unknown	Unknown	Unknown	
112	Haft tal (4)	32	17	41	53	49	16	I	Unknown	Unknown	Unknown	Unknown	
113	Jadeh sorkhu	32	24	37	55	9	24	I	Granodiorite and andesite/Tertiary	Hydrothermal	Pyrenean	Unknown	
114	Jahan Shir	32	47	44	55	38	53	I	Limestone, shale, and sandstone/ Triassic	Unknown	Unknown	Unknown	
115	Khak-e-sorkhe sadr Abad	32	56	27	53	55	50	I	Unknown	Unknown	Unknown	Unknown	
116	Khoshumi (Iron and Alum)	32	23	20	55	11	3	I	Schist and gneiss/Mesozoic	Hydrothermal	Pyrenean	Unknown	
117	Khoshumi (Pyrite)	32	23	54	55	11	3	I	Schist and gneiss/Mesozoic	Hydrothermal	Pyrenean	Unknown	

118	Kuh-e-Dozardalou 1	32	30	55	55	30	31	I	Schist and amphibolite/Upper Precambrian–Lower Cambrian	Volcanosedimentary	Pan-African	Unknown
119	Kuh-e-Dozardalou 2	32	30	0	55	41	49	I	Unknown	Unknown	Unknown	Unknown
120	Kuh-e-Mayoon	32	13	21	55	25	49	I	Granite, schist, and amphibolite/ Upper Precambrian–Lower Cambrian	Volcanic	Pan-African	Unknown
121	Metamorphic Pyrite	32	30	33	55	10	43	I	Unknown	Unknown	Unknown	Unknown
122	Moghestan	32	0	40	55	9	49	I	Unknown	Unknown	Unknown	Unknown
123	Nadushan	32	0	59	53	28	13	I	Andesite/Eocene	Hydrothermal	Pyrenean	Unknown
124	Neybaz	32	30	33	55	10	43	I	Granite, schist, and amphibolite/ Upper Precambrian–Lower Cambrian	Volcanic	Pan-African	Unknown
125	North of Allah Abad	32	46	41	55	19	0	I	Schist and amphibolite/Upper Precambrian–Lower Cambrian	Volcanosedimentary	Pan-African	Unknown
126	North of Allah abad	32	46	41	55	19	0	I	Unknown	Unknown	Unknown	Unknown
127	Poshteh sorkh (Alum)	32	19	48	55	24	51	I	Granite and schist/Upper Precambrian–Lower Cambrian	Volcanosedimentary	Pan-African	Unknown
128	Poshteh sorkh (Pyrite)	32	19	48	55	25	50	I	Granite and schist/Upper Precambrian–Lower Cambrian	Volcanosedimentary	Pan-African	Unknown
129	Rig dam khomar	32	15	7	55	8	22	I	Sedimentary rocks/Paleozoic	Unknown	Unknown	Unknown
130	Senjed Abad	32	2	7	54	42	9	I	Metamorphic rocks/ Jurassic–Cretaceous	Unknown	Unknown	Unknown
131	Tigh Noab (1)	32	5	50	60	33	10	I	Unknown	Unknown	Unknown	Unknown
132	Tigh Noab (2)	32	6	0	60	23	0	I	Unknown	Unknown	Unknown	Unknown
133	Tigh Now Ab 1	32	6	0	60	23	0	I	Unknown	Unknown	Unknown	Unknown
134	Tigh Now Ab 2	32	6	29	60	34	28	I	Gabbro and diorite/Mesozoic	Hydrothermal	Unknown	Unknown
135	Zaj Poshteh	32	21	8	55	24	0	I	Granite, schist, and amphibolite/ Upper Precambrian–Lower Cambrian	Volcanic	Pan-African	Unknown
136	Zaman abad	32	40	18	55	27	43	I	Unknown	Unknown	Unknown	Unknown
137	Ali Abad	33	1	24	60	27	25	I	Unknown	Unknown	Unknown	Unknown

(continued)

(continued)

No.	Name	Lat D°	Lat M	Lat S″	Long D°	Long M	Long S″	Size	Host rock/age	Genetic/ mineralization	Orogenic phases	Morphology	Elem para
138	Boshruyeh 1	33	22	11	57	21	14	I	Shale, sandstone, and limestone/ Upper Paleozoic	Unknown	Unknown	Unknown	
139	Boshruyeh 2	33	41	21	57	5	16	I	Shale, sandstone, and limestone/ Upper Paleozoic	Unknown	Unknown	Unknown	
140	Buteh Alam	33	34	29	53	45	53	S	Terrigenous carbonate rocks/Triassic	Hydrothermal	Unknown	Vein	
141	Chah Ali Khan	33	43	26	52	59	11	I	Andesite/Upper Eocene; tuff/ Middle–Upper Eocene	Hydrothermal	Pyrenean	Unknown	
142	Chah Sefid	33	40	1	53	27	48	S	Quartz keratophyre/Cretaceous	Volcanosedimentary	Late Cimmerian	Lens, vein	
143	Dazak	33	22	9	60	4	11	I	Limestone and dolomite/Cretaceous	Unknown	Unknown	Unknown	
144	Gorgab 1	33	58	39	52	25	47	I	Granodiorite/Lower Eocene, Metamorphics/Paleozoic	Skarn	Pyrenean	Lens	
145	Gorgab 2	33	58	47	52	27	5	I	Granodiorite and diorite/Eocene; limestone/Lower Cretaceous; schist/Paleozoic	Skarn	Pyrenean	Lens	
146	Hosein Abad	33	9	43	53	3	11	I	Diorite–quartz diorite/Oligo-Miocene; andesite/Eocene	Hydrothermal	Post-Pyrenean	Vein	
147	Khalu Heydar	33	23	7	53	44	15	S	Quartzite, schist, and dolomitized marble/Upper Proterozoic	Skarn	Pan-African	Lens	
148	Khomein	33	42	16	49	43	13	I	Limestone and dolomite/Lower Cretaceous	Volcanosedimentary	Late Cimmerian	Unknown	
149	Kuh-e-dom	33	59	51	52	49	24	S	Porphyry granodiorite/Late Eocene; limestone/Upper Cretaceous; phyllite/Paleozoic	Skarn	Post-Pyrenean	Lens, disseminated, Massive	
150	Lakab	33	20	30	53	26	17	I	Marble and schist/Upper Precambrian–Lower Cambrian	Hydrothermal	Post-Pyrenean	Vein	
151	Mahallat	33	57	21	50	23	23	I	Sedimentary rocks/Upper Paleozoic	Unknown	Unknown	Unknown	
152	Mohammad abad	33	13	49	53	41	31	I	Limestone/Upper Cretaceous, Schist and marble/Upper Proterozoic	Hydrothermal	Pan-African	Veinlet	
153	Monavad	33	4	11	60	20	50	I	Volcanics and limestone/Cretaceous	Unknown	Unknown	Unknown	

154	Northern Chah Palang	33	5	34	54	23	10	S	Metamorphic, volcanosedimentary rocks/Upper Precambrian–Lower Cambrian	Metamorphic	Pan-African	Stratiformbed	
155	Piyuk	33	20	24	53	44	2	I	Marble and schist/Upper Precambrian–Lower Cambrian	Hydrothermal	Unknown	Vein	Mn
156	Qayen	33	53	33	59	25	31	I	Andesite/Eocene	Volcanic	Pyrenean	Unknown	
157	Shams Abad	33	48	0	49	43	0	S	Lime–argillic/Lower Cretaceous	Volcanosedimentary	Late Cimmerian	Massive	
158	Soheil-e-Pakuh	33	10	16	53	1	12	I	Diorite–quartz diorite/Oligo-Miocene; andesite/Eocene	Volcanic	Post-Pyrenean	Vein	Mn
159	Ahangaran	34	7	7	48	59	20	S	Limestone, dolomite, and shale/Cretaceous; slate/Jurassic	Sedimentary	Late Cimmerian	Lens	
160	Anarak-Khur	34	10	37	55	23	39	I	Limestone and dolomite/Cretaceous	Unknown	Unknown	Unknown	
161	Asad abad	34	51	47	48	5	40	S	Metarhyolite, meta-andesite, limestone, and marble/Triassic	Volcanosedimentary	Early Cimmerian	Lens	
162	Baba Ali	34	56	5	48	9	32	S	Metavolcanics (acidic to intermediate), metasubvolcanics and marble/Jurassic	Volcanosedimentary and metamorphic	Early Cimmerian	Lens	
163	Chenare Olya	34	52	0	45	5	0	I	Metavolcanics (acidic to intermediate), metasubvolcanics and marble/Jurassic	Volcanosedimentary and metamorphic	Early Cimmerian	Lens	
164	Dasht-e-Mami tak	34	28	53	60	10	13	I	Tuff/Eocene	Volcanic	Pyrenean	Unknown	
165	Dehkhazal	34	53	24	47	50	56	S	Alkali rhyolite and limestone/Triassic	Volcanosedimentary and metamorphic	Early Cimmerian	Lens	
166	Dehzaman	34	58	0	57	50	0	S	Schist, phyllite, dolomite, and limestone/Upper Paleozoic–Lower Triassic	Volcanics	Hercynian	Lens	
167	Doroud	34	3	0	48	50	22	I	Granitoid/Mesozoic	Volcanosedimentary	Early Cimmerian	Unknown	
168	Fajrab	34	44	29	49	31	55	I	Tuff/Eocene; Limestone/Jurassic	Magmatic	Pyrenean	Unknown	
169	Farzoneh	34	30	52	60	35	13	I	Granite/Eocene; garnet skarn, amphibolite skarn, and shale/Paleozoic	Volcanosedimentary	Pyrenean	Unknown	

(continued)

(continued)

No.	Name	Lat D°	Lat M	Lat S″	Long D°	Long M	Long S″	Size	Host rock/age	Genetic/ mineralization	Orogenic phases	Morphology	Elem para
170	Galali	34	51	2	47	54	17	S	Metarhyolite, meta-andesite, limestone, and marble/Triassic	Volcanosedimentary and metamorphic	Early Cimmerian	Lens	
171	Hamehkasi	34	53	0	47	50	0	S	Limestone and volcanics/ Triassic–Jurassic	Volcanosedimentary	Early Cimmerian	Lens	
172	Hehho	34	41	19	60	24	47	I	Metamorphic rocks/ Paleozoic–Triassic	Unknown	Unknown	Unknown	
173	Hezarkhani	34	55	40	47	49	0	S	Metarhyolite, meta-andesite, limestone, and marble/Triassic	Volcanosedimentary	Early Cimmerian	Lens	
174	Jowzan	34	14	9	48	56	13	S	Metarhyolite and limestone/Jurassic	Volcanosedimentary	Early Cimmerian	Lens	
175	Khalt abad	34	11	36	59	41	29	I	Unknown	Hydrothermal	Pyrenean	Unknown	
176	Khosroabad	34	57	0	47	47	0	S	Metarhyolite, meta-andesite, limestone, and marble/Triassic	Volcanosedimentary and metamorphic	Early Cimmerian	Lens	
177	Kuh-e-Ahan	34	24	40	57	22	59	S	Limestone and dolomite/Jurassic	Unknown	Late Cimmerian	Unknown	
178	Kuh-e-Neybid	34	10	45	59	42	25	I	Andesite/Eocene	Hydrothermal	Pyrenean	Unknown	
179	Kuh-e-Taleb 1	34	28	57	60	25	26	S	Granite/Eocene; skarn, shale, and siltstone/Paleozoic	Volcanosedimentary	Pyrenean	Lens	
180	Kuh-e-Taleb 2	34	29	28	60	27	23	S	Granite/Eocene; skarn, shale, and siltstone/Paleozoic	Volcanosedimentary	Pyrenean	Lens	
181	Molhem Darreh	34	45	40	48	26	24	I	Granitoid/Mesozoic	Hydrothermal	Unknown	Unknown	
182	Niyasar	34	13	40	51	8	55	S	Diabase/Post-Eocene; limestone/ Eocene	Volcanic	Pyrenean	Lens	
183	Nok-e-Khorous	34	28	40	60	27	15	I	Granite/Eocene; skarn, shale, and siltstone/Paleozoic	Volcanosedimentary	Early Cimmerian	Lens	
184	Pahnehbur	34	55	54	48	10	35	S	Metarhyolite and limestone/Jurassic	Volcanosedimentary	Early Cimmerian	Lens	
185	Pey	34	58	3	60	29	1	I	Andesite/Eocene	Hydrothermal	Pyrenean	Unknown	
186	Poshteh(1)	34	41	19	60	23	16	I	Unknown	Unknown	Unknown	Unknown	
187	Poshteh(2)	34	41	23	60	23	51	I	Unknown	Unknown	Unknown	Unknown	
188	Ravanj	34	11	3	50	45	58	I	Granitoid/Oligo-Miocene; tuff/Eocene	Magmatic	Post-Pyrenean	Unknown	

189	Raveh	34	11	47	50	20	8	I	Limestone and dolomite/Cretaceous	Unknown	Unknown	Unknown
190	Sangan	34	28	0	60	25	0	S	Granite/Eocene; garnet skarn, amphibolite skarn, shale, and siltstone/Paleozoic	Volcanosedimentary, metasomatism	Pyrenean	Lens
191	Shizen 1	34	45	0	60	19	34	I	Andesite/Eocene; skarn and siltstone/ Paleozoic–Lower Triassic	Volcanosedimentary	Pyrenean	Unknown
192	Shizen 2	34	44	28	60	17	56	I	Andesite/Eocene; skarn and siltstone/ Paleozoic–Lower Triassic	Volcanosedimentary	Pyrenean	Unknown
193	South of Mahabad	34	18	24	59	37	0	I	Unknown	Hydrothermal	Pyrenean	Unknown
194	Southeast of Mahabad	34	17	19	59	41	29	I	Unknown	Hydrothermal	Pyrenean	Unknown
195	Bajestan	35	27	35	57	53	39	I	Volcanics/Mesozoic	Hydrothermal	Unknown	Unknown
196	Chalu	35	34	30	54	21	9	S	Andesite/Eocene	Volcanosedimentary	Pyrenean	Lens
197	Charmaleh Bala-va-Paein	35	1	21	47	46	32	S	Metarhyolite, meta-andesite, limestone, and marble/Triassic	Volcanosedimentary	Early Cimmerian	Lens
198	Divar	35	14	35	60	23	50	I	Andesite/Tertiary	Volcanic	Pyrenean	Unknown
199	Juzan	35	15	36	48	17	24	I	Unknown	Unknown	Unknown	Unknown
200	Kahriz now	35	10	5	60	3	25	I	Granitoid/Tertiary; limestone/ Cretaceous	Hydrothermal	Pyrenean	Unknown
201	Kuh-e-Sormeh Ali	35	10	14	47	18	31	I	Volcanics and limestone/Jurassic	Skarn	Laramide	Unknown
202	Mohammad Abad	35	24	42	57	23	11	I	Tuff and andesite/Eocene	Hydrothermal	Pyrenean	Unknown
203	Namegh	35	23	31	58	50	27	I	Diorite–granodiorite/Upper Eocene–Lower Oligocene	Hydrothermal	Pyrenean	Unknown
204	North of Semnan	35	35	0	53	25	0	S	Andesite and dacite/Tertiary	Volcanic	Pyrenean	Massive
205	Northwest of Tanureh	35	24	41	58	36	2	S	Diorite–granodiorite/Upper Eocene–Lower Oligocene	Hydrothermal	Pyrenean	Massive
206	Panj Kuh	35	47	22	54	20	0	S	Syenite and volcanics/ Eocene–Oligocene	Magmatic	Pyrenean	Lens
207	Siahbala	35	14	55	60	22	25	I	Andesite/Tertiary	Volcanic	Pyrenean	Unknown
208	Tanurcheh	35	23	51	58	37	38	S	Diorite and granodiorite/Upper Eocene–Lower Oligocene	Hydrothermal	Pyrenean	Massive
209	Tohen Abad	35	16	3	48	52	28	I	Sedimentary rocks/Oligo-Miocene	Unknown	Unknown	Unknown

(continued)

(continued)

No.	Name	Lat D°	Lat M	Lat S″	Long D°	Long M	Long S″	Size	Host rock/age	Genetic/ mineralization	Orogenic phases	Morphology	Elem para
210	Alam Kuh	36	25	54	50	57	48	I	Shale, sandstone, and limestone/ Upper Paleozoic–Jurassic	Sedimentary	Laramid	Stratiformbed, lens	
211	Amir abad (Gharavol Khaneh)	36	37	0	47	20	0	I	Intermediate to basic volcanics and subvolcanics/Oligo-Miocene	Volcanosedimentary	Post-Pyrenean	Lens	
212	Arjin	36	27	0	49	0	0	S	Dolomite/Lower Cambrian	Volcanosedimentary	Pan-African	Lens	
213	Bichaghchi	36	37	0	46	48	0	I	Shale, slate, and rhyolite/Precambrian	Sedimentary	Pan-African	Stratiformbed	
214	Biveh zhan	36	3	11	59	21	20	S	Shale and limestone/Upper Paleozoic–Lower Triassic	Sedimentary	Hercynian	Stratiformbed	
215	Chahar Tagh	36	28	0	47	16	0	S	Dolomite and schist/Upper Precambrian	Sedimentary	Pan-African	Lens	Iron, Zn, Pb, Ba
216	Eskand	36	37	25	48	45	36	I	Granite, granodiorite, and volcanics/ Eocene	Skarn	Pyrenean	Vein	
217	Ghalicheh Bolagh	36	48	0	47	53	0	S	Granite, metavolcanic, and shale/ Upper Precambrian	Volcanosedimentary	Pan-African	Lens	
218	Goljook	36	12	0	48	4	0	S	Dolomite/Upper Precambrian–Lower Cambrian	Sedimentary	Pan-African	Lens	
219	Incheh	36	37	0	48	24	0	I	Tuff/Eocene	Hydrothermal	Pyrenean	Unknown	
220	Kavand	36	42	16	48	8	42	S	Dolomite/Upper Precambrian	Sedimentary	Pan-African	Stratiformbed	
221	Kuh-e-baba (Alikandi)	36	46	0	47	14	0	S	Andesite and subvolcanic/ Post-Oligo-Miocene	Volcanic	Post-Pyrenean	Lens	
222	Kuh-e-Haraz	36	20	32	54	45	39	I	Sedimentary rocks/Paleogene	Unknown	Unknown	Unknown	
223	Mirjan	36	43	30	47	57	30	S	Dolomite/Upper Precambrian	Sedimentary	Pan-African	Stratiformbed	
224	Morvariyeh	36	34	0	48	33	0	S	Granite and granodiorite/Eocene–Oligocene; andesite/Eocene	Magmatic	Post-Pyrenean	Massive	
225	North of Rud-e-Alamut	36	33	39	50	30	56	I	Shale, sandstone, and limestone/ Lower Paleozoic	Unknown	Unknown	Unknown	
226	Northwest of Kahoor Charan synclinal	36	24	11	49	30	39	I	Andesite/Tertiary	Hydrothermal	Pyrenean	Unknown	
227	Northwestern Arjin	36	25	43	48	39	29	I	Dolomite/Upper Precambrian	Sedimentary	Pan-African	Lens	

228	Shah Bolagh	36	38	41	48	5	30	S	Dolomite and shale/Upper Precambrian–Lower Cambrian	Sedimentary	Pan-African	Stratiformbed
229	Shahrak	36	22	0	47	32	0	M	Andesite, rhyolite, diorite, and limestone/Miocene	Volcanic (flow magnetiti?)	Post-Pyrenean	Stratiformbed
230	Shahsavar	36	27	34	50	45	0	I	Shale, sandstone, and limestone/Lower Paleozoic	Unknown	Unknown	Unknown
231	Soltaniyeh	36	36	0	48	49	0	I	Granitoid/Oligo-Miocene	Hydrothermal	Post-Pyrenean	Unknown
232	Sorkhedizaj	36	35	0	48	50	0	S	Andesite, granodiorite to tonalite/Oligo-Miocene	Magmatic	Post-Pyrenean	Massive
233	Southwest of Arjin	36	23	24	48	41	35	I	Dolomite/Upper Precambrian	Sedimentary	Pan-African	Lens
234	Takab 1	36	22	15	47	31	5	I	Granitoid/Tertiary; phyllite and schist/Jurassic–Cretaceous	Hydrothermal	Pyrenean	Unknown
235	Takab 2	36	43	45	47	57	30	I	Granitoid/Precambrian	Hydrothermal	Pan-African	Unknown
236	West of Arjin	36	0	32	48	32	5	I	Dolomite/Upper Precambrian	Sedimentary	Pan-African	Lens
237	Zafarabad	36	12	0	46	57	59	S	Basic igneous rocks and limestone/Permian	Volcanosedimentary	Hercynian	Lens
238	Zanjan	36	31	0	48	29	8	I	Sandstone and shale/Precambrian	Unknown	Unknown	Unknown
239	Zardabiyeh	36	11	6	55	7	37	I	Andesite/Eocene	Magmatic	Pyrenean	Unknown
240	Aghbolagh	37	9	5	45	7	0	I	Granite and dolomite/Precambrian	Magmatic	Pan-African	Lens
241	Balestan	37	10	0	45	20	0	S	Granite, metavolcanic, shale, and dolomite/Upper Precambrian	Volcanosedimentary	Pan-African	Lens
242	Guran	37	9	5	48	54	23	S	Sandstone and shale/Permian	Sedimentary	Hercynian	Stratiformbed
243	Masuleh	37	9	0	48	59	0	I	Volcanosedimentary rocks/Upper Paleozoic–Lower Triassic	Volcanosedimentary	Hercynian	Unknown
244	Tatav Rud	37	12	59	49	7	45	I	Limestone/Cretaceous	Skarn	Pyrenean	Unknown
245	Amirchi	38	49	19	46	42	23	I	Volcanics and volcanoclastics/Eocene; limestone/Cretaceous	Skarn	Pyrenean	Unknown
246	Biruz	38	52	0	46	50	30	I	Acidic volcanic rocks and limestone/Upper Cretaceous–Lower Paleocene	Skarn	Laramide	Lens
247	Bizarin	38	40	16	46	47	26	I	Granodiorite/Oligo-Miocene; limestone/Cretaceous	Skarn	Post-Pyrenean	Unknown

(continued)

(continued)

No.	Name	Lat D°	Lat M	Lat S″	Long D°	Long M	Long S″	Size	Host rock/age	Genetic/ mineralization	Orogenic phases	Morphology	Elem para
248	Eskandian	38	0	0	44	40	0	S	Metamorphic rocks bearing amphibolite and granitoid/ Post-Cretaceous	Skarn	Pyrenean	Lens	
249	Mazraeh	38	38	47	47	4	11	I	Volcanics and volcanoclastics/ Eocene; limestone/Cretaceous	Skarn	Pyrenean	Unknown	
250	North of Bizarin	38	41	26	46	47	16	I	Granodiorite/Oligo-Miocene; limestone/Cretaceous	Skarn	Post-Pyrenean	Unknown	
251	Sarghin	38	29	0	47	10	20	I	Latite and ignimbrite/Eocene	Volcanic	Pyrenean	Disseminated	
252	Vayan	38	55	30	46	47	41	I	Intrusive rocks/Tertiary; volcanic, marble, and limestone/Upper Cretaceous	Skarn	Laramide	Lens	
253	Zaklike	38	33	0	47	4	0	S	Volcanic and intrusive rocks/ Eocene–Oligocene; sedimentary rocks/Cretaceous	Skarn	Pyrenean	Lens	
254	Kalat Naser Qaen							I	Schist and gabbro/Upper Paleozoic	Volcanosedimentary	Hercynian	Lens	
255	Neizar							S	Schist and limestone/Upper Paleozoic–Lower Triassic	Sedimentary	Hercynian	Stratiformbed	
256	Sheikhab							S	Andesite/Tertiary	Volcanic	Pyrenean	Lens	

[a] *S* Smal, *I* Index, *L* Large

Complete list of manganese mineral deposits and indications of Iran along with their detailed specifications

No.	Name	Lat D°	Lat M	Lat S″	Long D°	Long M	Long S″	Size	Host rock/age	Genetic/mineralization	Orogenic phases	Morphology	Elem para
1	Abband	28	54	23	54	24	29	S	Radiolarite chert/Upper Cretaceous	Sedimentary or hydrothermal	Laramide	Vein	Fe, Si
2	Abdol Abad	36	23	0	50	8	0	S	Andesite and tuff/Eocene	Hydrothermal	Post-Pyrenean	Vein	
3	Abdollah Abad	36	22	6	50	3	24	S	Volcanic/Eocene	Hydrothermal	Post-Pyrenean	Vein	
4	Aghcheh Mazar-e-Eshtehard	35	40	0	50	10	0	I	Dacite, andesite/Eo-Oligocene	Volcanosedimentary	Pyrenean	Vein	
5	Arat kuh	35	20	0	51	21	30	S	Dacite/Eocene–Oligocene	Volcanosedimentary	Pyrenean	Disseminated	
6	Asad	35	44	0	57	9	38	S	Volcanic and sedimentary rocks (from Kashmar ophiolite complex)/Upper Cretaceous–Paleocene	Volcanosedimentary	Laramide	Vein	
7	Bagh Ghareh	35	11	35	59	31	34	S	Tuff and andesite/Eocene; limestone/Cretaceous	Volcanosedimentary	Pyrenean	Vein	
8	Bardash Khaneh	36	19	0	57	52	0	S	Basalt, spilite, and limestone/Upper Cretaceous	Volcanosedimentary	Laramide	Lens	
9	Bazni (Baznin)	33	15	0	52	34	22	S	Dacite and andesite/Eocene	Hydrothermal	Pyrenean	Lens	
10	Benavid	32	39	0	53	9	32	S	Carbonaceous silica volcanic/Upper Cretaceous–Lower Paleogene	Volcanosedimentary	Pyrenean	Stratiformbed, lens	Fe
11	Benesporte	35	44	31	57	13	37	S	Volcanic and sedimentary rocks (from Kashmar ophiolite complex)/Upper Cretaceous–Paleocene	Volcanosedimentary	Laramide	Stratiformbed	
12	Chah Anjir	29	48	29	53	45	57	S	Dolomitic limestone, limestone, and chert/Cretaceous	Volcanosedimentary	Laramide	Stratiformbed, vein	
13	Chah Sefid	33	40	0	53	27	0	S	Andesite and andesitic dacite/Eocene	Hydrothermal	Pyrenean	Vein	Cu, As, Ba
14	Chahbashi	32	32	58	53	12	20	M	Marble and schist/Lower Cretaceous	Volcanosedimentary	Pyrenean	Vein	Fe
15	Chahgabri Sar-e-Kavir	35	24	1	54	14	48	S	Mica-schist/Upper Precambrian–Lower Cambrian	Hydrothermal	Pyrenean	Lens	Fe

(continued)

(continued)

No.	Name	Lat D°	Lat M	Lat S″	Long D°	Long M	Long S″	Size	Host rock/age	Genetic/ mineralization	Orogenic phases	Morphology	Elem para
16	Chahpalang	33	5	34	54	23	17	S	Dolomite, marble, schist, and metavolcanics/Upper Precambrian–Lower Cambrian	Volcanosedimentary	Pan-African	Stratiformbed	Fe, Mn
17	Chaytelvar	37	17	18	47	37	43	I	Rhyolite, trachyte, andesite, and tuff/ Lower Miocene	Hydrothermal	Post-Pyrenean	Disseminated	
18	Debeklu (Amir Abad)	37	39	43	46	53	43	S	Andesite and tuff/Oligo-Miocene	Volcanic	Post-Pyrenean	Vein	S, P, Fe
19	East of Rudkhaneh Shur	35	28	46	50	58	50	S	Dacite, andesite/Eocene	Volcanosedimentary	Pyrenean	Vein	
20	Eastern Torkemani	33	13	15	54	4	1	S	Dolomite/Upper Precambrian–Lower Cambrian	Telethermal	Pan-African	Vein	Fe
21	Gel-e-gonegu	33	4	25	53	2	10	S	Limestone and porphyritic volcanics/ Cretaceous	Volcanosedimentary	Laramide	Massive	
22	Ghazalcheh Maragheh	37	15	16	46	53	20	S	Tuff/Eocene–Oligocene	Hydrothermal	Pyrenean	Vein	
23	Ghupuz	37	27	50	46	50	0	I	Conglomerate/Upper Miocene	Hydrothermal	Post-Pyrenean	Veinlet, vein	
24	Goloujeh	37	39	27	47	9	0	I	Volcanics and carbonaceous volcanic rocks/Miocene	Volcanic	Post-Pyrenean	Vein	
25	Guvanich	27	56	45	60	36	44	S	Tuff, andesite, shale, and limestone/ Upper Cretaceous	Volcanosedimentary	Laramide	Vein	Fe
26	Hasan Robat	33	24	0	50	50	0	M	Dolomite and marble/Precambrian	Sedimentary	Hercynian	Lens, vein	Fe
27	Heneshk	30	47	30	53	16	0	M	Schist, phyllite, and metavolcanic/ Upper Paleozoic	Sedimentary	Hercynian	Lens, vein	Fe, Mn, Ba
28	Hojib	35	31	22	50	8	44	I	Dacite, andesite/Eo-Oligocene	Volcanosedimentary	Pyrenean	Vein	
29	Idehlu Chughani	37	11	13	46	51	36	I	Shale and sandstone/Jurassic and limestone, dolomite limestone, and quartz keratophyre/Miocene	Hydrothermal	Post-Pyrenean	Vein	S, P
30	Janbahan	37	53	43	46	47	59	S	Conglomerate and sandstone/ Miocene	Hydrothermal	Post-Pyrenean	Disseminated, veinlet	
31	Khalifeh Kamal	37	17	26	47	18	30	I	Olivine basaltic andesite, tuff, and dolomite limestone/Lower Miocene	Volcanic	Post-Pyrenean	Vein	S, P, Fe

32	Khalu Heydar	33	22	37	53	40	0	S	Quartzite schist and dolomitized marble/Upper Proterozoic	Skarn	Pan-African	Lens	Fe, Mn
33	Kolujeh	37	39	27	47	9	0	I	Volcanics and carbonaceous volcanic rocks/Miocene	Hydrothermal	Post-Pyrenean	Vein	
34	Kuh-e-Baba Khalou	33	40	0	53	26	48	I	Andesite/Upper Eocene	Hydrothermal	Pyrenean	Vein	Fe, Ba
35	Kuh-e-bababozorgi	33	40	36	53	27	34	S	Andesite/Upper Eocene	Hydrothermal	Pyrenean	Vein	
36	Kuh-e-darbid	32	30	42	54	28	25	S	Carbonaceous rocks/Upper Jurassic–Lower Cretaceous	Hydrothermal	Mid-Cimmerian	Lens	
37	Kuh-e-Zard	33	10	11	53	0	0	I	Limestone and dolomite/Cretaceous	Unknown	Unknown	Unknown	
38	Lyak (Niak)	35	40	20	50	24	48	S	Dacite/Eocene–Oligocene	Volcanosedimentary	Pyrenean	Vein	Fe, Cu, Pb
39	Malayer (Ahmadroghani)	34	11	48	48	55	34	M	Limestone/Cretaceous	Volcanosedimentary	Laramide	Stratiformbed, vein	S, P
40	Manamin	37	16	17	48	20	54	I	Tuff and andesite/Eocene–Oligocene	Hydrothermal	Post-Pyrenean	Vein	Pb
41	Mohammad Ali Khan	35	15	0	50	57	20	S	Volcanics/Paleogene	Volcanosedimentary	Pyrenean	Vein	Fe, Mn
42	Mourchekhort	33	15	0	51	15	0	M	Limestone/Lower Cretaceous; siltstone, shale, and sandstone/Jurassic	Hydrothermal	Mid-Cimmerian	Vein	Fe, Cu
43	Muteh	33	40	30	50	40	38	S	Schist/Upper Precambrian–Lower Cambrian	Hydrothermal	Pan-African	Lens	Fe
44	Narigan	31	47	3	55	41	2	M	Carbonaceous rocks and carbonaceous detrital and rhyolite/Upper Precambrian	Volcanosedimentary or Volcanic	Pan-African	Lens	Fe
45	Neyriz	29	10	38	54	14	22	S	Shale, radiolarite, red silic, and limestone/Jurassic–Cretaceous	Volcanosedimentary	Late Cimmerian	Stratiformbed	
46	Northwest Kuh-e-Kafardogh	35	12	29	59	26	25	I	Tuff, andesite/Eocene; limestone/Cretaceous	Volcanosedimentary	Pyrenean	Vein	Fe, Cu
47	Nugh	35	8	0	59	36	43	S	Tuff and andesite/Tertiary	Volcanic	Pyrenean	Vein	
48	Qaleh Gang	33	10	11	53	0	0	S	Limestone bearing quartz and shale bearing mica/Upper Cretaceous–Lower Paleogene	Hydrothermal	Laramide	Small lens	
49	Qom	34	25	9	50	45	26	I	Tuff and tuffaceous sandstone and limestone and volcanics/Eocene	Volcanosedimentary	Pyrenean	Stratiformbed	Fe

(continued)

(continued)

No.	Name	Lat D°	Lat M	Lat S″	Long D°	Long M	Long S″	Size	Host rock/age	Genetic/ mineralization	Orogenic phases	Morphology	Elem para
50	Robatkarim	35	29	11	50	58	51	M	Volcanics and pyroclastics/Eocene	Volcanosedimentary	Pyrenean	Stratiformbed	
51	Rudkhaneh	33	53	29	52	23	51	I	Limestone/Oligo-Miocene	Sedimentary	Post-Pyrenean	Stratiformbed	
52	Salamrud	35	18	0	56	54	0	S	Basalt, spilite, and limestone/Upper Cretaceous	Volcanosedimentary	Laramide	Vein	Fe
53	Sangaz (Sargaz)	27	55	34	56	40	0	S	Volcanic and sedimentary rocks/ Upper Jurassic–Upper Triassic	Massive	Late Cimmerian	Vein	Fe
54	Sarmak	30	26	0	53	40	0	S	Siltstone, sandstone, and limestone/ Upper Jurassic–Lower Cretaceous	Hydrothermal	Mid-Cimmerian	Vein	Fe
55	Sefid Kuh 1	34	54	3	57	5	10	S	Limestone, shale, and sandstone/ Jurassic	Unknown	Unknown	Unknown	
56	Sefid Kuh 2	34	54	0	57	8	23	S	Limestone, shale, and sandstone/ Jurassic	Unknown	Unknown	Unknown	
57	Sefid Kuh 3	34	56	13	57	9	40	I	Limestone, shale, and sandstone/ Jurassic	Unknown	Unknown	Unknown	
58	Separab	33	7	28	53	0	58	S	Limestone and porphyritic volcanics/ Cretaceous	Volcanosedimentary	Laramide	Massive	Fe, Cu, Pb
59	Shams Abad	33	48	0	49	43	0	S	Limestone and shale/Lower Cretaceous	Volcanosedimentary	Late Cimmerian	Massive	Fe
60	Shemsh Abad	33	51	0	50	9	3	I	Limestone and dolomite/Cretaceous; sandstone and shale/Jurassic	Unknown	Unknown	Unknown	Fe
61	Sorkhshad	33	33	34	53	36	28	S	Andesite/Eocene	Hydrothermal	Pyrenean	Vein, veinlet	Fe
62	Venarch	34	25	8	50	45	18	L	Tuff, tuffaceous sandstone, limestone, and volcanics/Eocene	Volcanosedimentary	Pyrenean	Stratiformbed	S, P
63	Viladarreh	38	12	50	48	2	52	I	Tuff/Pliocene; sandstone and siltstone/Upper Miocene	Volcanic	Post-Pyrenean	Disseminated, veinlet	Cu
64	Western Kuh-e-dom	33	57	41	52	40	31	I	Volcanics/Eocene; limestone/Lower Cretaceous	Volcanic	Pyrenean	Veinlet	Mn, Fe
65	Zereshlu	37	18	22	47	17	50	I	Andesite/Lower Miocene	Volcanic	Post-Pyrenean	Vein	

Complete list of chromite mineral deposits and indications of Iran along with their detailed specifications

No.	Name	Lat D°	Lat M	Lat S″	Long D°	Long M	Long S″	Size	Host rock/age	Genetic/mineralization	Orogenic phases	Morphology	Elem para
1	Gootich	26	54	8	59	0	45	I	Dunite, serpentinite, and pyroxenite/ Upper Cretaceous	Magmatic	Laramide	Lens	
2	Hari Duk	26	43	15	60	10	3	I	Dunite, serpentinite, and pyroxenite/ Upper Cretaceous	Magmatic	Laramide	Lens	
3	Mir Abad	26	54	8	59	15	0	I	Dunite, serpentinite, and pyroxenite/ Upper Cretaceous	Magmatic	Laramide	Lens	
4	Moter Abad	26	48	15	59	7	30	I	Dunite, serpentinite, and pyroxenite/ Upper Cretaceous	Magmatic	Laramide	Lens	
5	Remeshk	26	44	3	59	12	15	S	Dunite, serpentinite, and pyroxenite/ Upper Cretaceous	Magmatic	Laramide	Lens	
6	Amir	27	29	29	57	13	50	M	Dunite, serpentinite, and pyroxenite/ Upper Cretaceous	Magmatic	Laramide	Lens	
7	Badafshan Bala	27	35	57	57	12	15	I	Dunite, serpentinite, and pyroxenite/ Upper Cretaceous	Magmatic	Laramide	Lens	
8	East of ghallat	27	16	21	57	15	59	I	Dunite, serpentinite, and pyroxenite/ Upper Cretaceous	Magmatic	Laramide	Lens	
9	East of Kuh-e-Mishbadam	27	47	51	57	34	5	S	Dunite, serpentinite, and pyroxenite/ Upper Cretaceous	Magmatic	Laramide	Lens	
10	Faryab	27	27	49	57	2	42	L	Dunite, serpentinite, and pyroxenite/ Upper Cretaceous	Magmatic	Laramide	Lens	Cu
11	Ghallat	27	16	33	57	16	13	I	Dunite, serpentinite, and pyroxenite/ Upper Cretaceous	Magmatic	Laramide	Lens	
12	Gunich	27	58	1	60	34	27	S	Dunite, serpentinite, and pyroxenite/ Upper Cretaceous	Magmatic	Laramide	Lens	
13	Heydari	27	35	8	57	3	13	I	Dunite, serpentinite, and pyroxenite/ Upper Cretaceous	Magmatic	Laramide	Lens	
14	Jaghin	27	11	21	57	21	44	I	Dunite, serpentinite, and pyroxenite/ Upper Cretaceous	Magmatic	Laramide	Lens	

(continued)

(continued)

No.	Name	Lat D°	Lat M	Lat S″	Long D°	Long M	Long S″	Size	Host rock/age	Genetic/ mineralization	Orogenic phases	Morphology	Elem para
15	Kuh Sorkh	27	50	0	57	30	0	S	Dunite, serpentinite, and pyroxenite/ Upper Cretaceous	Magmatic	Laramide	Lens	
16	Kuh-e-Cheshmeh	27	36	57	57	3	50	S	Dunite, serpentinite, and pyroxenite/ Upper Cretaceous	Magmatic	Laramide	Lens	
17	Kuh-e-Garuk	27	47	34	57	34	35	I	Dunite, serpentinite, and pyroxenite/ Upper Cretaceous	Magmatic	Laramide	Lens	
18	Kuh-e-GolKahan	27	38	44	57	10	44	I	Dunite, serpentinite, and pyroxenite/ Upper Cretaceous	Magmatic	Laramide	Lens	
19	Kuh-e-Goltahan	27	38	44	57	10	44	S	Dunite, serpentinite, and pyroxenite/ Upper Cretaceous	Magmatic	Laramide	Lens	
20	Kuh-e-Mishbadam	27	47	43	57	21	49	S	Dunite, serpentinite, and pyroxenite/ Upper Cretaceous	Magmatic	Laramide	Lens	
21	Kuh-e-Sorkh	27	33	47	57	3	50	S	Dunite, serpentinite, and pyroxenite/ Upper Cretaceous	Magmatic	Laramide	Lens	
22	Kuh-e-Sorkh	27	50	0	57	30	0	I	Dunite, serpentinite, and ;pyroxenite/ Upper Cretaceous	Magmatic	Laramide	Lens	
23	Kuh-e-Tisa	27	0	24	57	35	12	I	Dunite, serpentinite, and pyroxenite/ Upper Cretaceous	Magmatic	Laramide	Lens	
24	Kutak	27	52	42	57	37	21	I	Dunite, serpentinite, and pyroxenite/ Upper Cretaceous	Magmatic	Laramide	Lens	
25	Nader	27	8	57	57	23	4	S	Dunite, serpentinite, and pyroxenite/ Upper Cretaceous	Magmatic	Laramide	Lens	
26	Northeast of Dastgerd fault	27	23	13	57	23	6	S	Dunite, serpentinite, and pyroxenite/ Upper Cretaceous	Magmatic	Laramide	Lens	
27	Reza	27	25	31	57	27	41	S	Dunite, serpentinite, and pyroxenite/ Upper Cretaceous	Magmatic	Laramide	Lens	
28	Shahin	27	23	17	57	24	33	S	Dunite, serpentinite, and pyroxenite/ Upper Cretaceous	Magmatic	Laramide	Lens	
29	Shahryar-e-Shahin	27	14	28	57	14	36	M	Dunite, serpentinite, and pyroxenite/ Upper Cretaceous	Magmatic	Laramide	Lens	

30	SouthofJaghin	27	11	4	57	21	17	I	Dunite, serpentinite, and pyroxenite/ Upper Cretaceous	Magmatic	Laramide	Lens
31	Southwest of Rudan fault	27	24	18	57	22	36	S	Dunite, serpentinite, and pyroxenite/ Upper Cretaceous	Magmatic	Laramide	Lens
32	Surakh	27	45	41	57	22	3	I	Dunite, serpentinite, and pyroxenite/ Upper Cretaceous	Magmatic	Laramide	Lens
33	Shahin 1	27	22	38	57	27	30		Ophiolite complex/Upper Paleozoic–Triassic	Magmatic	Laramide	Lens
34	Shahin 2	27	22	46	57	24	18		Ophiolite complex/Upper Paleozoic–Triassic	Magmatic	Early Cimmerian	Lens
35	Shahin 3	27	23	27	57	24	27		Ophiolite complex/Upper Paleozoic–Triassic	Magmatic	Early Cimmerian	Lens
36	Shahin 4	27	23	18	57	22	51		Ophiolite complex/Upper Paleozoic–Triassic	Magmatic	Early Cimmerian	Lens
37	South West Of Khosro Abad	27	18	2	57	21	31		Ophiolite complex/Upper Paleozoic–Triassic	Magmatic	Early Cimmerian	Lens
38	North of Kuh-e-Zendan	27	26	2	57	3	0		Ophiolite complex/Upper Paleozoic–Triassic	Magmatic	Early Cimmerian	Lens
39	Abdasht	28	23	0	56	45	0	M	Dunite, serpentinite, and pyroxenite/ Upper Cretaceous	Magmatic	Early Cimmerian	Lens
40	bagh Borj	28	33	14	57	3	3	I	Dunite, serpentinite, and pyroxenite/ Upper Cretaceous	Magmatic	Laramide	Lens
41	Esfandagheh	28	22	0	56	44	0	M	Dunite, serpentinite, and pyroxenite/ Upper Cretaceous	Magmatic	Laramide	Lens
42	ShahGhabel	28	26	57	54	4	36	S	Dunite, serpentinite, and pyroxenite/ Upper Cretaceous	Magmatic	Laramide	Lens
43	SheikhAli	28	13	28	56	49	13	S	Dunite, serpentinite, and pyroxenite/ Upper Cretaceous	Magmatic	Laramide	Lens
44	Soluo	28	37	45	57	3	4	S	Dunite, serpentinite, and pyroxenite/ Upper Cretaceous	Magmatic	Laramide	Lens
45	Bagh-e-Baluch	29	25	0	54	14	10	S	Dunite, serpentinite, and pyroxenite/ Upper Cretaceous	Magmatic	Laramide	Lens

(continued)

(continued)

No.	Name	Lat D°	Lat M	Lat S″	Long D°	Long M	Long S″	Size	Host rock/age	Genetic/ mineralization	Orogenic phases	Morphology	Elem para
46	Barandaz	29	30	0	54	5	0	I	Dunite, serpentinite, and pyroxenite/ Upper Cretaceous	Magmatic	Laramide	Lens	
47	Chah Jangal	29	33	14	60	3	0	I	Dunite, serpentinite, and pyroxenite/ Upper Cretaceous	Magmatic	Laramide	Lens	
48	Chenar-e-Mohseni	29	25	40	54	14	10	I	Dunite, serpentinite, and pyroxenite/ Upper Cretaceous	Magmatic	Laramide	Lens	
49	Darvazeh	29	24	40	54	7	45	I	Dunite, serpentinite, and pyroxenite/ Upper Cretaceous	Magmatic	Laramide	Lens	
50	East of Chah Bereshi	29	36	51	60	12	30	I	Dunite, serpentinite, and pyroxenite/ Upper Cretaceous	Magmatic	Laramide	Lens	
51	Golchah	29	22	20	54	9	20	I	Dunite, serpentinite, and pyroxenite/ Upper Cretaceous	Magmatic	Laramide	Lens	
52	Khajeh Jamal	29	25	30	54	7	0	I	Dunite, serpentinite, and pyroxenite/ Upper Cretaceous	Magmatic	Laramide	Lens	
53	KhajehJamali	29	57	30	53	45	30	I	Dunite, serpentinite, and pyroxenite/ Upper Cretaceous	Magmatic	Laramide	Lens	
54	Kuh-e-khul	29	36	14	59	59	42	I	Dunite, serpentinite, and pyroxenite/ Upper Cretaceous	Magmatic	Laramide	Lens	
55	Kuh-e-Shur Shirin 1	29	49	4	60	9	6	S	Dunite, serpentinite, and pyroxenite/ Upper Cretaceous	Magmatic	Laramide	Lens	
56	Kuh-e-Shur Shirin 2	29	45	51	60	9	23	I	Dunite, serpentinite, and pyroxenite/ Upper Cretaceous	Magmatic	Laramide	Lens	
57	Mohammad Saleh	29	16	5	54	11	15	I	Dunite, serpentinite, and pyroxenite/ Upper Cretaceous	Magmatic	Laramide	Lens	
58	Shuvin	29	42	20	60	7	35	I	Dunite, serpentinite, and pyroxenite/ Upper Cretaceous	Magmatic	Laramide	Lens	
59	Siah Jagun	29	42	20	60	10	18	S	Dunite, serpentinite, and pyroxenite/ Upper Cretaceous	Magmatic	Laramide	Lens	
60	Tang-e-Hana	29	24	30	54	11	0	S	Dunite, serpentinite, and pyroxenite/ Upper Cretaceous	Magmatic	Laramide	Lens	

61	tang-e-Hasan	29	24	30	54	10	0	S	Dunite, serpentinite, and pyroxenite/ Upper Cretaceous	Magmatic	Laramide	Lens
62	Aludeh Asagi 1	30	37	18	60	12	19	S	Dunite, serpentinite, and pyroxenite/ Upper Cretaceous	Magmatic	Laramide	Lens
63	Aludeh Asagi 2	30	37	18	60	13	52	I	Dunite, serpentinite, and pyroxenite/ Upper Cretaceous	Magmatic	Laramide	Lens
64	Aludeh Asagi 3	30	37	15	60	10	0	I	Dunite, serpentinite, and pyroxenite/ Upper Cretaceous	Magmatic	Laramide	Lens
65	Hipoo	30	39	0	60	10	49	I	Dunite, serpentinite, and pyroxenite/ Upper Cretaceous	Magmatic	Laramide	Lens
66	Kuh-e-Yalan	30	47	48	59	52	9	I	Dunite, serpentinite, and pyroxenite/ Upper Cretaceous	Magmatic	Laramide	Lens
67	Mordar Kuh	30	50	36	60	0	33	S	Dunite, serpentinite, and pyroxenite/ Upper Cretaceous	Magmatic	Laramide	Lens
68	Nehbandan	31	42	20	60	11	57	I	Dunite, serpentinite, and pyroxenite/ Upper Cretaceous	Magmatic	Laramide	Lens
69	Chah-e-Panjsar	32	27	47	60	17	30	I	Gabbro, dunite, serpentinite, and pyroxenite/Upper Cretaceous	Magmatic	Laramide	Lens
70	Mazraeh	32	16	45	60	4	30	I	Dunite, serpentinite, and pyroxenite/ Upper Cretaceous	Magmatic	Laramide	Lens
71	Chah Palang	33	6	29	53	4	18	I	Dunite, serpentinite, and pyroxenite/ Upper Cretaceous	Magmatic	Laramide	Lens
72	Espar	33	8	39	53	3	47	I	Dunite, serpentinite, and pyroxenite/ Upper Cretaceous	Magmatic	Laramide	Lens
73	Hoseyn Abad 1	33	8	55	53	2	52	I	Dunite, serpentinite, and pyroxenite/ Upper Cretaceous	Magmatic	Laramide	Lens
74	Hoseyn Abad 2	33	7	50	53	2	40	I	Dunite, serpentinite, and pyroxenite/ Upper Cretaceous	Magmatic	Laramide	Lens
75	Soheil-e-Pakuh	33	9	43	53	0	57	S	Dunite, serpentinite, and pyroxenite/ Upper Cretaceous	Magmatic	Laramide	Lens
76	Sorajeh	33	7	36	53	4	17	S	Dunite, serpentinite, and pyroxenite/ Upper Cretaceous	Magmatic	Laramide	Lens

(continued)

(continued)

No.	Name	Lat D°	Lat M	Lat S″	Long D°	Long M	Long S″	Size	Host rock/age	Genetic/ mineralization	Orogenic phases	Morphology	Elem para
77	Surand (Shurab)	33	19	36	60	4	50	I	Dunite, serpentinite, and pyroxenite/ Upper Cretaceous	Magmatic	Laramide	Lens	
78	Zillan (South of Zeidan)	33	10	59	60	9	34	S	Dunite, serpentinite, and pyroxenite/ Upper Cretaceous	Magmatic	Laramide	Lens	
79	Aghsu	35	45	24	59	14	40	S	Dunite, serpentinite, and pyroxenite/ Upper Cretaceous	Magmatic	Laramide	Lens	
80	Ajisur1	35	43	38	59	20	0	I	Dunite, serpentinite, and pyroxenite/ Upper Cretaceous	Magmatic	Laramide	Lens	
81	Ajisur2	35	41	7	59	19	30	I	Dunite, serpentinite, and pyroxenite/ Upper Cretaceous	Magmatic	Laramide	Lens	
82	Ajisur3	35	41	7	59	22	30	I	Dunite, serpentinite, and pyroxenite/ Upper Cretaceous	Magmatic	Laramide	Lens	
83	Amolgu	35	53	48	58	32	52	I	Dunite, serpentinite, and pyroxenite/ Upper Cretaceous	Magmatic	Laramide	Lens	
84	Ghoflavaren	35	40	24	59	6	57	I	Dunite, serpentinite, and pyroxenite/ Upper Cretaceous	Magmatic	Laramide	Lens	
85	Godar-e-Kutaku 1	35	44	3	59	27	0	I	Dunite, serpentinite, and pyroxenite/ Upper Cretaceous	Magmatic	Laramide	Lens	
86	Godar-e-Kutaku 2	35	47	13	59	26	29	I	Dunite, serpentinite, and pyroxenite/ Upper Cretaceous	Magmatic	Laramide	Lens	
87	Kajulingu	35	46	34	59	32	45	I	Dunite, serpentinite, and pyroxenite/ Upper Cretaceous	Magmatic	Laramide	Lens	
88	Kushk Kar 1	35	43	38	59	20	0	I	Dunite, serpentinite, and pyroxenite/ Upper Cretaceous	Magmatic	Laramide	Lens	
89	Kushk Kar 2	35	37	27	59	0	0	I	Dunite, serpentinite, and pyroxenite/ Upper Cretaceous	Magmatic	Laramide	Lens	
90	Mohammad Hasan	35	39	12	59	19	0	I	Dunite, serpentinite, and pyroxenite/ Upper Cretaceous	Magmatic	Laramide	Lens	
91	Nahagsat	35	47	12	59	15	0	I	Dunite, serpentinite, and pyroxenite/ Upper Cretaceous	Magmatic	Laramide	Lens	

92	Robat Sefid	35	24	44	59	58	30	I	Dunite, serpentinite, and pyroxenite/ Upper Cretaceous	Magmatic	Laramide	Lens
93	Sarab-e-Robat Sefid	35	46	22	59	21	30	I	Dunite, serpentinite, and pyroxenite/ Upper Cretaceous	Magmatic	Laramide	Lens
94	Sirsar1	35	26	54	59	26	39	I	Dunite, serpentinite, and pyroxenite/ Upper Cretaceous	Magmatic	Laramide	Lens
95	Sirsar2	35	28	7	59	29	57	I	Dunite, serpentinite, and pyroxenite/ Upper Cretaceous	Magmatic	Laramide	Lens
96	Sirsar3	35	27	47	59	29	17	I	Dunite, serpentinite, and pyroxenite/ Upper Cretaceous	Magmatic	Laramide	Lens
97	Tak Cheshmeh Mohammad Hoseyn	35	24	41	59	18	0	I	Dunite, serpentinite, and pyroxenite/ Upper Cretaceous	Magmatic	Laramide	Lens
98	Tang-e-Hasan Ali Shah	35	39	44	59	16	15	I	Dunite, serpentinite, and pyroxenite/ Upper Cretaceous	Magmatic	Laramide	Lens
99	Ziarat	35	43	38	59	27	20	I	Dunite, serpentinite, and pyroxenite/ Upper Cretaceous	Magmatic	Laramide	Lens
100	Agalgu	36	17	36	58	6	30	I	Ophiolite complex/Upper Cretaceous	Magmatic	Laramide	Lens
101	Avaz1	36	20	36	57	59	0	I	Dunite, serpentinite, and pyroxenite/ Upper Cretaceous	Magmatic	Laramide	Lens
102	Avaz2	36	18	11	57	58	30	I	Dunite, serpentinite, and pyroxenite/ Upper Cretaceous	Magmatic	Laramide	Lens
103	Bujandar	36	30	42	57	6	30	I	Dunite, serpentinite, and pyroxenite/ Upper Cretaceous	Magmatic	Laramide	Lens
104	Daran-e-Akvand	36	19	24	57	40	15	I	Dunite, serpentinite, and pyroxenite/ Upper Cretaceous	Magmatic	Laramide	Lens
105	Darreh Bid	36	23	41	57	34	40	S	Dunite, serpentinite, and pyroxenite/ Upper Cretaceous	Magmatic	Laramide	Lens
106	Darreh Parand	36	23	9	57	37	40	S	Dunite, serpentinite, and pyroxenite/ Upper Cretaceous	Magmatic	Laramide	Lens
107	Darvish1	36	18	14	58	5	0	I	Dunite, serpentinite, and pyroxenite/ Upper Cretaceous	Magmatic	Laramide	Lens

(continued)

(continued)

No.	Name	Lat D°	Lat M	Lat S″	Long D°	Long M	Long S″	Size	Host rock/age	Genetic/ mineralization	Orogenic phases	Morphology	Elem para
108	Darvish2	36	16	37	58	10	0	I	Dunite, serpentinite, and pyroxenite/ Upper Cretaceous	Magmatic	Laramide	Lens	
109	Dejangal	36	22	1	57	38	15	I	Dunite, serpentinite, and pyroxenite/ Upper Cretaceous	Magmatic	Laramide	Lens	
110	Denow1	36	19	23	57	49	0	I	Dunite, serpentinite, and pyroxenite/ Upper Cretaceous	Magmatic	Laramide	Lens	
111	Denow2	36	25	49	57	29	9	I	Dunite, serpentinite, and pyroxenite/ Upper Cretaceous	Magmatic	Laramide	Lens	
112	Forumad	36	32	50	56	50	24	S	Dunite, serpentinite, and pyroxenite/ Upper Cretaceous	Magmatic	Laramide	Lens	
113	Gaft	36	33	42	56	49	18	S	Dunite, serpentinite, and pyroxenite/ Upper Cretaceous	Magmatic	Laramide	Lens	
114	Gandevir	36	35	48	56	38	24	I	Dunite, serpentinite, and pyroxenite/ Upper Cretaceous	Magmatic	Laramide	Lens	
115	Gazavir	36	36	14	56	38	0	I	Dunite, serpentinite, and pyroxenite/ Upper Cretaceous	Magmatic	Laramide	Lens	
116	Gharyeh NurAbad	36	17	42	58	5	10	S	Dunite, serpentinite, and pyroxenite/ Upper Cretaceous	Magmatic	Laramide	Lens	
117	Ghasem Darvish1	36	17	44	58	15	0	I	Dunite, serpentinite, and pyroxenite/ Upper Cretaceous	Magmatic	Laramide	Lens	
118	Ghasem Darvish2	36	17	27	58	15	8	I	Dunite, serpentinite, and pyroxenite/ Upper Cretaceous	Magmatic	Laramide	Lens	
119	Golbini1	36	33	9	56	32	21	I	Dunite, serpentinite, and pyroxenite/ Upper Cretaceous	Magmatic	Laramide	Lens	
120	Golbini2	36	32	19	56	32	37	I	Dunite, serpentinite, and pyroxenite/ Upper Cretaceous	Magmatic	Laramide	Lens	
121	Hoseyn Abad	36	17	56	58	18	46	I	Dunite, serpentinite, and pyroxenite/ Upper Cretaceous	Magmatic	Laramide	Lens	
122	Hulangu Shir	36	25	15	57	8	0	I	Dunite, serpentinite, and pyroxenite/ Upper Cretaceous	Magmatic	Laramide	Lens	

123	Jangal	36	17	33	58	15	35	I	Dunite, serpentinite, and pyroxenite/ Upper Cretaceous	Magmatic	Laramide	Lens
124	Kalat-e-Sabz1	36	33	12	56	31	0	I	Dunite, serpentinite, and pyroxenite/ Upper Cretaceous	Magmatic	Laramide	Lens
125	Kalat-e-Sabz2	36	35	21	56	34	48	I	Dunite, serpentinite, and pyroxenite/ Upper Cretaceous	Magmatic	Laramide	Lens
126	Khurab1	36	21	48	57	49	30	I	Dunite, serpentinite, and pyroxenite/ Upper Cretaceous	Magmatic	Laramide	Lens
127	Khurab2	36	24	3	57	50	18	I	Dunite, serpentinite, and pyroxenite/ Upper Cretaceous	Magmatic	Laramide	Lens
128	Kuh-e-Sive	36	21	17	57	41	30	I	Dunite, serpentinite, and pyroxenite/ Upper Cretaceous	Magmatic	Laramide	Lens
129	Kuh-e-Surj	36	13	8	58	25	0	I	Dunite, serpentinite, and pyroxenite/ Upper Cretaceous	Magmatic	Laramide	Lens
130	Miandasht	36	26	57	56	0	0	I	Dunite, serpentinite, and pyroxenite/ Upper Cretaceous	Magmatic	Laramide	Lens
131	Mir Hoseyn	36	19	24	57	40	15	I	Dunite, serpentinite, and pyroxenite/ Upper Cretaceous	Magmatic	Laramide	Lens
132	Mir Mahmud	36	31	36	56	41	25	S	Dunite, serpentinite, and pyroxenite/ Upper Cretaceous	Magmatic	Laramide	Lens
133	Moghestan 2	36	31	20	57	10	53	I	Dunite, serpentinite, and pyroxenite/ Upper Cretaceous	Magmatic	Laramide	Lens
134	Moghestan1	36	32	53	56	28	30	I	Dunite, serpentinite, and pyroxenite/ Upper Cretaceous	Magmatic	Laramide	Lens
135	Morvarid Dor	36	35	50	56	46	0	I	Dunite, serpentinite, and pyroxenite/ Upper Cretaceous	Magmatic	Laramide	Lens
136	MourAli Abad	36	16	36	59	32	37	I	Ophiolite/Upper Paleozoic	Magmatic	Late Cimmerian	Lens
137	Mur1	36	28	58	56	50	0	I	Dunite, serpentinite, and pyroxenite/ Upper Cretaceous	Magmatic	Laramide	Lens
138	Mur2	36	33	21	57	5	22	I	Dunite, serpentinite, and pyroxenite/ Upper Cretaceous	Magmatic	Laramide	Lens

(continued)

(continued)

No.	Name	Lat D°	Lat M	Lat S″	Long D°	Long M	Long S″	Size	Host rock/age	Genetic/ mineralization	Orogenic phases	Morphology	Elem para
139	Najm Abad	36	5	38	57	59	6	I	Dunite, serpentinite, and pyroxenite/ Upper Cretaceous	Magmatic	Laramide	Lens	
140	Nezam Abad	36	17	26	58	1	0	I	Dunite, serpentinite, and pyroxenite/ Upper Cretaceous	Magmatic	Laramide	Lens	
141	Rak-e-Kamis	36	27	18	57	5	0	I	Dunite, serpentinite, and pyroxenite/ Upper Cretaceous	Magmatic	Laramide	Lens	
142	Sandana	36	27	0	57	0	15	I	Dunite, serpentinite, and pyroxenite/ Upper Cretaceous	Magmatic	Laramide	Lens	
143	Sarvar-e-Sabzevar	36	33	42	56	42	44	S	Dunite, serpentinite, and pyroxenite/ Upper Cretaceous	Magmatic	Laramide	Lens	
144	Seltesh Abad	36	33	1	56	43	30	I	Dunite, serpentinite, and pyroxenite/ Upper Cretaceous	Magmatic	Laramide	Lens	
145	Shuryab	36	21	36	57	48	38	I	Dunite, serpentinite, and pyroxenite/ Upper Cretaceous	Magmatic	Laramide	Lens	
146	Shuryab1	36	15	21	58	23	23	I	Dunite, serpentinite, and pyroxenite/ Upper Cretaceous	Magmatic	Laramide	Lens	
147	Shuryab2	36	13	44	58	22	0	I	Dunite, serpentinite, and pyroxenite/ Upper Cretaceous	Magmatic	Laramide	Lens	
148	SoodKhar	36	23	27	57	10	57	I	Dunite, serpentinite, and pyroxenite/ Upper Cretaceous	Magmatic	Laramide	Lens	
149	Sukhteh Dar	36	50	39	56	34	48	I	Dunite, serpentinite, and pyroxenite/ Upper Cretaceous	Magmatic	Laramide	Lens	
150	Sur-e-Ali Abad	36	30	49	56	52	0	I	Dunite, serpentinite, and pyroxenite/ Upper Cretaceous	Magmatic	Laramide	Lens	
151	Tak Cheshm	36	13	22	58	17	15	I	Dunite, serpentinite, and pyroxenite/ Upper Cretaceous	Magmatic	Laramide	Lens	
152	Tak Khaneh Hoseyn 1	36	11	21	58	14	30	I	Dunite, serpentinite, and pyroxenite/ Upper Cretaceous	Magmatic	Laramide	Lens	

153	Tak Khaneh Hoseyn 2	36	12	10	58	11	27	I	Dunite, serpentinite, and pyroxenite/ Upper Cretaceous	Magmatic	Laramide	Lens
154	Zarvand	36	15	45	58	23	0	I	Dunite, serpentinite, and pyroxenite/ Upper Cretaceous	Magmatic	Laramide	Lens
155	Tachakan	37	3	13	57	51	36	I	Dunite, serpentinite, and pyroxenite/ Upper Cretaceous	Magmatic	Laramide	Lens
156	Gheshlagh Khoy	38	39	30	44	47	34	S	Dunite, serpentinite, and pyroxenite/ Upper Cretaceous	Magmatic	Laramide	Lens
157	Abkuh							S	Ophiolite complex/Upper Paleozoic–Triassic	Magmatic	Hercynian	Lens
158	Anbar-e-Forumad							S	Dunite, serpentinite, and pyroxenite/ Upper Cretaceous	Magmatic	Late Cimmerian	Lens
159	BaghChenar							S	Dunite, Serpentinite, and pyroxenite/ Upper Cretaceous	Magmatic	Laramide	Lens
160	Barantin							S	Dunite, serpentinite, and pyroxenite/ Upper Cretaceous	Magmatic	Laramide	Lens
161	Chah Gardun							S	Dunite, serpentinite, and pyroxenite/ Upper Cretaceous	Magmatic	Laramide	Lens
162	Chah Zar							S	Dunite, serpentinite, and pyroxenite/ Upper Cretaceous	Magmatic	Laramide	Lens
163	Chahdust							I	Dunite, serpentinite, and pyroxenite/ Upper Cretaceous	Magmatic	Laramide	Lens
164	Cheshmehbid							S	Dunite, serpentinite, and pyroxenite/ Upper Cretaceous	Magmatic	Laramide	Lens
165	Dahaneh Shir							S	Dunite, serpentinite, and pyroxenite/ Upper Cretaceous	Magmatic	Laramide	Lens
166	DehBid-e-Sirjan							S	Dunite, serpentinite, and pyroxenite/ Upper Cretaceous	Magmatic	Laramide	Lens
167	Gharyeh Bang							S	Dunite, serpentinite, and pyroxenite/ Upper Cretaceous	Magmatic	Laramide	Lens

(continued)

(continued)

No.	Name	Lat D°	Lat M	Lat S″	Long D°	Long M	Long S″	Size	Host rock/age	Genetic/mineralization	Orogenic phases	Morphology	Elem para
168	Godar-e-Kabutary-e-DolatAbad							S	Dunite, serpentinite, and pyroxenite/Upper Cretaceous	Magmatic	Laramide	Lens	
169	Jamp							I	Dunite, serpentinite, and pyroxenite/Upper Cretaceous	Magmatic	Laramide	Lens	
170	Join							S	Dunite, serpentinite, and pyroxenite/Upper Cretaceous	Magmatic	Laramide	Lens	
171	Kalateh Sabz va Garmeh							S	Dunite, serpentinite, and pyroxenite/Upper Cretaceous	Magmatic	Laramide	Lens	
172	Khash							S	Dunite, serpentinite, and pyroxenite/Upper Cretaceous	Magmatic	Laramide	Lens	
173	Kheir Abad							S	Dunite, serpentinite, and pyroxenite/Upper Cretaceous	Magmatic	Laramide	Lens	
174	Kuh-e-Shah							S	Dunite, serpentinite, and pyroxenite/Upper Cretaceous	Magmatic	Laramide	Lens	
175	Mehr-e-Sabzevar							S	Dunite, serpentinite, and pyroxenite/Upper Cretaceous	Magmatic	Laramide	Lens	
176	Miand ehy Fariman							S	Ophiolite complex/Upper Paleozoic–Triassic	Magmatic	Late Cimmerian	Lens	
177	Mokhtar Abad							S	Dunite, serpentinite, and pyroxenite/Upper Cretaceous	Magmatic	Laramide	Lens	
178	Nipeyk-e-Sharghi							S	Dunite, serpentinite, and pyroxenite/Upper Cretaceous	Magmatic	Laramide	Lens	
179	North of Abbas Abad							S	Dunite, serpentinite, and pyroxenite/Upper Cretaceous	Magmatic	Laramide	Lens	
180	Nosrat Abad							S	Dunite, serpentinite, and pyroxenite/Upper Cretaceous	Magmatic	Laramide	Lens	
181	Nur Abad							S	Dunite, serpentinite, and pyroxenite/Upper Cretaceous	Magmatic	Laramide	Lens	
182	SangSefid							S	Dunite, serpentinite, and pyroxenite/Upper Cretaceous	Magmatic	Laramide	Lens	

183	Sarkhun	S	Dunite, serpentinite, and pyroxenite/ Upper Cretaceous	Magmatic	Laramide	Lens
184	Sefidab	S	Dunite, serpentinite, and pyroxenite/ Upper Cretaceous	Magmatic	Laramide	Lens
185	Shahrud `s Chromite	S	Dunite, serpentinite, and pyroxenite/ Upper Cretaceous	Magmatic	Laramide	Lens
186	Zarijak	S	Dunite, serpentinite, and pyroxenite/ Upper Cretaceous	Magmatic	Laramide	Lens
187	Ziv Fariman	S	Ophiolite complex/Upper Paleozoic–Triassic	Magmatic	Late Cimmerian	Lens

Complete list of wolframite mineral deposits and indications of Iran along with their detailed specifications

No.	Name	Lat D°	Lat M	Lat S″	Long D°	Long M	Long S″	Size	Host rock/age	Genetic/ mineralization	Orogenic phases	Morphology	Elem para
1	Chah Kalab	31	58	0	59	31	20	M	Granite/Upper Jurassic; marble and schist/ Jurassic	Skarn	Late Cimmerian	Vein	Cu, Pb, Zn
2	Southern Chahpalang	32	57	17	54	11	46	S	Granite/Post-Jurassic (Eocene); shale, sandstone/Triassic–Jurassic	Hydrothermal	Late Cimmerian	Vein, lens	Cu, Ni, Au, Bi
3	Bamsar	33	42	33	49	9	47	I	Phyllite and granodiorite and limestone/ Jurassic	Skarn	Late Cimmerian	Disseminated	Sn
4	Nazem Abad	33	40	0	49	17	0	S	Quartz diorite, schist, hornfels/Cretaceous	Hydrothermal	Laramide	Vein	Sn

Complete list of copper mineral deposits and indications of Iran along with their detailed specifications

No.	Name	Lat D°	Lat M	Lat S″	Long D°	Long M	Long S″	Size	Host rock/age	Genetic/ mineralization	Orogenic phases	Morphology	Elem para
1	Ab Bahry	29	24	0	56	35	0	S	Sedimentary rocks and conglomerate/ Upper Eocene	Hydrothermal	Post-Pyrenean	Vein	
2	Abad-e-pabagh	30	24	0	54	59	40	I	Diorite and andesite/Paleogene	Hydrothermal	Post-Pyrenean	Vein	
3	Abbas Abad (Narbolaghi)	35	8	50	50	32	0	S	Andesite and tuff/Eocene	Hydrothermal	Post-Pyrenean	Vein	
4	Abbas Abad (Zanjan)	36	28	10	49	9	10	S	Volcanic rocks/Eocene	Hydrothermal	Post-Pyrenean	Vein	Au
5	Abbas Abad-e-Aynalu	38	53	0	46	50	30	S	Granite/Upper Eocene–Lower Oligocene; volcanic rocks/Eocene; limestone/Upper Cretaceous	Skarn	Post-Pyrenean	Vein	Au, Pb
6	Abdar	30	30	40	54	56	0	I	Porphyritic intrusion/Oligo-Miocene; porphyrite andesite/Eocene	Porphyry	Post-Pyrenean	Vein	Au
7	Abtalkhun	29	48	55	56	4	31	S	Andesite, basaltic andesite, and pyroclastic/Eocene	Hydrothermal	Post-Pyrenean	Vein	
8	Abyaneh	33	31	0	51	31	0	S	Volcanic rocks/Mesozoic	Hydrothermal	Post-Pyrenean	Vein	
9	Adia bagh	30	24	0	54	59	40	I	Volcanic rocks/Eocene	Hydrothermal	Post-Pyrenean	Vein	
10	Afin	33	34	17	59	45	48	I	Granite/Tertiary; limestone/Cretaceous	Skarn	Post-Pyrenean	Vein	
11	Agh Darreh	36	30	40	49	10	0	I	Andesite and tuff/Eocene	Hydrothermal	Post-Pyrenean	Vein	
12	Agha baba sang	38	35	32	46	46	55	I	Andesite/Eocene	Hydrothermal	Post-Pyrenean	Vein	Au, Ag
13	Aghamira	38	49	30	46	39	0	I	Granodiorite/Upper Eocene–Lower Oligocene; porphyry andesite and dacite/Eocene; limestone/Upper Cretaceous	Hydrothermal	Post-Pyrenean	Vein	Au
14	Akhavaniyeh	35	38	0	50	40	0	I	Andesite and tuff/Eocene	Hydrothermal	Post-Pyrenean	Vein	
15	Ali Abad	36	26	0	49	11	30	I	Andesite and tuff/Eocene	Hydrothermal	Post-Pyrenean	Vein	
16	Ali Abad	31	39	0	53	51	0	L	Porphyry granite/Eocene; limestone and shale/Jurassic	Hydrothermal	Post-Pyrenean	Vein	
17	Ali Bolagh	36	27	20	49	10	25	I	Granodiorite/Upper Eocene–Lower Oligocene	Hydrothermal	Post-Pyrenean	Vein	

(continued)

(continued)

No.	Name	Lat D°	Lat M	Lat S″	Long D°	Long M	Long S″	Size	Host rock/age	Genetic/ mineralization	Orogenic phases	Morphology	Elem para
18	Ali Javad	38	39	5	46	55	33	S	Monzonite/Oligo-Miocene; andesite and basalt/Quaternary	Hydrothermal	Post-Pyrenean	Vein	Au
19	Allah Abad	29	39	50	56	42	20	I	Microdiorite/Oligo-Miocene; volcanic rocks and pyroclastic/Eocene	Hydrothermal	Post-Pyrenean	Vein	
20	Alvand	36	18	46	49	10	50	I	Granite and granodiorite/Upper Eocene–Lower Oligocene	Hydrothermal	Post-Pyrenean	Vein	Au
21	Amirabad	30	2	0	56	7	0	I	Andesite/Eocene	Hydrothermal	Post-Pyrenean	Vein	
22	Anar	30	34	52	55	3	41	I	Limestone and sandstone/Jurassic	Hydrothermal	Post-Pyrenean	Vein	
23	Andab Jadid	38	20	0	47	14	0	I	Granite/Eocene–Oligocene; tuff and limestone/Cretaceous	Skarn	Post-Pyrenean	Vein	Fe, Mo
24	Anjel	33	34	55	59	8	41	I	Andesite/Eocene	Hydrothermal	Post-Pyrenean	Vein	
25	Anjerd-e-olya	38	40	20	46	55	12	I	Granite/Oligocene; andesite/Eocene; limestone/Cretaceous	Skarn	Post-Pyrenean	Vein	Mo, Pb-Zn, Au, Ag
26	Ardiz	29	58	0	55	48	0	I	Granite/Oligo-Miocene; rhyolite and dacite/Eocene	Hydrothermal	Post-Pyrenean	Vein	
27	Arousan	33	42	17	52	57	26	I	Epeirogenic carbonate deposits (sandstone, conglomerate, and limestone)/Upper Cretaceous	Epithermal	Post-Pyrenean	Vein	
28	Arzandeh	34	46	40	60	14	2	I	Rhyolite and andesite lava/Eocene	Hydrothermal	Post-Pyrenean	Vein	
29	Arzandeh	35	1	0	60	1	0	I	Breccia, andesite, and sandstone/Upper Cretaceous	Hydrothermal	Post-Pyrenean	Vein	
30	Asheghlu	34	54	0	49	37	0	S	Rhyolite and dacite/Eocene	Hydrothermal	Post-Pyrenean	Vein	
31	Asiadive	36	26	40	56	28	0	S	Andesite/Eocene	Hydrothermal	Post-Pyrenean	Vein	
32	Astamal	38	42	0	46	24	30	S	Porphyritic dacite, andesite, and acidic tuff/Eocene; limestone/Cretaceous	Porphyry	Post-Pyrenean	Vein	Mo, Pb-Zn
33	Astamal cheshmeh ghan	38	44	18	46	23	48	I	Granite/Upper Oligocene–Lower Miocene; volcanic rocks, porphyry andesite, limestone/Eocene	Skarn	Post-Pyrenean	Vein	

34	Avan	38	45	47	46	23	11	I	Diorite/Upper Eocene–Lower Oligocene; limestone/Cretaceous	Skarn	Post-Pyrenean	Vein	
35	Aveh	36	18	53	50	36	4	I	Basalt and Andesite/Eocene	Hydrothermal	Post-Pyrenean	Vein	
36	Aveh Darreh	36	22	0	50	38	0	I	Crystalline limestone and mica-schist/ Jurassic	Hydrothermal	Post-Pyrenean	Vein	
37	Aynalu	38	52	30	46	49	0	I	Intrusion/Upper Eocene–Lower Oligocene; limestone and andesite/ Cretaceous	Hydrothermal	Post-Pyrenean	Vein	Mo
38	Bab Konuj	29	18	0	57	15	30	I	Granite/Oligo-Miocene; volcanic rocks/ Eocene	Porphyry	Post-Pyrenean	Vein	
39	Bab Nam	29	7	0	57	20	25	I	Tonalite and quartz diorite/Oligo-Miocene; volcanics/Eocene	Porphyry	Post-Pyrenean	Vein	
40	Bad Mish	28	32	0	58	24	0	I	Granite/Oligo-Miocene; limestone and volcanics/Upper Cretaceous	Hydrothermal	Post-Pyrenean	Vein	
41	Badamu	30	20	0	56	45	20	I	Basic dyke and volcanic rocks/ Paleogene	Hydrothermal	Post-Pyrenean	Vein	
42	Bagh Borji 1	28	32	52	57	1	59	S	Granite/Oligo-Miocene; limestone and volcanics/Upper Cretaceous	Hydrothermal	Post-Pyrenean	Vein	
43	Bagh Borji 2	28	33	41	57	0	55	I	Granite/Oligo-Miocene; dacite and rhyolite and andesite/Eocene	Hydrothermal	Post-Pyrenean	Vein	
44	Bagh Chamak	29	11	27	58	16	51	I	Dacite and rhyolite and pyroclastic/ Eocene	Hydrothermal	Post-Pyrenean	Vein	
45	Bagh Ghorogh	33	35	32	53	48	30	S	Andesite/Eocene	Epithermal	Post-Pyrenean	Vein	
46	Bagh khoshk	29	50	2	56	59	41	L	Tonalite and quartz diorite and diorite/ Oligo-Miocene; andesite and pyroclastic/Eocene	Hydrothermal	Post-Pyrenean	Vein	
47	Bagh-e-Alu	36	30	0	56	0	0	I	Andesite and sandstone and madstone/ Eocene	Hydrothermal	Post-Pyrenean	Vein	
48	Bagheraei	29	28	0	57	12	20	I	Granite/Oligo-Miocene; dacite and rhyolite/Eocene	Hydrothermal	Post-Pyrenean	Vein	
49	Bahr-e-Aseman	29	1	0	57	25	0	I	Andesite and pyroclastic/Eocene	Hydrothermal	Post-Pyrenean	Vein	

(continued)

(continued)

No.	Name	Lat D°	Lat M	Lat S″	Long D°	Long M	Long S″	Size	Host rock/age	Genetic/mineralization	Orogenic phases	Morphology	Elem para
50	Baluje	38	45	15	46	38	0	I	Andesite/Eocene; marbleization limestone/Cretaceous	Porphyry	Post-Pyrenean	Vein	Fe
51	Band-e-bagh	29	54	0	56	2	0	I	Microdiorite/Oligo-Miocene; volcanics and pyroclastic/Eocene	Hydrothermal	Post-Pyrenean	Vein	
52	Band-e-mamezar 1	29	53	2	55	54	19	I	Microdiorite/Oligo-Miocene; volcanics and pyroclastic/Eocene	Hydrothermal	Post-Pyrenean	Vein	Pb, Zn
53	Band-e-mamezar 2	29	54	0	55	54	9	I	Granodiorite/Oligo-Miocene; volcanics and pyroclastic/Eocene	Porphyry	Post-Pyrenean	Vein	
54	Bar Kuh	34	13	40	59	44	15	I	Porphyry lava and tuff/Eocene	Hydrothermal	Post-Pyrenean	Vein	
55	Barghzar	38	52	0	46	45	0	I	Porphyrite dacite, trachyte, and ignimbrite/Eocene–Oligocene	Hydrothermal	Post-Pyrenean	Vein	Pb-Zn
56	Barik	33	36	46	59	24	11	I	Schist, amphibolite, gneiss, and andesite/Upper Cretaceous (?)	Unknown	Post-Pyrenean	Vein	
57	Barikab	36	16	20	49	19	0	S	Andesite, sandstone, and mudstone/Eocene	Hydrothermal	Post-Pyrenean	Vein	
58	Barin	29	55	32	56	3	31	I	Dacite, rhyolite, and pyroclastic/Eocene	Hydrothermal	Post-Pyrenean	Vein	
59	Barmolk	38	35	0	46	10	0	S	Monzonite/Oligo-Miocene; andesite/Eocene	Hydrothermal	Post-Pyrenean	Vein	
60	Barsaei	36	47	9	49	47	9	I	Trachyte, trachyandesite, and breccia tuff/Oligo-Miocene	Hydrothermal	Post-Pyrenean	Vein	Zn, Pb, Mo, Ni, Co, Bi
61	Batu	35	17	0	57	19	0	S	Shale/Jurassic	Hydrothermal	Post-Pyrenean	Vein	
62	Baychebagh	36	51	59	47	17	16	S	Trachyte, trachyandesite, and breccia tuff/Oligo-Miocene	Hydrothermal	Post-Pyrenean	Vein	Zn, Pb, Mo, Ni, Co, Bi
63	Beh Abad	31	54	0	56	5	0	I	Rhyolite, dacite, and andesite/Eocene; diorite/Tertiary; limestone/Mesozoic	Hydrothermal	Post-Pyrenean	Vein	
64	Beyarjomand	36	0	16	55	53	23	I	Volcanic and pyroclastic rocks/Eocene	Hydrothermal	Post-Pyrenean	Vein	Pb
65	Bid Siah	28	47	0	58	0	0	I	Ophiolite complex/Upper Cretaceous	Unknown	Post-Pyrenean	Vein	

66	Bidak	27	53	47	57	44	51	I	Basic intrusion and metamorphic/Precambrian	Hydrothermal	Post-Pyrenean	Vein	
67	Bidkhan	29	37	0	56	31	0	S	Porphyrite quartz diorite/Oligo-Miocene; hornfels and volcanosedimentary/Eocene	Porphyry	Post-Pyrenean	Vein	
68	Bidshahi	29	40	38	56	9	20	I	Andesite and tracy andesite/Eocene; granite and granodiorite/Oligocene	Hydrothermal	Post-Pyrenean	Vein	
69	Bivarzin	36	41	0	49	35	0	I	Pyroxene andesite/Eocene	Hydrothermal	Post-Pyrenean	Vein	
70	Bizarin	38	42	0	46	47	0	I	Monzonite/Eocene; shale and limestone/Cretaceous	Skarn	Post-Pyrenean	Vein	Au
71	Bolboli	29	34	0	56	15	30	S	Diorite/Early Eocene; andesite, tuff, and limestone/Eocene	Skarn	Post-Pyrenean	Vein	
72	Bon Kuh	29	36	17	56	53	43	I	Andesite and andesitic tuff/Eocene	Porphyry	Post-Pyrenean	Vein	
73	Bondar-e-baghu	29	56	16	55	58	20	M	Porphyrite quartz diorite/Oligo-Miocene; andesite/Eocene	Porphyry	Post-Pyrenean	Vein	
74	Bondar-e-hanza	29	20	0	57	15	0	S	Porphyrite quartz diorite/Oligo-Miocene; volcanic rocks/Eocene	Porphyry(?)	Post-Pyrenean	Vein	Zn-Pb-Au
75	Boragh Band Kuh	33	24	0	50	33	13	I	Schist/Upper Precambrian	Hydrothermal	Post-Pyrenean	Vein	
76	Borjak	35	21	0	57	52	0	S	Volcanic rocks, sandstone, limestone, and schist/Eocene	Hydrothermal	Post-Pyrenean	Vein	Zn-Pb-Fe
77	Bulagh	37	20	0	48	25	30	I	Limestone, dolomite, shale/Jurassic–Triassic	Hydrothermal	Post-Pyrenean	Vein	
78	Buzan Abad 1	33	56	2	59	28	2	I	Andesite, basalt/Paleogene	Hydrothermal	Post-Pyrenean	Vein	
79	Buzan Abad 2	33	54	49	59	24	37	I	Andesite, basalt/Paleogene	Hydrothermal	Post-Pyrenean	Vein	
80	Buzan Abad 3	33	54	16	59	24	27	I	Andesite, basalt/Paleogene	Hydrothermal	Post-Pyrenean	Vein	
81	Buzan Abad 4	33	53	35	59	24	47	I	Andesite, basalt/Paleogene	Hydrothermal	Post-Pyrenean	Vein	Fe
82	Buzan Abad 5	33	52	54	59	25	26	I	Rhyolite, dacite, andesite/Paleogene	Hydrothermal	Post-Pyrenean	Vein	
83	Chah Darvish	36	18	32	56	19	53	I	Andesite and tuff/Eocene	Hydrothermal	Post-Pyrenean	Vein	Au, Ag
84	Chah Dollat	28	40	30	60	41	0	I	Diorite/Eocene; sandstone, limestone/unknown	Hydrothermal	Post-Pyrenean	Vein	

(continued)

(continued)

No.	Name	Lat D°	Lat M	Lat S″	Long D°	Long M	Long S″	Size	Host rock/age	Genetic/mineralization	Orogenic phases	Morphology	Elem para
85	Chah Farakh	35	28	39	54	18	8	I	Volcanic rocks and limestone/Eocene	Hydrothermal	Post-Pyrenean	Vein	Pb, Zn
86	Chah Farsakh	35	23	0	54	17	0	I	Andesite/Eocene	Hydrothermal	Post-Pyrenean	Vein	Pb, Zn
87	Chah Firouzeh	30	23	38	55	1	2	S	Porphyrite quartz diorite/Oligo-Miocene; andesite/Eocene	Porphyry	Post-Pyrenean	Vein	Turquoise
88	Chah Gorbeh	33	25	8	53	45	0	I	Schist, andesite, and tuff/Eocene	Hydrothermal	Post-Pyrenean	Vein	
89	Chah Jamal	32	24	15	55	9	9	I	Amphibolite and gneiss and volcanic rocks/Upper Cretaceous–Lower Tertiary	Hydrothermal	Post-Pyrenean	Vein	
90	Chah Jireh	33	58	47	52	9	34	I	Schist/Upper Precambrian	Hydrothermal	Post-Pyrenean	Vein	Pb-Zn
91	Chah Kharbozeh	33	28	51	53	47	54	I	Schist/Upper Precambrian	Hydrothermal	Post-Pyrenean	Vein	Pb-Zn-Ni
92	Chah Khatab (Nadushan)	31	50	0	53	20	0	S	Basic extrusive/Tertiary; marble and schist/Jurassic	Skarn	Post-Pyrenean	Vein	
93	Chah Mileh	33	26	0	53	48	30	I	Ultramafic rocks, schist and limestone/Upper Proterozoic–Lower Cambrian	Hydrothermal	Post-Pyrenean	Vein	Pb-Zn-Co-Ni
94	Chah raji	30	0	50	56	46	34	I	Andesite lava/Eocene	Hydrothermal	Post-Pyrenean	Vein	
95	Chah Shirin	35	17	49	54	9	1	S	Andesite/Eocene; limestone and dolomite/Cretaceous	Hydrothermal	Post-Pyrenean	Vein	
96	Chah shur	30	13	0	55	51	30	I	Porphyrite diorite/Oligo-Miocene; andesite/Eocene	Hydrothermal	Post-Pyrenean	Vein	
97	Chah Sodnow	32	16	20	55	23	41	I	Shale and sandstone/Jurassic	Hydrothermal	Post-Pyrenean	Vein	W, Au, Ni, Bi
98	Chah Toni	32	13	38	58	44	58	I	Andesite/Eocene–Oligocene	Hydrothermal	Post-Pyrenean	Vein	
99	Chahak	35	21	20	60	10	40	I	Porphyrite diorite/Paleogene	Hydrothermal	Post-Pyrenean	Vein	
100	Chahar Gonbad	29	35	30	56	11	0	L	Quartz diorite/Miocene–Pliocene; volcanic and sedimentary rocks/Eocene	Porphyry	Post-Pyrenean	Vein	
101	Chahdust	28	40	30	60	41	0	I	Porphyrite diorite/Oligo-Miocene; volcanics and pyroclastic/Eocene	Porphyry	Post-Pyrenean	Vein	

102	Chahmesi	30	24	30	55	10	0	M	Porphyrite diorite/Oligo-Miocene; porphyrite andesite and megaporphyrite tracy andesite/Eocene	Hydrothermal	Post-Pyrenean	Vein	
103	Chahpalang	32	59	0	54	11	30	S	Shale and sandstone/Jurassic	Hydrothermal	Post-Pyrenean	Vein	W, Au, Ni, Bi
104	Chamtal (Agha Ali)	38	42	0	46	10	50	I	Granodiorite/Upper Eocene–Lower Oligocene; limestone and volcanoclastic/Cretaceous	Skarn	Post-Pyrenean	Vein	
105	Chargar	36	31	0	49	10	30	I	Granite/Upper Eocene–Lower Oligocene; crystalline limestone, and Skarn/Cretaceous	Skarn	Post-Pyrenean	Vein	Fe
106	Chari	30	31	30	56	28	0	I	Sandstone, tuff, and limestone/Tertiary	Hydrothermal	Post-Pyrenean	Vein	
107	Chehel kureh	30	17	0	60	8	0	S	Sandstone, silt, shale, and limestone/Early Eocene	Hydrothermal	Post-Pyrenean	Vein	
108	Cherarlu	36	20	30	46	26	50	I	Andesite and tuff/Eocene	Hydrothermal	Post-Pyrenean	Vein	
109	Chergerd	36	26	0	49	5	0	I	Andesite and tuff/Eocene	Hydrothermal	Post-Pyrenean	Vein	
110	Cheshmeh Gaz	35	21	53	57	32	9	S	Andesite/Eocene	Hydrothermal	Post-Pyrenean	Vein	
111	Cheshmeh Khezr 1	30	18	33	55	38	59	I	Volcano sedimentary rocks/Eocene	Hydrothermal	Post-Pyrenean	Vein	
112	Cheshmeh Khezr 2	30	19	5	55	38	26	S	Volcano sedimentary rocks/Eocene	Hydrothermal	Post-Pyrenean	Vein	
113	Cheshmeh Khezr 3	30	19	43	55	37	58	I	Andesite, basalt, and trachyandesite/Upper Eocene	Hydrothermal	Post-Pyrenean	Vein	
114	Cheshmeh Mir	35	23	10	57	38	0	I	Volcanics and sedimentary/Upper Cretaceous	Hydrothermal	Post-Pyrenean	Vein	
115	Cheshmeh Sefid	30	31	20	56	28	40	I	Andesite/Eocene	Hydrothermal	Post-Pyrenean	Vein	
116	Cheshmeh Shur	29	44	9	56	34	26	I	Andesite and porphyrite lava/Eocene	Hydrothermal	Post-Pyrenean	Vein	
117	Chizeh	36	33	45	49	4	0	S	Andesite and tuff/Eocene	Hydrothermal	Post-Pyrenean	Vein	
118	Choon	35	46	1	57	23	39	I	Sandstone and shale/Cretaceous	Hydrothermal	Post-Pyrenean	Vein	
119	Chunt	35	48	44	57	39	10	S	Porphyrite andesite/Eocene	Hydrothermal	Post-Pyrenean	Vein	
120	Daghar Dagh	38	49	11	46	51	16	I	Monzonite/Oligo-Miocene; andesite and limestone/Cretaceous	Hydrothermal	Post-Pyrenean	Vein	
121	Dahaneh	36	31	25	49	7	25	I	Andesite/Eocene	Hydrothermal	Post-Pyrenean	Vein	Fe

(continued)

(continued)

No.	Name	Lat D°	Lat M	Lat S″	Long D°	Long M	Long S″	Size	Host rock/age	Genetic/mineralization	Orogenic phases	Morphology	Elem para
122	Dahaneh Bonehzar	32	53	38	55	43	40	I	Syenite, diorite, monzonite/Oligo-Miocene	Hydrothermal	Post-Pyrenean	Vein	
123	Dahaneh Siah	35	26	10	57	27	30	I	Melaphyre/Tertiary	Hydrothermal	Post-Pyrenean	Vein	
124	Daman Jala	36	24	0	56	24	0	I	Andesite/Eocene	Hydrothermal	Post-Pyrenean	Vein	
125	Damaneh	29	12	20	57	12	0	I	Porphyrite quartz diorite/Oligo-Miocene; andesite, trachyandesite, and tuff/Eocene	Porphyry	Post-Pyrenean	Vein	
126	Dar Alu	29	25	20	57	6	20	L	Porphyrite intrusion/Oligo-Miocene; agglomerate and pyroclastic/Eocene	Porphyry	Post-Pyrenean	Vein	
127	Dar Bidoo	30	29	0	55	7	30	I	Granodiorite and gabbro/Oligo-Miocene; andesite, rhyolite, and dacite/Eocene	Hydrothermal	Post-Pyrenean	Vein	
128	Dar Hamzeh	28	45	24	57	51	49	I	Porphyrite granitoid/Oligo-Miocene; andesite, trachyandesite, and dacite/Eocene	Porphyry	Post-Pyrenean	Vein	Pyrite
129	Dar khomreh	28	45	0	57	51	0	I	Granite/Oligo-Miocene; rhyolite and dacite/Eocene	Hydrothermal	Post-Pyrenean	Vein	
130	Darbini (Bidsorkh)	29	31	30	57	20	0	I	Porphyrite diorite and quartz diorite/Oligo-Miocene; andesite/Eocene	Porphyry	Post-Pyrenean	Vein	
131	Darestan	35	27	0	54	36	30	S	Granodiorite/Post-Eocene; breccia tuff and andesite/Mid-Eocene	Hydrothermal	Post-Pyrenean	Vein	
132	Darreh Mes	33	15	30	52	36	0	I	Graphite schist/Upper Precambrian–Lower Cambrian	Hydrothermal	Post-Pyrenean	Vein	Au, Mo
133	Darreh Zar	29	53	0	55	54	0	L	Quartz microdiorite/Miocene; andesite, tuff/Eocene	Porphyry	Post-Pyrenean	Vein	Au
134	Darreh Zereshk	31	32	0	53	55	0	L	Granodiorite and porphyrite diorite/Eocene–Oligocene; andesite, tuff/Eocene; limestone/Cretaceous	Porphyry, skarn	Post-Pyrenean	Vein	
135	Dasht-e-Lut	28	45	26	57	48	12	I	Rhyolite and dacite/Eocene	Hydrothermal	Post-Pyrenean	Vein	

136	Dasht-e-Ravan	28	51	0	58	3	30	I	Rhyolite and dacite and tuff/Eocene	Hydrothermal	Post-Pyrenean	Vein	
137	Dashtu	27	41	30	57	3	0	I	Phyllite and schist/Jurassic	Unknown	Post-Pyrenean	Vein	
138	Dastgerd	29	57	17	56	11	56	I	Dacite/Early Oligocene	Hydrothermal	Post-Pyrenean	Vein	
139	Davayataghy	36	29	0	49	10	20	I	Andesite and rhyolite/Oligo-Miocene; metamorphic rocks/Precambrian	Hydrothermal	Post-Pyrenean	Vein	
140	Deh Kalan	27	44	36	57	32	54	I	Marble, phyllite, and schist/Jurassic	Skarn	Post-Pyrenean	Vein	
141	Deh Siah Khan	29	59	0	55	58	0	I	Quartz monzonite and quartz diorite/ Oligo-Miocene	Porphyry	Post-Pyrenean	Vein	
142	Dehlarz	29	49	33	56	11	10	I	Granite and granodiorite/Early Oligocene; volcanosedimentary rocks/Eocene	Hydrothermal	Post-Pyrenean	Vein	
143	Dehsiahan 1	29	59	0	55	58	20	S	Quartz monzonite/Oligo-Miocene; andesite/Eocene	Hydrothermal	Post-Pyrenean	Vein	
144	Dehsiahan 2	29	59	19	56	3	11	M	Volcanics/Eocene; sandstone and marl/Cretaceous	Hydrothermal	Post-Pyrenean	Vein	
145	Delichay	35	40	0	52	32	0	I	Andesite and trachyandesite/Eocene	Hydrothermal	Post-Pyrenean	Vein	
146	Desk	28	58	0	58	6	20	I	Granite/Oligo-Miocene; rhyolite, Dacite/Eocene	Hydrothermal	Post-Pyrenean	Vein	
147	Divanehdar	36	23	0	57	43	0	I	Andesite and serpentinite rocks/Upper Cretaceous	Hydrothermal	Post-Pyrenean	Vein	Au
148	Dizehjin	36	39	30	49	5	15	S	Granodiorite/Post-Eocene; andesite/ Eocene	Hydrothermal	Post-Pyrenean	Vein	
149	Do zard akhtar	29	40	0	56	51	20	S	Pyroclastic/Eocene	Hydrothermal	Post-Pyrenean	Vein	
150	Dochahu	33	20	40	54	32	0	I	Red bed and evaporites/ Triassic–Jurassic	Hydrothermal	Post-Pyrenean	Vein	
151	Dodahaneh	36	31	25	49	7	25	I	Tuff/Eocene	Hydrothermal	Post-Pyrenean	Vein	
152	Domu	33	21	48	54	43	37	I	Porphyry andesite, tuff, sandstone/ Tertiary	Hydrothermal	Post-Pyrenean	Vein	
153	Doroh	32	18	45	60	30	5	S	Andesite and sandstone/Eocene	Hydrothermal	Post-Pyrenean	Vein	
154	Dowran	36	2	0	50	54	20	S	Pyroclastic/Eocene	Hydrothermal	Post-Pyrenean	Vein	

(continued)

(continued)

No.	Name	Lat D°	Lat M	Lat S″	Long D°	Long M	Long S″	Size	Host rock/age	Genetic/ mineralization	Orogenic phases	Morphology	Elem para
155	Durkan 1	27	10	40	57	54	32	I	Basaltic lava and limestone/Upper Cretaceous	Massive sulfide	Post-Pyrenean	Vein	
156	Durkan 2	27	11	13	57	52	7	I	Basaltic lava and limestone/Upper Cretaceous	Massive sulfide	Post-Pyrenean	Vein	
157	Durkan 3	27	12	17	57	52	7	I	Basaltic lava and limestone/Upper Cretaceous	Massive sulfide	Post-Pyrenean	Vein	
158	Durkan 4	27	12	17	57	51	13	I	Phyllite and schist/Cretaceous	Hydrothermal	Post-Pyrenean	Vein	
159	East of Bajgan	27	34	53	57	34	16	I	Limestone and dolomite/Cretaceous	Massive sulfide	Post-Pyrenean	Vein	
160	East of Gesh Miran 1	27	7	24	57	56	31	I	Ophiolite complex/Upper Cretaceous	Massive sulfide	Post-Pyrenean	Vein	
161	East of Howz Mohammad Hasan	32	21	18	58	57	46	I	Tuff/Eocene	Hydrothermal	Post-Pyrenean	Vein	
162	East of Jiroft fault	27	13	38	58	0	45	I	Limestone and dolomite/Cretaceous	Unknown	Post-Pyrenean	Vein	
163	East of Kakhk	32	35	21	58	52	34	I	Limestone and dolomite/Cretaceous	Unknown	Post-Pyrenean	Vein	
164	East of Khorram Shahi 1	28	57	32	57	40	6	I	Granite/Oligo-Miocene; rhyolite, andesite, and dacite/Eocene	Hydrothermal	Post-Pyrenean	Vein	
165	East of Kuh-e-Bidmeshk	31	56	46	59	38	24	I	Rhyolite and dacite/Tertiary	Hydrothermal	Post-Pyrenean	Vein	
166	East of Kuh-e-Bokikan	26	47	19	58	15	52	I	Granite/Mesozoic; limestone/ Cretaceous	Hydrothermal	Post-Pyrenean	Vein	
167	East of Kuh-e-Durkhan	27	8	53	57	55	45	I	Ophiolite complex/Upper Cretaceous	Unknown	Post-Pyrenean	Vein	
168	East of Kuh-e-Kalat	27	42	54	57	52	16	I	Ophiolite complex and limestone/ Upper Cretaceous	Hydrothermal	Post-Pyrenean	Vein	
169	East of Kuh-e-Madvar	30	31	45	54	57	30	I	Porphyrite diorite/Oligo-Miocene; andesite/Eocene	Porphyry	Post-Pyrenean	Vein	
170	East of Kutak	27	53	29	57	43	20	I	Limestone and dolomite/Cretaceous	Unknown	Post-Pyrenean	Vein	
171	East of Min Abad	31	37	15	60	9	51	I	Limestone, shale/Paleozoic	Unknown	Post-Pyrenean	Vein	

172	East Of Qalleh Sardar	29	24	19	58	6	34	I	Granite and granodiorite/Early Oligo-Miocene; porphyry andesite and andesibasalt/Eocene	Hydrothermal	Post-Pyrenean	Vein	
173	East of Sangoubal	37	3	16	49	6	25	I	Andesite/Tertiary	Hydrothermal	Post-Pyrenean	Vein	
174	Ekhvaniyeh	35	38	0	50	40	0	S	Andesite/Eocene	Hydrothermal	Post-Pyrenean	Vein	
175	Esmail Abad	36	30	40	49	8	20	I	Andesite/Eocene	Hydrothermal	Post-Pyrenean	Vein	
176	Estakhr-e-Abyari	35	14	50	60	21	20	I	Andesite and tuff/Eocene	Hydrothermal	Post-Pyrenean	Vein	Pb
177	Eypak	35	38	30	50	18	0	S	Lava and tuff/Tertiary	Hydrothermal	Post-Pyrenean	Vein	
178	Faryab 1	27	27	8	57	2	24		Basalt/Upper Cretaceous	Massive sulfide	Post-Pyrenean	Vein	
179	Faryab 2	27	27	49	57	2	15		Basalt/Upper Cretaceous	Massive sulfide	Post-Pyrenean	Vein	
180	Faryab 3	27	28	13	57	1	57		Granite/Eocene–Oligo-Miocene; tuff/Eocene	Hydrothermal	Post-Pyrenean	Vein	
181	Fatemeh Ali Shah	33	12	0	52	47	0	I	Granodiorite porphyry/Tertiary; sandstone and shale/Lower Jurassic; Dolomite/Triassic	Hydrothermal	Post-Pyrenean	Vein	
182	Feraydoun	32	49	30	50	58	0	I	Schist/Upper Precambrian	Hydrothermal	Post-Pyrenean	Vein	
183	Firuz Abad	36	30	0	56	36	0	S	Pyroxene andesite and volcanic breccia/Eocene	Hydrothermal	Post-Pyrenean	Vein	
184	Firuz kuh	35	52	0	52	42	0	I	Volcanic rocks/Eocene	Hydrothermal	Post-Pyrenean	Vein	
185	Gach-e-kenom	36	12	0	56	9	0	I	Basalt and andesite/Eocene; limestone, shale, and sandstone/Cretaceous	Hydrothermal	Post-Pyrenean	Vein	
186	Galeh	38	47	0	47	7	20	I	Granite and andesite/Upper Eocene–Lower Oligocene	Hydrothermal	Post-Pyrenean	Vein	Mo, W, (Au)
187	Gashulig	27	11	37	57	53	25	I	Ophiolite complex and limestone/ Upper Cretaceous	Hydrothermal	Post-Pyrenean	Vein	
188	Gavkhosb	36	56	0	49	30	0	I	Andesite/Eocene	Hydrothermal	Post-Pyrenean	Vein	
189	Gazu	33	12	20	57	23	30	S	Andesite/Eocene	Hydrothermal	Post-Pyrenean	Vein	
190	Gehdij	29	55	0	56	3	0	I	Volcanics/Eocene	Hydrothermal	Post-Pyrenean	Vein	
191	Gelmandeh	32	40	0	55	27	0	I	Andesite/Eocene; schist and phyllite/ Upper Precambrian	Hydrothermal	Post-Pyrenean	Vein	

(continued)

(continued)

		Lat			Long								
No.	Name	D°	M	S″	D°	M	S″	Size	Host rock/age	Genetic/ mineralization	Orogenic phases	Morphology	Elem para
192	Gerdu Kulu	29	7	0	57	15	0	I	Porphyritic granite/Oligo-Miocene; volcanic and pyroclastic rocks/ Eocene	Hydrothermal	Post-Pyrenean	Vein	
193	Gesh Kendar	27	53	47	57	43	1	I	Diabase, shale, limestone/Mesozoic	Hydrothermal	Post-Pyrenean	Vein	
194	Ghara Darreh	36	26	0	49	33	0	I	Basic volcanic/Eocene	Hydrothermal	Post-Pyrenean	Vein	
195	Ghareh Darreh	38	50	48	46	21	42	I	Gabbro and serpentinite/Cretaceous	Volcanic	Post-Pyrenean	Vein	
196	Ghar-e-Kaftar	33	8	51	58	26	44	I	Granite porphyry/Oligo-Miocene	Hydrothermal	Post-Pyrenean	Vein	
197	Gheshlag	36	55	0	48	50	30	S	Tuff/Eocene	Hydrothermal	Post-Pyrenean	Vein	
198	Gheshlagh-e-Yazdikan	38	24	0	44	47	30	I	Andesite and basalt/Quaternary	Hydrothermal	Post-Pyrenean	Vein	
199	Ghezal Darreh Shah Khani	36	35	0	49	30	0	I	Andesite and basalt/Eocene	Hydrothermal	Post-Pyrenean	Vein	
200	Ghohrud (Chahak)	33	28	0	51	47	0	I	Andesite/Eocene	Hydrothermal	Post-Pyrenean	Vein	
201	Ghomeshlu	35	41	0	50	37	0	S	Andesite/Eocene	Hydrothermal	Post-Pyrenean	Vein	Fe
202	Ghutur Souey	38	18	40	47	49	40	I	Granite/Upper Eocene–Lower Oligocene; limestone/Pre-Tertiary	Skarn	Post-Pyrenean	Vein	Au, Ag
203	Giravan	27	51	0	60	50	20	I	Ophiolite complex/Upper Cretaceous	Unknown	Post-Pyrenean	Vein	
204	Golestan	34	38	30	50	24	0	I	Tuff/Eocene	Hydrothermal	Post-Pyrenean	Vein	
205	Golicheh	36	46	0	48	41	30	I	Andesite, porphyrite trachyandesite/ Eocene	Hydrothermal	Post-Pyrenean	Vein	
206	Gomush Ulan	38	35	0	46	43	0	S	Granodiorite/Eocene–Oligocene; limestone/Cretaceous	Skarn	Post-Pyrenean	Vein	
207	Gonabad	34	26	0	58	32	20	I	Granite/Jurassic, schist/Upper Proterozoic	Hydrothermal	Post-Pyrenean	Vein	
208	Googhar	27	51	0	60	50	20	I	Sandstone and limestone/Miocene; sandstone tuff conglomerate/ Eocene	Hydrothermal	Post-Pyrenean	Vein	
209	Gourva	34	7	31	55	2	17	I	Granodiorite/Jurassic; schist and marble/Upper Proterozoic	Hydrothermal	Post-Pyrenean	Vein	

210	Gowdal	38	36	49	47	9	9	S	Intrusive/Upper Eocene–Lower Oligocene; andesite and limestone/ Cretaceous	Hydrothermal	Post-Pyrenean	Vein
211	Gowdar siah	29	27	0	57	27	0	S	Granite porphyry, dolerite, and basalt and porphyries dyabase/ Oligo-Miocene; andesite and pyroxene Andesite/Eocene	Hydrothermal	Post-Pyrenean	Vein
212	Gowdar-e-siah	33	57	1	52	48	21	I	Schist and marble/Upper Precambrian	Hydrothermal	Post-Pyrenean	Vein
213	Gowd-e-Kolvari	30	36	0	55	0	0	S	Porphyrite diorite, porphyrite quartz diorite/Oligo-Miocene; andesite/ Eocene	Porphyry	Post-Pyrenean	Vein
214	Gowd-e-Konarak	29	55	40	55	46	20	I	Conglomerate, sandstone, and limestone/Miocene	Hydrothermal	Post-Pyrenean	Vein
215	Gud-e-Morad	33	24	37	53	31	28	S	Metamorphic rocks/Upper Proterozoic	Hydrothermal	Post-Pyrenean	Vein
216	Gugher	29	29	11	56	26	17	S	Andesite to trachyandesite/Eocene	Hydrothermal	Post-Pyrenean	Vein
217	Gumush Dash	35	39	40	50	35	0	I	Volcanics and sedimentary/Upper Cretaceous	Hydrothermal	Post-Pyrenean	Vein
218	Gur Ali Esmayel	29	50	40	55	53	40	I	Diorite/Oligo-Miocene; andesite, tuff, and agglomerate/Eocene	Porphyry	Post-Pyrenean	Vein
219	Gurcheh berenj	33	50	0	54	2	0	I	Limestone/Oligo-Miocene	Skarn	Post-Pyrenean	Vein
220	Gurkhan	36	15	30	56	17	0	I	Andesite/Eocene	Hydrothermal	Post-Pyrenean	Vein
221	Guruh	29	22	0	57	18	30	I	Diorite and granodiorite/Oligo-Miocene; volcanics and pyroclas-tic/Eocene	Porphyry	Post-Pyrenean	Vein
222	Haji koshteh	29	41	0	60	50	0	I	Diorite/Oligo-Miocene; andesite, tuff, and agglomerate/Eocene	Porphyry	Post-Pyrenean	Vein
223	Halal Abad	36	19	0	49	23	30	I	Andesite/Eocene	Hydrothermal	Post-Pyrenean	Vein
224	Hamid Abad	34	12	10	59	47	42	I	Andesite, tuff/Eocene	Hydrothermal	Post-Pyrenean	Vein
225	Hamireh	35	43	0	57	12	0	I	Limestone and dolomite/Lower Cretaceous	Hydrothermal	Post-Pyrenean	Vein
226	Hammami	36	26	30	56	19	0	S	Volcanic rocks/Eocene	Hydrothermal	Post-Pyrenean	Vein

(continued)

(continued)

		Lat			Long								
No.	Name	D°	M	S″	D°	M	S″	Size	Host rock/age	Genetic/ mineralization	Orogenic phases	Morphology	Elem para
227	Hanza	29	20	0	57	14	30	S	Diorite/Oligo-Miocene; andesite, tuff, and agglomerate/Eocene	Porphyry	Post-Pyrenean	Vein	
228	Hararan	29	26	0	56	45	0	S	Limestone and dolomite/Cretaceous	Hydrothermal	Post-Pyrenean	Vein	Turquoise
229	Harzehvil	36	45	0	49	26	31	I	Tuffaceous shale and detrital tuffite(andesite)/Eocene	Hydrothermal	Post-Pyrenean	Vein	
230	Hasan Jir	26	41	19	58	55	42	I	Gabbro and diorite/Upper Cretaceous	Massive sulfide	Post-Pyrenean	Vein	
231	Hayun Bazar	29	38	54	60	54	11	I	Porphyrite diorite, diorite, porphyrite quartz diorite, and quartz diorite/ Oligo-Miocene	Porphyry	Post-Pyrenean	Vein	
232	Hendi Kandi	38	45	0	46	37	9	I	Granitoid and granophyre/Upper Eocene–Lower Oligocene	Hydrothermal	Post-Pyrenean	Vein	Pb-Zn
233	Hesar	36	29	0	49	9	0	I	Basalt/Eocene	Hydrothermal	Post-Pyrenean	Vein	
234	Heydar Abad	31	1	54	56	5	26	I	Limestone and shale/Cambrian	Hydrothermal	Post-Pyrenean	Vein	
235	Heydar Abad	31	6	50	60	3	45	S	Phyllite and schist/Upper Cretaceous–Lower Tertiary	Unknown	Post-Pyrenean	Vein	
236	Hoseyn Abad	33	41	20	49	25	0	I	Rhyolite, dacite, and tuff/Eocene; shale and dolomite/Cretaceous	Hydrothermal	Post-Pyrenean	Vein	
237	Hoseyn Abad	33	8	22	53	3	11	I	Andesite and tuff/Eocene	Hydrothermal	Post-Pyrenean	Vein	
238	Hoseyn Abad	29	50	1	55	59	46	S	Porphyrite quartz diorite and diorite/ Oligo-Miocene; andesite, dacite, pyroclastic, and sedimentary rocks/ Late Eocene	Porphyry (?)	Post-Pyrenean	Vein	
239	Howz-e-Dagh	33	2	11	58	22	35	I	Andesite and sandstone/Paleogene	Hydrothermal	Post-Pyrenean	Vein	
240	Howz-e-Raeis	33	7	54	58	15	58	I	Tuff and trachyandesite/Eocene	Hydrothermal	Post-Pyrenean	Vein	
241	Iju	30	33	0	54	57	20	L	Porphyrite diorite and quartz diorite/ Oligo-Miocene	Porphyry	Post-Pyrenean	Vein	
242	Ish Pash	27	45	20	59	17	0	I	Limestone and dolomite/Cretaceous	Unknown	Post-Pyrenean	Vein	
243	Izoseroy	32	53	5	55	43	40	I	Tuff/Eocene	Hydrothermal	Post-Pyrenean	Vein	

244	Jafary	33	32	30	54	9	0	I	Metamorphic rocks and limestone/Tertiary	Hydrothermal	Post-Pyrenean	Vein	
245	Jahan Abad	35	29	25	60	35	55	I	Porphyry dacite, andesite, and tuff/Eocene	Hydrothermal	Post-Pyrenean	Vein	Pb
246	Jamal Abad	29	22	30	56	31	0	I	Granodiorite and quartz diorite/Oligo-Miocene; volcanosedimentary/Eocene	Porphyry	Post-Pyrenean	Vein	Au
247	Jaru (Eshtehard)	35	41	0	50	33	25	I	Andesite/Eocene	Hydrothermal	Post-Pyrenean	Vein	
248	Jirandeh	36	42	57	49	48	51	I	Volcanic rocks/Eocene	Hydrothermal	Post-Pyrenean	Vein	
249	Jirvan	27	1	35	60	42	0	I	Gabbro and diorite/Mesozoic	hydrothermal	Post-Pyrenean	Vein	
250	Jonak	28	46	0	58	4	0	I	Granite/Oligo-Miocene; rhyolite, dacite, and andesite/Eocene	Hydrothermal	Post-Pyrenean	Vein	
251	Juiband	38	37	9	46	58	6	I	Granodiorite/Oligo-Miocene; andesite and tuff/Upper Cretaceous–Middle Eocene; volcanosedimentary/Eocene	Porphyry	Post-Pyrenean	Vein	Fe-Pb-Zn
252	Junas	27	24	28	57	41	56	I	Limestone and dolomite/Cretaceous	Unknown	Post-Pyrenean	Vein	
253	Jushin	29	4	5	57	37	30	I	Andesitic lava/Eocene	Hydrothermal	Post-Pyrenean	Vein	
254	Kaboutar Kuh	34	7	42	58	54	56	I	Limestone and dolomite/Jurassic	Hydrothermal	Post-Pyrenean	Vein	
255	Kader	30	37	57	54	45	38	S	Porphyrite diorite/Miocene; volcanic rocks/Eocene	Porphyry	Post-Pyrenean	Vein	
256	Kahnouj	30	12	50	55	42	2	I	Limestone and sandstone/Mid-Eocene; andesite, basalt/Upper Eocene	Hydrothermal	Post-Pyrenean	Vein	
257	Kajeh 1	34	10	35	57	44	40	I	Rhyolite, dacite, and green tuff/Paleogene	Hydrothermal	Post-Pyrenean	Vein	
258	Kajeh 2	34	9	22	57	47	5	I	Granite/Eocene; andesite, basalt/Paleogene	Hydrothermal	Post-Pyrenean	Vein	
259	Kakhk	34	8	23	58	38	48	I	Shale/Jurassic	Hydrothermal	Post-Pyrenean	Vein	
260	Kal Kolhie	30	18	55	59	52	20	I	Sedimentary rocks/Upper Jurassic	Hydrothermal	Post-Pyrenean	Vein	
261	Kalarud	36	48	0	48	29	30	I	Porphyrite diorite and granodiorite/Upper Eocene–Lower Oligocene	Hydrothermal	Post-Pyrenean	Vein	

(continued)

(continued)

No.	Name	Lat D°	Lat M	Lat S″	Long D°	Long M	Long S″	Size	Host rock/age	Genetic/ mineralization	Orogenic phases	Morphology	Elem para
262	Kalateh Ahani	34	6	0	58	50	0	I	Andesite/Eocene	Hydrothermal	Post-Pyrenean	Vein	
263	Kalat-e-Rajab	32	9	44	60	3	11	I	Gabbro and diorite/Upper Cretaceous	Massive sulfide	Post-Pyrenean	Vein	
264	Kalut	35	21	5	54	13	42	S	Granite and granodiorite/Oligo-Miocene; andesite, trachyandesite, dacite, and rhyodacite/Eocene	Hydrothermal	Post-Pyrenean	Vein	Pb-Zn
265	Kalut Boland	35	19	39	54	17	52	S	Granite/Oligo-Miocene; andesite and tuff/Eocene	Hydrothermal	Post-Pyrenean	Vein	
266	Kalutechah (Anarak)	33	1	0	55	22	30	I	Quartzite/Upper Cretaceous–Lower Paleocene	Hydrothermal	Post-Pyrenean	Vein	
267	Kamadurn (Sardab)	29	10	30	57	10	30	I	Granite/Oligo-Miocene; rhyolite and dacite/Eocene	Hydrothermal	Post-Pyrenean	Vein	
268	Kan Mes	33	24	13	53	33	9	S	Andesite/Eocene; limestone/Cretaceous	Hydrothermal	Post-Pyrenean	Vein	
269	Kanif 1	32	25	24	60	19	8	I	Shale and sandstone/Triassic–Jurassic	Hydrothermal	Post-Pyrenean	Vein	Au
270	Kanif 2	32	26	53	60	18	2	I	Dolomite and limestone/Upper Devonian–Lower Carboniferous	Hydrothermal	Post-Pyrenean	Vein	
271	Karreh	28	42	20	58	12	0	I	Granite/Oligo-Miocene; rhyolite, dacite/Eocene	Hydrothermal	Post-Pyrenean	Vein	
272	Kazgoon	33	22	58	60	0	48	I	Metavolcanic and marble/Upper Precambrian	Skarn	Post-Pyrenean	Vein	Pb, Au
273	Kbir Kuh-e-Bozorg	34	15	41	59	40	51	I	Rhyolite and dacite/Tertiary	Hydrothermal	Post-Pyrenean	Vein	
274	Keyghal	38	37	0	46	41	44	I	Microdiorite, diorite, and granodiorite/ Upper Eocene; andesitic tuff and pyroxene andesite and quartz andesite/Upper Cretaceous to Mid-Eocene	Porphyry	Post-Pyrenean	Vein	
275	Khalat Abad	34	15	24	59	40	57	I	Limestone/Cretaceous	Hydrothermal	Post-Pyrenean	Vein	
276	Khalifehlu	36	17	0	49	14	0	S	Andesite/Tertiary	Hydrothermal	Post-Pyrenean	Vein	
277	Khaliran	32	46	22	59	1	36	I	Schist, marble, and quartzite/ Precambrian	Hydrothermal	Post-Pyrenean	Vein	

278	Khalkhal	37	27	25	48	26	0	I	Tuff/Miocene	Hydrothermal	Post-Pyrenean	Vein		
279	Khaneghah	31	2	0	51	18	0	S	Porphyrite Granodiorite/Oligo-Miocene; limestone, shale, and quartzite/Triassic–Jurassic	Hydrothermal	Post-Pyrenean	Vein		
280	Khanuk	30	44	0	56	46	30	I	Porphyrite diorite and quartz diorite/Miocene	Hydrothermal	Post-Pyrenean	Vein		
281	Kharaghan	35	27	30	49	49	10	I	Dolomite and limestone/Cretaceous	Skarn	Post-Pyrenean	Vein		
282	Kharanjoo	37	19	0	46	31	30	I	Andesite, pyroclastic/Eocene	Hydrothermal	Post-Pyrenean	Vein		
283	Kharvanagh	38	32	30	46	15	30	I	Granodiorite/Oligo-Miocene; sedimentary rocks/Paleocene	Hydrothermal	Post-Pyrenean	Vein		
284	Khatibi	33	34	9	59	45	39	I	Volcanics and intrusive/Jurassic	Hydrothermal	Post-Pyrenean	Vein		
285	Kheibar Kuh-e-Bozorg 1	34	13	47	59	42	31	I	Shale/Lower Paleozoic; acidic to basic tuff/Eocene	Hydrothermal	Post-Pyrenean	Vein		
286	Kheibar Kuh-e-Bozorg 2	34	17	10	59	42	13	I	Granite/Eocene; shale, sandstone/Paleogene	Hydrothermal	Post-Pyrenean	Vein		
287	Kheibar Kuh-e-Kuchak	34	10	40	59	42	23	I	Dacite, andesite, basalt, pyroclastic/Paleogene	Hydrothermal	Post-Pyrenean	Vein		
288	Kheyr Abad	26	42	42	59	26	0	I	Dolomite and limestone/Cretaceous	Unknown	Post-Pyrenean	Vein		
289	Khonindizaj	38	40	32	46	47	0	I	Andesite/Tertiary (Eocene)	Hydrothermal	Post-Pyrenean	Vein		
290	Khoshumi	32	24	13	55	9	13	S	Syenite, diorite, and monzonite/Oligo-Miocene	Hydrothermal	Post-Pyrenean	Vein		
291	Khouresht	36	13	55	49	41	0	I	Volcanic rocks/Eocene	Hydrothermal	Post-Pyrenean	Vein		
292	Khut	31	53	0	53	42	30	S	Granite/Oligo-Miocene; volcanic rocks/Eocene	Porphyry	Post-Pyrenean	Vein		
293	Kol-e-firuzeh	33	40	5	57	17	36	I	Andesite and tuff/Eocene	Hydrothermal	Post-Pyrenean	Vein		
294	Komeri	38	7	0	48	41	30	I	Granite/Tertiary; limestone/Cretaceous	Skarn	Post-Pyrenean	Vein		Fe
295	Konchirud	33	22	0	54	44	0	I	Conglomerate/Oligocene	Unknown	Post-Pyrenean	Vein		
296	Koppeh Halval	33	46	59	53	32	52	S	Andesite, diorite, and microdiorite/Lower Eocene; carbonate rocks/Cretaceous	Hydrothermal and Skarn	Post-Pyrenean	Vein		

(continued)

(continued)

No.	Name	Lat D°	Lat M	Lat S″	Long D°	Long M	Long S″	Size	Host rock/age	Genetic/ mineralization	Orogenic phases	Morphology	Elem para
297	Kordkandy	36	51	0	48	27	30	S	Granite/Upper Eocene–Lower Oligocene	Hydrothermal	Post-Pyrenean	Vein	Mo, W, Zn, Pb, Ag, Bi, Fe, Au
298	Koroied Paien	30	1	50	56	46	36	I	Andesite, rhyolite, dacite, and tuff/Eocene	Hydrothermal	Post-Pyrenean	Vein	
299	Kuh Firouzeh	33	33	40	57	15	40	I	Limestone and conglomerate/Lower Paleogene	Hydrothermal	Post-Pyrenean	Vein	
300	Kuh Panj	29	51	49	56	4	17	S	Granodiorite and quartz diorite/ Oligo-Miocene; andesite, pyroclastic, breccia, and conglomerate/Eocene	Porphyry	Post-Pyrenean	Vein	
301	Kuh Zar (Baghu)	35	26	30	54	38	45	S	Granodiorite/Upper Eocene; volcanics/ Mid-Eocene	Hydrothermal	Post-Pyrenean	Vein	Au, Turquoise
302	Kuh-e-Abdol Abad	35	18	5	60	14	25	I	Rhyolite, dacite, and tuff/Tertiary	Hydrothermal	Post-Pyrenean	Vein	
303	Kuh-e-Bakhazar	34	27	38	60	27	23	I	Andesite/Eocene	Hydrothermal	Post-Pyrenean	Vein	
304	Kuh-e-Dom II	33	55	24	55	51	57	I	Granitoid/Upper Eocene; carbonate rocks/Cretaceous	Hydrothermal and Skarn	Post-Pyrenean	Vein	
305	Kuh-e-Dom IV	33	55	40	52	50	0	I	Limestone and conglomerate/Lower Paleogene	Hydrothermal	Post-Pyrenean	Vein	
306	Kuh-e-Durkhan	27	9	28	57	54	11	I	Ophiolite complex/Upper Cretaceous	Unknown	Post-Pyrenean	Vein	
307	Kuh-e-Hajin	30	2	9	56	6	43	I	Porphyrite diorite/Upper Miocene–Pliocene	Porphyry	Post-Pyrenean	Vein	
308	Kuh-e-Hanza 1	29	19	43	57	12	30	I	Granodiorite, porphyrite granodiorite and porphyrite diorite/Oligo-Miocene; andesite, andesibasalt, and pyroclastic/Mid-Eocene	Hydrothermal	Post-Pyrenean	Vein	
309	Kuh-e-Hanza 2	29	24	3	57	8	35	I	Granite, granodiorite, and diorite/ Oligo-Miocene; volcanics/Eocene	Hydrothermal	Post-Pyrenean	Vein	
310	Kuh-e-Hayan Jam	35	4	30	60	18	30	I	Andesite/Eocene	Hydrothermal	Post-Pyrenean	Vein	

311	Kuh-e-Jaj	32	25	20	60	21	45	I	Granite/Tertiary; metamorphic rocks/ Cretaceous–Paleocene	Hydrothermal	Post-Pyrenean	Vein
312	Kuh-e-Jana 1	31	13	2	60	20	55	I	Limestone dolomite/Cretaceous	Hydrothermal	Post-Pyrenean	Vein
313	Kuh-e-Janja 2	31	12	58	6	21	47	I	Porphyrite hornblend Granite/ Oligocene; volcanic, pyroclastic, and limestone/Eocene	Hydrothermal	Post-Pyrenean	Vein
314	Kuh-e-Kalat	27	42	8	57	52	39	I	Ophiolite complex/Upper Cretaceous	Unknown	Post-Pyrenean	Vein
315	Kuh-e-Kamar Sabz	34	56	23	60	29	1	I	Andesite/Eocene	Hydrothermal	Post-Pyrenean	Vein
316	Kuh-e-Kamarkhid	34	5	49	58	40	16	I	Shale and sandstone/Jurassic	Hydrothermal	Post-Pyrenean	Vein
317	Kuh-e-Lar 1	29	42	32	60	59	24	I	Intrusive/Tertiary; carbonate rocks and filish/Paleocene	Skarn	Post-Pyrenean	Vein
318	Kuh-e-Lar 2	29	43	27	60	53	46	I	Porphyry granodiorite/Oligo-Miocene; dacite and pyroclastic/Eocene	Porphyry	Post-Pyrenean	Vein
319	Kuh-e-Maadan	28	33	30	58	0	0	I	Andesite, dacite, and tuff/Eocene	Hydrothermal	Post-Pyrenean	Vein
320	Kuh-e-Malou	35	18	20	61	2	50	I	Intermediate lava and pyroclastic/ Eocene	Hydrothermal	Post-Pyrenean	Vein
321	Kuh-e-masahim	30	22	58	55	11	24	I	Porphyrite quartz diorite/Oligo-Miocene; andesite/Pliocene	Porphyry	Post-Pyrenean	Vein
322	Kuh-e-Neybid	34	10	35	59	42	33	I	Andesite/Eocene	Hydrothermal	Post-Pyrenean	Vein
323	Kuh-e-Pasakuh	34	58	22	59	58	3	I	Limestone and volcanics/Cretaceous	Hydrothermal	Post-Pyrenean	Vein
324	Kuh-e-Rouz	33	23	42	54	54	25	S	Andesite and tuff/Eocene	Hydrothermal	Post-Pyrenean	Vein
325	Kuh-e-Rud	33	39	0	51	25	0	I	Andesite/Tertiary	Hydrothermal	Post-Pyrenean	Vein
326	Kuh-e-Sangmes	33	11	0	52	45	20	S	Quartzite and phyllite/Precambrian	Hydrothermal	Post-Pyrenean	Vein
327	Kuh-e-SangSiah	29	40	0	56	51	30	I	Tuff and breccia/Miocene; dacite and dasitic tuff/Eocene	Hydrothermal	Post-Pyrenean	Vein
328	Kuh-e-sarhangi	34	50	0	57	16	0	S	Sandstone, limestone, and dolomite/ Upper Paleozoic	Hydrothermal	Post-Pyrenean	Vein
329	Kuh-e-Sefid	31	51	36	60	13	20	I	Andesite to basalt/Oligo-Miocene; alkali-rich tuff/Eocene	Unknown	Post-Pyrenean	Vein
330	Kuh-e-Sefid Donbeh	32	54	19	55	24	8	I	Syenite, diorite, and monzonite/ Tertiary; volcanics/Eocene	Hydrothermal	Post-Pyrenean	Vein

(continued)

(continued)

No.	Name	Lat D°	Lat M	Lat S″	Long D°	Long M	Long S″	Size	Host rock/age	Genetic/ mineralization	Orogenic phases	Morphology	Elem para
331	Kuh-e-Shotoran	34	18	55	58	59	11	I	Granite/Tertiary	Hydrothermal	Post-Pyrenean	Vein	
332	Kuh-e-Siah Targi	30	35	7	60	27	42	I	Andesite/Eocene; sedimentary rocks/ Paleogene	Hydrothermal	Post-Pyrenean	Vein	
333	Kuh-e-Sorkheh	31	56	37	58	43	28	I	Phyllite and schist/Jurassic	Unknown	Post-Pyrenean	Vein	
334	Kuh-e-Spid Kanarun	27	18	6	57	21	35	I	Ophiolite complex/Upper Cretaceous	Unknown	Post-Pyrenean	Vein	
335	Kuh-e-Tak Khord	34	18	38	59	37	40	I	Granite/Eocene; andesite and basalt/ Paleogene	Hydrothermal	Post-Pyrenean	Vein	
336	Kuh-e-Taleb	34	29	13	60	29	21	I	Granite/Tertiary; limestone/Triassic	Hydrothermal	Post-Pyrenean	Vein	
337	Kuh-e-Yalan	31	5	59	59	33	14	I	Shale and limestone/Cretaceous; dolomite/Jurassic; dolomite, limestone, and sandstone/ Precambrian	Hydrothermal	Post-Pyrenean	Vein	
338	Kuh-e-Zolfaghar	35	21	15	61	9	20	I	Andesite/Eocene	Hydrothermal	Post-Pyrenean	Vein	
339	Kuhian	36	46	30	48	52	0	I	Trachyandesite/Eocene	Hydrothermal	Post-Pyrenean	Vein	Fe
340	Lab-e-Kal	36	27	50	56	32	0	I	Volcanics/Tertiary	Hydrothermal	Post-Pyrenean	Vein	
341	Laibid	33	27	13	50	40	2	I	Porphyrite trachyandesite, basalt/Lower Eocene	Hydrothermal	Post-Pyrenean	Vein	Ni, Co
342	Lalehzar F3	29	24	0	56	53	30	S	Granodiorite/Oligo-Miocene; volcanics/Eocene	Porphyry	Post-Pyrenean	Vein	
343	Lalehzar F4	29	22	0	56	47	30	S	Granite/Oligo-Miocene; rhyolite, dacite, and tuff/Eocene	Porphyry	Post-Pyrenean	Vein	
344	Lar	29	41	21	60	53	36	S	Intrusive/Tertiary; carbonate rocks and filish/Paleocene	Porphyry	Post-Pyrenean	Vein	Au, Mo
345	Laymordeh	26	32	14	58	40	23	I	Basaltic lava/Upper Cretaceous	Volcano- sedimentary	Post-Pyrenean	Vein	
346	Limiar	36	23	20	50	18	0	I	Volcanics/Tertiary	Hydrothermal	Post-Pyrenean	Vein	
347	Lubin Zardeh	36	41	0	49	6	0	I	Andesite/Eocene	Hydrothermal	Post-Pyrenean	Vein	
348	Maadan (Karun Rud)	31	43	20	50	53	0	I	Basic extrusive/Oligo-Miocene	Hydrothermal	Post-Pyrenean	Vein	

349	Maadan-e-Bozorg	36	25	0	56	27	0	S	Andesite and serpentine/Eocene	Hydrothermal	Post-Pyrenean	Vein		
350	Maadan-e-Rughani	31	15	49	59	13	25	I	Shale, limestone, and dolomite/Lower Cretaceous	Volcano-sedimentary	Post-Pyrenean	Vein		
351	Madgh Kahriz	31	10	26	60	20	16	I	Volcanoclastics sedimentary and pyroclastic/Cretaceous	Massive sulfide	Post-Pyrenean	Vein		
352	Mahallati	27	10	49	58	2	12	I	Sedimentary rocks/Paleozoic	Hydrothermal	Post-Pyrenean	Vein		
353	Mahdi Abad	31	30	30	55	10	0	S	Phyllite and schist/Upper Cretaceous–Lower Tertiary	Unknown	Post-Pyrenean	Vein		
354	Mahgalui	27	45	30	59	42	40	I	Ophiolite complex/Upper Cretaceous	Hydrothermal	Post-Pyrenean	Vein		
355	Mahmud Abad	38	45	40	46	48	29	I	Andesite and subvolcanic/Eocene–Oligocene; limestone and shale/Cretaceous	Hydrothermal	Post-Pyrenean	Vein		Pb, Zn, Au, Ag
356	Mahrud	32	11	13	60	47	24	I	Dolomite and limestone/Upper Devonian–Lower Carboniferous	Hydrothermal	Post-Pyrenean	Vein		
357	Makki	26	57	19	60	47	42	S	Basaltic, andesite, and spilite from ophiolite complex/Upper Cretaceous	Massive sulfide	Post-Pyrenean	Vein		
358	Malek Gori	30	34	16	60	37	50	I	Trachyandesite and pyroxene andesite/Eocene	Hydrothermal	Post-Pyrenean	Vein		
359	Mardanal (Mardan Qom)	38	50	0	46	33	0	S	Porphyrite granodiorite/Eocene–Oligocene; volcanoclastics sedimentary/Cretaceous to Eocene	Vein	Post-Pyrenean	Vein		Fe
360	Mary	37	0	20	48	28	30	I	Andesite and tuff/Paleogene	Hydrothermal	Post-Pyrenean	Vein		Pb, Fe
361	Mashar Tighdar	31	6	6	60	30	43	I	Granite and granodiorite/Oligo-Miocene; sandstone, siltstone, and limestone	Porphyry	Post-Pyrenean	Vein		
362	Mazraeh	38	39	0	47	4	0	M	Monzonite to granite/Eocene–Oligocene; tuff, and ignimbrite/Eocene; limestone/Upper Cretaceous	Skarn	Post-Pyrenean	Vein		Fe, Au
363	Mazraeh sadat	30	26	30	55	35	0	I	Porphyrite diorite/Mid- to Upper Miocene	Porphyry	Post-Pyrenean	Vein		

(continued)

(continued)

No.	Name	Lat D°	Lat M	Lat S″	Long D°	Long M	Long S″	Size	Host rock/age	Genetic/ mineralization	Orogenic phases	Morphology	Elem para
364	Mazraehmirha	32	26	0	54	16	0	I	Shale and sandstone/ Jurassic–Cretaceous	Hydrothermal	Post-Pyrenean	Vein	
365	Mazraeh-ye-Haji Hassan	32	32	52	47	2	30	I	Amphibolite and gneiss/Precambrian	Hydrothermal	Post-Pyrenean	Vein	
366	Mesgar (Mazagar)	38	49	0	46	44	0	I	Monzonite/Oligo-Miocene	Hydrothermal	Post-Pyrenean	Vein	
367	Meskani	33	19	0	53	28	0	M	Trachyandesite and basalt/Eocene	Hydrothermal	Post-Pyrenean	Vein	
368	Meyduk (Lachah)	30	25	0	55	10	0	L	Porphyrite diorite/Miocene; andesite/ Eocene	Porphyry	Post-Pyrenean	Vein	Mo, Au
369	Milsefid and mohammadabad	31	57	0	53	44	0	I	Intermediate intrusive/Tertiary; limestone and dolomite/Permian	Hydrothermal	Post-Pyrenean	Vein	
370	Mir Hoseyn	36	20	30	57	41	30	I	Dacite/Eocene	Hydrothermal	Post-Pyrenean	Vein	
371	Mirkhash	33	7	50	58	16	0	I	Andesite/Eocene; limestone/Cretaceous	Hydrothermal	Post-Pyrenean	Vein	
372	Moein Abad	33	21	21	60	6	21	I	Metavolcanic and marble/Upper Proterozoic	Skarn	Post-Pyrenean	Vein	Fe
373	Mogh Mohammad	27	22	50	57	45	54	I	Gabbro and diorite/Mesozoic; ophiolite complex and limestone/Upper Cretaceous	Hydrothermal	Post-Pyrenean	Vein	
374	Moghestan	32	35	49	55	12	4	I	Porphyrite andesite and tuff/Unknown	Hydrothermal	Post-Pyrenean	Vein	
375	Mohammad Abad	27	54	52	57	53	7	I	Phyllite and schist/Upper Cretaceous–Lower Tertiary	Unknown	Post-Pyrenean	Vein	
376	Morghak-e-Bam	29	9	50	57	20	30	S	Porphyrite andesite/Paleogene	porphyry	Post-Pyrenean	Vein	
377	Mour	33	7	50	60	34	28	I	Limestone/Jurassic	Hydrothermal	Post-Pyrenean	Vein	
378	Naft Abad	33	2	0	59	52	0	I	Granite/Tertiary; limestone/Cretaceous	Skarn	Post-Pyrenean	Vein	
379	Nahroo (Joze)	30	32	40	54	44	25	I	Andesite and tuff/Eocene	Hydrothermal	Post-Pyrenean	Vein	
380	Najm Abad	34	12	26	58	51	4	I	Rhyolite and dacite/Eocene Limestone and dolomite/Cretaceous	Hydrothermal	Post-Pyrenean	Vein	
381	Namesh	29	16	15	57	21	0	I	Pyroxene andesite/Eocene	Hydrothermal	Post-Pyrenean	Vein	
382	Naran	29	12	0	57	8	0	I	Andesite and pyroclastic/Eocene	Hydrothermal	Post-Pyrenean	Vein	

383	Narbulaghi	32	45	30	55	45	30	I	Limestone and dolomite/ Triassic–Cretaceous	Unknown	Post-Pyrenean	Vein
384	Narp Allah Abad	29	42	0	56	39	0	I	Porphyry quartz/Oligo-Miocene; volcanosedimentary rocks/Eocene	Hydrothermal	Post-Pyrenean	Vein
385	Nasr	35	31	0	59	29	0	I	Andesite and tuff/Eocene	Hydrothermal	Post-Pyrenean	Vein
386	Nasr Abad	31	45	58	53	52	0	S	Rhyolite and dacite/Tertiary	Hydrothermal	Post-Pyrenean	Vein
387	Nehbandan	31	32	21	60	9	23	I	Andesite/Eocene	Hydrothermal	Post-Pyrenean	Vein
388	Neyshabur	36	29	40	58	23	0	I	Pyroxene andesite/Eocene	Hydrothermal	Post-Pyrenean	Vein
389	Nish Tafoun	34	27	30	60	2	50	I	Granite and limestone/Jurassic	Skarn	Post-Pyrenean	Vein
390	Norhwest of Kuh-e-Mishbadam	27	49	25	57	24	24	I	Ophiolite complex/Upper Cretaceous	Unknown	Post-Pyrenean	Vein
391	North East Of Ghaem Abad	32	21	32	58	58	5	I	Diabase/Upper Cretaceous	Hydrothermal	Post-Pyrenean	Vein
392	North East Of Gheshmiran	27	9	6	57	55	45	I	Basaltic pillow lava and limestone/ Cretaceous	Massive sulfide	Post-Pyrenean	Vein
393	North East Of Hedar Abad	31	8	18	60	8	54	I	Volcanoclastics sedimentary and pyroclastic/Cretaceous	Massive sulfide	Post-Pyrenean	Vein
394	North East Of Kalateh Mir	32	18	8	59	50	6	I	Gabbro and diorite/Upper Cretaceous	Massive sulfide	Post-Pyrenean	Vein
395	North East Of Khoramshahi	28	58	30	57	44	10	I	Microgranite and microgranodiorite/ Oligocene	Porphyry	Post-Pyrenean	Vein
396	North East Of Tahrud	29	21	8	58	12	41	I	Dacite and rhyolite and pyroclastic/ Eocene	Hydrothermal	Post-Pyrenean	Vein
397	North of Asad Abad	32	58	41	60	10	18	I	Basalt/Tertiary; ophiolite complex/ Cretaceous	Hydrothermal	Post-Pyrenean	Vein
398	North of Bad Afshan-e-Bala	27	37	54	57	12	34	I	Phyllite and schist/Cretaceous	Unknown	Post-Pyrenean	Vein
399	North Of Bagh Chamak	29	16	21	58	16	51	I	Dacite and rhyolite and pyroclastic/ Eocene	Hydrothermal	Post-Pyrenean	Vein
400	North of Chahar Gonbad	29	45	34	56	18	45	I	Granite/Oligo-Miocene; rhyolite, dacite/Eocene	Hydrothermal	Post-Pyrenean	Vein

(continued)

(continued)

		Lat			Long					Genetic/	Orogenic		
No.	Name	D°	M	S″	D°	M	S″	Size	Host rock/age	mineralization	phases	Morphology	Elem para
401	North of Choghondarsar Shomali	36	19	0	56	16	0	I	Basalt/Eocene; limestone, shale, and sandstone/Cretaceous	Hydrothermal	Post-Pyrenean	Vein	
402	North Of Darzin	29	18	55	58	9	44	I	Dacite and rhyolite and pyroclastic/ Eocene	Hydrothermal	Post-Pyrenean	Vein	
403	North of Deh Kahan 1	27	43	47	57	31	4	I	Phyllite and schist/Jurassic	Unknown	Post-Pyrenean	Vein	
404	North Of Deh Kahan 2	27	44	27	57	31	4	I	Phyllite and schist/Cretaceous	Hydrothermal	Post-Pyrenean	Vein	
405	North Of Deh Kahan 3	27	44	52	57	32	17	I	Basalt/Upper Cretaceous	Massive sulfide	Post-Pyrenean	Vein	
406	North of Espily	36	57	33	49	52	54	I	Trachyandesite/Eocene	Hydrothermal	Post-Pyrenean	Vein	
407	North of Kal-e-Sorkh	35	22	5	58	40	21	I	Andesite and tuff/Eocene	Hydrothermal	Post-Pyrenean	Vein	
408	North of Khounik	33	33	31	59	10	21	I	Andesite/Eocene	Hydrothermal	Post-Pyrenean	Vein	
409	North Of Kuh-e-Azghand 1	35	18	8	58	51	16	I	Granite and granodiorite/Tertiary; andesite and tuff/Eocene	Hydrothermal	Post-Pyrenean	Vein	
410	North Of Kuh-e-Azghand 2	35	19	46	58	48	37	I	Granite and granodiorite/Tertiary; andesite and tuff/Eocene	Hydrothermal	Post-Pyrenean	Vein	
411	North of Kuh-e-Bidkhan	29	37	1	56	30	46	I	Tuff and andesite/Eocene	Hydrothermal	Post-Pyrenean	Vein	
412	North of Kuh-e-Cheheltan	29	48	14	56	20	56	I	Andesite and rhyolite/Eocene; granite and granodiorite/Triassic–Jurassic	Hydrothermal	Post-Pyrenean	Vein	
413	North of Kuh-e-K alleh Gav	30	1	46	56	46	34	I	Tuff/Eocene	Hydrothermal	Post-Pyrenean	Vein	
414	North of Kuh-e-Lalehzar	29	26	29	56	41	56	I	Granite/Oligo-Miocene; rhyolite and dacite/Eocene	Hydrothermal	Post-Pyrenean	Vein	
415	North of Kuh-e-Movlai	27	26	44	57	27	59	I	Limestone and dolomite/Cretaceous	Unknown	Post-Pyrenean	Vein	
416	North of Ramun	29	3	50	57	29	11	I	Diorite and volcanosedimentary/ Oligo-Miocene; volcanics and pyroclastics/Eocene	Hydrothermal	Post-Pyrenean	Vein	
417	North of Rud-e-Jamali	32	29	10	59	30	58	I	Tuff/Eocene	Hydrothermal	Post-Pyrenean	Vein	

418	North of Sabzevaran	28	47	34	57	48	26	I	Granite/Oligo-Miocene; andesite and tuff/Eocene	Hydrothermal	Post-Pyrenean	Vein
419	North of Sorkh Kuh	33	17	58	58	24	6	I	Rhyolite and dacite and tuff/Eocene; shale and dolomite/Cretaceous	Hydrothermal	Post-Pyrenean	Vein
420	Northeast of Asfaran	36	18	47	50	35	49	I	Ophiolite complex/Upper Paleozoic–Lower Mesozoic	Hydrothermal	Post-Pyrenean	Vein
421	Northeast of Choghondarsar Shomali	36	18	20	56	17	0	I	Basalt and tuff/Eocene	Hydrothermal	Post-Pyrenean	Vein
422	Northeast of Godar-e-Sangbor	30	1	43	56	51	16	I	Tuff/Eocene	Hydrothermal	Post-Pyrenean	Vein
423	Northeast of Kakhk	32	36	49	58	56	10	I	Andesite/Eocene	Hydrothermal	Post-Pyrenean	Vein
424	Northeast of Kal-e-Sorkh	35	22	30	58	42	14	I	Granite and granodiorite/Post-Eocene; andesite and tuff/Eocene	Hydrothermal	Post-Pyrenean	Vein
425	Northeast of Kuh-e-Durkhan	27	11	12	57	52	7	I	Ophiolite complex/Upper Cretaceous	Unknown	Post-Pyrenean	Vein
426	Northeast of Kuh-e-Kam Sefid	27	5	1	58	6	40	I	Ophiolite complex/Upper Cretaceous	Unknown	Post-Pyrenean	Vein
427	Northeast of Kuh-e-Sar Sefidal	35	21	49	58	46	40	I	Andesite and tuff/Eocene	Hydrothermal	Post-Pyrenean	Vein
428	Northeast of Namakzar	30	14	27	54	56	52	I	Granite/Oligo-Miocene; rhyolite and dacite/Eocene	Hydrothermal	Post-Pyrenean	Vein
429	Northern Cheheltan	29	51	30	56	24	0	I	Diorite and volcanosedimentary/Oligo-Miocene; volcanics and pyroclastics/Eocene	Hydrothermal	Post-Pyrenean	Vein
430	Northwest of Chahar Gonbad	29	38	6	56	5	56	I	Granite/Oligo-Miocene	Hydrothermal	Post-Pyrenean	Vein
431	Northwest of Kuh-e-Bokikan	26	45	0	58	24	53	I	Ophiolite complex and limestone/Upper Cretaceous	Unknown	Post-Pyrenean	Vein
432	Northwest of Kuh-e-Jeyran	26	40	0	58	44	23	I	Gabbro and diorite/Cretaceous	Hydrothermal	Post-Pyrenean	Vein

(continued)

(continued)

No.	Name	Lat D°	Lat M	Lat S″	Long D°	Long M	Long S″	Size	Host rock/age	Genetic/ mineralization	Orogenic phases	Morphology	Elem para
433	Northwest of Sar Chah Shur	32	21	18	58	52	40	I	Syenite, diorite, monzonite/Tertiary	Hydrothermal	Post-Pyrenean	Vein	
434	Northwest of Shahr-e-Babak	30	15	8	55	26	5	I	Andesite and pyroclastic/Eocene	Hydrothermal	Post-Pyrenean	Vein	
435	Northwest of Zabl	27	27	46	57	2	39	S	Biotite granite/Unknown	Skarn	Post-Pyrenean	Vein	
436	Northwest of Zahedan	29	33	8	60	31	23	I	Rhyodacite, andesite hornfels/ Post-Eocene	Porphyry	Post-Pyrenean	Vein	%46 Cu
437	Now Darreh	36	20	0	59	27	0	I	Volcanics and sedimentary rocks/ Eocene	Hydrothermal	Post-Pyrenean	Vein	
438	Nowbahar	35	2	8	60	19	30	I	Volcanics and sedimentary rocks/Upper Cretaceous	Hydrothermal	Post-Pyrenean	Vein	
439	Nowchun	29	55	40	55	51	15	L	Granite, porphyry diorite, and Rhyodacite/Oligo-Miocene; andesite and pyroclastic/Eocene	Hydrothermal	Post-Pyrenean	Vein	
440	Nowdeh	35	43	0	57	15	0	I	Limestone and dolomite/Jurassic	Hydrothermal	Post-Pyrenean	Vein	
441	Padeh	35	29	15	61	4	50	I	Andesite/Eocene	Hydrothermal	Post-Pyrenean	Vein	Zn, Pb, Au
442	Pakoo	33	10	40	53	1	21	I	Schist/Upper Precambrian	Hydrothermal	Post-Pyrenean	Vein	Ni, Sb
443	Palangi	30	18	33	55	38	59	M	Porphyry diorite/Early Eocene	Porphyry	Post-Pyrenean	Vein	
444	Pariz	29	52	10	55	47	0	I	Andesite, tuff, and agglomerate/Eocene	Hydrothermal	Pyrenean	Vein	
445	Parsan	29	59	20	56	10	0	I	Granodiorite and diorite/Oligo-Miocene; volcanics/Eocene	Porphyry	Post-Pyrenean	Vein	
446	Pas Qaleh	35	50	0	51	30	0	I	Andesite, trachyandesite, and tuff/ Eocene	Hydrothermal	Pyrenean	Vein	Pb, Au
447	Patiyar	33	19	57	53	47	23	S	Limestone and dolomite/Cretaceous	Hydrothermal	Pyrenean	Vein	
448	Pay Negin	29	20	0	56	54	0	S	Granodiorite/Oligo-Miocene; volcanics/Eocene	Hydrothermal	Post-Pyrenean	Vein	
449	Pey Godar-e-Sorkh	32	30	30	60	20	0	I	Metamorphosed limestone, quartzite and schist/Jurassic	Hydrothermal	Late Cimmerian	Vein	

450	Pinavand	33	5	0	52	0	0	I	Granodiorite to diorite porphyry and volcanite pyroclastite/Eocene	Hydrothermal	Pyrenean	Vein	Zn, Pb,
451	Piran	29	57	0	55	51	28	S	Volcanics and pyroclastics/Eocene	Hydrothermal	Post-Pyrenean	Vein	
452	Pirouzy	33	34	17	54	4	0	I	Limestone and dolomite/Cretaceous	Hydrothermal	Laramide	Vein	
453	Piruzaki (Bazman)	27	50	40	59	54	30	S	Ophiolite complex/Upper Cretaceous	Unknown	Laramide	Vein	
454	Pish Kuh	34	7	17	55	4	8	I	Granite limestone/Jurassic	Skarn	Late Cimmerian	Vein	
455	Posht Kam	27	37	18	57	13	28	I	Ophiolite complex/Upper Cretaceous	Unknown	Laramide	Vein	
456	Poshteh	34	41	3	60	21	15	I	Phyllite and schist/Jurassic	Hydrothermal	Pyrenean	Vein	
457	Prayjan	36	16	21	50	56	0	S	Volcanosedimentary rocks/Eocene	Hydrothermal	Pyrenean	Vein	
458	Qaleh Arab	32	36	0	51	6	0	S	Ophiolite complex/Cretaceous	Unknown	Unknown	Vein	
459	Qaleh Asgar	29	28	0	56	41	20	I	Andesite and pyroclastic/Eocene	Hydrothermal	Pyrenean	Vein	
460	Qaleh Gazik	36	14	30	56	8	0	I	Sandstone and limestone/Upper Paleozoic–Lower Jurassic	Hydrothermal	Late Cimmerian	Vein	
461	Qaleh Narp	29	42	0	56	41	30	I	Microdiorite, andesite, pyroclastic, sandstone, and conglomerate/Upper Eocene	Hydrothermal	Pyrenean	Vein	
462	Qaleh Now	33	44	30	49	58	0	I	Tonalite/Post-Miocene; limestone/Oligo-Miocene	Skarn	Post-Pyrenean	Vein	Co
463	Qaleh Sorb	29	42	0	56	41	30	S	Granite/Oligo-Miocene; rhyolite and dacite/Eocene	Hydrothermal	Post-Pyrenean	Vein	
464	Qaleh Zangian	27	47	8	57	29	51	I	Ophiolite complex/Cretaceous	Unknown	Unknown	Vein	
465	Qaleh Zari	31	49	43	58	55	15	M	Andesite, dacite, and basalt/Eocene	Hydrothermal	Pyrenean	Vein	
466	Qamsar	33	45	0	51	23	0	I	Tonalite/Post-Miocene; limestone/Oligo-Miocene	Skarn	Post-Pyrenean	Vein	
467	Qanat Darreh	29	12	43	57	15	26	I	Volcanics and pyroclastics/Eocene	Hydrothermal	Post-Pyrenean	Vein	
468	Qanat-e-Marvan	29	20	0	56	46	30	I	Sandstone and shale/Paleocene	Unknown	Unknown	Vein	
469	Qayen	33	53	10	59	25	39	I	Limestone/Lower Cretaceous	Hydrothermal	Late Cimmerian	Vein	
470	Qebleh	33	28	13	53	26	13	S	Andesite and tuff/Eocene	Hydrothermal	Pyrenean	Vein	
471	Qolleha	31	24	0	59	20	30	I	Limestone with phyllitic intercalate/Jurassic	Hydrothermal	Pyrenean	Vein	

(continued)

(continued)

No.	Name	Lat D°	Lat M	Lat S″	Long D°	Long M	Long S″	Size	Host rock/age	Genetic/ mineralization	Orogenic phases	Morphology	Elem para
472	Rashid Abad	28	58	5	57	39	40	I	Granodiorite, quartz diorite, and diorite porphyry/Oligo-Miocene; volcanite and pyroclastite/Eocene	Porphyry	Post-Pyrenean	Vein	
473	Rashid Abad	33	9	46	58	21	42	I	Ultramafic rocks/Upper Precambrian	Epithermal	Pan-African	Vein	
474	Rashidabad	37	5	11	48	22	49	S	Andesite and rhyodacite/ Eocene–Oligocene	Hydrothermal	Pyrenean	Vein	Ni, Au.Ag
475	Rashkan	26	51	0	59	36	0	I	Ophiolite complex and limestone/ Upper Cretaceous	Unknown	Unknown	Vein	Zn, Pb
476	Rasur	33	23	30	53	43	20	I	Ultramafic rocks/Upper Precambrian	Epithermal	Pan-African	Vein	
477	Razan (Mazraeh)	35	27	0	49	11	0	S	Rhyolite, dacite, and tuff/Eocene	Hydrothermal	Pyrenean	Vein	
478	Razgah	38	5	10	47	20	10	I	Granite/Eocene; tuff and limestone/ Cretaceous	Hydrothermal	Pyrenean	Vein	Fe
479	Razi Abad Madin	28	58	0	57	41	10	I	Granodiorite/Oligo-Miocene; rhyolite, dacite, and andesite/Eocene	Porphyry	Post-Pyrenean	Vein	
480	Remeshk 1	26	45	32	58	37	39	I	Ophiolite complex and limestone/ Upper Cretaceous	Unknown	Unknown	Vein	
481	Remeshk 2	26	49	11	58	37	2	I	Ophiolite complex and limestone/ Upper Cretaceous	Unknown	Unknown	Vein	
482	Revesht	33	43	0	49	10	0	I	Granite/Jurassic	Hydrothermal	Mid-Cimmerian	Vein	
483	Rigane Bam	28	13	49	58	47	26	I	Sedimentary/Eocene	Hydrothermal	Pyrenean	Vein	
484	Robat-e-Kureh	35	3	35	60	18	10	I	Andesite and basalt/Eocene	Hydrothermal	Pyrenean	Vein	
485	Roniz (Kohneh mes)	29	22	0	53	39	0	I	Andesite and pyroclastic/Eocene	Hydrothermal	Post-Pyrenean	Vein	
486	Rudab	28	43	0	58	7	15	I	Tuff and pyroclastic/Eocene	Hydrothermal	Pyrenean	Vein	
487	Rud-e-shalang	29	38	0	56	6	30	S	Andesite, pyroclastic, and limestone/ Eocene	Hydrothermal	Post-Pyrenean	Vein	
488	Rud-e-tangu	29	39	30	56	8	30	S	Granodiorite/Oligo-Miocene; andesite basalt with tuff/Eocene	Hydrothermal	Post-Pyrenean	Vein	
489	Rughareh Chah (Chah Maadan)	35	54	0	55	47	30	I	Rhyolite, dacite, and tuff/Eocene	Hydrothermal	Pyrenean	Vein	

490	Sahl Abad	32	18	22	59	54	56	I	Schist and phyllite/Precambrian	Hydrothermal	Pan-African	Vein	
491	Sahlegu	33	25	0	53	25	58	I	Limestone/Cretaceous; shale/Jurassic	Epithermal	Late Cimmerian	Vein	
492	Saleh Dareh	38	39	32	46	56	20	I	Monzonite/Oligo-Miocene; andesite/Eocene	Hydrothermal	Post-Pyrenean	Vein	
493	Sang Siani	26	46	37	58	27	51	I	Volcanics/Mesozoic	Unknown	Unknown	Vein	
494	Sang-e-Asq	29	28	0	57	8	0	I	Shale and pyroclastic/Paleogene	Hydrothermal	Pyrenean	Vein	
495	Sang-e-siyat	29	40	40	56	46	30	I	Porphyrite diorite/Oligo-Miocene	Porphyry	Post-Pyrenean	Vein	
496	Sanjedeh	35	9	50	60	19	45	I	Granite/Oligo-Miocene; andesite and tuff/Eocene	Hydrothermal	Post-Pyrenean	Vein	Fe
497	Sar Lakhlu	38	18	40	47	22	30	I	Porphyrite monzonite/Oligo-Miocene; andesite, tuff, and limestone/Paleocene–Eocene	Porphyry	Post-Pyrenean	Vein	Zn-Pb, Ag, Mo
498	Sar Shirin	35	20	0	54	15	0	I	Andesite and volcanic rocks/Eocene	Hydrothermal	Pyrenean	Vein	
499	Sara	30	26	40	55	8	20	S	Porphyrite quartz diorite/Oligo-Miocene; andesite/Eocene	Porphyry	Post-Pyrenean	Vein	
500	Sarbala	32	51	30	55	44	30	I	Andesitic to dacitic lava and tuff/Paleogene	Hydrothermal	Pyrenean	Vein	
501	Sarband	36	48	0	54	59	0	I	Monzonite/Eocene–Oligocene; volcanoclastic/Eocene	Hydrothermal	Pyrenean	Vein	
502	Sar-cheshmeh	29	56	40	55	52	20	L	Granodiorite/Oligo-Miocene; andesite/Eocene	Porphyry	Post-Pyrenean	Vein	
503	Sarduiyeh	29	12	9	57	16	33	I	Porphyrite diorite/Oligo-Miocene; andesite and pyroclastic/Eocene	Porphyry	Post-Pyrenean	Vein	
504	Sar-e-bagh	29	59	3	56	3	4	S	Granodiorite, porphyrite granodiorite, and porphyrite quartz diorite/Oligo-Miocene; porphyrite andesite, tuffy sandstone/Eocene	Porphyry	Post-Pyrenean	Vein	%0.46 Cu
505	Sar-e-Godarsorkh	33	19	0	54	57	0	I	Conglomerate/Paleocene; schist and serpentinite/Upper Cretaceous	Hydrothermal	Laramide	Vein	Ni, Co, Pb, Zn

(continued)

(continued)

No.	Name	Lat D°	Lat M	Lat S″	Long D°	Long M	Long S″	Size	Host rock/age	Genetic/ mineralization	Orogenic phases	Morphology	Elem para
506	Sar-e-kuh	29	55	50	55	46	0	L	Granodiorite, porphyrite granodiorite, and porphyrite quartz diorite/ Oligo-Miocene; volcanosedimentary rocks/Eocene	Porphyry	Post-Pyrenean	Vein	
507	Sargowd	29	10	0	57	13	30	I	Porphyrite quartz diorite/Oligo-Miocene; andesite/Eocene	Hydrothermal	Post-Pyrenean	Disseminated	
508	Sarkavir-e-Semnan (Sarkalateh Mehran 1)	35	21	8	54	12	33	S	Granodiorite/Oligo-Miocene; andesite and pyroclastics/Eocene	Hydrothermal	Post-Pyrenean	Vein	
509	Sarkavir-e-Semnan (Sarkalateh Mehran 2)	35	21	38	54	11	13	S	Granodiorite/Oligo-Miocene; andesite and pyroclastics/Eocene	Hydrothermal	Post-Pyrenean	Vein	
510	Sarmeshk	29	24	25	57	9	30	S	Granodiorite/Oligo-Miocene; pyroclastics/Eocene	Porphyry	Post-Pyrenean	Disseminated	
511	Sarzen	27	17	18	57	47	18	I	Peridotite and serpentinite/Upper Cretaceous	Unknown	Unknown	Unknown	
512	Sebarz	33	22	0	53	36	0	S	Porphyritic lava and tuff/Tertiary	Hydrothermal	Pyrenean	Vein	
513	Segheynow	30	29	0	54	59	30	S	Porphyrite diorite/Oligo-Miocene; andesite/Eocene	Porphyry	Post-Pyrenean	Disseminated	
514	Sehchangi	32	32	0	58	3	0	S	Ophiolitic complex/Cretaceous	Hydrothermal	Laramide	Unknown	
515	Senj (Doorvan)	36	2	0	50	54	20	S	Porphyrite andesite/Eocene	Hydrothermal	Pyrenean	Vein	
516	Senjedeh	37	20	0	48	13	40	I	Andesite and tuffaceous sandstone/ Eocene	Hydrothermal	Pyrenean	Vein	Pb, Zn, Ag
517	Senjedu	33	10	30	52	50	0	I	Limestone and dolomite/Cretaceous	Hydrothermal	Laramide	Vein	
518	Seridun-e-shomali	29	56	40	55	54	0	I	Andesite and dacite/Eocene	Hydrothermal	Post-Pyrenean	Vein, disseminated	
519	Seyed Mohammad	35	2	25	60	25	15	I	Andesite/Eocene	Hydrothermal	Pyrenean	Veinlet	
520	Shah Ali Beiglu	37	22	0	48	8	0	I	Intrusion and volcanic rocks/Eocene	Hydrothermal	Pyrenean	Vein	
521	Shahmil	37	3	0	48	4	0	I	Andesite/Eocene	Hydrothermal	Pyrenean	Vein	
522	Shalang	29	37	54	56	4	26	I	Andesite and dacite/Eocene	Hydrothermal	Post-Pyrenean	Vein, disseminated	

523	Shams Abad	33	48	0	49	43	40	I	Tuffaceous–sedimentary/Mid–Upper Eocene	Hydrothermal	Pyrenean	Stratiformbed	
524	Shams Abad II	33	25	52	53	25	39	I	Lava flow and andesite/Eocene	Volcanic	Pyrenean	Vein	
525	Sharif Abad	34	8	9	51	10	25	S	Basic intrusive and limestone/ Cretaceous	Hydrothermal	Laramide	Veinlet	
526	Sheikh Ahmad	28	45	50	59	56	0	I	Basic intrusive/Eocene; limestone and volcanic/Upper Cretaceous	Massive sulfide	Pyrenean	Vein, disseminated	
527	Sheikh Ahmad Lu Borazin	38	42	30	46	47	0	I	Basic volcanicand andesite/Upper Eocene	Hydrothermal	Pyrenean	Vein	
528	Sheikh Ali	28	9	0	56	46	20	S	Volcanosedimentary rocks from ophiolite complex/Upper Cretaceous	Massive sulfide	Laramide	Lens	
529	Shekar Abad	37	22	0	47	35	0	I	Andesite/Eocene	Hydrothermal	Pyrenean	Vein	Zn-Pb
530	Shekarab	33	40	34	55	20	40	I	Tuff and andesite/Eocene	Hydrothermal	Pyrenean	Veinlet	
531	Shekasteh-e-Chah Sad	34	25	56	58	33	5	I	Granite/Eocene; andesite and basalt/ Paleogene	Hydrothermal	Pyrenean	Vein	
532	Shekastehsabze	33	2	30	58	22	0	I	Dyke, shale, and sandstone/Jurassic	Hydrothermal	Late Cimmerian	Vein	
533	Shirkkuh	26	59	11	58	18	4	I	Ophiolitic complex, limestone/ cretaceous	Unknown	Unknown	Unknown	
534	Shirvaneh	34	48	40	46	59	0	S	Granite/Tertiary	Hydrothermal	Pyrenean	Veinlet	
535	Shonig	31	56	30	58	53	9	I	Limestone, dolomite, shale/ Paleozoic–Triassic	Unknown	Unknown	Unknown	
536	Shoveh	29	43	20	60	12	0	I	Sedimentary rocks/Paleogene; peridotite and serpentinite/Upper Cretaceous	Unknown	Unknown	Unknown	
537	Shovin	29	40	0	60	4	36	I	Granite, plagiogranite, tonalite/Upper Cretaceous	Hydrothermal	Unknown	Lens	
538	Shurab	33	34	0	58	3	0	I	Andesite/Neogene	Hydrothermal	Pyrenean	Veinlet	
539	Shurak	32	17	0	60	2	52	S	Rhyolite, tuff, and dacite/Eocene	Hydrothermal	Pyrenean	Unknown	
540	Shurak	33	12	6	58	27	46	I	Rhyolite and dacite/Eocene	Hydrothermal	Pyrenean	Veinlet	

(continued)

(continued)

No.	Name	Lat D°	Lat M	Lat S″	Long D°	Long M	Long S″	Size	Host rock/age	Genetic/mineralization	Orogenic phases	Morphology	Elem para
541	Siah Bala	35	14	50	60	22	35	I	Granite/Oligo-Miocene; andesite and tuff/Eocene	Hydrothermal	Post-Pyrenean	Veinlet	
542	Siah Ghara Ghan	27	48	30	57	35	30	I	Basic intrusive and metamorphic rocks/Cretaceous	Hydrothermal	Laramide	Lens	
543	Siah Gharaghan	27	48	30	57	35	30	I	Basic intrusive and metamorphics/Upper Cretaceous	Hydrothermal	Pan-African	Lens	
544	Siah Jekul	29	53	0	60	18	20	I	Granodiorite/Oligo-Miocene; volcanics/Eocene	Porphyry	Post-Pyrenean	Disseminated	
545	Siahkalan	38	37	0	46	44	40	S	Porphyrite monzonite/Oligo-Miocene; andesite, tuff, and limestone/Cretaceous (Paleocene–Eocene?)	Porphyry	Post-Pyrenean	Lens, disseminated, veinlet	Mn, Pb-Zn, Au.Ag
546	Siahkuh	28	31	22	56	45	47	I	Porphyrite quartz diorite/Oligo-Miocene; andesite and pyroclastics/Eocene	Hydrothermal	Post-Pyrenean	Disseminated	
547	Silicat-e-Neyshabur (Adibpur)	36	44	0	58	22	30	I	Basalt and tuff/Eocene	Hydrothermal	Pyrenean	Veinlet	
548	Silijerd	35	9	17	50	16	29	I	Granite/Upper Eocene–Lower Oligocene	Hydrothermal	Pyrenean	Vein	
549	Simiya	36	23	20	50	18	0	I	Volcanics and pyroclastics/Eocene; shale and limestone/Triassic	Hydrothermal	Pyrenean	Veinlet	
550	Sin Abad	29	12	10	57	16	0	I	Diorite/Oligo-Miocene;pyroclastic/Eocene	Hydrothermal	Post-Pyrenean	Veinlet, disseminated	
551	Sirjend (Khanehzar)	35	5	30	50	13	0	I	Andesite and limestone/Upper Cretaceous	Hydrothermal	Laramide	Vein	
552	Sistanak 1	33	37	54	59	36	23	I	Volcanics and intrusive/Jurassic	Hydrothermal	Mid-Cimmerian	Vein	
553	Sistanak 2	33	37	14	59	38	27	I	Basalt and andesite/Paleogene	Hydrothermal	Pyrenean	Vein, veinlet	
554	Soheil	33	15	0	53	1	30	I	Andesite/Eocene; limestone dolomite/Jurassic	Hydrothermal	Pyrenean	Veinlet	
555	Soleymani	27	49	44	57	41	38	I	Ophiolite complex/Upper Cretaceous	Unknown	Unknown	Unknown	

556	Sorkh Darreh	35	45	30	58	15	0	S	Andesite and quartz porphyry/ Tertiary(?) (Upper Cretaceous)	Hydrothermal	Unknown	Vein	
557	Sorkha (Ghalcheh)	29	20	30	57	27	20	I	Ophiolite complex/Upper Cretaceous	Unknown	Unknown	Unknown	
558	Sorkhpay	35	52	0	57	32	30	I	Granite and tuff/Eocene	Hydrothermal	Pyrenean	Unknown	
559	Soungun	38	42	0	46	43	0	L	Porphyrite micromonzonite/Miocene; limestone/Cretaceous	Porphyry, skarn	Post-Pyrenean	Lens, disseminated	Fe, Mo
560	South of Chahar Gonbad	29	32	9	56	7	58	I	Tuff/Eocene	Hydrothermal	Pyrenean	Veinlet	
561	South of Halil Rud	28	49	27	57	12	58	I	Granite/Oligo-Miocene; rhyolite and dacite/Eocene	Hydrothermal	Post-Pyrenean	Vein	
562	South of Kal-e-Azghand	35	18	16	58	51	12	I	Andesite and tuff/Eocene	Hydrothermal	Pyrenean	Unknown	
563	South of Kuh-e-Panj	29	52	9	56	3	54	I	Tuff/Eocene	Hydrothermal	Post-Pyrenean	Disseminated	
564	South of Kusheh1	35	24	20	58	31	48	I	Andesite/Tertiary	Hydrothermal	Pyrenean	Unknown	
565	South of Kusheh2	35	24	16	58	31	38	I	Subvolcanic rocks/Eocene–Oligocene; andesite/Eocene	Hydrothermal	Pyrenean	Vein	
566	South of Lakar Kuh	31	2	26	57	6	43	I	Andesite/Tertiary	Hydrothermal	Unknown	Unknown	
567	South of Mahabad	34	18	36	59	37	59	I	Microsyenite and red limestone/ Tertiary	Hydrothermal	Pyrenean	Vein	
568	South of Mahallati	27	6	23	58	6	40	I	Phyllite and schist/Cretaceous	Unknown	Unknown	Unknown	
569	South of Rayen	29	28	14	57	26	33	I	Tuff and andesite/Eocene	Hydrothermal	Post-Pyrenean	Unknown	
570	South of Seh Farsakh	31	17	54	60	2	24	I	Rhyolite, dacite, and limestone/Eocene	Hydrothermal	Pyrenean	Veinlet	
571	South of Siah Rud	36	44	57	49	34	30	I	Andesite/Eocene	Hydrothermal	Pyrenean	Vein, veinlet, disseminated	
572	South of Toshab	29	40	55	56	34	56	I	Tuff/Eocene	Hydrothermal	Pyrenean	Veinlet	
573	South west Of Chah Akhteh 1	35	13	54	57	10	13	I	Andesite/Paleogene	Hydrothermal	Pyrenean	Vein	
574	South west Of Chah Akhteh 2	35	13	30	57	12	41	I	Andesite, tuff/Eocene; granite, granodiorite/Tertiary	Hydrothermal	Pyrenean	Lens	
575	South West of Darreh Garm	30	0	49	56	0	0	I	Volcanosedimentary rocks/Eocene	Hydrothermal	Pyrenean	Vein	

(continued)

(continued)

No.	Name	Lat D°	Lat M	Lat S″	Long D°	Long M	Long S″	Size	Host rock/age	Genetic/ mineralization	Orogenic phases	Morphology	Elem para
576	South West Of Deh Tahir	30	9	11	55	50	19	I	Monzonite, granosyenite/Tertiary	Hydrothermal	Pyrenean	Vein	
577	South West Of Estin	31	17	50	60	2	30	I	Basaltic pillow lava/Cretaceous	Massive sulfide	Laramide	Lens	
578	South West Of Kalateh-ye-Bala	34	9	52	58	55	16	I	Shale, sandstone/Jurassic	Hydrothermal	Pyrenean	Vein	
579	Southeast of Hamand	32	26	38	59	1	14	I	Rhyolite, dacite, and tuff/Eocene	Hydrothermal	Pyrenean	Unknown	
580	Southeast of Kuh-e-Panj	29	48	47	56	7	58	I	Tuff/Eocene	Hydrothermal	Pyrenean	Veinlet	
581	Southeast of Mour	32	39	33	59	40	13	I	Basalt and andesite/Tertiary	Hydrothermal	Pyrenean	Unknown	
582	Southeast of Shahr-e-Babak	29	54	19	55	15	18	I	Sedimentary rocks/Paleogene	Unknown	Unknown	Unknown	
583	Southern Cheheltan	29	44	0	56	25	20	I	Tuff and andesite/Eocene	Hydrothermal	Pyrenean	Unknown	
584	Southwest of Bardsir	29	52	25	56	24	3	I	Porphyrite diorite and porphyrite quartz diorite/Oligo-Miocene; andesite and pyroclastites/Eocene	Porphyry	Post-Pyrenean	Disseminated	
585	Southwest of Bashagard fault	26	47	14	58	28	1	I	Ophiolite complex and limestone/ Upper Cretaceous	Unknown	Unknown	Unknown	
586	Southwest of Chah Akhteh 1	35	13	58	57	10	14	I	Andesite/Tertiary	Hydrothermal	Pyrenean	Unknown	
587	Southwest of Chah Akhteh 2	35	13	34	57	12	41	I	Andesite/Tertiary	Hydrothermal	Pyrenean	Unknown	
588	Southwest of Choghondarsar Shomali	36	17	0	56	16	10	I	Andesite/Eocene	Hydrothermal	Pyrenean	Vein, veinlet, disseminated	
589	Southwest of Gorazi	35	1	55	60	5	0	I	Andesite and tuff/Eocene	Hydrothermal	Pyrenean	Veinlet	
590	Southwest of Halatun	26	40	40	58	48	38	I	Ophiolite complex and limestone/ Upper Cretaceous	Unknown	Unknown	Unknown	
591	Southwest of Hengam	32	4	19	59	10	0	I	Andesite/Eocene	Hydrothermal	Pyrenean	Unknown	

592	Southwest of Khousf	32	44	3	58	41	8	I	Basalt/Tertiary; ophiolite complex/ Cretaceous	Hydrothermal	Pyrenean	Unknown	
593	Southwest of Kuh-e-Gir 1	26	43	51	58	47	43	I	Ophiolite complex and limestone/ Upper Cretaceous	Unknown	Unknown	Unknown	
594	Southwest of Kuh-e-Gir 2	26	45	16	58	48	20	S	Ophiolite complex and limestone/ Upper Cretaceous	Unknown	Unknown	Unknown	
595	southwest of Kuh-e-Govag	27	57	17	57	21	40	I	Phyllite and schist/Jurassic	Unknown	Unknown	Unknown	
596	Southwest of Mour	32	39	49	59	28	6	I	Peridotite and serpentinite/Upper Cretaceous	Unknown	Unknown	Unknown	
597	Southwest of Sar Chah Shur	32	13	28	58	41	52	I	Andesite and tuff/Eocene; schist and phyllite/Precambrian	Hydrothermal	Pyrenean	Veinlet	
598	Southwest of Sorkh Kuh	33	13	18	58	21	38	I	Shale/Triassic–Jurassic; metamorphic rocks and limestone/Triassic Jurassic	Hydrothermal	Unknown	Disseminated	
599	Southwest of Tanoureh	35	3	30	60	7	30	I	Andesite/Eocene	Hydrothermal	Pyrenean	Vein	
600	Surakh-e-mar 1	29	18	0	57	26	0	I	Diorite and porphyrite diorite/ Oligo-Miocene; andesite/Eocene	Porphyry	Post-Pyrenean	Disseminated	
601	Surakh-e-mar 2	29	10	15	57	23	0	I	Porphyrite quartz diorite/Oligo-Miocene; quartzite, hornfels, and volcanosedimentary/Eocene	Porphyry	Post-Pyrenean	Disseminated	
602	Takht	29	36	0	56	16	30	S	Tuff/Eocene	Hydrothermal	Pyrenean	Veinlet	
603	Takht Chahar Gonbad	29	36	0	56	13	0	I	Diorite/Oligo-Miocene; volcanics and limestone/Eocene	Hydrothermal	Post-Pyrenean	Veinlet	
604	Takht-e-baneh	29	35	15	56	12	15	S	Limestone and tuff/Eocene	Porphyry	Post-Pyrenean	Disseminated	
605	Takht-e-Chaman	35	10	30	50	31	10	I	Chlorite schist/Precambrian	Chlorite schist/ Precambrian	Chlorite schist/ Precambrian	Chlorite schist/ Precambrian	
606	Takieh-e-Taleghan	36	10	20	50	36	30	I	Volcanic and volcanoclastic/Eocene	Hydrothermal	Pyrenean	Veinlet	
607	Taknar	35	27	0	57	48	0	S	Marble and schist/Upper Precambrian	Hydrothermal	Pan-African	Veinlet, disseminated	Ni, Co, U, Au, Ag

(continued)

(continued)

		Lat			Long								
No.	Name	D°	M	S″	D°	M	S″	Size	Host rock/age	Genetic/mineralization	Orogenic phases	Morphology	Elem para
608	Talarji	33	26	20	54	2	30	I	Limestone and schist/Upper Proterozoic–Lower Cambrian	Hydrothermal	Pan-African	Veinlet,	Au
609	Taleh Seiah	33	23	0	54	57	0	S	Porphyrite diorite/Eocene; red and evaporite bed/Jurassic	Hydrothermal	Pyrenean	Veinlet	
610	Talkheh	33	20	8	53	35	0	S	Marble and schist/Upper Precambrian	Hydrothermal	Pan-African	Veinlet, disseminated	Ni, Co, U, Au, Ag
611	Talle Maadan	29	22	0	56	57	40	I	Diorite, granite, tuff/Eocene	Hydrothermal	Pyrenean	Veinlet	Au
612	Talmesi	33	22	41	53	27	30	M	Andesite/Eocene	Hydrothermal	Pyrenean	Veinlet, vein, disseminated	Au
613	Tamam Deh	31	49	41	60	32	23	I	Granite/Tertiary; marble/Cretaceous	Hydrothermal	Pyrenean	Vein	
614	Tang-e-Chenar	31	22	30	54	22	0	I	Basalt/Tertiary	Hydrothermal	Pyrenean	Unknown	
615	Tangel Chamani	32	54	0	55	43	40	I	Gabbro and Diorite/Mesozoic	Hydrothermal	Pyrenean	Unknown	
616	Tappeh Seyfollah	36	31	0	49	31	0	I	Granite/Eocene Andesitic tuff/Eocene	Hydrothermal	Pyrenean	Vein, veinlet	
617	Tarik Darreh	35	29	55	60	42	30	I	Diorite/Post-Jurassic, Shale, Sandstone/Jurassic	Hydrothermal	Late Cimmerian	Vein	
618	Tarkin	33	18	0	51	54	0	I	Porphyritic lava and Tuff/Tertiary	Hydrothermal	Pyrenean	Vein, veinlet	
619	Tek Tekeh (Kayaz)	33	59	30	52	42	30	S	Recrystallization marble/Upper Precambrian–Lower Cambrian	Hydrothermal	Pan-African	Vein	
620	Tezelgahai	31	57	51	58	54	48	I	Basalt/Cretaceous	Massive sulfide	Laramide	Lens	
621	Tezerg (Tizark)	30	34	20	55	3	40	I	Granodiorite and gabbro/Oligo-Miocene; andesitic lava/Eocene	Hydrothermal	Post-Pyrenean	Vein	
622	Tigh NowAb	32	6	29	60	34	28	S	Tuff/Eocene	Hydrothermal	Pyrenean	Veinlet	
623	Tirkuh	30	36	0	55	0	10	I	Granite/Oligo-Miocene	Hydrothermal	Post-Pyrenean	Veinlet	
624	Titoeyeh	29	22	0	56	29	0	I	Andesite and agglomerate/Eocene	Hydrothermal	Post-Pyrenean	Vein	
625	Torkamani	33	13	0	54	1	0	I	Metamorphic rocks and ulterabasic rocks/Upper Precambrian	Hydrothermal	Pan-African	Vein	
626	Tovazari	33	24	14	54	15	34	I	Granodiorite/Upper Jurassic; schist, marble/Proterozoic	Hydrothermal	Late Cimmerian	Vein	

627	Tut	32	32	50	54	25	22	I	Andesite/Eocene	Hydrothermal	Pyrenean	Vein	
628	Uros Morghi	29	9	30	57	10	30	I	Granite, tuff/Eocene; andesite, dacite, rhyolite/Eocene	Hydrothermal	Pyrenean	Vein, veinlet	
629	Uzun Darreh	37	28	0	48	4	0	I	Limestone/Cretaceous; marble/Jurassic	Unknown	Unknown	Vein	
630	Vafez	34	50	30	49	23	0	S	Porphyritic lava and tuff/Tertiary (Eocene)	Hydrothermal	Pyrenean	Lens	
631	Valimohammad	35	9	30	48	24	30	I	Granite, tuff/Eocene	Hydrothermal	Pyrenean	Lens	
632	Varche Jajarm	37	5	30	57	15	23	I	Porphyry intrusive/Eocene; limestone, tuff/Cretaceous	Skarn	Pyrenean	Lens	Fe
633	Varzaj 1	33	37	50	59	22	20	I	Andesite basalt/Paleogene	Hydrothermal	Pyrenean	Vein and veinlet	
634	Varzaj 2	33	36	45	59	24	37	I	Andesite basalt/Paleogene	Hydrothermal	Pyrenean	Vein and veinlet	
635	Varzaj 3	33	37	10	59	22	59	S	Andesite basalt/Paleogene	Hydrothermal	Pyrenean	Vein and veinlet	
636	Varzaj 4	33	37	58	59	24	23	i	Andesite basalt/Paleogene	Hydrothermal	Pyrenean	Vein and veinlet	
637	Veshnaveh	34	15	0	51	0	0	S	Andesite/Eocene	Hydrothermal	Pyrenean	Veinlet	
638	West of Bahadrak	27	44	19	57	17	3	I	Phyllite and schist/Jurassic	Unknown	Unknown	Unknown	
639	West of Bahr-e-Aseman 1	29	3	14	57	14	41	I	Granite/Oligo-Miocene; rhyolite and dacite/Eocene	Hydrothermal	Post-Pyrenean	Unknown	
640	West of Bahr-e-Aseman 2	29	5	56	57	27	0	I	Andesite and tuff/Eocene	Hydrothermal	Post-Pyrenean	Unknown	
641	West of Dozad	28	51	16	59	43	19	I	Granite/Oligo-Miocene; rhyolite and dacite/Eocene	Hydrothermal	Post-Pyrenean	Unknown	
642	West Of Kalateh-ye-Bala	34	11	21	58	53	19	I	Dacite, andesite, basalt, pyroclastic/Paleogene	Hydrothermal	Pyrenean	Vein	
643	West of Kal-e-Azghand	35	19	55	58	48	35	I	Andesite/Eocene	Hydrothermal	Post-Pyrenean	Vein and veinlet	
644	West of Kuh-e-Ahurak	29	35	34	56	45	0	I	Tuff/Eocene	Hydrothermal	Post-Pyrenean	Veinlet	
645	West of Kuh-e-Govag	27	59	15	57	19	8	I	Ophiolite complex and limestone/Upper Cretaceous	Unknown	Unknown	Unknown	
646	West of Kuh-e-Halatun	26	41	12	58	52	16	I	Ophiolite complex and limestone/Upper Cretaceous	Unknown	Unknown	Unknown	

(continued)

(continued)

No.	Name	Lat D°	Lat M	Lat S″	Long D°	Long M	Long S″	Size	Host rock/age	Genetic/ mineralization	Orogenic phases	Morphology	Elem para
647	West of Kuh-e-Jeyran	26	44	12	58	40	3	I	Ophiolite complex and limestone/ Upper Cretaceous	Unknown	Unknown	Unknown	
648	West of Kuh-e-Lalehzar	29	22	50	56	31	33	I	Granite/Oligo-Miocene; rhyolite and dacite/Eocene	Hydrothermal	Post-Pyrenean	Unknown	
649	West of Kuh-e-Madvar	30	37	1	54	44	1	I	Rhyolite and dacite/Eocene	Hydrothermal	Pyrenean	Veinlet	
650	West of Kuh-e-Mishbadam	27	47	27	57	22	44	I	Gabbro, diorite/Mesozoic	Hydrothermal	Laramide	Unknown	
651	West of Lachah	30	25	56	55	2	48	I	Rhyolite and dacite/Eocene	Hydrothermal	Pyrenean	Veinlet	
652	West of Sarduiyeh	29	12	9	57	12	11	I	Granite/Oligo-Miocene; andesite/ Eocene	Hydrothermal	Post-Pyrenean	Vein, disseminated	
653	West to northwest of Khaliran	32	41	27	58	56	48	I	Pillow lava, pyroxenite/Upper Cretaceous	Hydrothermal	Pyrenean	Unknown	
654	Yakhab	33	58	38	52	11	19	I	Shale and sandstone/Jurassic; limestone/Triassic	Hydrothermal	Unknown	Unknown	
655	Yakhab	36	31	55	49	41	0	I	Andesite/Eocene	Hydrothermal	Pyrenean	Lens, massive	
656	Yamakhan	36	35	0	49	5	30	I	Andesite/Eocene	Hydrothermal	Pyrenean	Lens, massive	
657	Yamghan	36	35	0	49	5	30	S	Granite/Eocene–Oligocene; andesitic tuff/Eocene	Hydrothermal	Pyrenean	Vein, veinlet	
658	YazBanoo	34	12	17	59	48	19	I	Volcanosedimentary rocks/Upper Cretaceous	Hydrothermal	Laramide	Vein	
659	Zali Bulagh (Kuh panj)	35	10	0	50	36	0	I	Andesite/Paleogene	Hydrothermal	Laramide	Lens	
660	Zamin-e-Hoseyn	29	7	0	57	15	0	I	Granite/Oligo-Miocene; rhyolite and dacite/Eocene	Hydrothermal	Post-Pyrenean	Unknown	
661	Zanaghlu (Zanaglu)	35	29	0	57	36	0	S	Andesite/Paleogene	Hydrothermal	Laramide	Lens	
662	Zand Abad	38	37	0	46	56	0	S	Granite/Upper Eocene–Lower Oligocene; volcanics/Paleogene; limestone/Cretaceous	Skarn	Post-Pyrenean	Lens	Au, Pb

663	Zandiyeh	30	27	57	55	34	22	I	Granodiorite/Jurassic; sandstone, siltstone, quartzite/Tertiary	Porphyry	Mid-Cimmerian	Disseminated	
664	Zangzin	29	10	0	57	12	0	I	Granodiorite and diorite/Oligo-Miocene; volcanics/Eocene	Porphyry	Post-Pyrenean	Disseminated	
665	Zarmak-e-delijan (Atash Kuh)	33	52	5	50	37	20	S	Intrusive and volcanic rocks	Hydrothermal	Mid-Cimmerian	Vein	
666	Zarrin Khani	36	33	30	49	36	50	S	Basalt and tuff/Eocene	Hydrothermal	Pyrenean	Veinlet	
667	Zarrineh Rekab	38	40	10	46	45	0	I	Andesite and subvolcanic/ Eocene–Oligocene	Hydrothermal	Pyrenean	Disseminated	Ag
668	Zaruiyeh	30	28	46	54	58	51	I	Volcanics and sedimentary/Eocene	Hydrothermal	Pyrenean	Vein	
669	Zaveh Olia	30	28	18	54	58	52	I	Rhyolite and dacite/Eocene	Hydrothermal	Pyrenean	Veinlet	
670	Zavork	29	1	0	57	44	30	I	Porphyrite rhyolite/Eocene	Hydrothermal	Pyrenean	Vein, veinlet	
671	Zebra Kuh 1	34	52	42	57	27	43	S	Rhyolite, dacite, tuff/Paleogene	Hydrothermal	Pyrenean	Vein	
672	Zereshlu	38	18	0	47	17	40	I	Dyke and intrusive rocks/Unknown	Porphyry	Unknown	Vein	
673	Ziaran	36	8	49	50	33	32	I	Andesite/Eocene	Hydrothermal	Pyrenean	Lens	Fe
674	Ziarat	27	25	27	57	46	23	I	Limestone, phyllite/Cretaceous	Massive sulfide	Laramide	Lens	

Complete list of lead and zinc mineral deposits and indications of Iran along with their detailed specifications

No.	Name	Lat D°	Lat M	Lat S″	Long D°	Long M	Long S″	Size	Host rock/age	Genetic/ mineralization	Orogenic phases	Morphology	Elem para
1	Ab Bagh Koruyeh	31	54	0	51	45	0	I	Limestone/Cretaceous	Hydrothermal	Laramide	Unknown	
2	Chah Ahan (Kal-e-Sabz)	34	15	30	57	5	40	S	Shale, sandstone, and limestone/ Mid-Jurassic	Sedimentary	Mid-Cimmerian	Disseminated, vein	Cu, Fe
3	Ab Bagh-e-Koruyeh Shahr Reza	31	46	19	51	48	7	I	Limestone/Lower Cretaceous	Mississippi Valley Type	Laramide	Unknown	
4	Ab Bid	30	53	55	56	58	55	I	Limestone and dolomite/Permo-Triassic	Sedimentary	Hercynian–Early Cimmerian	Unknown	
5	Ab Chamutash-e-Shahrud	36	30	30	54	41	0	S	Limestone/Triassic–Jurassic	MVT	Early Cimmerian	Vein, veinlet	
6	Abbas Abad (Kerman)	31	32	45	56	17	58	I	Dolomite/Permo-Triassic	Mississippi Valley Type	Hercynian	Vein	
7	Abdol Abad	34	9	23	56	31	5	I	Dolomite/Mid-Triassic	Sedimentary	Laramide	Vein, veinlet, disseminated	
8	Abgarm	35	39	45	53	11	20	S	Sandy limestone/Jurassic	Hydrothermal	Early Cimmerian	Vein	
9	Abgarm	35	39	45	53	11	20	s	Sandstone limestone/Jurassic	Hydrothermal	Late Cimmerian	Vein	Ba
10	Abheydar	31	56	36	56	4	29	S	Dolomite and limestone/Mid-Triassic	Hydrothermal	Early Cimmerian	Vein	Ba, F
11	Abkamar	34	38	14	50	3	0	I	Pyroclastics/Eocene; carbonates/ Upper Cretaceous	Hydrothermal	Pyrenean	Unknown	
12	Abolhasani	35	21	50	54	38	30	S	Andesite/Eocene–Oligocene	Hydrothermal	Pyrenean	Vein	Ba
13	Abolhasani	35	21	50	54	38	30	s	Andesite/Eocene–Oligocene	Hydrothermal	Pyrenean	Vein	Cu
14	Abyek-e-Eshtehard 1	36	19	0	50	32	0	I	Tuff and pyroclastics/Tertiary	Hydrothermal	Pyrenean	Unknown	
15	Abyek-e-Eshtehard 2	35	38	0	50	19	0	I	Volcanics and pyroclastics/ Tertiary(Eocene)	Volcanic	Pyrenean	Unknown	Cu
16	Aghuzaki	36	44	0	50	35	0	I	Limestone/Mid-Permian	Hydrothermal	Hercynian	Unknown	
17	Ahangaran	34	10	11	48	59	58	M	1- Dolomite/Lower Cretaceous; 2- slate/Jurassic	Mississippi Valley Type (MVT) and SEDEX	Early Cimmerian	Stratiformbed, lens	Cu, Ba, Ag
18	Ahmad Abad	31	58	0	55	54	0	S	Limestone and dolomite/Upper Triassic	Mississippi Valley Type	Early Cimmerian	Lens	
19	Ahou	34	34	3	50	2	46	I	Pyroclastics/Eocene; carbonates/ Upper Cretaceous	Hydrothermal	Pyrenean	Unknown	

20	Ahovanu	36	13	7	54	10	54	S	Limestone/Mid-Jurassic	Sedimentary	Mid-Cimmerian	Disseminated	
21	Ahuan Gordel Kuh	34	11	31	49	25	15	I	Dolomite and limestone/Cretaceous	Epithermal	Pyrenean	Vein	Sb, Ag
22	Ai Qalehsi	36	20	24	47	22	30	S	Limestone and sandstone, tuff/Oligo-Miocene	Hydrothermal	Post-Pyrenean	Vein	Au
23	Akhlamad	36	41	0	58	50	0	S	Sandstone and shale/Jurassic	Unknown	Unknown	Unknown	
24	Akhory	35	17	12	54	27	21	I	Dolomite/Devonian	Hydrothermal	Pyrenean	Disseminated	
25	Akhory	35	17	12	54	27	21	I	Dolomite/Devonian	Hydrothermal	Pyrenean	Disseminated	
26	Akhtarchi	33	47	0	50	15	0	I	Dolomite limestone/Lower Cretaceous	Mississippi Valley Type (MVT)	Laramide	Unknown	
27	Akuhi	35	50	0	52	31	0	I	Limestone/Cretaceous	Sedimentary	Late Cimmerian	Unknown	
28	Akuhi	35	50	0	52	31	0	I	Limestone/Cretaceous	Sedimentary	Late Cimmerian	Unknown	
29	Alam Kandy	36	45	32	47	15	40	M	Schist, amphibolite, and marble/Upper Precambrian–Lower Cambrian	SEDEX	Pan-African	Lens, vein	
30	Alamut	33	47	0	50	28	0	I	Sandstone and carbonates/Lower Cretaceous	Hydrothermal	Pyrenean	Vein	
31	Alavijeh	33	4	0	51	7	0	I	Limestone/Lower Cretaceous	Mississippi Valley Type (MVT)	Laramide	Unknown	
32	Ali Abad va Dasht-e-Baft	28	56	0	56	30	0	S	Limestone/Devonian	Hydrothermal	Early Cimmerian	Vein	
33	Amin Abad	35	35	30	51	30	20	I	Andesite/Eocene	Volcanic	Pyrenean	Unknown	
34	Amirnan	36	11	53	50	32	11	I	Limestone, volcanics, and pyroclastics/Paleocene–Eocene	Hydrothermal	Pyrenean	Vein, lens	Ba
35	Amrudak	28	43	15	60	55	0	S	Meta-phyllitic shale/Eocene	Hydrothermal	Post-Pyrenean	Vein	
36	Anabu	35	48	9	53	3	55	S	Carbonaceous rocks/Jurassic	Hydrothermal	Laramide	Vein, disseminated	
37	Anabu	35	48	9	53	3	55	s	Carbonates/Jurassic	Hydrothermal	Laramide	Vein, disseminated	Cu, Ba
38	Analou	35	57	0	55	58	30	I	Metamorphic rocks and carbonates/Cretaceous	Hydrothermal	Pyrenean	Unknown	Cu, Ba

(continued)

(continued)

No.	Name	Lat			Long			Size	Host rock/age	Genetic/ mineralization	Orogenic phases	Morphology	Elem para
		D°	M	S″	D°	M	S″						
39	Anaru (Qulleh Anaru)	35	18	29	54	29	28	S	1- Gray limestone/Devonian (Upper Paleozoic); 2- dolomite, limestone/Triassic	Hydrothermal	Early Cimmerian	Disseminated, vein	
40	Anaru (Qulleh Anaru)	35	18	29	54	29	28	s	Gray limestone/Devonian; dolomite and limestone/Triassic	Hydrothermal	Late Cimmerian	Vein, disseminated	
41	Anguran	36	37	40	47	24	20	L	Schist, dolomite, and marble/Upper Precambrian–Lower Cambrian	Massive sulfide	Pan-African	Lens	Cd
42	Anjerd	38	40	13	46	55	10	S	Monzonite/Eocene–Oligocene; tuff and limestone/Cretaceous	Skarn	Pyrenean	Lens	Cu, Ba
43	Anjireh	31	43	41	54	13	54	S	Dolomite and limestone/Lower Cretaceous	Mississippi Valley Type	Laramide	Vein	Fe
44	Anjireh Kuhpayeh	33	12	57	52	2	54	I	Sandstone and carbonates/Lower Cretaceous	Mississippi Valley Type	Laramide	Vein	
45	Anjireh Tiran	32	44	43	51	7	30	S	Limestone, marl and marl limestone/ Lower Cretaceous	Mississippi Valley Type (MVT)	Laramide	Lens, vein	Cu
46	Arask-e-Damghan (Cheshmeh Naanai)	36	9	6	54	0	23	S	Limestone, dolomitic limestone/ Upper Jurassic	Sedimentary	Late Cimmerian	Vein	
47	Arous kuh	35	58	0	53	3	0	I	Shale, sandstone, and limestone/ Jurassic	Hydrothermal	Pyrenean	Unknown	
48	Arpachay	36	47	0	47	58	0	S	Sandy tuff and Acidic tuff/ Oligo-Miocene	Hydrothermal	Post-Pyrenean	Vein	Cu
49	Arreh	32	31	45	51	36	36	I	Sandy dolomite/Lower Cretaceous	Mississippi Valley Type (MVT)	Laramide	Unknown	
50	Arusan	33	24	13	54	26	49	I	Schist and marble/Upper Proterozoic–Lower Paleozoic	Hydrothermal	Pan-African(?)	Vein, stratiformbed	Ag
51	Arzuyeh (Esfandagheh 2)	28	37	50	56	22	40	I	Basic dyke/Upper Triassic; calc and dolomite marble and green schist/ Devonian	Hydrothermal	Early Cimmerian	Vein	

52	Asagi	30	35	37	60	11	16	S	Andesite and dacite/Oligo-Miocene	Hydrothermal	Post-Pyrenean	Vein, disseminated	Cu, Fe
53	Asaran	35	51	19	53	16	55	S	Massive limestone/Upper Cretaceous	Sedimentary	Laramide	Veinlet	
54	Asbeh Kosh	34	46	13	50	20	45	I	Tuff and Andesite/Eocene	Hydrothermal	Pyrenean	Lens	Ba, Cu
55	Ashtiyan 1	34	32	11	50	0	48	S	Pyroclastics/Eocene	Hydrothermal	Pyrenean	Unknown	
56	Ashtiyan 2	34	30	48	50	4	7	I	Pyroclastics/Eocene	Hydrothermal	Pyrenean	Unknown	
57	Ashtiyan 3	34	33	57	50	2	48	S	Dolomite and limestone/Cretaceous	Hydrothermal	Pyrenean	Vein	
58	Atash Kuh-e-Mahallat	33	53	56	50	38	0	I	Thick-bedded limestone and dolomite/Cretaceous	Unknown	Unknown	Unknown	
59	Avan	36	30	32	50	28	39	I	Volcanosedimentary/Triassic	Hydrothermal	Early Cimmerian	Vein	
60	Aynalou	35	57	0	55	58	30	I	Metamorphic and carbonate rocks/Cretaceous	Hydrothermal	Pyrenean	Unknown	
61	Ayrijeh	33	46	28	49	45	7		Limestone/Lower Cretaceous	MVT	Laramide	Vein	
62	Baba Qolleh	33	33	20	49	59	40	S	Dolomite and limestone/Lower Cretaceous	MVT	Laramide	Vein, veinlet	Cu
63	Baba Rais	34	18	48	48	57	16	I	Quartzite/Jurassic	Mississippi Valley Type (MVT)	Laramide	Vein, veinlet	Cu, Fe
64	Baba Sheikh	33	29	41	50	16	0	S	Thick-bedded limestone/Cretaceous	Hydrothermal	Laramide	Vein, veinlet	
65	Baey	35	8	38	60	11	15	I	Limestone/Triassic	Hydrothermal	Pyrenean	Vein	
66	Bafgh	31	55	0	55	54	0	I	Dolomite/Triassic	Hydrothermal	Early Cimmerian	Vein	
67	Bagh Bahadoran	32	28	0	51	10	0	I	Sandstone and carbonates/Lower Cretaceous	Mississippi Valley Type	Laramide	Vein	
68	Bagh Bakhshi	34	16	29	59	44	9	I	Volcanics and pyroclastics/Tertiary	Hydrothermal	Pyrenean	Unknown	
69	Bagh Ghorogh	33	35	38	53	48	49	I	Dolomite and limestone/Upper Cretaceous	Hydrothermal	Pyrenean	Vein	
70	Bagh Kumeh	32	31	30	51	36	30	I	Sandstone and carbonates/Lower Cretaceous	Mississippi Valley Type	Laramide	Vein	
71	Bagher Abad	32	17	0	51	16	0	I	Dolomite, limestone/Cretaceous; shale, sandstone/Jurassic	MVT	Laramide	Vein	
72	Bahram Abad	30	48	37	56	52	39	S	Limestone/Cretaceous	Hydrothermal	Unknown	Vein	

(continued)

(continued)

No.	Name	Lat D°	Lat M	Lat S″	Long D°	Long M	Long S″	Size	Host rock/age	Genetic/ mineralization	Orogenic phases	Morphology	Elem para
73	Bahram Taj	32	19	0	53	30	32	S	Limestone, shale, and sandy tuff/ Upper Proterozoic–Lower Paleozoic	SEDEX	Pan-African	Vein	
74	Bajgan	31	27	52	55	53	45	I	Limestone/Triassic–Jurassic	Sedimentary	Cimmerian	Vein	
75	Bala Kuh	37	10	26	48	48	5	I	Limestone/Upper Devonian–Lower Permian	Hydrothermal	Hercynian	Vein, lens, Disseminated	
76	Balam	33	25	0	49	55	0	I	Limestone/Lower Cretaceous	MVT	Laramide	Vein	
77	Balamcheh Ardestan	33	3	53	51	58	6	I	Dolomitic limestone/Lower Cretaceous	MVT	Laramide	Vein	
78	Bam Kamar	31	15	0	59	17	0	I	Carbonates/Cretaceous	Hydrothermal	Early Cimmerian	Vein	
79	Band-e-Bagh	29	53	35	56	1	55	I	Andesite/Mid-Eocene	Hydrothermal	Pyrenean	Vein, veinlet	
80	Band-e-Gel II	33	21	11	55	1	12	I	Limestone and shale/Lower Cretaceous	Hydrothermal	Pyrenean	Vein	Cu, Au, Ag
81	Baneh	34	5	18	49	23	19	S	Thickness layers limestone/Lower Cretaceous	Mississippi Valley Type (MVT)	Laramide	Veinlet	
82	Barik Ab	36	16	46	49	18	21	S	Breccia tuff/Eocene	Hydrothermal	Pyrenean	Vein	Cu, Au
83	Basari	34	5	57	49	15	50	S	Gray limestone and marl/Upper Cretaceous	Mississippi Valley Type (MVT)	Laramide	Veinlet	Ag, Fe
84	Betu Neyshbur	36	26	0	58	17	0	I	Andesite/Tertiary	Volcanic	Pyrenean	Unknown	
85	Bibi Shahrbanoo	35	36	0	51	28	0	I	Dolomite/Precambrian	Volcanic	Pan-African (?)	Unknown	
86	Bidu	34	52	0	57	22	0	S	Thick-bedded to massive dolomite/ Upper Precambrian–Lower Cambrian	Sedimentary	Pan-African	Vein	
87	Bikhtang	34	13	37	48	56	2	S	Dolomite and limestone/Lower Cretaceous	Mississippi Valley Type (MVT)	Laramide	Stratiformbed	Cu, Fe
88	Binaloud	36	6	35	56	32	10	I	Dolomite/Triassic; shale/Jurassic	MVT	Mid-Cimmerian	Vein	
89	Bivarzan	36	42	0	49	34	35	S	Massive dolomite and sandy limestone/Permian	Hydrothermal	Early Cimmerian	Vein	
90	Boghose	33	36	4	57	14	30	I	Reef limestone/Jurassic	MVT	Late Cimmerian	Vein	

91	Bolazoom	33	57	0	53	8	0	I	Volcanics and pyroclastics/Eocene	Hydrothermal	Pyrenean	Vein	
92	Bolmocheh	33	3	53	51	58	6	S	Thick-bedded limestone/Lower Cretaceous	Mississippi Valley Type (MVT)	Laramide	Veinlet	
93	Bon Anar	31	57	39	55	54	6	S	Limestone and dolomite/Permo-Triassic	Sedimentary	Hercynian	Vein	
94	Bonidar	34	10	52	49	8	0	S	Breccia sandstone/Lower Cretaceous	Mississippi Valley Type (MVT)	Laramide	Vein	Cu, Au
95	Borj Band Kuh	33	24	0	50	33	30	I	Limestone/Lower Cretaceous	MVT	Laramide	Vein	
96	Borzoo	34	10	49	59	43	24	I	Volcanics and pyroclastics/Tertiary	Hydrothermal	Pyrenean	Unknown	
97	Boz Kamar	35	23	46	54	33	45	I	Breccia volcanic/Eocene–Oligocene	Hydrothermal	Pyrenean	Vein	
98	Bozbarreh-ye-AmmarLu	36	44	0	49	34	25	S	Massive dolomite and sandy limestone/Permian	Hydrothermal	Early Cimmerian	Vein	Cu, Ag, Cd
99	Bozeh	35	23	48	54	33	45	I	Lava and breccia volcanics/Eocene	Hydrothermal	Pyrenean	Vein	
100	Bozgoush	33	26	33	54	43	16	I	Limestone/Lower Cretaceous	Mississippi Valley Type	Laramide	Unknown	
101	Buteh Gaz	35	28	25	60	48	45	I	Intrusive and volcanic rocks/Tertiary; carbonates/Triassic	Hydrothermal	Pyrenean	Vein	
102	Chadorgi	33	20	37	50	34	48	S	Limestone/Upper Cretaceous	Mississippi Valley Type	Laramide	Vein, veinlet	Ba, Cu
103	Chah Allah	30	29	50	56	21	20	I	Limestone/Jurassic	Sedimentary	Early–Middle Cimmerian	Unknown	
104	Chah Farakh	35	28	56	54	18	53	S	Tephrite/Eocene	Hydrothermal	Pyrenean	Vein	Cu
105	Chah Gaz	29	31	5	55	2	20	I	Schist/Triassic–Jurassic	Massive sulfide	Late Cimmerian	Vein	
106	Chah Gorbeh	33	25	3	53	44	48	I	Marbles/Upper Precambrian–Lower Cambrian	Hydrothermal	Pyrenean	Vein	Ag
107	Chah Kharbozeh	33	28	51	53	47	54	I	Limestone and limestone conglomerate/Lower Cretaceous	Hydrothermal	Laramide–Pyrenean	Vein, veinlet	Cu
108	Chah Makan	30	22	30	56	34	10	I	Dolomite/Permo-Triassic	Sedimentary	Early Cimmerian	Unknown	
109	Chah Mesi	29	25	29	55	10	20	I	Andesite and basalt/Mid-Eocene	Hydrothermal	Pyrenean	Vein	Cu, Au
110	Chah Mileh	33	25	42	53	48	41	I	Schist and marble/Upper Proterozoic–Lower Paleozoic	Hydrothermal	Pan-African (?) Eocene (?)	Vein	

(continued)

(continued)

No.	Name	Lat D°	Lat M	Lat S″	Long D°	Long M	Long S″	Size	Host rock/age	Genetic/ mineralization	Orogenic phases	Morphology	Elem para
111	Chah Musa	35	29	0	54	53	30	S	Andesite, trachyandesite, dacite/ Eocene	Hydrothermal	Pyrenean	Vein	Ba, Cu
112	Chah Shirin	35	21	0	54	5	57	I	Dolomite and limestone/Devonian	Hydrothermal	Pyrenean	Veinlet, vein	Cu
113	Chah Sorb	34	4	0	56	40	0	S	Limestone and dolomite/Mid-Triassic	Mississippi Valley Type (MVT)	Late Cimmerian	Vein, veinlet, disseminated	
114	Chah Sormeh	34	14	20	48	54	25	I	Limestone/Lower Cretaceous	MVT	Laramide	Vein	
115	Chah Talkh	33	18	32	50	33	36	S	Limestone/Upper Cretaceous	Hydrothermal	Laramide	Vein	Ba, Cu
116	Chah Talkh (Saeid Abad)	29	8	14	55	26	6	M	Limestone/Upper Cretaceous	Hydrothermal	Pyrenean	Vein	Ba, Cu
117	Chahak	34	47	0	50	18	36	S	Tuff and andesite/Eocene	Hydrothermal	Pyrenean	Lens	
118	Chahak	34	47	0	50	18	36	I	Tuff/Eocene	Hydrothermal	Pyrenean	Vein	
119	Chahar Darreh-ye-Damghan	36	22	24	54	18	20	S	Limestone/Triassic	Hydrothermal	Early Cimmerian	Disseminated	Cu
120	ChaharDeh Damghan	36	22	24	54	18	20	I	Vermicular limestone/Triassic	Hydrothermal	Early Cimmerian	Vein	
121	Chahbad	35	22	40	54	17	45	I	Dolomite and limestone/Upper Paleozoic	Hydrothermal	Pyrenean	Disseminated	
122	Chahmir	31	39	0	56	52	30	S	Black shale and tuff/Upper Precambrian–Lower Cambrian	Volcanosedimentary/ SEDEX	Pan-African	Stratiformbed, lens	
123	Chahriseh (Lapalang)	33	0	32	52	8	43	S	Thick-bedded limestone/Upper Cretaceous	Mississippi Valley Type (MVT)	Laramide	Vein, veinlet	
124	Chahriseh (Lapalangchi)	33	0	17	52	2	43	S	Dolomite/Mid-Triassic	Mississippi Valley Type (MVT)	Laramide	Veinlet, disseminated	
125	Chahriseh (Qassabi)	33	0	40	52	8	17	S	Thick-bedded limestone/Upper Cretaceous	Mississippi Valley Type (MVT)	Laramide	Vein	
126	Cham Darreh	33	17	0	51	50	0	S	Sandstone and carbonates/Lower Cretaceous	Mississippi Valley type	Laramide	Vein	
127	Changarzeh	33	16	0	51	56	0	S	Dolomite limestone and limestone/ Lower Cretaceous	Mississippi Valley Type (MVT)	Laramide	Vein	

128	Changureh	36	13	0	49	26	0	S	Andesite, andesibasalt, and tuff/Eocene	Hydrothermal	Pyrenean	Unknown	
129	Chara Malek	33	43	0	49	36	0	I	Dolomite and limestone/Upper Cretaceous	Mississippi Valley Type (MVT)	Laramide	Unknown	
130	Chara Molk	33	43	0	49	36	30	I	Sandstone and carbonates/Lower Cretaceous	Mississippi Valley Type	Laramide	Vein	
131	Charderagi	33	21	40	50	34	0	I	Limestone/Lower Cretaceous	MVT	Laramide	Vein	
132	Chari	29	36	0	57	5	0	I	Pyroclastics/Eocene	Hydrothermal	Pyrenean	Vein	
133	Charkat Zanguli	36	17	0	53	45	0	I	Sedimentary rocks/Upper Paleozoic–Lower Triassic	Hydrothermal	Pyrenean	Unknown	
134	Cheft	33	31	18	54	11	11	I	Schist, marble/Precambrian	Hydrothermal	Pyrenean	Vein	
135	Chehr Abad	36	51	0	47	51	54	I	Marl/Oligo-Miocene	Hydrothermal	Post-Pyrenean	veinlet, Disseminated	
136	Chekab	32	43	53	51	8	54	S	Limestone/Lower Cretaceous	Mississippi Valley Type (MVT)	Laramide	Vein	Ag, Cd
137	ChekCheku	32	22	37	54	21	45	S	Thick-bedded to massive limestone, Shale and sandstone/Lower Cretaceous	Mississippi Valley Type (MVT)	Laramide	Stratiformbed, vein	
138	Cheshmeh Adineh	33	23	0	50	43	0	I	Sandy dolomite/Lower Cretaceous	Mississippi Valley Type (MVT)	Laramide	Unknown	
139	Cheshmeh Akhury	35	42	0	52	47	0	I	Sandstone and shale/Jurassic	Mississippi Valley Type	Laramide	Unknown	
140	Cheshmeh Firuzeh	31	56	18	55	53	8	I	Dolomite and limestone/Triassic–Jurassic	Sedimentary	Early Cimmerian	Vein	Cu
141	Cheshmeh Hafez	35	24	57	54	44	59	S	Andesitic breccia tuff/Eo-Oligocene	Hydrothermal	Post-Pyrenean	Vein, veinlet	
142	Cheshmeh Sadegh Ali	34	1	46	49	13	12	I	Dolomite and limestone/Lower Cretaceous	Mississippi Valley Type (MVT)	Laramide	Disseminated	Ba
143	Cheshmeh Sang	33	2	26	53	10	19	I	Dolomite/Mid-Triassic	Hydrothermal	Pyrenean	Disseminated	
144	Cheshmeh Sefid	30	31	20	56	28	40	I	Shale/Jurassic; dolomite and limestone/Triassic	Hydrothermal	Early–Middle Cimmerian	Vein, veinlet	F, Ba
145	Cheshmeh Sefid	34	1	0	49	14	0	S	Limestone and dolomite/Upper Cretaceous	Mississippi Valley Type	Laramide	Vein	Cu

(continued)

(continued)

No.	Name	Lat D°	Lat M	Lat S″	Long D°	Long M	Long S″	Size	Host rock/age	Genetic/ mineralization	Orogenic phases	Morphology	Elem para
146	Cheshmehvar	33	24	0	49	55	30	I	Limestone, dolomite/Lower Cretaceous	MVT	Laramide	Vein	
147	Chichaklou	36	26	46	47	26	0	I	Shale and dolomite/Upper Precambrian–Lower Cambrian	Volcano Sedimentary	Pan-African	Vein	Au, Ag
148	Chika Buh	33	52	24	52	59	12	I	Tuffite/Eocene; limestone/Cretaceous	Hydrothermal	Pyrenean	Disseminated	
149	Chirak	33	22	42	57	15	58	I	Limestone/Carboniferous	Sedimentary	Hercynian	Unknown	
150	Chogha Siah	33	25	0	49	55	0	I	Dolomite and limestone/Upper Cretaceous	Mississippi Valley Type (MVT)	Laramide	Vein	
151	Choghart Karim Abad	31	59	25	56	1	42	I	Limestone, dolomite/Triassic	Hydrothermal	Early Cimmerian	Vein	
152	Chomalu	38	7	8	48	11	42	S	Andesitic and rhyolitic tuff/Eocene	Hydrothermal	Pyrenean	Vein, disseminated	Cu
153	Chugha Soukhteh	33	7	13	50	38	3	I	Limestone/Cretaceous	Mississippi Valley Type (MVT)	Laramide	Unknown	Cu
154	Chumul	36	43	45	49	38	55	S	Limestone/Upper Jurassic	Hydrothermal	Late Cimmerian	Lens	Ba
155	Dahagh Arabestan	33	2	22	50	47	18	I	Limestone/Lower Cretaceous	MVT	Laramide	Vein	
156	Dahaneh Bagh-ha	31	43	50	56	19	40	I	Dolomite and dolomitic limestone/ Triassic	Hydrothermal	Early Cimmerian	Vein	
157	Dahaneh Pahne Kamar	32	31	58	58	26	24	I	Shale/Upper Triassic	Hydrothermal	Pyrenean	Vein	
158	Dahaneh Sabz	33	2	12	52	9	41	S	Thick-bedded and gray limestone/ Lower Cretaceous	Mississippi Valley Type (MVT)	Laramide	Veinlet	
159	Dahaneh Shur	31	56	0	56	4	45	I	Limestone, dolomite, and shale/ Jurassic–Triassic	Hydrothermal	Pyrenean	Unknown	
160	Dahaneh Tangool	32	27	20	57	20	0	I	Shale, limestone, sandstone/Triassic	Hydrothermal	Pyrenean	Vein	
161	Dahaneh-Pahneh Kamar (Birjand)	32	31	58	58	26	24	I	Shale/Upper Triassic	Hydrothermal	Early Cimmerian	Vein	
162	Dar Dahaneh	33	23	46	50	35	0	I	Dolomite/Lower Cretaceous	MVT	Laramide	Vein	
163	Darband	35	44	55	53	21	17	S	Limestone/Upper Cretaceous	Sedimentary	Laramide	Vein, veinlet	

164	Darband Hendeh	33	21	0	49	58	0	I	Limestone, dolomite/Lower Cretaceous	MVT	Laramide	Vein	
165	Darestan	35	26	5	54	39	25	I	Tuff/Eocene	Hydrothermal	Pyrenean	Vein	
166	Darin	33	21	0	49	58	0	I	Dolomite and limestone/Upper Cretaceous	Mississippi Valley Type (MVT)	Laramide	Vein	
167	Darin Tabas	33	34	0	56	11	30	S	Massive dolomite and dolomite limestone/Ordovician	Hydrothermal	Caledonian-Hercynian	Vein	
168	Darreh	33	52	38	51	18	54	I	Andesitic lava and pyroclastics/Eocene	Hydrothermal	Pyrenean	Vein	
169	Darreh Amrud (Qohrud)	33	38	47	51	23	37	I	Granodiorite/Post-Eocene; acidic tuff/Eocene	Hydrothermal	Pyrenean	Veinlet, disseminated	
170	Darreh Arezoo	33	0	16	50	42	0	I	Sandstone and carbonates/Lower Cretaceous	Mississippi Valley Type	Laramide	Vein	Fe
171	Darreh Bayat	34	14	52	48	55	4	S	Dolomite and limestone/Lower Cretaceous	Mississippi Valley Type (MVT)	Laramide	Lens	
172	Darreh Bid	32	58	38	50	44	50	I	Dolomite and limestone/Lower Cretaceous	Mississippi Valley Type (MVT)	Laramide	Unknown	
173	Darreh Farah	33	23	0	49	55	0	I	Limestone/Lower Cretaceous	MVT	Laramide	Vein	
174	Darreh Garmeh	33	25	0	49	56	0	I	Limestone, dolomite/Lower Cretaceous	MVT	Laramide	Vein	
175	Darreh Harzu	33	0	16	50	42	0	I	Sandy dolomite/Lower Cretaceous	Mississippi Valley Type (MVT)	Laramide	Unknown	
176	Darreh Kamran	36	28	0	49	25	30	I	Tuff/Eocene	Unknown	Unknown	Unknown	Fe
177	Darreh Molki	34	3	19	49	25	9	I	Thickness layers limestone/Lower Cretaceous	Mississippi Valley Type (MVT)	Laramide	Vein	
178	Darreh Noghreh	33	31	41	50	14	17	S	Tuff and pyroclastics/Cretaceous	Mississippi Valley Type	Laramide	Vein, veinlet	Ag
179	Darreh Rasul	34	11	8	49	3	42	I	Dolomite limestone and silica/Lower Cretaceous	Mississippi Valley Type (MVT)	Laramide	Stratiformbed	
180	Darreh Sharif	31	8	0	55	34	0	I	Limestone and dolomite/Lower Cretaceous	Mississippi Valley Type	Laramide	Vein	
181	Darreh Varandoun	33	38	47	51	23	37	I	Dacite/Post-Eocene; metamorphosed silica tuff/Eocene	Hydrothermal	Pyrenean	Disseminated	

(continued)

(continued)

No.	Name	Lat D°	Lat M	Lat S″	Long D°	Long M	Long S″	Size	Host rock/age	Genetic/ mineralization	Orogenic phases	Morphology	Elem para
182	Darreh Zanjir	31	43	0	54	13	0	S	Limestone/Lower Cretaceous	Hydrothermal	Pyrenean	Disseminated	
183	Darreh Zarreh	31	12	10	56	41	27	I	Dolomite/Permo-Triassic	Sedimentary	Early Cimmerian	Vein	
184	Darz Abad	34	25	32	58	9	57	S	Volcanics/Tertiary; sedimentary rocks/Paleozoic	Hydrothermal	Pyrenean	Unknown	
185	Dashtak	33	2	0	50	42	0	I	Shale/Lower Jurassic	Mississippi Valley Type (MVT)	Laramide	Unknown	
186	Dasht-e-Varamin	35	12	0	51	50	0	I	Unknown	Unknown	Unknown	Unknown	
187	Dastgerd	36	19	0	50	18	0	I	Basalt/Tertiary	Volcanic	Pyrenean	Unknown	Fe
188	Deh Asghar	31	46	30	56	12	0	S	Limestone, limestone dolomite, shale, and sandstone/Permo-Triassic	MVT	Early Cimmerian	Vein, veinlet	
189	Deh Chuneh	34	6	0	49	6	0	I	Limestone and dolomite/Upper Cretaceous	Mississippi Valley Type	Laramide	Unknown	
190	Deh Siahan	30	0	10	55	59	30	I	Tuff/Eocene	Hydrothermal	Pyrenean	Unknown	
191	Deh Sufian	35	49	30	53	23	45	I	Limestone/Triassic	Mississippi Valley Type	Early Cimmerian	Veinlet	
192	Dehagh	33	2	8	50	47	20	S	Sandstone and carbonates/Lower Cretaceous	Mississippi Valley Type	Laramide	Vein	
193	Dehreza	28	57	16	61	5	56	I	Shale and sandstone/Eocene	Hydrothermal	Post-Pyrenean	Vein	
194	Derasheh	33	23	0	50	35	0	I	Sandstone and carbonates/Lower Cretaceous	Mississippi Valley Type	Laramide	Vein	
195	Digar Kuh	31	2	0	56	56	0	I	Gypsum, gypsy marl, limestone/ Jurassic	Hydrothermal	Pyrenean	Unknown	
196	Doab	36	0	0	53	4	0	I	Sandstone and shale/Jurassic	Hydrothermal	Pyrenean	Unknown	
197	Dour	33	18	14	50	39	30	I	Dolomite and limestone/Lower Cretaceous	Mississippi Valley Type (MVT)	Laramide	Unknown	
198	Dowlat Abad	35	46	0	52	42	0	I	Pyroclastics/Tertiary	Hydrothermal	Pyrenean	Unknown	
199	Duna-Elika	36	9	55	51	26	45	M	Crystalline limestone/Permian– Triassic	MVT	Early Cimmerian	Vein	

200	Dush kharat	33	6	33	50	26	36	S	Dolomite and limestone/Cretaceous	Mississippi Valley Type	Laramide	Vein, veinlet, and lens	
201	East of Kachuieh	32	26	0	51	12	0	I	Sandstone and carbonates/Lower Cretaceous	Mississippi Valley Type	Laramide	Vein	
202	East of Kuh-e-Piadehrow	29	47	19	57	53	18	I	Pyroclastics/Eocene	Hydrothermal	Pyrenean	Vein	Cu
203	Ejilku	36	55	0	47	5	0	I	Andesitic lava/Oligo-Miocene	MVT	Post-Pyrenean	Vein	
204	Emarat	33	51	0	49	36	0	L	Thin-bedded limestone, shale, marl and chert/Cretaceous	Mississippi Valley Type (MVT)	Laramide	Vein, lens	
205	Eram-e-Bozorg	35	51	30	53	14	30	S	Thickness layers limestone/Upper Cretaceous	Sedimentary	Laramide	Vein	Cu, Ba, Ag
206	Eram-e-Kochak	35	51	19	53	16	18	S	Thickness layers limestone/Upper Cretaceous	Sedimentary	Laramide	Vein	
207	Espily	36	55	42	49	52	45	S	Limestone/Jurassic	Hydrothermal	Mid-Cimmerian	Vein	
208	Farah Abad	31	38	30	53	46	0	S	Massive to bedded limestone/Lower Cretaceous	Hydrothermal	Laramide	Vein	Cu, Ba
209	Farak (Parak)	31	51	30	55	44	0	S	Silica dolomite and limestone/Upper Precambrian–Lower Cambrian	Volcanosedimentary	Pan-African	Vein	
210	Fars Manesh (Vaymand)	34	53	52	49	49	28	S	Tuff/Eocene	Hydrothermal	Pyrenean	Vein	
211	Farvaz	34	14	18	48	56	12	S	Dolomite, limestone/Lower Cretaceous	MVT	Laramide	Lens	
212	Faskhoud	33	16	0	51	56	0	S	Sandstone and carbonates/Lower Cretaceous	Mississippi Valley Type	Laramide	Vein	
213	Feyz Abad	32	44	0	52	59	0	I	Volcanics and pyroclastics/Tertiary	Hydrothermal	Pyrenean	Vein	
214	Firuzkuh 1	35	48	0	52	35	0	I	Pyroclastics/Tertiary	Hydrothermal	Pyrenean	Unknown	
215	Firuzkuh 2	35	47	0	52	52	0	I	Pyroclastics/Tertiary	Hydrothermal	Pyrenean	Unknown	
216	Gadok Firuzkuh	35	50	0	52	54	0	I	Limestone and dolomite/Permo-Triassic	Hydrothermal	Late Cimmerian	Vein	
217	Gaduk-e-Firuzkuh	35	50	0	52	54	0	I	Dolomite and limestone/Permo-Triassic	Hydrothermal	Early Cimmerian	Vein and veinlet	

(continued)

(continued)

No.	Name	Lat D°	Lat M	Lat S″	Long D°	Long M	Long S″	Size	Host rock/age	Genetic/ mineralization	Orogenic phases	Morphology	Elem para
218	Gandi	35	19	15	54	38	45	S	Andesite/Eocene–Oligocene	Hydrothermal	Pyrenean	Vein	
219	Gara Gheshlagh	35	38	0	50	17	0	I	Volcanics and pyroclastics/Tertiary	Volcanic	Pyrenean	Unknown	
220	Garadu	34	15	0	57	6	0	S	Dolomite/Mid-Devonian, Massive to thick-bedded limestone/Permian	Hydrothermal	Early Cimmerian	Vein, disseminated	
221	Gardaneh	32	0	0	51	0	0	I	Sandy dolomite/Lower Cretaceous	Mississippi Valley Type (MVT)	Laramide	Unknown	
222	Gardaneh Dolasoul	32	56	0	51	3	0	I	Sandstone and carbonates/Lower Cretaceous	Mississippi Valley Type	Laramide	Vein	
223	Gardaneh Rokh	32	23	0	51	8	0	I	Sandstone and carbonates/Lower Cretaceous	Mississippi Valley Type	Laramide	Vein	
224	Gardaneh Shir (Gol Mouniai)	33	2	0	52	13	29	S	Dolomite limestone and dolomite/ Mid-Triassic	Mississippi Valley Type (MVT)	Laramide	Veinlet, vein	
225	Garuki	34	46	0	50	14	30	1	Andesite/Eocene	Hydrothermal	Pyrenean	Vein	
226	Gavart	36	25	5	51	5	49	S	Diorite and syenite/Tertiary; carbonates/Paleozoic	Hydrothermal	Pyrenean	Unknown	
227	Gazaldar 1	33	47	0	49	46	0	I	Sandstone and carbonates/Lower Cretaceous	Mississippi Valley Type	Laramide	Vein	
228	Gazaldar 2	33	45	8	49	46	37	I	Sandstone and carbonates/Lower Cretaceous	Mississippi Valley Type	Laramide	Vein	
229	Geloangah	36	10	0	51	20	0	I	Tuff and pyroclastics/Tertiary	Hydrothermal	Pyrenean	Unknown	Ba, Cu
230	Geruki	34	46	0	50	14	40	I	Tuff and andesite/Eocene	Hydrothermal	Pyrenean	Lens	
231	Ghadir	31	37	10	55	59	5	I	Limestone and dolomite/Permo-Triassic	Sedimentary	Early Cimmerian	Vein	
232	Ghahestan	34	22	0	57	15	0	I	Marl, limestone/Jurassic	MVT	Late Cimmerian	Vein	
233	Ghahr Rud	33	38	47	51	23	37	I	Rhyolitic tuff/Eocene	Hydrothermal	Pyrenean	Vein	
234	Ghahyaz Ardestan	33	2	21	52	26	30	I	Limestone, shale, tuff/Eocene	Mississippi Valley Type	Laramide	Vein	
235	Ghajreh	35	56	0	51	11	0	I	Pyroclastic/Eocene	Hydrothermal	Pyrenean	Vein	
236	Ghara Boltagh	33	12	0	50	15	0	I	Sandstone and carbonates/Lower Cretaceous	Mississippi Valley Type	Laramide	Vein	

237	Ghara Gheshlagh	35	38	0	50	17	0	I	Volcanics and pyroclastics/Tertiary	Volcanic	Pyrenean	Unknown	
238	Ghara Sarjedgal	33	49	0	49	33	0	I	Dolomite and limestone/Upper Cretaceous	Mississippi Valley Type (MVT)	Laramide	Unknown	
239	Gharah Darreh	36	27	0	49	33	0	I	Andesite/Eocene	Unknown	Unknown	Unknown	
240	Ghare Golkan	34	8	30	56	32	50	S	Dolomite/Mid-Triassic	Sedimentary	Laramide	Disseminated, veinlet	
241	Ghar-e-Baba Jaber	33	40	0	50	28	0	I	Sandstone and carbonates/Lower Cretaceous	Mississippi Valley Type	Laramide	Vein	Cu
242	Ghar-e-Morvarik	33	53	44	49	37	54	S	Sandstone and conglomerate/Lower Cretaceous	Mississippi Valley Type (MVT)	Laramide	Vein	
243	Ghar-e-Palangi	32	27	50	57	19	50	I	Dolomite/Lower–Mid- Triassic	Hydrothermal	Early Cimmerian	Unknown	
244	Ghar-e-Sar-e-jadeh Kal	33	46	52	49	35	6	I	Thick-bedded limestone/Lower Cretaceous	Mississippi Valley Type (MVT)	Laramide	Disseminated	
245	Ghar-e-Shir Ali	34	9	30	57	8	12	I	Shale, carbonaceous marl, and sandstone/Lower Jurassic	Sedimentary	Early Cimmerian	Vein, veinlet	
246	Gharmee	33	45	0	49	48	0	I	Sandstone and carbonates/Lower Cretaceous	Mississippi Valley Type	Laramide	Vein	Ba
247	Gharyeh Seh	33	28	15	51	30	47	S	Thick-bedded and gray limestone/ Lower Cretaceous	Mississippi Valley Type (MVT)	Laramide	Veinlet, disseminated	
248	Ghasem Abad	33	54	0	49	44	0	I	Sandstone and carbonates/Lower Cretaceous	Mississippi Valley Type	Laramide	Vein	
249	Ghassabi	33	0	40	52	8	17	I	Limestone/Upper Cretaceous	MVT	Laramide	Vein	
250	Ghazan	33	41	50	51	22	43	I	Andesitic tuff/Eocene	Hydrothermal	Pyrenean	Vein	
251	Ghazi Darreh	33	26	0	50	41	0	I	Sandstone and carbonates/Lower Cretaceous	Mississippi Valley Type	Laramide	Vein	
252	Ghazi Kolayeh	36	9	0	50	23	0	I	Volcanics and pyroclastics/Tertiary	Volcanic	Pyrenean	Unknown	
253	Ghehestan	34	22	42	57	14	22	I	Dolomite and limestone/Jurassic	Mississippi Valley Type	Late Cimmerian	Stratiformbed	
254	Ghelich kandi	35	38	37	49	50	26	I	Volcanics and pyroclastics/Tertiary	Volcanic	Pyrenean	Unknown	
255	Gheshleghiri	33	23	30	50	36	0	I	Dolomite/Lower Cretaceous	MVT	Laramide	Vein	
256	Gheslinsar	36	23	0	50	10	0	I	Pyroclastics/Tertiary; carbonates/ Jurassic	Volcanic	Pyrenean	Unknown	

(continued)

(continued)

No.	Name	Lat D°	Lat M	Lat S″	Long D°	Long M	Long S″	Size	Host rock/age	Genetic/ mineralization	Orogenic phases	Morphology	Elem para
257	Ghezel Dar	33	47	0	49	46	0	I	Limestone/Lower Cretaceous	MVT	Laramide	Vein	
258	Ghezel Qaleh	37	13	18	48	15	15	S	Acidic to intermediate tuff/Eocene	Hydrothermal	Pyrenean	Vein	Cu
259	Ghjareh	35	56	0	51	11	0	I	Tuff and pyroclastics/Tertiary	Volcanic	Pyrenean	Unknown	
260	Gholatu	31	36	30	56	27	15	I	Dolomite and limestone/Permo-Triassic	Sedimentary	Early Cimmerian	Vein	
261	Ghole Bozi	32	31	54	58	22	20	I	Dolomite and limestone/Triassic	Hydrothermal	Pyrenean	Vein	
262	Gholuri	31	35	46	56	6	8	I	Limestone and dolomite/Permo-Triassic	Sedimentary	Early Cimmerian	Vein	
263	Ghor-e-Sefid	32	32	0	58	19	52	I	Dolomite, limestone/Triassic	Hydrothermal	Pyrenean	Vein	
264	Ghoshrow	34	16	55	56	35	48	I	Limestone and dolomite/Jurassic	Mississippi Valley Type	Late Cimmerian	Vein	
265	Ghuch Kuhi	34	36	30	57	9	0	S	Dolomite/Mid-Devonian	Sedimentary	Caledonian	Vein	
266	Gijar Kuh	31	52	53	56	7	27	S	Dolomite/Triassic	MVT	Early Cimmerian	Vein, veinlet	
267	Giutangeh	36	6	7	53	55	22	S	Dolomite and limestone/Jurassic (Malm)	Mississippi Valley Type	Late Cimmerian	Vein, veinlet	
268	Givtangeh Damghan	36	5	50	53	55	45	I	Dolomite and limestone/Jurassic	Hydrothermal	Late Cimmerian	Vein	
269	God Abbas Abad	33	27	8	54	30	11	I	Limestone/Lower Cretaceous	Hydrothermal	Pyrenean	Vein	
270	Godar Sorkh	32	22	0	57	16	0	I	Shale and limestone/Triassic	Sedimentary	Early Cimmerian	Unknown	
271	Godar-e-Dour	33	22	30	50	45	16	I	Sandstone and carbonates/Lower Cretaceous	Mississippi Valley Type	Laramide	Vein	
272	Godar-e-Pouneh	36	11	21	59	48	1	I	Limestone, sandstone, and shale/ Jurassic	Hydrothermal	Late Cimmerian	Vein, veinlet	
273	Godar-e-Sorkh	33	19	21	54	47	14	I	Limestone/Lower Cretaceous	Mississippi Valley Type	Laramide	Unknown	
274	Godar-e-Sorkh	33	23	3	57	21	12	S	Siltstone and dolomite/Lower–Mid-Triassic	Hydrothermal	Early Cimmerian	Vein	
275	Gojar	31	35	1	56	25	55	S	Limestone and dolomite/Jurassic	Mississippi Valley Type (MVT)	Laramide	Vein	
276	Gol Boneh	33	49	0	50	30	30	I	Sandstone and carbonates/Lower Cretaceous	Mississippi Valley Type	Laramide	Vein	

277	Gol Tappeh 2	33	50	0	49	48	0	I	Sandstone and carbonates/Lower Cretaceous	Mississippi Valley Type	Laramide	Vein	Ba, Cu
278	Golestan	34	38	35	50	30	0	I	Tuff and andesite/Eocene	Hydrothermal	Pyrenean	Lens	
279	Golestani (Delbar)	35	57	23	55	58	47	S	Dolomite/Jurassic	Hydrothermal	Laramide	Vein	
280	Golharan	32	58	37	50	44	50	S	Limestone/Cretaceous; shale and sandstone/Jurassic	MVT	Laramide	Vein and veinlet	
281	Golpar Abad (Darreh Cheshmeh Gandomeh)	34	9	46	49	3	35	I	Slate and quartzite/Lower Cretaceous	Mississippi Valley Type (MVT)	Laramide	Vein	
282	Golzard	33	26	30	49	49	46	I	Phyllite, calc phyllite, quartz phyllite/Jurassic	Mississippi Valley Type (MVT)	Laramide	Stratiformbed, vein	
283	Gondehdar	31	48	53	55	44	30	I	Black dolomite/Permo-Triassic	Sedimentary	Early Cimmerian	Vein	
284	Gonharan	32	59	4	50	41	34	S	Marl/Mid-Cretaceous	Mississippi Valley Type (MVT)	Laramide	Vein, veinlet	Fe
285	Goor Delbandan	31	39	30	56	22	40	I	Dolomite, limestone/Cretaceous	Hydrothermal	Laramide	Vein	
286	Goor Ravar	31	21	4	56	16	45	S	Dolomitic limestone/Cretaceous; shale, sandstone, siltstone/Permo-Triassic	Hydrothermal	Late Cimmerian	Vein	Ba, Fe
287	Goor Vafadari	31	21	15	55	55	5	I	Shale, dolomite, pyroclastic/Precambrian	Sedex	Pan-African	Lens	
288	Gordel Kuh	34	11	31	49	25	15	S	Dolomite, limestone, and quartzite/Lower Cretaceous	Mississippi Valley Type (MVT)	Laramide	Vein	
289	Gor-e-Ravar	31	21	40	56	16	45	S	Dolomite/Permo-Triassic	Sedimentary	Hercynian	Vein	
290	Gorgab	33	59	53	52	26	38	I	Schist/Paleozoic	Hydrothermal	Pyrenean	Stratiformbed	Cu, Au, Ag
291	Gorgab 3	33	59	19	52	27	5	I	Schist/Paleozoic	Hydrothermal	Pyrenean	Stratiformbed	
292	Gorgan	33	29	0	50	4	0	I	Sandstone and carbonates/Lower Cretaceous	Mississippi Valley Type	Laramide	Vein	
293	Gourestani	34	7	25	56	33	6	S	Dolomite/Mid-Triassic	Sedimentary	Laramide	Veinlet, disseminated	
294	Gourt Chalous	36	29	0	51	9	0	I	Limestone/Lower Cretaceous	Hydrothermal	Late Cimmerian	Vein	

(continued)

(continued)

No.	Name	Lat D°	Lat M	Lat S″	Long D°	Long M	Long S″	Size	Host rock/age	Genetic/ mineralization	Orogenic phases	Morphology	Elem para
295	Gousheh Kamar	34	37	58	57	11	27	S	Porphyry andesite/unknown; dolomite/Devonian	Hydrothermal	Unknown	Vein, veinlet	
296	Govar	31	21	52	56	16	21	S	Dolomite/Permo-Triassic	Sedimentary	Early Cimmerian	Vein	
297	Gowd-e-Abbas	33	26	17	54	30	50	S	Schist and marble/Upper Cambrian–Lower Ordovician	Hydrothermal	Caledonian	Vein	
298	Gowd-e-Vafadary	31	21	15	55	55	5	I	Dolomite/Permo-Triassic	Sedimentary	Early Cimmerian	Vein	
299	Gurcheh Berenj II	34	6	37	54	44	18	I	Limestone/Oligo-Miocene	Hydrothermal	Post-Pyrenean	Vein	Eu, Nd, Ag
300	Gurt-e-Kolardasht	36	25	40	51	8	0	S	Porphyrite quartz monzonite/Tertiary; diabase dike/Pre-Tertiary; limestone/Upper Paleozoic	Hydrothermal	Post-Pyrenean	Vein, veinlet	
301	Gushkharat	33	6	33	50	26	36	S	Silica limestone/Lower Cretaceous	Mississippi Valley Type (MVT)	Laramide	Unknown	
302	Haft Savaran	33	34	39	49	55	57	S	Sandstone and shale/Lower Cretaceous	Hydrothermal	Late Cimmerian	Vein, veinlet	
303	Haft Tal-e-Aghda	32	16	40	53	28	16	I	Diorite and dolerite/Post-Cambrian; shale and sandstone/Cambrian	Hydrothermal	Pan-African	Vein	Ba
304	Hafther	32	16	0	53	28	0	S	Shale, sandstone, and limestone/ Lower to Upper Cambrian	Hydrothermal	Pan-African	Vein	
305	Haji Abad-e-Zarrin	32	53	0	54	58	0	I	1- Shale and sandstone/Jurassic; 2-limestone/Cretaceous	Hydrothermal	Pyrenean	Unknown	
306	Hasan Abad	31	33	0	53	51	0	S	Gray dolomite and limestone/ Cretaceous	Hydrothermal	Laramide	Vein	
307	Hasham	32	17	0	53	34	0	I	Shale, sandstone, and limestone/ Lower to Upper Cambrian	Hydrothermal	Pan-African	Unknown	
308	Hashiu	31	35	39	56	4	9	I	Limestone/Permo-Triassic	Sedimentary	Early Cimmerian	Vein	
309	Hashtenjan	34	6	35	56	32	10	S	Dolomite/Mid-Triassic	Sedimentary	Late Cimmerian	Vein	
310	Hesar	35	24	20	58	46	36	S	Volcanics/Tertiary	Hydrothermal	Pyrenean	Vein	
311	Hezar Bisheh	33	22	42	50	38	42	I	Sandstone and carbonates/Lower Cretaceous	Mississippi Valley Type	Laramide	Vein	

312	Hoseyn Abad	33	38	11	49	50	42	S	Limestone and shale/Jurassic	MVT	Late Cimmerian	Vein	
313	Hoseyn Abad Chaplagh	36	29	0	51	9	0	I	Limestone and dolomite/Permo-Triassic	Hydrothermal	Post-Pyrenean	Vein	
314	Howz Riz	32	36	43	58	1	20	I	Andesite and dacite/Eocene	Hydrothermal	Pyrenean	Unknown	
315	Howzanak	36	12	0	50	33	0	I	Volcanics and pyroclastics/Tertiary	Hydrothermal	Pyrenean	Unknown	
316	Howz-e-Raeis	32	37	0	58	1	0	I	Andesite and dacite/Tertiary	Volcanic	Pyrenean	Vein	
317	Howz-e-Sefid	32	21	8	54	13	27	S	Shale, sandstone, limestone, and dolomite/Lower Cretaceous	Mississippi Valley Type (MVT)	Laramide	Vein, lens, stratiformbed	
318	Iran Kuh Complex Mine	32	28	0	51	31	0	M	Shale and limestone/Cretaceous, shale/Jurassic	MVT	Laramide	Lens, vein	
319	Irankuh	32	31	45	51	36	36	I	Sandstone, limestone/Lower Cretaceous	MVT	Laramide	Unknown	
320	Jalayer	34	10	0	49	14	0	I	Dolomite, limestone/Lower Cretaceous; shale, sandstone/Jurassic	MVT	Laramide	Vein	
321	Jarou	35	41	0	50	35	0	I	Andesite/Tertiary	Volcanic	Pyrenean	Unknown	
322	Jigery-e-Paborj	32	57	0	51	4	0	I	Sandy dolomite/Lower Cretaceous	Mississippi Valley Type (MVT)	Laramide	Unknown	
323	Jouzvar	34	53	0	50	10	30	I	Pyroclastics limestone and limestone/Eocene	Hydrothermal	Pyrenean	Vein	Ba, Cu, Fe
324	Jouyband	38	37	17	46	58	51	I	Marble, reef limestone/Upper Cretaceous	Hydrothermal	Pyrenean	Vein	
325	Juband	38	37	17	46	58	51	I	Marl, reef limestone/Upper Cretaceous	Hydrothermal	Pyrenean	Vein	
326	Jubariun	36	15	57	54	59	0	S	Gypsiferous marl/Eocene	Hydrothermal	Pyrenean	Veinlet	
327	Jutan	36	21	0	50	29	0	I	Tuff/Eocene	Hydrothermal	Post-Pyrenean	Unknown	
328	Kabu	31	35	31	56	3	30	I	Dolomite and limestone/Permo-Triassic	Sedimentary	Early Cimmerian	Vein	
329	Kabutar Kuh	34	7	58	58	55	6	S	Sedimentary rocks/Jurassic	Mississippi Valley Type	Laramide	Vein	
330	Kah Kuh	33	35	30	51	6	30	S	Sandstone and carbonates/Lower Cretaceous	Mississippi Valley Type	Laramide	Vein	

(continued)

(continued)

No.	Name	Lat D°	Lat M	Lat S″	Long D°	Long M	Long S″	Size	Host rock/age	Genetic/ mineralization	Orogenic phases	Morphology	Elem para
331	Kahang	32	20	0	53	24	0	I	Diorite and dolerite/Post-Cambrian; shale and sandstone/Cambrian	Hydrothermal	Pan-African	Vein	
332	Kalar	34	8	20	56	31	35	S	Dolomite/Triassic	Hydrothermal	Laramide	Vein	
333	Kalardasht	36	29	23	51	4	55	I	Limestone/Triassic	Hydrothermal	Late Cimmerian	Vein	
334	Kalardustak	34	19	20	50	47	18	S	Tuff and andesite/Eocene	Hydrothermal	Pyrenean	Lens	
335	Kall Sorkh	33	52	0	50	41	0	I	Sandy dolomite/Lower Cretaceous	MVT	Laramide	Vein	
336	Kallak	35	48	0	51	3	0	I	Volcanics and pyroclastics/ Tertiary(Eocene)	Volcanic	Pyrenean	Unknown	
337	Kalleh	36	11	17	54	10	34	S	Dolomite/Devonian	Sedimentary	Hercynian	Vein, veinlet, disseminated	
338	Kalout-e-Ahani	33	54	0	58	34	0	I	Andesite/Tertiary	Hydrothermal	Pyrenean	Vein	
339	KamarAllah	31	39	10	56	11	58	I	Dolomite/Permo-Triassic	Mississippi Valley Type	Early Cimmerian	Vein	
340	KamarBarAb	31	39	48	56	11	40	I	Dolomite/Permo-Triassic	Sedimentary	Hercynian	Vein	
341	Kandeh Nader	33	24	0	50	33	30	I	Sandstone and carbonates/Lower Cretaceous	Mississippi Valley Type	Laramide	Vein	
342	Kang Savar	33	15	0	50	51	0	I	Sandstone and carbonates/Lower Cretaceous	Mississippi Valley Type	Laramide	Vein	
343	Kangarmaza	33	45	49	49	36	30	I	Sandstone and carbonates/Lower Cretaceous	Mississippi Valley Type	Laramide	Vein	
344	Kardan	35	56	30	50	51	50	I	Andesite/Eocene	Hydrothermal	Pyrenean	Vein	
345	Kardodeh (Chah-e-Alighorban)	32	30	52	58	18	18	I	Dolomite and limestone/Triassic	Hydrothermal	Early Cimmerian	Vein	
346	Karevangah	31	32	40	56	19	31	S	Dolomite/Permo-Triassic	MVT	Early Cimmerian	Vein	
347	Karvand 1	33	28	37	51	32	46	S	Dolomite/Paleozoic	Unknown	unknown	Lens	
348	Karvand 2	33	29	10	51	34	4	S	Dolomite/Paleozoic	Unknown	unknown	Lens	
349	Kashmar	35	20	0	58	41	0	S	Granitoid and volcanics/Tertiary	Hydrothermal	Pyrenean	Vein	
350	Katook	29	12	0	57	21	0	I	Andesite/Tertiary	Hydrothermal	Pyrenean	Vein	
351	Kavosh 1	33	38	10	56	12	15	I	Massive dolomite and dolomite limestone/Ordovician	Hydrothermal	Caledonian	Vein	Cu

352	Kelisheh	33	32	56	50	0	5	S	Shale and sandstone/Jurassic	Hydrothermal	Late Cimmerian	Vein, veinlet	
353	Khan Abad	33	40	0	49	48	0	I	Sandstone and carbonates/Lower Cretaceous	Mississippi Valley Type	Laramide	Vein	Cu
354	Khaneh Sormeh	32	42	53	51	14	38	S	Dolomite and limestone/Lower Cretaceous	Mississippi Valley Type (MVT)	Laramide	Vein	Cu
355	Khangah Ghoncheh Khoran	36	26	0	49	21	0	S	Olivine diabase, porphyrite monzonite/Oligocene; volcanosedimentary/Eocene	Hydrothermal	Post-Pyrenean	Vein	
356	Khangeh Savar	33	15	22	50	50	42	I	Limestone/Lower Cretaceous	MVT	Laramide	Vein	
357	Khanj	36	13	0	59	30	0	I	Sandstone and shale/Jurassic	Unknown	Unknown	Unknown	
358	Khanjar Reshm	35	18	14	54	32	40	S	Limestone/Cretaceous	Hydrothermal	Pyrenean	Veinlet, disseminated	
359	Kharestan	28	30	0	60	50	0	I	Limestone/Cretaceous	Hydrothermal	Pyrenean	Vein	
360	Kherad Abad	34	1	0	58	39	0	I	Shale and Sandstone/Jurassic	Hydrothermal	Pyrenean	Vein	
361	Khormatu	31	35	46	56	4	54	I	Dolomite and limestone/Permo-Triassic	Sedimentary	Early Cimmerian	Vein	
362	Khormayou	34	8	35	57	8	45	I	Shale, carbonaceous marl, and sandstone/Lower Jurassic	Hydrothermal	Early Cimmerian	Vein, veinlet	
363	Khorram Abad	31	24	45	55	58	20	I	Dolomite/Permo-Triassic	Sedimentary	Early Cimmerian	Vein	
364	Khougan 1	33	51	0	50	7	0	I	Sandy dolomite/Lower Cretaceous	Mississippi Valley Type (MVT)	Laramide	Unknown	
365	Khougan 2	33	51	0	50	9	0	I	Sandstone and carbonates/Lower Cretaceous	Mississippi Valley Type	Laramide	Vein	
366	Khuni Kalkafi	33	26	31	54	12	9	I	Schist and marble/Upper Proterozoic–Lower Paleozoic	Hydrothermal	Pan-African	Stratiformbed	
367	Khuonj-e-Chupanan	33	22	19	54	56	11	I	Limestone/Lower Cretaceous	Hydrothermal	Late Cimmerian	Vein	
368	Khusf	32	32	0	58	16	0	I	Andesite and dacite/Tertiary	Hydrothermal	Pyrenean	Vein	
369	Klongeh	33	33	20	49	53	30	I	Sandstone and carbonates/Lower Cretaceous	Mississippi Valley Type	Laramide	Vein	
370	Kohnehvar	35	41	0	52	40	30	I	Pyroclastics/Eocene	Hydrothermal	Pyrenean	Unknown	

(continued)

(continued)

No.	Name	Lat D°	Lat M	Lat S″	Long D°	Long M	Long S″	Size	Host rock/age	Genetic/ mineralization	Orogenic phases	Morphology	Elem para
371	Kojarestan	33	44	22	49	40	51	I	Thick-bedded limestone/Lower Cretaceous	Mississippi Valley Type (MVT)	Laramide	Disseminated	
372	Kolah Bid	34	5	18	49	43	19	S	Breccia sandstone/Lower Cretaceous	Mississippi Valley Type (MVT)	Laramide	Vein	Ba
373	Kolah Darvazeh (Iran Kuh)	32	31	10	51	34	0	S	Dolomite and limestone/Lower Cretaceous	Mississippi Valley Type (MVT)	Laramide	Vein	
374	Kooreh	33	43	0	49	36	30	I	Limestone/Lower Cretaceous	MVT	Laramide	Vein	
375	Koppeh Motallebi	32	44	0	51	11	0	I	Limestone/Lower Cretaceous	Mississippi Valley Type (MVT)	Laramide	Unknown	
376	Kour Cheshmeh Damghan	36	22	0	54	29	0	I	Limestone/Mesozoic	Hydrothermal	Unknown	Unknown	
377	Kuh Gabri	31	43	30	54	16	0	I	Limestone/Lower Cretaceous	Hydrothermal	Laramide	Vein	
378	Kuh Sang Panbeh	31	33	31	56	29	42	I	Dolomite, volcanic, and pyroclastic/ Infra-Cambrian	SEDEX	Pan-African	Vein	
379	Kuh Siah	31	55	37	56	5	26	I	Limestone, dolomite/Triassic	Hydrothermal	Early Cimmerian	Vein	
380	Kuh Sormeh	28	29	0	52	18	0	M	Dolomitic limestone and sandy dolomite/Upper Permian–Lower Triassic	Mississippi Valley Type (MVT)	Hercynian	Vein	
381	Kuh-e- Alighorban1	32	33	34	58	25	0	S	Limestone and dolomite/Upper Triassic	Sedimentary	Early Cimmerian	Vein	
382	Kuh-e- Alighorban2	32	31	38	58	26	17	S	Limestone and dolomite/Upper Triassic	Sedimentary	Early Cimmerian	Vein	
383	Kuh-e-Anar	28	31	30	60	52	0	I	Andesite and dacite/Miocene	Hydrothermal	Post-Pyrenean	Vein	
384	Kuh-e-Anar (Seniu)	28	37	0	60	52	30	S	Shale and sandstone/Eocene	Hydrothermal	Post-Pyrenean	Vein	
385	Kuh-e-Atabak	34	26	4	58	31	37	S	Pyroclastics/Eocene	Hydrothermal	Pyrenean	Vein	
386	Kuh-e-Banan	31	25	14	56	16	32	I	Dolomite/Permo-Triassic	Sedimentary	Hercynian	Vein	
387	Kuh-e-Baroo	33	2	0	50	45	0	I	Sandstone and carbonates/Lower Cretaceous	Mississippi Valley Type	Laramide	Vein	

388	Kuh-e-Bashm	35	50	0	53	27	0	I	Shale, sandstone, and limestone/Jurassic	Hydrothermal	Pyrenean	Unknown	
389	Kuh-e-Bidu	30	10	0	56	43	0	I	Limestone/Cretaceous	Unknown	Laramide	Unknown	
390	Kuh-e-Bozkoosh	34	20	30	49	24	30	I	Limestone and dolomite/Upper Cretaceous	Mississippi Valley Type	Laramide	Unknown	
391	Kuh-e-Cheshmeh Zardab	32	30	50	58	26	13	I	Dolomite and limestone/Triassic; limestone/Cretaceous	Hydrothermal	Pyrenean	Vein	
392	Kuh-e-Dandaneh	32	35	0	54	26	0	I	Shale and limestone/Permian	Hydrothermal	Hercynian	Unknown	
393	Kuh-e-Garmab	32	31	10	58	20	26	I	Limestone and dolomite/Upper Triassic	Sedimentary	Early Cimmerian	Vein	
394	Kuh-e-Gerd (Kalleh gerdoo)	31	35	28	56	3	26	S	Limestone and dolomite/Middle Triassic	Mississippi Valley Type	Early Cimmerian	Vein	
395	Kuh-e-Ghara	33	42	30	49	48	0	I	Sandstone and carbonates/Lower Cretaceous	Mississippi Valley Type	Laramide	Vein	
396	Kuh-e-Hengam	34	7	58	58	55	6	S	Intrusive/Tertiary; shale, sandstone, and carbonates/Jurassic	Hydrothermal	Pyrenean	Unknown	
397	Kuh-e-Jamal	33	22	37	57	18	22	I	Dolomite and limestone/Lower Triassic	Hydrothermal	Mid-Cimmerian	Vein	
398	Kuh-e-Kaftar	30	29	50	56	21	30	I	Limestone and dolomite/Cretaceous	Mississippi Valley Type	Laramide	Vein	
399	Kuh-e-Kolah Ghazi	34	11	0	49	26	0	S	Limestone and dolomite/Upper Cretaceous	Mississippi Valley Type	Laramide	Vein	Cu
400	Kuh-e-Masahim	30	18	6	55	19	13	I	Andesite and pyroclastics/Quaternary	Hydrothermal	Post-Pyrenean	Vein	
401	Kuh-e-Mazar	38	33	12	46	10	48	I	Syenite/Post-Oligocene; breccia dacite (or diorite)/Oligocene	Hydrothermal	Post-Pyrenean	Vein	Cu
402	Kuh-e-Nayband 1	32	26	34	57	20	17	I	Shale and limestone/Triassic	Sedimentary	Early Cimmerian	Unknown	
403	Kuh-e-Noghreh	33	25	0	49	49	0	I	Dark brown sandstone/Lower Cretaceous	Mississippi Valley Type (MVT)	Laramide	Vein	
404	Kuh-e-Qaleh	31	35	26	56	3	26	S	Limestone and dolomite/Middle Triassic	Mississippi Valley Type	Early Cimmerian	Vein	
405	Kuh-e-Rose	33	23	38	54	54	5	I	Limestone/Lower Cretaceous	Mississippi Valley Type	Laramide	Unknown	

(continued)

(continued)

No.	Name	Lat D°	Lat M	Lat S″	Long D°	Long M	Long S″	Size	Host rock/age	Genetic/ mineralization	Orogenic phases	Morphology	Elem para
406	Kuh-e-Rudbar	35	40	56	53	10	30	I	Carbonates/Jurassic	Hydrothermal	Pyrenean	Unknown	
407	Kuh-e-Sangpanbeh	31	33	31	56	29	42	I	Dolomite/Upper Precambrian–Lower Cambrian	Volcanosedimentary	Pan-African	Vein	
408	Kuh-e-Sarchah	32	43	42	51	28	34	I	Sandstone and carbonates/Lower Cretaceous	Mississippi Valley Type	Laramide	Vein	
409	Kuh-e-Siah	34	9	28	57	8	19	S	Dolomite and limestone/Jurassic	Mississippi Valley Type	Late Cimmerian	Lens	
410	Kuh-e-Sorb (Sormeh Kuh)	35	40	50	53	9	30	S	Limestone and dolomite/Jurassic	Hydrothermal	Pyrenean	Veinlet, disseminated	
411	Kuh-e-Tappehha-e-Godar-e-Dour	33	22	0	50	44	30	I	Sandstone and carbonates/Lower Cretaceous	Mississippi Valley Type	Laramide	Vein	
412	Kuh-e-Yazdan (Southern Faskhud)	33	14	51	52	1	5	S	Gray sandy limestone/Lower Cretaceous	Mississippi Valley Type (MVT)	Laramide	Veinlet	Ag, Co, As
413	Kuh-e-Zir Sorkh	33	2	0	50	49	0	I	Sandstone and carbonates/Lower Cretaceous	Mississippi Valley Type	Laramide	Vein	
414	Kuhbanam	31	25	14	56	16	32	I	Dolomite/Permo-Triassic	Sedimentary	Early Cimmerian	Vein	
415	KuhZar	33	50	24	49	50	32	I	Limestone and sandstone/Lower Cretaceous	Mississippi Valley Type (MVT)	Laramide	Vein	
416	Kukueyeh	30	23	10	56	32	50	I	Dolomite/Permo-Triassic	Sedimentary	Early Cimmerian	Unknown	
417	Kur Cheshmeh	36	22	0	54	29	0	I	Limestone/Upper Jurassic	Hydrothermal	Mid-Cimmerian	Veinlet	Cu, Ba
418	Kuranak Taleghan	36	12	0	50	33	0	I	Limestone, volcanics, and pyroclastic/Paleocene–Eocene	Hydrothermal	Pyrenean	Vein, lens	
419	Kushk	31	45	30	55	45	0	L	Black shale/Upper Precambrian–Lower Cambrian	Massive Sulfide	Pan-African	Lens	
420	La Palang	33	0	32	52	8	43	I	Limestone/Upper Cretaceous	MVT	Laramide	Vein	
421	La Palangchi	33	0	17	52	2	43	I	Dolomite/Mid-Triassic	MVT	Laramide	Vein	
422	Lak	35	34	0	49	56	30	I	Andesite/Eocene	Hydrothermal	Pyrenean	Vein	
423	Lakan	33	41	52	49	43	48	S	Marl, limestone, and dolomite limestone/Cretaceous	Mississippi Valley Type (MVT)	Laramide	Vein, disseminated	

424	Lakhouni	33	0	40	52	13	45	S	Limestone/Lower Cretaceous	Mississippi Valley Type (MVT)	Laramide	Veinlet	F, Ba
425	Lali	36	2	36	53	55	30	S	Dolomite and limestone/ Devonian–Carboniferous	Hydrothermal	Hercynian	Veinlet, disseminated	
426	Larestan	35	44	30	54	4	15	S	Limestone and dolomite/Upper Paleozoic (Devonian)	Hydrothermal	Laramide	Veinlet, disseminated	
427	Lashkary	31	36	33	56	3	50	I	Limestone and dolomite/Permo-Triassic	Sedimentary	Early Cimmerian	Vein	
428	Laybid	33	26	0	50	42	0	I	Sandstone and carbonates/Lower Cretaceous	Mississippi Valley Type	Laramide	Vein	Ag, Ba, Cu
429	Leilan	33	42	49	49	51	50	S	Limestone/Lower Cretaceous	Mississippi Valley Type (MVT)	Laramide	Veinlet, disseminated	Cu
430	Look Siah	31	55	37	56	5	26	I	Limestone, dolomite, and shale/ Jurassic–Triassic	Hydrothermal	Pyrenean	Unknown	
431	Luk-e-Gabri (Mazraeh Qebleh)	31	43	30	54	16	50	I	Limestone and dolomite/Lower Cretaceous	Mississippi Valley Type	Laramide	Vein	
432	Maaden-e-Sorb (North of Bam)	29	13	0	58	14	0	I	Limestone/Cretaceous	Sedimentary	Late Cimmerian	Unknown	
433	Magasu	31	21	2	56	8	8	I	Limestone/Permo-Triassic	Sedimentary	Early Cimmerian	Vein	
434	Mahabad	34	17	9	59	42	30	I	Volcanics and pyroclastics/Tertiary	Hydrothermal	Pyrenean	Unknown	Ba
435	Mahdi Abad	31	29	0	55	1	30	L	Shale, limestone, and dolomite/Lower Cretaceous	Mississippi Valley Type (MVT)	Laramide	Massive	
436	Mahdi Shahr	35	45	0	53	19	0	I	Sandstone and shale/Middle Jurassic	Hydrothermal	Pyrenean	Vein	Cu, Ba
437	Mahmoud Abad	31	51	52	56	9	0	I	Limestone, dolomite, and shale/ Jurassic–Triassic	Hydrothermal	Pyrenean	Unknown	
438	Mahram Taj	32	19	0	53	30	33	I	Dolomite, shale, sandstone/Upper Cambrian–Lower Cambrian	SEDEX	Pan-African	Lens	
439	Malayer	34	10	0	49	14	0	I	Limestone and dolomite/Upper Cretaceous	Mississippi Valley Type	Laramide	Veinlet	
440	Maneshk	36	14	0	59	34	0	I	Granite and metamorphic rocks/ Upper Paleozoic–Lower Triassic	Hydrothermal	Hercynian	Unknown	

(continued)

(continued)

No.	Name	Lat D°	Lat M	Lat S″	Long D°	Long M	Long S″	Size	Host rock/age	Genetic/ mineralization	Orogenic phases	Morphology	Elem para
441	Mansour Abad (Amr Abad)	31	36	30	53	46	30	S	Massive and gray limestone/Lower Cretaceous	Mississippi Valley Type	Laramide	Lens and vein	
442	Maravand	33	34	17	51	24	58	S	Limestone and sandstone/Upper Triassic	Hydrothermal	Early Cimmerian	Vein	
443	Marjan Abad 1	36	47	12	49	29	30	S	Siltstone, carbonaceous siltstone, and sandstone/Upper Triassic–Lower Jurassic	Hydrothermal	Pyrenean	Unknown	
444	Marjan Abad 2	36	46	12	49	27	30	S	Tuff/Tertiary; carbonaceous rocks/ Jurassic	Hydrothermal	Pyrenean	Vein	
445	Mashhad	36	12	0	59	18	0	I	Sandstone and shale/Jurassic	Unknown	Unknown	Unknown	
446	Mazraeh Sadr	31	46	0	53	26	0	S	Andesite and dacite/Tertiary	Hydrothermal	Pyrenean	Vein	
447	Mazraeh Selow	34	9	44	56	32	10	I	Dolomite/Mid-Triassic	Sedimentary	Late Cimmerian	Vein, disseminated	
448	Mazraehnow (Zavareh)	32	22	50	53	30	32	S	Diorite and dolerite/Post-Cambrian; dolomite, shale, and sandstone/ Cambrian	Hydrothermal	Pan-African	Lens	
449	Mehraneh	34	0	50	56	31	15	S	Dolomite/Mid-Triassic	Sedimentary	Late Cimmerian	Vein	Cu, Ag, Cd
450	Mehrjerd	32	15	34	53	51	31	S	Andesite and tuff/Eocene	Hydrothermal	Post-Pyrenean	Vein	
451	Meimeh(Kahru)	33	25	24	51	60	19	S	Bedded limestone/Lower Cretaceous	Mississippi Valley Type (MVT)	Laramide	Veinlet	
452	Meyduk	30	25	0	55	10	0	S	Andesite and pyroclastics/Eocene	Hydrothermal	Pyrenean	Vein	
453	Miankuhi	31	57	0	55	53	42	I	Dolomite and limestone/ Triassic–Jurassic	Sedimentary	Early Cimmerian	Vein	F, Ba
454	MilaKuh	35	59	30	53	47	47	S	Massive and red-layered carbonaceous/Permian	Hydrothermal	Late Cimmerian	Veinlet	
455	Milandar	33	17	34	51	55	10	S	Yellow limestone/Mid-Triassic	Mississippi Valley Type (MVT)	Laramide	Vein, veinlet	
456	Mobarak	32	50	0	51	8	0	I	Limestone/Lower Cretaceous	MVT	Laramide	Vein	

457	Moghestan	32	35	56	55	11	0	I	Shale, sandstone, and dolomite/ Lower Cretaceous	Mississippi Valley Type	Laramide	Stratiformbed	
458	Moshir	29	21	0	56	44	0	I	Carbonates/Cretaceous	Hydrothermal	Post-Pyrenean	Vein	Cu
459	MostafaLu (1, 2)	36	33	40	49	18	57	I	Trachyte and andesite/Eocene	Hydrothermal	Pyrenean	Vein	
460	Mozd Abad	34	1	0	58	39	0	I	Carbonates/Cretaceous; shale, and sandstone/Jurassic	Hydrothermal	Pyrenean	Vein	Cu, Ag
461	Muchan	33	50	40	49	36	37	S	Limestone/Lower Cretaceous	Mississippi Valley Type (MVT)	Laramide	Vein, disseminated	
462	Mujan	34	29	35	50	10	45	S	Limestone, sandstone, tuff/Eocene	Hydrothermal	Pyrenean	Vein	
463	Nabidar	34	10	52	49	8	0	I	Silicic sandstone/Lower Cretaceous	MVT	Laramide	Vein	
464	Najm Abad	34	10	24	58	54	27	I	1- Intrusive rocks/Tertiary, 2-shale, sandstone, and carbonates/ Jurassic	Hydrothermal	Pyrenean	Unknown	Ag
465	Nakhlak	33	33	52	53	50	19	L	Limestone and dolomite/Upper Cretaceous	Hydrothermal	Pyrenean	Vein	
466	Namar	36	5	0	52	3	0	S	Limestone and tuff/Eocene	Hydrothermal	Pyrenean	Vein	
467	Namegh Kashmar	35	24	0	58	48	0	I	Volcanic/Eocene	Hydrothermal	Pyrenean	Vein	
468	Naser Abad (Abadeh)	31	50	0	53	53	0	S	Gray dolomite/Upper Triassic	Hydrothermal	Early Cimmerian	Vein	Cu, Ba
469	Naser Abad 1 (Amol)	36	24	0	51	29	0	M	Dolomitization limestone/Upper Paleozoic (Carboniferous)	Hydrothermal	Hercynian	Vein, veinlet	
470	Naser Abad 2	36	23	46	51	27	49	M	Dolomitization limestone/Upper Paleozoic (Carboniferous)	Hydrothermal	Hercynian	Vein, veinlet	
471	Nasr Abad	31	46	50	53	54	50	S	Gray dolomite/Upper Triassic	Hydrothermal	Early Cimmerian	Disseminated	
472	Nayband	32	22	0	57	47	0	I	Andesite and dacite/Tertiary	Hydrothermal	Pyrenean	Vein	
473	Negin	33	36	59	54	31	42	I	Carbonates/Paleozoic	Hydrothermal	Pyrenean	Vein	
474	Negooiem	33	40	0	54	30	0	I	Limestone/Lower Cretaceous	Mississippi Valley Type	Laramide	Unknown	
475	Neignan 1	34	22	5	57	23	18	S	Dolomite and limestone/Mid-Jurassic	Mississippi Valley Type (MVT)	Late Cimmerian	Vein	
476	Neygnan 2	34	22	25	57	19	33	S	Dolomite and limestone/Mid-Jurassic	MVT	Late Cimmerian	Vein	

(continued)

(continued)

		Lat			Long								
No.	Name	D°	M	S″	D°	M	S″	Size	Host rock/age	Genetic/ mineralization	Orogenic phases	Morphology	Elem para
477	Neyzar	33	18	47	50	34	33	I	Sandstone and carbonates/Lower Cretaceous	Mississippi Valley Type	Laramide	Vein	
478	Niarak	36	33	5	49	24	54	I	Trachytic to basaltic tuff/Eocene	Hydrothermal	Pyrenean	Vein, veinlet	
479	Niaz Morq	33	31	40	51	29	30	S	Limestone and sandstone/Upper Triassic	Mississippi Valley Type (MVT)	Early Cimmerian	Veinlet	
480	Nimhoor	33	27	0	51	4	0	I	Sandstone and carbonates/Lower Cretaceous	Mississippi Valley Type	Laramide	Vein	
481	Nineh	34	2	17	50	32	7	I	Limestone and dolomite/Upper Cretaceous	Mississippi Valley Type	Laramide	Vein	
482	North of Bam	29	13	0	58	14	0	S	Pyroclastics/Eocene	Hydrothermal	Pyrenean	Vein	
483	North of Sar Chah Shur	32	21	18	58	55	12	I	Andesite/Tertiary	Hydrothermal	Pyrenean	Vein	Fe
484	OrasKuh	35	55	48	53	51	13	S	Limestone and dolomite/Permian	Hydrothermal	Early Cimmerian	Unknown	
485	Ouran-e-Bozorg	35	50	48	53	14	32	S	Limestone and dolomite/Cretaceous	Hydrothermal	Late Cimmerian	Vein, veinlet	
486	Ouran-e-Kouchak	35	50	4	53	17	1	S	Limestone and dolomite/Cretaceous	Hydrothermal	Late Cimmerian	Vein, veinlet	
487	Oushk2	32	57	25	57	28	13	S	Reef limestone/Upper Triassic	Hydrothermal	Early Cimmerian	Vein, veinlet	
488	Ozbak Kuh	34	39	17	57	7	19	S	Porphyry andesite/unknown; dolomite, and dolomite limestone/Devonian	Hydrothermal	Unknown	Stratiformbed, vein	
489	Ozun darreh	37	27	0	48	10	0	I	Andesite/Eocene	Hydrothermal	Pyrenean	Vein	
490	Pachy Miana	36	3	46	53	16	9	S	Limestone/Triassic	Hydrothermal	Early Cimmerian	Vein, lens	
491	Pahneh Kamar	32	32	32	58	25	46	I	Limestone/Upper Triassic	Hydrothermal	Early Cimmerian	Disseminated	Ba
492	Panjsar	33	21	34	51	40	0	S	Thick-bedded and gray limestone/ Lower Cretaceous	Mississippi Valley Type (MVT)	Laramide	Veinlet, disseminated	
493	Parachan	36	16	0	50	56	0	I	Volcanic pyroclastic/Eocene	Hydrothermal	Pyrenean	Vein	
494	Parmagasu	35	24	15	54	36	45	I	Andesite/Eocene	Hydrothermal	Pyrenean	Veinlet	
495	Pasar	36	58	0	49	5	0	S	Intrusive rocks/Tertiary; phyllite and schist/unknown	Hydrothermal	Pyrenean	Unknown	

496	Pashtuk	36	41	15	47	37	30	S	Amphibolite and marble/Upper Precambrian–Lower Cambrian	Massive Sulfide	Pan-African	Vein	
497	Pay Negin	29	25	29	56	56	5	I	Andesite and basalt/Eocene	Hydrothermal	Pyrenean	Vein, veinlet	
498	Paychamtu va Barfakeh	36	31	37	54	40	10	S	Limestone/Jurassic	Hydrothermal	Late Cimmerian	Veinlet, lens	
499	Pazanou	32	22	54	50	37	0	I	Sandstone and carbonates/Lower Cretaceous	Mississippi Valley Type	Laramide	Vein	Cu
500	PeyKuh (Oushk1)	32	57	58	57	28	13	S	Reef limestone/Upper Triassic	Hydrothermal	Early Cimmerian	Veinlet	
501	Piazkesh	36	48	0	50	40	0	S	Limestone and dolomite/Triassic	Hydrothermal	Early Cimmerian	Veinlet	
502	Pinavand	33	32	0	51	38	0	I	Dolomite and limestone, schist/ Lower Cambrian(?)	Mississippi Valley Type	Laramide	Vein, disseminated	Cu, Fe
503	Pirhaji	33	23	0	50	42	0	I	Sandstone and carbonates/Lower Cretaceous	Mississippi Valley Type	Laramide	Vein	
504	Pirhaji	34	17	9	57	34	0	S	Dolomite and limestone/Permo-Triassic	Mississippi Valley Type	Late Cimmerian	Vein	Ba, Cu
505	Pish Kuh	34	7	17	55	4	8	I	Limestone and dolomite/Cretaceous	Hydrothermal	Unknown	Massive, veinlet, disseminated	
506	PolSefid	33	21	0	49	58	0	I	Limestone/Lower Cretaceous	MVT	Laramide	Vein	
507	Posht-e-Asiab Hendeh	33	23	0	50	0	49	I	Sandstone and carbonates/Lower Cretaceous	Mississippi Valley Type	Laramide	Vein	
508	Posht-e-Badam	32	58	0	55	22	0	I	Metamorphic rocks/Upper Cretaceous–Lower Tertiary	Hydrothermal	Pyrenean	Unknown	
509	Potkestan	32	29	0	54	30	0	I	Limestone and dolomite/Permian	Hydrothermal	Hercynian	Unknown	
510	Qaleh	34	42	14	57	13	24	S	Limestone and dolomite/ Carboniferous	Hydrothermal	Caledonian	Vein	
511	Qaleh Arab (Hoseyn Abad)	32	38	12	50	59	55	S	Schist, gneiss, and metamorphic volcanics/Precambrian	Hydrothermal	Pan-African	Vein	
512	Qaleh Chah	33	35	30	58	7	24	S	Shale/Lower Jurassic	Hydrothermal	Pyrenean	Vein	
513	Qaleh Madar	34	42	14	57	13	24	S	Limestone/Carboniferous	Mississippi Valley Type	Late Cimmerian	Vein, veinlet	
514	Qaleh-ye-Esfandiar	35	42	30	51	3	30	I	Limestone/Eocene	Hydrothermal	Pyrenean	Vein	

(continued)

(continued)

No.	Name	Lat D°	Lat M	Lat S″	Long D°	Long M	Long S″	Size	Host rock/age	Genetic/mineralization	Orogenic phases	Morphology	Elem para
515	Qanat-e-Marvan	29	21	5	56	45	54	S	Andesite, andesibasalt, and tuff/Mid-Oligocene–Lower Miocene	Hydrothermal	Post-Pyrenean	Vein, veinlet	
516	Qaramilk	33	43	0	49	36	30	I	Shale/Lower Jurassic	MVT	Late Cimmerian	Vein	
517	Qez An	33	41	50	51	22	43	I	Andesitic lava and pyroclastics/Eocene	Hydrothermal	Pyrenean	Vein	Cu
518	Qolleh Kaftarha	35	30	37	54	54	32	S	Porphyrite dacite/Eocene– Oligocene	Hydrothermal	Pyrenean	Vein, veinlet	Ba, Mn, Cu
519	Qomeishlu	32	3	0	51	30	0	I	Limestone/Lower Cretaceous	Mississippi Valley Type (MVT)	Laramide	Unknown	
520	Qomsheh	33	4	0	51	55	0	S	Sandstone and carbonates/Lower Cretaceous	Mississippi Valley Type	Laramide	Vein	
521	Ramu	33	45	0	54	34	0	I	Carbonate/Paleocene	Hydrothermal	Pyrenean	Disseminated	
522	Rana Abad	33	23	0	49	55	0	I	Limestone/Lower Cretaceous	MVT	Laramide	Vein	
523	Rashid Abad	37	6	25	48	23	24	I	Volcanics and volcanoclastics/Eocene	Hydrothermal	Pyrenean	Disseminated	F
524	Rashk and Mashkuh	36	29	23	51	4	55	S	Limestone/Triassic	Hydrothermal	Early Cimmerian	Vein	Cu, Ag
525	Ravanj 1	34	10	0	50	43	0	M	Dolomite and limestone/Mid-Cretaceous	Hydrothermal	Pyrenean	Lens	
526	Ravanj 2	34	10	40	50	43	31	S	Limestone/Lower Cretaceous	Hydrothermal	Post-Pyrenean	Vein	
527	Ravar	31	38	30	57	0	2	I	Dolomite/Mid-Triassic	Sedimentary	Early Cimmerian	Vein	
528	Razan	33	52	29	49	34	32	I	Thick-bedded limestone/Lower Cretaceous	Mississippi Valley Type (MVT)	Laramide	Vein	
529	Reghe	33	48	0	57	14	0	I	Reef limestone/Jurassic	MVT	Late Cimmerian	Vein	
530	Reihan	31	7	25	56	43	40	I	Dolomite/Permo-Triassic	Sedimentary	Early Cimmerian	Vein	Ba
531	Reskan	34	33	26	50	9	16	I	Tuff/Eocene	Hydrothermal	Pyrenean	Vein, veinlet	
532	Reza Abad	35	51	36	53	31	8	S	Limestone/Upper Cretaceous	Sedimentary	Laramide	Veinlet, disseminated	
533	Rezabarag	35	49	40	53	28	18	S	Gray limestone/Upper Cretaceous	Sedimentary	Laramide	Veinlet	
534	Rezavand	34	11	12	48	59	40	I	Limestone and dolomite/Upper Cretaceous	Mississippi Valley Type	Laramide	Stratiformbed	Cu

535	Rig Kalaghi	31	53	39	56	6	31	S	Massive dolomite/Permian	Hydrothermal	Early Cimmerian	Vein	Cu
536	Rivasar	34	46	45	50	19	45	S	Tuff and andesite/Eocene	Hydrothermal	Pyrenean	Lens	Cu
537	Rizab-e-Maryam	33	48	28	53	11	8	I	Red sandy conglomerate/Eocene–Oligocene; tuffite/Eocene	Hydrothermal	Pyrenean	Disseminated	
538	Robai	35	22	0	54	27	45	I	Andesite and dacite/Eocene	Hydrothermal	Pyrenean	Vein	
539	Robat (Arreh Gijeh)	33	46	0	49	46	0	L	Limestone and marl/Lower Cretaceous	Mississippi Valley Type (MVT)	Laramide	Stratiformbed, lens	Fe, Au, Ag, Cu
540	Robat (Takht-e-Hoseyn)	33	45	0	49	46	37	S	Limestone/Lower Cretaceous	Mississippi Valley Type (MVT)	Laramide	Vein, disseminated	
541	Robat 1	33	45	21	49	46	48	S	Limestone/Lower Cretaceous	MVT	Laramide	Vein	
542	Robat 2	33	45	57	49	48	27	S	Limestone/Lower Cretaceous	MVT	Laramide	Vein	
543	Robat 3	33	46	10	49	48	14	S	Limestone/Lower Cretaceous	MVT	Laramide	Vein	
544	Robat Anarak-e-Posht-e-Badam	33	1	0	55	9	0	I	Limestone/Cretaceous	Mississippi Valley Type (MVT)	Laramide	Unknown	
545	Robat Paien	33	46	10	49	47	30	I	Limestone/Lower Cretaceous	MVT	Laramide	Vein	
546	Robat Zilli 2	33	32	2	50	13	12	S	Sandstone and carbonates/Lower Cretaceous	MVT	Laramide	Vein	
547	Robat-e-Bala	33	44	46	49	47	30	I	Sandstone and carbonates/Lower Cretaceous	Mississippi Valley Type	Laramide	Vein	
548	Robat-e-Kerman	30	21	35	56	35	1	I	Dolomite and limestone/Triassic–Jurassic	Sedimentary	Early Cimmerian	Vein	
549	Robat-e-Zilli 1	33	30	8	50	13	51	I	Sandstone and carbonates/Lower Cretaceous	Mississippi Valley Type	Laramide	Vein	
550	Roft	29	39	0	56	43	0	I	Pyroclastics/Eocene	Hydrothermal	Pyrenean	Vein	
551	Roghani	33	44	30	49	35	0	I	Sandstone/Lower Jurassic	Unknown	Unknown	Unknown	Fe
552	Roman	33	39	0	53	22	0	I	Sandstone and andesite basalt/Eocene–Oligocene	Hydrothermal	Pyrenean	Vein	
553	Rutshun (Esfandagheh 1)	28	35	55	56	18	5	I	Basic dike, calc, and dolomite marble and green schist/Devonian	Hydrothermal	Early Cimmerian	Vein	
554	Sadegh Abad	31	44	0	54	20	30	S	Limestone/Cretaceous	Sedimentary	Laramide	Vein	

(continued)

(continued)

No.	Name	Lat D°	Lat M	Lat S″	Long D°	Long M	Long S″	Size	Host rock/age	Genetic/ mineralization	Orogenic phases	Morphology	Elem para
555	Sadegh Abad	32	38	55	50	50	2	I	Sandstone and carbonates/Jurassic	Mississippi Valley Type	Laramide	Vein	
556	Sadr Abad	31	42	0	54	27	0	I	Limestone/Lower Cretaceous	Hydrothermal	Laramide	Vein	
557	Saghand	32	29	0	55	14	0	I	Limestone/Cretaceous	Unknown	Laramide	Unknown	
558	Said Abad	34	20	0	57	26	0	I	Marl, limestone/Jurassic	MVT	Late Cimmerian	Vein	
559	Saied Abad	35	8	55	60	11	55	I	Pyroclastics/Tertiary; carbonates/ Triassic	Hydrothermal	Pyrenean	Vein	
560	Saki Bala	33	49	36	49	50	49	S	Sandstone and shale/Lower Jurassic	Mississippi Valley Type (MVT) and SEDEX	Laramide	Vein	Cu, Fe
561	Saki Paien	33	48	0	49	48	0	S	Limestone/Lower Cretaceous	MVT	Laramide	Vein	
562	Saleh Peighambar (Bagh-e-Bandkuh)	33	24	53	50	33	17	S	Orbitolina limestone/Mid-Cretaceous	Hydrothermal	Laramide	Vein	
563	Salek	33	42	20	49	41	30	I	Sandstone and carbonates/Lower Cretaceous	Mississippi Valley Type	Laramide	Vein	
564	Sang Sayyad	29	39	40	56	45	30	I	Andesite/Upper Eocene	Hydrothermal	Pyrenean	Unknown	
565	Sang-e-Sefid	33	54	0	49	39	0	I	Sandstone and carbonates/Lower Cretaceous	Mississippi Valley Type	Laramide	Vein	
566	Sangkar	35	20	40	54	27	50	S	Gray limestone/Devonian (Upper Paleozoic)	Hydrothermal	Pyrenean	Vein	
567	Sangoubal	37	3	0	49	5	0	I	Carbonates/Paleozoic	Volcanic	Hercynian	Unknown	
568	Sangsar	35	45	0	53	19	0	I	Shale, sandstone, and limestone/ Jurassic	Hydrothermal	Pyrenean	Unknown	
569	Sara	29	13	0	55	12	0	I	Dolomite/Devonian	Unknown	Caledonian	Unknown	
570	Sarbisheh-ye-fulad Mahalleh	36	8	0	53	38	30	S	Limestone/Upper Devonian–Lower Carboniferous	Hydrothermal	Hercynian	Disseminated	Cu, Ag
571	Sarchelnoo	32	24	56	57	22	58	I	Shale and limestone/Triassic	Sedimentary	Early Cimmerian	Unknown	
572	Sarkahnow	28	47	6	61	8	15	I	Shale and sandstone/Eocene	Hydrothermal	Post-Pyrenean	Vein	
573	Sarlash	35	48	16	53	3	49	S	Limestone/Upper Cretaceous	Sedimentary	Laramide	Vein	
574	Sartakht-e-Bahram	33	36	0	57	14	0	I	Shale and limestone/Jurassic	MVT	Laramide	Vein	

575	Sarv	33	47	0	49	42	0	I	Sandstone and carbonates/Lower Cretaceous	Mississippi Valley Type	Laramide	Vein	
576	Savad Kuh	33	25	0	49	56	0	I	Limestone/Lower Cretaceous	MVT	Laramide	Vein	
577	Savojbolagh	36	2	0	50	45	0	I	Carbonates/Jurassic	Hydrothermal	Pyrenean	Unknown	
578	Se	33	24	0	51	37	0	S	Sandstone and carbonates/Lower Cretaceous	Mississippi Valley Type	Laramide	Vein	
579	Sechah	28	45	50	56	30	35	I	Basic dike, calc, and dolomite marble and green schist/Devonian	Hydrothermal	Early Cimmerian	Vein	
580	Sechangi	32	32	43	58	2	52	I	Andesite and dacite/Paleogene	Hydrothermal	Pyrenean	Vein	
581	Sedar	31	35	24	56	2	8	I	Limestone and dolomite/Permo-Triassic	Sedimentary	Early Cimmerian	Vein	
582	Sefid Khal	32	44	0	51	13	0	I	Limestone/Lower Cretaceous	Mississippi Valley Type (MVT)	Laramide	Unknown	
583	Seh Changi	32	32	23	58	2	10	S	Andesite and dacite/Tertiary (Eocene)	Volcanic	Pyrenean	Vein	Cu, Ba, Au
584	Sekarnab	36	18	0	50	19	0	I	Volcanics and volcanoclastics/ Tertiary Carbonates/Paleozoic	Volcanic	Pyrenean	Unknown	
585	Semnan	35	48	0	53	18	0	I	Shale, sandstone, and limestone/ Jurassic	Hydrothermal	Pyrenean	Unknown	
586	Senjetoo (Deh Asgar)	31	46	23	56	12	31	S	Dolomite/Permo-Triassic	Hydrothermal	Early Cimmerian	Lens, veinlet	Cu
587	Shah Ali Beiglu (Babin)	37	21	24	48	10	8	S	Andesitic tuff and tuffaceous sandstone/Eocene	Hydrothermal	Pyrenean	Vein	
588	Shahmirzad	35	43	55	53	16	30	S	Chert, limestone/Upper Cretaceous	Sedimentary	Laramide	Vein	
589	Shakhab	36	2	16	53	54	0	S	Massive limestone, dolomitic limestone/ Devonian–Carboniferous	Hydrothermal	Hercynian	Vein, veinlet, disseminated	
590	Shakin	35	54	0	49	20	0	M	Limestone and dolomite/Permian	Hydrothermal	Hercynian	Veinlet	
591	Shams Abad	33	48	0	49	44	0	I	Sandstone and carbonates/Lower Cretaceous	Mississippi Valley Type	Laramide	Vein	
592	Sharif Abad	34	6	0	51	12	0	S	Andesite and limestone/Eocene	Skarn	Pyrenean	Lens, vein	
593	Sheykh Ahmad	28	43	15	60	55	0	I	Shale/Eocene	Hydrothermal	Post-Pyrenean	Vein	

(continued)

(continued)

No.	Name	Lat D°	Lat M	Lat S″	Long D°	Long M	Long S″	Size	Host rock/age	Genetic/ mineralization	Orogenic phases	Morphology	Elem para
594	Shilandar	36	49	20	48	41	30	S	Tuff and tuffite/Eocene (Karaj formation)	Hydrothermal	Pyrenean	Vein	Ba, Ag
595	Shourabeh AbbasAbad	33	25	35	54	33	56	S	Schist and marble/Upper Proterozoic–Lower Paleozoic	Hydrothermal	Pan-African	Vein	Cu
596	Shur Cheshmeh (Shurab)	36	2	51	53	52	55	S	Limestone, dolomitic limestone/ Devonian–Carboniferous	Hydrothermal	Hercynian	Veinlet	F
597	Shurab	33	34	33	58	34	33	S	Dacite/Paleogene; shale/Lower Jurassic	Hydrothermal	Pyrenean	Veinlet, disseminated	Ba
598	Siah Cheshmeh	36	18	16	54	40	24	S	Limestone/Mid–Upper Permian	Hydrothermal	Hercynian	Veinlet, disseminated	
599	Siah Darreh	36	31	5	54	4	55	I	Limestone/Upper Jurassic	Hydrothermal	Mid-Cimmerian	Veinlet	
600	Siah Kamar	31	12	56	59	3	27	I	Andesite and pyroclastics/Tertiary	Hydrothermal	Pyrenean	Vein	
601	Siah Kuh	32	47	0	53	56	10	I	Limestone/Precambrian	Hydrothermal	Pan-African	Vein	
602	Siahoo	32	57	0	51	1	0	I	Sandstone and carbonates/Lower Cretaceous	Mississippi Valley Type	Laramide	Vein	
603	Sibzar	34	37	0	57	10	30	S	Thick-bedded to massive gray dolomite/Upper Precambrian–Lower Cambrian	Hydrothermal	Hercynian	Vein	
604	Sineh Kuh	36	12	14	54	12	50	I	Limestone/Devonian	Hydrothermal	Hercynian	Vein	
605	Sira	36	2	0	51	8	0	I	Pyroclastics/Eocene	Hydrothermal	Pyrenean	Unknown	
606	Soltan Abad	34	18	12	48	54	25	S	Dolomite and limestone/Lower Cretaceous	Mississippi Valley Type (MVT)	Laramide	Lens	
607	Soltan Abad	34	53	0	49	37	0	I	Pyroclastics/Eocene	Volcanic	Pyrenean	Unknown	Cu, Ag, Sb
608	Somagh	36	23	0	49	21	0	I	Tuff and andesite/Eocene	Hydrothermal	Pyrenean	Vein	
609	Sorb	33	19	22	54	46	59	I	Limestone/Lower Cretaceous	Hydrothermal	Pyrenean	Vein	
610	Sormeh Chal	34	14	20	48	54	25	S	Limestone and dolomite/Upper Cretaceous	Mississippi Valley Type	Laramide	Stratiformbed	Cu, F
611	Soumarubar	37	13	0	48	47	0	S	Limestone/Mid–Upper Permian	Hydrothermal	Pyrenean	Vein	
612	Sourmagh	36	32	0	49	21	0	I	Tuff and andesite/Eocene	Hydrothermal	Pyrenean	Vein	

613	South East Of Kuh-e Kolahbid	34	11	19	48	59	50	S	Limestone/Cretaceous	MVT	Laramide	Vein, veinlet	
614	South Of Abkhorak 1	34	15	57	57	6	31	S	Limestone and dolomite/Triassic	Hydrothermal	Mid-Cimmerian	Vein	
615	South Of Abkhorak 2	34	15	24	57	6	11	S	Limestone and dolomite/Triassic	Hydrothermal	Mid-Cimmerian	Vein	
616	South Of Abkhorak 3	34	14	43	57	6	2	S	Limestone and dolomite/Triassic	Hydrothermal	Mid-Cimmerian	Vein	
617	South Of Abkhorak 4	34	14	3	57	6	2	S	Limestone and dolomite/Triassic	Hydrothermal	Mid-Cimmerian	Vein	
618	South of Bailaq 1	33	33	0	49	59	49	I	Sandstone and carbonates/Lower Cretaceous	MVT	Laramide	Vein	
619	South of Bailaq 2	33	45	0	49	46	37	I	Sandstone and carbonates/Lower Cretaceous	MVT	Laramide	Vein	
620	South of Dadkin	33	17	10	54	57	6	I	Limestone/Mesozoic	Hydrothermal	Pyrenean	Vein	
621	South of Klongeh	33	42	0	49	53	0	I	Sandstone and carbonates/Lower Cretaceous	Mississippi Valley Type	Laramide	Vein	
622	Southeast of Kal-e-Sorkh	33	52	0	50	41	0	I	Sandstone and carbonates/Lower Cretaceous	Mississippi Valley Type	Laramide	Vein	
623	Southwest of Lachah	30	24	11	55	9	41	I	Andesite and pyroclastics/Eocene	Hydrothermal	Pyrenean	Vein	
624	Suk	33	43	31	49	41	6	I	Thick-bedded limestone/Lower Cretaceous	Mississippi Valley Type (MVT)	Laramide	Vein, veinlet	
625	Surakh-e-Tarik	32	16	0	54	25	0	I	Limestone and dolomite/Lower Cretaceous	Sedimentary	Laramide	Stratiformbed, vein	
626	SynehKuh	36	12	14	54	12	50	S	Limestone/Devonian	Hydrothermal	Hercynian	Veinlet, disseminated	Cu, Fe
627	Tabas	33	32	0	56	9	0	I	Shale, limestone, and dolomite/Upper Triassic	Skarn	Early Cimmerian	Vein	
628	Tafresh	34	38	18	50	3	7	I	Dolomite, limestone/Cretaceous	Hydrothermal	Pyrenean	Vein	
629	Taft	31	43	45	54	10	50	S	Shale, marl, and massive limestone/ Lower Cretaceous	Sedimentary	Laramide	Vein, veinlet	
630	Taj Kuh	31	26	15	55	57	14	S	Limestone dolomite/Permo-Triassic	Mississippi Valley Type (MVT)	Early Cimmerian	Stratiformbed	
631	Tajareh	35	14	0	49	41	32	I	Carbonates/Oligo-Miocene	Hydrothermal	Post-Pyrenean	Unknown	

(continued)

(continued)

No.	Name	Lat D°	Lat M	Lat S″	Long D°	Long M	Long S″	Size	Host rock/age	Genetic/mineralization	Orogenic phases	Morphology	Elem para
632	Takht Kuh	33	45	0	50	45	0	I	Sandstone and carbonates/Lower Cretaceous	Mississippi Valley Type	Laramide	Vein	
633	Takhtabi Zanjani	36	31	45	54	40	55	I	Dolomitic limestone/Jurassic	Hydrothermal	Pyrenean	Vein	
634	Takht-e-Payzendegani	36	32	5	54	40	55	S	Massive and large-layered limestone/Upper Jurassic	Hydrothermal	Late Cimmerian	Veinlet	
635	Takieh	33	51	22	49	38	30	I	Marl, limestone/Lower Cretaceous	MVT	Laramide	Vein	
636	Taknar	35	22	0	57	47	0	I	Volcanics/Tertiary; schist/Precambrian	Hydrothermal	Pyrenean	Stratiformbed	
637	Taleh Dizak, Darreh Farah, Cheshmehvar, Darreh Garmeh	33	23	0	49	58	30	I	Dolomite and limestone/Lower Cretaceous	Mississippi Valley Type (MVT)	Laramide	Disseminated	
638	Talkhab	34	13	10	56	33	5	S	Dolomite/Triassic	Sedimentary	Late Cimmerian	Vein, disseminated	
639	Tambali	33	49	0	49	32	0	I	Limestone, dolomite/Lower Cretaceous	MVT	Laramide	Vein	
640	Tangal-e-Ghar	31	38	46	56	2	23	S	Limestone and dolomite/Permo-Triassic	Mississippi Valley Type (MVT)	Early Cimmerian	Lens	
641	Tangeh	35	25	27	54	35	15	S	Volcanics and pyroclastics/Eo-Oligocene	Hydrothermal	Pyrenean	Vein	
642	Tangol-e-Bala	33	19	14	54	47	9	I	Dolomite and limestone/Lower Cretaceous	Hydrothermal	Laramide	Vein	
643	Tangol-e-Khoriti	34	50	0	57	15	0	I	Dolomite and limestone/Upper Precambrian	Hydrothermal	Pyrenean	Vein	
644	Tanureh	35	20	30	54	26	15	I	Limestone and dolomite/Devonian	Hydrothermal	Pyrenean	Vein	
645	Tappeh Sorkh (yazd)	31	45	6	56	16	6	S	Dolomitic limestone and dolomite/Permo-Triassic	MVT	Early Cimmerian	Vein, lens	
646	Tappeh Tagh (Bidoo)	34	51	56	57	20	50	S	Dolomite and phyllite/Upper Precambrian; igneous body/Post-Precambrian	Hydrothermal	Early Cimmerian	Vein	

647	Tar	33	26	4	51	43	47	I	Sandstone/Upper Triassic; basaltic dike and partial of limestone/Mid-Triassic	Hydrothermal	Early Cimmerian	Veinlet, disseminated	
648	Taraz	31	22	0	56	27	39	S	Dolomite/Permo-Triassic	Sedimentary	Early Cimmerian	Vein	
649	Taraz Ravar	31	22	0	56	27	39	S	Limestone and dolomite/Lower Cretaceous	Mississippi Valley Type	Laramide	Vein	
650	Tashud	34	16	55	56	35	48	I	Dolomite/Mid-Triassic	Sedimentary	Early Cimmerian	Vein	
651	Teymoor Mahallat	33	27	0	51	4	0	I	Shale, limestone, sandstone/Triassic	MVT	Laramide	Vein	
652	Tikan 1	33	21	40	50	34	49	I	Shale and marl/Mid-Cretaceous	Mississippi Valley Type (MVT)	Laramide	Unknown	Fe, Cu
653	Tikan 2	33	22	58	50	36	46	I	Shale and marl/Mid-Cretaceous	Mississippi Valley Type	Laramide	Unknown	
654	Tobe Asli	36	6	0	53	55	0	I	Shale, sandstone, limestone, and dolomite/Jurassic	Hydrothermal	Pyrenean	Vein	
655	Tochal	35	35	0	51	41	0	I	Andesite/Eocene	Volcanic	Pyrenean	Unknown	
656	Tomburli	33	31	13	49	34	7	S	Sandstone/Jurassic	Mississippi Valley Type (MVT)	Laramide	Vein	Ba
657	Torogh	33	19	52	51	47	26	S	Sandstone/Lower Jurassic	Hydrothermal	Early Cimmerian	Veinlet, lens	
658	Toureh	34	2	0	49	15	30	I	Limestone and dolomite/Upper Cretaceous	Mississippi Valley Type	Laramide	Unknown	Cu
659	Tout	32	31	0	54	26	0	I	Dolomite/Upper Precambrian–Lower Cambrian	Hydrothermal	Pan-African	Unknown	F
660	Tuyehdarvar	36	1	20	53	47	27	S	Limestone and shale limestone/Jurassic	Mississippi Valley Type	Laramide	Veinlet, vein, Lens	Polymetal
661	Valucheh	36	17	0	53	46	0	I	Sedimentary rocks/Upper Paleozoic	Unknown	Hercynian	Unknown	
662	Varrin 1	34	5	48	50	25	45	I	Limestone and dolomite/Upper Cretaceous	Mississippi Valley Type	Laramide	Lens	
663	Vejin Bala	32	44	24	51	8	57	S	Limestone/Lower Cretaceous	Mississippi Valley Type (MVT)	Laramide	Vein	Ag
664	Vejin Paien	32	43	35	51	9	35	S	Limestone/Lower Cretaceous	Mississippi Valley Type (MVT)	Laramide	Vein	

(continued)

(continued)

No.	Name	Lat D°	Lat M	Lat S″	Long D°	Long M	Long S″	Size	Host rock/age	Genetic/ mineralization	Orogenic phases	Morphology	Elem para
665	VishanTekiyeh	33	51	22	49	38	30	S	Limestone/Lower Cretaceous	Mississippi Valley Type (MVT)	Laramide	Vein, lens	
666	West of Agricheh	33	46	0	49	59	0	I	Sandstone and carbonates/Lower Cretaceous	Mississippi Valley Type	Laramide	Vein	
667	West of Kachuieh	32	17	42	51	9	43	S	Sandstone and carbonates/Lower Cretaceous	Mississippi Valley Type	Laramide	Vein	
668	West of Kalot-e-Chapdouni	32	36	49	51	58	39	I	Sandstone and carbonates/ Triassic–Jurassic	Mississippi Valley Type	Laramide	Vein	
669	West of Klongeh	33	33	0	49	52	0	I	Sandstone and carbonates/Lower Cretaceous	Mississippi Valley Type	Laramide	Vein	
670	Yazd	34	22	0	57	19	0	I	Dolomite and limestone/Mid-Jurassic	MVT	Late Cimmerian	Vein	
671	Yurt Baba	36	31	20	54	37	10	S	Massive and large-layered Limestone/ Upper Jurassic	Hydrothermal	Late Cimmerian	Veinlet	
672	Zafreh	32	55	8	52	16	24	S	Dolomite and limestone/Lower Cretaceous	Mississippi Valley Type (MVT)	Laramide	Vein, veinlet	
673	Zaghoo	31	46	13	56	14	56	S	Dolomite and limestone/Permo-Triassic	Mississippi Valley Type (MVT)	Early Cimmerian	Lens	
674	Zah	34	26	21	52	22	20	S	Trachyandesite/Lower Eocene	Hydrothermal	Pyrenean	Vein	
675	Zajegan-e-Sofla	36	16	46	49	25	12	S	Breccia tuff/Eocene	Hydrothermal	Pyrenean	Vein, disseminated	
676	Zakaria	36	1	30	53	52	37	S	Crystalline limestone/ Devonian–Carboniferous	Hydrothermal	Hercynian	Veinlet, disseminated	F
677	Zali	36	21	16	54	4	5	S	Sandstone/Jurassic (Lias)	Sedimentary	Early Cimmerian	Veinlet, disseminated	Ba
678	Zangoulu	33	3	30	50	42	0	I	Sandstone and carbonates/Lower Cretaceous	Mississippi Valley Type	Laramide	Vein	
679	Zar Abad	36	29	0	50	27	0	I	Basalt/Tertiary; carbonates/ Cretaceous	Hydrothermal	Pyrenean	Unknown	
680	Zar Shokuh	35	21	40	54	37	40	I	Andesite and tuff/Eocene– Oligocene	Hydrothermal	Pyrenean	Vein	

681	Zarankab	38	39	29	46	44	32	I	Unknown	Hydrothermal	Unknown	Vein	
682	Zarbisheh	36	6	0	53	24	0	I	Limestone, dolomite/Permo-Triassic	Hydrothermal	Late Cimmerian	Vein	
683	Zarigan (Zireh khaman)	32	3	0	55	33	0	S	Intrusive and volcanic rocks/Upper Precambrian–Lower Cambrian	Hydrothermal	Pan-African	Stratiformbed, vein	
684	Zarrin Dasht	35	36	0	52	38	0	I	Pyroclastics/Eocene; carbonates/ Cretaceous	Hydrothermal	Pyrenean	Unknown	
685	Zaru-e-Abarghu	31	44	40	53	24	35	S	Granodiorite/Oligo-Miocene; thick- to mid-bedded limestone/ Lower Cretaceous	Hydrothermal	Post-Pyrenean	Vein	
686	Zarvan	35	46	0	52	46	0	I	Sandstone and shale/Jurassic	Mississippi Valley Type	Laramide	Unknown	
687	Zefreh	32	55	8	52	16	24	I	Dolomite/Mid-Triassic	Mississippi Valley Type	Laramide	Vein	Cu, Ag, Au, Co
688	Zeh Abad	36	28	0	49	25	0	S	Porphyrite monzonite/Oligocene; tuff/Eocene	Hydrothermal	Post-Pyrenean	Vein	Cu
689	Zendegan	33	25	0	49	56	0	I	Limestone/Cretaceous	MVT	Laramide	Vein	
690	Zeresh Amiriyeh	33	32	0	50	0	0	I	Sandstone and carbonates/Lower Cretaceous	Mississippi Valley Type	Laramide	Vein	Cu
691	Zeresh Kuh	35	21	40	54	37	40	S	Andesite/Eocene–Oligocene	Unknown	Unknown	Vein, veinlet	
692	Zizgan	34	31	30	50	10	35	S	Tuff/Eocene	Hydrothermal	Pyrenean	Vein	
693	Zour Abad	35	26	0	60	51	0	I	Intrusive and volcanic rocks/Tertiary; carbonates/Triassic	Hydrothermal	Pyrenean	Vein	Ba

Complete list of bauxite mineral deposits and indications of Iran along with their detailed specifications

No.	Name	Lat D°	Lat M	Lat S″	Long D°	Long M	Long S″	Size	Host rock/age	Genetic/mineralization	Orogenic phases	Morphology	Elem para
1	Abgarm-e-Qazvin	35	30	0	49	30	0	S	Limestone/Permian	Surface Alteration	Early Cimmerian	Lens	
2	Aghajari	36	34	22	46	36	0	S	Limestone and sandstone/Permo-Triassic	Surface Alteration	Late Cimmerian	Lens	
3	Aineh varzan-Delichay	35	45	0	52	15	0	I	Dolomite/Triassic; limestone/Permian	Surface Alteration	Early Cimmerian	Stratiformbed	
4	Alamut	35	24	49	50	35	20	S	Black shale and sandstone/Jurassic; limestone and dolomite/Permo-Triassic	Surface Alteration	Early Cimmerian	Stratiformbed	
5	Aliabad-e-Karaftu	36	15	0	47	0	0	S	Limestone and dolomite/Permo-Triassic	Surface Alteration	Late Cimmerian	Stratiformbed	
6	Alibalta-va-Kordkandi	36	42	24	46	42	30	S	Dolomite/Triassic; quartzitic sandstone/Permian	Surface Alteration	Late Cimmerian	Stratiformbed	
7	Bazargan	30	27	49	57	1	34	S	Dolomite, shale, and sandstone/Permo-Triassic	Surface Alteration	Late Cimmerian	Lens	
8	Bolboluiyeh	30	12	0	57	18	0	S	Shale and sandstone/Upper Triassic;dolomite/Mid-Triassic	Surface Alteration	Early Cimmerian	Stratiformbed	
9	Chagharlu	36	20	0	46	45	0	S	Dolomite, marl, sandstone, and limestone/Permian	Surface Alteration	Late Cimmerian	Stratiformbed	
10	Chapu	36	59	35	46	58	39	S	Dolomite/Triassic; limestone/Permian	Surface Alteration	Late Cimmerian	Stratiformbed	
11	Chekcheku	32	15	0	54	22	0	S	Shale and sandstone/Jurassic; dolomite and limestone/Permian	Surface Alteration	Mid-Cimmerian	Lens	
12	Dasht-e-Zar	31	15	0	51	15	0	S	Shale, sandstone, and dolomite/Permian	Surface Alteration	Late Cimmerian	Lens	
13	Durak	31	48	30	50	44	30	S	Limestone and dolomite/Triassic; dolomite Permo-Carboniferous	Surface Alteration	Late Cimmerian	Stratiformbed	
14	Ganow	36	4	5	53	48	29	I	Shale and sandstone/Jurassic; dolomite and limestone/Triassic	Surface Alteration	Mid-Cimmerian	Stratiformbed	
15	Gheshlagh	36	45	0	55	15	0	S	Carbonaceous rocks/Triassic; limestone and siltstone/Permian	Surface Alteration	Late Cimmerian	Stratiformbed	
16	Golcharmu-Sarighomeish	36	28	27	46	29	42	I	Limestone/Permian	Surface Alteration	Hercynian	Stratiformbed	
17	Haft Cheshmeh	31	45	0	50	15	0	S	Dolomite and shale/Triassic; limestone/Permian	Surface Alteration	Late Cimmerian	Stratiformbed	

18	Hengam	28	24	0	52	23	0	S	Marl limestone/Upper Cretaceous	Surface Alteration	Laramide	Stratiformbed
19	Jajarm	37	2	30	56	33	0	S	Shale and sandstone/Jurassic; dolomite/Triassic; limestone/Carboniferous	Surface Alteration	Mid-Cimmerian	Lens
20	Kal-e-Jafar Agha (Jahan Abad)	36	49	7	56	7	8	S	Shale and sandstone/Jurassic; dolomite and limestone/Triassic	Surface Alteration	Mid-Cimmerian	Stratiformbed
21	Khatb-e-Maragheh(Khatib)	37	9	0	46	26	0	S	Limestone and shale/Upper Permian	Surface Alteration	Late Cimmerian	Stratiformbed
22	Kombolu	35	59	30	53	53	20	S	Tuff/Eocene; limestone/Jurassic	Surface Alteration	Pyrenean	Stratiformbed
23	Nilchian Dopelan	31	55	0	50	30	0	S	Limestone and dolomite/Permo-Triassic	Surface Alteration	Late Cimmerian	Stratiformbed
24	North of Hoseyn Abad	36	40	3	45	54	0	I	Limestone/Permian	Surface Alteration	Late Cimmerian	Stratiformbed
25	North of Yazd	32	0	0	54	45	0	S	Shale and sandstone/Jurassic; dolomite and limestone/Permian	Surface Alteration	Mid-Cimmerian	Lens
26	Nowruz Abad	36	49	0	46	46	0	S	Dolomite/Triassic; limestone/Permian	Surface Alteration	Late Cimmerian	Stratiformbed
27	Reza Abad	36	0	49	53	33	20	S	Shale and sandstone/Jurassic; dolomite and limestone/Triassic	Surface Alteration	Mid-Cimmerian	Stratiformbed
28	Sadr Abad	31	50	0	53	45	0	S	Limestone, shale, and sandstone/Upper Triassic–Lower Jurassic; limestone/Mid-Triassic	Surface Alteration	Early Cimmerian	Lens
29	Sarchaveh	35	0	0	49	30	0	S	Carbonaceous rocks/Permo-Triassic	Surface Alteration	Early Cimmerian	Lens
30	Sarfaryab	30	45	0	50	37	0	S	Limestone and marl/Upper Cretaceous	Surface Alteration	Laramide	Lens
31	Shah Bolagh	35	45	0	53	25	0	S	Shale and sandstone/Jurassic; dolomite/Triassic	Surface Alteration	Early Cimmerian	Lens
32	Shahmirzad	35	47	41	53	21	12	S	Shale and sandstone/Jurassic; dolomite and limestone/Triassic	Surface Alteration	Early Cimmerian	Stratiformbed
33	Siahrudbar and Shirin Abad	36	38	0	55	1	30	S	Conglomerate and sandstone/Jurassic; dolomite and limestone/Triassic	Surface Alteration	Mid-Cimmerian	Stratiformbed and lens
34	Sorkh Hesar	35	4	22	60	14	40	S	Limestone and dolomite/Mid–Upper Triassic	Surface Alteration	Early Cimmerian	Stratiformbed
35	Southwest of Kalijeh	36	39	35	45	53	12	I	Limestone/Permian	Surface Alteration	Late Cimmerian	Stratiformbed
36	Taveh Qoran	36	30	0	46	0	0	I	Limestone and dolomite/Permo-Triassic	Surface Alteration	Late Cimmerian	Stratiformbed
37	Zan	35	37	55	52	3	20	I	Limestone/Jurassic	Surface Alteration	Mid-Cimmerian	Stratiformbed

Complete list of gold mineral deposits and indications of Iran along with their detailed specifications

No.	Name	Lat D°	Lat M	Lat S″	Long D°	Long M	Long S″	Size	Host rock/age	Genetic/ mineralization	Orogenic phases	Morphology	Elem para
1	Abbas Abad Aynalu	38	53	0	46	50	30	I	Granite/Upper Eocene–Lower Oligocene; volcanic rocks/Eocene; limestone/Cretaceous	Skarn	Pyrenean	Vein	Cu
2	Abdar	30	18	0	55	19	0	I	Altered porphyry intrusive/Oligo-Miocene	Porphyry	Post-Pyrenean	Disseminated, stockwork	Cu, Pb, Ag
3	Agh Darreh	36	40	0	47	10	0	S	Limestone/Miocene	Epithermal	Pyrenean	Vein (lens)	Sb, Ag
4	Aghamira	38	49	30	46	39	0	I	Andesite/Eocene	Hydrothermal	Pyrenean	Vein	Cu(Ag)
5	Aghuran	37	24	30	47	30	30	S	Rhyodacite and trachyandesite/Eocene	Epithermal	Pyrenean	Vein	
6	Ali Abad-e-Musavi	35	30	25	49	58	14	I	Volcanic and pyroclastic rocks/Eocene	Hydrothermal	Pyrenean	Vein	
7	Ali javad	38	39	5	46	55	33	S	Monzonite/Oligo-Miocene, andesite, and basalt/Quaternary	Hydrothermal	Post-Pyrenean	Disseminated	Cu
8	Alvand	36	18	46	49	10	50	I	Granite and granodiorite/Upper Eocene–Lower Oligocene	Hydrothermal	Pyrenean	Vein	Cu, Fe, Zn, Pb, Ag, As
9	Alvand-e-Hamedan	34	44	0	48	28	30	I	Granite and hornfels/Upper Cretaceous–Paleocene; schist/ Upper Jurassic–Lower Cretaceous	Hydrothermal	Laramide	Disseminated and vein	Ag
10	Alvand-e-Hamedan	34	44	0	48	28	30	I	Alluvial/Quaternary	Placery	Post-Pyrenean	Placer	
11	Anarak	32	45	0	55	12	0	S	Granite, metavolcanics, and marble/ Upper Proterozoic	Hydrothermal	Pan-African	Lens	Cu, Pb, Zn, W, Bi, As, Fe, Ag
12	Anguran Chay	36	39	28	47	28	45	M	Alluvial/Quaternary	Sedimentary	Post-Pyrenean	Placer	Ag, W, Pb, Zn
13	Anjedan	32	57	46	50	20	7	I	Shale, Marl, and limestone/Cretaceous	Hydrothermal	Laramide	Vein	Pb, Ba
14	Anjerd-e-Olya	38	40	20	46	55	12	I	Granite/Oligocene; andesite/Eocene; limestone/Cretaceous	Skarn	Post-Pyrenean	Vein and veinlet	Cu, Ag
15	Arabshah-e-Garus	36	24	0	47	20	10	I	Schist and marble/Precambrian	Epithermal	Pyrenean	Vein and veinlet	As, Ba
16	Arghash (Cheshmeh Zard)	35	52	30	58	36	20	M	Granite/Eo-Oligocene; volcanics/ Eocene	Epithermal	Pyrenean	Vein	
17	Armudagh	37	46	0	47	59	0	I	Andesite and tuff/Miocene	Hydrothermal	Post-Pyrenean	Vein	Cu, Ag
18	Asagi	30	37	25	60	12	6	I	Andesite and dacite/Oligocene–Eocene	Hydrothermal	Pyrenean	Vein and veinlet	Cu, Fe, Ag

19	Ashin	33	31	46	53	24	46	I	Quartz keratophyre, plagiogranite and gabbro/Eocene	Hydrothermal	Pyrenean	Vein, veinlet, lens	Cu
20	Astaneh	33	50	8	49	19	0	S	Microgranodiorite/Cretaceous	Hydrothermal	Pyrenean	Vein	
21	Astaneh	33	52	0	49	20	0	S	Alluvial/Quaternary	Placery	Post-Pyrenean	Placer	
22	Ay Qalehsi	36	20	0	47	22	0	S	Limestone and tuff/Oligo-Miocene	Epithermal	Post-Pyrenean	Vein	Zn, Pb
23	Bab Kahnuj	29	18	0	57	15	30	I	Granite/Oligo-Miocene; volcanics/ Eocene	Porphyry	Post-Pyrenean	Vein	Cu
24	Baghdeh							I	Granite and granodiorite/Oligo-Miocene; volcanics and pyroclastics/Eocene	Hydrothermal	Pyrenean	Vein	Cu
25	Bagheraey	28	46	0	57	11	0	I	Andesitic lava/Eocene	Hydrothermal	Pyrenean	Vein	Cu
26	Barika (Alut)							S	Metavolcanics/Upper Cretaceous	Hydrothermal	Laramide	Vein and veinlet	Ba
27	Bayche Bagh	36	52	15	47	18	40	S	Hydrobreccia andesite and porphyry dacite/Oligo-Miocene	Catathermal	Post-Pyrenean	Vein	Cu, Zn, Pb, Ag (Polymetal)
28	Bibi shahr banu	35	19	0	51	54	0	S	Andesite/Eocene	Mesothermal	Pyrenean	Vein	Pb, Cu
29	Bid Mohammad Hasan	35	24	0	54	25	0	S	Volcanoclastics/Mid-Eocene	Hydrothermal	Pyrenean	Vein	Cu
30	Bolboli	29	34	0	56	15	30	I	Diorite/Post-Eocene; limestone and tuff/Eocene	Skarn	Pyrenean	Vein, veinlet, disseminated	Cu, Fe
31	Borjak	35	19	18	57	41	46	I	Rhyolite/Paleozoic; schist/Precambrian	Hydrothermal	Hercynian	Massive, lens	Cu, Fe, Ag, Pb, Zn
32	Buteh Alam	33	34	35	53	45	0	I	Porphyry granite/Eocene; limestone and sandstone/Post-Triassic	Hydrothermal	Pyrenean	Vein	Fe, Pb
33	Chah Ali Khan	33	23	0	53	8	0	I	Andesite, basalt, tuff, and tuffite/ Eocene	Hydrothermal	Pyrenean	Vein	Cu, Bi
34	Chah Gaz	29	30	45	55	2	0	S	Granodiorite, schist, and limestone/ Permo-Triassic	Massive sulfide	Late Cimmerian	Vein	Cu, Pb, Zn
35	Chah Gir	33	21	0	53	28	0	I	Andesite/Upper Eocene	Hydrothermal	Pyrenean	Veinlet	Cu, Pb, Ag
36	Chah mesi	30	24	30	55	10	0	S	Porphyry diorite/Oligo-Miocene; andesite porphyry, megaporphyry trachyandesite/Eocene	Hydrothermal	Post-Pyrenean	Vein	Cu, Pb, Zn, Ag

(continued)

(continued)

		Lat			Long								
No.	Name	D°	M	S″	D°	M	S″	Size	Host rock/age	Genetic/ mineralization	Orogenic phases	Morphology	Elem para
37	Chah mileh	33	26	0	53	48	30	I	Schist and limestone/Upper Proterozoic–Lower Cambrian	Hydrothermal	Pan-African	Lens, vein	Pb, Zn, Cu, W, Ag, Sb, Ni
38	Chah musa	35	30	0	54	55	0	S	Granite and diorite/Upper Eocene–Lower Oligocene; volcanics/ Eocene	Hydrothermal	Pyrenean	Vein	Cu, Ag
39	Chah Zaghu	32	0	31	59	17	35	I	Granite/Tertiary; carbonate, silicic vein/Cretaceous	Hydrothermal	Pyrenean	Vein	Cu
40	Chahar Gonbad	29	35	30	56	11	0	S	Quartz diorite/Early Miocene; volcanosedimentary rocks/Eocene	Porphyry	Post-Pyrenean	Vein, disseminated	CuPb, Zn, Ag
41	Chah-e-chahar nafari	31	49	30	59	58	0	I	Listwanite and limestone/Upper Cretaceous–Lower Paleogene	Hydrothermal	Laramide	Disseminated	Cu, Pb, Zn, Mo, Fe
42	Chal							I	Volcanics and pyroclastics/Eocene	Hydrothermal	Pyrenean	Vein	
43	Chalpu (Kuh Sorkh)	35	38	0	58	27	0	S	Sedimentary rocks and pyroclastics/ Eocene	Epithermal	Pyrenean	Vein	As, Sb, (Hg)
44	Chamtal (Agha Ali)	38	42	0	46	10	50	I	Granodiorite/Upper Eocene–Lower Oligocene; volcanoclastic rocks and limestone/Cretaceous	Skarn	Pyrenean	Vein	Cu, Fe
45	Chant	35	43	30	57	26	0	I	Sedimentary rocks/Paleocene	Hydrothermal	Pyrenean	Vein	Cu
46	Chare Darreh	38	49	40	46	20	20	I	Granitic rocks/Oligocene	Hydrothermal	Pyrenean	Vein and veinlet	
47	Chargerd	36	26	30	49	5	0	I	Tuff and andesite/Eocene	Hydrothermal	Pyrenean	Vein (lens)	Ag, Cu
48	Cheshmeh Ghan	38	44	16	46	23	48	I	Limestone, shale, volcanic rocks/ Cretaceous; granite Ordoubad/ Eocene	Skarn	Pyrenean	Vein and veinlet	Cu, Fe
49	Cheshmeh Hafez	35	24	57	54	44	59	I	Granite and granodiorite/Upper Eocene–Lower Oligocene; andesite, dacite, rhyolite, and tuff/Eocene	Hydrothermal	Pyrenean	Vein	Cu
50	Cheshmeh Sefid	35	21	0	54	41	20	I	Andesite and dacite/Eocene	Hydrothermal	Pyrenean	Vein and veinlet	Cu
51	Cheshmeh Shurab	33	26	40	53	22	10	I	Serpentinite/Upper Proterozoic–Lower Paleozoic	Hydrothermal	Pan-African	Vein	Cu, Ni, Ag

52	Chichalku	36	25	0	47	22	0	I	Schist and marble/Precambrian	Epithermal	Pyrenean	Vein and veinlet	Zn, Pb
53	Damaneh	29	12	20	57	12	0	I	Porphyrite quartz diorite/Oligo-Miocene; andesite, trachyandesite, and tuff/Eocene	Porphyry	Post-Pyrenean	Disseminated	Cu
54	Dar Bagh	27	9	0	57	48	0	I	Andesite/Eocene–Oligocene	Hydrothermal	Post-Pyrenean	Vein	Cu
55	Dar Bagh	33	39	50	48	16	40	I	Pillow lava, tuff, and limestone/Paleozoic	Volcanic	Hercynian	Vein	
56	Darestan	35	27	0	54	36	30	S	Granodiorite/Early Eocene, breccia tuff and andesite/Mid-Eocene	Hydrothermal	Pyrenean	Vein	Cu
57	Dargiaban	29	7	30	60	57	30	S	Locogranite/Post-Eocene; schist/Eocene	Hydrothermal	Pyrenean	Vein	Ag, As, Sb
58	Darreh Hamzeh	28	54	0	57	51	0	S	Porphyry granitoid/Oligo-Miocene; andesite, trachyandesite, and dacite/Eocene	Porphyry	Post-Pyrenean	Vein	Cu, Mo, Ag, Fe (very low)
59	Darreh Zereshk	31	33	30	53	51	0	S	Granodiorite/Oligo-Miocene; andesite, trachyte, sandstone, and limestone/Tertiary	Epithermal	Post-Pyrenean	Vein	Cu, Ag
60	Dashkasan (Sari Dagh)	35	12	0	48	5	0	L	Dacite and rhyodacite/Pliocene; limestone/Lower Miocene	Epithermal	Post-Pyrenean	Vein	Sb, As, Ag, Hg
61	Dastgerd	27	10	16	57	12	48	I	Alluvial/Quaternary	Placery	Post-Pyrenean	Placer	
62	Divaneh Dar	36	0	0	57	27	0	I	Andesite and basalt/Eocene	Hydrothermal	Pyrenean	Vein and veinlet	Cu
63	Dizehjine	36	39	30	49	5	15	S	Granodiorite/Post-Eocene; andesite/Eocene	Hydrothermal	Pyrenean	Vein and veinlet	Cu, Pb, Zn
64	Do Zarad Akhtar	29	40	0	56	51	20	I	Pyroclastics/Eocene	Hydrothermal	Pyrenean	Vein	Cu, Ag
65	Dowlat Abad	29	9	33	57	13	9	I	Granodiorite/Oligo-Miocene; volcanosedimentary/Eocene	Hydrothermal	Post-Pyrenean	Vein and veinlet	Cu
66	Dozal	38	51	10	46	14	30	I	Diorite and granodiorite/Oligo-Miocene	Hydrothermal	Post-Pyrenean	Vein	
67	Dustbeyglu	38	33	30	47	33	55	S	Intrusions/Oligo-Miocene; andesite/Eocene	Epithermal	Pyrenean	Vein and veinlet	Polymetal
68	Eshtovin	38	53	0	46	22	0	I	Granite and granodiorite/Upper Eocene–Lower Oligocene	Hydrothermal	Pyrenean	Vein	

(continued)

(continued)

No.	Name	Lat D°	Lat M	Lat S″	Long D°	Long M	Long S″	Size	Host rock/age	Genetic/ mineralization	Orogenic phases	Morphology	Elem para
69	Gandy	35	19	0	54	38	0	S	Tuff and sedimentary rocks/Eocene	Hydrothermal	Pyrenean	Vein and veinlet	
70	Garoos	38	19	30	46	1	39	I	Marl, shale, and sandstone/Upper Miocene	Sedimentary	Post-Pyrenean	Massive	
71	Gavkamar	37	18	28	48	16	35	I	Monzonite/Post-Eocene; green tuff/ Eocene	Hydrothermal (epithermal)	Pyrenean	Vein	Ag, Sb, As
72	Gharachilar	38	50	48	46	23	0	S	Granite and granodiorite/Upper Eocene–Lower Oligocene	Hydrothermal	Pyrenean	Vein	cu, Mo(Ag), W
73	Ghare Chy	38	51	0	46	23	0	I	Granite and granodiorite/Upper Eocene–Lower Oligocene	Hydrothermal	Pyrenean	Vein	
74	Ghasam Abad	33	48	0	52	23	0	I	Volcanosedimentary/Eocene	Hydrothermal	Pyrenean	Vein	Pb, W, Cu, Ag
75	Gheichagh	37	41	3	47	2	47	I	Andesite/Oligocene	Hydrothermal	Pyrenean	Vein	Cu, Zn, Pb, Ag
76	GomushUlan	38	35	0	46	43	0	I	Granodiorite/Eocene– Oligocene; limestone/Cretaceous	Hydrothermal	Pyrenean	Vein	Cu(Pb, Ag)
77	Gorgab 3	33	59	40	52	28	0	I	Dike of granodiorite porphyry, dacite, and andesite/Upper Eocene; schist/ Paleozoic	Hydrothermal	Pyrenean	Vein	Pb, Zn, Cu, Ag
78	Gowd	32	55	0	54	29	0	I	Marble and schist/Upper Precambrian	Hydrothermal	Pan-African	Vein	Zn, Pb (Cu, Ag, Hg)
79	Gowd-e-Morad	33	24	37	53	31	28	S	Metamorphic rocks/Upper Proterozoic	Hydrothermal	Pan-African	Lens	Polymetal
80	Gowzal Bologh	36	27	0	46	37	0	I	Granite and granodiorite/Tertiary	Hydrothermal	Pyrenean	Vein	Cu
81	Gudarzi	32	51	0	53	12	0	I	Listwanite/Upper Cretaceous	Hydrothermal	Laramide	Vein	Cu
82	Gurcheh berenj	33	50	0	54	2	0	I	Limestone/Oligo-Miocene	Skarn	Post-Pyrenean	Vein and veinlet	Cu, Pb, Ba, Ag
83	Gurva	34	0	0	54	36	29	I	Granodiorite/Upper Jurassic; schist and marble/Proterozoic	Hydrothermal	Pan-African	Vein	Cu, Pb, Zn, Co, Mo
84	Halil Rud	28	54	0	57	6	0	I	Young alluvial/Quaternary	Placery	Post-Pyrenean	Placer	
85	Helmesi	37	41	0	47	31	0	I	Rhyodacite and andesite/ Eocene–Miocene	Hydrothermal	Pyrenean	Disseminated	Cu
86	Hengaran	32	4	55	59	14	0	I	Granite/Tertiary; listwanite/Upper Cretaceous	Hydrothermal	Pyrenean	Vein	

87	Hyred	31	56	22	59	12	10	S	Granite, granodiorite, and tuff/Upper Cretaceous	Hydrothermal	Pyrenean	Vein	Cu
88	Jushin	29	4	5	57	37	30	I	Andesite/Eocene	Hydrothermal	Pyrenean	Vein	Cu
89	Kakiyeh	35	21	0	54	40	0	I	Tuff and sedimentary rocks/Eocene	Hydrothermal	Pyrenean	Vein and veinlet	Cu
90	Kal Kafi	33	24	0	54	14	0	S	Schist/Upper Precambrian; microgranite, quartz monzosyenite porphyry/Eocene	Porphyry	Pyrenean	Disseminated	Cu, Mo
91	Kalateh Teymour	35	25	5	58	20	40	S	Granite and granodiorite/Early Eocene; volcanics/Eocene	Hydrothermal	Pyrenean	Vein	
92	Kavand	36	37	30	48	8	30	S	Dolomite/Upper Precambrian	Hydrothermal	Pyrenean	Vein (lens)	Fe, Ag, W
93	Keyghal	38	37	0	46	41	44	I	Microdiorite, diorite, and granodiorite/Upper Eocene; andesitic tuff, pyroxene andesite, and quartz andesite/Upper Cretaceous–Mid-Eocene	Porphyry	Pyrenean	Disseminated and veinlet	Cu, Pb, Zn
94	Khalifehlu	36	19	0	49	14	0	S	Rhyolite, andesite, and trachyte/Eocene	Epithermal mesothermal	Pyrenean	Vein and veinlet	Cu, Ag, Pb, Zn
95	Khangoul	39	13	43	44	7	24	S	Ophiolite melange/Upper Cretaceous	Hydrothermal	Laramide	Vein	
96	Kharvana (Kharvanagh)	38	32	30	46	15	30	S	Granodiorite/Oligo-Miocene; sedimentary rocks/Paleocene	Epithermal	Post-Pyrenean	Disseminated and vein	Cu
97	Khoin Rud	38	41	45	46	36	10	S	Monzodiorite porphyry/Oligo-Miocene; andesite and trachyte/Eocene	Hydrothermal (epithermal)	Post-Pyrenean	Disseminated	(Ag, Cu, As, Sb, Hg)
98	Khuni	33	26	35	54	12	8	S	Dolomite and dolomitic marble/Upper Precambrian; volcanics/Eocene	Hydrothermal	Pan-African	Vein	Cu, Zn, Pb, Ag
99	Khunik	32	23	31	59	7	12	S	Microdiorite and andesitic dike, latite–andesite/Early Eocene	Hydrothermal	Pyrenean	Vein	
100	Kilisa kandy	38	49	38	46	10	0	I	Volcanics and pyroclastics/Oligo-Miocene	Hydrothermal	Post-Pyrenean	Vein and veinlet	
101	Koruyan	36	8	0	46	6	0	S	Metamorphic rocks/Cretaceous	Hydrothermal	Laramide	Vein	
102	Kuh Zar (Baghu)	35	26	30	54	38	45	S	Granodiorite/Upper Eocene; volcanic rocks/Mid-Eocene	Mesothermal	Pyrenean	Vein	Cu, turquoise

(continued)

(continued)

No.	Name	Lat D°	Lat M	Lat S″	Long D°	Long M	Long S″	Size	Host rock/age	Genetic/ mineralization	Orogenic phases	Morphology	Elem para
103	Kuh Zar (Baghu)	35	26	30	54	38	45	S	Alluvial/Quaternary	Placery	Post-Pyrenean	Placer	
104	Kuh Zar (Torbat-e-Heydariyeh)	35	21	58	58	54	20	S	Granite and granodiorite/Upper Eocene–Lower Oligocene	Epithermal	Pyrenean	Vein	Cu, Ag
105	Kuh-e-Dom (Gharghar)	33	59	39	52	51	30	S	Granite and granodiorite/Oligocene; lava and pyroclastic/Eocene	Hydrothermal	Pyrenean	Vein	Cu, Ag
106	Kuh-e-Janja	31	12	58	60	21	48	I	Porphyry hornblende granite/Oligocene; volcanics, pyroclastics, and limestone/Eocene	Hydrothermal	Pyrenean	Veinlet	Cu, Ag, Mo
107	Kuh-e-Lar	29	41	21	60	53	36	S	Intrusive body/Tertiary; carbonaceous and flysch/Paleocene	Porphyry	Post-Pyrenean	Disseminated	Cu, Mo
108	Kuh-e-Mearaji	33	10	49	54	11	29	I	Dolomite and limestone/Permian	Hydrothermal	Hercynian	Vein	Pb, Zn
109	Kuh-e-Zar Arak	33	51	0	49	49	0	I	Sandy dolomite, sandstone, and conglomerate/Lower Cretaceous	Hydrothermal	Late Cimmerian	Vein	Pb, As, Ag, Co
110	Kuhiyan	36	46	30	48	52	0	S	Granite/Oligocene; tuff/Eocene	Hydrothermal	Pyrenean	Vein and veinlet	
111	Latala	30	25	51	55	10	57	I	Volcanics/Eocene	Hydrothermal	Pyrenean	Vein	Cu
112	Maranjab	34	15	0	51	48	0	I	Granodiorite and volcanics/Eocene	Hydrothermal	Pyrenean	Vein	Pb, Cu
113	Mardan Qom (Mirdanal)	38	50	0	46	33	0	S	Granodiorite porphyry/Eocene–Oligocene; volcanosedimentary/Cretaceous–Eocene	Hydrothermal	Pyrenean	Vein	Cu(Ag), Au
114	Masjed Daghi	38	52	30	45	56	10	I	Volcanics and pyroclastics/Oligo-Miocene	Hydrothermal	Post-Pyrenean	Vein	Cu
115	Mazraeh	38	39	0	47	4	0	S	Granite/Eocene–Oligocene; tuff and ignimbrite/Eocene; limestone/Upper Cretaceous	Skarn	Pyrenean	Vein (lens)	Cu(Ag), Au
116	Mazraeh Seyed Ali	32	13	22	59	50	40	S	Silica/Cretaceous	Epithermal	Laramide	Vein	
117	Meskani	33	19	0	53	28	0	S	Trachyandesite and basalt/Eocene	Hydrothermal	Pyrenean	Vein, veinlet	Cu, Ni, Co, Ag
118	Muteh	33	34	0	50	45	0	L	Schist/Upper Precambrian–Lower Cambrian	Hydrothermal	Pan-African	Vein and veinlet	Cu, Fe

119	Nabijan	38	46	30	46	48	30	S	Quartz monzodiorite/Oligo-Miocene; marl, and limestone/Cretaceous	Hydrothermal	Post-Pyrenean	Vein, veinlet, lens	Zn, Pb, Fe, Sb, Hg, Cu
120	Nadushan	31	48	51	53	21	3	S	Tuff and tuffite/Eocene; dolomite limestone and limestone/Paleozoic	Hydrothermal	Pyrenean	Lens	Polymetal
121	Naibandan	38	46	30	46	48	30	S	Monzodiorite porphyry/Oligo-Miocene; volcanic and limestone/Eocene	Skarn	Post-Pyrenean	Vein	Zn, Pb, Fe
122	Noghdouz	38	23	0	47	19	0	S	Granodiorite/Early Eocene	Hydrothermal	Pyrenean	Vein	Ag, Cu
123	North of Bazman	28	17	30	60	1	12	S	Volcanics and pyroclastics/Miocene	Hydrothermal	Post-Pyrenean	Vein, veinlet, disseminated	Ag
124	North of Kamkueyh	30	24	0	55	9	0	I	Porphyry dike/Early Eocene; volcanic rocks/Eocene	Hydrothermal	Pyrenean	Vein	
125	Olang-e-Firuz Kuh	35	25	0	60	52	30	I	Diorite and shale/Lower Jurassic	Hydrothermal	Mid-Cimmerian	Vein	
126	Oshar	38	52	20	45	47	0	I	Rhyolite/Eocene	Hydrothermal	Pyrenean	Vein	
127	Pahnavar (Nowjemehr)	38	49	35	46	10	43	I	Intermediate intrusive rocks/Miocene; volcanics/Oligo-Miocene	Hydrothermal	Post-Pyrenean	Vein	
128	Pas Qaleh	35	52	30	51	30	0	S	Andesite, trachyandesite, and tuff/Eocene	Hydrothermal	Pyrenean	Stratiformbed, lens	Polymetal
129	Pice of gold Eastern Torkemani	33	14	0	54	1	0	I	Listwanite bearing gold or conglomerate/Pliocene	Unknown	Pan-African	A piece of gold with weight 60 gr.	Sb
130	Pir Bolaghi	38	52	0	46	22	50	I	Granite and granodiorite/Upper Eocene–Lower Oligocene	Hydrothermal	Pyrenean	Vein and veinlet	
131	Qaleh (West of Asad abad)	29	42	0	56	39	0	I	Volcanics and sedimentary/Eocene	Hydrothermal	Pyrenean	Vein	
132	Qaleh Narp	29	42	0	56	41	30	I	Andesite, pyroclastics, microdiorite, sandstone, and conglomerate/Late Eocene,	Hydrothermal	Pyrenean	Vein	Cu, Ag
133	Qaleh Sardar va Talbur	34	4	0	52	12	0	I	Volcanics and quartzite veins/Eocene	Hydrothermal	Pyrenean	Vein	Pb, Cu
134	Qaleh Zary	31	49	43	58	55	15	S	Andesite, dacite, and basalt/Eocene	Hydrothermal	Pyrenean	Vein	Cu, Fe, Pb, Zn
135	Qaleh-jugh	35	18	30	58	56	8	S	Volcanosedimentary/Eocene	Hydrothermal	Pyrenean	Vein	

(continued)

(continued)

No.	Name	Lat D°	Lat M	Lat S″	Long D°	Long M	Long S″	Size	Host rock/age	Genetic/ mineralization	Orogenic phases	Morphology	Elem para
136	Qebleh	33	27	0	53	26	0	I	Trachyandesite/Eocene; limestone/ Lower Cretaceous	Hydrothermal	Pyrenean	Veinlet	Cu, Ag
137	Rashid Abad	37	6	0	48	22	0	S	Andesite and rhyodacite/Eo-Oligocene	Hydrothermal	Pyrenean	Vein and veinlet	Cu, Fe(Ag)
138	Razligh	38	8	0	47	32	0	I	Andesite, dacite, and rhyolitic tuff/ Eocene	Hydrothermal	Pyrenean	Vein	
139	Revesht (Hoseyn Abad)	33	42	30	49	11	30	I	Granite/Jurassic	Hydrothermal	Mid-Cimmerian	Vein	W, Cu
140	Sabzevaran	28	37	10	57	56	30	I	Granodiorite/Oligocene	Hydrothermal	Post-Pyrenean	Vein	
141	Sahlehgou	33	23	25	53	27	30	I	Limestone/Cretaceous; shale/Jurassic	Hydrothermal	Late Cimmerian	Vein	Cu, Ag
142	Samagh	37	6	0	48	22	0	I	Granodiorite/Post-Eocene; andesite and dacite porphyry/Eocene	Hydrothermal	Pyrenean	Vein and veinlet	Zn, Pb, Cu, Ag
143	Sarajiyeh	33	59	12	52	4	14	I	Volcanics/Eocene	Hydrothermal	Pyrenean	Vein	Pb, W, Cu, Ag
144	Sar-Cheshmeh	29	56	40	55	52	20	S	Granodiorite/Miocene; andesite/ Eocene	Porphyry	Post-Pyrenean	Disseminated	Cu, Mo, Ag
145	Sar-e-Kuh	29	55	50	55	46	0	S	Granodiorite/Oligo-Miocene;	Hydrothermal	Post-Pyrenean	Veinlet, disseminated	Cu
146	Sarha	35	23	0	54	35	0	I	Breccia tuff and tuff/Eocene	Hydrothermal	Pyrenean	Vein	
147	Sarikhanlu	38	33	33	47	33	57	S	Volcanics and pyroclastics/Eocene	Hydrothermal	Pyrenean	Vein and veinlet	
148	Sarnow	30	13	0	54	56	0	I	Quartz diorite porphyry/Oligocene; volcanics/Eocene	Hydrothermal	Pyrenean	Vein, veinlet, disseminated	Cu
149	Senjedeh	37	20	0	48	13	0	S	Andesite and sandy tuff/Eocene	Hydrothermal	Pyrenean	Vein	Polymetal
150	Sharaf Abad-Hizeh jan	38	37	7	46	29	52	S	Monzodiorite/Mio-Pliocene; trachyandesite and tuff/Eocene	Hydrothermal	Post-Pyrenean	Vein	Cu, Co, Zn, Pb
151	Shekarab	33	40	22	55	20	0	I	Tuff, andesite/Eocene	Hydrothermal	Pyrenean	Veinlet	Cu, Mo, Ag
152	Shend-e-Mohammad	35	32	24	58	4	34	S	Rhyolite/Post-Jurassic; shale and sandstone/Jurassic	Hydrothermal	Late Cimmerian	Vein	Sb, Ag
153	Shurchah-e-Almasaki							I	Granodiorite/Oligocene; siltstone and sandstone/Eocene	Hydrothermal	Pyrenean	Vein	Sb, As

154	Siah Jangal	28	46	28	61	13	28	I	Porphyry diorite/Oligocene; shale, sandstone/Eocene	Hydrothermal	Post-Pyrenean	Vein, veinlet	Fe, Mn, Pb
155	Siah Rud	38	46	21	46	12	48	I	Volcanic rocks, sandstone, shale, and tuff/Oligo-Miocene	Volcanic	Post-Pyrenean	Vein	
156	Siasetorgi	30	35	18	60	26	44	S	Granodiorite and hornfels/ Eocene–Oligocene	Hydrothermal	Pyrenean	Vein	Cu
157	Solado Gol							I	Serpentinite/Upper Cretaceous	Hydrothermal	Laramide	Veinlet	
158	Soltan Baba	36	41	24	49	5	36	I	Acidic intrusions/Oligo-Miocene; volcanics/Eocene	Hydrothermal	Pyrenean	Vein	Pb, Zn, Cu, Ag
159	Sorfeh Paeyn	29	19	38	57	19	29	I	Andesite and pyroclastics/Eocene	Hydrothermal	Post-Pyrenean	Vein	Cu
160	South of Tabriz	38	3	19	46	16	56	I	Rhyolite, dacite, and tuff/Eocene	Hydrothermal	Pyrenean	lens	
161	Sungoun	38	41	30	46	42	20	S	Porphyry micromonzonite/Miocene; limestone/Cretaceous	Porphyry, skarn	Post-Pyrenean	Lens Disseminated	Cu(Mo, Pb)
162	Taknar	35	21	55	57	46	24	S	Granite/Precambrian; slate and phyllite/Precambrian	Massive sulfide	Pan-African	Massive	Cu, Pb, Zn, Au (Ag)
163	Talarji	33	26	20	54	2	30	I	Schist and limestone/Upper Proterozoic–Lower Cambrian	Hydrothermal	Pan-African	Veinlet	Cu, Pb
164	Talkhe	33	20	8	53	35	0	S	Schist, marble/Upper Precambrian–Lower Cambrian	Hydrothermal	Pan-African	Veinlet, lens	Ni, Co, U, Au, Ag
165	Talhe (Cheshmeh Talhe)	34	2	42	52	49	14	I	Andesite and acidic tuff/Eocene	Hydrothermal	Pyrenean	Vein	Sr, Pb(Ba)
166	Tall-e-Siah	33	23	0	54	55	48	S	Porphyry diorite/Eocene; red bed and evaporites/Jurassic	Hydrothermal	Pyrenean	Veinlet	Cu, Pb, Ba
167	Talmesi	33	22	40	53	27	30	S	Trachyandesite and basalt/Eocene	Hydrothermal	Pyrenean	Vein, veinlet, disseminated	Cu, Ni, Co, Ag
168	Tar							I	Silica iron lens and sandy limestone dolomite/Mid-Triassic	Hydrothermal	Early Cimmerian	Vein, veinlet, disseminated	Pb, Zn, Ba, Cu
169	Tarik Darreh	35	19	30	60	39	0	S	Diorite/Post-Jurassic; shale and sandstone/Jurassic	Hydrothermal	Early Cimmerian	Vein	W
170	Tickmedash	37	49	0	47	56	0	I	Rhyodacite/Oligo-Miocene	Hydrothermal	Pyrenean	Vein	Fe, W, Ag

(continued)

(continued)

No.	Name	Lat D°	Lat M	Lat S″	Long D°	Long M	Long S″	Size	Host rock/age	Genetic/ mineralization	Orogenic phases	Morphology	Elem para
171	Torghabeh (Daghestan)	36	19	0	59	25	0	S	Granodiorite/Upper Paleozoic–Lower Triassic	Hydrothermal	Late Cimmerian	Vein	Ag, Cu, Fe, W
172	Tovazari	33	25	24	54	21	9	I	Granodiorite/Upper Jurassic; schist and marble/Proterozoic	Hydrothermal	Late Cimmerian	Vein	Cu, Co, Ni
173	Turah							I	Serpentinite/Upper Cretaceous	Hydrothermal	Laramide	Veinlet	
174	Tuzgi							I	Hornfels/Tertiary	Hydrothermal	Pyrenean	Vein	Sb, As
175	Tuzlar	36	50	0	47	28	0	S	Andesite/Oligo-Miocene	Hydrothermal	Post-Pyrenean	Vein	Cu, Ag
176	Urus Morghi	29	9	30	57	10	30	I	Granite/Oligo-Miocene; rhyolite, dacite, and andesite/Eocene	Hydrothermal	Pyrenean	Vein	Cu
177	Yaromghiyeh	38	41	0	45	5	0	I	Mica-schist, hornblende schist, gabbro, and pyroxenite/Upper Cretaceous–Lower Paleogene	Hydrothermal	Post-Pyrenean	Lens	
178	Yousoflu-Noghduz	38	23	30	47	19	2	I	Alkali granite, monzogranite porphyry/ Oligocene	Epithermal	Post-Pyrenean	Lens	Cu
179	Zaglik	38	36	22	47	20	51	S	Granite and granodiorite/Oligocene; andesite and tuff/Eocene	Hydrothermal (Epithermal)	Pyrenean	Vein	
180	Zanjireh	38	21	0	45	17	0	I	Marl, shale, and sandstone/Upper Miocene	Placery	Post-Pyrenean	Massive	
181	Zarmehr	35	14	11	58	58	41	S	Alluvial/Quaternary	Sedimentary	Post-Pyrenean	Placer	
182	Zarrin	32	40	30	54	37	30	S	Granite/Pre-Cretaceous; metamorphosed shale and sandstone/ Jurassic	Hydrothermal	Late Cimmerian	Vein	W
183	Zarrin Khoy	38	42	0	45	6	0	I	Gneiss, schist, and mica-schist/Upper Cretaceous–Lower Paleogene	Hydrothermal	Post-Pyrenean	Vein	
184	Zarrine-rekab	38	40	0	46	45	0	S	Andesite and subvolcanics/ Eocene–Oligocene	Epithermal	Pyrenean	Disseminated	Cu, Zn, Pb, Ag
185	Zarshuran	36	43	0	47	8	0	L	Marble and black shale/Precambrian	Epithermal	Pyrenean	Disseminated, veinlet	Sb, As, Hg, Zn

186	Zartarasht	28	12	18	57	12	40	S	Volcano-green schist/Devonian	Hydrothermal	Hercynian	Vein, veinlet, . lens	
187	Zeh Abad	36	28	0	49	25	10	S	Intrusive rocks/Eo-Oligocene; volcanoclastics and volcanics/ Eocene	Hydrothermal	Pyrenean	Vein	Zn, Pb, Ag
188	Zenooz	38	35	30	45	49	38	I	Rhyolite, dacite, and tuff/Eocene	Hydrothermal	Pyrenean	Lens	
189	Ziyakuh	36	59	24	50	8	34	S	Volcanics/Mesozoic	Hydrothermal	Unknown	Vein	

Complete list of antimony, arsenic, and mercury mineral deposits and indications of Iran along with their detailed specifications

No.	Name	Lat D°	Lat M	Lat S″	Long D°	Long M	Long S″	Size	Host rock/age	Genetic/ mineralization	Orogenic phases	Morphology	Type
1	Ferdows (Shurab)	33	34	54	58	6	49	I	Shale/Jurassic; rhyolite and andesite/Tertiary	Hydrothermal	Pyrenean	Vein	Sb
2	Khuni	33	33	34	53	23	8	I	Argillized rocks/Mid-Eocene; tuff/Eocene	Hydrothermal	Pyrenean	Vein	Hg
3	Nakhlak	33	32	42	53	49	14	I	Limestone and sandstone/Triassic and Upper Cretaceous	Hydrothermal	Laramide	Vein	Hg
4	Pateyar	33	19	57	53	47	23	S	Red sandstone/Oligo-Miocene; schist and dolomite/Upper Precambrian–Lower Cambrian	Telethermal	Post-Pyrenean	Vein	Sb
5	Saleh Abad	33	0	28	53	54	27	I	Silicified keratophyre, phyllite/Cretaceous	Hydrothermal	Laramide	Vein	Hg
6	Torkemani	33	13	43	54	1	52	S	Marble/Upper Precambrian–Lower Cambrian	Hydrothermal	Pyrenean	Vein and veinlet	Sb
7	Gonabad	34	10	13	58	59	56	I	Granite/Tertiary	Hydrothermal	Pyrenean	Vein	Sb
8	Posht-e-Kalleh Nignan	34	20	49	57	22	13	S	Volcanic rocks/Tertiary; limestone, shale, marble, and sandstone/Upper Jurassic; volcanics/Tertiary	Hydrothermal	Pyrenean	Vein	Sb
9	Dashkasan (Sarigona)	35	12	0	48	5	0	L	Porphyry microgranite and microgranodiorite, dacite, and rhyodacite/Pliocene–Pleistocene	Epithermal	Post-Pyrenean	Vein	Sb
10	Chalpu Kuh-e-Sorkh	35	37	30	58	27	30	S	Tuff and marl and sandstone/Eocene	Epithermal	Pyrenean	Vein, veinlet, disseminated	Sb
11	Ghosun (Kashmar)	35	29	0	58	18	0	I	Flysch/Cretaceous–Eocene	Hydrothermal	Pyrenean	Vein	Sb
12	Kuh-e-Sorkh	35	29	39	58	23	0	I	Tuff/Tertiary	Hydrothermal	Post-Pyrenean	Vein	Sb
13	Shend-e-Mahmud	35	32	24	58	4	34	S	Rhyolite/Upper Eocene–Lower Oligocene	Epithermal	Pyrenean	Vein	Sb
14	Shurab	35	33	0	58	4	0	S	Rhyolite/Upper Eocene–Lower Oligocene	Epithermal	Pyrenean	Vein, veinlet	Sb
15	TakSanguta	35	37	15	58	8	45	I	Sandstone and tuff/Paleogene	Hydrothermal	Pyrenean	Vein	Sb
16	Taktanur (Kalateh Choubak)	35	35	40	58	20	30	S	Andesite and tuff/Eocene	Hydrothermal	Pyrenean	Veinlet	As

17	West of Torbat-e-Heydariyeh 1	35	17	19	58	34	53	I	Rhyolite, andesite, and tuff/Eocene	Hydrothermal	Pyrenean	Vein	As
18	West of Torbat-e-Heydariyeh 2	35	20	39	58	32	47	I	Rhyolite, andesite, and tuff/Eocene	Hydrothermal	Pyrenean	Vein	As
19	Agh Darreh	36	40	38	47	0	20	I	Andesite/Miocene; sedimentary/Oligo-Miocene	Hydrothermal	Post-Pyrenean	Vein	Sb
20	Balar Ghani	36	43	21	47	6	14	I	Schist and marble/Precambrian; volcanic/Tertiary	Hydrothermal	Post-Pyrenean	Vein	Sb
21	Ghareh Zagh	36	39	11	46	49	17	I	Schist and volcanics and sedimentary rocks/Lower Cambrian; andesite, and limestone/Eocene	Hydrothermal	Post-Pyrenean	Disseminated	Hg
22	MoghanLu	36	38	0	48	47	53	S	Gneiss and granite/Post-Miocene; limestone and sandstone and shale/Plio-Quaternary	Epithermal	Pyrenean	Vein	Sb
23	Zarshuran	36	43	0	47	8	0	I	Marble and black shale/Precambrian	Epithermal	Pyrenean	Veinlet, disseminated	As
24	Zarshuran	36	42	5	47	7	39	S	Andesite/Eocene	Hydrothermal	Pyrenean	Unknown	Sb
25	Kelisa Kandi	38	51	0	46	10	0	I	Limestone/Oligo-Miocene	Hydrothermal	Post-Pyrenean	Vein	As
26	Valilu	38	21	40	46	50	20	S	Contact of volcanics and sedimentary rocks/Oligo-Miocene	Hydrothermal	Post-Pyrenean	Vein, veinlet, disseminated	As
27	Khangoli	39	3	15	44	23	11	S	Limestone and serpentinite/Upper Cretaceous	Hydrothermal	Pyrenean	Vein, veinlet, disseminated	Hg
28	Arsenic in the Torbat-e-Jam							I	Unknown	Hydrothermal	Unknown	Vein	As
29	Sefidabeh							I	Unknown	Unknown	Unknown	Unknown	Sb
30	SeQaleh							S	Rhyolite/Upper Eocene–Lower Oligocene	Epithermal	Pyrenean	Vein	Sb
31	Torbat-e-Jam Antimony							I	Volcanics/Tertiary; sedimentary rocks/Permian	Hydrothermal	Pyrenean	Vein and veinlet	Sb

Complete list of kaolin mineral deposits and indications of Iran along with their detailed specifications

No.	Name	Lat D°	Lat M	Lat S″	Long D°	Long M	Long S″	Size	Host rock/age	Genetic/ mineralization	Orogenic phases	Morphology	Elem para
1	Ghazan daghi	36	6	0	49	34	0	L	Rhyolite/Paleocene	Surface Alteration	Pyrenean	Massive	Alunite
2	Azarguyuni	36	41	0	47	15	0	L	Andesite/Oligo-Miocene	Surface Alteration	Post-Pyrenean	Unknown	Badeleite
3	Blueteluk	38	34	45	45	48	45	L	Andesite and carbonates/ Tertiary	Surface Alteration	Post-Pyrenean	Lens	Ball clay
4	Chah kular robat khan	33	21	8	56	3	18	L	Sedimentary rocks/ Permo-Triassic	Sedimentary	Hercynian	Lens	Boaxite
5	Dasht-e-Kalat	35	30	0	53	39	0	L	Sedimentary rocks/Eocene	Surface Alteration	Post-Pyrenean	Stratiformbed	Boaxite and fire clay
6	Kaboutar Kuh	34	13	0	58	58	0	S	Limestone and shale/Jurassic	Surface Alteration	Laramide	Massive	Clay
7	Ab-garm Mahallat	34	2	0	50	30	0	L	Sedimentary rocks/Jurassic	Surface Alteration	Post-Pyrenean	Stratiformbed	Clay
8	Bagh Siah	34	25	15	58	34	44	L	Volcanics/Eocene	Surface Alteration	Laramide	Vein and veinlet	Clay
9	Maki Kashmar	35	35	0	58	22	0	S	Rhyodacite/Eocene	Surface Alteration	Pyrenean	Lens	Clay
10	Shurjeh bolagh	35	31	0	53	50	0	S	Volcanics/Eocene	Surface Alteration	Pyrenean	Lens	Clay
11	Vijeh	31	22	0	52	46	14	L	Sedimentary/Permo-Triassic	Sedimentary	Hercynian	Stratiformbed	Clay
12	Zilig	38	31	30	46	56	0	S	Sedimentary Rocks/Pliocene	Surface Alteration	Pyrenean	Massive	Clay
13	Sahah bolagh	36	39	55	48	5	0	S	Granite/Precambrian	Surface Alteration	Pan-African	Massive	Feldspar
14	Istisu	36	0	0	47	0	0	S	Volcanics, tuff, and sandstone/ Tertiary	Surface Alteration	Post-Pyrenean	Massive	Fire clay
15	Kandaj	34	52	0	50	2	0	I	Acidic tuff/Eocene	Surface Alteration	Post-Pyrenean	Stratiformbed	Fire clay
16	Kandi Bad Ab	35	20	0	54	40	0	L	Volcanics/Eocene–Oligocene	Surface Alteration	Pyrenean	Massive	Fire Clay
17	Kavir16	31	21	51	52	35	3	L	Sedimentary rocks/ Permo-Triassic	Sedimentary	Hercynian	Stratiformbed	Fire Clay
18	Kuh-e-Atabak 3	34	26	8	58	36	49	I	Tuff/Eocene	Surface Alteration	Post-Pyrenean	Unknown	Green clay
19	Zoghal Choob	34	6	45	58	27	20	I	Granite/Eocene	Surface Alteration	Post-Pyrenean	Unknown	Green clay
20	Ab-Torsh	36	21	0	49	32	0	S	Volcanics/Eocene	Surface Alteration	Pyrenean	Vein	Illite
21	Kusk Nosrat	35	7	13	50	56	0	L	Acidic volcanics/Eocene	Surface Alteration	Post-Pyrenean	Massive	Illite
22	Nikouiyeh	36	17	0	49	32	0	S	Acidic volcanics/Tertiary	Surface Alteration	Post-Pyrenean	Massive	Illite

23	South of Atabak Kuh	34	25	15	58	34	44	S	Tuff/Eocene	Surface Alteration	Post-Pyrenean	Lens	Illite
24	Niagh	36	22	0	50	4	0	M	Volcanics/ Neogene–Quaternary	Surface Alteration	Post-Pyrenean	Lens	Laterite
25	Hajib	35	32	0	50	2	0	M	Rhyolite/Eocene	Surface Alteration	Pyrenean	Stratiformbed	Yellow clay
26	Abadeh	33	9	32	52	47	1	S	Rhyolite to dacite/Eocene	Surface Alteration	Post-Pyrenean	Lens to Stratiformbed	
27	Abak	37	30	0	47	30	0	I	Volcanics/Tertiary	Surface Alteration	Post-Pyrenean	Unknown	
28	Abdol Abad	36	10	48	50	9	30	S	Andesite/Tertiary	Surface Alteration	Post-Pyrenean	Stratiformbed, massive	
29	Abdollah	31	15	12	52	43	31	L	Sedimentary rocks/ Permo-Triassic	Surface Alteration	Pyrenean	Unknown	
30	Alanghiyeh	36	11	20	49	36	30	S	Rhyolite/Paleocene	Surface Alteration	Pyrenean	Massive	
31	Amir Nan	36	12	1	50	32	9	I	Rhyolite, andesite, and tuff/ Eocene	Surface Alteration	Post-Pyrenean	Unknown	
32	Azadi	31	21	51	52	35	30	S	Claystone, sandstone, and clay-bearing Fe/ Permian–Triassic	Sedimentary	Hercynian	Stratiformbed	
33	Bibijanlu	38	20	0	48	45	0	S	Unknown	Surface Alteration	Post-Pyrenean	Massive	
34	Chah Bid	32	26	0	56	15	0	L	Sedimentary rocks/ Permo-Triassic	Sedimentary	Hercynian	Stratiformbed	
35	Chapu Shahindezh	36	59	35	46	58	39	S	Sedimentary rocks/ Permo-Triassic	Sedimentary	Hercynian	Stratiformbed	
36	Cheshmeh Shotoran	33	26	10	56	11	22	M	Sedimentary rocks/ Permo-Triassic	Sedimentary	Hercynian	Unknown	
37	Doplan	31	13	2	51	43	20	L	Sedimentary rocks/ Permo-Triassic	Sedimentary	Hercynian	Stratiformbed	
38	Drehjazin	35	40	0	53	38	0	L	Acidic Volcanics/Eocene	Surface Alteration	Pyrenean	Stratiformbed	
39	East of Rudkhaneh Shur	35	28	32	50	59	30	I	Rhyolite, andesite, and tuff/ Eocene	Surface Alteration	Post-Pyrenean	Unknown	
40	Esteghlal	31	15	8	52	43	25	S	Shale, sandstone, sandy limestone/Devonian	Sedimentary	Hercynian	Stratiformbed	

(continued)

(continued)

No.	Name	Lat D°	Lat M	Lat S″	Long D°	Long M	Long S″	Size	Host rock/age	Genetic/ mineralization	Orogenic phases	Morphology	Elem para
41	Gavazn	37	44	0	48	15	0	L	Sedimentary rocks/Neogene	Surface Alteration	Post-Pyrenean	Unknown	
42	Ghezel Gheshlagh	34	52	0	49	40	0	I	Rhyolite and tuff/Eocene	Surface Alteration	Pyrenean	Stratiformbed	
43	Isisu 1	36	42	0	46	58	0	L	Sedimentary rocks/Neogene	Surface Alteration	Post-Pyrenean	Massive	
44	Isisu 2	36	59	0	46	58	0	L	Sedimentary rocks/Neogene	Surface Alteration	Post-Pyrenean	Massive	
45	Kamblu	35	43	50	54	12	0	M	Feldspathic rocks/Tertiary	Surface Alteration	Post-Pyrenean	Unknown	
46	Kavir 2							L	Sedimentary rocks/ Permo-Triassic	Sedimentary	Hercynian	Stratiformbed	
47	Kavir 5							L	Sedimentary rocks/ Permo-Triassic	Sedimentary	Hercynian	Stratiformbed	
48	Kavir17	31	15	8	52	39	0	S	Sedimentary rocks/ Permo-Triassic	Sedimentary	Hercynian	Stratiformbed	
49	Kuh-e-Atabak 1	34	25	2	58	35	0	I	Tuff/Tertiary	Surface Alteration	Post-Pyrenean	Unknown	
50	Kuh-e-Atabak 2	34	24	34	58	36	31	I	Tuff/Eocene	Surface Alteration	Post-Pyrenean	Unknown	
51	Kuh-e-Hengam	34	13	0	58	55	0	I	Granite/Eocene	Surface Alteration	Post-Pyrenean	Unknown	
52	Kushkak	35	33	0	49	49	0	L	Volcanic Rocks/Eocene	Surface Alteration	Post-Pyrenean	Massive	
53	Maneshkeh	36	50	0	47	10	0	L	Andesite/Oligo-Miocene	Surface Alteration	Post-Pyrenean	Massive	
54	Nowrouz Abad	37	27	0	46	39	0	S	Sedimentary rocks/ Miocene–Paleocene	Sedimentary	Hercynian	Stratiformbed	
55	Parandak	35	23	5	50	41	32	L	Acidic volcanics/Eocene	Surface Alteration	Pyrenean	Stratiformbed	
56	Posht-e-Semirom	31	12	16	51	41	55	L	Sedimentary rocks/Cretaceous	Sedimentary	Late Cimmerian	Stratiformbed	
57	Qazvin	36	7	0	49	37	0	L	Andesite porphyry/Tertiary	Surface Alteration	Post-Pyrenean	Stratiformbed	
58	Robat-e-Kham	31	23	10	56	21	22	S	Shale, dolomite, and limestone/Cretaceous	Sedimentary	Late Cimmerian	Stratiformbed	
59	Sang Rud	36	28	6	49	40	0	S	Sandstone and limestone/ Jurassic	Sedimentary	Early Cimmerian	Stratiformbed	
60	Shahindezh	36	41	38	46	42	48	S	Sedimentary rocks/ Permo-Triassic	Sedimentary	Hercynian	Stratiformbed	

61	Surtejin	35	32	43	49	3	40	L	Volcanics/Neogene-Quaternary	Surface Alteration	Post-Pyrenean	Massive
62	Tavakkol Abad	35	26	37	50	38	20	L	Acidic volcanics/Eocene	Surface Alteration	Pyrenean	Lens
63	Zajkan (Qazvin)	36	19	40	49	31	5	S	Lava and tuff/Eocene	Surface Alteration	Pyrenean	Vein
64	Zonous 1	38	33	34	45	48	15	L	Rhyolitic tuff/Eocene	Surface Alteration	Pyrenean	Lens
65	Zonous 2	38	34	0	45	48	45	I	Rhyolitic tuff/Eocene	Surface Alteration	Pyrenean	Unknown

Complete list of phosphate mineral deposits and indications of Iran along with their detailed specifications

No.	Name	Lat D°	Lat M	Lat S″	Long D°	Long M	Long S″	Size	Host rock/age	Genetic/mineralization	Orogenic phases	Morphology	Elem para
1	Ahar	38	15	0	46	30	0	I	Sedimentary rock/Neogene	Sedimentary	Post-Pyrenean	Unknown	
2	Bafgh 1	31	46	6	55	26	12	I	Metamorphic/Precambrian	Magmatic	Pan-African	Unknown	Fe
3	Bibi Shahrbanou	35	38	10	51	33	41	I	Limestone and andesite/Eocene	Sedimentary	Pyrenean	Unknown	
4	Chalmish	35	54	0	53	4	0	L	Sedimentary rocks/Devonian	Sedimentary	Hercynian	Stratiformbed	
5	Dahouiyeh	30	47	0	56	43	0	I	Sedimentary rocks/Ordovician	Sedimentary	Caledonian	Stratiformbed	
6	Dalir	36	20	0	51	5	0	L	Sedimentary rocks/Cambrian	Sedimentary	Pan-African	Stratiformbed	
7	Deh Zoka	30	24	30	56	46	30	I	Sedimentary rocks/Devonian	Sedimentary	Hercynian	Stratiformbed	
8	Dehdasht	30	56	30	51	20	0	I	Unknown	Unknown	Unknown	Unknown	
9	Dehmolla	36	21	25	54	41	0	L	Sedimentary rocks/Devonian	Sedimentary	Hercynian	Stratiformbed	
10	Dogonbadan	30	34	34	51	13	58	I	Unknown	Unknown	Unknown	Unknown	
11	East of Cheshmeh godar sorkh	30	21	13	56	47	54	I	Shale and dolomite/Upper Precambrian–Lower Cambrian	Sedimentary	Pan-African	Stratiformbed	
12	Esfordi	31	47	0	55	38	0	L	Igneous/Upper Precambrian	Magmatic	Pan-African	Lens	
13	Firouz Abad Keykuh	36	17	30	51	22	30	L	Sedimentary rocks/Cambrian	Sedimentary	Pan-African	Unknown	
14	Gadoukdogol	35	51	0	52	55	0	L	Sedimentary rocks/Devonian	Sedimentary	Hercynian	Stratiformbed	
15	Gasak	30	38	50	57	15	30	I	Sedimentary rocks/Cambrian	Sedimentary	Hercynian	Stratiformbed	
16	Gazestan	31	39	45	55	57	0	I	Igneous and volcanic rocks, sandstone, shale, limestone, and dolomitic limestone/ Precambrian–Cambrian	Magmatic	Pan-African	Lens	Fe
17	Ghalghanlu	36	53	0	46	22	30	I	Sedimentary rocks/Cambrian	Sedimentary	Pan-African	Stratiformbed	
18	Hoseyn Abad	34	40	54	48	11	12	I	Phyllite and schist/Triassic–Jurassic	Hydrothermal	Laramide	Unknown	
19	Jajarm	37	5	0	56	30	0	I	Shale, sandstone, and limestone/ Upper Cretaceous	Sedimentary	Late Cimmerian	Unknown	
20	Jeyroud	35	59	30	51	30	0	L	Sedimentary rocks/Devonian	Sedimentary	Hercynian	Stratiformbed	
21	Kalat-e-Nadery	36	50	10	59	36	39	I	Shale, sandstone, and limestone/ Upper Cretaceous	Sedimentary	Late Cimmerian	Unknown	
22	Kalmard	35	25	0	56	10	20	L	Sedimentary rocks/Ordovician	Sedimentary	Caledonian	Stratiformbed	

23	Kasil	36	1	35	51	24	0	L	Sedimentary rocks/Devonian	Sedimentary	Hercynian	Unknown	
24	Kasokochek	39	20	27	44	51	25	I	Sedimentary rock/Upper Paleozoic	Sedimentary	Hercynian	Stratiformbed	
25	Kuh Tizi	30	27	56	57	15	7	I	Sedimentary rocks/Devonian	Sedimentary	Hercynian	Stratiformbed	
26	Kuh Zangu	30	22	30	56	47	30	I	Sedimentary rocks/Devonian	Sedimentary	Hercynian	Stratiformbed	
27	Kuh-e-Lakh	33	1	53	53	59	56	I	Dolomite and limestone/Cretaceous	Sedimentary	Late Cimmerian	Stratiformbed	
28	Kuh-e-Lar	30	30	45	50	45	51	L	Sedimentary rocks/ Eocene–Oligocene	Sedimentary	Pyrenean	Stratiformbed	
29	Kuh-e-Nil	31	43	0	50	40	0	L	Sedimentary rocks/Oligocene	Sedimentary	Pyrenean	Stratiformbed	
30	Kuh-e-Pabdeh 1	32	6	49	50	1	36	I	Dolomite and limestone/Cretaceous	Sedimentary	Late Cimmerian	Stratiformbed	
31	Kuh-e-Pabdeh 2	32	27	16	49	45	26	I	Dolomite and limestone/Cretaceous	Sedimentary	Late Cimmerian	Stratiformbed	
32	Kuh-e-Pabdeh 3	32	27	16	49	45	36	I	Dolomite and limestone/Cretaceous	Sedimentary	Late Cimmerian	Stratiformbed	
33	Kuh-e-Shurab	35	52	27	54	6	20	I	Sedimentary rocks/Devonian	Sedimentary	Hercynian	Stratiformbed	
34	Kuh-Kumeh	30	35	0	50	43	0	L	Sedimentary rocks/ Eocene–Oligocene	Sedimentary	Pyrenean	Stratiformbed	
35	KuhSefid	31	15	25	49	49		L	Unknown/Eocene	Sedimentary	Pyrenean	Unknown	
36	Kuhzangi	36	22	0	54	12	0	I	Sedimentary rocks/Devonian	Sedimentary	Hercynian	Unknown	
37	Lakak	30	57	16	50	8	13	S	Shale, dolomite, and limestone/ Cretaceous	Sedimentary	Late Cimmerian	Stratiformbed	
38	Lakeh Siah	31	47	3	55	42	7	L	Intrusion/Upper Precambrian	Magmatic	Pan-African	Vein	Fe
39	Lalun	35	59	0	51	37	0	L	Sedimentary rocks/Devonian	Sedimentary	Hercynian	Stratiformbed	
40	Mahi Rud	32	16	18	60	44	40	I	Sedimentary rocks/Cretaceous	Sedimentary	Late Cimmerian	Stratiformbed	
41	Mishdovan	31	5	48	55	31	12	I	Granitoid/Upper Precambrian–Lower Cambrian	Magmatic	Pan-African	Lens	
42	Morgdar	36	21	15	54	32	0	L	Sedimentary rocks/Devonian	Sedimentary	Hercynian	Lens	
43	Nir Synclin	30	34	31	50	45	50	I	Shale, dolomite, and limestone/ Cretaceous	Sedimentary	Late Cimmerian	Stratiformbed	
44	Northeast of Cheshmeh gaz	30	26	28	56	43	45	I	Shale and dolomite/Precambrian	Sedimentary	Pan-African	Stratiformbed	
45	Northeast of Qasr-e-Shirin	34	45	33	45	45	45	I	Shale, sandstone, and limestone/ Upper Cretaceous	Sedimentary	Late Cimmerian	Unknown	

(continued)

(continued)

No.	Name	Lat D°	Lat M	Lat S″	Long D°	Long M	Long S″	Size	Host rock/age	Genetic/ mineralization	Orogenic phases	Morphology	Elem para
46	Pa-Ghaleh	35	55	0	53	10	0	L	Sedimentary rocks/Devonian	Sedimentary	Hercynian	Lens	
47	Robat-e-Karabil	37	14	43	56	14	0	I	Shale, sandstone, and limestone/ Upper Cretaceous	Sedimentary	Late Cimmerian	Unknown	
48	Seyed Kandi	36	52	0	48	15	0	L	Dolomite/Cambrian	Sedimentary	Pan-African	Stratiformbed	
49	Shahin Dezh	36	55	0	46	47	0	I	Sedimentary rock/Precambrian	Sedimentary	Pan-African	Unknown	
50	Sheikh Abad	30	54	32	50	50	57	I	Shale, dolomite and limestone/ Cretaceous	Sedimentary	Late Cimmerian	Stratiformbed	
51	Shemshak	36	0	0	51	33	41	S	Limestone and dolomite/Cambrian	Sedimentary	Pan-African	Stratiformbed	
52	Sheykh Habil	30	45	0	50	50	0	L	Sedimentary rocks/ Cretaceous–Eocene	Sedimentary	Pyrenean	Stratiformbed	
53	Southeast of Khoy	38	30	0	45	0	0	I	Sedimentary rock/Neogene	Sedimentary	Post-Pyrenean	Unknown	
54	Taghdis-e-Chenareh	32	57	0	48	20	0	L	Sedimentary rocks/Upper Cretaceous	Sedimentary	Laramide	Lens	
55	Taghdis-e-Khartang	28	18	0	51	55	0	I	Sedimentary rocks/Paleocene	Sedimentary	Laramide	Stratiformbed	
56	Taghdis-e-Kuh Namak	28	15	0	51	40	0	L	Sedimentary rocks/Paleocene	Sedimentary	Laramide	Lens	Fe
57	Taghdis-e-rit	32	6	0	48	15	0	L	Sedimentary rocks/Upper Cretaceous	Sedimentary	Laramide	Lens	
58	Taleghan	36	7	30	50	45	0	I	Sedimentary rocks/Jurassic	Sedimentary	Pan-African	Stratiformbed	
59	Vali Abad Shalzi	36	17	30	51	17	30	L	Sedimentary rocks/Cambrian	Sedimentary	Pan-African	Vein	
60	Zarigan	32	3	33	55	30	56	M	Intrusion/Upper Precambrian	Magmatic	Pan-African	Disseminated	

Complete list of bentonite mineral deposits and indications of Iran along with their detailed specifications

No.	Name	Lat D°	Lat M	Lat S″	Long D°	Long M	Long S″	Size	Host rock/age	Genetic/mineralization	Orogenic phases	Morphology	Elem para
1	Ab Bandamou	32	59	21	55	31	0	I	Tuff/Oligo-Miocene	Surface Alteration	Post-Pyrenean	Stratiformbed	
2	Abbas Abad	35	21	11	59	13	0	I	Volcanics and tuff/Upper Eocene	Surface Alteration	Pyrenean	Stratiformbed	
3	Aghol Salari	32	27	50	55	9	34	I	Tuff/Oligo-Miocene	Surface Alteration	Post-Pyrenean	Stratiformbed	
4	Alajugh	36	22	30	47	11	15	I	Limestone, marl, and rhyolite/Eocene	Surface Alteration	Pyrenean	Unknown	
5	Aliabad	35	56	14	55	30	0	I	Tuff and volcanics/Oligo-Miocene	Surface Alteration	Post-Pyrenean	Stratiformbed	
6	Aliyan	35	55	54	54	15	0	I	Tuff/Eocene	Unknown	Unknown	Unknown	
7	Anarestan	35	47	0	54	10	0	S	Pyroclastics/Mid-Eocene	Surface Alteration	Pyrenean	Lens	
8	Arhan	36	42	0	49	21	0	I	Tuff/Oligo-Miocene	Surface Alteration	Post-Pyrenean	Stratiformbed	
9	Avaj	35	34	41	49	14	0	I	Tuff/Oligo-Miocene	Surface Alteration	Post-Pyrenean	Lens	
10	Bostagh1	33	42	32	58	34	0	I	Tuff and volcanics/Oligo-Miocene	Hydrothermal	Post-Pyrenean	Vein	
11	Bostagh2	33	58	21	58	39	20	I	Tuff and volcanics/Oligo-Miocene	Hydrothermal	Post-Pyrenean	Vein	
12	Brosir	29	51	0	56	9	0	I	Tuff/Eocene	Surface Alteration	Pyrenean	Stratiformbed, lens	
13	Chah Kam	34	5	0	56	4	0	S	Sedimentary rocks and volcanics/ Oligo-Miocene	Surface Alteration	Post-Pyrenean	Stratiformbed	
14	Darya-Belik	35	45	0	52	32	0	I	Tuff/Mid-Eocene	Surface Alteration	Pyrenean	Stratiformbed	
15	Deh Amshhad	34	49	0	49	46	0	I	Tuff/Oligo-Miocene	Surface Alteration	Post-Pyrenean	Stratiformbed	
16	Deh Hoseyn-e-Kolangi	32	30	0	59	58	0	I	Volcanics/Upper Eocene	Surface Alteration	Pyrenean	Lens	
17	Dizaran	31	25	0	53	5	0	S	Sedimentary rocks/Oligo-Miocene	Sedimentary	Unknown	Stratiformbed	
18	Eliatu	35	33	59	59	29	16	S	Volcanics and tuff/Eocene	Surface Alteration	Unknown	Unknown	
19	Fajan (Kilan 1)	35	30	0	52	10	0	M	Tuff/Mid-Eocene	Surface Alteration	Pyrenean	Lens	
20	Galou	33	38	0	59	50	0	I	Tuff/Upper Eocene	Surface Alteration	Pyrenean	Stratiformbed	
21	Gazalak	36	11	0	50	31	40	I	Tuff and volcanics/Oligo-Miocene	Hydrothermal	Post-Pyrenean	Vein	
22	Gel Abir	36	20	0	48	19	0	I	Tuff/Oligo-Miocene	Surface Alteration	Post-Pyrenean	Stratiformbed	
23	Gelkan	33	59	0	56	59	0	S	Limestone, marl, andesite, and tuff/ Oligo-Miocene	Surface Alteration	Post-Pyrenean	Stratiformbed	

(continued)

(continued)

No.	Name	Lat D°	Lat M	Lat S″	Long D°	Long M	Long S″	Size	Host rock/age	Genetic/ mineralization	Orogenic phases	Morphology	Elem para
24	Gelkan	35	22	0	60	22	0	I	Sedimentary rocks/Upper Eocene	Sedimentary	Pyrenean	Unknown	
25	GhashGha Jan	35	48	0	48	0	0	I	Tuff/Oligo-Miocene	Surface Alteration	Post-Pyrenean	Stratiformbed	
26	Ghazalcheh	34	46	0	49	52	0	L	Sedimentary rocks/Oligo-Miocene	Surface Alteration	Post-Pyrenean	Stratiformbed	
27	Golabir	36	2	0	48	13	0	I	Tuff/Oligo-Miocene	Surface Alteration	Pyrenean	Unknown	
28	Gonabad	34	22	0	58	9	0	M	Volcanics and tuff/Upper Eocene	Surface Alteration	Pyrenean	Massive	
29	Gooiech	35	33	30	52	21	40	S	Tuff/Mid-Eocene	Surface Alteration	Pyrenean	Stratiformbed	
30	Goosh mir	33	38	0	58	25	0	I	Tuff and volcanics/Upper Eocene	Surface Alteration	Unknown	Unknown	
31	Gouyad	31	38	47	60	4	50	I	Tuff/Upper Eocene	Surface Alteration	Pyrenean	Stratiformbed	
32	Halab	36	18	0	47	26	0	I	Limestone, marl, andesite, and tuff/ Oligo-Miocene	Surface Alteration	Post-Pyrenean	Unknown	
33	Hesamiyeh	33	4	0	58	22	0	S	Volcanics/Upper Eocene	Surface Alteration	Pyrenean	Lens	
34	Kalateh	36	21	20	54	7	30	I	Sedimentary rocks/Eocene	Surface Alteration	Pyrenean	Unknown	
35	Kalishom	36	42	19	49	57	38	S	Tuff/Mid-Eocene	Hydrothermal	Pyrenean	Lens	
36	Keyvan	35	32	0	52	10	20	I	Tuff/Eocene	Surface Alteration	Pyrenean	Stratiformbed	
37	Khushab	35	25	20	58	9	0	S	Volcanics and tuff/Upper Eocene	Surface Alteration	Pyrenean	Massive	
38	Kilan (Fajan)	35	22	2	52	50	0	S	Tuff/Mid-Eocene	Hydrothermal	Pyrenean	Lens	
39	Kuh-e-Bazman	27	50	0	60	15	0	I	Andesite and tuff/Oligo-Miocene	Surface Alteration	Post-Pyrenean	Lens	
40	Kuh-e-Ghozghalan	35	36	0	49	11	0	I	Tuff/Eocene	Surface Alteration	Pyrenean	Stratiformbed	
41	Kuseh-Lar	36	29	0	48	22	0	I	Tuff/Mid-Eocene	Surface Alteration	Pyrenean	Unknown	
42	Manian	28	57	0	53	12	0	S	Sedimentary rocks/Oligo-Miocene	Sedimentary	Post-Pyrenean	Stratiformbed	
43	Mehrejan	33	35	57	55	9	10	L	Tuff and marl/Mid-Eocene	Surface Alteration	Pyrenean	Lens	
44	Merikeh	33	54	0	58	4	0	I	Volcanics/Oligo-Miocene	Surface Alteration	Post-Pyrenean	Lens	
45	Meydan-e-Tir-e-41	35	33	0	51	32	0	I	Sedimentary rocks/Eocene	Sedimentary	Post-Pyrenean	Lens	
46	Moemen Abad	35	33	18	53	13	51	S	Volcanics and sedimentary rocks/ Oligo-Miocene	Surface Alteration	Post-Pyrenean	Stratiformbed	
47	Najaf Abad	32	40	0	51	21	0	I	Sedimentary rocks/Lower Eocene	Surface Alteration	Pyrenean	Stratiformbed	
48	Nashtafan (Kalatehha)	35	17	0	59	46	0	S	Volcanics and tuff/Upper Eocene	Surface Alteration	Pyrenean	Stratiformbed	

49	Niasar	33	59	10	51	6	40	S	Tuff/Mid-Eocene	Surface Alteration	Pyrenean	Stratiformbed
50	Nish Tafoon	33	45	48	58	47	6	I	Tuff/Eocene	Surface Alteration	Pyrenean	Stratiformbed
51	Ortu-Cheshmeh (Orghusheh)	37	5	0	58	16	0	I	Volcanics/Eocene	Surface Alteration	Pyrenean	Stratiformbed
52	Oshar	38	52	20	45	47	0	S	Tuff/Eocene	Surface Alteration	Pyrenean	Stratiformbed
53	Our Of	30	4	0	55	55	0	I	Volcanics/Middle Eocene	Surface Alteration	Pyrenean	Stratiformbed
54	Ourfasa	30	40	30	53	55	0	I	Unknown	Unknown	Unknown	Unknown
55	Oyaghlou	38	43	10	45	13	0	I	Tuff/Oligo-Miocene	Surface Alteration	Post-Pyrenean	Unknown
56	Panjeh Sofla	36	4	0	47	26	0	I	Limestone, marl, andesite, and Tuff/ Oligo-Miocene	Surface Alteration	Post-Pyrenean	Unknown
57	Parvan Avoj	35	35	0	49	11	0	I	Tuff/Eocene	Surface Alteration	Pyrenean	Stratiformbed
58	Reshm	35	16	0	54	30	0	S	Pyroclastics/Mid-Eocene	Surface Alteration	Pyrenean	Stratiformbed
59	Saghand	32	34	0	55	17	0	S	Tuff and volcanics/Oligo-Miocene	Surface Alteration	Post-Pyrenean	Lens
60	Sangsar	35	42	0	53	50	0	S	Sedimentary rocks/Oligo-Miocene	Surface Alteration	Post-Pyrenean	Stratiformbed
61	Sar Darya	28	34	0	61	10	0	I	Andesite and tuff/Oligo-Miocene	Surface Alteration	Post-Pyrenean	Stratiformbed
62	Sepideh (Ardekan)	30	17	0	51	48	0	I	Volcanics/Oligo-Miocene	Surface Alteration	Post-Pyrenean	Stratiformbed
63	Siah Kuh	34	43	0	52	51	0	S	Tuff and volcanics/Mid-Eocene	Surface Alteration	Pyrenean	Lens
64	South of Zanjan	36	39	0	48	30	0	I	Tuff/Mid-Eocene	Surface Alteration	Pyrenean	Unknown
65	Susanvar	35	17	20	54	32	51	S	Pyroclastics/Mid-Eocene	Surface Alteration	Pyrenean	Lens
66	Tab	36	18	0	48	3	0	I	Tuff/Oligo-Miocene	Surface Alteration	Post-Pyrenean	Stratiformbed
67	Takab	36	28	20	47	11	15	I	Tuff/Oligo-Miocene	Surface Alteration	Post-Pyrenean	Stratiformbed
68	Tashtab	33	39	17	55	16	19	S	Tuff/Eocene	Hydrothermal	Post-Pyrenean	Vein
69	West of Ferdows	34	0	0	58	1	23	I	Tuff/Eocene	Surface Alteration	Pyrenean	Stratiformbed
70	Zahab	31	52	0	60	9	50	I	Tuff/Upper Eocene	Surface Alteration	Pyrenean	Lens
71	Zanjan-Bijar	36	37	0	48	27	0	I	Tuff/Mid-Eocene	Surface Alteration	Pyrenean	Unknown
72	Zanjan-Mianeh	36	35	0	48	25	0	I	Tuff/Eocene	Surface Alteration	Pyrenean	Unknown
73	Zarnigh (Mehrban)	38	12	0	47	7	0	I	Tuff/Paleocene–Eocene	Surface Alteration	Unknown	Stratiformbed
74	Zarrin	32	42	57	54	36	47	M	Tuff and volcanics/Oligo-Miocene	Surface Alteration	Post-Pyrenean	Stratiformbed
75	Zarrinabad	36	25	0	48	20	0	S	Volcanics/Quaternary	Sedimentary	Unknown	Unknown

Complete list of magnesite mineral deposits and indications of Iran along with their detailed specifications

		Lat			Long								
No.	Name	D°	M	S″	D°	M	S″	Size	Host rock/age	Genetic/ mineralization	Orogenic phases	Morphology	Elem para
1	Afin1	33	30	40	59	47	35	S	Serpentinite and peridotite/Upper Cretaceous–Lower Paleocene	Surface Alteration	Post-Laramide	Lens	
2	Afin2	33	32	0	59	44	21	S	Serpentinite and peridotite/Upper Cretaceous–Lower Paleocene	Surface Alteration	Post-Laramide	Lens	
3	Afzal Abad	31	52	0	60	16	0	I	Serpentinite and peridotite/Upper Cretaceous–Lower Paleocene	Surface Alteration	Post-Laramide	Lens	
4	Ali Abad	33	0	46	60	22	55	I	Serpentinite and peridotite/Upper Cretaceous–Lower Paleocene	Surface Alteration	Post-Laramide	Lens	
5	Ali Gol	31	47	36	60	10	23	I	Serpentinite and peridotite/Upper Cretaceous–Lower Paleocene	Surface Alteration	Post-Laramide	Lens	
6	Asad Abad	31	52	58	60	16	8	I	Serpentinite and peridotite/Upper Cretaceous–Lower Paleocene	Surface Alteration	Post-Laramide	Lens	
7	Bandan	31	13	30	60	59	55	I	Serpentinite and peridotite/Upper Cretaceous–Lower Paleocene	Surface Alteration	Post-Laramide	Lens	
8	Borj Ali Khan	32	11	53	60	19	9	I	Serpentinite and peridotite/Upper Cretaceous–Lower Paleocene	Surface Alteration	Post-Laramide	Lens	
9	Borj Mohammad 1	33	35	2	59	28	46	I	Serpentinite and peridotite/Upper Cretaceous–Lower Paleocene	Surface Alteration	Post-Laramide	Lens	
10	Borj Mohammad 2	33	36	0	59	28	12	I	Serpentinite and peridotite/Upper Cretaceous–Lower Paleocene	Surface Alteration	Post-Laramide	Lens	
11	Chah Chocho	31	38	55	60	10	30	I	Serpentinite and peridotite/Upper Cretaceous–Lower Paleocene	Surface Alteration	Post-Laramide	Lens	
12	Chah Khu	32	22	30	60	1	6	S	Serpentinite and peridotite/Upper Cretaceous–Lower Paleocene	Surface Alteration	Post-Laramide	Lens	
13	Chah Mondad	31	8	24	59	9	4	I	Serpentinite and peridotite/Upper Cretaceous–Lower Paleocene	Surface Alteration	Post-Laramide	Lens	
14	Chah Palang 1	33	0	56	53	3	40	I	Serpentinite and peridotite/Upper Cretaceous–Lower Paleocene	Surface Alteration	Post-Laramide	Lens	
15	Chah Siah	32	22	54	59	43	34		Serpentinite and peridotite/Upper Cretaceous–Lower Paleocene	Surface Alteration	Post-Laramide	Lens	

16	Chahar Chah Mohammad Rahim	29	50	39	60	10	9	I	Serpentinite and peridotite/Upper Cretaceous–Lower Paleocene	Surface Alteration	Post-Laramide	Lens
17	Cheshmeh Lagury– Kasrab	32	11	35	60	0	58	S	Serpentinite and peridotite/Upper Cretaceous–Lower Paleocene	Surface Alteration	Post-Laramide	Lens
18	Chorogh	28	31	0	60	0	30	I	Serpentinite and peridotite/Upper Cretaceous–Lower Paleocene	Surface Alteration	Post-Laramide	Lens
19	Dahan Siahu	32	16	10	60	17	45	I	Serpentinite and peridotite/Upper Cretaceous–Lower Paleocene	Surface Alteration	Post-Laramide	Lens
20	Deh Reza	32	38	1	60	15	42	I	Serpentinite and peridotite/Upper Cretaceous–Lower Paleocene	Surface Alteration	Post-Laramide	Lens
21	East of Ali Abad	32	46	31	60	15	37	S	Serpentinite and peridotite/Upper Cretaceous–Lower Paleocene	Surface Alteration	Post-Laramide	Lens
22	East of Bidmeshk	31	54	51	59	40	56	I	Serpentinite and peridotite/Upper Cretaceous–Lower Paleocene	Surface Alteration	Post-Laramide	Lens
23	East of Khuni	33	31	12	53	21	24	I	Serpentinite and peridotite/Upper Cretaceous–Lower Paleocene	Surface Alteration	Post-Laramide	Lens
24	East of Neh fault	31	24	45	60	9	8	I	Serpentinite and peridotite/Upper Cretaceous–Lower Paleocene	Surface Alteration	Post-Laramide	Lens
25	Eastakhr–e-Abyari	32	5	24	60	7	39	I	Serpentinite and peridotite/Upper Cretaceous–Lower Paleocene	Surface Alteration	Post-Laramide	Lens
26	East-Northeast of Jamal Abad	32	38	42	60	15	42	S	Serpentinite and peridotite/Upper Cretaceous–Lower Paleocene	Surface Alteration	Post-Laramide	Lens
27	Esfordi	31	45	22	55	40	49	I	Serpentinite and peridotite/Upper Cretaceous–Lower Paleocene	Surface Alteration	Post-Laramide	Lens
28	Espiky- Farmaj	32	41	2	59	20	31	S	Serpentinite and peridotite/Upper Cretaceous–Lower Paleocene	Surface Alteration	Post-Laramide	Lens
29	GarmTamam Ab	32	11	13	60	0	29	S	Serpentinite and peridotite/Upper Cretaceous–Lower Paleocene	Surface Alteration	Post-Laramide	Lens
30	Gazdez -va- Shurab	32	23	15	60	2	33	I	Serpentinite and peridotite/Upper Cretaceous–Lower Paleocene	Surface Alteration	Post-Laramide	Lens

(continued)

(continued)

No.	Name	Lat D°	Lat M	Lat S″	Long D°	Long M	Long S″	Size	Host rock/age	Genetic/ mineralization	Orogenic phases	Morphology	Elem para
31	Gomenj	33	19	41	60	5	21	S	Serpentinite and peridotite/Upper Cretaceous–Lower Paleocene	Surface Alteration	Post-Laramide	Lens	
32	Holur Tappeh	38	45	16	60	9	28		Serpentinite and peridotite/Upper Cretaceous–Lower Paleocene	Surface Alteration	Post-Laramide	Lens	
33	Hoseyn Abad	33	2	42	53	4	18	I	Serpentinite and peridotite/Upper Cretaceous–Lower Paleocene	Surface Alteration	Post-Laramide	Lens	
34	Howz-e-Sefid-e-Arabkhaneh	32	30	32	59	29	31	S	Serpentinite and peridotite/Upper Cretaceous–Lower Paleocene	Surface Alteration	Post-Laramide	Lens	
35	Jamal Abad	32	36	16	60	16	35	I	Serpentinite and peridotite/Upper Cretaceous–Lower Paleocene	Surface Alteration	Post-Laramide	Lens	
36	Kalat-e-Akhundan	32	38	19	59	37	6		Serpentinite and peridotite/Upper Cretaceous–Lower Paleocene	Surface Alteration	Post-Laramide	Lens	
37	Kalat-e-Ali Mohammad	32	43	46	59	27	18	S	Serpentinite and peridotite/Upper Cretaceous–Lower Paleocene	Surface Alteration	Post-Laramide	Lens	
38	Kalat-e-Manzarieh	32	49	45	59	8	3		Serpentinite and peridotite/Upper Cretaceous–Lower Paleocene	Surface Alteration	Post-Laramide	Lens	
39	Kal-e-Hedied	31	59	43	60	11	44	I	Serpentinite and peridotite/Upper Cretaceous–Lower Paleocene	Surface Alteration	Post-Laramide	Lens	
40	Kaliran	32	29	22	59	54	25	I	Serpentinite and peridotite/Upper Cretaceous–Lower Paleocene	Surface Alteration	Post-Laramide	Lens	
41	Kalut-e-Soleiman	32	32	35	59	27	47	I	Serpentinite and peridotite/Upper Cretaceous–Lower Paleocene	Surface Alteration	Post-Laramide	Lens	
42	Kanift	32	23	20	60	22	15	I	Serpentinite and peridotite/Upper Cretaceous–Lower Paleocene	Surface Alteration	Post-Laramide	Lens	
43	Khakshuran	31	41	35	60	13	6	I	Serpentinite and peridotite/Upper Cretaceous–Lower Paleocene	Surface Alteration	Post-Laramide	Lens	
44	Khuni	33	31	5	53	24	10	I	Serpentinite and peridotite/Upper Cretaceous–Lower Paleocene	Surface Alteration	Post-Laramide	Lens	
45	Khunik	30	25	0	60	35	0	I	Serpentinite and peridotite/Upper Cretaceous–Lower Paleocene	Surface Alteration	Post-Laramide	Lens	

46	Kuh-e-Aqavile 1	33	28	46	59	54	34	I	Serpentinite and peridotite/Upper Cretaceous–Lower Paleocene	Surface Alteration	Post-Laramide	Lens
47	Kuh-e-Aqavile 2	33	25	38	59	57	55	S	Serpentinite and peridotite/Upper Cretaceous–Lower Paleocene	Surface Alteration	Post-Laramide	Lens
48	Kuh-e-Aqavile 3	33	24	27	59	58	52	1	Serpentinite and peridotite/Upper Cretaceous–Lower Paleocene	Surface Alteration	Post-Laramide	Lens
49	Kuh-e-Azik	32	59	25	60	16	30	I	Serpentinite and peridotite/Upper Cretaceous–Lower Paleocene	Surface Alteration	Post-Laramide	Lens
50	Kuh-e-Cheshmeh Zarbi	33	32	8	59	36	8	S	Serpentinite and peridotite/Upper Cretaceous–Lower Paleocene	Surface Alteration	Post-Laramide	Lens
51	Kuh-e-Dabil	31	46	47	60	14	41	I	Serpentinite and peridotite/Upper Cretaceous–Lower Paleocene	Surface Alteration	Post-Laramide	Lens
52	Kuh-e-jaj	32	26	0	60	23	0	I	Serpentinite and peridotite/Upper Cretaceous–Lower Paleocene	Surface Alteration	Post-Laramide	Lens
53	Kuh-e-KamarSiah	32	40	49	60	15	31	I	Serpentinite and peridotite/Upper Cretaceous–Lower Paleocene	Surface Alteration	Post-Laramide	Lens
54	Kuh-e-Naharjan	32	32	49	59	29	44	I	Serpentinite and peridotite/Upper Cretaceous–Lower Paleocene	Surface Alteration	Post-Laramide	Lens
55	Kuh-e-Ranje	33	6	7	53	4	35	S	Serpentinite and peridotite/Upper Cretaceous–Lower Paleocene	Surface Alteration	Post-Laramide	Lens
56	Kuh-e-Ratook	32	35	8	60	15	30	I	Serpentinite and peridotite/Upper Cretaceous–Lower Paleocene	Surface Alteration	Post-Laramide	Lens
57	Kuh-e-Zenzal	32	24	15	60	16	37	I	Serpentinite and peridotite/Upper Cretaceous–Lower Paleocene	Surface Alteration	Post-Laramide	Lens
58	Lakuh-e-Garmeh 1	31	39	40	60	13	7	I	Serpentinite and peridotite/Upper Cretaceous–Lower Paleocene	Surface Alteration	Post-Laramide	Lens
59	Lakuh-e-Garmeh 2	31	40	48	60	12	53	I	Serpentinite and peridotite/Upper Cretaceous–Lower Paleocene	Surface Alteration	Post-Laramide	Lens
60	Lakuh-e-Garmeh 3	31	41	36	60	11	47	I	Serpentinite and peridotite/Upper Cretaceous–Lower Paleocene	Surface Alteration	Post-Laramide	Lens

(continued)

(continued)

No.	Name	Lat D°	Lat M	Lat S″	Long D°	Long M	Long S″	Size	Host rock/age	Genetic/ mineralization	Orogenic phases	Morphology	Elem para
61	Maarufan	32	19	28	60	1	7	S	Serpentinite and peridotite/Upper Cretaceous–Lower Paleocene	Surface Alteration	Post-Laramide	Lens	
62	Mahi Rud	32	16	18	60	44	40	I	Serpentinite and peridotite/Upper Cretaceous–Lower Paleocene	Surface Alteration	Post-Laramide	Lens	
63	Mazraeh	32	16	13	60	5	6	I	Serpentinite and peridotite/Upper Cretaceous–Lower Paleocene	Surface Alteration	Post-Laramide	Lens	
64	Neskhandeh Kuh	31	36	16	60	12	25	I	Serpentinite and peridotite/Upper Cretaceous–Lower Paleocene	Surface Alteration	Post-Laramide	Lens	
65	North Khunik	31	26	19	60	9	47	I	Serpentinite and peridotite/Upper Cretaceous–Lower Paleocene	Surface Alteration	Post-Laramide	Lens	
66	North of Gomenj	33	18	55	60	5	19	I	Serpentinite and peridotite/Upper Cretaceous–Lower Paleocene	Surface Alteration	Post-Laramide	Lens	
67	North of Jamal Abad 1	32	39	5	60	13	43	S	Serpentinite and peridotite/Upper Cretaceous–Lower Paleocene	Surface Alteration	Post-Laramide	Lens	
68	North of Jamal Abad 2	32	39	47	60	13	2	S	Serpentinite and peridotite/Upper Cretaceous–Lower Paleocene	Surface Alteration	Post-Laramide	Lens	
69	North of Kanaft	32	27	15	60	18	45	I	Serpentinite and peridotite/Upper Cretaceous–Lower Paleocene	Surface Alteration	Post-Laramide	Lens	
70	North of Tamamdeh	31	59	28	60	27	18	I	Serpentinite and peridotite/Upper Cretaceous–Lower Paleocene	Surface Alteration	Post-Laramide	Lens	
71	NorthEast of Jamal Abad	32	39	21	60	14	45	S	Serpentinite and peridotite/Upper Cretaceous–Lower Paleocene	Surface Alteration	Post-Laramide	Lens	
72	Northeast of Kuh-e-ratook	32	35	50	60	16	10	I	Serpentinite and peridotite/Upper Cretaceous–Lower Paleocene	Surface Alteration	Post-Laramide	Lens	
73	Northeast of Rud-e-Marghzar	32	3	6	60	7	49	S	Serpentinite and peridotite/Upper Cretaceous–Lower Paleocene	Surface Alteration	Post-Laramide	Lens	
74	Northeast of Tabas	32	49	24	60	15	0	S	Serpentinite and peridotite/Upper Cretaceous–Lower Paleocene	Surface Alteration	Post-Laramide	Lens	
75	Northern AfzalAbad	31	59	20	60	14	41	S	Serpentinite and peridotite/Upper Cretaceous–Lower Paleocene	Surface Alteration	Post-Laramide	Lens	

76	Northwest of Afzal Abad	31	53	50	60	12	23	I	Serpentinite and peridotite/Upper Cretaceous–Lower Paleocene	Surface Alteration	Post-Laramide	Lens
77	NorthWest of Yadmaro	33	30	52	59	29	44	I	Serpentinite and peridotite/Upper Cretaceous–Lower Paleocene	Surface Alteration	Post-Laramide	Lens
78	Noughab	32	21	58	59	42	8	S	Serpentinite and peridotite/Upper Cretaceous–Lower Paleocene	Surface Alteration	Post-Laramide	Lens
79	Omar	29	58	9	59	54	38	I	Serpentinite and peridotite/Upper Cretaceous–Lower Paleocene	Surface Alteration	Post-Laramide	Lens
80	Pelleh Sabz	32	19	51	60	24	0	I	Serpentinite and peridotite/Upper Cretaceous–Lower Paleocene	Surface Alteration	Post-Laramide	Lens
81	Pirbid	38	34	13	60	14	32		Serpentinite and Peridotite/Upper Cretaceous–Lower Paleocene	Surface Alteration	Post-Laramide	Lens
82	Rezgh-e-Soleyman	32	33	14	59	33	48	S	Serpentinite and peridotite/Upper Cretaceous–Lower Paleocene	Surface Alteration	Post-Laramide	Lens
83	Rud-e-Marghzar	32	5	5	60	21	52	I	Serpentinite and peridotite/Upper Cretaceous–Lower Paleocene	Surface Alteration	Post-Laramide	Lens
84	Rud-e-Torshab1	32	10	0	60	2	52	S	Serpentinite and peridotite/Upper Cretaceous–Lower Paleocene	Surface Alteration	Post-Laramide	Lens
85	Rud-e-Torshab2	32	7	26	60	7	40	S	Serpentinite and peridotite/Upper Cretaceous–Lower Paleocene	Surface Alteration	Post-Laramide	Lens
86	SarLord	31	52	5	60	13	16	S	Serpentinite and peridotite/Upper Cretaceous–Lower Paleocene	Surface Alteration	Post-Laramide	Lens
87	Seh Farsakh	31	22	54	60	2	29	I	Serpentinite and peridotite/Upper Cretaceous–Lower Paleocene	Surface Alteration	Post-Laramide	Lens
88	Shah Dezh	31	35	2	60	10	15	I	Serpentinite and peridotite/Upper Cretaceous–Lower Paleocene	Surface Alteration	Post-Laramide	Lens
89	Shahrakht	33	22	58	60	0	38	I	Serpentinite and peridotite/Upper Cretaceous–Lower Paleocene	Surface Alteration	Post-Laramide	Lens
90	Shandul	30	50	3	60	8	32	I	Serpentinite and peridotite/Upper Cretaceous–Lower Paleocene	Surface Alteration	Post-Laramide	Lens
91	Shir Kuhak	31	41	2	60	11	3	S	Serpentinite and peridotite/Upper Cretaceous–Lower Paleocene	Surface Alteration	Post-Laramide	Lens

(continued)

(continued)

No.	Name	Lat D°	Lat M	Lat S″	Long D°	Long M	Long S″	Size	Host rock/age	Genetic/mineralization	Orogenic phases	Morphology	Elem para
92	Shovin	29	42	27	60	6	25	I	Serpentinite and peridotite/Upper Cretaceous–Lower Paleocene	Surface Alteration	Post-Laramide	Lens	
93	South Heidar Abad	31	1	38	60	5	3	I	Serpentinite and peridotite/Upper Cretaceous–Lower Paleocene	Surface Alteration	Post-Laramide	Lens	
94	South Khunik	31	24	49	60	9	18	I	Serpentinite and peridotite/Upper Cretaceous–Lower Paleocene	Surface Alteration	Post-Laramide	Lens	
95	South of Kuh-e-Sefid	31	49	28	60	12	30	I	Serpentinite and peridotite/Upper Cretaceous–Lower Paleocene	Surface Alteration	Post-Laramide	Lens	
96	South of Separab	33	1	13	53	3	53	M	Serpentinite and peridotite/Upper Cretaceous–Lower Paleocene	Surface Alteration	Post-Laramide	Vein	
97	Southeast of Afin	33	24	47	59	59	31	I	Serpentinite and peridotite/Upper Cretaceous–Lower Paleocene	Surface Alteration	Post-Laramide	Lens	
98	Southern Afzal Abad	31	50	30	60	16	30	S	Serpentinite and peridotite/Upper Cretaceous–Lower Paleocene	Surface Alteration	Post-Laramide	Lens	
99	Southwest of Ab Garm	32	53	18	60	16	29	S	Serpentinite and peridotite/Upper Cretaceous–Lower Paleocene	Surface Alteration	Post-Laramide	Lens	
100	Southwest of Hengaran	32	6	36	59	11	27	I	Serpentinite and peridotite/Upper Cretaceous–Lower Paleocene	Surface Alteration	Post-Laramide	Lens	
101	Southwest of Khuboo	33	31	20	59	29	31	S	Serpentinite and peridotite/Upper Cretaceous–Lower Paleocene	Surface Alteration	Post-Laramide	Lens	
102	Southwest of Kuh-e-ratook	32	35	10	60	15	27	S	Serpentinite and peridotite/Upper Cretaceous–Lower Paleocene	Surface Alteration	Post-Laramide	Lens	
103	Tak Siah	31	24	8	60	3	41	S	Serpentinite and peridotite/Upper Cretaceous–Lower Paleocene	Surface Alteration	Post-Laramide	Lens	
104	Tang-e-Omar	29	46	57	59	55	3	I	Serpentinite and peridotite/Upper Cretaceous–Lower Paleocene	Surface Alteration	Post-Laramide	Lens	
105	Torshab	29	38	47	59	58	32	I	Serpentinite and peridotite/Upper Cretaceous–Lower Paleocene	Surface Alteration	Post-Laramide	Lens	

106	Torshak-e-Mohammadi	31	56	10	60	12	18	S	Serpentinite and peridotite/Upper Cretaceous–Lower Paleocene	Surface Alteration	Post-Laramide	Lens
107	West of Gur Chang	32	52	0	60	14	18	S	Serpentinite and peridotite/Upper Cretaceous–Lower Paleocene	Surface Alteration	Post-Laramide	Lens
108	West of Heidar Abad	31	2	32	60	1	48	I	Serpentinite and peridotite/Upper Cretaceous–Lower Paleocene	Surface Alteration	Post-Laramide	Lens
109	Zahed	31	52	41	60	9	37	I	Serpentinite and peridotite/Upper Cretaceous–Lower Paleocene	Surface Alteration	Post-Laramide	Lens
110	Zek	31	43	6	59	55	1	I	Serpentinite and peridotite/Upper Cretaceous–Lower Paleocene	Surface Alteration	Post-Laramide	Lens

Complete list of silica mineral deposits and indications of Iran along with their detailed specifications

No.	Name	Lat D°	Lat M	Lat S″	Long D°	Long M	Long S″	Size	Host rock/age	Genetic/ mineralization	Orogenic phases	Morphology	Elem para
1	Abdar	34	14	11	48	28	42	S	Phyllite and slate/Jurassic	Metamorphic	Late Cimmerian	Lens, vein	
2	Ardehin	36	13	0	49	1	0	S	Sandstone/Lower Cambrian	Sedimentary	Pan-African	Stratiformbed	
3	Azadkuh (Jajrud)	35	46	55	51	42	20	S	Quartzite/Ordovician	Sedimentary	Pan-African	Stratiformbed	
4	Azandarian	34	30	0	48	40	0	S	Garnet schist and hornfels/Jurassic	Metamorphic	Laramide	Lens	
5	Baba raeis	34	19	22	48	57	23	S	Slate and schist/Jurassic	Metamorphic	Late Cimmerian	Vein, lens	
6	BarudAb-e-Nahavand	34	26	11	48	22	30	S	Marble and schist/Jurassic	Metamorphic	Late Cimmerian	Vein	
7	Chand Sin	36	5	0	49	12	0	S	Quartzite/Upper Cambrian	Sedimentary	Pan-African	Stratiformbed	
8	Chashm	35	55	55	53	15	20	S	Sandstone and conglomerate/Jurassic	Sedimentary	Early Cimmerian	Vein	
9	Chenaruiyeh	30	14	51	56	45	6	S	Sandstone and limestone/Cretaceous	Sedimentary	Late Cimmerian	Stratiformbed	
10	Cheshin	36	3	0	49	4	58	S	Sandstone/Lower Cambrian	Sedimentary	Pan-African	Stratiformbed	
11	Cheshmeh Paein-e-Arak	33	55	55	49	15	0	S	Schist and phyllite/Jurassic	Metamorphic	Late Cimmerian	Vein	
12	Cheshmeh Pahn-e-Nanj	34	27	0	48	36	18	S	Mica-schist/Jurassic	Metamorphic	Pyrenean	Stratiformbed to lens	
13	Chidarreh (Tahmasb Abad)	36	22	54	48	39	26	S	Quartzite/Upper Cambrian	Sedimentary	Pan-African	Stratiformbed	
14	Dahoiyeh	30	47	38	56	44	0	S	Quartzite/Ordovician	Sedimentary	Pan-African	Stratiformbed	
15	Dalmeh	32	25	40	54	19	0	S	Quartzite and sandstone/Cambrian	Sedimentary	Pan-African	Stratiformbed	
16	Darbeand-e-Shahmirzad	35	43	52	53	15	37	S	Quartzitic sandstone/Mid-Jurassic	Sedimentary	Early Cimmerian	Stratiformbed	
17	Darkaj, Hashish, Darandun	30	46	0	56	6	0	S	Limestone and sandstone/Lower Paleozoic	Sedimentary	Pan-African	Stratiformbed	
18	Darreh Zanjir-e-Taft	31	40	17	54	12	10	S	Sandstone and shale and limestone/ Lower Cretaceous	Sedimentary	Late Cimmerian	Lens	
19	Dashlijeh	36	13	48	48	49	12	S	Sandstone/Ordovician	Sedimentary	Pan-African	Stratiformbed	
20	Dorud	33	27	31	49	0	0	S	Rudist and radiolarite limestone/ Upper Cretaceous	Sedimentary	Laramide	Stratiformbed	
21	Emamzadeh Hashem	35	46	38	51	59	47	S	Quartzite/Ordovician	Sedimentary	Pan-African	Stratiformbed	

22	Firuz Abad-Nazvol	34	12	30	48	37	30	S	Granodiorite/Tertiary; schist/Jurassic	Hydrothermal	Pyrenean	Vein
23	Ghani Abad-e-Rey	35	34	55	51	26	40	S	Quartzite/Cambrian; limestone and dolomite/Lower Cambrian	Sedimentary	Pan-African	Stratiformbed
24	Ghanlu Chopoghlu	36	11	14	48	51	54	L	Quartzite/Lower Cambrian	Sedimentary	Pan-African	Stratiformbed
25	Gharajeh Fid	35	56	0	49	14	0	S	Quartzite/Ordovician	Sedimentary	Pan-African	Stratiformbed
26	Ghasem Abad-e-Latian	35	48	0	51	38	0	S	Top quartzite in upper red sandstone/Cambrian	Sedimentary	Pan-African	Stratiformbed
27	Gheid Ali	33	34	22	49	27	23	S	Phyllite, schist, quartzite, and marble/Precambrian	Hydrothermal	Late Cimmerian	Vein
28	Ghermez Abad	35	55	12	49	15	0	S	Quartzite/Ordovician	Sedimentary	Pan-African	Stratiformbed
29	Gholou Shajaroo	34	22	30	48	52	30	I	Quartzite/Ordovician	Metamorphic	Early Cimmerian	Vein
30	Havasku (Khavasku)	32	20	23	53	40	16	S	Quartzite/Ordovician	Sedimentary	Pan-African	Stratiformbed
31	Hiroud	35	49	0	53	56	20	S	Sandstone and quartzite/Jurassic	Sedimentary	Hercynian	Stratiformbed
32	Hovire	35	43	0	52	18	0	S	Quartzite/Ordovician	Sedimentary	Pan-African	Stratiformbed
33	Jozvan	35	12	36	48	17	24	S	Schist and phyllite/Jurassic	Metamorphic	Late Cimmerian	Vein
34	Kabutarak- Deh Jalal	36	18	36	48	42	28	S	Sandstone/Lower Cambrian	Sedimentary	Pan-African	Stratiformbed
35	Kamareh	34	37	30	48	37	30	S	Pegmatite/Upper Cretaceous; mica-schist bearing garnet and andalusite schist/Jurassic	Hydrothermal	Laramide	Lens
36	Kandhallan	34	37	30	48	52	30	I	Phyllite and schist/Jurassic	Metamorphic	Early Cimmerian	Vein
37	Kanhelal	34	30	0	48	45	0	S	Garnet schist and wollastonite and mica-schist/Jurassic	Metamorphic	Late Cimmerian	Vein
38	KhairAbad-Seyed Shahab	34	22	30	48	41	52	S	Slate and phyllite and schist/Jurassic	Metamorphic	Late Cimmerian	Vein
39	Khankuk	30	44	0	56	46	0	S	Quartzite/Ordovician	Sedimentary	Pan-African	Stratiformbed
40	Kikleh, Achmizan, Payhan, Sibar, Davijeh	34	5	0	48	52	30	S	Granodiorite/Tertiary; schist/Jurassic	Hydrothermal	Pyrenean	Vein
41	Kisehjin	35	49	48	49	4	48	S	Quartzite/Ordovician	Sedimentary	Pan-African	Stratiformbed
42	Kolushe gerd	34	21	18	48	49	34	S	Slate, phyllite, and schist/Jurassic	Metamorphic	Late Cimmerian	Vein

(continued)

(continued)

No.	Name	Lat D°	M	S″	Long D°	M	S″	Size	Host rock/age	Genetic/ mineralization	Orogenic phases	Morphology	Elem para
43	Kowli Kosh	30	46	55	53	8	45	S	Quartzite and schist/Paleozoic	Metamorphic	Hercynian	Stratiformbed	
44	Kuvank (Gueichin)	36	5	0	49	10	0	S	Quartzite/Lower Cambrian	Sedimentary	Pan-African	Stratiformbed	
45	Lalun	35	58	55	51	34	40	S	Limestone/Cambrian	Sedimentary	Pan-African	Stratiformbed	
46	Larestan	35	37	55	54	0	0	S	Sandstone and shale and limestone/ Jurassic	Sedimentary	Early Cimmerian	Stratiformbed	
47	Markasiduneh	32	13	42	53	39	54	S	Quartzitic sandstone/Upper Jurassic	Sedimentary	Late Cimmerian	Stratiformbed	
48	Mazarin	33	38	0	50	20	0	S	Mica-schist and gneiss/Paleozoic	Metamorphic	Late Cimmerian	Vein	
49	Mighan	36	36	37	54	56	0	S	Quartzite/Upper Cambrian	Sedimentary	Pan-African	Stratiformbed	
50	Mizvaj	36	22	10	50	2	14	S	Dolomite and limestone/Triassic	Hydrothermal	Hercynian	Vein	
51	MobarakAbad, Tappeh Faraj, Mosha	35	47	0	52	0	0	S	Layered quartzite and red sandstone/ Cambrian	Sedimentary	Pan-African	Stratiformbed	
52	Nadushan-e-Hafther	32	18	31	53	28	30	S	Quartzite/Ordovician	Sedimentary	Pan-African	Stratiformbed	
53	Nazkhatun	34	30	10	48	45	0	S	Schist/Jurassic	Metamorphic	Late Cimmerian	Vein	
54	Nir	31	29	24	54	7	32	S	Granite and granodiorite/Jurassic	Hydrothermal	Post-Pyrenean	Massive	
55	Northeast of Kal-e-Nari	35	22	19	59	58	19	I	Sandstone and shale/Paleogene	Sedimentary	Pyrenean	Stratiformbed	
56	Ordaklu-Malian	34	12	30	48	47	30	S	Slate, phyllite, and schist/Jurassic	Metamorphic	Late Cimmerian	Vein to lens	
57	Paridar Abolhasan-Biaghan-e-Bala	34	4	0	44	26	30	S	Slate, phyllite, and schist/Jurassic	Metamorphic	Late Cimmerian	Lens	
58	Patiyeh	34	5	0	48	55	0	S	Slate and phyllite and schist/Jurassic	Hydrothermal	Laramide	Vein	
59	Qezeljeh	36	6	0	48	58	48	S	Quartzite/Upper Cambrian	Sedimentary	Pan-African	Stratiformbed	
60	RigSefid	30	3	49	55	8	8	S	Alluvial/Quaternary	Sedimentary	Post-Pyrenean	Massive	
61	Robat Morad-e-Arak	33	36	0	50	16	0	S	Sandy conglomerate/Tertiary	Sedimentary	Pyrenean	Vein	
62	Saranza	35	43	12	52	48	0	S	Shale, marl, and sandstone/Lower Jurassic	Sedimentary	Early Cimmerian	Stratiformbed	
63	Senjetou-Mansour Abad	32	1	45	54	44.	10	S	Quartzite/Ordovician	Sedimentary	Pan-African	Stratiformbed	
64	Shabjereh	31	2	0	56	14	0	S	Quartzite/Ordovician	Sedimentary	Pan-African	Stratiformbed	

65	Shah Bolaghi	36	9	0	48	55	48	S	Quartzite/Upper Cambrian	Sedimentary	Pan-African	Stratiformbed
66	Shalvar	36	22	12	48	45	0	M	Quartzite/Upper Cambrian	Sedimentary	Pan-African	Stratiformbed
67	Shineh In	36	4	0	49	12	0	S	Quartzite/Upper Cambrian	Sedimentary	Pan-African	Stratiformbed
68	Shurab Niglejeh	36	25	12	48	41	24	S	Quartzite/Upper Cambrian	Sedimentary	Pan-African	Stratiformbed
69	South of Kuh-e-Dochahi	35	54	7	59	53	49	I	Sandstone and volcano metamorphic and schist/Paleozoic	Metamorphic	Late Cimmerian	Vein
70	Tasbandy	34	34	51	48	47	40	S	Schist/Jurassic	Metamorphic	Late Cimmerian	Vein
71	Urneh	36	3	0	49	11	0	S	Quartzite/Upper Cambrian	Sedimentary	Pan-African	Stratiformbed
72	Vali Mahd	35	7	30	48	22	30	S	Schist and phyllite/Jurassic	Metamorphic	Late Cimmerian	Vein
73	Vasleh	35	17	24	48	22	30	S	Schist and phyllite/Jurassic	Metamorphic	Late Cimmerian	Vein
74	Vir-Algazir	36	17	39	48	52	12	S	Quartzite/Upper Cambrian	Sedimentary	Pan-African	Stratiformbed
75	Yaghub Abad	34	45	0	48	34	40	S	Schist and phyllite/Jurassic	Metamorphic	Late Cimmerian	Vein
76	Yusbashi	36	22	55	49	28	40	S	Andesite/Eocene	Hydrothermal	Pyrenean	Vein
77	Zagheh	34	52	30	48	17	30	S	Schist/Jurassic	Metamorphic	Late Cimmerian	Unknown

Complete list of fluorite mineral deposits and indications of Iran along with their detailed specifications

		Lat			Long					Genetic/	Orogenic		
No.	Name	D°	M	S″	D°	M	S″	Size	Host rock/age	mineralization	phases	Morphology	Elem para
1	Abbas Abad	33	12	58	52	2	46	I	Limestone and dolomite/Mid-Triassic	Hydrothermal	Early Cimmerian	Vein, Lens	
2	Atashkuh	33	51	0	50	40	0	I	Limestone and dolomite/Mid-Triassic	Hydrothermal	Early Cimmerian	Vein, lens	
3	Kalateh Ashrafu	32	52	50	54	58	50	I	Limestone and dolomite/Mid-Triassic	Hydrothermal	Early Cimmerian	Vein, lens	
4	Kamar Mehdi	33	1	0	56	31	15	S	Limestone and dolomite/Mid-Triassic	Hydrothermal	Early Cimmerian	Vein, lens	
5	Kamar Zard	33	6	0	56	32	0	S	Limestone and dolomite/Mid-Triassic	Hydrothermal	Early Cimmerian	Vein, lens	
6	Kamshecheh	32	56	0	51	51	0	S	Limestone and dolomite/Mid-Triassic	Hydrothermal	Early Cimmerian	Vein, lens	Pb, Ba
7	Kouppeh Ashrafou	32	52	50	54	58	50	I	Limestone and dolomite/Mid-Triassic	Hydrothermal	Early Cimmerian	Vein, lens	
8	Laalkan	36	47	8	47	24	10	S	Crystalline carbonate rocks/Paleozoic	Hydrothermal	Early Cimmerian	Vein, lens	
9	Northwest of Kerman	30	25	5	56	58	5	I	Limestone, dolomite, and shale/Cretaceous	Hydrothermal	Unknown	Unknown	
10	Northwest of Saqqez	36	18	36	46	11	39	I	Limestone, dolomite, and shale/Cretaceous	Hydrothermal	Unknown	Unknown	
11	Qahrabad Soleyman							S	Limestone and dolomite/Mid-Triassic	Hydrothermal	Early Cimmerian	Vein, lens	
12	Sheshrud-e-Palam							S	Limestone and dolomite/Mid-Triassic	Hydrothermal	Early Cimmerian	Vein, lens	
13	Tabas	33	10	39	56	22	11	S	Limestone and dolomite/Mid-Triassic	Hydrothermal	Early Cimmerian	Vein, lens	
14	Pachy Miana	36	3	45	53	11	0	S	Limestone and dolomite/Mid-Triassic	Sedimentary	Early Cimmerian	Stratiformbed	Pb

Complete list of barite mineral deposits and indications of Iran along with their detailed specifications

No.	Name	Lat D°	Lat M	Lat S″	Long D°	Long M	Long S″	Size	Host rock/age	Genetic/ mineralization	Orogenic phases	Morphology	Elem para
1	Abyk	36	7	0	50	31	40	S	Andesitic tuff/Eocene	Hydrothermal	Pyrenean	Vein, massive	
2	Aghcheh Mazar	35	38	0	50	53	0	S	Pyroclastics/Eocene	Hydrothermal	Pyrenean	Vein, stratiformbed	
3	Aghzigan Amereh	34	41	0	50	10	30	S	Andesite and tuff/Eocene	Hydrothermal	Pyrenean	Lens, vein	
4	Ahmad Abad	35	23	0	49	41	30	S	Tuff/Eocene	Hydrothermal	Pyrenean	Vein	Mn
5	Ahwarak	36	18	0	50	22	0	S	Tuff/Eocene	Volcanic	Pyrenean	Vein	
6	Ali Naghieh	36	11	3	49	38	0	S	Unknown	Hydrothermal	Pyrenean	Vein	Pb
7	Amrudak	36	18	0	50	32	30	S	Tuff/Eocene	Hydrothermal	Pyrenean	Vein	Pb, Zn
8	Ardahin	35	13	30	49	48	30	S	Tuff/Eocene	Hydrothermal	Pyrenean	Lens, vein	
9	Ardekan	32	24	0	54	20	0	M	Limestone and dolomite (marble)/ Devonian–Permian	Hydrothermal	Hercynian	Vein, lens	
10	Ardemin	35	15	30	49	47	0	S	Tuff/Eocene	Hydrothermal	Pyrenean	Vein	
11	Armiun	35	31	0	49	45	0	S	Tuff/Eocene; middle-bedded limestone/Paleocene	Hydrothermal	Pyrenean	Lens	
12	Arzandeh	34	27	30	60	30	20	I	Rhyolite and dacite/Eocene	Hydrothermal	Pyrenean	Vein	
13	Asgar Abad	35	11	0	51	47	0	S	Tuff/Eocene	Hydrothermal	Pyrenean	Vein	
14	Atanak	36	9	30	50	27	0	S	Tuff/Eocene	Hydrothermal	Pyrenean	Vein	Pb
15	Azadkin	35	16	0	50	7	0	S	Tuff/Oligocene	Hydrothermal	Pyrenean	Vein	
16	Baba Hemmat	34	54	30	51	58	30	S	Trachyte and andesite/Eocene	Hydrothermal	Pyrenean	Vein	
17	Bagher Abad	34	53	0	51	14	0	S	Diorite and granodiorite/Post-Eocene; tuff/Eocene	Hydrothermal	Pyrenean	Vein	
18	Bagher Abad-e-Qom	34	57	0	50	52	0	S	Andesite and tuff/Eocene	Hydrothermal	Pyrenean	Vein	Cu, Fe
19	Bagh-e-Zendan	35	35	0	51	26	0	S	Unknown	Unknown	Unknown	Vein	
20	Barik	33	36	2	59	24	12	I	Tuff and andesite/Eocene	Hydrothermal	Pyrenean	Unknown	
21	Barkah	34	5	57	59	44	3	I	Rhyolite and dacite/Eocene	Hydrothermal	Pyrenean	Vein	
22	Bidhend 1	34	17	0	50	46	58	I	Tuff and andesite/Eocene	Hydrothermal	Pyrenean	Vein	
23	Bidhend 2	34	22	0	50	49	30	S	Andesite and tuff/Eocene	Hydrothermal	Pyrenean	Lens	

(continued)

(continued)

No.	Name	Lat D°	Lat M	Lat S″	Long D°	Long M	Long S″	Size	Host rock/age	Genetic/ mineralization	Orogenic phases	Morphology	Elem para
24	Bighan	34	21	0	50	35	3	S	Andesite and tuff/Eocene	Hydrothermal	Pyrenean	Vein	
25	Blueband	35	17	0	50	2	30	S	Tuff/Oligocene	Hydrothermal	Pyrenean	Vein	
26	Chah Shirin	35	23	20	54	2	0	S	Andesite/Eocene; limestone, dolomite/Cretaceous	Hydrothermal	Late Cimmerian	Vein	
27	Chahmes	32	13	0	54	44	0	S	Limestone and shale and sandstone/ Jurassic	Hydrothermal	Late Cimmerian	Lens	
28	Chalambar	35	31	0	49	47	30	S	Tuff/Eocene	Hydrothermal	Unknown	Unknown	Pb
29	Chamanak	34	32	30	50	21	0	S	Andesite and tuff/Eocene	Hydrothermal	Pyrenean	Vein	
30	Chanjan	38	16	0	45	42	30	S	Flysch/Cretaceous	Hydrothermal	Pyrenean	Vein	
31	Chary Abtorsh	31	15	0	55	44	0	M	Reefy limestone/Lower Cretaceous; shale and	Epithermal	Pyrenean	Vein	Pb, F
32	Chenar Vardeh	35	7	30	50	41	0	S	Tuff/Eocene	Hydrothermal	Pyrenean	Vein, lens	
33	Dasht-e-Deh	32	12	0	54	45	0	S	Limestone/Jurassic	Hydrothermal	Late Cimmerian	Lens, vein	
34	Dasht-e-Kavir	32	30	0	57	15	0	I	Sandstone and limestone/Triassic	Hydrothermal	Pyrenean	Unknown	
35	Dasht-e-Kavir	34	52	0	51	55	0	S	Trachyte and andesite/Eocene	Hydrothermal	Pyrenean	Vein	
36	Dastgerd	36	19	6	50	18	5	I	Basalt/Tertiary	Hydrothermal	Pyrenean	Vein	
37	Deh Khatib	34	6	16	59	43	50	I	Andesite and dacite/Eocene	Unknown	Unknown	Unknown	
38	Deh Shur	35	38	0	50	52	0	S	Andesite/Oligocene	Unknown	Unknown	Vein	
39	Dorbid	32	7	16	54	24	41	S	Limestone, dolomite, shale and quartzitic sandstone/Cretaceous	Hydrothermal	Late Cimmerian	Vein, lens	
40	Dorreh-ye-Kashan	33	53	0	51	18	0	L	Volcanics and limestone/Oligo-Miocene	Hydrothermal	Post-Pyrenean	Lens to strati-formbed massive	
41	Duna	36	9	55	51	20	45	S	Limestone/Permian–Triassic	Hydrothermal	Late Cimmerian	Vein	
42	East of Borzoo	34	16	35	59	43	40	I	Tuff and andesite/Eocene	Hydrothermal	Pyrenean	Vein	
43	Ebdal-e-Samadi	35	39	0	46	36	0	M	Andesibasalt, limestone, and shale/ Cretaceous	Hydrothermal	Laramide	Vein to massive	
44	Elit	36	20	0	51	6	0	L	Sandstone/Upper Precambrian	Hydrothermal	Pyrenean	Vein, lens	Pb, Zn

45	Esfarayen	36	33	47	50	36	0	I	Tuff/Eocene	Hydrothermal	Pyrenean	Vein	
46	Faskhoud	32	29	20	51	59	15	S	Limestone/Cretaceous	Hydrothermal	Pyrenean	Vein	
47	Gachsar	35	49	0	51	28	0	S	Tuff/Eocene	Hydrothermal	Pyrenean	Vein	
48	Galleh Siah	34	22	40	50	50	30	S	Shale and tuff/Eocene; limestone/ Cretaceous	Hydrothermal	Pyrenean	Vein	
49	Garmab	35	19	0	53	47	0	S	Limestone/Cretaceous	Hydrothermal	Pyrenean	Vein	
50	Gavan Kuh	33	45	0	49	48	0	S	Limestone/Cretaceous	Unknown	Unknown	Vein	
51	Gelak	35	14	30	50	4	30	S	Tuff/Eocene	Hydrothermal	Pyrenean	Vein, lens	
52	Ghara Chay	34	53	32	49	42	11	I	Tuff and andesite/Eocene	Hydrothermal	Pyrenean	Vein	
53	Ghareh Cheytan	36	0	0	60	45	0	S	Unknown	Unknown	Unknown	Vein	
54	Ghargh Abad	34	29	0	50	8	30	S	Andesite and tuff/Eocene	Hydrothermal	Pyrenean	Vein	
55	Ghasem Abad	35	1	30	49	48	0	S	Tuff/Eocene	Unknown	Unknown	Vein	
56	Ghazi Kalayeh	36	8	38	50	38	38	I	Basalt/Tertiary	Hydrothermal	Pyrenean	Vein	
57	Gonabad	34	25	20	58	31	35	I	Tuff and andesite/Eocene	Hydrothermal	Pyrenean	Vein	
58	Haft Her	32	10	0	53	45	0	M	Limestone, sandy limestone/ Precambrian	Hydrothermal	Hercynian	Vein, lens	Pb, F
59	Haji Abad	33	9	0	54	47	0	S	Carbonaceous rocks/Lower Cretaceous; carbonate rocks/ Lower Paleozoic	Volcanosedi- mentary	Unknown	Stratiformbed, lens, veinlet	Pb
60	Hemat Abad	35	8	46	60	11	26	I	Sandstone and limestone/Triassic	Hydrothermal	Pyrenean	Vein	
61	Hoseyn Abad	36	33	27	49	14	8	S	Volcanics/Tertiary	Hydrothermal	Pyrenean	Vein	
62	Hudeh	32	15	0	53	37	0	S	Dolomite, shale, and limestone/Upper Precambrian–Lower Cambrian	Hydrothermal	Pan-African	Vein	Pb, Cu
63	Ipak	35	38	18	50	17	45	S	Pyroclastics/Eocene	Hydrothermal	Pyrenean	Vein	
64	Ivanak	36	20	0	50	50	0	S	Tuff/Eocene	Hydrothermal	Pyrenean	Vein	
65	Jasb	34	10	0	50	51	0	M	Andesite and tuff/Eocene	Hydrothermal	Post-Pyrenean	Vein	
66	Kaboutar Kuh	34	7	45	58	53	35	I	Limestone, dolomite, and sandstone/ Jurassic	Hydrothermal	Pyrenean	Vein	
67	Kalateh	34	34	10	60	6	23	I	Limestone and dolomite/Cretaceous	Hydrothermal	Pyrenean	Vein	
68	Khaman	30	29	30	56	28	30	M	Volcanics/Eocene	Hydrothermal	Pyrenean	Vein, stratiformbed	

(continued)

(continued)

No.	Name	Lat D°	Lat M	Lat S″	Long D°	Long M	Long S″	Size	Host rock/age	Genetic/ mineralization	Orogenic phases	Morphology	Elem para
69	Kamar Kajgan	34	18	6	50	48	37	I	Tuff and andesite/Eocene	Hydrothermal	Pyrenean	Vein	
70	Kamshecheh	32	56	0	51	51	0	L	Dolomite and limestone/Mid-Triassic	Hydrothermal	Early Cimmerian	Vein, lens	
71	Karat	34	33	57	60	32	56	I	Limestone, dolomite, and sandstone/ Jurassic	Hydrothermal	Pyrenean	Vein	
72	Kelak-e-Karaj	35	47	0	51	7	0	S	Tuff and shale/Eocene	Volcanic	Pyrenean	Vein	
73	Khureh Abad	31	3	40	52	51	15	S	Limestone and schist/Lower Mesozoic	Hydrothermal	Early Cimmerian	Vein, Stratiformbed	
74	kordin	34	53	36	60	13	5	I	Rhyolite and dacite/Eocene	Hydrothermal	Pyrenean	Vein	
75	Kubi Lark	36	32	30	50	0	30	S	Tuff/Eocene	Hydrothermal	Pyrenean	Vein	
76	Kuh-e-Arzandeh 1	35	0	45	60	0	40	I	Tuff and andesite/Eocene	Hydrothermal	Pyrenean	Vein	Pb
77	Kuh-e-Arzandeh 2	34	45	0	60	12	27	I	Rhyolite and dacite/Eocene	Hydrothermal	Pyrenean	Vein	
78	Kuh-e-Arzandeh 3	34	55	50	60	10	51	I	Rhyolite and dacite/Eocene	Hydrothermal	Pyrenean	Vein	
79	Kuh-e-Bahereh	34	27	54	60	29	1	I	Rhyolite and dacite/Eocene	Hydrothermal	Pyrenean	Vein	
80	Kuh-e-Bazd 1	35	9	15	60	21	39	I	Andesite/Eocene	Hydrothermal	Pyrenean	Vein	
81	Kuh-e-Bazd 2	35	7	37	60	23	43	I	Rhyolite and dacite/Eocene	Hydrothermal	Pyrenean	Vein	
82	Kuh-e-Cheshmeh zebri	33	36	13	59	30	38	I	Diorite, monzonite, and syenite/ Tertiary	Hydrothermal	Pyrenean	Unknown	
83	Kuh-e-Gijkin	36	31	0	50	23	0	S	Tuff/Eocene	Hydrothermal	Pyrenean	Vein	
84	Kuh-e-Hendi	34	49	51	50	14	39	I	Tuff and andesite/Eocene	Hydrothermal	Pyrenean	Vein	
85	Kuh-e-kafar Rah	35	55	0	51	46	0	S	Unknown	Unknown	Unknown	Vein	
86	Kuh-e-Kamar Sabz	34	58	20	60	27	23	I	Rhyolite and dacite/Eocene	Hydrothermal	Pyrenean	Vein	
87	Kuh-e-Khashchal	36	31	30	50	30	0	S	Tuff/Eocene	Hydrothermal	Pyrenean	Vein	
88	Kuh-e-Mahyar	33	45	35	59	8	23	I	Andesite/Eocene	Hydrothermal	Pyrenean	Vein	
89	Kuh-e-Mianjani	35	4	47	60	26	7	I	Rhyolite and dacite/Eocene	Hydrothermal	Pyrenean	Vein	
90	Kuh-e-Nahour	34	32	40	59	57	13	I	Limestone/Cretaceous	Hydrothermal	Pyrenean	Vein	
91	Kuh-e-Nakhjiv	35	7	9	60	22	30	I	Rhyolite and dacite/Eocene	Hydrothermal	Pyrenean	Vein	
92	Kuh-e-Padeshahan	35	24	30	60	4	25	I	Limestone and dolomite/Cretaceous	Hydrothermal	Pyrenean	Vein	

93	Kuh-e-Pasakuh	34	58	39	59	58	5	I	Tuff and andesite/Eocene	Hydrothermal	Pyrenean	Vein	
94	Kuh-e-she Pestan	33	12	2	60	14	50	I	Volcanics/Cretaceous	Hydrothermal	Late Cimmerian	Vein	
95	Kuh-e-Sorkheh Hesar	35	6	29	60	16	7	I	Sandstone and limestone/Upper Paleozoic	Hydrothermal	Pyrenean	Vein	
96	Kuh-e-Tah darreh	34	57	30	60	4	28	I	Rhyolite and dacite/Eocene	Hydrothermal	Pyrenean	Vein	
97	Kuh-e-Tajdar	35	1	30	60	21	18	I	Rhyolite and dacite/Eocene	Hydrothermal	Pyrenean	Vein	
98	Kuh-e-Zard	33	10	0	54	57	0	S	Massive limestone/Lower Cretaceous	Hydrothermal	Late Cimmerian	Vein, lens, veinlet	
99	Ladinar Chahmish	33	14	0	51	52	30	S	Limestone/Cretaceous	Hydrothermal	Late Cimmerian	Vein	
100	Lar	35	48	0	51	53	0	S	Tuff/Eocene	Hydrothermal	Pyrenean	Vein, lens	
101	Lielestan	36	9	0	51	8	0	S	Tuff/Eocene	Hydrothermal	Pyrenean	Vein	
102	Lirestan	37	11	30	50	4	30	S	Unknown	Hydrothermal	Pyrenean	Vein	
103	Manzariyeh-ye-Qom	34	53	0	50	57	30	S	Andesite/Eocene	Hydrothermal	Post-Pyrenean	Vein	
104	Mehdi Abad	31	29	0	55	1	30	M	Limestone, dolomite, and sandstone/ Lower Cretaceous	Hydrothermal	Late Cimmerian	Lens	
105	Nabi Abad	34	13	35	59	45	30	I	Tuff and andesite/Eocene	Hydrothermal	Pyrenean	Vein	
106	Nayeh	34	38	30	50	15	0	I	Tuff and andesite/Eocene	Hydrothermal	Pyrenean	Vein	
107	Nigh	34	4	0	51	16	0	S	Volcanics and pyroclastics/Eocene	Hydrothermal	Pyrenean	Vein	
108	Nineh-ye-Qom	34	3	30	50	31	0	S	Tuff/Eocene	Hydrothermal	Pyrenean	Vein	
109	Niyuk	32	24	30	54	36	30	S	Limestone and dolomite/ Jurassic–Cretaceous	Hydrothermal	Late Cimmerian	Vein to lens	Pb, F
110	North of Gazorkan	36	33	0	50	30	0	I	Shale, sandstone, and limestone/ Precambrian	Hydrothermal	Pyrenean	Vein	
111	Nowbaran	35	1	30	49	44	0	S	Tuff/Eocene; middle-bedded limestone/Paleocene	Hydrothermal	Pyrenean	Lens	
112	Nowbaran 1	36	1	21	50	55	53	I	Tuff and volcanics/Eocene	Hydrothermal	Pyrenean	Vein	
113	Nowdaran	36	11	30	50	28	30	S	Tuff/Eocene	Hydrothermal	Pyrenean	Vein	
114	Nusha	36	36	30	50	32	20	S	Tuff/Eocene	Hydrothermal	Pyrenean	Vein	
115	Ostad Bakharno	34	51	23	60	15	0	I	Rhyolite and dacite/Eocene	Hydrothermal	Pyrenean	Vein	
116	Parachan	36	17	0	50	51	0	S	Tuff/Eocene	Hydrothermal	Pyrenean	Vein	
117	Parkeh	36	21	0	50	6	0	S	Unknown	Hydrothermal	Pyrenean	Stratiformbed	

(continued)

(continued)

No.	Name	Lat D°	Lat M	Lat S″	Long D°	Long M	Long S″	Size	Host rock/age	Genetic/ mineralization	Orogenic phases	Morphology	Elem para
118	Poshteh	35	23	0	53	37	10	S	Medium- to thick-bedded limestone/ Upper Jurassic	Hydrothermal	Late Cimmerian	Vein	
119	Qez Gholleh	36	19	0	50	18	30	S	Unknown	Hydrothermal	Pyrenean	Vein	
120	Qom	34	4	0	50	30	0	S	Andesite and tuff/Eocene	Hydrothermal	Pyrenean	Lens	
121	Rasour	33	22	30	53	24	5	I	Trachyandesite and volcanosedimentary rocks/Eocene	Hydrothermal	Pyrenean	Vein and lens	
122	Razaghan	35	20	0	49	57	0	S	Tuff/Eocene	Hydrothermal	Pyrenean	Vein	Pb
123	Renj-e-Vejan	36	5	0	52	4	0	S	Unknown	Unknown	Unknown	Vein	
124	Robat	35	1	33	60	18	45	I	Rhyolite and dacite/Eocene	Hydrothermal	Pyrenean	Vein	
125	Rudkhaneh-ye-Ali Zan	36	1	24	50	51	3	I	Tuff and volcanics/Eocene	Hydrothermal	Pyrenean	Vein	Pb
126	RudShur-Shahdasht	35	39	14	50	46	15	S	Limestone and andesite/Eocene	Hydrothermal	Pyrenean	Vein	
127	Safar Abad	34	52	0	50	54	30	S	Andesite/Oligo-Miocene	Hydrothermal	Post-Pyrenean	Vein	F
128	Salafchegan	34	29	3	50	24	23	I	Tuff and andesite/Eocene	Hydrothermal	Pyrenean	Vein	Pb, F
129	Sangan Paien	34	27	6	60	29	1	I	Rhyolite and dacite/Eocene	Hydrothermal	Pyrenean	Lens	
130	Sangobal	37	3	16	49	5	45	I	Limestone/Jurassic	Hydrothermal	Late Cimmerian	Vein	
131	Sefid Abad	35	8	35	60	11	55	I	Sandstone and limestone/Triassic	Hydrothermal	Pyrenean	Vein	
132	Sefid Kuh-e-Khur	34	52	13	60	11	49	I	Rhyolite and dacite/Eocene	Hydrothermal	Pyrenean	Vein	
133	Sepidarak	36	5	10	50	40	50	S	Tuff/Eocene	Hydrothermal	Pyrenean	Vein	
134	Seyed Mohammad	35	2	25	60	23	45	I	Rhyolite and dacite/Eocene	Hydrothermal	Pyrenean	Vein	
135	Shah Alamdar	36	15	0	49	59	30	S	Rhyolite, trachyte, and tuff/Eocene	Hydrothermal	Pyrenean	Vein	
136	Shalamzar	36	5	10	50	40	50	S	Tuff/Eocene	Hydrothermal	Pyrenean	Vein	Pb
137	Shekarnab	36	18	14	50	18	30	I	Basalt/Tertiary	Hydrothermal	Pyrenean	Vein	Cu
138	Shirin Bolagh	34	22	0	50	34	30	S	Shale and tuffaceous sandstone/ Eocene	Hydrothermal	Pyrenean	Vein	
139	Siah Kuh	34	57	0	51	58	30	S	Trachyte and andesite/Eocene	Hydrothermal	Pyrenean	Vein	
140	Siah Lakh	34	52	13	60	18	35	I	Rhyolite and dacite/Eocene	Hydrothermal	Pyrenean	Vein	
141	Siavashan	34	28	55	50	8	24	I	Tuff and andesite/Eocene	Hydrothermal	Pyrenean	Vein	

142	Soltan Abad	34	36	0	50	14	30	S	Tuff/Eocene	Hydrothermal	Pyrenean	Vein, stratiformbed	
143	South of Abbas Abad	34	5	35	59	5	35	I	Limestone, dolomite, and sandstone/ Jurassic	Hydrothermal	Pyrenean	Vein	Pb
144	South of Siah Rud	36	45	7	49	34	30	I	Sandstone and limestone/Upper Paleozoic	Hydrothermal	Pyrenean	Vein	
145	Souzan	34	26	23	60	0	38	I	Limestone and dolomite/Cretaceous	Hydrothermal	Pyrenean	Vein	
146	Takht-e-Chaman	35	9	30	50	23	30	S	Tuff/Eocene	Hydrothermal	Pyrenean	Vein	
147	Tall-e-Siah	33	15	0	54	47	0	M	Limestone/Cretaceous	Hydrothermal	Late Cimmerian	Lens, stratiformbed	
148	Tamasha	35	39	0	51	50	0	S	Limestone/Paleocene	Epithermal	Pyrenean	Vein	
149	Tang-e-Duzan (Tang-e-Bid)	33	3	39	49	57	30	S	Limestone/Cretaceous	Hydrothermal	Late Cimmerian	Vein	
150	Tappeh Sorkh-e-Bijegan	34	6	0	50	46	0	S	Limestone and tuff/Eocene	Hydrothermal	Pyrenean	Vein, massive	
151	Tayersar	37	9	30	50	5	0	S	Unknown	Hydrothermal	Pyrenean	Vein	
152	Teikhur	36	10	15	50	29	20	S	Tuff/Eocene	Hydrothermal	Pyrenean	Vein	
153	Tighchal	36	18	30	50	32	0	S	Tuff/Eocene	Hydrothermal	Pyrenean	Lens	
154	Vandar	36	32	0	50	30	0	S	Rhyolite, trachyte, and tuff/Eocene	Hydrothermal	Pyrenean	Vein, lens	Pb
155	Varan	34	10	0	50	55	0	L	Andesite and tuff/Eocene	Hydrothermal	Post-Pyrenean	Vein	
156	Varbon	36	32	0	50	30	0		Schist and sandstone/Precambrian	Hydrothermal	Unknown	Vein	
157	Varedeh	33	56	31	59	15	48	I	Andesite/Eocene	Hydrothermal	Pyrenean	Vein	Pb
158	Varsan	34	28	10	50	15	50	S	Trachytic tuff/Eocene	Hydrothermal	Pyrenean	Vein	Cu, F
159	Yarak	36	18	0	50	47	19	I	Shale, sandstone, and limestone/ Precambrian	Hydrothermal	Pyrenean	Vein	Pb
160	Yark	36	25	52	50	48	24	I	Tuff and basalt/Eocene	Hydrothermal	Pyrenean	Vein	
161	Yekeh Kuh	34	34	44	60	27	23	I	Phyllite and dacite/Eocene	Hydrothermal	Pyrenean	Vein	Pb
162	Zar1andZar2	34	2	0	50	50	0	S	Andesite/Oligo-Miocene	Hydrothermal	Post-Pyrenean	Vein	

Complete list of Mica mineral deposits and indications of Iran along with their detailed specifications

No.	Name	Lat D°	Lat M	Lat S″	Long D°	Long M	Long S″	Size	Host rock/age	Genetic/ mineralization	Orogenic phases	Morphology	Elem para
1	Amlash	37	4	0	50	8	20	S	Dyke lamprophyre/Eocene	Magmatic	Pyrenean	Vein	
2	Arzanfud	34	45	0	48	35	0	I	Metamorphic rock/Jurassic	Muscovite: pegmatitic	Laramide	Vein	Sodic feldspar
3	Bug	29	14	0	61	9	0	I	Granitoid/Eocene–Oligocene	Pegmatitic	Pyrenean	Vein	
4	Chah Zard	34	3	44	54	34	0	I	Schist/Upper Precambrian–Lower Cambrian	Metamorphic	Pan-African	Vein	
5	Chahgaz	29	31	5	55	2	20	S	Granite, schist, and limestone/ Permo-Triassic	Pegmatitic	Hercynian	Vein	
6	Ching Kalagh	36	10	0	59	36	0	I	Granite/Paleozoic	Pegmatitic	Hercynian	Vein	Sodic and potasic feldspar
7	Deh Gheiby	36	12	0	59	53	0	I	Granite/Paleozoic	Pegmatitic	Hercynian	Vein	
8	Deh Gheshlagh	36	4	0	59	42	0	S	Granite/Paleozoic	Pegmatitic	Hercynian	Vein	Feldspar
9	Dehnow	33	26	0	49	54	0	I	Granodiorite/Upper Pre–Mid-Cretaceous	Pegmatitic	Laramide	Vein	
10	Dehnow Asadollah Khan	34	43	0	48	41	0	S	Metamorphic rock/Jurassic	Metamorphic	Laramide	Vein	Sodic feldspar
11	Gharah Bagh	38	4	21	45	2	14	S	Alkali granite, phlogopite bearing pyroxenite and hornfels/Upper Cretaceous–Lower Paleocene	Metasomatic	Laramide	Vein	Apatite

12	Gheshlagh	36	5	51	59	40	39	S	Granite and pegmatite/Paleozoic	Pegmatitic	Hercynian	Vein	
13	Jandagh	34	5	11	54	34	32	S	Mica-schist, amphibolite, and crystalline limestone/Precambrian	Metamorphic pegmatitic	Pan-African	Vein	
14	Khajeh Morad	36	8	0	59	41	0	I	Granite/Paleozoic	Magmatic	Hercynian	Vein	
15	Khonardeh	33	47	49	49	13	0	I	Granodiorite/Upper Cretaceous	Pegmatitic	Laramide	Vein	
16	Kuh-e-Kafari	35	43	38	55	18	21		Mica-schist/Precambrian	Metamorphic	Pan-African	Lens	
17	Mangavy	34	41	0	48	43	0	I	Metamorphic rock/Jurassic	Metamorphic	Laramide	Vein	Sodic feldspar
18	Molla Bagh (Masouleh)	37	8	26	49	0	52	S	Granite and pegmatitic/Paleozoic; limestone, shale, and sandstone/ Jurassic	Pegmatitic	Late Cimmerian	Vein	
19	Molla Taleb	33	31	0	49	38	0	S	Granodiorite/Pre–Mid-Cretaceous	Pegmatitic	Laramide	Vein	
20	Southeast of Jandagh	33	59	44	54	40	36	I	Schist/Upper Precambrian–Lower Cambrian	Pegmatitic	Pan-African	Vein	
21	Yaromghiyeh	38	40	13	45	1	13	S	Granite, mica-schist, phlogopite bearing pyroxenite and hornfels/ Upper Cretaceous–Lower Paleocene	Metasomatism	Post-Pyrenean	Lens	Feldspar
22	Zahran	38	42	0	45	6	0	S	Granite, mica-schist, phlogopite bearing pyroxenite and hornfels/ Tertiary	Pegmatitic	Laramide	Lens	
23	Zaman Abad	34	38	0	48	41	0	S	Pegmatite Upper Cretaceous–Lower Paleocene; schist/Jurassic	Magmatic	Laramide	Massive	
24	Zanjanbar	33	42	0	51	33	0	I	Granodiorite/Oligo-Miocene; limestone/Paleozoic	Metasomatism (Skarn)	Post-Pyrenean	Massive	

Complete list of talc mineral deposits and indications of Iran along with their detailed specifications

No.	Name	Lat D°	Lat M	Lat S″	Long D°	Long M	Long S″	Size	Host rock/age	Genetic/mineralization	Orogenic phases	Morphology	Elem para
1	Agh Bolagh	33	44	27	49	1	5	S	Limestone and schist and phyllite/Triassic	Metasomatic	Late Cimmerian	Vein	
2	Baghestan	34	14	38	58	27	34	I	Limestone/Paleogene	Metasomatic	Pyrenean	Vein	
3	Bahabad-e-Jandagh	34	1	54	54	32	16	S	Crystalline schist and limestone/Upper Precambrian–Lower Cambrian	Metasomatic	Pan-African	Vein	
4	Bandan 1	31	18	43	60	40	9	I	Ophiolite complex and limestone/Cretaceous	Metasomatic	Laramide	Unknown	
5	Bandan 2	31	20	21	60	40	9	I	Ophiolite complex and limestone/Cretaceous	Metasomatic	Laramide	Unknown	
6	Bavaki	33	34	43	49	20	6	I	Limestone and schist/Triassic	Metasomatic	Late Cimmerian	Vein	
7	Chah Chvchu	31	40	50	60	13	25	I	Turbidites/Upper Cretaceous	Metasomatic	Late Cimmerian	Vein	
8	Chah Ebrahim Zahra Jandagh	34	7	0	55	3	0	S	Crystalline limestone and schist/Upper Precambrian–Lower Cambrian	Metasomatic	Pan-African	Vein	
9	Chah Hajat 1	31	33	26	60	30	19	I	Ophiolite complex/Upper Cretaceous	Metasomatic	Laramide	Unknown	
10	Chah Hajat 2	31	32	53	60	31	16	S	Ophiolite complex/Upper Cretaceous	Metasomatic	Laramide	Unknown	
11	Chah Rostam	31	35	20	60	25	16	I	Ophiolite complex/Upper Cretaceous	Metasomatic	Laramide	Unknown	
12	Chah Safahu	34	4	0	54	35	0	S	Crystalline schist and limestone/Upper Precambrian–Lower Cambrian	Metasomatic	Pan-African	Vein	
13	Chah Zard-e-Jandagh	34	2	35	54	35	0	S	Crystalline schist/Upper Precambrian–Lower Cambrian	Metasomatic	Pan-African	Vein	
14	Chahar Cheshmeh	36	16	22	59	31	8	S	Ophiolite complex/Upper Paleozoic	Metasomatic	Hercynian	Vein	
15	Cheshmeh Shapour	33	1	39	55	2	4	S	Sedimentary rocks/Upper Cretaceous–Tertiary	Sedimentary	Pyrenean	Stratiformbed	
16	DarehhSukhteh	33	36	5	49	17	58	I	Limestone and schist/Triassic	Metasomatic	Late Cimmerian	Vein	
17	Darreh Hadavaneh	33	32	17	49	12	45	I	Limestone and schist and phyllite/Triassic	Metasomatic	Late Cimmerian	Vein	
18	Darreh Pahn	26	37	0	57	37	0	S	Flysch/Miocene, shale, and sandstone/Eocene	Metasomatic	Post-Pyrenean	Vein	

19	Darreh Shiran (Darizhan Bala)	33	31	37	49	15	43	I	Limestone and schist/Triassic	Metasomatic	Late Cimmerian	Vein
20	Deh Haji	33	31	0	49	19	0	S	Limestone, dolomite, and schist/ Triassic	Metasomatic	Late Cimmerian	Vein
21	Deh Mousa	33	34	0	49	20	0	S	Dolomite and schist/Triassic	Metasomatic	Late Cimmerian	Vein
22	Digelan-e-Fariman	35	35	0	59	40	0	S	Ultramafic rocks/Upper Cretaceous	Metasomatic	Laramide	Vein
23	East of Anarak	33	19	8	53	48	41	I	Crystalline schist/Upper Precambrian–Lower Cambrian	Metasomatic	Pan-African	Vein
24	East of Nari	35	32	11	59	42	49	I	Ophiolite complex/Upper Cretaceous	Metasomatic	Laramide	Vein
25	Easth of Kuh-e-Malekdun	31	24	25	60	35	19	I	Sandstone, limestone, and dolomite/ Triassic	Metasomatic	Laramide	Unknown
26	Ferdows	33	45	21	57	17	36	I	Rhyolite and dacite/Tertiary; limestone/Jurassic	Hydrothermal	Pyrenean	Vein
27	Gheshlagh	38	39	0	44	47	20	S	Amphibolite, mica-schist, and ultramafic rocks/Upper Cretaceous	Hydrothermal	Laramide	Vein
28	Gonabad	34	2	15	59	1	5	I	Dolomite and limestone/Paleogene	Metasomatic	Pyrenean	Vein
29	Gorji DarrehSy	36	46	0	47	17	0	S	Ultramafic rocks/Precambrian	Metasomatic	Pan-African	Lens
30	Gushe (Mohsenebne Ali)	33	42	0	49	1	0	S	Schist, metamorphosed limestone, and volcanic rocks/Triassic	Metasomatic	Late Cimmerian	Vein
31	Hola Heydar	33	22	25	53	43	14	I	Phyllite and schist/Precambrian	Metasomatic	Pan-African	Vein
32	Jandagh 1	34	4	0	54	36	0	S	Phyllite and schist/Precambrian	Metasomatic	Pan-African	Vein
33	Jandagh 2	34	4	30	54	36	0	S	Phyllite and schist/Precambrian	Metasomatic	Pan-African	Vein
34	Khanghah	38	39	20	44	47	50	S	Amphibolite, mica-schist, and ultramafic rocks/Upper Cretaceous	Metasomatic	Laramide	Vein
35	Lakuh-e-Sefid	31	31	6	60	13	27	I	Turbidites/Cretaceous	Metasomatic	Late Cimmerian	Vein
36	Mansh	27	4	50	61	11	45	S	Ophiolite complex/Upper Cretaceous	Metamorphic and metasomatic	Laramide	vein
37	Masod Abad	33	31	0	49	18	0	S	Limestone, dolomite/Permian	Hydrothermal and metasomatic	Late Cimmerian	Vein
38	Mesgar Abad	35	34	5	52	0	14	S	Talc schist, andesite, and tuff/Eocene	Metasomatic	Pyrenean	Vein

(continued)

(continued)

No.	Name	Lat D°	Lat M	Lat S″	Long D°	Long M	Long S″	Size	Host rock/age	Genetic/mineralization	Orogenic phases	Morphology	Elem para
39	Moluk	34	9	40	49	4	10	S	Limestone, dolomite, and schist/Jurassic	Metasomatic	Laramide	Vein	
40	Mozdabad	34	1	50	59	2	30	I	Dolomite and limestone/Paleogene	Metasomatic	Pyrenean	Vein	
41	Nain-Zavar	33	3	37	51	21	3	I	Limestone/Cretaceous	Sedimentary	Laramide	Stratiformbed	
42	Nehbandan	31	41	34	60	13	21	I	Turbidites/Upper Cretaceous	Metasomatic	Late Cimmerian	Vein	
43	Norh of Tamam Deh 1	31	55	58	60	28	1	I	Rhyolite and dacite/Tertiary	Hydrothermal	Pyrenean	Vein	
44	Norh of Tamam Deh 2	31	56	30	60	27	40	I	Rhyolite and dacite/Tertiary	Hydrothermal	Pyrenean	Vein	
45	Norh of Tamam Deh 3	31	58	47	60	27	18	I	Rhyolite and dacite/Tertiary	Hydrothermal	Pyrenean	Vein	
46	Norh of Tamam Deh 4	31	59	40	60	27	2	I	Rhyolite and dacite/Tertiary	Hydrothermal	Pyrenean	Vein	
47	North of Patiyar	33	11	29	53	55	22	I	Phyllite and schist/Precambrian	Metasomatic	Pan-African	Vein	
48	Northeast of Anarak	33	21	35	53	43	45	I	Crystalline schist/Upper Precambrian–Lower Cambrian	Metasomatic	Pan-African	Vein	
49	Northwest of Deh Garw	31	58	55	60	12	15	I	Phyllite and flysch/Upper Cretaceous	Metasomatic	Laramide	Vein	
50	Patiyar	33	19	2	53	48	6	I	Phyllite and schist/Precambrian	Metasomatic	Pan-African	Vein	
51	Sagh	34	57	30	60	4	28	S	Ultramafic rocks/Paleozoic	Metasomatic	Pan-African	Vein	
52	Salehabad	33	2	36	49	45	21		Limestone/Cretaceous	Metasomatic	Pyrenean	Vein	
53	Seh Farsakh	31	16	5	60	2	49	I	Andesite, phyllite, and schist/Tertiary	Hydrothermal	Pyrenean	Vein	
54	Shovin	29	41	56	60	3	35	I	Ophiolite complex/Upper Cretaceous	Metasomatic	Laramide	Lens	
55	Sira	36	7	0	51	8	0	S	Shale, sandstone, and dolomite/Triassic	Hydrothermal	Pan-African	Vein, veinlet, lens	
56	Sivakh Bolagh	36	46	0	47	15	0	S	Ultramafic rocks/Precambrian	Metasomatic	Pan-African	Lens	
57	South of Bandan fault	31	34	3	60	25	54	I	Ophiolite complex/Upper Cretaceous	Metasomatic	Laramide	Unknown	
58	South of Kuh-e-Malekdun	31	19	2	60	40	24	I	Ophiolite complex and limestone/Cretaceous	Metasomatic	Laramide	Unknown	
59	Southwest of Arusan	34	5	54	55	1	58	I	Crystalline limestone and schist/Upper Precambrian–Lower Cambrian	Metasomatic	Pan-African	Vein	

60	Tamam Deh	31	49	54	60	28	6	I	Rhyolite and dacite/Tertiary	Hydrothermal	Pyrenean	Vein
61	Tanurdar	33	38	44	49	3	54		Limestone and schist/Triassic	Metasomatic	Late Cimmerian	Vein
62	Tidar	33	36	0	49	13	0	S	Limestone, dolomite, and schist/ Triassic	Metasomatic	Late Cimmerian	Vein
63	Vardeh	35	13	46	50	1	49	S	Talc schist, andesite, and tuff/ Oligocene	Metasomatic	Pyrenean	Vein
64	West of Alam Kandy	36	47	0	47	17	0	S	Ultramafic rocks/Precambrian	Metasomatic	Pan-African	Lens
65	Zailik	38	40	54	46	6	25	I	Rhyolite and dacite/Eocene, limestone/ Cretaceous	Metasomatic	Pyrenean	Vein

Complete list of polymetal deposits and indications of Iran along with their detailed specifications

No.	Name	Lat D°	Lat M	Lat S″	Long D°	Long M	Long S″	Size	Host rock/age	Genetic/ mineralization	Orogenic phases	Morphology	Elem para
1	Baychebagh	36	51	59	47	18	16	S	Hydrobreccia and acidic tuff and porphyrite Dacite/Oligo-Miocene	Hydrothermal	Post-Pyrenean	Vein	Cu, Zn, Pb, Mo, Ni, Co, Bi, Au
2	Barikab	36	16	49	49	18	21		Acidic Tuff and intrusive granitoid rocks/Eocene–Oligocene	Hydrothermal	Pyrenean	Vein	
3	Chah Gaz	29	30	45	55	2	0	I	Granodiorite, schist, and limestone/ Permo-Triassic	Massive sulfide	Cimmerian	Vein	Cu, Pb, Zn, Au
4	Chah kalab	31	58	0	59	31	20	S	Basic extrusive/Tertiary; marble and schist/Jurassic	Skarn	Pyrenean	Vein	W, Cu, Zn, Pb
5	Chah Musa	35	30	0	54	55	0	S	Granite and diorite/Upper Eocene–Lower Oligocene	Hydrothermal	Pyrenean	Vein	Cu, Pb, Au
6	Chah Shureh	33	23	20	53	41	30	S	Schist and ultramafic/Upper Precambrian	Hydrothermal	Pan-African	Vein	Ni, Cu, Co, Zn, Pb, Au, Ag
7	Chahpalang	32	56	59	54	11	33	S	Shale and sandstone/Jurassic; schist/ Upper Proterozoic	Hydrothermal	Late Cimmerian	Vein	Cu, W, Au, Ni, Bi
8	Cheshmeh Chah Sefid	33	36	30	53	30	0	I	Schist/Precambrian; quartz keratophyre/Cretaceous	Hydrothermal	Pyrenean	Vein	Cu, Ni
9	Cheshmeh Hafez	35	24	57	54	44	59	S	Andesite and tuff/Eocene	Hydrothermal	Post-Pyrenean	Vein	Cu, Pb, Au, Zn

10	Cheshmeh Karim	33	23	30	53	32	20	S	Schist and marble/Upper Precambrian	Hydrothermal	Pan-African	Lens	Cu, Ni
11	Cheshmeh Shurab	33	26	40	53	32	10	S	Serpentinite/Upper Precambrian	Hydrothermal	Pan-African	Vein	Cu, Ni
12	Gharechilar	38	50	25	46	23	30	S	Granite and granodiorite/Upper Eocene–Lower Oligocene	Hydrothermal	Pyrenean	Vein	Cu, Mo, (Au), W
13	Jameni	33	26	35	54	13	40	S	Chlorite-quartz-sericite schist/ Precambrian; monzonite porphyry/Eocene	Skarn	Pyrenean	Lens	Cu, Ag, Mo
14	Kalkafi	33	24	0	54	14	0	L	Schist/Upper Precambrian; granite/ Eocene	Porphyry	Pyrenean	Disseminated	Cu, Mo, Au
15	Kharestan	28	42	0	60	54	0	S	Trachyte, trachyandesite, and breccia tuff/Oligo-Miocene	Hydrothermal	Post-Pyrenean	Vein	Cu, Pb, Zn, Ag
16	Khuni	33	26	35	54	12	8	S	Dolomite and dolomitic marble/ Upper Precambrian; volcanics/ Eocene	Pb-Zn: MVT, Au: Hydrothermal	Pan-African	Vein	Cu, Pb, Au
17	Qamsar	33	45	0	51	23	0	I	Tonalite/Post-Miocene; limestone/ Oligo-Miocene	Skarn	Post-Pyrenean	Vein	Cu, Co
18	Rasur	33	23	30	53	43	20	I	Ultramafic rocks/Upper Precambrian	Hydrothermal	Pan-African	Vein	Cu, Ni, Au, Ag
19	Taknar	35	22	0	57	46	0	S	Granite and volcanic rocks, slate phyllite/Precambrian	Massive sulfide	Pan-African	Stratiformbed	Cu, Zn, Pb, Au, Ag

Complete list of sulfur mineral deposits and indications of Iran along with their detailed specifications

No.	Name	Lat D°	Lat M	Lat S″	Long D°	Long M	Long S″	Size	Host rock/Age	Genetic/ Mineralization	Orogenic Phases	Morphology	Elem Para
1	Berkeh Seflin	26	42	16	55	13	12	S	Sandstone and shale/Oligocene, Limestone/Eocene	Hydrothermal	Post-Pyrenean	Vein, veinlet, and lens	
2	Bostaneh	26	36	16	54	49	13	S	Limestone, sandstone, siltstone, dolomite, and gypsum/Upper Proterozoic-Lower Paleozoic	Hydrothermal	Post-Pyrenean	Vein	
3	Dastkuh Basaeidu	26	39	16	55	17	6	S	Gypsum, limestone, and sandstone/ Precambrian	Hydrothermal	Post-Pyrenean	Disseminated, lens	
4	Poljar	26	48	50	55	12	18	I	Shale, marl, sandstone, and gypsum/ Precambrian	Hydrothermal	Post-Pyrenean	Vein	
5	Qeshm	26	55	22	56	4	12	S	Shale, marl, sandstone, and gypsum/ Precambrian	Hydrothermal	Post-Pyrenean	Vein	
6	Hormoz	27	5	30	56	28	30	I	Shale, marl, sandstone, and gypsum/ Precambrian	Hydrothermal	Post-Pyrenean	Vein	
7	Khamir	27	2	47	55	41	32	S	Quartz porphyry/Upper Proterozoic– Lower Paleozoic	Hydrothermal	Post-Pyrenean	Vein, veinlet, and Lens	
8	Khatuneh	27	5	34	54	10	46	I	Shale, marl, sandstone, and gypsum/ Precambrian	Hydrothermal	Post-Pyrenean	Vein	
9	Kormusen	27	30	0	54	30	0	S	Limestone and gypsum/Miocene	Hydrothermal	Post-Pyrenean	Stratiformbed and lens	
10	Behbahan	30	4	5	51	23	50	S	Limestone and gypsum/Miocene	Hydrothermal	Post-Pyrenean	Stratiformbed and lens	
11	Dokuh	32	28	38	48	40	0	I	Limestone and gypsum/Miocene	Hydrothermal	Post-Pyrenean	Stratiformbed and lens	
12	Delazin	35	24	32	53	22	49	S	Gypsum and limestone/Miocene	Hydrothermal	Post-Pyrenean	Disseminated, vein, and veinlet	
13	Haji Abad	35	23	55	53	29	20	S	Sandstone, marl, and gypsum/ Miocene	Hydrothermal	Post-Pyrenean	Stratiformbed	
14	Kuh-e-Sorkh	35	24	32	53	9	2	I	Gypsum and limestone/Miocene	Hydrothermal	Post-Pyrenean	Vein	
15	Semnan	35	32	27	53	31	20	S	Sandstone, marl, and gypsum/ Miocene	Hydrothermal	Post-Pyrenean	Lens	

Complete list of turquoise mineral deposits and indications of Iran along with their detailed specifications

No.	Name	Lat			Long			Size	Host rock/age	Genetic/ mineralization	Orogenic phases	Morphology	Elem para
		D°	M	S″	D°	M	S″						
1	Chah Ebrahim Zahra	34	5	58	55	2	56	I	Schist, marble, and quartzite/Upper Proterozoic	Hydrothermal	Pyrenean	Vein and veinlet	Ag
2	Chahposhti	34	6	31	55	1	38	I	Schist, marble, and quartzite/Upper Proterozoic	Hydrothermal	Pyrenean	Vein and veinlet	
3	Neignan	34	23	19	57	22	25	I	Dacitic volcanic and andesite/Upper Eocene	Hydrothermal	Pyrenean	Vein and veinlet	
4	Baghu	35	26	30	54	38	45	S	Andesite/Eocene	Hydrothermal	Pyrenean	Vein and veinlet	Au
5	Neyshabur	36	27	20	58	25	24	S	Andesite/Eocene	Hydrothermal	Pyrenean	Vein and veinlet	Au

Complete list of turquoise mineral deposits and indications of Iran along with their detailed specifications

No.	Name	Lat D°	Lat M	Lat S″	Long D°	Long M	Long S″	Size	Host rock/age	Genetic/ mineralization	Orogenic phases	Morphology	Elem para
1	Ghareh Aghaj	37	46	0	44	48	30	M	Ultramafic and mafic rocks/Precambrian (probably Cretaceous–Paleocene)	Magmatic	Pan-African (Pyrenean)	Massive, Placery	
2	Kahnouj	27	41	0	57	47	0	M	Ultramafic and mafic rocks/Cretaceous	Magmatic	Pan-African	Massive	

Complete list of titanium mineral deposits and indications of Iran along with their detailed specifications

No.	Name	Lat D°	Lat M	Lat S″	Long D°	Long M	Long S″	Size	Host rock/age	Genetic/ mineralization	Orogenic phases	Morphology	Elem para
1	Ghareh Aghaj	37	46	0	44	48	30	M	Ultramafic and mafic rocks/Precambrian (probably Cretaceous–Paleocene)	Magmatic	Pan-African (Pyrenean)	Massive, placery	
2	Kahnouj	27	41	0	57	47	0	M	Ultramafic and mafic rocks/Cretaceous	Magmatic	Pan-African	Massive	

Complete list of strontium mineral deposits and indications of Iran along with their detailed specifications

		Lat			Long								
No.	Name	D°	M	S″	D°	M	S″	Size	Host rock/age	Genetic/ mineralization	Orogenic phases	Morphology	Elem para
1	Bebahan	30	41	0	50	24	0	S	Gypsum, marl, and limestone/ Lower Miocene	Sedimentary	Post-Pyrenean	Stratiformbed	
2	Gurecheh Berenj	34	6	39	54	3	14	S	Shale, mudstone, sandstone, and limestone/Oligo-Miocene	Hydrothermal	Post-Pyrenean	Vein	
3	Malek Abad	34	35	4	52	24	0	S	Gypsumferous marl/Oligo-Miocene	Sedimentary and hydrothermal	Post-Pyrenean	Stratiformbed	
4	Talheh	34	30	20	52	39	31	S	Shale, mudstone, sandstone, and limestone/Oligo-Miocene	Hydrothermal	Post-Pyrenean	Stratiformbed	Zn
5	North Of Daryaye Namak 1	34	58	43	51	58	50	S	Gypsumferous marl/Oligo-Miocene	Sedimentary	Post-Pyrenean	Stratiformbed	
6	North Of Daryaye Namak 2	34	59	32	52	0	0	S	Gypsumferous marl/Oligo-Miocene	Sedimentary	Post-Pyrenean	Stratiformbed	
7	Mesgar Abad	35	10	0	51	43	0	S	Gypsumferous marl/Oligo-Miocene	Sedimentary	Post-Pyrenean	Stratiformbed	
8	Nazar Abad	35	9	49	51	45	0	S	Gypsumferous marl/Oligo-Miocene	Sedimentary	Post-Pyrenean	Stratiformbed	

Complete list of sulfate sodium mineral deposits and indications of Iran along with their detailed specifications

No.	Name	Lat D°	Lat M	Lat S″	Long D°	Long M	Long S″	Size	Host rock/age	Genetic/ mineralization	Orogenic phases	Morphology	Elem para
1	Abbas Abad	35	13	0	52	48	0	S	Clay and silt/Pleistocene	Evaporite	Post-Pyrenean	Surface stratiformbed	
2	Agha Beyk	35	11	0	53	24	0	S	Clay and silt/Pleistocene	Evaporite	Post-Pyrenean	Surface stratiformbed	
3	Asgar Abad	35	2	0	51	46	0	S	Clay and silt/Pleistocene	Evaporite	Post-Pyrenean	Surface stratiformbed	
4	Belgheitas	35	14	0	51	30	0	S	Clay and silt/Pleistocene	Evaporite	Post-Pyrenean	Surface stratiformbed	
5	Chah Jires	33	48	0	58	7	0	I	Clay and silt/Pleistocene	Evaporite	Post-Pyrenean	Surface stratiformbed	
6	Chamburak	35	16	0	51	29	0	S	Clay and silt/Pleistocene	Evaporite	Post-Pyrenean	Surface stratiformbed	
7	Dasht-e-Azadegan	31	43	0	48	28	0	I	Clay and silt/Pleistocene	Evaporite	Post-Pyrenean	Surface stratiformbed	
8	Deh Naghsh	35	4	0	52	32	0	S	Clay and silt/Pleistocene	Evaporite	Post-Pyrenean	Surface stratiformbed	
9	Dehnamak	35	15	0	52	44	0	S	Clay and silt/Pleistocene	Evaporite	Post-Pyrenean	Surface stratiformbed	
10	DehShureh	35	16	0	51	53	0	S	Clay and silt/Pleistocene	Evaporite	Post-Pyrenean	Surface stratiformbed	
11	Dolat Abad-e-Kharabeh	34	37	11	51	8	0	S	Clay and silt/Pleistocene	Evaporite	Post-Pyrenean	Surface stratiformbed	
12	Fasha Fouieyeh	35	12	0	51	26	0	S	Clay and silt/Pleistocene	Evaporite	Post-Pyrenean	Surface stratiformbed	
13	Gol Ara	34	12	0	51	17	0	S	Clay and silt/Pleistocene	Evaporite	Post-Pyrenean	Surface stratiformbed	
14	Hesar Gely	34	57	0	51	18	0	S	Clay and silt/Pleistocene	Evaporite	Post-Pyrenean	Surface stratiformbed	
15	Hows-e-Soltan	35	3	0	50	58	0	I	Clay and silt/Pleistocene	Evaporite	Post-Pyrenean	Surface stratiformbed	
16	Ijdanak	35	8	0	51	18	0	S	Clay and silt/Pleistocene	Evaporite	Post-Pyrenean	Surface stratiformbed	
17	Jahan Abad-e-Varamine	35	6	0	51	52	0	S	Clay and silt/Pleistocene	Evaporite	Post-Pyrenean	Surface stratiformbed	
18	Karim Abad	34	47	0	51	48	0	S	Clay and silt/Pleistocene	Evaporite	Post-Pyrenean	Surface stratiformbed	
19	Kavir-e-Mighan	34	8	0	49	52	0	I	Clay and silt/Pleistocene	Evaporite	Post-Pyrenean	Surface stratiformbed	
20	Lasjerd	35	23	0	53	5	0	S	Clay and silt/Pleistocene	Evaporite	Post-Pyrenean	Surface stratiformbed	
21	Manzariyeh	34	51	0	50	46	0	S	Clay and silt/Pleistocene	Evaporite	Post-Pyrenean	Surface stratiformbed	
22	Mighan	34	9	6	49	47	32	S	Clay and silt/Pleistocene	Evaporite	Post-Pyrenean	Surface stratiformbed	
23	Mohammad Abad	35	22	0	51	30	0	S	Clay and silt/Pleistocene	Evaporite	Post-Pyrenean	Surface stratiformbed	
24	Najm Abad	35	53	0	50	25	0	S	Clay and silt/Pleistocene	Evaporite	Post-Pyrenean	Surface stratiformbed	
25	Nurrod in Abad	35	5	0	52	26	0	S	Clay and silt/Pleistocene	Evaporite	Post-Pyrenean	Surface stratiformbed	

(continued)

(continued)

No.	Name	Lat D°	Lat M	Lat S″	Long D°	Long M	Long S″	Size	Host rock/age	Genetic/mineralization	Orogenic phases	Morphology	Elem para
26	Qaleh HasanAli Rain	29	31	0	58	18	0	S	Clay and silt/Pleistocene	Evaporite	Post-Pyrenean	Surface stratiformbed	
27	Sagzi	32	32	0	51	58	0	S	Clay and silt/Pleistocene	Evaporite	Post-Pyrenean	Surface stratiformbed	
28	Shokr Abad	34	59	0	51	43	0	S	Clay and silt/Pleistocene	Evaporite	Post-Pyrenean	Surface stratiformbed	
29	South west of Varamine	35	12	0	51	20	0	S	Clay and silt/Pleistocene	Evaporite	Post-Pyrenean	Surface stratiformbed	
30	West of Safaiyeh	35	12	0	52	44	0	I	Clay and silt/Pleistocene	Evaporite	Post-Pyrenean	Surface stratiformbed	
31	Yusef Abad	34	59	30	51	25	0	S	Clay and silt/Pleistocene	Evaporite	Post-Pyrenean	Surface stratiformbed	
32	Zarrin Abad	36	3	0	54	37	0	S	Clay and silt/Pleistocene	Evaporite	Post-Pyrenean	Surface stratiformbed	

Complete list of pumice deposits and indications of Iran along with their detailed specifications

No.	Name	Lat D°	Lat M	Lat S″	Long D°	Long M	Long S″	Size	Host rock/age	Genetic/ mineralization	Orogenic phases	Morphology	Elem para
1	Gazaneh	35	55	0	52	15	0	M	Pumice with trachyandesite composition/Neogene	Volcanic	Post-Pyrenean	Lens	
2	Karaf	35	59	0	52	13	0	L	Pumice/Quaternary with andesite composition	Volcanic	Post-Pyrenean	Lens	
3	Malar	35	58	40	52	14	14	M	Pumice with trachyte composition/Quaternary	Volcanic	Post-Pyrenean	Stratiformbed	
4	Rineh	35	54	0	52	6	0	S	Pumice with trachyandesite composition/Neogene	Volcanic	Post-Pyrenean	Lens and layered	
5	Esfandagheh	37	53	57	46	28	40	M	Ignimbrite/Neogene	Volcanic	Post-Pyrenean	Stratiformbed	
6	Iranagh	37	54	51	46	34	8	M	Ignimbrite/Neogene	Volcanic	Post-Pyrenean	Stratiformbed	
7	Avaj	38	5	0	44	40	0	M	Basalt and andesite/Quaternary	Volcanic	Post-Pyrenean	Stratiformbed	
8	Sar Eine	38	9	0	48	3	0	M	Latite basalt and andesite/Plio-Quaternary	Volcanic	Post-Pyrenean	Stratiformbed	
9	Velheriz	38	15	20	47	35	45	L	Pumice with trachyandesite composition/Neogene	Volcanic	Post-Pyrenean	Lens	

Complete list of perlite deposits and indications of Iran along with their detailed specifications

		Lat			Long								
No.	Name	D°	M	S″	D°	M	S″	Size	Host rock/age	Genetic/ mineralization	Orogenic phases	Morphology	Elem para
1	Abak	37	21	51	47	26	58	M	Tuff/Eocene	Volcanic	Pyrenean	Stratiformbed	
2	Ajami	37	27	51	47	26	58	M	Tuff/Eocene	Volcanic	Pyrenean	Stratiformbed	
3	Ashlagh Chay	37	26	47	47	36	2	M	Tuff/Eocene	Volcanic	Pyrenean	Stratiformbed	
4	Sefid Khaneh	37	41	47	47	52	15	L	Tuff/Miocene	Volcanic	Post-Pyrenean	Stratiformbed	
5	Shirin Bolagh	37	27	51	47	38	3	M	Tuff/Eocene	Volcanic	Pyrenean	Stratiformbed	

Complete list of ocher deposits and indications of Iran along with their detailed specifications

		Lat			Long								
No.	Name	D°	M	S″	D°	M	S″	Size	Host rock/age	Genetic/mineralization	Orogenic phases	Morphology	Elem para
1	Hormoz	27	5	0	56	27	0	L	Sedimentary and volcanic rock/Upper Precambrian–Lower Cambrian	Surface Alteration	Pan-African	Lens	
2	Nadoushan	32	2	0	53	34	0	S	Clay, silite sandstone and shale bearing iron/Lower Triassic	Surface Alteration	Early Cimmerian	Lens	

Complete list of halite mineral deposits and indications of Iran along with their detailed specifications

No.	Name	Lat D°	Lat M	Lat S″	Long D°	Long M	Long S″	Size	Host rock/age	Genetic/ mineralization	Orogenic phases	Morphology	Elem para
1	AbouMousa	25	53	15	55	1	12	S	Gypsum, acidic to intermediate lava/Upper Precambrian–Lower Cambrian	Evaporite	Pan-African	Stratiformbed	
2	Ammarlu	36	26	11	58	30	40	S	Salt, gypsum, and sulfurous marl/Miocene; tuff and pyroclastics/ Eocene–Oligocene	Evaporite	Post-Pyrenean	Stratiformbed	
3	Angouran	27	16	9	55	17	42	S	Gypsum, acidic to intermediate lava/Upper Precambrian–Lower Cambrian	Evaporite	Pan-African	Stratiformbed	
4	Chupanlu (Duzlakh)							I	Marl and shale/Miocene	Evaporite	Post-Pyrenean	Stratiformbed	
5	Farour	26	17	44	54	30	0	S	Gypsum, acidic to intermediate lava/Upper Precambrian–Lower Cambrian	Evaporite	Pan-African	Stratiformbed	
6	Gachin	27	5	24	55	55	28	I	Gypsum, dolomite limestone, shale limestone, and rhyolite/Upper Precambrian–Lower Cambrian	Evaporite	Pan-African	Stratiformbed	
7	Gachin	27	7	32	55	55	0	S	Gypsum, acidic to intermediate lava/Upper Precambrian–Lower Cambrian	Evaporite	Pan-African	Stratiformbed	
8	Gavkhuni`s Swamp	32	7	5	52	51	42	I	Lake salt/Quaternary	Evaporite	Post-Pyrenean	Stratiformbed	
9	Ghalat (Kalat)	27	18	35	56	4	14	S	Gypsum, acidic to intermediate lava/Upper Precambrian–Lower Cambrian	Evaporite	Pan-African	Stratiformbed	

10	Hormoz	27	3	13	54	53	6	I	Gypsum, acidic to intermediate lava/Upper Precambrian–Lower Cambrian	Evaporite	Pan-African	Stratiformbed
11	Hoseyn Abad-e-Dehnamak	35	19	25	52	37	28	S	Marl and shale/Neogene	Evaporite	Post-Pyrenean	Stratiformbed
12	Howz-e-Soltan	35	0	0	51	0	0	I	Lake salt/Quaternary	Evaporite	Post-Pyrenean	Stratiformbed
13	Khoy`s Playa Lacks	33	50	11	55	30	20	I	Lake salt/Quaternary	Evaporite	Post-Pyrenean	Stratiformbed
14	Kuh-e-Namak-e-Qom							S	Marl, gypsum, and halite/Oligo-Miocene	Evaporite	Post-Pyrenean	Stratiformbed
15	Lark	26	51	26	56	21	36	S	Gypsum, acidic to intermediate lava/Upper Precambrian–Lower Cambrian	Evaporite	Pan-African	Stratiformbed
16	Melheh							S	Marl, gypsum, and halite/Neogene	Evaporite	Post-Pyrenean	Stratiformbed
17	Nahand	38	16	4	46	28	38	I	Marl and shale/Miocene	Evaporite	Post-Pyrenean	Stratiformbed
18	Qeshm	26	37	4	55	29	36	S	Gypsum, acidic to intermediate lava/Upper Precambrian–Lower Cambrian	Evaporite	Pan-African	Stratiformbed
19	Rah Rahak	35	18	55	52	18	16	S	Gypsum and salt/Eocene	Evaporite	Pyrenean	Stratiformbed
20	Siahoo	27	46	20	56	17	27	S	Gypsum, acidic to intermediate lava/Upper Precambrian–Lower Cambrian	Evaporite	Pan-African	Stratiformbed
21	Sialak	35	22	10	52	14	1	S	Marl/Miocene	Evaporite	Post-Pyrenean	Stratiformbed

Complete list of feldspar and feldspathoid mineral deposits and indications of Iran along with their detailed specifications

No.	Name	Lat D°	Lat M	Lat S″	Long D°	Long M	Long S″	Size	Host rock/age	Genetic/ mineralization	Orogenic phases	Morphology	Elem para
1	Agh Bolagh	36	48	0	47	14	0	S	Granite/Upper Precambrian	Pegmatitic	Pan-African	Vein	
2	Anarak	32	0	24	55	40	19	I	Granite/Precambrian	Pegmatitic	Pan-African	Unknown	
3	Bamrud-e-Qayen	33	41	0	60	16	0	S	Granite/Tertiary	Pegmatitic	Pyrenean	Lens and vein	
4	Bazman Darkiaban								Alkali granite/Post-Triassic	Pegmatitic	Mid-Cimmerian	Massive	
5	Bid Akhvid	31	37	0	53	55	0	S	Granite/Upper Jurassic	Pegmatitic	Late Cimmerian	Vein	
6	Bozgush	37	47	19	47	29	19	S	Nepheline syenite/Eocene-Oligocene	Pegmatitic	Pyrenean	Massive	
7	Chador Malou	32	18	38	55	41	45	I	Granite/Precambrian	Pegmatitic	Pan-African	Unknown	
8	Choghaie	34	11	0	48	38	0	S	Granodiorite/Upper Cretaceous	Pegmatitic	Laramide	Vein	
9	Damak-e-Ali Abad							S	Granite/Upper Jurassic	Pegmatitic	Late Cimmerian	Vein	
10	Dehgan	33	55	40	48	47	15	S	Granite and pegmatic/ Jurassic–Cretaceous	Pegmatitic	Laramide	Vein	
11	Dehnow	34	31	0	48	37	0	S	Granodiorite/Upper Cretaceous	Pegmatitic	Laramide	Vein	
12	Dowran	36	42	0	48	23	0	S	Granite/Upper Precambrian	Pegmatitic	Pan-African	Massive	
13	Duzakh Darreh	32	32	4	55	42	44	I	Granite/Precambrian	Pegmatitic	Pan-African	Unknown	
14	Emarat	36	13	38	52	27	56	I	Unknown	Unknown	Unknown	Unknown	
15	Ganjnameh	34	41	0	48	11	0	S	Granodiorite/Upper Cretaceous	Pegmatitic	Laramide	Vein	
16	Gatehdeh (Taleghan)	36	10	0	51	15	0	S	Rhyolite and tuff/Eocene	Volcanic	Pyrenean	Vein	
17	Gazdun	31	33	23	53	40	2	S	Granite/Upper Jurassic	Pegmatitic	Late Cimmerian	Vein	
18	Gharahdash	37	14	0	47	12	0	S	Rhyolite/Upper Precambrian	Pegmatitic	Pan-African	Vein	
19	Gheshlagh	36	2	0	59	36	0	S	Granite and pegmatic/ Paleozoic–Triassic	Pegmatitic	Early Cimmerian	Stratiformbed	
20	Ghorgho Shomalou	36	47	27	47	22	7	I	Granite/Precambrian	Pegmatitic	Pan-African	Vein	
21	Haji Beig	33	54	11	48	50	31	S	Granite and pegmatic/ Jurassic–Cretaceous	Pegmatitic	Laramide	Vein	
22	Jahan Shir	32	48	42	55	43	18	I	Granite/Precambrian	Pegmatitic	Pan-African	Unknown	

23	Jandagh	33	58	39	54	36	47	I	Granite/Mesozoic	Pegmatitic	Unknown	Unknown	
24	Kaboutar Lan	33	51	45	48	56	43	S	Granodiorite/Jurassic–Cretaceous	Pegmatitic	Laramide	Vein	
25	Kaleibar	38	48	0	46	59	18	I	Nepheline syenite/Eocene–Oligocene	Pegmatitic	Pyrenean	Massive	
26	Khaf	35	2	0	59	32	0	S	Granite/Tertiary	Pegmatitic	Pyrenean	Vein	Mica
27	Khalil Abad							S	Granite/Upper Jurassic	Pegmatitic	Late Cimmerian	Vein	
28	Khoshkehdar							S	Granite and pegmatic/Upper Cretaceous	Pegmatitic	Laramide	Vein	
29	Marzian	33	34	28	49	26	52	S	Granite and pegmatic/ Jurassic–Cretaceous	Pegmatitic	Laramide	Vein	
30	Masuleh	37	21	33	49	5	25	S	Diorite and gabbro/Triassic	Magmatic	Early Cimmerian	Vein	Mica
31	Mir Akhor	36	37	0	47	45	0	S	Granite/Upper Precambrian	Surface Alteration	Pan-African	Massive	
32	Moghanlu	36	42	0	47	52	0	S	Granite/Upper Precambrian	Magmatic	Pan-African	Massive	
33	Morassaee	36	32	0	47	54	0	S	Granite and rhyolite/Upper Precambrian	Magmatic	Pan-African	Massive	
34	Northeast of Bafgh	31	57	30	55	30	0	S	Granite/Upper Precambrian	Pegmatitic	Pan-African	Vein	
35	Qezelcheh	36	47	0	47	14	0	S	Granite/Upper Precambrian	Magmatic	Pan-African	Vein	
36	Razgah	37	47	47	47	26	15	S	Nepheline syenite/Eocene–Oligocene	Magmatic	Pyrenean	Massive	
37	Robat-e-Posht-e-Badam	32	52	30	55	7	0	S	Granite/Upper Precambrian	Pegmatitic	Pan-African	Vein	
38	Sabalan	38	15	0	47	44	34	I	Rhyolite/Tertiary	Volcanic	Post-Pyrenean	Unknown	
39	Sangan(Dorduy)	34	34	0	60	27	0	S	Granite/Upper Eocene	Pegmatitic	Pyrenean	Vein	
40	Southwest of Mehriz	31	27	15	54	11	20	I	Granite/Mesozoic	Unknown	Unknown	Unknown	
41	Valilou	38	19	5	46	51	25	I	Rhyolite/Tertiary	Volcanic	Pyrenean	Unknown	
42	Varzaghan	38	13	38	47	6	51	I	Rhyolite/Tertiary	Volcanic	Pyrenean	Unknown	
43	Zaman Abad	32	55	50	55	42	9	I	Granite/Precambrian	Unknown	Pan-African	Unknown	
44	Zaman Abad	34	38	0	48	41	0	S	Granodiorite/Upper Cretaceous	Pegmatitic	Laramide	Vein	Mica

Complete list of coal deposits and indications of Iran along with their detailed specifications

		Lat			Long								
No.	Name	D°	M	S″	D°	M	S″	Size	Host rock/age	Genetic/ mineralization	Orogenic phases	Morphology	Elem para
1	Fath Abad	29	47	51	57	6	55	I	Shale and sandstone/Jurassic	Sedimentary	Early–Middle Cimmerian	Stratiformbed	
2	Badamoo	30	28	18	56	36	18	I	Shale and sandstone/Jurassic	Sedimentary	Early–Middle Cimmerian	Stratiformbed	
3	Darreh Gaz	30	53	55	56	50	31	I	Shale and sandstone/Jurassic	Sedimentary	Early Cimmerian	Stratiformbed	
4	Neyestan	30	55	13	57	9	2	I	Shale and sandstone/Jurassic	Sedimentary	Early–Middle Cimmerian	Stratiformbed	
5	Parvar Deh	30	41	35	56	58	21	I	Shale and sandstone/Jurassic	Sedimentary	Early Cimmerian	Stratiformbed	
6	Pabdana	31	12	48	56	20	31	S	Shale and sandstone/Jurassic	Sedimentary	Early–Middle Cimmerian	Stratiformbed	
7	Robat-e-Kham	31	15	40	56	19	43	I	Shale and sandstone/Jurassic	Sedimentary	Early–Middle Cimmerian	Stratiformbed	
8	Nayband 1	32	22	0	57	10	0	S	Limestone, shale, and sandstone/Upper Triassic	Sedimentary	Early Cimmerian	Stratiformbed	
9	Nayband 2	32	34	33	57	23	47	I	Limestone, shale, and sandstone/Upper Triassic	Sedimentary	Early Cimmerian	Stratiformbed	
10	Nayband Area1	32	15	0	57	11	58	I	Limestone, shale, and sandstone/Upper Triassic	Sedimentary	Early Cimmerian	Stratiformbed	
11	Nayband Area2	32	10	0	57	19	8	I	Limestone, shale, and sandstone/Upper Triassic	Sedimentary	Early Cimmerian	Stratiformbed	
12	Shekasteh Abshar	32	20	49	57	26	10	I	Limestone, shale, and sandstone/Upper Triassic	Sedimentary	Early Cimmerian	Stratiformbed	
13	Borj Kachal	33	23	5	56	11	36	I	Limestone, shale, and sandstone/Upper Triassic	Sedimentary	Early Cimmerian	Stratiformbed	
14	East of Kuh-e-Jamal	33	21	39	57	19	21	I	Limestone, shale, and sandstone/Upper Triassic	Sedimentary	Early Cimmerian	Stratiformbed	
15	Faskhoud	33	16	29	51	56	34	I	Limestone, shale, and sandstone/Upper Triassic	Sedimentary	Early Cimmerian	Stratiformbed	
16	Hashashegh	33	29	30	51	34	4	I	Limestone, shale, and sandstone/Upper Triassic	Sedimentary	Early–Middle Cimmerian	Stratiformbed	
17	Kal-e-Shur	33	5	32	56	1	56	I	Limestone, shale, and sandstone/Upper Triassic	Sedimentary	Early Cimmerian	Stratiformbed	

18	Mazinow	33	10	0	56	13	5	I	Limestone, shale, and sandstone/Upper Triassic	Sedimentary	Early–Middle Cimmerian	Stratiformbed
19	Parvardeh	33	40	0	57	15	0	S	Limestone, shale, and sandstone/Triassic–Jurassic	Sedimentary	Early Cimmerian	Stratiformbed
20	Tarak	33	58	37	56	32	36	I	Limestone, shale, and sandstone/Upper Triassic	Sedimentary	Early–Middle Cimmerian	Stratiformbed
21	Torogh	33	19	27	51	49	33	I	Limestone, shale, and sandstone/Upper Triassic	Sedimentary	Early–Middle Cimmerian	Stratiformbed
22	Nesa Bala	34	4	56	51	20	39	I	Limestone, shale, and sandstone/Upper Triassic	Sedimentary	Early Cimmerian	Stratiformbed
23	Alborz	35	59	19	51	25	28	S	Limestone, shale, and sandstone/Upper Triassic	Sedimentary	Early–Middle Cimmerian	Stratiformbed
24	Cheshmeh Gol	35	29	30	60	44	20	I	Limestone, shale, and sandstone/Upper Triassic	Sedimentary	Early Cimmerian	Stratiformbed
25	Gol Banou	35	20	45	59	57	30	I	Limestone, shale, and sandstone/Upper Triassic	Sedimentary	Early Cimmerian	Stratiformbed
26	Jam	35	40	54	54	5	17	I	Limestone, shale, and sandstone/Upper Triassic	Sedimentary	Early–Middle Cimmerian	Stratiformbed
27	Kalut-e-Sefid	35	16	0	60	20	5	I	Limestone, shale, and sandstone/Upper Triassic	Sedimentary	Early Cimmerian	Stratiformbed
28	Kuh-e-Cheshmeh Rodangi	35	15	20	60	18	5	I	Limestone, shale, and sandstone/Upper Triassic	Sedimentary	Early Cimmerian	Stratiformbed
29	Shemshak2	35	57	16	51	41	55	S	Limestone, shale, and sandstone/Upper Triassic	Sedimentary	Early–Middle Cimmerian	Stratiformbed
30	Siankuhi Agh Darband	35	59	0	60	56	0	S	Limestone, shale, and sandstone/Upper Triassic	Sedimentary	Early Cimmerian	Stratiformbed
31	Abyek	36	19	31	50	8	12	I	Limestone, shale, and sandstone/Upper Triassic	Sedimentary	Early Cimmerian	Stratiformbed
32	Abyek To Ziaran1	36	3	48	50	36	15	I	Limestone, shale, and sandstone/Upper Triassic	Sedimentary	Early–Middle Cimmerian	Stratiformbed
33	Abyek To Ziaran2	36	5	39	50	36	55	I	Limestone, shale, and sandstone/Upper Triassic	Sedimentary	Early–Middle Cimmerian	Stratiformbed

(continued)

(continued)

No.	Name	Lat D°	Lat M	Lat S″	Long D°	Long M	Long S″	Size	Host rock/age	Genetic/ mineralization	Orogenic phases	Morphology	Elem para
34	Alnghayeh	36	9	51	49	35	49	I	Limestone, shale, and sandstone/Upper Triassic	Sedimentary	Early–Middle Cimmerian	Stratiformbed	
35	Asbavin	36	40	8	49	39	3	I	Limestone, shale, and sandstone/Upper Triassic	Sedimentary	Early–Middle Cimmerian	Stratiformbed	
36	Baba Hafez	36	14	30	54	6	27	I	Limestone, shale, and sandstone/Upper Triassic	Sedimentary	Early–Middle Cimmerian	Stratiformbed	
37	Bayandor	36	30	12	48	27	38	I	Limestone, shale, and sandstone/Upper Triassic	Sedimentary	Early–Middle Cimmerian	Stratiformbed	
38	Binzheh	36	48	12	49	14	39	I	Limestone, shale, and sandstone/Upper Triassic	Sedimentary	Early–Middle Cimmerian	Stratiformbed	
39	Chal	36	10	0	51	51	0	S	Limestone, shale, and sandstone/Upper Triassic	Sedimentary	Early–Middle Cimmerian	Stratiformbed	
40	Damghan	36	4	5	53	48	29	I	Limestone, shale, and sandstone/Upper Triassic	Sedimentary	Early–Middle Cimmerian	Stratiformbed	
41	Derakht-e-Toot	36	8	10	59	23	50	I	Limestone, shale, and sandstone/Upper Triassic	Sedimentary	Early–Middle Cimmerian	Stratiformbed	
42	East of Abrendan	36	24	40	54	5	28	I	Limestone, shale, and sandstone/Upper Triassic	Sedimentary	Early–Middle Cimmerian	Stratiformbed	
43	Elika	36	5	27	51	18	54	S	Limestone, shale, and sandstone/Upper Triassic	Sedimentary	Early–Middle Cimmerian	Stratiformbed	
44	Feshneh	36	1	44	50	46	24	I	Limestone, shale, and sandstone/Upper Triassic	Sedimentary	Early–Middle Cimmerian	Stratiformbed	
45	Ghatar	36	57	58	46	30	36	I	Limestone, shale, and sandstone/Upper Triassic	Sedimentary	Early–Middle Cimmerian	Stratiformbed	
46	Ghatar Paien Taraz	36	55	8	46	32	15	I	Limestone, shale, and sandstone/Upper Triassic	Sedimentary	Early–Middle Cimmerian	Stratiformbed	
47	Gheshlagh	36	59	18	55	15	0	S	Limestone, shale, and sandstone/Upper Triassic	Sedimentary	Early–Middle Cimmerian	Stratiformbed	
48	Goland rud	36	24	0	51	54	0	I	Limestone, shale, and sandstone/Upper Triassic	Sedimentary	Early–Middle Cimmerian	Stratiformbed	

49	Jozchal	36	51	4	55	14	54	I	Limestone, shale, and sandstone/Upper Triassic	Sedimentary	Early–Middle Cimmerian	Stratiformbed
50	Kolariz	36	18	41	54	23	23	I	Limestone, shale, and sandstone/Upper Triassic	Sedimentary	Early–Middle Cimmerian	Stratiformbed
51	Kalavi	36	19	0	54	40	0	S	Limestone, shale, and sandstone/Upper Triassic	Sedimentary	Early–Middle Cimmerian	Stratiformbed
52	Kaleshtar1	36	49	20	49	24	38	I	Limestone, shale, and sandstone/Upper Triassic	Sedimentary	Early–Middle Cimmerian	Stratiformbed
53	Kaleshtar2	36	48	34	49	25	26	I	Limestone, shale, and sandstone/Upper Triassic	Sedimentary	Early–Middle Cimmerian	Stratiformbed
54	Kar Sang	36	16	0	52	17	0	S	Limestone, shale, and sandstone/Upper Triassic	Sedimentary	Early–Middle Cimmerian	Stratiformbed
55	Kharsang	36	21	49	54	8	13	I	Limestone, shale, and sandstone/Upper Triassic	Sedimentary	Early–Middle Cimmerian	Stratiformbed
56	Kuh-e-Farbas	36	22	30	54	44	0	I	Limestone, shale, and sandstone/Upper Triassic	Sedimentary	Early–Middle Cimmerian	Stratiformbed
57	Kuh-e-Kahkeshan	36	45	7	54	6	0	I	Limestone, shale, and sandstone/Upper Triassic	Sedimentary	Early–Middle Cimmerian	Stratiformbed
58	Kuh-e-Khandaghi	36	30	15	48	30	12	I	Limestone, shale, and sandstone/Upper Triassic	Sedimentary	Early–Middle Cimmerian	Stratiformbed
59	Kuh-e-Labansar	36	24	39	54	22	39	I	Limestone, shale, and sandstone/Upper Triassic	Sedimentary	Early–Middle Cimmerian	Stratiformbed
60	Kuh-e-Petal	36	23	6	54	45	40	I	Limestone, shale, and sandstone/Upper Triassic	Sedimentary	Early–Middle Cimmerian	Stratiformbed
61	Kuh-e-Sangi	36	21	12	54	14	9	I	Limestone, shale, and sandstone/Upper Triassic	Sedimentary	Early–Middle Cimmerian	Stratiformbed
62	Kuh-e-Tamuz	36	23	38	54	19	39	I	Limestone, shale, and sandstone/Upper Triassic	Sedimentary	Early–Middle Cimmerian	Stratiformbed
63	Kuh-e-Tasher	36	28	38	51	48	9	I	Limestone, shale, and sandstone/Upper Triassic	Sedimentary	Early–Middle Cimmerian	Stratiformbed

(continued)

(continued)

No.	Name	Lat D°	Lat M	Lat S″	Long D°	Long M	Long S″	Size	Host rock/age	Genetic/ mineralization	Orogenic phases	Morphology	Elem para
64	Kuh-e-Varjeh1	36	19	8	54	25	23	I	Limestone, shale, and sandstone/Upper Triassic	Sedimentary	Early–Middle Cimmerian	Stratiformbed	
65	Kuh-e-Varjeh2	36	27	21	54	25	4	I	Limestone, shale, and sandstone/Upper Triassic	Sedimentary	Early–Middle Cimmerian	Stratiformbed	
66	North of Kuh-e-Cheshmeh Ali	36	21	30	54	9	36	I	Limestone, shale, and sandstone/Upper Triassic	Sedimentary	Early–Middle Cimmerian	Stratiformbed	
67	North of Zarband	36	9	44	52	21	9	I	Limestone, shale, and sandstone/Upper Triassic	Sedimentary	Early–Middle Cimmerian	Stratiformbed	
68	Pishsar	36	14	34	54	4	27	I	Limestone, shale, and sandstone/Upper Triassic	Sedimentary	Early–Middle Cimmerian	Stratiformbed	
69	Pymod	36	26	29	51	53	39	I	Limestone, shale, and sandstone/Upper Triassic	Sedimentary	Early–Middle Cimmerian	Stratiformbed	
70	Sang Rud	36	36	0	49	49	0	S	Limestone, shale, and sandstone/Upper Triassic	Sedimentary	Early–Middle Cimmerian	Stratiformbed	
71	Sefid Darak1	36	11	1	50	33	48	I	Limestone, shale, and sandstone/Upper Triassic	Sedimentary	Early–Middle Cimmerian	Stratiformbed	
72	Sefid Darak2	36	3	5	50	33	48	I	Limestone, shale, and sandstone/Upper Triassic	Sedimentary	Early–Middle Cimmerian	Stratiformbed	
73	Sefid Darak3	36	10	16	50	34	18	I	Limestone, shale, and sandstone/Upper Triassic	Sedimentary	Early–Middle Cimmerian	Stratiformbed	
74	Sefid Kuh1	36	15	48	54	5	49	I	Limestone, shale, and sandstone/Upper Triassic	Sedimentary	Early–Middle Cimmerian	Stratiformbed	
75	Sefid Kuh2	36	16	51	54	9	9	I	Limestone, shale, and sandstone/Upper Triassic	Sedimentary	Early–Middle Cimmerian	Stratiformbed	
76	Seyed	36	4	5	59	33	41	I	Limestone, shale, and sandstone/Upper Triassic	Sedimentary	Early–Middle Cimmerian	Stratiformbed	
77	Shemshak1	36	0	32	51	30	30	I	Limestone, shale, and sandstone/Upper Triassic	Sedimentary	Early–Middle Cimmerian	Stratiformbed	
78	South of Deh Dushab	36	28	4	50	11	39	I	Limestone, shale, and sandstone/Upper Triassic	Sedimentary	Early–Middle Cimmerian	Stratiformbed	

79	South of Nesa Bala	36	3	38	51	22	30	I	Limestone, shale, and sandstone/Upper Triassic	Sedimentary	Early–Middle Cimmerian	Stratiformbed
80	Takht	36	38	37	54	56	39	I	Limestone, shale, and sandstone/Upper Triassic	Sedimentary	Early–Middle Cimmerian	Stratiformbed
81	Tarz	36	23	26	54	48	19	S	Limestone, shale, and sandstone/Upper Triassic	Sedimentary	Early–Middle Cimmerian	Stratiformbed
82	Tash	36	34	0	54	40	43	S	Limestone, shale, and sandstone/Upper Triassic	Sedimentary	Early–Middle Cimmerian	Stratiformbed
83	Tazareh	36	25	8	54	23	15	S	Limestone, shale, and sandstone/Upper Triassic	Sedimentary	Early–Middle Cimmerian	Stratiformbed
84	Vaz Paien	36	22	15	52	9	53	S	Limestone, shale, and sandstone/Upper Triassic	Sedimentary	Early–Middle Cimmerian	Stratiformbed
85	Yakhchal	36	30	40	49	31	21	I	Limestone, shale, and sandstone/Upper Triassic	Sedimentary	Early–Middle Cimmerian	Stratiformbed
86	Zirab	36	0	0	53	3	32	S	Limestone, shale, and sandstone/Upper Triassic	Sedimentary	Early–Middle Cimmerian	Stratiformbed
87	Zoghan	36	23	30	54	20	6	I	Limestone, shale, and sandstone/Upper Triassic	Sedimentary	Early–Middle Cimmerian	Stratiformbed
88	Chevan Chai	37	27	25	46	10	30	I	Limestone, shale, and sandstone/Upper Triassic	Sedimentary	Early–Middle Cimmerian	Stratiformbed
89	Chobal	37	3	41	49	15	0	I	Limestone, shale, and sandstone/Upper Triassic	Sedimentary	Early–Middle Cimmerian	Stratiformbed
90	Gavmishlar Daghi	37	14	11	46	15	40	I	Limestone, shale, and sandstone/Upper Triassic	Sedimentary	Early–Middle Cimmerian	Stratiformbed
91	Gheimas Khan	37	27	34	46	5	25	I	Limestone, shale, and sandstone/Upper Triassic	Sedimentary	Early–Middle Cimmerian	Stratiformbed
92	Govareh Paien	37	20	18	46	17	32	I	Limestone, shale, and sandstone/Upper Triassic	Sedimentary	Early–Middle Cimmerian	Stratiformbed
93	Kuh-e-Goy Dagh	37	16	45	46	17	18	I	Limestone, shale, and sandstone/Upper Triassic	Sedimentary	Early–Middle Cimmerian	Stratiformbed
94	Kuh-e-Goy Posht1	37	27	44	46	10	33	I	Limestone, shale, and sandstone/Upper Triassic	Sedimentary	Early–Middle Cimmerian	Stratiformbed

(continued)

(continued)

No.	Name	Lat			Long			Size	Host rock/age	Genetic/mineralization	Orogenic phases	Morphology	Elem para
		D°	M	S″	D°	M	S″						
95	Kuh-e-Goy Posht2	37	29	11	46	10	28	I	Limestone, shale, and sandstone/Upper Triassic	Sedimentary	Early–Middle Cimmerian	Stratiformbed	
96	Lali Daghi	37	15	40	46	17	22	I	Limestone, shale, and sandstone/Upper Triassic	Sedimentary	Early–Middle Cimmerian	Stratiformbed	
97	Orojeh1	37	13	35	46	21	46	I	Limestone, shale, and sandstone/Upper Triassic	Sedimentary	Early–Middle Cimmerian	Stratiformbed	
98	Orojeh2	37	7	11	46	27	31	I	Limestone, shale, and sandstone/Upper Triassic	Sedimentary	Early–Middle Cimmerian	Stratiformbed	
99	Taft	37	8	14	46	26	35	I	Limestone, shale, and sandstone/Upper Triassic	Sedimentary	Early–Middle Cimmerian	Stratiformbed	

Complete list of borax mineral deposits and indications of Iran along with their detailed specifications

No.	Name	Lat			Long			Size	Host rock/age	Genetic/ mineralization	Orogenic phases	Morphology	Elem para
		D°	M	S″	D°	M	S″						
1	Ghareh Aghaj	36	56	20	47	29	20	I	Marl/Upper–Mid-Miocene	Evaporation	Post-Pyrenean	Vein	
2	Ghareh Gol	36	34	15	47	50	0	S	Conglomerate, clay, and sandstone/Pliocene	Evaporation	Post-Pyrenean	Vein	
3	Mianj	36	55	45	47	27	40	I	Conglomerate, clay, and sandstone/Pliocene	Evaporation	Post-Pyrenean	Vein	

Complete list of alunite mineral deposits and indications of Iran along with their detailed specifications

No.	Name	Lat D°	Lat M	Lat S″	Long D°	Long M	Long S″	Size	Host rock/age	Genetic/ mineralization	Orogenic phases	Morphology	Elem para
1	Hasan Abad	36	36	30	49	12	0	S	Tuff/Mid-Eocene	Hydrothermal	Early Pyrenean	Stratiformbed	
2	Sirdan	36	39	0	49	11	0	I	Tuff/Paleogene	Hydrothermal	Early Pyrenean	Lens	
3	Youzbashi Chay	36	23	0	49	26	0	I	Tuff/Middle Eocene	Hydrothermal	Early Pyrenean	Massive	
4	Zaj Kandi	36	57	0	48	22	0	I	Sandy tuff/Eocene	Hydrothermal	Early Pyrenean	Massive	
5	Zajkan	36	22	0	49	25	0	S	Tuff/Middle Eocene	Hydrothermal	Early Pyrenean	Lens	
6	Zaylik-Ghalan Dar	38	34	30	47	6	30	I	Tuff/Mid–Upper Eocene	Hydrothermal	Early Pyrenean	Massive	

References

Aghanabati A (1998) Jurassic stratigraphy of Iran, vol 1. Geological Survey of Iran, Tehran

Ahmad ebn-e Omar ebn-e Rasteh, Al-aalagh Al-nafiseh translated by Hossein Qare-Chanlou (1986)

Ahmadi-khalaji A (2007) Study of petrology and petrofabric of intrusive bodies and contact metamorphism in Boroujerd area. PhD thesis, Tehran University, Tehran

Aistov L, Melnikov B, Krivyakin B, Morozov L (1984) Geology of the Khur area (Central Iran), Explanatory text of the Khur quadrangle map 1:250,000. Geological Survey of Iran, Tehran

Alavi Naeini et al (1976) Geological map of Takab, with the scale of 1:250,000. Geological Survey of Iran

Alavi NM (1979a) The Virian ophiolite complex and surrounding rocks, Geological Randsch, Dtsch 68(1):334–341, ISSN 0016–7835

Alavi NM (1991) Sedimentary and structural characteristics of the Paleo–Tethys remnants in North Eastern Iran. Bull Geol Soc Am 103:983–992

Alavi NM (1994) Tectonic of the Zagros, orogenic belt of Iran, new data and interpretation. Tectonophysics 299:211–238

Al-e Taha B (1996) 'Sanat-e Tahiye-ye Tootia dar Iran-e Bastan' ('Tutia' Industry of Ancient Iran), Faslname-ye Kerman (In Farsi)

Alipour K (1993) History of earth sciences and mining in Iran. Geological Survey of Iran, Tehran (In Farsi)

Alipour-asl (1998) Evaluation of Titanium in illmentie-apatite-bearing ultramafic-mafic intrusive bodies in north of Ghara-ajagh. MSc thesis, Shahid Beheshti University, Tehran

Almasian M (1998) Tectonic of Anarak Area (Central Iran). PhD thesis, Islamic Azad University, 164 p

Ataloo (1998) Study of Anjerd copper deposit, M.Sc. Thesis, Tarbiyat Moaalem University

Bazin D, Hübner H (1969) Copper deposits in Iran. Geological Survey of Iran, report no. 13, 225 p

Berberian M (1984) Structural evolution and tectonics of the Iranian plateau, a plate tectonic approach, Bogdanov, N.A., Tesizy, 27-Ymezhdunarodnyy geologicheskiy congress, Abstracts. In: 27th international Geological Congress, vol 27, no 3, 140 p. Geological Survey Iran, Tehran

Berberian M, King GCP (1981) Towards a paleogeography and tectonic evolution of Iran. Can J Earth Sci 18(2):210–265

Berthier F, Billiault JP, Halbronn B, Maurizot P (1974) Etude stratigraphique, pétrologique et structurale de la region de Khorramabad (Zagros, Iran), S.L., thesis Doct. 3e Cycle, Spec. Geol. Appl., Unvi. Sci. Med., pp 1–307

Berthoud T (1979) Etude par l'analyse de traces et la filiation enter mineral de cuivre et object archeologique du Moyen- Orient (iveme et Illeme millenaire avant notre ere), Fra., V 30, 14 p, Thesis, Sci. Phys., Paris 6/1976

M. Ghorbani, *The Economic Geology of Iran: Mineral Deposits and Natural Resources*, Springer Geology, DOI 10.1007/978-94-007-5625-0,

Bolourchi, Halalat (1995) On Choghart area; Russian experts; Code 6, Archive 3 NISCO
Boulin J (1991) Hercynian and Eo Cimmerian events in Afghanistan and adjoining regions. Tectonophysics 148:253–258
bp Website (2012) Natural gas reserves, 2 July 2012, Retrieved from http://www.bp.com/sectiongenericarticle800.do?categoryId=9037178&contentId=7068624
Chardon J (1956) Expedition diary; translated by Abbasi M. Amir Kabir Publications, Tehran
Coleman H (1984) Iran and the Iranian Civilization (tr. to Persian by Hassan Anousheh)
Daliran F (1990) The magnetite-apatite deposit of Mishdovan. Unpublished PhD thesis, München University
Eftekhar Nezhad J (1975a) Brief history and structural development of Azarbaijan. Geological Survey of Iran, Internal report no.8, 9 p
Eftekhar Nezhad J (1975b) Brief description of tectonic history and structural development of Azarbaijan, Field excursion guide, no. 2. Note A Sym. Geodynamic of southeast Asia, Tehran, pp 469–478
Eftekharnejad J (1980) Tectonic classification of Iran based on the pattern of sedimentary basins. Bull Pet Soc 82 (In Farsi)
Etemadi M (1997) Study of petrography and geochemistry of ultramafic-mafic and metamorphic rocks in southwest of Bandar-abbas. MSc thesis, Shahid Beheshti University
Feyzi F (2005) Copper mineralization in Taknar area. PhD thesis, Islamic Azad University
Gabriel A (1934) Durch Persiens Wüsten. Strecker and Schroedr, Stoccarda, 390 p
Ghorbani M (1990) Exploration of mineral deposits in Abadeh area, Iran's Industrial minerals and rocks of Iran Co
Ghorbani M (1991a) Bentointe and its geologic properties in Iran. In: 10th earth sciences conference, GSI, Tehran
Ghorbani M (1991b) Observation of montmorillonite clay mineral group: properties, usages and resources in Iran. 1st mineralogy seminar of Iran. BoAli University, Hamedan
Ghorbani M (1991c) Prospecting and exploration of bentonite deposits of Iran and their geological issues. MSc thesis, Shahid Beheshti University
Ghorbani M (1993a) Kaolinite mineralization modeling in Iran. In: 2nd seminar of mineralogy and crystallography of Iran, Science and Industry University
Ghorbani M (1993b) Mineral distribution map of Iran, colleague, GSI Publication, Treatise on the Geology of Iran
Ghorbani M (1994a) Bentonite and Zeolite in Iran, GSI Publication, Treatise on the Geology of Iran, 147 p
Ghorbani M (1994b) An introduction to Takab metallogeny state. In: 3rd conference of Iran geology society
Ghorbani M (1994c) Iron metallogeny in Iran. Internal seminar of science Faculty of Tehran University
Ghorbani M (1994d) Silica in Iran. Geological Survey of Iran Publication, Treatise on the Geology of Iran, 127 p
Ghorbani M (1994e) Kaolinite and fireproof clay minerals in Iran. Geological Survey of Iran Publication, Treatise on the Geology of Iran, 171 p
Ghorbani M (1995a) Antimony, arsenic and mercury deposits of Iran. Geological Survey of Iran Publication, Treatise on the Geology of Iran, 132 p
Ghorbani M (1995b) Iron deposits in Iran. Geological Survey of Iran Publication, Treatise on the Geology of Iran, 213 p.
Ghorbani M (2001) Magmatism in Takab Region. Earth Sci Seas J, SBU
Ghorbani M (2005a) Pyroclastic rocks in Alborz and their relation with bentonite deposits. In: 1st conference of Alborz and Caspian Sea Marginal Regions Earth Sciences
Ghorbani M (2005b) Mining industry. In: Encyclopedia IRANICA
Ghorbani M (2005c) Geological setting of Iran within Alpine-Himalayan framework. Structure, Tectonics and Ore Mineralization Processes (STOMP), Townsville
Ghorbani M (2005d) Copper mineralization of Iran. In: Encyclopedia IRANICA
Ghorbani M (2005e) Classification of metallogenic and mining provinces, belt and zones of Iran. In: 6th international conferences of Middle East, UAE

Ghorbani M (2005f) Alborz Zone or Alborz geology state and its mineralization potential. In: 1st conference of Alborz and Caspian Sea Marginal Regions Earth Sciences

Ghorbani M (2005g) An introduction to geology and mineralization in Bafgh Area, Iran, STOMP, Structure, Tectonics and Ore Mineralization Processes, 2005, Townsville, Australia

Ghorbani M (2005i) History of mining activity in Iran. In: Encyclopedia IRANICA

Ghorbani M (2005j) Lead-Zinc mineralization of Iran. In: Encyclopedia IRANICA

Ghorbani M (2005k) Metallogenic and mineralization phases of Iran. In: 6th international conferences of Middle East, UAE

Ghorbani M, Mozaffarzadeh R, Iranmanesh M (2005) Geological zone or province of Alborz and its mineralization potentials, Proceedings of The First Conference of Geoscience on Alborz area and Caspian Sea, November 2005, Islamic Azad University, Lahijan, p 17–38

Ghorbani M (2006b) All you need to know about copper deposits in Iran. Sang va Maadan Q 1:12–26

Ghorbani M (2009a) The economic importance of Caspian Sea and its energy sources, Energy Newcom (7)

Ghorbani M (2009b) Dimension stones: rich reserves but poor exports, Stone-Minemagzine (IMH), pp 8–20

Ghorbani M (2009c) Antimony, arsenic and mercury as strategic metals in industries and key elements in exploration geology of epithermal gold deposits. Sang va Maadan Q 13:4–15

Ghorbani M (2009d) Barite: a strategic mineral for petroleum industry, Stone-Minemagzine, (IMH), pp 4–25

Ghorbani M (2010b) Aluminum and its sources, Stone-Minemagzine, (IMH), pp 4–24

Ghorbani M (2011) Magmatism and alteration, and their association with mineralization along Ahar – Jolfa (Arasbaran) axis. Research-Scientific Journal, Islamic Azad University, Science and Research Department

Ghorbani M, Iranmanesh M (2012) Titanium; a light element with magnificent application in industry. Sang va Maadan Q 20:22–31

Ghorbani M, Momenzadeh M (1994) Mineralization phases of Iran. In: 13th congregation of earth scientists. Geological Survey of Iran (In Farsi)

Ghorbani M, Momenzadeh M (1995) Explanation of mineralization phases in Iran. In: Proceedings of the 13th conference on GeoScience held in Geological Survery of Iran

Ghorbani M, Mostafavi F (1995) Study of iron deposits in northwest of Iran, Ajin Co

Ghorbani M, Ojaghi B (2003) Exploration report on Galai Baba-ali iron deposit, Sabanour Co

Ghorbani M, Tajbakhsh P (2010) Bentonite; its resources and geological setting in Iran. In: 20th general meeting of the international mineralogical association (IMA2010), Budapest

Ghorbani M, Tajbakhsh P, Khoi N (2000) Lead-zinc deposits in Iran. Geological Survey of Iran, 512 p

Ghorbani M, Mozafarzadeh R, Mousavi-pak N (2009) Chrome and chromite, strategic metal. Sang va Maadan Q 11

Girshman R (1934) Fouilles de Sialk, prés de Kashan, 1933, 1934, 1937. Librairie Orientaliste Paul Geuthner, Paris (in two volumes)

Girshman R (1954) Iran: from the earliest times to the Islamic conquest. Penguin Books, Harmondsworth

Haghipour A (1972) Hercynian movement in Iran. Int Geol Cong, Abstr Cong Geol Int., No. 24, 79 p

Haghipour A (1974) Etude géologique de la région de Biabanak-Bafgh (Iran central), Pétrologie et tectonique du socle precambrien et de sa couverture, France. PhD thesis, Sci Nat, Gronoble University, 403 p

Haghipour A (1977) Etude geologique de la région de Biabanak-Bafgh (Iran central), with coloured map. Geological Survey Iran, report no. 34

Haghipour A, Farooqui M, Amidi SM, Aghanabati A (1979) The destructive Tabas earthquake of September 16/1978, East Iran, A preliminary report. Proc Annu Eng Geol Soil Eng Symp 17:1–12

Hajali-lou B (2000) Metallogeny of tertiary in western Alborz-Azerbaijan (Miyaneh-Siyahroud) with specific view on Hashtjin area. PhD thesis, Shahid Beheshti University

Halalat H, Bolourchi MH (1994) Geology of Iran: phosphate. Geological Survey of Iran, Tehran, 360 p

Hamdi B (1989) Stratigraphy and paleontology of the Late Precambrian to Early Cambrian in the Alborz mountains, northern Iran, report no. 59
Hamdi B (1995) Precambrian and Cambrian sedimentary rocks of Iran. Treatise on the Geology of Iran, Ministry of Mines and Metals
Hedin S (1910) Zu land nach Indien, Bd. I, II. S.A. Brockhaus, Leipzig
Helmi F (2000) A depiction of halite and, potash, and their distribution in Iran. Geological Survey of Iran, Tehran
Hillier JA Mills AJ, White R (2000) World Mineral Statistics, (1994–1998). British Geological Survey, Keyworth, 300 p
Hosseini B (2011) Petrology and petrogenesis of Paleozoic metamorphic rocks in Sanadaj-Sirjan zone and their relation to mineralization of Iron. MSc thesis, Shahid Beheshti University
Houshmand-Zadeh A (1997) Magmatism and metamorphism of Iran, Course materials taught at Shahid Beheshti University
Houshmand-Zadeh A, Berberian M (1972) Implication of plate tectonics for the Sanandaj-Sirjan zone (Northeastern border of Zagros). Geological Survey of Iran, internal report, 6 p
Houshmand-Zadeh A, Posht-kouhi M (1996) Metamorphic map of Iran 1:1,000,000, Treatise on the Geology of Iran
Houshmand-Zadeh A, Alavi-naini M, Haghipour A (1978) The evolution of geologic phenomenon in Toroud area (from Precambrian to present day), report H5. Geological Survey of Iran
Index mundi (2012) World crude oil consumption by year, 1 July 2012. Retrieved from http://www.indexmundi.com/energy.aspx
Innocenti F, Manetti P, Mazzuoli R, Pasquare G, Villari L, Thorpe R (eds) (1982) Anatolia and North-Western Iran. Andesites and Orogenic Andesites and Related Rocks, GBR. Wiley, Chichester, pp 327–349, 73 Ref, ISBN
Kassiani-Avval J (2004) Dimension stones atlas of Iran. Ministry of Mines and Metals
Keary travel diary, Giambi Keary, Giovanni Francesco; translated by Nakhjavani A, Golrang A. East Azerbaijan Bureau of Culture and Arts, Franklin publication, 1969
LeFond SJ (ed) (1975) Industrial rocks and minerals/Bauxite raw materials, vol 1, 5th edn. SME-AIME, New York
Lindenberg H, Gorler K, Ibbeken G (1983) Stratigraphy, structure and orogenic evolution of the Sabzevar zone in the area of Gorgan- Khorasan, NE Iran, Geological Survey of Iran, ISSN 0075–0484, report no. 51. Freie Universität Berlin Institute of Geological Sciences, Dtsch, pp 119–143
Maghsoudi A, Rahmani M, Rashidi B (2005) Gold deposits and indications of Iran. Research manual for Students of Earth Science (In Persian)
Masoudi F (1997) Contact metamorphism and pegmatite development in the region SW of Arak, Iran. PhD thesis, Department of Earth Sciences, University of Leeds
Metz A (1985) Islamic Civilization in 4th Hijri Century (tr. To Persian by Alireza Zakavati Qaragozlou A)
Memenzadeh M (1981) Distribution and metallogeny within Ahar 1:250,000 Quadragle, Geological Survery of Iran
Minoreski V (1963) Expedition diary of Abudolaf Khazraji; translated by Abolfazl Tabatabayi. Farhang-e Iranzamin, Tehran
Moein M (1976) 'Farhang-e Farsi' (A Persian dictionary), Entesharat-e Amir Kabir
Moein-Vaziri H (1996) An introduction to the Magmatism in Iran, Tarbiyat Moaalem University publication
Momen-Zadeh M (1976) Stratabound lead–zinc ores in the Lower Cretaceous and Jurassic sediments in the Malayer–Esfahan district (west central Iran). PhD thesis, Heidelberg, 300 p
Momen-Zadeh M (2004) Metallic mineral resources of Iran, mined in ancient times – A brief review; with contribution from Ali Haji-Soltan and Mahsa Momensade. Zarneh Research Institute, Tehran
Momenzadeh M, Ghorbani M, Moradi M (1995) Deposits of iron in Hamedan, Ajin Co
Momen-Zadeh M, Heidari E (1995) Ore–hydrocarbon resources and alkaline magmatism of Late Proterozoic–Early Cambrian in Iran, genetic interpretation. Carbonate Evapor 10(1):79–88

Momen-Zadeh M, Walther HW (1984) Mineral deposits and metallogenic epochs in the area of the Geotraverse Iran, A review. N Jb Geol Palaeont Abh 168(2/3):468–478, Stuttgrat
Motiei H (1993) Stratigraphy of Zagros. In: Treatise on the geology of Iran. Iran Geological Survey, Tehran
Nabavi M (1976) An introduction to geology of Iran. Geological Survey of Iran, Tehran
NISCO (1977) Commercial and genetic type of manganese raw materials of Iran, V/O Technoexport USSR, Isfahan (unpublished), 179 p
Nogole-Sadat MAA (1978) Les zones de decrochement et les virgation structurales en Iran. Consequences des resultants de l`analyse structural de la région du Qom. Unpublished PhD thesis, France, 201 p
Nogol Sadat and Hushmandzadeh (1999) Exploration report of Sistan-Baluchestan Provice, Science and Technology Institute
OPEC (2012) Crude oil, 1 July 2012. Retrieved from http://www.opec.org/opec_web/en/press_room/180.htm
Parvin-pour F (2007) Porphyry copper deposits within Abdar-Dahaj area. MSc thesis, Islamic Azad University
Peyrovan H (1992) Petrographic, petrologic and geochemical studies of intrusive rocks north of Abhar and the association of plutonism in the area with mineralization. MSc thesis, Tarbiyat Moalem University
Ramezani J, Tucker R (2003) Petrogenesis and implications for Gondwana tectonics. Department of Earth and Planetary Sciences, Washington University, St. Louis
Roads atlas of Iran, Gitashenasi Cartographic and Geographic Institute (2004)
Romanko E, Momenzadeh M, Mogarovskiy VV, Kokorin YG (1985) The new celestite region in the eastern part of the Mediterranean Belt, Zagros, Southern Iran, Doklady Academii Nauk Tadzhikskoy S.S.R, vol 28, No. 4, 227–230
Sabzehei M (1974) Les mélanges ophiolotiques de la région d'Esfandagheh (Iran meridional), étude pétrographique et structurale. Thése Doct. Etat, Univ. Grenoble, 306 p
Sadeghi A (1999) Cretaceous stratigraphy in southern flank Alborz. PhD thesis, Shahid Beheshti University
Samani BA (1988) Metallogeny of the Precambrian in Iran. Precambrian Res 39(1–2):85–106
Sengör AMC (1991) Late Paleozoic and Mesozoic tectonic evolution of the Middle Eastern Tethysides: implications of the Paleozoic geodynamics of the Tethyan realm, IGCP Project 276, Newsletter No. 2, 149 p
Seyed Emami K, Brants A, Bozorgnia F (1971) Stratigraphy of the Cretaceous rocks southeast of Esfahan, in contribution to the paleontology and stratigraphy of Iran. Geological Survey Iran, report no. 20, pp 5–40
Shahriyari M (1986) Karstic bauxite deposits with carbonate floor. Faculty of Technology, Tehran University, Tehran
Sheikhi-kahrizaki (2006) Ophiolitic rocks in Khajeh-jamali area (Shiraz) and associated chromite deposits, Shahroud University
Shemirani A (1988) Cretaceous in Iran, course materials. Shahid Beheshti University, Tehran
Smirnov VI (1983) Geology of Mineral Reserves; translated by K. Alipour. University Printing Center
Soheili S (2004) Bauxite deposits of Iran. MSc thesis, Islamic Azad University
Stöcklin J (1961) Lagoon formation and salt domes in East Iran. Bull Iran Pet Inst 3:29–46
Stöcklin J (1962) Note on the geology of the Iran- Pakistan border area. Geological Survey Iran, Geological Note No. 1, 15 p
Stöcklin J (1968) Structural history and tectonics of Iran; a review. Am Assoc Pet Geol Bull 52(7):1229–1258
Stöcklin J (1974) Evolution of the continental margins bounding a former Southern Tethys. In: Geology of continental margins. Springer, Berlin, pp 873–887, BIBL. 2P, 5Illus. U.N. Geological Survey Institute
Stöcklin J, Eftekharnezhad J, Hushmandzadeh A (1972) Central Lut reconnaissance, East Iran. Geological Survey Iran, report no. 22, 62 p

STOMP (2005) An introduction to geology and mineralization of Bafgh area, Iran. Structure, Tectonics and ore mineralization (STOMP), Townsville

Takin M (1971) Geological history and tectonics of Iran, a discussion of continental drift in the Middle East since the Early Mesozoic. Geological Survey of Iran, internal report

Tarkhani S (2010) Sanadaj-Sirjan magmatism within Saghez area. PhD thesis, Islamic Azad University

Tehran Padir (2002) Exploration report of Astamal area, Bureau of industry and mines of East Azerbaijan Province

Torkamani E (1997) Petrologic study of intrusive rocks north of Abhar-Khoramdarreh. MSc thesis, Shahid Beheshti University

Travaux 13th International Symposium of ICSOBA "Development of Aluminium Industry in Asia", (2000) Iran Itok Engineering, Technology and Contracting Company

Vatanpour H (1998) Study of mineralogy and geochemistry of chromites in northwest of Sabzevar. PhD thesis, Shahid Beheshti University

Wulff HE (1966) The traditional crafts of Persia: their development, technology and influence on Eastern and Western civilizations. MIT Press, Cambridge, MA

Zarei (1992) Petrography and geochemistry of volcanic rocks north of Abhar, M.Sc. Thesis, Tarbiyat Moaalem University

Zavosh M (1976) Mineralogy in ancient Iran, vol 2. Institute for Humanities and Cultural Studies, Tehran

Zonenshain LP, Le Pichon X (1986) Deep basins of the Black sea and Caspian Sea as remand of Mesozoic back- arc basins. Tectonophysics 124:181–211

Knowledge Sources

Abdalian S (1953) Le tremblement de terre de Toroud, en Iran, An. Nature (Paris) 81(3222): 314–319

Adabi MH, Ager DV (1997) Late Jurassic brachiopods from northeast Iran. Palaeontology 40(Part 2):355–362

Adabi MH, Moussavi-Harami R (1989) Geomorphology of eastern Kopet-Dagh Basin. In: Popoli MH (ed) International Geographical Seminar, Mashhad, Iran, vol 1, pp 87–104

Adams TD, Bourgeois F (1967) Asmari biostratigraphy. Geological and exploration division, report no. 1074. IOOC, Iran

Ager DV (1988) Mesozoic Turkey as part of Europe. In: Audley-Charles MG, Hallam A (eds) Gondwana and Tetheys, Special publication no. 37. Geological Society, London, pp 241–245

Afshar-Harb A (1970) Geology of Sarakhs area and Khangiran gas field. Geological Division, Exploration and Product Group, N.I.O.C, Tehran, pp 1–17

Afshar-Harb A (1979) The stratigraphy, tectonics and petroleum geology of the Kopeh-Dagh region, northern Iran. Unpublished PhD thesis, Imperial College, London, 316 p

Afshin Y (1994) Rivers of Iran. Ministry of Energy

Aghanabati A (2004) Geology of Iran. GSI Publication, Tehran

Aghanabati SA, Ghorbani M (2011) Metallogenic map of the middle east. Geological Survey of Iran

Aghanabati SA (1986) Lecture notes on the geology of Iran. Faculty of Sciences, Ahwaz University, Ahwaz (In Farsi)

Aghanabati SA (1992) Map of metamorphic rocks of Iran (1:2,500,000). Geological survey of Iran (Unpublished)

Alavi NM (1972) Etude géologique de la région de Djam, report no. 23. Geological Survey of Iran, 293 p

Alavi TN (1977) Geology and petrography in the ophiolite range NW of Sabzevar (Khorassan/Iran) with special regard to metamorphism and genetic relation in an ophiolite suite, H.T.9, S.l, pp 1–47

Alavi TN (1979b) Ophiolitic rocks complex in Iran; results and discussable issues. Geological Survey of Iran, Tehran

Alavi NM (1993) Paleozoic stratigraphy of Iran

Alavi NM (2009) An introduction to stratigraphy of Iran. Geological Survey of Iran, Tehran

Alavi NM, Bolourchi MH (1982) Geology of Takab- Saein Qaleh, report no. 50. Geological Survey of Iran, Tehran

Alavi NM, Bolourchi MH, Navaee I (1973) Explanatory text of the Maku quadrangle map 1:250,000; No. A1. Geological Survey of Iran, 47 p

Alavi NM (1980) Tectonostratigraphic evolution of the Zagrosides in Iran. Geology 8:144–149

Alipour K (1988) Geology of ore deposits. University Publication Centre, Tehran (In Farsi)

M. Ghorbani, *The Economic Geology of Iran: Mineral Deposits and Natural Resources*, Springer Geology, DOI 10.1007/978-94-007-5625-0,

Alipour-asl (2003) Europium (REE); prospecting of monazite in alluvial sediments south of Marvast. Geological Survey and Mining Exploration of Iran, Tehran

Alleman F, Peters T (1972) The ophiolite- radiolarite belt of the north- Oman mountains. Eclogae Geol Helv 65:657–698

Amidi SM (1977) Etude géologique de la région du Natanz– Surk (Iran Central) stratigraphie et pétrographie. PhD thesis, Geological Survey of Iran, report no. 42, 316 p

Amidi M et al (1978) Geological quadrangle map of Mianeh, 1:250,000, Geological survey of Iran, no. C3

Amidi SM, Emami MH, Michel R (1984) Alkaline character of Eocene volcanism in the middle part of Central Iran and its geodynamic situation. Geol Randsch 73(3):917–932

Anaraki H (1969) Chromite in Iran

Andrusov (1896) On the geological investigation in the course of the summer 1895 in the government of Baku and on the eastern littoral of the Caspian. Transact Soc Natur St Petersb Sect Geol Miner, vol 24 (In Russian)

Annells RN, Arthurton RS, Bazley RA, Davies RG (1975) Explanatory text of the Qazvin and Rasht quadrangle map, 1:250,000, Geological Survey of Iran, no. E3, E4

Ansari HJ (1965) Geology of Southern Meyaneh basin in Azarbayjan, Iran. Pet Geol Bull 49(1)

Arian Zamin Co. (2010a) Prospective exploration report on Abdar-Dahaj axis

Arian Zamin Co. (2010b) Prospective exploration report on Jebal-e Barez area

Armstead HCH (1973) Geothermal energy, Review of research and development, Third impression (1977). Unesco, France, 185 p

Armstead HCH (1983) Geothermal energy, 2nd edn. Wiley, New York

Armstrong FC (1981) Genesis of uranium and gold- bearing Precambrian quartz- pebble, conglomerates. US Geological Survey professional paper. US Geological Survey, Tehran, 1161 p

Arnorsson S (1983) Chemical equilibrium in Icelandic geothermal systems, implications for chemical geothermometry investigations. Geothermics 12:119–128

Asch W, Asch D (1914) The Silicate of chemistry and commerce. Constable, London

Assereto R (1966a) Geological map of upper Djadjerud and Lar valleys, Ist Geological University Milano. Series G Publ, Ital., no. 232, pp 1–86

Assereto R (1966b) Explanatory note on the geological map of Upper Djadjerud and Lar valleys (Central Elburz, Iran), Milan University, Ist Geological Publication Series G. no 232, 86 p

Barthel F, Hahn L (1985) Sedimentary uranium occurrences in Eastern Europe with special reference to sandstone formations, in, geological environments of sandstone type, uranium deposits, TECDOC-328. IAEA, Vienna, pp 51–68

Bates R, Jackson J (1980) Glossary of geology-Virginia, 2nd edn. American Geological Institute, Alexandria, 751 p

Baumann A, Spies O, Lensch G (1984) Strontium isotopic composition of post-ophiolitic Tertiary volcanics between Kashmar, Sabzevar and Quchan /NE Iran, Ch. G. plate tectonics, orogeny and mineralization in the Iranian fold belt, Report of a German-Iranian Research Program 1977–1978, Univ. Saarlandes, Saarbrucken, Federal Republic of Germany, Neues Jahrb fuer geologie and palaeontaologie. Abhan Dlungen 168(2–3):409–416

Belov AA, Gatinskiy YG, Mossakovskiy AA (1985) The Indosinides of Eurasia. Geotectonics 19(6):446–459

Belov AA et al (1986) A precis on pre-Alpine tectonic history of Tethyan Paleooceans. Tectonophysics 127:197–211

Berberian M (1976) An explanatory note on the first seismotectonics map of Iran, A seismotectonic review of the Country, in contribution to the seismotectonic of Iran (Part II). Geological of Iran, report no. 43, 518 p

Berberian M (1977) Against the rigidity of the "Lut Block", Iran. Geological Mineral Research Organization report. no. 40

Berberian F (1981) Petrogenesis of Iranian plutons. A study of the Natanz and Bazman intrusive complexes. PhD thesis, University of Cambridge, England, 300 p

Berberian M (1983a) Continental deformation of Iran Plateau. Geological survey of Iran, report no 52 (In Farsi)

Berberian M (1983b) The southern Caspian; a compressional depression floored by a trapped modified oceanic crust. Can J Earth Sci 20(2):163–183
Berberian M (1995) Master blind thrust faults hidden under the Zagros Fold active basement tectonics and surface morphotectonics. Tectonophysics 241:193–224
Berberian M, Hamdi B (1977) First discovery of Ordovician beds and conodonts in the slightly metamorphosed rocks of Kuh-Agh Baba, Maku quadrangle map, Azarbaijan. Geological Survey of Iran, internal report, 7 p
Berberian F, Berberian M (1981) Tectono-plutonic episodes in Iran. In: Zagros, Hidu Kush and Himalaya Geodynamic Evolution, American Geophysical Union, Geodynamic Series 3, pp 5–32
Berberian M, Manougian M (1997) A review of the history of universe and earth in Iranvij Sazman-e Pajouheshi Iranzamin. Bonyad-e Neishabour, Tehran
Berberian M, Houshmandzadeh A, Lotfi M (1973) Deformation phases and related metamorphism in the Gorgan area. Geological Survey of Iran, Internal report, 14 p
Berbrian F, Billiault JP, Maurizot P (1974) Preliminary explanatory text of the geology of Deh Sard and Khalor with some remarks on the metamorphic complexes and the tectonics of the area (Two geological maps 1:100,000 from the Haji Abad quadrangle map). Geological Survey of Iran, Internal report, 60 p
Berning JR, Cooke SA, Hiemstra SA, Hoffman U (1976) Rossing uranium deposit, southwest Africa. Econ Geol 71:351–368
Bibliography of geological literature on Iran (1995) Geological Survey of Iran, vol 1, Author Index, 764 p
Bonskowski R (2007) Coal news and markets, Energy Information Administration (EIA)
Bordet P (1975) Volcanisme, plutonisme et orogenese Tertiares en Iran du NW, le massif du Sahand, In IIIE. Reun. Annu. Sci. Terre, Montpellier, Montpellier University of Science and Technology Langueddoc, I.G.A.L., Paris, 57 p
Borna B (1986) Prospective exploration report on antimony deposits in Anarak, Boshrouyeh, Ferdow and Kashmar areas. Geological Survey of Iran, Tehran
Borna B (1991) Exploration report on Shourab antimony deposit, and preliminary prospecting of Shand-mahmoud, She-halgheh and Hesamiyeh gold-bearing antimony deposits in Ferdows area. Geological Survey of Iran, Tehran
Boroumandi (1982) Exploration report on Sangan Iron Mine. Geological Survey of Iran, Tehran
Boudier F, Nicolas A, Ceuleneer G (1989) De l'accretion océanique á la convergence, le cas de l'ophiolite d'Oman. Bull Soc Géol France 8:221–230
Boyle RW (1979) The geochemistry of gold and its deposits. Geol Surv Con Bull 280:584
Boyle RW (1982) Geochemical prospecting for thorium and uranium deposits. Elsevier Scientific Publishing Company, Amsterdam, 493 p
Bradossy G (1993) Carboniferous to Jurassic bauxite and laterite as paleoclimatic indicators
Braud J (1970) Les formations du Zagros dans la region de Kermanshah (Iran) et leurs rapports structuraux. Acad Sci Paris CR Ser D Geol Sk map 271(15):1241–1244
Braud J (1987) La suture du Zagros au niveaude Kermanshah (Kurdestan, Iranian), évolution géodynamique, magmatique et structurale. Fara DA (1983) nonpaging, 30 cm, Thesis. Doct., Tectonic. Paris 11/1987, 110 p
Braud J (1990) Explanatory text of the Bakhtaran (Kermanshah) quadrangle map, 1:250,000; no. C6. Geological Survey of Iran, Tehran
Braud J, Bellon H (1975) Sur le domaine metamorphique du Zagros (Zone de Sanandaj-Sirjan) au niveau de Kermanshah-Hamadan (Iran). Eclogae Geologicae Helvetiae
Braud J, Ricou LE (1971) L'accicent du Zagros ou main thrust, un charriage et un coulissement. Acad Sci Paris, CR Ser D 272(2):203–206
BRGM (1978a) Eastern Iran project, Zone IV. Mining reconnaissance, report no. 5, Qayen area, text of the report. Geological and Mineral Survey of Iran, Tehran, 248 p
BRGM (1978b) Eastern Iran Project, Zone IV. Mining reconnaissance, report no. 7, Shahrakht area, Text of the report. Geological and Mineral Survey of Iran, 165 p
British Geological Survey (1999) World mineral statistics 1993–97: production, exports, imports, Keyworth, Nottingham

British Geological Survey (2002) World mineral statistics, 1997–2001, Production, exports, imports. British Geological Survey, Keyworth, Nottingham, Printed by Saxon Graphics Ltd., Derby, 285 p

Brobst DA (1983) In: Lefond SJ (ed) Industrial minerals and rocks/barium minerals, vol 1, 5th edn. AIME, New York, pp 485–501

Brosses R, MoeinVaziri H (1980) Le volcanisme du Kouh -e-Tchah- e- Shah, au nord du Makran (Iran), Vl. 69, no. 1

Brunn JH (1960) Mise en place et différentiation de l'association pluto- volcanique du cortége ophiolitique. Rev Géo Phys et Géol Dyn S.2–3:115–132

Buhne E (1928) Bericht uber die untersuchung der ezgruben uon karadjadagh, report in Geological Survey, Iran Library, 5 p

Bunteharth G (1984) Geothermics, an introduction. Springer, Berlin/Heidelberg, 139 p

Burnol L (1968) Contribution a l'etudedes gisements de plomb et zinc de L'Iran. Geological Survey of Iran, report no. 1, 113 p

Byers HG, Alexander LT, Holmes RS (1935) The composition and constitution of the colloids of the great soil groups. U.S. Department Agriculture Technical Bulletin, Beltsville, 484 p

Cartier E (1971) Die geologie des Unteren Chalus Tales Zentral-Alborz Iran. Mitt. Geol.Inst. Eidgen.Tech. Hochsch. Univ. Zuerich, Neve folge.ch, no. 164, pp 1–133

Central Bank of the Islamic Republic of Iran (various years) Annual review of the economy of Iran; Available from: http://www.cbi.ir/publications/. Accessed Aug 24 2004

Central Bank of the Islamic Republic of Iran (various years) Economic report and balance sheet. Available from: http://www.cbi.ir/publications/. Accessed Aug 24 2004

Central Bank of the Islamic Republic of Iran (various years) Economic trends and indices. Available from: http://www.cbi.ir/publications/. Accessed Aug 24 2004

Chalenko J (1975) Introduction a la paléogeographie de l'Elbourz meridional (Iran), Tectonophysics; Netherl. In: International symposium on recent crustal movements. 5. Proceedings, Zurich, 1974. Imperial College of Science, London, G.B., vol 29, no 1–4, pp 411–420

Cherven Victor B (1986) Tethys-marginal sedimentary basins in western Iran. Geological Society of American Bulletin, ISSN 0016–7606, USA, Da. 1986–05, vol 97/no 5, pp 516–522

Christian L (1996) Paleozoic opach and facies reconstructions of Iran and the Arabian Peninsula. Geosci Sci Q J 20, Summer Geological Survey of Iran

Chromite deposits in Torbat-heydarieh – Fariman area, Arian Zamin Co. (2010)

Church WR (1972) Ophiolite: its definition, origin as oceanic crust, and mode of emplacement in orogenic belts, with special reference to the Appalachians. Department of Energy, Mines, Resources, Canada Publication, vol 42, pp 71–85

Church SG (1985) Genetic interpretation of lead- isotopic data from the Columbia River Basalt Group, Oregon, Washington and Idaho. Bull Geol Soc Am 96:676–690

Clark GC, Davies R, Hamzehpour B, Jones CR, Ghorashizadeh M, Hamidi B, Navai N (1975) Bandar-e-Pahlavi, Geological Quadrangle map of Iran; 1:250,000. Geological Survey of Iran, no. D3

Clark GC, Davies R, Hamzehpour B, Jones C, Ghorashizadeh M, Hamidi B, Navai N (1977) Explanatory text of the Bandar-e-Pahlavi quadrangle map 1:250,000; no D3, 205 p

Climatic Atlas of Iran (1965) Tehran University, Institute of Geography

Comprehensive Geography of Iran (1987) Ministry of Education publication

Complementary exploration report on iron deposits in Hamedan and Kurdistan (2008) Sabanour Co.

Conrad G, Conrad J, Girod M (1977) Les formations continentaux Tertiares et Quaternaires du block du Lout (Iran). Importance du plutonisme et du volcanisme; MEM. Hors-Ser., Soc. Geol. Fr.; DA. (1977) no. 8, pp 53–75

Crawford AR (1972) Iran, continental drift and plate tectonics. In: 24th international Geological Congress, Montreal, vol 3, pp 106–112

Crawford AR (1977) A summary of isotopic age data for Iran, Pakistan and India. Mem Hors Ser Geol Fri; no. 8, pp 251–260

CRU Monitor (2003) Steelmaking raw materials, 3 p

Dahlkamp FJ (1989) Classification scheme for uranium deposits, state of the art review in metallogenesis of uranium deposits. International Atomic Energy Agency, Vienna, pp 1–32

Dahlkamp FJ (1993) Uranium-ore deposits. Springer, Berlin/Heidelberg, 449 p
Darehshouri BF, Kasraian N (1998) Nature of Iran. Rowzaneh Kar Publication, Tehran
Davies RG, Jones CR, Hamzepour B, Clark GC (1972) Geology of the Masuleh, Sheet 1:100,000; Northwest Iran. Geological Survey of Iran, report no. 24, 110 p
Davoudzadeh M (1969) Geologie und petrography des Gebietes Nordlich von Nain, Zentral– Iran, Mitt. Geol. Inst. Eidg. Tech. Hochsch. Univ. Zuerich, Che, no. 98, pp 1–40
Davoudzadeh M (1972) Geology and petrography of the area North of Nain, Central Iran. Geological Survey of Iran, report no. 14, 92 p
Davoudzadeh M, Soffel H, Schmidt K (1981) On the rotation of Central– East Iran microplate. N Jb Geol Plaent Mh 1981(3):180–192
De Vivo B (1984) Uranium geochemistry, mineralogy, geology, exploration and resources. The Institution of Mining and Metallurgy, London, 194 p
Dedual E (1967) Zur Geologie des mittleren und unteren Karaj- Tales, Zentral- Elburz (Iran), Zuerich, Univ., Geol. Inst. Eidgenoess. Tech. Hochsh, Hochsh., Inst., Mitt., no. 76, 123 p
Dehghani GA, Markis J (1983) The gravity field and crustal structure of Iran. Geodynamic project (geotraverse) in Iran. Geological Survey of Iran, report no. 51, pp 51–68
Dellenbach J (1964) Contribution et l'etude géologique de la region situee a l'est du Tehran (Iran). Fac. Sic. Univ, Strasbourg, 117 p
Dewey JF, Pittman WC, Ryan WFB, Bonnin J (1973) Plate tectonics and the evolution of the Alpine system. Geol Soc Am Bull 84:3137–3180
Didon J, Gemain YM (1976) Le Sabalan, volcan Plio–Quaternaire de L'Azerbaijan oriental (Iran), étude géologique et pétrographique de ledifice et de son environnement regional. Thése de 3esme Cycle, Univ. Grenoble, 304 p
Difenbach W, Davoudzadeh M, Alavi Tehrani N, Lensch G (1986) Paleozoic ophiolites in Iran: geology, geochemistry and geodynamic implication. Ofioliti 11(3):305–338
Dimitrijevic M, Djokovic I (1973) Geological map of Kerman region, scale 1:500,000, Geological Survey of Iran, 1 sheet
Diyakonov MM (1967) 'Tarikh-e Iran-e Bastan'(History of ancient Iran); translated by: Rouhi Arbab; Bongah-e tarjome va nashr-e Ketab, Tehran
Dykstra JD, Birnie R (1979) Segmentation of the quaternary subduction zone under the Baluchistan region of Pakistan and Iran. In: Geodynamic of Pakistan. Elite Publishers, Karachi, pp 319–325
Economides M, Ungemach JP (1987) Applied geothermics. Wiley, New York, 235 p
Eftekhar Nezhad J (1991) Geodynamic significance of recent discoveries of ophiolites and late Paleozoic rocks in NE. Iran (including Kopet-Dagh). Abh. Geol. B.A. Wien, pp 89–110
Eftekharnejad J (1973) Notes on plate tectonics of Iran and formation of sedimentary basins in the east of the country. J Fac Eng Tehran Univ 26 (In Farsi)
Eftekharnejad J (1996) Lecture notes on the geology of Iran and its neighbouring countries. Faculty of Earth Sciences, Shabid Beheshti University (In Persian)
Eftekharnejad J, Behroozi A (1991) New finds on the Ophiolitic and Late Paleozoic rocks of north-eastern Iran (including Kopet-Dagh) and their geodynamic significance. J Earth Sci (Geological Survey of Iran) 1 (In Farsi)
Eftekhar Nezhad J, Nabavi M, Valeh N (1965) Geology of Tarom-Talesh area. Geological Survey of Iran, note no.16 with map 1:100,000 (unpublished), pp 1–29
Eftekhar Nezhad J, Samimi M, Arshadi S (1985) Explanatory text of the Pishin quadrangle map, 1:250,000; no. M14, Geological Survey of Iran
Elder S (1957) Lexicon of stratigraphy, Southwest Iran, Unpublished report, Iranian Oil Operating Companies
Emami MH (1981) Geologie de la region du Qom–Aran (Iran) contribution a l'étude dynamique et géochemique du volcanism Tertiaire de l'Iran Central. PhD thesis, 489 p
Emami MH (1991) Magmatic activity and mineral exploration. J Earth Sci (Geological Survey of Iran) 1 (In Farsi)
Emami MH (2000) Magmatic activities in Iran. Geological Survey of Iran, Tehran
Emami MH, Mir Mohammad Sadeghi M, Omrani SJ (1993) Magmatic map of Iran, 1:1,000,000. Geological Survey of Iran, Tehran

Engalenc M (1968) Geologie, geomorphologie, hydrogeologie de la region du Tehran (Iran). These en sciences, Montpellier, 180 p
Erni A (1931) Decouverte du Bathonian fossilifere dans l'Elbourz. Ecol Geol Helv 24(2):165–166
Etminan H (1977) Le porphyre cuprifere de Sar Cheshmeh (Iran), Role des phases fluids dans les mecanismes d`alteration et de mineralization, Science de la terre, Mem, France, no. 34, pp 1–249
Evans AM (1993) Ore geology and industrial minerals. Blackwell, Oxford
Exploration for ore deposits in Kerman region (1973) Report no. Yu/53-1973. Geological Survey of Iran, 218 p
Exploration report on Astamal area (2002) Tehran Padir Co.
Exploration report on Chodar-chay copper deposit (2009) Arian Zamin Co.
Exploration report on Galali – Baba-ali mines (2003) Sabanour Co.
Falcon NL (1961) Major earth–flexuring in the Zagros mountain of southwest Iran. Q J Geol Soc London 117(Prt.4 No. 637) obs –with discussion 1961 Proceedings. Geological Survey of Iran no. 1591, pp 129–130
Falcon NL (1967a) Equal area of Gondwana and Laurasia. Nature (London) 213(5076):580–581
Falcon NL (1967b) The geology of northeast margin of the Arabian basement shield. Adv Sci 24(119):31–42
Falcon NL (1974) Southern Iran; Zagros mountains, in Mesozoic-Cenozoic orogenic belts. Data for orogenic studies; Alpine-Himalayan Orogenes Geological Society of London, Special Publication no. 4, pp 199–211
Farhoudi G (1978) A comparison of Zagros geology to island arcs. J Geol USA 86(3):323–334
Fauvelet E, Eftekhar Nezhad J (1990) Explanatory text of the Qayen map, 1:250,000; no. K7. Geological Survey of Iran, Tehran, 315 p
Foley DC, Foreman NE (1995) Middle East source rock equivalents on other continents. Geosci Sci Q J 4(15&16)
Forster RP (1993) Gold metallogeny and exploration. Blackie and Son Ltd., London, 432 p
Forster H, Kermani A (1991) Petrographic, mineralogical and geochemical investigations of Sangan iron ore deposit, North eastern Iran. Geology Survey of Iran, Tehran
Forster H, Larsen G (1991) The Para genesis of inclusions in salt diapers in southern Iran. In: Proceedings of third mining symposium, Iran, vol 2
Forster H, Fesefeldt K, Kuersten M (1972) Magmatic and orogenic evolution of the Central Iranian volcanic zone. International Geological Congress, Abstract, no. 24, 41 p
Geodynamic Project (1983) Geotraverse in Iran, final report. Geological Survey of Iran, no. 51
Ghani M, Khalatbari M, Jan-nesari MR (2010) The importance of position of Afchang supra-ophiolitic series within Sabzevar ophiolites
Ghasemipour (1976) Study of economic geology of Sangan iron deposit in Khorasan Province. Barite Co., Tehran
Ghazvini E (2002) Exploration of natural oil seepage. Bureau of Industry and Mine of Kermanshah Province, Tehran
Ghorbani M (1992a) Observation of eastern Iran metallogeny. In: 1st seminar of geology of Eastern Iran
Ghorbani M (1992b) 'Tarikh-e ma'dankari va fonoon an dar Iran-e Bastan' (History of mining and its techniques in ancient Iran). In: The 1st symposium of Iranology, Iranology Foundation
Ghorbani M (1992c) Dibache-i bar zaminshenasi-e eqtesadi Iran (An introduction to economic geology of Iran). National Geological Data Center, Geological Survey of Iran, Tehran
Ghorbani M (1999a) Chemical properties of magmatic rocks of Iran, colleague. Geological Survey of Iran Publication, 2v, 1466 p
Ghorbani M (1999b) Copper deposits in Iran. Geological Survey of Iran Publication, Tehran, 421 p
Ghorbani M (1999c) Introduction of required resources of steel industry in NW Iran. In: Steel industry of Iran symposium, Esfahan
Ghorbani M (1999d) Perlite and lightweight aggregate. Geological Survey of Iran, Tehran

Ghorbani M (1999e) Petrology study of tertiary – quaternary magmatic rocks of Takab Region. In: 4th conference of Iran geology society, Shiraz
Ghorbani M (1999f) Petrological investigations of tertiary-quaternary magmatic rocks and their metallogeny in Takab area. PhD thesis (Unpublished), Faculty of Earth Sciences, Shahid Beheshti University (In Farsi)
Ghorbani M (2000a) Perlite and pomis in Iran. Geological Survey of Iran Publication, Treatise on the Geology of Iran, 114 p
Ghorbani M (2000b) Lead- Zinc resources of Iran. Geological Survey of Iran Publication, 480 p
Ghorbani M (2000c) Mineralization in copper metallogeny states in Iran. In: 19th earth sciences conference. Geological Survey of Iran, Tehran
Ghorbani M (2000d) Mineralization and distribution of lead-zinc reserves in Iran. In: 19th earth science congregation. Geological Survey of Iran, Tehran
Ghorbani M (2000e) Mineralization phases, distribution and mineralogical characteristics of lead and zinc deposits in Iran. In: 19th Geological Congress. Geological Survey of Iran, Tehran
Ghorbani M (2000f) The origin of mineralization phases and relative advantages of copper in Iran. In: 19th Geological Congress. Geological Survey of Iran, Tehran
Ghorbani M (2000g) Copper metallogenic map of Iran. Zariaban Exploration Company, Tehran
Ghorbani M (2000h) Mineral distribution atlas of Iran (Arianzamin)
Ghorbani M (2002a) Economic geology of Iran. Geological Survey of Iran Publication, 700 p
Ghorbani M (2002b) Metallogeny regions of Iran. In: 21st Earth Sciences conference. Geological Survey Iran, Tehran
Ghorbani M (2002c) An introduction to economic geology of Iran. National Geological Data Center. Geological Survey of Iran, Tehran, 695 p
Ghorbani M (2003a) The position of mining industry and metallogenic provinces of Iran. Khaneh Maadan Iran J
Ghorbani M (2003b) Observation of Bafgh region geology. In: 1st specialized seminar of Payam-e-Nour University
Ghorbani M (2003c) Gemology history of Iran and its role in Persian poems. In: 2nd seminar of Iranology Institute
Ghorbani M (2003d) Metamorphism in Takab Region. Earth Sci Seas J, SBU
Ghorbani M (2003e) Study of magmatism and alteration in Ahar Region and their relationship with mineralization. In: 21st earth sciences conference, Geological Survey of Iran
Ghorbani M (2003f) Scientific edition of geochemistry basics. Payam-e-Nour University, Urmia
Ghorbani M (2004a) Volcanology basics with a view on Iran volcanoes, Pars (arian zamin) geology research center, 356 p
Ghorbani M (2004b) Precious minerals (gems) and rocks and their place in Iran, Pars (arian zamin) geology research, 396 p
Ghorbani M (2004c) Petrologic characteristics of magmatic rocks in alteration zones and their association with mineralization within Ahar-Jolfa axis, Research project at Shahid Beheshti University
Ghorbani M (2005h) Geology of energy resources. Payam-e-Nour University, Urmia
Ghorbani M (2006a) Irreversible environmental degradation: lessons from past, South Africa, pp 12–26
Ghorbani M (2007a) Economic geology of natural and mineral resources of Iran, Pars (arian-zamin), 492 p
Ghorbani M (2007b) Lead- zinc reserves of Iran capacities and requirements, Stone-Minemagzine' (IMH)
Ghorbani M (2007c) All you need to know about Iron deposits in Iran. Sang va Maadan Q 3
Ghorbani M (2007d) Geology of energy-producing raw materials. Payam-nour University, Urmia
Ghorbani Gh (2007e) Magmatism in Toroud area. PhD thesis, Shahid Beheshti University
Ghorbani M (2007f) Gold mining of Iran, Stone-Minemagzine' (IMH)
Ghorbani M (2007g) Coal: a forgotten prized mineral, Stone-Minemagzine' (Imh)
Ghorbani M (2007h) Refractory material (esp. Magnesite), Stone-Minemagzine, (IMH); Ghorbani M (2007) Refractories with specific view on Magnesite; Sang va Maadan Q 5:Autumn
Ghorbani M (2008a) Gold in Iran. International Geological Congress, Oslo

Ghorbani M (2008b) Copper in Iran. International Geological Congress, Oslo
Ghorbani M (2008c) Gemestone of Iran, Stone-Minemagzine, (IMH)
Ghorbani M (2008d) Phosphate; more valuable than what we thought. Sang va Maadan Q 7
Ghorbani M (2008e) Industrial soils. Stone-Minemagzine (IMH)
Ghorbani M (2008f) Manganese: a strategic element for steel industry. Stone-Minemagzine (IMH)
Ghorbani M (2008g) Economic geology of ore deposits and indication of Iran, Pars (arian-zamin), 639 p
Ghorbani M (2008h) All you need to know about Gold deposits in Iran. Sang va Maadan Q
Ghorbani M (2008i) Time and place distribution of phosphate deposits of Iran
Ghorbani M (2010a) Talc; an industrial mineral with developing application. Sang va Maadan Q 15:4–13
Ghorbani M (2010c) Mica, an industrial and strategic mineral in electronic industry, Stone-Minemagzine, (IMH), pp 4–14
Ghorbani M (2012a) Geology of Iran. ArianZamin Publication, Tehran
Ghorbani M (2012b) Rare Earth Elements (REE); rare metals but strategic in the third millennium. Sang va Maadan Q 20:4–14
Ghorbani M (2013) Metallogeny atlas for Iran deposits
Ghorbani M, Parvinpour F (2001) Iranian mining position in the world (standing). In: The 20th symposium on geosciences. Geological Survey of Iran, Tehran
Ghorbani M, Parvinpour F (2002) History of mining and its techniques in ancient Iran. In: The 1st symposium of Iranology, Iranology Foundation
Ghorbani M, Mavahebi N, Ansari M (2004) Relative advantages of mining industry in Iran
Gibbs A (1974) Geology and genesis of the Bafq lead-zinc deposit. Iran Inst Min Metall Iran Sect Bull 85:205–220
Gillet S (1967) Le Néogène de la région de Moghan (Azerbaidjan Iranien) d`après la macrofaune: Athens. Praktika les Akademias Athenon 42:57–78
Girod M, Conrad G (1976) Les formations volcaniques recentes du sud de l`Iran (Kouh-e-Shahsasvaran), Donneés pétrologiques preliminaries, implications structurales. Bull. Volcano; Ital., Da. 1975–1976, 39(4):495–511, Abstr. Angl.; BIBL. 1P. 1/2; 5 Illus.; 33 Anal Cent Geol Geophys, C.N.R.S., Montpellier, France
Girshman (1976) From beginning to rise of Islam; translated by Moein M., Bongah-e Tarjome va Nashr-e Ketab
Glennie KW (2000) Cretaceous tectonic evolution of Arabia eastern plate margin, a tale of two oceanic. In: Middle East models of Jurassic/Cretaceous carbonates systems. SEMP (Society for Sedimentary Geology) Special Publication no. 69, pp 9–20
Glennie KW et al (1973) Late Cretaceous nappes in Oman mountain and their geologic evolution. AAPG Bull 57:5–27
Golubiatnikov D (1904) Main results of the geological works on the Apsheron Peninsula in 1903. Bull Geol Commun 3(1) (In Russian)
Greene EW, Duke JB (1962) Selective froth flotation of ultra fine, minerals or slimes. Trans SME/AIME 223:1389–1395
Grim RE (1968) Clay mineralogy, 2nd edn. McGraw Hill Book Company, New York, 569 p
Grim RE, Guven N (1978) Bentonites: geology, mineralogy, properties and uses, Developments in sedimentology. Elsevier, New York
Gunn FA, Morris HH (1965) Delaminate domestic sedimentary clay products and method of preparation Thereof, US Patent
Gupta VJ, Harsh K (1980) Geothermal resources, an energy alternative. Elsevier Scientific Publishing Company, New York, 219 p
Hadding A (1923) Ein rontgenographische Method Kristalline und Kryptokristalline Substanzen zu Identifizieren. Z Krist 58:108–112
Haghipour A (1981) Precambrian in Central Iran. Bull Iran Pet Inst 81:1–17
Haghipour A (1989) Geologic Map of Iran 1:2,500,000. Geological Survey of Iran
Haghipour A, Pelissier G (1968) Geology of the Posht-e-Badam Saghand area, (East- central Iran). Geological Survey Iran, note no. 48, p 144, 51 Figs., 3 Pls., Map (Unpublished)

Haghipour A, Valeh N, Pelissier G, Davoudzadeh M, Bolourchi MH, Houshmandzadeh A, Sabzehei M, Stöcklin J, Sluiter W, Aghanabati A (1977) Explanatory text of the Ardekan quadrangle map, 1:250,000, Geological Quadrangle map of Iran: No. H8. Geological Survey of Iran, 114 p
Hajian J (1996) Paleocene-Eocene deposit in Iran
Hallam A (1988) A re-evaluation of Jurassic eustasy in the light of new data and revised Exxon curve. In: Wilgus CK, Hastings BS, Kendall CGSTC, Posamentier HW, Ross CA, Van Wagoner (eds) Sea-level changes-an integrated approach, vol 42, Society of Economic Palaeontologists and Mineralogists Special Publication. Everest Geotech, Houston, pp 261–273
Hamdi B, Barasier MD, Zhiwen J (1989) Earliest skeletal fossils from Precambrian/Cambrian boundary strata, Alborz mountains. Geol Mag 126(3):283–289
Haq BV, Hardenbol J, Vall PR (1987) Chronology of fluctuating sea levels since the Triasic. Science 235:1156–1167
Harben PW, Bates RL (1990) Topic of barite. In: Industrials minerals geology and world deposits. Metal Bulletin Plc, London, pp 10–18
Harris RA (1992) Pre –collisional extension in the BND orogen as a model for the tectonic evolution of Oman–type ophiolites. In: Parson LM, Mourton BJ, Browning P (eds) Ophiolites and their modern oceanic analogues, Geological Society Special Publication no. 60. Institute of Oceanographic Sciences, Godalming
Hassanzadeh J (1993) Metallogenic and tectonomagmatic events in the SE sector of Cenozoic, active Continental margin of Central Iran (Shahr-e-Babak area), Kerman province. PhD thesis, California, 204 p
Hooper R, Baron I, Hatcher R, Agah S (1994) The development of the southern Tethyan margin in Iran after the break up of Gondwana-implications for the Zagros hydrocarbon province. Sci Q J 13
Houshmand-Zadeh A (1969) Métamorphism et granitization du massif Chapedony (Iran central). Thesis, Univ. Grenoble, 242 p
Houshmand-Zadeh A (1977) Ophiolites of south Iran and their genesis problems. Geological Survey, Tehran, Unpublished
Houshmand-Zadeh A (1998) Geology of the basement of Central Iran
http//www.usbm
http://minerals.usgs.gov/minerals/pubs/commodity/Barite
http://www.bgs.ac.uk
http://www.platts.com/Coal/.>AnalyticSolutions>COALdat,or>Newsletters>Coal Outlook
Huber H (1978a) Geological map of Iran, 1 sheet. Geological Survey of Iran, Tehran
Huber H (1978b) Tectonic map of Iran, 1 sheet. Geological Survey of Iran, Tehran
Huber H (1978c) Geological map of Iran, 1:1,000,000; with explanatory note. North-east Iran. Exploration and Production Affairs, Tehran, 1 sheet
Huckriede R (1961) Gebietes zwischen Kerman und Sangan, Beith. Geol Jahrb 1 map S1:197 p
Huckriede R (1962) Kerman, mesolithic, historical geology. Quaternary 51
Huckriede R, Kuersten M, Venzlaff H (1962) Zur Geologie des Gebietes Zwischen Kerman und Sagand (Iran) Beith. Geol Jahb 51:24–26
Huttrer GW (1994) The status of world geothermal power production. In: Proceeding of world geothermal congress, 1995, vol 1, pp 3–14
IAEA (1982) Vein- Type and similar uranium deposits in rocks younger than Proterozoic. IAEA, Vienna, 391 p
IAEA (1986a) Correlation of uranium geology between South America and Africa. IAEA, Vienna, 475 p
IAEA (1986b) (INTURGEO) US- DOE National Uranium Resource Evaluation program (NURE), Uranium occurrence reports, database compiled by the IAEA from Appendix. C of the NURE quadrangle reports series PG/F- and GJQ- published from 1978 through 1982 (Available on magnetic media), 5000 p
Iannicelli J (1976) New developments in magnetic separation. Trans Magn MAG-12:436–443
Innocenti F, Mazzuoli R, Pasquare G, Radicati D, Villari L, Brozolo F (1976) Evolution of the volcanism in the area of interaction between the Arabian, Anatolian and Iranian plates (Lake Van, Eastern, Turkey). J Volcano Geotherm 1(2):103–112

International Lead and Zinc Study Group (ILZSG) (2003) Company reports. Government Agencies, AME

International Lead and Zinc Study Group (ILZSG) (2006) Lead and zinc statistics, Company reports, Government Agencies, AME

Iran Geodynamic report no. 52 (1983) Geological Survey of Iran

Irvine TN, Baragar WRA (1971) A guide to the classification of the common volcanic rocks. Can J Earth Sci 8:523–548

Jacob KH, Quittmeyer RC (1979) The Makran region of Pakistan and Iran trench-arc system with active plate subduction. In: Geodynamics of Pakistan. Geological Survey of Pakistan, Quetta, pp 305–318

Jafarzadeh A (1981) Die magnetiterzlagerstaette Chador-malu in zentral Iran und ihre exploration, Dtsch. Thesis, Tech. Hochsc. Aachen/1981, 152 p

Jafari A (2005) Mountains of Iran

Jafarzadeh A, Ghorbani M, Pezeshkpour M (1996) Iron deposits of Iran

Jahangiri H (1999) Study of anomalies of Tungsten and Tin in Nezam-abad and Bamsar areas, Tungsten exploration project in Nezam-abad and Bamsar areas

James R (1968) Wairakei and Larderello, geothermal power systems compared, New Zealand. J Sci 11:706–719

James R (1976a) Possible serious effect of steam on water flow measurements. In: Proceeding of 2nd U.N. symposium on the development and use of geothermal resources, vol 3. U.S. Government Printing office, Washington, DC, pp 1703–1706

James R (1976b) Optimum well spacing for geothermal power. In: Proceeding of 2nd U.N. symposium on the development and use of geothermal resources, vol 3. U.S. Government Printing office, Washington, DC, pp 1681–1683

James P, Seals P (2004) Barite statistics (2002), U.S.G.S., Mineral Commodity, Summaries

Jung D, Keller J, Khorassani R, Marcks C, Baumann A, Horn P (1983) Petrology of the tertiary magmatic activity in the Northern Lut area, East of Iran. Geological Survey, Iran, ISSN 0075–0484, report no. 51, pp 285–336, Universität Hamburg Mineral. Petrogr. Inst., Hamburg 2000, Dtsch

Juteau T, Maury R (1999) The oceanic crust, from accretion to mantle recycling. Springer, Berlin/Heidelberg/New York/London/Paris/Tokyo/Hong Kong, 390 p

Kabudar-Ahang quadrangle map, 1,250,000; No. D5, Geological Survey of Iran

Kadinsky CK, Barazangi M (1982) Seismotectonics of Southern Iran, the Oman line. Tectonics 1(5): 389–412, Illus. Incl. Sk. Maps, 50 Ref., October 1982, Cornell University, Department of Geological Science, Contribution no. 729

Kalantari A (1987) Biofacies map of Kopet-Dagh region, Unpub. Map, N.I.O.C. Exploration and Production, Tehran, 1 sheet

Kazmin VG, Kulakov VV (1969) Quelques traits de la structure tectonique de l'Iran et de l'Afghanistan, Bull. Moskov, Ispytatelej prirody, Otd. Geol., SUN: 1969(4), 44(2):61–76

Kesler S (1992) Mineral resources, economics and the environment. McGraw-Hill, New York, 450 p

Khadem N (1968) Types of copper ore deposits in Iran, Geological Survey Iran internal report no. 11, Based on Khadem, 1964. In: CENTO symposium on mining geology and the base metals, Ankara, pp 101–115

Khoshjou A (1999) Barite deposits of Iran. Geological Survey of Iran, Tehran

Khoyi N, Ghorbani M, Tajbakhsh P (1992) Copper deposits of Iran. Geological Survey of Iran, Tehran

Khoyi N, Ghorbani M, Tajbakhsh P (1999) Copper ores of Iran. Geological Survey of Iran, Tehran

Kirk O (1977) Encyclopedia of chemical technology. Wiley, New York

Kluyver HM, Griffis RJ, Trivel R, Chance PN, Meixner HM (1983a) Explanatory text of the Lakar-Kuh quadrangle map, 1:250,000; No. J9. Geological Survey of Iran, Tehran

Kluyver HM, Tirrel R, Chance PN, Johns GW, Meixner HM (1983b) Explanatory text of the Naybandan quadrangle map, 1:250,000; No. J8. Geological Survey of Iran, Tehran

Kowsari S et al (1986) Study of Kahnuj deposit

Kowsari S (1991) Titanium: exploration geology and economic situation of Titanium in Iran and the world

Le Pichon X (1968) Sea floor spreading and continental drift. J Geophys Res 73(12):3661–3697

Lensch G, Mihm A, Alavi Tehrani N (1977) Geology and petrography of the ophiolite belt north of Sabzevar/Khorasan (Iran). N Jb Min Abh, Univ Saarlandes 131(2):156–178
Lescuyer J, Michel R, Riou R, Vivier G (1976) Etude géochimique du volcanisme Tertiaire de la region du Mianeh (Azerbaijan, Iran). Geol Alp Grenoble Fran 52:85–98
Lescuyer JL, Riou R (1976) Géologie de la région de Mianeh (Azerbaijan): Contribution a l`étude volcnanism Tertiaire, 1, Iran. Thesis 3EME, University of Grenoble, France, 232 p
Lindenberg H, Jacobshagen V (1983) Post Paleozoic geology of the Taknar zone and adjacent area (NE Iran Khorasan). Geological Survey of Iran, ISSN 0075–0484, report no. 51. Freie Universität Berlin Institute of Geological Sciences, Dtsch, pp 145–163
Lorenz G (1964) Die geologie des oberen Karaj-Tales (Zentral-Elburz), Iran, PhD dissertation, University of Zuerich, 113 p
Lotfi M (1982) Geological and geochemical investigation on the volcanogenic Pb, Zn, Cu, Sb ore mineralization in Shurab-Golechan and NW of Khur (Lut, East Iran). PhD thesis, University of Hamburg, 151 p
Lotfi M, Mir Mohammad Sadeghi M, Omrani SJ (1993) Mineral distribution map of Iran, 1:1,000,000. Geological Survey of Iran, Tehran
Madani M (1977) A study of the sedimentology, stratigraphy and regional geology of the Jurassic rocks of eastern Kopet-Dagh. NE Iran. Unpublished PhD thesis, Imperial College, London, 246 p
MAGCOBAR, Operations oil field products division dresser industries INC (1972) Drilling fluid engineering manual. Dresser Industries Inc., Houston
Maghsoudi A (2008) Study of heavy minerals and their geology in Sero area northwest of Iran. MSc thesis, Geo-Science Research Institute
Majidi B (1978) Etude pétrostructural de la région du Mashhad (Iran), Les problemes des métamorphites, serpentinites et granitoides Hercyniens, Thése Universite Scientifique et Medical de Granobel France, 277 p
Majidi B (1981) The ultrabasic lava flows of Mashhad, northeast Iran. Geol Mag Gbr 118(1):49–58, Pl. Geological Survey of Iran, Tehran
Majidi B (1983) The geochemistry of ultrabasic lava flows occurrences in NE of Iran. Geological Survey of Iran, report no. 51, pp 463–477
Matsumto R (1977) Marine gas hydrates in the Gulf of Oman, potential importance as natural gas sources. In: Proceedings of the first Iranian marine geology symposium with special reference to Oman Sea, ChahBahar, 7–9 May 1997
Maurizot P (1990a) Explanatory text of the Shahrakht quadrangle map, 1:250,000; No, L7. Geological Survey of Iran, Tehran, 141 p
Maurizot P (1990b) Explanatory text of the Gazik quadrangle map, 1:250,000; No, L8. Geological Survey of Iran, Tehran
McCall GJH (1985a) Explanatory text of the Fahnuj quadrangle map, 1:250,000; No. K14. Geological Survey of Iran, Tehran
McCall GJH (1985b) Explanatory text of the Taherie quadrangle map, 1:250,000; No. J14. Geological Survey of Iran, Tehran
McCall GJH (1995) Explanatory text of the Minab quadrangle map, 1:250,000; No. J13. Geological Survey of Iran, Tehran
McCall G, Simonian K (1985) The Makran project, a case history prospecting in areas of desert terrain. In: International conference/1985/Rabat GBR, London. Inst Minng Metallurgy, pp 31–42
Metallogenic Map of Anarak (1984) Technoexport Co.
Meyer C (1981) Ore- forming processes in earth history. In: Economic geology, 75th Anniversary volume, pp 6–41
Milos K (1984) Industrial Minerals and Rocks/Developments in Economic Geology, pp 142–147
Mines and Mining in Iran (2005) Public. Relations Department of Iranian Mines and Mining Industries, Development and Renovation Organization (IMIDRO)
Mineral Facts and Problems (1985) Bureau of mines, Washington, DC
Ministry of Energy (1973) Subterranean canals and dams in ancient Iran
Ministry of Energy (1998) Proceedings of Iran and Germany joint seminar on new energy sources

Ministry of Mines and Metals (various years) Annual Review of the Mining Sector. Information Bureau, Development and Planning Directorate, Ministry of Mines and Metals, Tehran

Moayyed M (2001) Petrologic study of Tertiary volcano-plutonic belt in western Alborz, Azerbaijan, Ph.D. Thesis, Shahid Beheshti University

Mohajer Ashjai A, Behzadi H, Berberian M (1975) Reflection on the rigidity of the Lut Block and recent crustal deformation in eastern Iran, Tectonophysics. NLD 25(3–4):281–301

Mohajjel M (1997) Structure and tectonic evolution of Paleozoic – Mesozoic, rock Sanandaj-Sirjan, Iran. PhD thesis, University of Wollongong Australia

Moinvaziri H (1985) Volcanism Tertiaire et Quaternair en Iran. PhD thesis d'Etat, Orsay, Fra., 285 p

Momen-Zadeh M (1989) The role of submarine volcanism in the formation of synsedimentary ores, a hypothetical approach. In: Boyle RW, Brown AC, Jefferson CW, Jewett EC, Kirkham RV (eds) The symposium on sediment-hosted stratiform copper deposits, special paper 36, pp 185–186

Momen-Zadeh M (1990) Saline deposits and alkaline magmatism: a genetic model. J Pet Geol 13(3):341–356

Momen-Zadeh M, Wauschkuhn A (1984) The Allahabad Fe deposit, an example of Fe, Pb, Turquoise occurrences in the Zeber Kuh–Sarhangi Mountains, SW Kashmar, NE- Iran. N Jb Geol Palaeont Abh 168(2–3):490–496

Momen-Zadeh M, Seddighi T (1989) A guide in detecting ancient gold mines in Iran. In: International symposium on old world archaeometallurgy, Heidelberg, Germany

Momen-Zadeh M, Vatandoust R (1997) Ancient mining and metallurgy in Iran, A brief review. In: International symposium on archaeometallurgy in central and western Asia, Tehran, Iran, pp 19–24

Momen-Zadeh M, Shafighi S, Rastad E, Amstutz GC (1979) Ahangaran lead – silver deposits, West Central Iran. Miner Deposita 14:323–341, Berlin

Momen-Zadeh et al (1987) Introducing Khoueeye – rood gold-mercury deposit from mining and archeological standpoints

Momen-Zadeh M, Keesmann I, Rashidnejad N, Hezarhkhani Z (1993) Archaeometallurgische spurensuche in Iran, forschungsmagazin der Johannes Gutenberg – Universitaet Meinz, 2/93, 9, pp 45–22

Momen-Zadeh M, Chegini N, Parziger H, Pernicka E, Stollner Th, Vatandoust R, Weissgerber G (2000) Preliminary report on archaeometallurgical investigations around the prehistoric site of Arisman, near Kashan, west Central Iran, Deutsches Archaeologisches Institute, Archaeologische Mitteilungen aus Iran und Turan, band 32, Sonderdruk

Mor F, Hormozi A, Ya'ghoubpour A (1996) 'Manabe'ma'dani az didgah-e eqtesadi va zistmohiti' (Mineral resources, economics and the environment). Vizhe Nashr, Tehran

Moritz RP, Lotfi M, Saup F (1993) The sedimentary rock–hosted gold deposit at Zarshuran, Iran, A preliminary fluid inclusion and sulfur isotope study. In: Second biennial meeting of the "Society for Geology Applied to Mineral Deposits", Goanda, Spain

Motiei H (1992) Stratigraphy of Zagros

Moussavi-Harami R, Brenner RL (1992) Geohistory analysis and petroleum reservoir characteristics of Lower Cretaceous (Neocomian) sandstones, eastern Kopet-Dagh Basin, northeastern Iran. Am Assoc Pet Geol Bull 76:1200–1208

Muller R, Walter R (1983) Geology of the Precambrian – Paleozoic Taknar Inlier north west of Kashmar, Khorasan Province (NE Iran). Geodynamic project (Geotraverse) in Iran. Geological Survey of Iran, report no. 51, pp 165–183

Nabavi M (1994) Climatological Map of Iran

Nabavi MH et al (1972) Geological quadrangle map of Yazd, 1:250,000; Geological Survey of Iran, No. H9. Geological Survey of Iran, Tehran

Nabavi MH et al (1978) Geological quadrangle map of Nain, 1:250,000; Geological Survey of Iran, No. G8. Geological Survey of Iran, Tehran

National Historical Atlas of Iran, National Cartographic Center (1999)

Natural gas information, IEA (2003) Oil Gas J

Nicholson K (1993) Geothermal fluids chemistry and exploration techniques. Springer, Berlin/ Heidelberg, 255 p

Nicolas A (1989) Structure of ophiolites and dynamics of oceanic lithosphere. Kluwer, Dordrecht, 367 p

Nicolas A, Peters TJ, Coleman RG (1990) Ophiolite genesis and evolution of the oceanic lithosphere: proceeding of the Ophiolite Conference, held in Muscat, Oman, 7–18 January 1990. Kluwer Academic Publishers, Dordrecht

Nogol A (1993) Tectonic map of Iran, treatise on the geology of Iran. Ministry of Mines and Metals, Tehran

Nogole-Sadat MAA (1993) Seismotectonic Map of Iran, scale, 1:1,000,000; Treatise on the geology of Iran. Geological Survey of Iran, Tehran

Nogole-Sadat MA, Almasian M (1993) Tectonic map of Iran. Geological Survey of Iran, Tehran

Novelli G (1972) Properties and possible use of montmorillonite in pharmaceutical and cosmetic preparation, Riv. Ital. E.ssenze, Profumi, Piante Off., Olliveg., Saponi, vol 54, pp 267–270

Nowroozi AA (1972) Focal mechanism of earthquakes in Persian, Turkey, West Pakistan and Afghanistan and plate tectonics of the Middle East. Seismd Soc Bull 62(3):823–850

Ohanian T (1983) The Birjand ophiolite, an intercontinental transform structure, Eastern Iran. Geological Survey of Iran, report no. 51, pp 239–245

Omrani J, Agard Ph., Whitechurch H, Benoit M, Prouteau G, Jolivet L (2007) Arc-magmatism and subduction history beneath the Zagros Mountains. A new report of adakites and geodynamic consequences, Iran

OPEC (2002) Annual statistical bulletin

OPEC Secretariat (2004) Oil and energy outlook to 2025

Papp JF (1985) Chromium. US. Bureau of Mines, Washington, DC

Patterson SH, Murray HH (1983) In: Lefond SJ (ed) Industrial minerals and rocks/gem materials, vol 1, 5th edn. AIME, New York, pp 585–651

Pelissier G, Bolourchi MH (1967) East Takab metamorphic complex. Geological Survey of Iran, Tehran (Unpublished)

Phillips KA, Hicks CJ (2001) Prospecting for barite, Circular No. 4, February 1981, revised 2001. Arizon Department of Mines and Mineral Resources, Phoenix

Pilgrim GE (1908) The geology of the Persian Gulf and adjoining portions of Persian and Arabia. Geol Surv India Mem 24:1–177

Platt's Coal Outlook (2006), pp 8–9

Rahimzadeh F (1994) Oligo-Miocene, Miocene, Pliocene

Razzagh Manesh B (1968) Metallogeny of copper and zinc of Taknar area, Khorasan Province, Thesis Rhein-Westph., 130 p

Research and Technology Research Institute (1999) Metallogenic report on Sistan-Baluchestan Province

Reyer D, Mohafez S (1972) A first contribution to the NIOC–ERAP agreements to the knowledge of Iranian geology. editions techni, Paris, pp 1–58

Ricou LE (1974) L`etude géologique de la région du Neyriz (Zagros Iranien) et l`évolution structurale des Zagrosides, S.L. PhD thesis, Es-Sci, University Paris-Sud cent, Orsay, 321 p

Rieben H (1935) Contribution de la geology de l'Azarbaijan Persan, Neuchatel. Thesis, Sci Nat, pp 1–44

Rieben H (1955) The geology of the Tehran plain. Am J Sci 253(11):617–639

Rieben H (1966) Geological observation on alluvial deposits in northern Iran. Geological Survey of Iran, report no. 9, 40 p

Robert LV (2002) Clay and Shale. U.S.G.S., Mineral Commodity, Summaries (2001)

Romanko E et al (1984) Outline of Metallogeny of Anarak Area (central Iran), Explanatory text to metallogenic map, 1:250,000. Geological Survey of Iran, report TE/No. 21, 136 p

Roskill (2000) The economics of clay. Roskill Information Services Ltd., London

Roskill (2001) The economics of bentonite. Roskill Information Services Ltd., London

Roskill (2003a) The economics of bentonite. Roskill Information Services Ltd., London

Roskill (2003b) The economics of Barite, 9th edn. Roskill Information Services Ltd., London

Ruttner A (1991) Geology of the Aghdarband Area (Kopet Dagh). NE-Iran Abh Geol B A 38:44–47

Ruttner A (1994) Southern borderland of Triassic Laurasia in north–east Iran. Sci Q J 12:28–35
Ruttner A, Nabavi MH, Hajian J (1968) Geology of the Shirgesht Area (Tabas area, East Iran). Geological Survey Iran, report no. 4, pp 5–133
Ruttner A et al (1984) The Pre–Liassic basement of the eastern Kopeh Dagh Range, Plate tectonics, orogeny and mineralization in the Iranian fold belt, Report of a German-Iranian research program 1977–1978, University of Saarlandes, vol 168, No. 2–3, pp 256–268
Rybach L (1981) Geothermal systems, conductive heat flow, geothermal anomalies. In: Rybach L, Maffler LJP (eds) Geothermal system, principles and histories. Wiley, Chichester, pp 3–36
Rybach L, Buntebarth G (1981) Heat generating radioelement in granitic magmas. J Volcan Geotherm Resour 10:395–404
Rybach L, Muffler LJP (1981) Geothermal system- principles and case histories. Wiley, Chichester/ New York/Brisbane/Toronto
Sabzehei M (1991) Esfandagheh ophiolites, a polygenic association of Palaeozoic ophiolitic fragment of diverse origins, regional cent. Kerman. In: Symposium on ophiolite genesis and evolution of oceanic lithosphere, Geological Survey Iran
Sabzehei M (1995) Layered ultramafic- mafic komatiitic lava flow and their bearing on the genesis of Iranian ophiolites. In: 30th International Geological congress, Bijing, China, Abstract
Sabzehei M, Roshanravan J, Amini B, Eshraghi SA, Alai Mahabadi S, Seraj M (1993) Geological map of the Neyriz quadrangle, H-11, 1:250,000. Geological Survey of Iran
Samani BA (1989a) Metallogenesis and ore provinces of Iran, In the contexts of plate tectonics, International Geological Congress 28/1989-07-09/Washington, USA, vol 3, 14 p
Samani BA (1989b) Uranium and thorium mineralization related to the Pan-African magmatism and alkaline metamorphism in Central Iran. International Geological Congress 28/1989-07-09/ Washington, USA, vol 3, pp 14–15
Samimi M, Movahed Aval M, Haghipour A (1966) New discoveries of phosphate in Iran, CENTO
Samani B (1995) Manganese deposits
seminar on field techniques for mineral investigation, held in Iran, Sep 13–Oct 1 1965. Central Treaty Organ, Office US Economic Coordinator, pp 145–156, Illus. (Incl.Geol. maps), Ankara
Schubert JK, Bottjer DJ (1995) Aftermath of the Permian-Triassic mass extinction event, paleoecology of Lower Triassic carbonates in the western USA. Pleogeogr Paleoclimatol Paleoecol 116:1–39
Schuster Waltser S (1985) Iran-e safavi az didgah-e safarname-ye oroupaiyan; translated by Verahram G. Amir Kabir Publications, Tehran
Sengör AMC, Kidd WSF (1978) Post- collisional tectonics of the Turkish- Iranian Plateau and a comparison with Tibet. Eos Trans Am Geophys Union 59(4), 323 p
Sengör AMC, Yilmaz Y (1979) The suture zone of Permo- Triassic Tethys, extent and tectonic history. Geol Soc Am Abstr Progr 11(7):513–514
Sengör AMC, Altiner D, Cin A, Ustaosmer T, Hsu K (1988) Origin and assembly of the Tethside orogenic collage at the expense of Gondwana Land. In: Audley-charles MG, Hallam A (eds) Gondwana and Tethys, vol 37. Special Publication Geological Society, London, pp 119–181
Sengör AMC, Robertson AHF, Searle MP, Ries AC (1990) A new model for the Late Paleozoic – Mesozoic tectonic evolution of Iran and implication for Oman, Special Publication No. 49. Geological Society, London, pp 773–786
Shahrabi M (1994) Seas and Lakes of Iran, treatise on the Geology of Iran. Ministry of Mines and Metals, Tehran
Shahrabi M et al (1985) Geological map of Urumiyeh, 1:250,000. Quadrangle Map of Iran No. B-3
Sharkovski M et al (1984) Geology of the Anarak area (central Iran), explanatory text of the Anarak quadrangle map 1:250000, report TE/No. 19. Geological Survey of Iran, Tehran
Sieger FE (1977) Zur Geologie des Nord–Alamut Gebietes (Central-Elbourz Iran), Eidgenosische Technische Hochschule (ETH) Zuerich, Thesis no. 6093, 161 p
Sjögren H (1891) Preliminära meddelanden om de kaukasiska naftafälten. Geol För Stockh Förh 13(3)
Smirnov VI (1976) Geology of mineral deposits. MIR Publishers, Moscow

Smirnov VI, Gin AI, Bury RM, Yakovlew Grigorilv GF (1983) Studies of mineral deposits. MIR Publishers, Moscow
Smith E (1970) Climatic change and the development of agriculture in the Zagros Mountains. In: American Quaternary Association, Meeting 1st, 125 p
Spies O, Lensh G, Miha A (1983) Geochemistry of the post–ophiolithic Tertiary volcanics between Sabzevar and Quchan/NE Iran. Geological Survey Iran, report no. 51, pp 247–265
Spies O, Lensh G, Miha A (1984) Petrology and geochemistry of the post–ophiolithic Tertiary volcanics between Sabzevar and Quchan /NE Iran, Neues Jahrb. Geol Palaeontol. Abh., ISSN 0077–7749, Dtsch, vol 168, report no. 51, pp 247–265
Stahl AF (1904) Die Erze. Des karadag in Persien, chemiker-zeitung 17(6):85–86
Stalder P (1971) Magmatism Tertiaire et subrecent enter Taleghan et Alamout Alborz, Central, Bullsuiss. Min Pet 51:1–138
Stampfli GM (1978) Etude geologique general de l'Elbourz oriental au sud Gonbad-e-Qabus, Iran NE, These Geneve, 328 p
Stampfli GM (1989) Tectonogenesis and evolution of a segment of the Cimmerides, the volcano sedimentary Triassic of Aghdarband (Kopeh Dagh, north–east Iran). In: Sengör AMD (ed) Tectonic evolution of the Tethyan Region. Kluwer, Dordrecht, pp 265–275
Statistical Center of Iran (1994) Statistical report on operative mines of Iran. Available from: http://www.sci.org.ir/. Accessed 20 Aug 2005
Statistical Center of Iran (2004) Statistical report on operative mines of Iran. Available from: http://www.sci.org.ir/. Accessed 20 Aug 2004
Stöcklin J (1960) General geology, East Alborz, vol 52, No. 2. Geological Survey of Iran
Stöcklin J (1964) New data on the Lower Paleozoic- Pre Cambrian of North Iran. Geological Survey Iran, Tehran, 32 p
Stöcklin J (1967) Tectonic map of Iran. Geological Survey of Iran, Tehran
Stöcklin J (1971) Stratigrafic Lexicon of Iran, Part I, Central, North and East Iran. Geological Survey of Iran, report no. 18, 384 p
Stöcklin J (1975) On the origin of ophiolite complexes in the southern Tethys region. Tectonophysics 25:303–322
Stöcklin J (1977) Structural correlation of the Alpine ranges between Iran and central Asia. Anonymous. Livre a la memoire de Albert F. de Lapparent (1905–1975) consacre aux Recherches geologiqes dans les chaines alpines de l`Asie du Sud-Ouest. Soc Geol Fr Mem Hors-Ser 8:333–353
Stöcklin J (1983) Himalayan orogeny and earth expansion. In: Carey SW (ed) Expanding earth symposium, Sydney, 1981, University of Tasmania, pp 119–130
Stöcklin J (1986) The Vendian – Lower Cambrian Salt Basins of Iran, Oman and Pakistan, Stratigraphy, Correlations, Paleogeography, Science de la terre, Memorie, 47 p
Stöcklin J, Eftekharnezhad J (1969) Explanatory text of the Zanjan quadrangle map, No. D4, 1:250,000. Geological Survey of Iran, Tehran, 61 p
Stöcklin J, Eftekharnezhad J, Hushmandzadeh A (1965) Geology of the Shotori range (Tabas area, East Iran), with 1:100,000 coloured map. Geological Survey Iran, report no. 3
Stöcklin J, Nabavi MH (1971) Explanatory text of the Boshruyeh quadrangle map, 1:250,000; No. J7. Geological Survey of Iran, Tehran, 50 p
Stöcklin J, Nabavi MH (1973) Tectonic Map of Iran, scale 1:2,500,000. Geological Survey Iran, Tehran Offset Press, report no. 31, 100 p
Stöcklin J, Setudehnia A (1972) Lexique Stratigraphique International, V. III, Asie 9B, Iran, 1. Iran central septentrional et oriental, 2. Iran du Sud-ouest, CNRS (Paris), vol 3, No. 9, 376 p
Stremme H (1911) On allophane, hallyosite and montmorillonite, Centr Mineral, Geol: 205–211
Sussli PE (1976) The geology of the lower Haraz valley area, Central Alborz, Iran. Geological Survey Iran, report no. 38, pp 1–116
Taghizadeh N (1966a, b, c, 1967a, b, 1968a, b) Reports of Prospecting in Baluchestan, Azarbayejan, Gonabad, Damghan. Geological Survey of Iran, internal reports
Takin M (1972a) An outline of the evidence for continental drift in the Iranian region before the early Mesozoic. Geological Survey of Iran, internal report, no. 89, 4 p

Takin M (1972b) Iranian geology and continental drift in the Middle East. Nature 235(5334):147–150
Takin M (1973) Geological history and tectonics of Iran in a discussion of continental drift in the Middle East. In: Semin. Geodyn. Himalaya reg.; Hyderabad, Hyderabad Natl Geophys. Res, pp 199–200
Taraz H (1969) Permo – Triassic section in Central Iran. Am Assoc Pet Geol Bull 53(3):688–693
Taraz H (1974) Geology of the Surmaq–Dehbid area, Abadeh region, Central Iran. Geological Survey of Iran, report no. 37, 148 p
Tarkian M, Lotfi M, Baumann A (1983) Tectonic, magmatism and the formation of mineral deposits in Central Lut (East Iran). Geodynamic Project (Geotraverse) in Iran, final report, pp 357–384
Tarkian M, Lotfi M, Baumann A (1984) Magmatic copper and lead-zinc ore deposits in the Central Lut (East Iran), Neues Jahrb. Geol. Palaeontol. Abh., ISSN 0077–7749, Dtsch, vol 168, No. 2–3, pp 497–523, University of Hamburg Mineralogical, Petrographic Institute, Hamburg 2000, Dtsch
Tatsch IH (1976) Geothermal depositions, origin, evolution and present characteristics, Second Printing, Tatsch Associates, 292 p
Taylor SR, McLennan SM (1985) The continental crust: its composition and evolution. Blackwell Scientific Publication, Oxford, 312 p
Ternet Y (1990) Explanatory text of the Taybad quadrangle map, 1:250,000; No, L6. Geological Survey of Iran, Tehran
Thayer TP (1960) Some critical differences between Alpine-type and stratiform peridotite- gabbro complexes, International Geological Congress, 21st Session, Copenhagen, vol 13, pp 247–259
The balance report on water resources of Iran, Office of Water Resource Research Organization (2005–2006)
The Saghand Region, U-Pb Geochronology, Central Iran
Thiele O, Alavi Naini M, Assefi R, Houshmandzadeh A, Seyed Emami SK, Zahedi M (1967) Geological map of the Golpaygan quadrangle, 1:250,000. Geological Survey of Iran, Geological Quadrangle Map of Iran No. E7
Thiele O, Alavi Naini M, Assefi R, Houshmandzadeh A, Seyed Emami SK, Zahedi M (1968) Explanatory text of the Golpaygan quadrangle map, 1:250,000. Geological Survey of Iran, Quadrangle Map E-7, pp 3–24
Tietze E (1879) Die Mineralreichtümer Persiens. Jb Geol Bundestonst 29:565–568
Tirrul R, Bell IR, Griffis R, Camp V (1983) The Sistan suture zone of eastern Iran. Geol Soc Am Bull 94(1):134–150
Ullman (1986) Encyclopedia of industrial chemistry, vol A7, 5th edn. VCH, New York, pp 109–136
United Nation (1997) Handbook of world mineral trade statistics. United Nation, New York/ Geneva [22B]
United Nations (1999a) Geology and mineral resources of Mongolia, vol 14. United Nations, New York
United Nations (1999b) Mineral resources assessment. Development and management series, vol 5
United Nations (2000) Handbook of world mineral trade statistics, year 1993–8. United Nations, New York, 380 p
United States Geological Survey (2003) Minerals and Materials Information (CD-ROM). United States Department of the Interior, Washington, DC
United States Geological Survey (2005) Mineral commodity summaries. United States Government Printing Office, Washington, DC, 197 p
United States Geological Survey (2006) Mineral industry surveys. United States Government Printing Office, Washington, DC
U.S. Census Bureau (2006)
Unterhossel F (1934) Die. Wichtigeren Erzvorkommen des persischen karadag- Gebrges. Erzmetall H 9:425–429
Valizadeh MV, Cantagrel JM (1975) Premieres donées radiometriques (K-Ar et Rb-Sr) sur les micas du complexe magmatique du mont Alvand pres Hamadan (Iran Occidental) Acad. Sci. Paris, C.R., Ser. D., vol 281, No. 15, pp 1083–1086

Vatan A, Yassini I (1969) Les grandes lignes de la géologie de l'Elbourz Central dans la région de Téhéran et la plaine de la Caspienne. Inst Franq Petrole 24(7–9):841–878
Velde B (1992) Introduction to caly minerals. Chapman & Hall, London
Weber L, Zsak G (1999) World mining data, vol 14. Federal Ministry of Economics and Labour of the Republic of Austria, Vienna, 231 p
Weber L, Zsak G (2005) World mining data, vol 20. Federal Ministry of Economics and Labour of the Republic of Austria, Vienna, 249 p
Weddige K (1983) New stratigraphic data on Devonian and Carboniferous formations from the Binalud and Ozbak-Kuh Mountains, NE Iran. Geological Survey of Iran, report no. 51, pp 105–117
Weekly Price Survey (2007) Listed prices in Platts Coal Outlook
White RS, Klitgord K (1976) Sediment deformation and plate tectonic in the Gulf of Oman. Earth Planet Sci Lett Netherl 32(2):199–209
White DE, Maffler LPJ, Truesdell AH (1971) Vapor dominated hydrothermal systems compared with hot Water systems. Econ Geol 66:75–97
Wiickmann L (1922) Zum Klima der Türkei, Bd. I: Luftdrick u. Windverhältnisse im östlichen Mittel- Meer. Bayerische Landeswetterwarte, München
Williams GJ (1973) Contributions to the geology of mineral resources in Iran. Geological Survey of Iran, report no 21, 73 p
Williams GJ, Movahed Aval M, Samimi N, Ghasempour R, Altschuler ZS, Hignett T (1968) Recent phosphate discoveries in Iran. Geological survey of Iran, report no. 10, pp 1–84
World Bureau of Metal Statistics (2003)
World Mineral Statistics (1968–1990) vol 1
World Mining Data/Vennia (2004) Minerals production, vol 19. Federal Ministry for Economy and Labour of the Republic of Austria, Vienna, 238 p
Wright WS (1965) Lead and zinc in Iran, US AID Mission to Iran (unpublished), report 231
www.eere.energy.gov
www.energysmartschools.gov
www.fsec.ucf.edu/index.htm
www.homepower.com
www.ises.org
www.scincedirect.com
www.solarenergy.net
www.solartoday.org
Yaghubpur A (1984) Mineralization in relation to tectonic environments in Iran, part of Tethyan region, Mezhdunarodnyj geologicheskij congress, 27/1984/Moskov, SUN, Moskov: Nauka, vol 6, 356 p., Teacher Training University, Tehran, Iran
Zahedi M (1990) Explanatory text of the Sanandaj quadrangle map, 1:250,000; No, Cs. Geological Survey of Iran, Tehran
Zavosh M (1968) Mineralogy in ancient Iran, vol 1. Institute for Humanities and Cultural Studies, Tehran
Zawvosh M (1996) Mineralogy in ancient Iran
Zonenshain LP, Kazmin MI, Lisistsin AP, Bogdanov YA, Baranov BV (1989) Tectonics of the Mid- Atlantic Ridge Rift Valley between the TAG and MARK areas (26°-24°N): evidence for vertical tectonism. Tectonophysics 159:1–23

Index

A

Alborz, 4, 47, 66, 80, 90, 200, 339, 535
Antimony, 85, 134, 136, 140–146, 202, 204, 217, 220, 224, 225, 231, 235–238, 240, 270, 279, 280, 282, 288, 291, 298, 301, 312, 320–321, 480, 481
Apatite, 81, 94, 149, 154, 159, 163, 197, 202, 207, 208, 210, 512
Arabian plate, 4, 5, 8, 47, 207
Arsenic, 71, 77, 85, 133, 134, 136, 141–146, 200, 202, 220, 222, 231, 235, 236, 240, 270, 279, 280, 291, 298, 319–320, 480, 481
Azerbaijan, 1, 2, 4, 7, 8, 20, 22, 26, 30, 51–52, 54, 60–62, 72, 80, 81, 85, 90, 91, 93, 95–97, 105, 115, 117, 125, 126, 153, 154, 163, 169, 170, 172–175, 180, 181, 188–190, 197, 226, 227, 266, 273, 276, 310, 317, 326, 327, 337, 339, 340, 342, 349, 351, 352

B

Barite, 82, 83, 86, 88, 91, 117–119, 131, 169–171, 200, 219, 235, 236, 243, 257, 298, 300, 304, 308, 312, 327, 505
Bauxite, 82–84, 124–128, 187, 298, 301, 304, 308, 312, 318–320, 466
Bentonite, 85, 181–188, 219, 261, 263, 264, 298, 300, 304, 308, 312, 326, 489
Bismuth, 136, 146, 147, 220, 224, 235, 284
Borax, 68, 74, 75, 77, 165, 298, 300, 304, 308, 312, 325–326, 541
Boron, 165, 235, 240, 325–326
Brines, 173

C

Caspian Sea, 1, 2, 8, 12, 16, 18–22, 24–26, 29–31, 50, 63, 66, 148, 266, 339, 340, 348
Celestine, 166–168, 225, 312
Chalcopyrite, 147, 149, 222, 270, 284, 285
Chromite, 84, 97–103, 146, 200, 220, 223, 257, 260, 292, 298, 302, 304, 308, 312, 314–317, 375, 387
Clay, 33, 34, 72, 75, 82, 124, 125, 161, 168, 178, 179, 188, 204, 298, 482, 483, 525, 526, 529, 541
Climate, 1–44, 165, 187, 250
Copper, 57, 65, 81, 101, 199, 298, 389
Cupellation, 76–77

D

Dacite, 52, 110, 204, 361
Diatomite, 85, 188, 298, 300, 327
Dimension stones, 195, 196, 328–330
Diorite, 90, 220, 356
Dolerite, 102, 293, 401, 444, 446, 452
Dolomite, 55, 81, 88, 207, 298, 356
Dunite, 99–102, 134, 220, 223, 291, 292, 375–387

E

Eastern Iran, 33, 47, 52, 60, 69, 82–85, 91, 94, 95, 111, 114, 115, 134, 135, 180, 181, 321, 350

M. Ghorbani, *The Economic Geology of Iran: Mineral Deposits and Natural Resources*, Springer Geology, DOI 10.1007/978-94-007-5625-0,

F

Faults, 6, 8, 40, 41, 47, 51–54, 62, 63, 100, 109, 117, 118, 133–138, 140, 142, 143, 172, 202–204, 208–210, 212, 216, 217, 222, 224, 225, 230, 231, 241, 242, 245, 248–250, 256, 257, 260–263, 266–268, 270–274, 279, 280, 283, 287, 288, 291, 294, 352, 356, 376, 377, 398, 422, 493, 516
Feldspar, 163, 164, 170–172, 180, 181, 187, 235, 236, 261, 263, 264, 279, 298, 300, 304, 309, 313, 327–328, 482, 512, 513, 532
Fireclay, 82, 84, 124, 177–180, 227, 235, 237, 298, 300, 304, 308, 312
Fluorite, 82, 83, 115, 117–119, 169, 202, 298, 322, 324, 504
Flysch zone, 52, 54, 161, 286–288, 292
Fossil fuels, 199, 298, 307, 330–332, 334, 336–343, 345

G

Gas, 25, 26, 75, 200, 294–295, 298, 330–332, 334–340, 342–348, 350
Gold, 65, 81, 104, 199, 298, 468
Gondwana, 2, 3, 5, 202, 207
Granite, 54, 80, 88, 203, 300, 357
Grano-diorite, 108, 164, 195, 204, 220, 222, 262, 267, 268, 276, 284, 362, 364, 367–370, 388–390, 392, 393, 395–397, 399–401, 403–409, 411–418, 420, 421, 424, 427, 437, 465, 468–478, 480, 501, 502, 505, 512, 513, 518, 519, 532, 533

I

Iranian Plateau, 1–7, 10, 16–22, 29, 34, 66
Iron, 57, 65, 80, 87–95, 200, 298, 356
Iron mineralization, 87, 88, 90, 91, 96, 117, 207, 210, 223, 238, 240, 256, 263, 271, 282, 286

K

Kaolin, 84, 85, 177–181, 325–326, 482
Kerogen, 335–336

L

Laurasia, 3, 5, 7, 45, 202
Lead, 66, 80, 95, 200, 298, 428
Limestone, 50, 81, 91, 204, 298, 356
Lut block, 47, 52, 140–141, 202, 286–288
Lut desert, 8, 10, 24, 34, 212

M

Magnesite, 84, 165–167, 298, 301, 304, 308, 312, 329–331, 492
Makran, 2, 24, 26, 53, 54, 102, 108, 110–113, 257, 286–288, 290, 293, 315
Manganese, 80–85, 88, 90, 92, 95–98, 137, 150, 200, 202, 203, 207, 208, 220, 222–224, 235, 236, 240, 243, 249, 253, 255, 256, 265, 268, 271, 292, 298, 301, 304, 308, 312, 314–316, 371
Maragheh, 6, 7, 85, 97, 342, 352, 372, 467
Marble, 66, 72, 102, 135, 151, 194–196, 212–217, 222, 226, 227, 232, 234, 252, 264, 274, 300, 357, 361, 364–367, 370–373, 388, 392, 394, 397, 400, 401, 404, 410, 423–425, 429, 430, 433, 435, 444, 445, 447, 455, 457, 459, 460, 468, 471–473, 478, 480, 481, 500, 501, 505, 518, 519, 521
Mercury, 68–70, 77, 85, 133, 134, 141–146, 202, 224, 235, 237, 238, 240, 241, 280, 282, 286, 288, 291, 298, 480
Metallurgy, 66, 74–77, 104
Mica, 161–164, 217, 226, 227, 232, 234, 298, 300, 304, 309, 313, 326, 327, 361, 371, 391, 478, 500–502, 512, 513, 515, 533
Molybdenum, 109–111, 139, 140, 147, 149, 150, 220, 222, 224, 253, 270, 271, 280, 282, 283, 291, 293, 298, 301, 316, 317
Monzonite, 109, 195, 220, 247, 262, 267, 268, 270, 284, 290, 356, 390, 392, 393, 395, 396, 405, 407, 409, 410, 414, 417, 420–422, 430, 444, 447, 465, 468, 472, 477, 508, 519
Mountain, 2, 52, 71, 81, 101, 224, 324

N

Nuclear energy, 346–348

O

Oil, 25, 26, 72, 73, 75, 160, 181, 192, 200, 294, 298, 299, 331, 334–348, 350
Oman Sea, 1, 8, 10, 16, 24–26, 29, 31, 32, 53, 148, 353

Ophiolite, 53–54, 63, 84, 100–102, 107, 137, 203, 204, 210, 212, 220, 257, 260, 356, 371, 377, 381, 383, 385–387, 392, 398–400, 406–411, 413, 415, 416, 419–423, 425, 426, 473, 514–516
Orogenic belt, 45, 52, 119, 200, 202, 243, 294

P
Perlite, 85, 189, 298, 300, 304, 309, 313, 328, 329, 528
Persian Gulf, 1, 2, 8, 10, 11, 16, 17, 21, 24–26, 29, 31–33, 72, 80, 173, 342, 343, 350
Phosphate, 55, 81–84, 148, 149, 153–161, 175, 197, 203, 205, 207, 209, 235, 236, 298, 301, 304, 308, 312, 322, 323, 486
Porphyry, 57, 80, 105, 203, 316, 357
Potash, 174
Pyrite, 118, 147, 149, 222, 224, 225, 270, 279, 283–286, 298, 362, 363, 396

Q
Quaternary sediments, 6

R
Rhenium, 321

S
Salt, 2, 71, 81, 88, 204, 298, 530
Sanandaj-Sirjan, 47, 51, 60–63, 82–84, 90, 92–96, 99, 100, 115, 133, 135, 138, 150, 160, 169, 175, 176, 190, 191, 195–197, 200, 202, 229, 230, 234, 237, 243, 249–257, 294
Sand, 14, 27, 37, 147, 177, 218, 275, 304, 309, 313, 334–336
Schist, 52, 100, 212, 301, 356
Seasons, 10, 12, 16–20, 24–28, 193
Serpentines, 146, 160, 161, 194, 292, 294, 298, 409
Silica, 81–83, 117, 118, 123, 161, 165, 174–177, 182, 236, 252, 253, 261, 271, 279, 285, 298, 300, 304, 308, 313, 356, 371, 420, 437, 439, 444, 474, 477, 500
Strontium, 166–168, 220, 222, 225, 326–327, 524
Sulfide, 81, 82, 84, 90, 104, 107, 110, 113, 114, 117–119, 122, 136, 140, 149–151, 209, 210, 224, 235, 236, 260, 291, 292, 398, 399, 402, 404, 409, 411, 412, 419, 422, 424, 427, 430, 433, 450, 455, 469, 477, 518, 519
Sulfur, 70–72, 120, 193, 241, 268, 270, 271, 279, 298, 335, 520, 530
Syenite, 90, 154, 155, 174, 262, 267, 268, 275, 276, 282, 283, 292, 356, 361, 362, 367, 396, 405, 407, 414, 440, 449, 508, 532, 533

T
Talc, 160–161, 235, 236, 257, 293–294, 298, 300, 514, 515, 517
Titanium, 147–149, 180, 197, 523
Tonalite, 109, 142, 232, 245, 247, 262, 265, 369, 391, 415, 419, 519
Tootia, 71, 75–76, 114
Tungsten, 82–84, 136, 141, 149–151, 224, 225, 249, 253, 255, 256, 271, 291, 293, 298

U
Uranium, 88, 146, 147, 152, 197, 209, 210, 298, 336, 348

V
Vermiculite, 154, 298, 301, 326

W
Water, 5, 18, 20, 23, 25–29, 31–34, 42, 43, 75, 153, 173, 174, 177, 189, 194

Z
Zagros, 5, 6, 8–15, 20–24, 26, 29, 32, 34, 42, 43, 47–50, 55, 56, 61–64, 66, 81, 82, 84, 86, 101, 114, 124–126, 147, 153–155, 158, 160, 166, 167, 172–174, 179, 191, 196, 200, 249, 250, 257, 273, 294, 295
Zinc silver, 82

MIX
Papier aus verantwortungsvollen Quellen
Paper from responsible sources
FSC® C105338

If you have any concerns about our products,
you can contact us on
ProductSafety@springernature.com

In case Publisher is established outside the EU,
the EU authorized representative is:
Springer Nature Customer Service Center GmbH
Europaplatz 3, 69115 Heidelberg, Germany

Printed by Libri Plureos GmbH
in Hamburg, Germany